Encyclopedia of Environmental Ethics and Philosophy

Encyclopedia of Environmental Ethics and Philosophy

VOLUME 2

JACKSON TO WRIGHT
APPENDICES, INDEX

J. Baird Callicott and Robert Frodeman
EDITORS IN CHIEF

MACMILLAN REFERENCE USA
A part of Gale, Cengage Learning

GALE
CENGAGE Learning™

Detroit • New York • San Francisco • New Haven, Conn • Waterville, Maine • London

Encyclopedia of Environmental Ethics and Philosophy

J. Baird Callicott and Robert Frodeman, Editors in Chief

For product information and technology assistance, contact us at
Gale Customer Support, 1-800-877-4253
For permission to use material from this text or product,
submit all requests online at **www.cengage.com/permissions**
Further permissions questions can be emailed to
permissionrequest@cengage.com

Library of Congress Cataloging-in-Publication Data

Encyclopedia of environmental ethics and philosophy / J. Baird Callicott, Robert Frodeman, editors in chief.
p. cm
Includes bibliographical references and index.
ISBN 978-0-02-866137-7 (set) — ISBN 978-0-02-866138-4 (vol 1) — ISBN 978-0-02-866139-1 (vol 2) — ISBN ISBN 978-0-02-866140-7 (ebook)
1. Environmental ethics. 2. Environmental sciences—Philosophy. I. Callicott, J. Baird. II. Frodeman, Robert.

GE42.E5318 2009
179'.1—dc22 2008027495

Gale
27500 Drake Rd.
Farmington Hills, MI 48331-3535

ISBN-13: 978-0-02-8661377 (set) ISBN-10: 0-02-866137-0 (set)
ISBN-13: 978-0-02-866138-4 (vol. 1) ISBN-10: 0-02-866138-9 (vol. 1)
ISBN-13: 978-0-02-866139-1 (vol. 2) ISBN-10: 0-02-866139-7 (vol. 2)

This title is also available as an e-book.
ISBN-13: 978-0-02-866140-7; ISBN-10: 0-02-866140-0
Contact your Gale sales representative for ordering information.

Printed in the United States of America
2 3 4 5 6 7 12 11 10 09

Editorial Board

Editorial and Production Staff

PROJECT EDITOR
Jason M. Everett

CONTRIBUTING EDITOR
Ken Wachsberger

EDITORIAL TECHNICAL SUPPORT
Mark Drouillard

MANUSCRIPT EDITORS
Judith Clinebell
William Kaufman
Christine Kelley
Eric Lowenkron
Alan Thwaits

PROOFREADERS
Melodie Monahan
Amy Unterburger

CAPTION WRITER
Jaclyn Setili

INDEXER
Katharyn Dunham

PRODUCT DESIGN
Pamela A.E. Galbreath

IMAGING
Lezlie Light

GRAPHIC ART
Pre-Press PMG

PERMISSIONS
Margaret Abendroth
Jennifer Altschul
Margaret Chamberlain-Gaston
Leitha Etheridge-Sims

COMPOSITION
Evi Seoud

MANUFACTURING
Rita Wimberley

DIRECTOR, NEW PRODUCT DEVELOPMENT
Hélène Potter

PUBLISHER
Jay Flynn

Contents

J

JACKSON, WES
1936–

Wesley Jackson, agricultural scientist, founder of the Land Institute, and proponent of perennial polyculture, was born on June 15, 1936, and grew up just outside Topeka, Kansas. Jackson pursued a variety of options for environmental activism during his career, developing one of the first environmental-studies texts, founding an alternative school, and developing an agricultural-research institution. His most significant contribution to environmental thought was the plant-breeding research at his institute, where he attempted to apply the idea of biomimicry (or nature as model) to science and agriculture.

Jackson earned his Ph.D. in plant genetics at North Carolina State University in 1967 and then moved to Kansas Wesleyan University to teach. While there, he transformed an introductory biology course into one titled Man and the Environment, and in 1971 he edited a textbook based on that course. Though he was raising student awareness of environmental issues, he was not satisfied with his impact.

In 1974 he took a sabbatical and, as part of the back-to-the-land movement, moved to a plot of land in Salina, Kansas. Eventually he and his then-wife Dana founded the Land Institute to create an ideal learning environment and to explore appropriate technology. He also explored environmental ethics. In 1976 he presented a paper called "Toward an Ecological Ethic," in which he lightheartedly suggested that Aldo Leopold would be the most important thinker in an ecological Bible. Yet in the end it was not in philosophy, but in the practical application of ideas, where Jackson's work proved most significant.

The foundations of his later work were presented in the 1978 article "Soil Loss and the Search for a Permanent Agriculture," published in *The Land Report*, the journal of the Land Institute. Soil erosion provided an immediate spur for Jackson to consider how the split between humanity and nature might be addressed. In this he followed in a line of agricultural conservationists who see preservation of soil and the wise use of it as important for human health. He suggested that a more sustainable agriculture would rely on the imitation of key aspects of natural systems. In nature, multiple plant species grow in a field, and the soil is not disrupted each year for planting. Hence, a more sustainable agriculture might also follow those traits.

He further developed his ideas about perennial polyculture and science in his books *New Roots for Agriculture* (1985 [1980]) and *Altars of Unhewn Stone: Science and the Earth* (1987). The Land Institute slowly transformed from a school into a research institute. In his work Jackson sought to develop an agriculture suited to his region, and so he studied the prairie ecosystems of the Great Plains as models. In his emphasis on using nature as a model, he was a pioneer in the field of biomimicry.

Jackson assigned practical and philosophical value to functions at the ecosystem level. Early on, a sense of holism motivated his proposal, but eventually he drew on the work of J. Stan Rowe and Arnold Schultz to argue more specifically that emergent properties on higher levels of organization require identification and respect. Jackson was less interested in valuing the ecosystem in itself than he was in preserving properties of the ecosystem. He believed that plants can be grown more efficiently when humans make use of the properties of the ecosystem.

Jackson looked at two key ecosystem-level properties, one being polyculture and the other being perennialism, which his research staff focused more on as years passed. Agriculture that uses perennials can protect a host of soil microorganisms and the interactions between them, which annual plowing disrupts. The Land Institute saw roots and what occurs in the soil as important, as shown by its logo. Their research assumed that what evolved did so because it developed a useful function in an ecosystem, and thus all aspects of an ecosystem can be worthy of study and preservation. In *The Land Report* through articles by himself and others, and perhaps more so through the photographs of Terry Evans, Jackson promoted an ethic of respect for the prairie (which had few champions when he began his work).

Mainstream agricultural research since World War II promoted reductionism in what it studied and valued, focusing on plants in relative isolation and on chemicals that could be applied to those plants. Little value was assigned to interactions between plants and soil, or to interactions among plants. Little value was assigned to farmers or to the knowledge that farmers developed. The knowledge developed by scientists was privileged, as was the goal of increasing production per acre, and the realization of this goal continually put more and more farmers out of work.

Jackson's work developed in response to these trends. He valued systems, not just parts in isolation. So when he used nature as a model, it was originally on the ecosystem level. Even when he focused more exclusively on perennialism, he did so in part to preserve soil interactions. He sought to respect farmers by having the Land Institute develop a presence in Matfield Green (a small town near Salina), and he made the case for protecting rural communities in his book *Becoming Native to this Place* (1994). Though farmers rarely contributed to his research, he hoped that farmers' knowledge could eventually play a role in a future sustainable agriculture, with farmers breeding species appropriate to their particular places.

Along with focusing on the practicalities of agriculture, Jackson also sculpted an epistemology different from that of most research. He did not believe that laboratory experiments were an effective means of gaining knowledge about nature, since results in the field were so different from results in experiments. For Jackson, reductionist experimental knowledge was of questionable utility and validity. Along with his frequent collaborator Wendell Berry, Jackson promoted a worldview cognizant of human ignorance, arguing that what humans do not know is much greater and more significant than what humans do know. He hoped that humans could find ways to benefit from natural processes, even if they did not understand how those processes functioned. He sought an agriculture that, as he phrased it, relied more on nature's wisdom and design, and less on human cleverness.

SEE ALSO *Agriculture; Berry, Wendell; Environmental Activism; Environmental Education; Environmental Philosophy: V. Contemporary Philosophy; Land Ethic; Regionalism; Shiva, Vandana; Sustainable Agriculture.*

BIBLIOGRAPHY

Filipiak, Jeffrey M. "Learning from the Land: Wendell Berry and Wes Jackson on Knowledge and Nature." Ph.D. diss. Ann Arbor: University of Michigan.

Heat-Moon, William Least. 1991. *PrairyErth (A Deep Map)*. Boston: Houghton Mifflin. Maps and Kansas petroglyphs drawn by the author.

Jackson, Wes. 1978. "Soil Loss and the Search for a Permanent Agriculture." *Land Report*, no. 4, February.

Jackson, Wes. 1979. "Toward an Ecological Ethic." In *Man and the Environment*, 3rd edition, ed. Wes Jackson, pp. 344–355. Dubuque, IA: William C. Brown.

Jackson, Wes. 1985. *New Roots for Agriculture*, new edition. Lincoln: University of Nebraska Press. First edition, 1980.

Jackson, Wes. 1987. *Altars of Unhewn Stone: Science and the Earth*. New York: North Point.

Jackson, Wes. 1994. *Becoming Native to This Place*. Lexington, KY: University Press of Kentucky.

Land Institute. Web site. Available from http://www.landinstitute.org

Thompson, Paul B. 1994. *The Spirit of the Soil: Agriculture and Environmental Ethics*. London: Routledge.

Jeffrey M. Filipiak

JAINISM

Jainism, which originated prior to 500 BCE in northeastern India, supports key ideas and practices that accord well with environmental ethics. Its cosmology states that soul (*jiva*) is found even in plants and the elements, and its rules of behavior advocate avoiding harm to all beings. Its monastic and lay leaders have advocated personal and societal life patterns that protect life in its myriad forms.

The oldest extant Jain text, the *Acaranga Sutra* (ca. 300 BCE), proclaims that "a wise person should not act sinfully towards the earth, nor cause others to act so, nor allow others to act so" (1:1.3; anon. 1968, p. 5), that one "should not kill, nor cause others to kill, nor consent to the killing of others" (1:3.2; anon. 1968, p. 31), and that "all breathing, existing, living, sentient beings should not be slain, nor treated with violence, nor abused, nor tormented, nor driven away. This is the pure, unchangeable, eternal law" (1:4.1; anon. 1968, p. 36). These passages outline the fundamental rule to be obeyed by all Jains: the observance of nonviolence (*ahimsa*). This exhortation extends not merely to behavior toward other humans but, as noted, also to behavior toward the earth itself, toward the elements (water, fire, and air), and toward plants and animals. This earliest text outlines several techniques for avoiding violence to living beings, including not wearing clothing produced in ways that unduly harm mobile living beings, such as fur or silk garments; not consuming meat, fish, or eggs; and not moving about excessively during the rainy season (this last practice helps one to avoid stepping on the many insects that proliferate when it rains).

Jains remember twenty-four great religious leaders, of whom the most recent was Mahavira Vardhamana. Early Jain textual and archaeological materials indicate that Mahavira Vardhamana most likely lived in the fifth century BCE and taught a fivefold discipline of nonviolence, truthfulness, not stealing, sexual restraint, and nonpossession. He was preceded by Parsvanath, a Tirthankar (enlightened ascetic) who lived around 800 BCE. Later texts extol these twenty-four great teachers' accomplishments, especially the *Adipurana* of Jinasena (ninth century), which tells detailed stories about Rsibha, the first Tirthankar, and his son Bharata, the first world ruler. Like his successors, this first teacher, as an expression of his deep commitment to nonviolence, eventually renounced all clothing and entered death in old age by refusing food.

Jain literature and philosophy parallels that of the other two great ancient traditions: Brahmanic Hinduism and Buddhism. All three traditions concern themselves with making sense of the human condition. For Hinduism, adherence to one's dharma (duty), observance of ritual, and prayerful reflection and meditation constitute the good life. Buddhists seek to understand the root causes of human suffering and to follow an eightfold path of ethical behavior and meditation. Jainism emphasizes the role of ethics in advancing along a fourteenfold path toward total liberation. All three traditions include detailed assessments of karma, particularly in the Samkhya texts of Hinduism, the Abhidharma texts of Buddhism, and the extensive commentaries on Umasvati's *Tattvartha Sutra* (ca. 400) in Jainism, along with the *Karmagranthas*, the *Pancasamgraha*, and the *Karmaprakrti*. These texts make clear that action (*karma*) taken in the present will leave a residue or seed (*samskara*, *vasana*, *bija*) that will bear fruit (*phala*) at some future time. Consequently, ethics must be assiduously observed to assure a propitious outcome to human endeavors.

Jain philosophy accords particularly well with thinking about the state of the material world. According to Jainism, as poetically expressed by Mahavira and encapsulated in the aphorisms of Umasvati, life forms pervade the universe. From time without beginning, masses of living entities, known as *jivas*, have operated reciprocally with matter (*dravya*, *karma*) through movement (*dharma*) in time (*kala*). The *Acaranga Sutra* warns that life is to be found even in the particles of the earth itself, and that to avoid accruing harmful karma that will ripen inauspiciously, monks and nuns must shun any abuse to living creatures, including plants and the soil itself. Water must be strained to avoid ingesting small bugs, and food must be eaten before sunset so as to avoid inadvertently harming anything. In some traditions, various unusual observances can be found for protecting the status of one's karma. For instance, in the Sthanakvasi Svetambara branch of Jainism, monks and nuns generally wear a mouth covering (*muhpatti*) to avoid harm to the air and the beings living in the air through breathing or speaking too forcefully. The covering also prevents one from inhaling bugs. Even laypeople wear the *muhpatti* on special occasions, such as during temple visits, during particular holidays, or just to increase one's awareness while at home. All Jains espouse vegetarianism. Although they acknowledge that harm is done by taking the lives of plants, this is seen as necessary for survival. Periodic fasting is universally observed by all Jains, the most notable being the fast of Parysan, which occurs a few weeks before the fall equinox. Fasting ensures that no life forms have been injured or killed to support one's own life.

Jains have developed a scrupulous regimen for deciding what livelihoods are most conducive to the observance of nonviolence. As early as the *Acaranga Sutra*, lists were developed prohibiting Jains from participating in specific occupations that kill or injure animals (*Acaranga Sutra* 1:1.6; anon. 1968, p. 12). Consequently, Jains will not participate in butchery, livestock rearing, or agricultural practices that abuse animals. This concern results in the shunning of perhaps less obvious forms of violence as well: "dealing in charcoal, selling timber, driving oxcarts, dealing in ivory, manufacturing or selling alcohol, dealing in poisons or weapons, burning fields, draining water, breeding destructive animals" (Jaini 1979, p. 172). Six professions are approved: government, writing, farming, education, commerce, and crafts. The most preferred occupation for Jains is commerce. By some estimates, Jains constitute the single wealthiest group within India today—a result in large part of generations of conscious choice of profession.

Jains have developed a profound ethical awareness and conscience in their worldview, which sees life forms as passing through a cycle of birth after birth and regards every human being as having spent time as an almost unimaginable array of other life forms, including animals and bacteria (referred to by Jains as *nigoda*). Knowing that they were once sheep or goats or cows, they take special care to protect all animals. Even kindling fire becomes problematic for observant Jains, on account of the physical pain created by the friction of generating a flame. Jain monks and nuns never light or extinguish lamps or cook food. Lay Jains often avoid overusing electricity, with some families eschewing air conditioning not for lack of wealth but out of concern not to steal from other life forms for the sake of one's own physical comfort.

Jains adhere to nonviolence to purify their own souls. Jainism does not escape anthropocentrism and in fact lauds human birth as a necessary prerequisite for the liberation of the soul. The ultimate goal, omniscience (*kevala*), exists beyond the concerns of birth, life, death, and rebirth. Jainism does not advocate love of nature in the sense that it might be practiced in New England, but preaches self-restraint and caution around nature. If one harms a being, that harm will return to hurt oneself. Jain literature, such as the story of Yashodhara, does not celebrate the beauty of plants and animals, but rather serves as a cautionary tale, warning its reader not to succumb to the violence and lust that runs rampant in nature (Chapple 2006, pp. 241–249).

Various Jains in the past several decades have taken leadership roles in calling attention to problems of pollution and environmental degradation. In 1949 Acarya Tulsi, a confidant and adviser to Mahatma Gandhi, promulgated a list of twelve vows, starting with a vow not to commit violence in any form, and ending with "I will do my best to avoid contributing to pollution" (Kumar and Prakash 1997, p. 71). L. M. Singhvi, a member of Parliament who also served on India's Supreme Court, published *The Jain Declaration on Nature*, which lists the core teachings (nonviolence, interdependence, the doctrine of manifold aspects, equanimity, and compassion) that constitute the foundation for a Jain ecological ethics. He reiterates that in Jain cosmology, life pervades the world, appearing as "earth-bodies, water-bodies, fire-bodies, air-bodies, vegetable-bodies, and mobile bodies ranging from bacteria, insects, worms, birds and larger animals to human beings, infernal beings, and celestial beings." Singhvi asserts that by applying the five traditional vows and practicing kindness to animals, vegetarianism, avoidance of waste, and charity, one can find "a viable route plan for humanity's common pilgrimage for holistic environmental protection, peace, and harmony" (Chapple 2002, p. 222).

The traditional worldview developed by the Jain community over the course of several centuries could not directly anticipate the environmental crises of the twenty-first century. It does, however, provide conceptual resources that might be marshaled and applied to specific problems as they arise. The complexity of environmental issues requires approaching each situation from a variety of perspectives. The Jain philosophy of many-sidedness (*anekanta*) can be instructive in this regard. John Cort has pointed out that one of the Jain environmental initiatives in India, planting trees to reforest the mountain that houses the renowned Satrunjaya temple complex, has prevented shepherds from grazing their sheep on lands once accessible in common (Chapple 2002, p. 89). Environmental justice, though perhaps enhanced by vegetarianism and kindness to animals and monastic communities that leave a negligible footprint on the earth's resources, requires a level of social and economic analysis broader than the simple observance of a moral code. Interpreting and applying the principles and practices of Jainism to the environmental problems of the early twenty-first century presents new challenges to this ancient faith.

Jains are well poised to make strategic business decisions to help protect the environment. In India they are well known for their ownership of major newspapers, steel companies, mining concerns, and insurance companies. As knowledge comes to light regarding the potentially devastating effects that global climate change will have on India, Jains hold positions of leadership and can provide an important voice for change. As the Himalayan glaciers continue to melt, some have estimated that by 2050 the Ganges River will go dry for a few months each year. Jain industrialists and jurists may support legislation that can slow the progress of global climate change. Relief agencies will need continued support from Jains as flooding and heat spikes create a need for emergency food and shelter. Jain engineering firms might also help design and implement water-catchment systems to compensate for lost river water. Jains are well positioned and hopefully will respond to the challenges posed by climate change.

SEE ALSO *Animal Ethics; Asian Philosophy; Buddhism; Environmental Justice; Global Climate Change; Hinduism; India and South Asia; Pollution; Vegetarianism.*

BIBLIOGRAPHY

Akalaṅka. 1999. *Biology of Jaina Treatise on Reals*. Varanasi, India: Parsvanatha Vidyapitha.

Anonymous. 1968 (1884). *Jaina Sutras*, Vol. 1: *Akaranga Sutra, Kalpa Sutra*, trans. Hermann Jacobi. New York: Dover.

Babb, Lawrence A. 1996. *Absent Lord: Ascetics and Kings in a Jain Ritual Culture*.Berkeley: University of California Press.

Chapple, Christopher. 1993. *Nonviolence to Animals, Earth, and Self in Asian Traditions*. Albany: State University of New York Press.

Chapple, Christopher, ed. 2002. *Jainism and Ecology: Nonviolence in the Web of Life*. Cambridge, MA: Center for the Study of World Religions, Harvard Divinity School.

Chapple, Christopher. 2006. "Inherent Value without Nostalgia: Animals and the Jaina Tradition." In *A Communion of Subjects: Animals in Religion, Science, and Ethics*, ed. Paul Waldau and Kimberley Patton, pp. 241–249. New York: Columbia University Press.

Dundas, Paul. 2002. *The Jains*, 2nd edition. London: Routledge.

Gandhi, S. L., ed. 1987. *Anuvrat Movement: A Constructive Endeavour towards a Nonviolence Multicultural Society.* Rajasmand, India: Anuvrat Vishva Bharati.

Glasenapp, Helmuth von. 1942. *Doctrine of Karma in Jain Philosophy*, trans. G. Barry Gifford. Bombay, India: Bai Vijibai Jivanlal Panalal Charity Fund.

Jaini, Padmanabh S. 1979. *The Jaina Path of Purification.* Berkeley: University of California Press.

Kumar, Muni Prashat, and Muni Lok Prakash. 1997. "Lokesh." *Anuvibha Reporter* 3, no. 1 (October–December).

Sāntisūri. 1950. *Santisurisvaraji's Jiva vicara Prakaranam*, trans. Jayant P. Thaker. Madras, India: Sri Jaina Siddhanta Society.

Tobias, Michael. 1991. *Life Force: The World of Jainism.* Berkeley, CA: Asian Humanities Press.

Umāsvāti. 1994. *That Which Is: Tattvartha Sutra*, trans. Nathmal Tatia. San Francisco: HarperCollins.

Waldau, Paul, and Kimberley Patton, eds. 2006. *A Communion of Subjects: Animals in Religion, Science, and Ethics.* New York: Columbia University Press.

Christopher Key Chapple

JAMIESON, DALE
1947–

Dale Jamieson was born in Sioux City, Iowa, on October 21, 1947, and received a doctorate in philosophy from the University of North Carolina in 1976. He is currently a professor of environmental studies and philosophy, an affiliated professor of law, and the director of environmental studies at New York University. He is the author of *Morality's Progress* (2002) and *Ethics and the Environment: An Introduction* (2008), the editor of *A Companion to Environmental Philosophy* (2001) and *Singer and His Critics* (1999), and the coeditor of *Reflecting on Nature* (1994) with Lori Gruen and *Readings in Animal Cognition* (1996) with Marc Bekoff.

Jamieson approaches environmental philosophy from a perspective that is "philosophically naturalist, morally consequentialist, and metaethically constructivist" (Jamieson 2002, p. vii). His contributions to the discipline have been so broad and deep as to defy easy summary, but the following aims to give a sense of some of the most prominent work.

First, Jamieson has published extensively on the ethics of the treatment and study of nonhuman animals, especially on cognitive ethology, animal cognition, and animal experimentation. In general, he defends an animal welfare approach that is based on a utilitarian ethic that places substantial value on individual liberty for both humans and animals. Perhaps his most famous essay in this area is "Against Zoos" (1986), which has been widely anthologized. In it he argues for two claims: Although there is something to be said for the usual defenses of zoos—for example, that they educate people about animals and assist in preserving endangered species—these defenses provide reasons for different types of zoos, and these reasons are in tension with each other; and, despite their positive aspects, all things considered zoos ought to be abolished because they deny liberty to individual animals, cause significant suffering in other respects, and "teach us a false ... [and also 'dangerous'] ... sense of our place in the world" (Jamieson 1986, p. 175).

Second, Jamieson has written a number of influential articles that critically examine, and reject, key environmental concepts such as ecosystem health and sustainability. In general, he believes that people should be wary of attempts to articulate environmental concerns through the invention of new quasi-scientific terms because "the environmental problems we face are not fundamentally scientific problems ... but [problems] in our institutions of governance, our systems of value, and our ways of knowing" (Jamieson 2002, p. 224). Instead, they should seek positive visions of ways to relate to animals and nature that have been absent from the Western tradition to this point.

Third, Jamieson has worked to undermine a number of important schisms in environmental philosophy, such as those between animal advocates and environmentalists (Jamieson 1998), between metaphysical realists and subjectivists (Jamieson 2003), and between those concerned with environmental justice between humans and those concerned with the human relationship to nature (Jamieson 1994).

Finally, Jamieson has been a pioneer of work on the ethical aspects of climate change. His many articles on the topic include the early paper "Ethics, Public Policy and Global Warming" (1992), which assails contemporary economics as a useful paradigm for understanding climate change; "Adaptation, Mitigation, and Justice" (2005), which argues that both adaptation and mitigation need to be addressed in a serious climate policy; "Ethics and Intentional Climate Change" (1996), which explores the moral constraints that should be imposed on any attempt to "geoengineer" the climate; and "When Utilitarians Should Be Virtue Theorists" (2007), which argues that, in order to confront the looming environmental crisis, utilitarians should embrace an uncompromising set of green virtues.

In addition to his substantial research effort, Jamieson has made major contributions to the development of the field. He coedited an early reader on the topic (Gruen and Jamieson 1994) and helped the discipline come of age with his monumental *A Companion to Environmental Philosophy* (2001). More generally, Jamieson has spent many decades acting as an effective bridge between mainstream academic philosophy and environmental issues, not only playing a significant role in encouraging the younger philosophers in the field but also explaining and defending the usefulness and integrity of applied ethics in general to a sometimes skeptical wider audience. On this topic, his essay "Is Applied Ethics Worth Doing?" (1988) is widely regarded as a classic.

SEE ALSO *Animal Ethics; Global Climate Change; Life: Respect/Reverence; Utilitarianism.*

BIBLIOGRAPHY

Bekoff, Marc, and Dale Jamieson, eds. 1996. *Readings in Animal Cognition.* Cambridge, MA: MIT Press.

Gruen, Lori, and Dale Jamieson, eds. 1994. *Reflecting on Nature: Readings in Environmental Philosophy.* New York: Oxford University Press.

Jamieson, Dale. 1986. "Against Zoos." In *In Defense of Animals*, ed. Peter Singer. Oxford, UK: Basil Blackwell. Reprinted in Jamieson 2002.

Jamieson, Dale. 1988. "Is Applied Ethics Worth Doing?" In *Applied Ethics and Ethical Theory*, ed. David M. Rosenthal and Fadlou Shehadi. Salt Lake City: University of Utah Press. Reprinted in Jamieson 2002.

Jamieson, Dale. 1992. "Ethics, Public Policy and Global Warming." *Science, Technology and Human Values* 17(2): 139-153. Reprinted in Jamieson 2002.

Jamieson, Dale. 1994. "Global Environmental Justice." In *Philosophy and the Natural Environment*, ed. Robin Attfield and Andrew Belsey. Cambridge, UK, and New York: Cambridge University Press. Reprinted in Jamieson 2002.

Jamieson, Dale. 1996. "Ethics and Intentional Climate Change." *Climatic Change* 33(3): 323–336.

Jamieson, Dale. 1998. "Animal Liberation Is an Environmental Ethic." *Environmental Values* 7(1): 41-57. Reprinted in Jamieson 2002.

Jamieson, Dale, ed. 1999. *Singer and His Critics.* Malden, MA: Blackwell.

Jamieson, Dale, ed. 2001. *A Companion to Environmental Philosophy.* Malden, MA: Blackwell.

Jamieson, Dale. 2002. *Morality's Progress.* Oxford, UK, and New York: Oxford University Press.

Jamieson, Dale. 2003. "Values in Nature." In *A Companion to Applied Ethics*, ed. R. G. Frey and Christopher Heath Wellman. Malden, MA: Blackwell. Reprinted in Jamieson 2002.

Jamieson, Dale. 2005. "Adaptation, Mitigation, and Justice." In *Perspectives on Climate Change: Science, Economics, Politics, Ethics*, ed. Walter Sinnott-Armstrong and Richard Howarth. Amsterdam: Elsevier.

Jamieson, Dale. 2007. "When Utilitarians Should Be Virtue Theorists." *Utilitas* 19: 160–183.

Jamieson, Dale. 2008. *Ethics and the Environment: An Introduction.* Cambridge, UK: Cambridge University Press."""

Stephen Gardiner

JAPAN

Encompassing some 3,000 islands in the Pacific Ocean in East Asia, Japan is one of the great economic success stories of the post–World War II era. Although it ranks tenth in the world in population (127,433,494 as estimated in 2007), it is the world's third-largest economy (behind the United States and China) by one measure (purchasing power parity) and the second-largest (behind the United States) by other yardsticks (real gross domestic product [GDP] and nominal GDP). Japan's economic might is all the more remarkable considering the starting point—the country was in economic ruins after its defeat in World War II but applied its collective energy and entrepreneurial ingenuity in achieving a spectacular recovery: near-miraculous annual growth rates of an average of 10 percent in the 1950s and 1960s, 5 percent in the 1970s, with declining but steady rates of growth since then. Home to the world's largest automaker—Toyota—and one of the world's most powerful media and electronics conglomerates—Sony—Japan has led the way among industrial powers in confronting the tradeoffs between economic growth and environmental protection. Toyota and Honda were among the first auto companies to offer hybrid vehicles, and their fleets of cars rank among the highest in fuel efficiency and the lowest in emissions. Japan hosted the 1997 conference that promulgated the Kyoto Protocols on climate change (to which it is a signatory).

JAPANESE RELIGION AND NATURE

Japanese attitudes toward nature and the environment have deep roots in the religious traditions that have shaped the country's cultural ethos. Japan's indigenous religion, Shinto, is a form of animistic nature worship, in which the divine "kami" (deities, spirits, or gods) are believed to reside in animals, trees, rice fields, and certain human beings. Viewed from a contemporary perspective, Shinto might be called an ecoholistic religion because not only sentient beings but also whole mountains and the land itself are the objects of worship. Every village once had its own Shinto shrines, many of which still exist today. Buddhism, Taoism, and Confucianism found their way into this religious context but were transformed and melded into native Japanese traditions.

Buddhism thrived in the Nara and Heian era (710–1180). The Buddhist monks Kukai (774–835) and Saicho (767–822) propounded the belief that "mountain, river, grass, and trees have attained Buddhahood." This thesis is different from the sentient/nonsentient dualism of original Buddhism and expresses the ecological continuity of beings. In the Kamakura era, Zen Buddhism thrived, represented by Dōgen Zenji (1200–1253) (Callicott 1994).

After a period of civil war, Japan was united. Thus began the Edo Era (1603–1867). Edo (today's Tokyo) was densely populated (roughly 1 million people resided there in the seventeenth century) and because the ruling class itself was not wealthy, the difference between rich and poor was small. Meat eating and land development were illegal. Edo society retained and even enriched the natural environment by its symbiotic human/nature interactions.

When Jesuit missionaries visited Japan, some major local rulers accepted Christianity, but later the missionaries were exiled, and Japan became a closed country with the exception of the admission of occasional foreign traders. The principal ideological opponent of Christianity was Hukan Fabian (1565–1621), the author of *Refuting Deus*. By then Buddhism had become the national religion; every family had both Shinto and Buddhist altars in its house. Further, Confucianism was adopted as a governmental ideology, and its implicitly ecological world view thus complemented the combined Buddhism and Shinto traditions of Japan.

Japanese Confucianism was founded by Kaibara Ekken and developed and practiced by Ogyu Sorai (1666-1728) and Ninomiya Sontoku. For Japanese Confucians, "heaven" ("heaven-earth-nature" or *ten-chi-sizen*) was a symbol of the natural environment. Although the social ethics of these thinkers was similar to today's European and North American utilitarianism, their environmental views were ecoholistic, giving serious consideration to nature's welfare or well-being. Sorai's ethicopolitical outlook respected heaven and the happiness of people; hence human happiness was embedded in an ecological worldview. Sorai's motto was "happiness of people and world peace."

When Japan was forced to open it economy and culture to Europe and North America in the 1850s, the goal of "rich country, strong army," based on European Enlightenment ideals, became the guiding spirit of the age. Japan's national independence became an overriding concern of public policy. Yet the leading philosophers such as Nakamura Keiu were originally Confucians. Keiu combined his own belief in a Confucian heaven with a Christian God. Uchimura Kanzo (1861–1930), a leading Christian thinker, argued for an agriculture-based (instead of industry-based) state, "small-countryism" and "non-warism," positions that ran against the tide of industrialization and imperialism. Contrary to other Enlightenment thinkers, they accepted utilitarianism in the social and ethical spheres while retaining a traditional Confucian view of nature.

Although Japan began to evolve into a capitalist-industrial society under the influence and pressure of the United States and European powers, the culture remained steeped in traditional Shinto, Confucian, and Buddhist thinking while critically evaluating and absorbing European ideas. Out of this confluence original Japanese philosophies have emerged. The so-called Kyoto School, led by Nishida Kitaro (1870–1945), created a nondualistic philosophy based on Zen Buddhism. Some of the members of this school argued against European and North American modernism. Their philosophies were mainly concerned with religion, aesthetics, and culture. Today the Kyoto School is reviving, but it has not yet given much attention to the global ecological crisis, despite the influence on global environmentalist thought by Nishida's friend Daisetz Teitaro Suzuki (1870–1966).

ENVIRONMENTAL CONCERNS SINCE WORLD WAR II

Japan's defeat in World War II drastically shifted people's attitudes and values toward modern anthropocentric concerns. Under the guidance of the occupation authority, traditional thought was suppressed and channeled toward European and North American modernism. Modern democratic values eclipsed traditional morals. The attack against traditional thought—now denigrated as the ideology of a feudal, class society—prevailed and become institutionalized.

Japan's economic success—attained through aggressive industrialization—brought with it domestic environmental degradation and, indeed, some environmental calamities that were subsequently ameliorated. The most notorious was Minamata disease, a neurological syndrome caused by severe mercury poisoning. It was caused by the methyl mercury in the effluent from the Chisso chemical factory from 1932 to 1968; this effluence bioaccumulated in the marine life of Japan's Minamata Bay and the Shiranui Sea. Those eating seafood from these waters were the principal victims; their misfortunes made the dangers of heavy-metals pollution well known to medical science.

More subtly destructive was the introduction of the rich lifestyle of affluent societies, which displaced the traditional symbiotic way of life. Mass consumption and mass abandonment replaced traditional recycling systems. Indeed, the industrialization of densely populated Asian countries, following the models of European and North American modernism, is a major force in the contemporary global ecological crisis.

Those Japanese thinkers who are disciples of European and North American modernism have paid scant attention to the contemporary environmental crisis. However, if Japanese thinkers honestly confront the global environmental crisis, criticize the modernism (including industrialization) that produced it, and revive the traditional philosophies with their rich heritage of environmental

Victim of Mercury Poisoning, Minamata, Japan. *A woman holds a victim of "Minamata Disease," or mercury poisoning, in Japan in 1973. Many of the victims of the disease suffer from physical deformities, such as the malformed hand of the girl shown. Between 1932 and 1968, Chisso Corporation, originally a Japanese fertilizer and carbicle company, dumped an estimated 27 tonnes of mercury compounds into Minamata Bay.* AP IMAGES.

ethics, then a distinctive environmental philosophy can emerge in Japan (Callicott 1994).

Since the 1960s Japan has seen the emergence of various green civil movements, sometimes spurred by overseas aid programs focused on environmental quality. Books concerning green movements in other parts of the world (such as Vandana Shiva's [2005]) are being translated and introduced to the Japanese public. In academic circles various research projects on the environment have arisen, including the founding of the Society for Studies on Entropy, a group that brings together physicists, economists, and environmentalists to discuss the nature of living systems, technology, and sustainable modes of economic growth (1994–2003).

SEE ALSO *Buddhism; Confucianism; Pollution; Utilitarianism.*

BIBLIOGRAPHY

Callicott, J. Baird. 1994. *Earth's Insights: A Multicultural Survey of Ecological Ethics from the Mediterranean Basin to the Australian Outback.* Berkeley: University of California Press.

Palmer, Joy A. 2001. *Fifty Key Thinkers on the Environment.* London: Routledge.

Shiva, Vandana. 2005. *Earth Democracy: Justice, Sustainability, and Peace.* Cambridge, MA: South End Press.

Society for Studies on Entropy. 1994–2003. Selected Papers, vols. 1–7. Available from http://entropy.ac/modules/mydownloads/viewcat.php?cid=3

Tamanoi, Y.; A.Tsuchida; and T. Murota. 1984. "Towards an Entropic Theory of Economy and Ecology: Beyond the Mechanistic Equilibrium Approach." *Economie appliqué* 37: 279–294.

Tucker, Mary E. 1990. *Moral and Spiritual Cultivation in Japanese Neo-Confucianism: The Life and Thought of Kaibara Ekken (1630–1714).* Binghamton: State University of New York Press.

Watanabe, Kyoji. 1998. *Yukishi-Yo-no-Omokage (The Vestiges of the Society that Passed Away).* Fukuoka, Japan: Ashi-Shoho.

T. Yamauchi

Thanks are due to Ms. Karen Mather for review of an earlier draft.

JEFFERS, ROBINSON

1887–1962

Robinson Jeffers, born in Pittsburgh, Pennsylvania, on January 10, 1887, was a regional poet. His work is rooted in California, but it speaks of larger themes, such as the cycles of life, the cruelty of humans, the dance of death and renewal, and the ways in which culture and civilization can blind people to the beauty of the world. Jeffers published over fifteen volumes of poetry in his lifetime. He wrote epic narrative and lyrical poetry. He laid a foundation for bioregionalists and environmentalists concerned with place and human obligations to nature.

In September of 1914, Jeffers and his wife Una moved to Carmel, California. There, under the influence of the Big Sur coast and its people, Jeffers found his voice. His collection *Californians* (1916) described the Big Sur region and explored ideas about the decline and future of Western civilization. By 1919 Jeffers and his family had moved two miles south of Carmel to live on the windswept, rocky coast. Jeffers learned the craft of stonemasonry. With local granite, he built Tor House for his family residence and Hawk Tower, a four-story tower overlooking the Pacific. Jeffers wrote poetry in the morning. In the afternoon, he worked with stone and tended to the trees he had planted, a grove of over two thousand eucalyptus, cypress, and pine trees. His poetry and his life grew out of his attachment to this place along the Big Sur coast. It was a landscape of rock, stormy skies, kelp, sea lions, and intense, passionate people. Jeffers remarked that his poems grew like a plant out of particular places—a canyon; a promontory; a relationship of rock, water, wood, and grass.

Jeffers's poetry not merely described these places, but plunged into symbolic and metaphorical depth. The West was the place of the Pacific Ocean, but also of the end of

Robinson Jeffers, 1934. *Robinson Jeffers was an early twentieth-century poet known for his symbolic and metaphorical depictions of the natural world. Jeffers also developed the concept of "inhumanism," seeing the nonhuman as significant.* COURTESY OF THE BANCROFT LIBRARY, UNIVERSITY OF CALIFORNIA, BERKELEY.

western migration and the end of human civilization. In such collections of poetry as *Tamar and Other Poems* (1924), *Roan Stallion, Tamar, and Other Poems* (1925), *The Women at Point Sur* (1927), and *Give Your Heart to the Hawks and Other Poems* (1933), Jeffers developed themes that would remain consistent throughout his work. He believed that poetry should be of beauty and of larger things; poetry was a way to deepen our awareness of the nature of the world—the flux, the beauty, the cycle of death and rebirth. His long, epic narratives are laden with violence, murder, and incest. These narratives are indebted to Greek tragedy and are based on stories and people from the Big Sur region. The characters are rough, violent, passionate, and driven to destruction. Jeffers despaired of human cruelty and self-centeredness. For Jeffers, it was civilization that blinded people to the beauty and divine nature of the world. His poetry is a reminder and a tribute to the natural world. His lines are filled with hawks, water, rocks, cypress trees, pelicans, and horses. His poetry brings attention to the natural world and asks that people participate in it.

Jeffers developed the concept of inhumanism, in contrast to the idea of humanism. Inhumanism represents a shift from seeing humans as significant to seeing the nonhuman as significant. Inhumanism is a move away from egocentrism to an acknowledgment of the divine beauty of the world. People must overcome self-centeredness to keep their integrity and appreciate nature. In "The Answer," Jeffers writes, "Integrity is wholeness, the greatest beauty is / Organic wholeness, the wholeness of life and things, the divine beauty of the universe. Love that, not man / Apart from that, or else you will share man's pitiful confusions, or drown in despair when his days darken." For Jeffers, we must move away from an obsession with self to love things greater than ourselves. To do otherwise is to suffer pain and confusion.

Loren Eiseley, in the foreword to *Not Man Apart* (1965), shows how Jeffers, his poetry, and the Big Sur coast are all connected. Eiseley writes, "The sea-beaten coast, the fierce freedom of its hunting hawks, possessed and spoke through [Jeffers]. It was one of the most uncanny and complete relationships between a man and his natural background that I know in literature" (p. 23). The rocky, windy coast of Big Sur carved Jeffers and his poetry.

SEE ALSO *Deep Ecology; Regionalism.*

BIBLIOGRAPHY

Eiseley, Loren. 1965. Foreword. In *Not Man Apart: Lines from Robinson Jeffers*, ed. David Ross Brower. San Francisco: Sierra Club.

Jeffers, Robinson. 1916. *Californians.* New York: Macmillan Co.

Jeffers, Robinson. 1924. *Tamar and Other Poems.* New York: P. G. Boyle.

Jeffers, Robinson. 1925. *Roan Stallion, Tamar, and Other Poems.* New York: Boni and Liveright.

Jeffers, Robinson. 1927. *The Women at Point Sur.* New York: Boni and Liveright.

Jeffers, Robinson. 1933. *Give Your Heart to the Hawks and Other Poems.* New York: Random House.

Jeffers, Robinson. 2001. *The Selected Poetry of Robinson Jeffers.* Stanford, CA: Stanford University Press.

Karman, James. 1995. *Robinson Jeffers: Poet of California*, rev. edition. Brownsville, OR: Storyline Press.

Lisa Gerber

JUDAISM

Judaism's teachings, including those on environmental matters, derive their authority from being found in, traced to, or implied by the Torah. In Hebrew, *torah* means "instruction." As a proper name, it refers primarily to the laws divinely revealed in the Pentateuch (the first five books of the Bible). Secondarily, it refers to the Hebrew Bible and its teachings as a whole. These include, above all, belief in an all-powerful, all-knowing God who has created the world (Gen. 1:1–2:4), revealed a definitive set of laws for the Jewish people (Ex. 20:1–23:33, 25:1–31:17; Lev. 1:1–7:38, 11:1–24:9, 25:1–26:2, 27:1–34; Num. 27:8–11, 28:1–30:17, 34:50–35:34, 36:7–9; Deut. 4:15–19, 5:6–18,12:1–26:19), and promised to reward obedience to those laws with communal prosperity and punish disobedience with political, social, and environmental ruin (Lev. 26:2–45; Deut. 11:10–17, 28:1–69, 29:9–30:20). Except for laws that spell out some minimum standards of moral decency which are meant to be appropriate for any political society (Gen. 9:3–7; Sacks 1990, 66), the Torah's laws are not directly intended for non-Jews, though they do invite prudent emulation (Deut. 4:6–7).

Central to the Torah's laws are the institutions of the Sabbatical and Jubilee Years (Sacks 1990, 94–114). Their general purpose is to protect and foster community life in the Promised Land (ancient Israel) insofar as its flourishing depends on each patriarchal family's ancestral farm. The Sabbatical laws (Ex. 21:2–6, 23:10–11; Lev. 25:1–7; Deut. 15:1–3, 24:19–22, 31:10–13) stipulate that every seventh year farms must lie fallow. Meanwhile spontaneous vegetation must be left unharvested and unfenced for any passers-by (including animals) to consume. Also, all debts must be forgiven. Finally, heads of families nation-wide must assemble to hear a public reading of the laws. The Jubilee laws (Lev. 25:8–55, with Ex. 21:2–6 and Deut. 15:12–18) add that farms can never be sold, only leased. All leases expire every fiftieth year, when each farm reverts to its ancestral owners. Simultaneously, all slaves must be freed, with one-on-one economic assistance to prevent their relapse into slavery. As for the theological and political importance of these laws, the prophet Jeremiah laments (Jer. 34:13–22) that it was neglect of the Sabbatical and Jubilee Years which brought God's punishment in the form of the Babylonian conquest and the resulting exile of the Jewish people from their land in 586 BCE.

THE ORAL LAW

Judaism's post-biblical legal and related writings are called the "Oral Torah" (or "Oral Law"). These extend the Pentateuch's "Written Law" to the circumstances of Jewish life in exile. Foremost among such writings are the Mishnah (a legal code compiled around 200 CE), the Talmud (containing the Mishnah with wide-ranging legal and homiletical controversy and commentary, called

Preparing for Sukkot. *An Ultra Orthodox Jew inspects willow branches for Sukkot, the fall harvest celebration when pious Jews thank God for the rains that irrigate farms, forests, oases, and rivers. Responding to critiques of being anti-environmental, several Revisionist proposals have been suggested in an effort to "green" Judaism.* DAVID SILVERMAN/GETTY IMAGES.

Gemara, compiled around 500 CE), and Midrash (contemporaneous compilations of brief homilies, or mini-sermons, loosely geared to biblical texts). There are also ongoing Teshuvot ("responses" by authorities in the Oral Law to specific legal inquiries), Bible translations (from Hebrew into vernacular languages) and, eventually, detailed biblical, legal, ethical, and theological commentaries. Also authoritative is Moses Maimonides' *Mishneh Torah* ("recapitulation of Torah," circa 1170–1180 CE), a codified digest of the Talmud whose invisible backdrop is Platonic-Aristotelian philosophy. The Oral Law's purely legal component is called Halakhah ("walking," i.e., rulings reached step-by-step). The Halakhah's original formulators were the Pharisees ("separatists," i.e., innovators), who also instituted synagogue worship, personal prayers, and ritual blessings to be recited over mundane activities such as eating. Its subsequent authorities are called Rabbis ("teachers" of the Oral Law).

Nowadays Jews are divided into Orthodox and non-Orthodox denominations, according to how strictly they adhere to the Oral Law. Demographically, most Jews are non-Orthodox—Reform Jews considering themselves only loosely or voluntarily bound to the Oral Law, Conservative Jews adhering somewhat more strictly. These differences show up in Judaism's discussions of environmental matters as follows. How well, it is asked, does the Oral Law address the unprecedented legal and moral issues posed by the environmental crisis of the twentieth and twenty-first centuries—the massive pollution of the air, water, and land as a side effect of the steadily accumulating advances of modern technology, with the resulting threats to the flourishing of plant, animal, and human life? Orthodox scholars find the Oral Law adequately equipped with precedents for meeting today's crisis. Among other things, they note its prohibitions against wanton or needless destruction (called by the Rabbis *bal tashchit*, "Don't destroy!") and against causing animals pain (called *tsa'ar ba'alei chayyim*, "animal suffering"), as well as to its Sabbatical and Jubilee provisions. In contrast, various non-Orthodox scholars have proposed radically revising—or *greening*—Judaism with a view to letting it speak to the crisis more directly. Their proposals have resulted in a variety of syncretistic arguments—innovative syntheses of Jewish and environmentalist views—shaped partly by traditional Jewish teachings, partly by how the religious implications of the crisis happen to be understood (or perhaps misunderstood) by prominent green authors, notably the historian Lynn T. White Jr. (1967).

WHITE, LEOPOLD, AND TOYNBEE VERSUS THE BIBLE

In "The Historical Roots of our Ecologic Crisis" (1967), White alleged that the source of the contemporary environmental crisis was Genesis 1:28—where God tells the first humans to "fill the earth and master it; and exercise dominion over the fish of the sea, the birds of the sky, and all the living things that creep on the earth." White's allegation echoed similar allegations by the forest ecologist Aldo Leopold, who protested against what he called "our Abrahamic concept of land"—by which he meant regarding the land "as a commodity belonging to us"—and who commented sarcastically that "Abraham knew exactly what the land was for: it was to drip milk and honey into Abraham's mouth" (Leopold 1949, viii, 204f.). White's and Leopold's allegations resonated further in the historian Arnold Toynbee (1972), who called for a return to pagan nature-worship, on the supposition that it would protect us against the environmental deterioration accompanying modern industrialization and commercialization, which the Bible (he asserted) had inspired. Neither White nor Leopold nor Toynbee, however, adduced much more than a biblical verse or two, cited out of context, to substantiate their far-reaching, if historically dubious, claims.

Nor did they examine other, more direct sources for the modern view of the relation between human beings and their natural environment. In particular, they overlooked the detailed arguments for the environmentally invasive project to "conquer nature ... for the relief of man's estate" and make human beings "like masters and owners of nature" which are found in the writings of the philosophical founders of modern technology, Francis Bacon and René Descartes, respectively (Yaffe 2001, 66 nn. 28–29, 70 n. 73). Bacon, for example, in his rhetorical appeal to the early chapters of Genesis to underwrite that project (Bacon 1963, 296–297), bypasses Genesis 1:28 altogether. He refers instead to the biblical description of the neediness of human beings after they had been expelled from the Garden of Eden for having acquired "knowledge of good and evil" (Gen. 2:9, 17, 3:1–7a) and could no longer look to God to supply their wants but had to rely entirely on their own initiative and industriousness (Gen. 3:7b, 17–19, 23). Here Bacon mentions, as precursors to his own project, figures of an "active" bent—Cain, the first tiller of the ground (Gen. 4:2, 17), and Cain's offspring, including the inventors of the musical and metalworking arts and the builders of the Tower of Babel (Gen. 4:21–22, 11:3–5)—whose very inventiveness the Bible criticizes as evidence of their ongoing rebellion against God (Gen. 4:10–14, 23–24, 11:5–9), as Bacon duly notes. For Bacon, the project to bring about "the limitless scientific mastery of nature and the technologizing of human life," being of strictly human origin, is not to be identified with the way of life mandated by the divine command set forth in Genesis 1:28, with which it is in direct competition (Kennington 2004, 5–6, 70, with 123–144 on Bacon's thoroughgoing influence on Descartes).

THE ENVIRONMENTAL CREDENTIALS OF JUDAISM

Jewish environmental ethics—as distinct from Judaism's traditional theological and ethical discussions—emerged once contemporary scholars saw a pressing need to defend Jewish texts and practices against the unsupported allegations by White and the others. In their various counterarguments, Orthodox and non-Orthodox scholars alike consider three interrelated questions—historical, ethical, and philosophical. Historically, they ask, what is Judaism's ecological track record? Ethically, they add, how should Jews as Jews face the crisis here and now? And, they wonder philosophically, does contemporary environmentalism fit, or fail to fit, with traditional Jewish piety? These questions are not easily separated. Moreover, they are often followed out all too incompletely, intermittently, and haltingly. Nor is there strict unanimity of answers. Also, as already stated, Orthodox and non-Orthodox scholars differ on whether their task is simply to recount Judaism's traditional teachings with an eye to today's environmental concerns, or to revise (or green) those teachings. Last but not least, the allegations by White and the others force scholars who would defend Judaism's ecological heritage to construe it in terms of "nature," a notion for which there is no exact equivalent in the Torah. As a result, Jewish green responses to White and the others often fall short in addressing the full issue at stake.

The shortcomings have to do with how the Torah's silence about "nature" and science correlates with its articulateness about "creation" and law. Generally speaking, either living things and their habitats are "natural" (that is, self-originating, self-maintaining, and scientifically intelligible in their own terms) or they are "created" (that is, originated, arranged, and provided for by a divine Creator). By assuming the second of these two incompatible alternatives, the Torah implicitly excludes the first (Maimonides 1963, 281–360). Its reticence about "nature" cannot be ascribed simply to a lack of scientific sophistication or an excess of mythological imagination, however, since the Torah, by its own lights, is neither "science" nor "myth"—biblical Hebrew lacks terms for these as well—but revealed law. The Torah's presenting things as created (rather than "natural") is tailored to its practical interest in the extra-human or supra-human support for law-abidingness. It spells out that support in a manner that is intelligible and persuasive to its immediate adherents. Thus, for example, its account of day two of the first week of creation, when God is said to separate the primordial watery chaos into "waters above" and "waters below" and to maintain that separation henceforth (Gen. 1:2, 6–8), is meant as a straightforward description of God's ongoing separating activity, and not, say, as a poetic description of a "natural" separation on the part of the waters themselves. The straightforward implication is that God could in principle abandon or otherwise modify his activity whenever appropriate for disciplinary or instructional purposes. Such purposes are evident, for example, in God's allowing the "waters above" to collapse into the "waters below" during Noah's flood (Gen. 6:17–8:19), and subsequently—in keeping with a post-flood covenant whereby God vows never to allow that collapse again and stipulates in return that human beings must henceforth govern their behavior by written law (Gen. 8:20–9:17)—in God's promising to provide seasonable rain as a reward for obeying the Sabbatical and Jubilee laws in particular and threatening to withhold the rain as a punishment for disregarding those laws (Lev. 26:3–5, 18–20; Deut. 11:10–17). In short, from the Torah's viewpoint, the stability and flourishing of human beings and their larger environment depend on the will of their Creator as revealed for the sake of fostering adherence to law, rather than on "nature." The conceptual and other shortcomings that show up in Jewish environmental ethics are, generally speaking, traceable to its following White and Leopold and Toynbee uncritically in ignoring the likelihood that the differences between the Torah, with its theological supposition that the earth and its inhabitants are creatures of God, and Baconian science, with its philosophical supposition that they are to be mastered by human beings, are, at bottom, irreconcilable (Spinoza 2004, 67–101).

THE ECOLOGICAL TRACK RECORD OF JUDAISM

Be that as it may, the main historical counterargument against White and the others (Ehrenfeld and Bentley 2001) is that Genesis 1:28 does not give humans unrestricted sway over nature, but only caretaking authority, or stewardship. Much theological support is found for this view. Biblically speaking, all the land—or earth (*arets* in Hebrew means both)—belongs to God (Ex. 9:29; Lev. 25:23; Deut. 10:14; Ps. 24:1). God is said to govern the earth providentially so as to reward law-abidingness and punish rebelliousness (Kay 2001; Allen 2001), especially by supplying or withholding rain in the manner already mentioned. *Sukkot*, the fall harvest festival, is also a water festival (Schaffer 2001), whose ritually displayed plant species—willow branches, date-palm fronds, myrtle twigs, citrons—celebrate ancient Israel's divinely-bestowed water abundance in its four ecologically distinct regions: riverine wetlands (willow), desert oases (palm), forested highlands (myrtle), and cultivated farmlands (citron). In addition, post-biblical Midrashim criticize the very notion that humans should be masters over all other creatures (Cohen 2001, 74ff.), since although they resemble angels in being created in God's image, they also resemble beasts in being procreated and mortal—and thereby susceptible to lawlessness.

Halakhic evidence for Judaism's environmental conscientiousness is also considerable. Deuteronomy 20:19–20's prohibition against destroying fruit trees for siegeworks during wartime serves as precedent for the Oral Law's *bal tashchit* prescriptions (Cohen 2001, 77ff.; Schwartz 2001). Ambiguity about whether the Torah's concern here is—in today's parlance—biocentric (that is, for the trees) or anthropocentric (that is, for their human beneficiaries) permeates Talmudic rulings as well. For example, according to the Mishnah (*Baba Kamma* 8:6), destroying any fruit tree is punishable by a fine. The tree's owner is exempt, however, on the presumption that no one willingly destroys one's own property needlessly. Here the Mishnah surrounds a biocentric principle (saving a tree for its own sake) with an anthropocentric restriction (accommodating the needs of its owner). Nor is this all. The Gemara (*Baba Kamma* 91b–92a) adds that the owner's need must be specific—the tree's unprofitability, say, or its inconvenience to others. Here the Gemara trims the aforementioned restriction (the owner's putative need) so as not to obstruct the basic principle (the tree's presumptive right to exist and flourish as a living creature). As in the foregoing, a continual need to reconcile anthropocentric and biocentric (or, as we shall see, zoocentric) considerations also guides post-Talmudic case-law deliberations in this and other matters—including questions of *tsa'ar ba'alei chayyim*, about which more shortly.

The evidence just cited shows that the allegations by White and Leopold and Toynbee are ill-informed about biblical and post-biblical Jewish law. Nevertheless it does not dispose of those allegations altogether. This would require showing, in addition, that the meaning of Genesis 1:28 is absorbed without remainder into the Torah's legal and moral prescriptions. On the contrary, in its immediate context Genesis 1:28 invites its reader to consider, if only for a moment, the possibility of a way of life which is not only completely free of legal restraints and oriented to the unimpeded human domination of the earth but, even so, backed by God's blessing (Sacks 2001, 153ff.)—as White and the others surmise—although or because the Torah goes on to replace the foregoing possibility with that of life under a divinely revealed law that is, in turn, both ethically and ecologically enlightened (Kass 2001, 384, 409). Here, perhaps, White and Leopold and Toynbee share some moral and intellectual high-ground with the Torah after all. That is, despite their unfounded resentment and unscholarly presumptions about the Torah, White and the others may have a viable, if one-sided, insight into the human-all-too-human starting-point that prompts its environmental teachings—namely, unrestricted human freedom and the environmental and other risks that would and do arise from it. Current defenders of Judaism's ecological track record would do well to explore this high-ground more fully.

FACING THE CONTEMPORARY CRISIS

Three noteworthy proposals would revise traditional Jewish teachings so as to incorporate contemporary environmental activism. Another noteworthy proposal would adjust contemporary environmental activism so as to incorporate traditional Jewish teachings. In addition, several Jewish activist organizations and projects are worth noting.

One revisionist proposal (Artson 2001, 161, 171) would extend the putative sacredness of Israel's environment to the entire earth. Israel's environment is sacred, it is argued, as the original location for obeying the Mitzvot (divine "commandments"). Because the Mitzvot remain binding on Jews in the Galut ("Exile"), that is, in post-biblical circumstances anywhere on the earth, it follows that the latter's environment must to that extent be sacred too. Meanwhile Israel's exclusivity is preserved, on the premise that the ultimate purpose of the Mitzvot is to live piously in the Land of Israel. This argument manifestly privileges *aliyah* ("ascent" or immigration to Israel). A difficulty with its premise, however, is that it discounts the intrinsic value of Mitzvot whose evident purpose is simple moral decency, regardless of location (Ex. 23:1–9; Lev. 19:13–18; Deut. 15:1–18, 22:6–7).

Another revisionist proposal (Troster 2001) would amalgamate Gaia worship with Judaism, on the premise that the earth ("Gaia") is God's creature. A difficulty here is the Torah's uncompromising prohibition of creature-worship (Ex. 20:3–6; Deut. 5:7–10 with 4:15–19; Wyschogrod 2001).

A third revisionist proposal (Benstein 2001) would offset the Mishnah's warning against being distracted from Torah study while looking at trees and fields (*Pirkei Avot* 3:7; Schwarzschild 2001), by supplementing the two traditional categories of Mitzvot—ethical obligations to fellow humans and ritual obligations to God—with a third category: obligations to one's natural environment. Fitting scientific ecology neatly into this new category, however, would depend on whether or not the aforementioned tension between "nature" and "creation" is resolvable within the purview of Jewish law (Jonas 2001).

A non-revisionist proposal (Rosenblum 2001) would educate environmental activists generally in the ecological benefits of the Sabbatical and Jubilee Years: Resting farms every seven years would reverse the soil-depleting effects of steady planting. Remitting debts would relieve economic pressures on poor farmers to exhaust their land's resources. Explicating the laws in public forums would foster communal environmental awareness. Returning farms to their original land-grant owners every half-century would forestall absentee land-accumulators who might abuse the land with impunity. Finally, freeing indentured servants and offering one-on-one economic assistance might keep them out of the impoverished underclasses, who are especially hard on the environment. If these policies seem utopian and unlikely to be implemented, they nevertheless provide a lodestar for activists and a measuring-stick for their successes.

Among Jewish activist organizations is COEJL (Coalition on the Environment and Jewish Life), an umbrella group that facilitates green projects in Jewish communities both locally, with activities such as nature-walks and guest-speakers, and nationally, with campaigns to purchase energy-efficient light-bulbs, and green political advocacy in general. Teva Learning Center (*teva* is the post-biblical Hebrew word for "nature") supplies green educational materials, teacher-training, and recreational opportunities for Jewish schools, camps, and youth groups. Shomrei Adamah ("Guardians of the Earth") is a popular label for Jewish green programs in various localities. Finally, the Heschel Center for Environmental Learning and Leadership is an educational and policy think-tank located in Israel. (For a longer list of organizations, see Waskow 2000, I:290–92.)

CONTEMPORARY ENVIRONMENTALISM AND JEWISH PIETY

The foregoing defenses against White and Leopold and Toynbee result in a philosophical quandary. On the one hand, revisionist proposals for greening Judaism risk forfeiting traditional Jewish teachings that are otherwise indispensable. On the other hand, traditionalist attempts to read Genesis 1:28 exclusively as a mission-statement for environmental stewardship ignore its surface connotation as a permission-slip (subsequently withdrawn) for environmental recklessness. This quandary is intellectually humbling, but not hopeless. Given that our present crisis is no mere academic exercise, we need not expect our philosophical reflections simply to underwrite this or that ecologically friendly view, whether traditional or revisionist, which we may have favored before we started thinking about it. The tension between environmental ethics and Jewish piety—between an independent-minded search for clarity about our ecological predicament, and reverent devotion to the teachings of the Torah—is perhaps unbridgeable. But one may come to grips with it by being either a philosophical inquirer open to Judaism's ecological insights, or a piously observant Jew open to those of philosophy.

Consider, in this regard, current controversies over animals' welfare. Prodded to some extent by White and Leopold and Toynbee, advocates of animal rights and vegetarianism wonder about Judaism's notion of *tsa'ar ba'alei chayyim.* At issue here is not whether Jewish teachings promote animals' welfare, but whether they do so as far as they might. Certainly the Torah prohibits, for example, working animals on the Sabbath (Ex. 20:10, 23:12; Deut. 5:14), overburdening draft-animals (Ex. 23:5), and muzzling oxen treading edible grain (Deut. 25:4). And the Talmud goes on to forbid, for example, dismembering live animals, hunting animals for sport, and leaving animals hungry while humans eat (Bleich 2001, 333–38). But, it is asked, what about possible cruelties inflicted when using animals in laboratory experiments or even slaughtering animals for meat (Levy 2001)? Would espousing animal rights not make Jewish teachings more up-to-date and consistent? As in cases of *bal tashchit,* however, there is a need to reconcile, in this case, animals' welfare with that of human beings. As for animal experimentation, Orthodox and non-Orthodox scholars alike would reply that humans need ongoing scientific medical research. Besides, Jewish teachings are designed to ennoble character (Bleich 2001, 349–52), not just regulate behavior. Hence pious Jews are expected to recognize, and refrain from, inflicting or endorsing cruelty and unbearable suffering anywhere, even or especially in laboratories (Bleich 2001, 344–49). As for animal slaughtering, rules for *kashrut* (dietary certification) specify butchering methods aimed at—and effective in—minimizing if not eliminating animals' pain (Bleich 2001, 338–44). Also, even though vegetarianism is permissible halakhically (Bleich 2001, 371–83), meat eating in conformity with *kashrut* restrictions remains desirable for theological reasons. On reflection, kosher land-animals, fish, and birds turn out to be "pure" (or "clean"; the Hebrew *tahor* means both) in the ecological sense that each as a species inhabits its own generic environment-sector—land, water, or air—as designated at the time of creation (Lev. 11 with Gen. 1; Kass 2001, 386–93, 397–403), none mimics in its motility genera outside its own, each has a "clean" or clearly delineated shape, and none is carnivorous (except that kosher fish may eat other kosher fish; Kass 2001, 408f. n. 29). Piously and self-consciously restricting meat-eating choices to these—following proper ritual slaughtering and after reciting appropriate blessings—reminds Jews versed in the Torah of the diverse and mind-engaging character of creation itself, therefore, and of Jews' longstanding obligations to its Creator.

SEE ALSO *Bible; Christianity; Leopold, Aldo; Stewardship; White, Lynn, Jr.*

BIBLIOGRAPHY

Allen, E. L. 2001. "The Hebrew View of Nature." In *Judaism and Environmental Ethics: A Reader*, ed. Martin D. Yaffe. Lanham, MD: Lexington Books.

Artson, Bradley Shavit. 2001. "Our Covenant with Stones: A Jewish Ecology of Earth." In *Judaism and Environmental Ethics: A Reader,* ed. Martin D. Yaffe. Lanham, MD: Lexington Books.

Bacon, Francis. 1963. *Advancement of Learning*. In *The Works of Francis Bacon*, ed. James Spedding, Robert Leslie Ellis, and Douglas Dennon Heath. 14 vols. Facsimile reprint, Stuttgart-Bad Constatt: Friedrich Frommann Verlag Günther Holzboog. Vol. 3.

Benstein, Jeremy. 2001. "'One, Walking and Studying ... ': Nature vs. Torah." In *Judaism and Environmental Ethics: A Reader*, ed. Martin D. Yaffe. Lanham, MD: Lexington Books.

Benstein, Jeremy. 2006. *The Way into Judaism and the Environment.* Woodstock, VT: Jewish Lights Publishing.

Bernstein, Ellen, ed. 1997. *Ecology and the Jewish Spirit: Where Nature and the Sacred Meet.* Woodstock, VT: Jewish Lights Publishing.

Bernstein, Ellen. 2005. *The Splendor of Creation.* Cleveland, OH: Pilgrim Press.

Bleich, J. David. 2001. "Judaism and Animal Experimentation: Vegetarianism and Judaism." In *Judaism and Environmental Ethics: A Reader,* ed. Martin D. Yaffe. Lanham, MD: Lexington Books.

Cohen, Jeremy. 2001. "On Classical Judaism and Environmental Crisis." In *Judaism and Environmental Ethics: A Reader,* ed. Martin D. Yaffe. Lanham, MD: Lexington Books.

Ehrenfeld, David, and Philip J. Bentley. 2001 "Judaism and the Practice of Stewardship." In *Judaism and Environmental*

Ethics: A Reader, ed. Martin D. Yaffe. Lanham, MD: Lexington Books.

Eisenberg, Evan. 1998. *The Ecology of Eden.* New York: Knopf.

Jonas, Hans. 2001. "Contemporary Problems in Ethics from a Jewish Perspective." In *Judaism and Environmental Ethics: A Reader,* ed. Martin D. Yaffe. Lanham, MD: Lexington Books.

Kass, Leon R. 2001. "Sanctified Eating." In *Judaism and Environmental Ethics: A Reader,* ed. Martin D. Yaffe. Lanham, MD: Lexington Books.

Kay, Jeanne. 2001. "Concepts of Nature in the Hebrew Bible." In *Judaism and Environmental Ethics: A Reader,* ed. Martin D. Yaffe. Lanham, MD: Lexington Books.

Kennington, Richard. 2004. *On Modern Origins: Essays in Early Modern Philosophy,* ed. Pamela Kraus and Frank Hunt. Lanham, MD: Lexington Books.

Leopold, Aldo. 1949. *A Sand County Almanac, Sketches Here and There.* New York: Oxford University Press.

Levy, Ze'ev. 2001. "Ethical Issues of Animal Welfare in Jewish Thought." In *Judaism and Environmental Ethics: A Reader,* ed. Martin D. Yaffe.Lanham, MD: Lexington Books.

Maimonides, Moses. 1963. *The Guide of the Perplexed,* trans. Shlomo Pines. Chicago: University of Chicago Press.

Rosenblum, Eric. 2001. "Is Gaia Jewish? Finding a Framework for Radical Ecology in Traditional Judaism." In *Judaism and Environmental Ethics: A Reader,* ed. Martin D. Yaffe. Lanham, MD: Lexington Books.

Sacks, Robert D. 1990. *Commentary on the Book of Genesis.* Lewiston, NY: Edwin Mellen Press.

Sacks, Robert D. 2001. "Commentary on the Book of Genesis, Chapter 1." In *Judaism and Environmental Ethics: A Reader,* ed. Martin D. Yaffe. Lanham, MD: Lexington Books.

Schaffer, Arthur. 2001. "The Agricultural and Ecological Symbolism of the Four Species of *Sukkot.*" In *Judaism and Environmental Ethics: A Reader,* ed. Martin D. Yaffe. Lanham, MD: Lexington Books.

Schwartz, Eilon. 2001. "*Bal Tashchit:* A Jewish Environmental Precept." In *Judaism and Environmental Ethics: A Reader,* ed. Martin D. Yaffe. Lanham, MD: Lexington Books.

Schwarzschild, Steven S. 2001. "The Unnatural Jew." In *Judaism and Environmental Ethics: A Reader,* ed. Martin D. Yaffe. Lanham, MD: Lexington Books.

Spinoza, Benedict. 2004. *Theologico-Political Treatise,* trans. Martin D. Yaffe. Newburyport, MA: Focus Publications

Tirosh-Samuelson, Hava, ed. 2002. *Judaism and Ecology: Created World and Revealed Word.* Cambridge, MA: Harvard University Press.

Toynbee, Arnold. 1972. "The Religious Background of the Present Environmental Crisis." *International Journal of Environmental Studies* 3: 141–146.

Troster, Lawrence. 2001. "Created in the Image of God: Humanity and Divinity in an Age of Environmentalism." In *Judaism and Environmental Ethics: A Reader,* ed. Martin D. Yaffe. Lanham, MD: Lexington Books.

Waskow, Arthur, ed. 2000. *Torah of the Earth: Exploring 4,000 Years of Ecology in Jewish Thought.* 2 vols. Woodstock, VT: Jewish Lights.

White, Lynn, Jr. 1967. "The Historical Roots of Our Ecologic Crisis." *Science* 155: 1203–1207.

Wyschogrod, Michael. 2001. "Judaism and the Sanctification of Nature." In *Judaism and Environmental Ethics: A Reader,* ed. Martin D. Yaffe. Lanham, MD: Lexington Books.

Yaffe, Martin D., ed. 2001. *Judaism and Environmental Ethics: A Reader.* Lanham, MD: Lexington Books.

Martin D. Yaffe

K

KOREA, NORTH AND SOUTH

The history of post–World War II Korea is a tale of two countries—or, more accurately, one country split in two. The two parts of this divided nation share a name (informally South Korea and North Korea, formally the Republic of Korea and the Democratic People's Republic of Korea, respectively), a landmass, a language, and a culture, but are bitterly divided by ideology, economic circumstances, and geopolitical pressures.

A Japanese colony until the end of World War II, Korea was divided into two administrative zones at the war's end: the Soviet Union occupied the territory north of the 38th parallel, and the United States was in charge of the area to the south. With the collapse of postwar reunification talks, the northern zone declared itself the Korean Democratic People's Republic on May 1, 1948. On June 25, 1950, North Korean troops invaded the southern zone with Soviet and Chinese backing, seeking to unify the country under a communist regime. United Nations troops, under U.S. leadership, fought off the invaders, and the war ended in a standoff in 1953, with the country indefinitely divided.

Since then the capitalist economy of South Korea has registered spectacular rates of growth, skyrocketing from a per capita gross domestic product (GDP) of $100 in 1963 to $10,000 in 1995; its economy now ranks tenth in the world as measured by nominal GDP. North Korea, meanwhile, has stagnated under its planned, centralized communist economy; it is one of the poorest countries in the world, ranking 156th in total GDP with a per capita GDP of roughly $1,000. Beginning in the 1990s, the country's ideologically rigid leadership began to make grudging initiatives in the direction of limited free-market experiments, especially in its tentative economic exchanges with South Korea.

SOUTH KOREAN ENVIRONMENTALISM

South Korea's transformation from a traditional agricultural society into a modern industrial economic powerhouse began in the 1960s. Until the 1970s and mid-1980s, pollution was the main environmental issue, and antipollution movements took the form of victim demands for damage compensation or farmers' and fishermen's protests concerning the siting of industrial facilities. Civil protests against industrial and urban pollution were not taken seriously, however, because of the government's headlong pursuit of economic development.

The Korean Pollution Research Institute (PRI, *Gonghae-Yonguhoe*), established in 1982 by progressive Christian clergymen, made determined efforts to support victims of pollution, to research various environmental problems, and to educate the public about such issues. The majority of the group's supporters were intellectuals and antigovernment activists under the leadership of Yeol Choi. The PRI allied itself with the nationalist-democratic movement, a merger that has led many analysts to view the PRI's environmentalism as part of the left. Yeol Choi later became copresident of the Korean Antipollution Movement Association (KAPMA, *Kongchuryon*), founded in 1988. Its ideology remained broadly leftist, demanding environmental justice combined with criticism of the monopolizing capitalist power.

But South Korean environmentalism was not monopolized by leftist activism. Quite apart from KAPMA and other environmental movements, the national forestation policy, a public campaign for tree planting, and the effort to preserve greenbelts in urban and suburban areas were initiated by the government as early as the 1950s and carried out from then through the time of President Park Chung-hee's regime (1961–1979) effectively enough to secure a comparatively solid ecological foundation for agricultural productivity and sound rural landscape at least in the southern part of the peninsula.

Notable ideological changes of the environmental movement began to occur in South Korea in the late 1980s, namely from leftist activism to a moderate professionalism, and from the radical antipollution movement to a realistic ecological conservationism. While KAPMA regarded huge industrial corporations and political rulers in South Korea as the main violators of ecological balance, Citizens' Coalition for Economic Justice (CCEJ, *Kyungshilyon*), YMCA, and YWCA, for example, criticized the radical strategy of the former antipollution activists.

The South Korean government began, though late, to keep pace with increasing public consciousness that the natural environment must not be sacrificed in pursuit of economic development and that economic value should be harmonized with environmental value through the legal and political systems. As the 1990s began, increasing pollution and environmental degradation were brought under control with higher efficiency on the basis of advanced scientific research and green technologies. The government's earlier repressive attitude toward antipollution activists changed to that of a more responsive as well as responsible and reasonable policy maker. The Ministry of Environment was inaugurated in January 1994, replacing the former National Environment Agency. New laws were enacted in the early 1990s to complement the existing legal measures for environmental policy, including the Environmental Conservation Law, the Fundamental Law for Environmental Policy, the Air Pollution Control Act, the Water Pollution Prevention Act, the Conciliation Law for Environmental Pollution Conflict, and others.

But a variety of hot-button environmental issues continued to arise in South Korea during the 1990s and after 2000. They took in general the form of civil protest against the government's development plans. But the real controversy arose often out of the conflicts between civil interest groups. In some cases, they became nationwide struggles, involving public opinion as well as the political and legal systems, to preserve ecologically valuable resources or endangered rare species of flora and fauna.

Protesters who tried to preserve wide mud flat regions along the west coast of the Korea Peninsula couldn't stop the Shihwa Lake project in Kyunggi Province and the Saemangeum Mud Flat project in Cholla Province. But central and local governments had to cancel the Dong River Dam project in Kangwon Province and the Buan Radioactive Waste Depot project in Cholla Province because of tenacious protests and legal battles to defend the natural landscape and ecological living conditions of the human and biotic inhabitants throughout a wide range of related areas. Protesters rallied against the Cheonseong-Keumjeong Mountain Tunnel project and express railway construction in Kyungsang Province to preserve the habitat of the Korean clawed salamander (*Onychodactylus fischeri*). Only following intervention by the South Korean president and the Supreme Court was tunnel construction able to continue after it waited three years for the final legal decision.

Other notable developments have fostered increased attention to environmental issues in South Korea: the sandy wind blowing every spring from Mongolia and China, the reduction of habitat areas for migratory birds, fish, and marine animals that inhabit or stop on and around the peninsula, and international projects for the protection and preservation of flora and fauna on the peninsula and in the Demilitarized Zone (DMZ) that separates the two Koreas. The United Nations Conference on Environment and Development (UNCED) in 1992 also motivated the Korean people to pay more attention to global environmental issues than they had before. A major contributor to the growing public awareness of environmental problems has been the bimonthly magazine *Green Criticism* (*Noksaek Pyongron*), founded in October 1991 by Jongchul Kim, a poet and professor of English literature. This magazine provides the public with copious information and critical essays on environmentalism, Deep Ecology, critiques of modern civilization, and thoughts on alternative ways of living and their practicability.

The ethical significance of the issues that arose in the 1990s stimulated academic research and philosophical discussions about the most fundamental themes of environmental ethics and the ecological sustainability of human civilization. The Korean Society for the Study of Environmental Philosophy was founded in 1995 and has published its journal, *Environmental Philosophy (Hwankyung Cheolhak)*, since 2002. Scholars and theologians have awakened anew to the ecological wisdom of the religious traditions of Asia (especially, Buddhism, Daoism, and worship of spirits) and in native Korean ideas of nature (e.g., Ch'i-ecology, worship of mountains, and cosmology of heaven-earth-human).

Noteworthy private institutions for the promotion of green culture include the Canaan Farming School (founded by Young-Gi Kim, presbyter, 1962), the Toji (Earth)

Foundation of Culture (founded by Kyung-ree Park, novelist, 1999), the Korea Green Foundation (founded by Yeol Choi, agriculturist, 2002), Green University (founded by Hoeick Chang, professor emeritus, physicist, 2003), and the World Life-Culture Forum (initiated by Jiha Kim, poet, 2003).

GROWING CONSCIOUSNESS IN NORTH KOREA

Since its founding, North Korea has been governed by one-ruler dictatorship, first Kim Il-sung till 1994, then his son Kim Jong-il, with the support of one political party system. It may be assumed that all of North Korean environmental policy has been directed or supported by the same centralized governing structure. The dominant environmental issue in the years before and after the Korean War was forestation policy. It aimed to restore the forest resources that were exploited during the Japanese colonization (1910–1945), and was carried out so effectively that timber production in the 1970s quadrupled that of the late 1940s. But North Korea has had to cope with other environmental issues since the 1950s, in particular, industrial pollution that began in the late 1950s; and the adverse effects of the agricultural development policies introduced in 1976 by Kim Il-sung.

North Korea's industrial development, which had begun earlier than in South Korea, was effective enough that its industry made up 74 percent of the whole national economy in the early 1970s, while it made up only 34 percent in 1956. This process of development must have raised problems of industry pollution and public health. Kim Il-sung himself emphasized continuously in his official speeches beginning in 1972 the seriousness of water, air, and land pollution caused by toxic materials that were emitted from factories and in the urban areas. Material evidence suggests that he didn't stop to promote administrative and legal measures against every kind of pollution. The Land Law, enacted in 1977, contains regulations for environmental protection both in industrial sites and in urban and rural areas. The Pollution Research Institute (later renamed Research Institute for Environmental Protection) was founded also in 1977. Three years later the People's Health Protection Law was enacted following the will of the Supreme People's Assembly to mobilize all the possible antipollution measures.

The Environmental Protection Law, enacted in 1986, introduced environmental protectionism as North Korea's state ideology together with socialism and communism.

An amendment to North Korea's National Constitution in 1992 gave priority to environmental protection over the growth of economic production. Many other new laws and enforcement regulations for environmental protection were continuously enacted almost every year until 1998. But the environmental situation in North Korea worsened during the 1990s as repeated floods and droughts brought disastrous famine, destruction of water supply facilities, a drop in agricultural productivity, and the spread of waterborne epidemics.

Analysts believe that ecological collapse must be, if not the primary, at least one of the main reasons for North Korea's agricultural collapse, economic stagnation, and failure to increase its energy supply in the 1990s, and that the primary reason for the ecological collapse was a false agricultural policy. Kim Il-sung, whose political goal was to realize a self-sufficient national economy, had introduced his Terraced Farm Policy (*Darakbat Campaign*) in 1976 to raise agricultural productivity. But this policy was executed so excessively that after ten years every hill and grassland in the provinces had been converted to plow lands, while the transformation of mountain tops and the devastation of forests had caused other forms of geographic and climate calamities, including earth erosion, aridity, a decrease of fertility, drying up of streams and groundwater, and diminished ecological habitats for flora and fauna species.

Fortunately, North Koreans have learned from their experiences. Since 1998 they have introduced new plans for securing water resources and new policies for environmental protection. A new long-term plan for forestation was started in 2000. The Floodgate Law, the National Land Planning Law, and the River Act were enacted for environmental protection, and systematic reservation plans for the preservation of biotic species were introduced in 2003. North Koreans are enlarging various cooperative measures and possibilities of exchange plans with South Korea not only for industry, agriculture, and energy supply but also in agriculture, forestation, and environmental conservation.

BIBLIOGRAPHY

Asia Environment Report, 1997–1998 1. 2000. Seoul: Ddanim.

Chang, Jaeyeon. 2005. *Environment in North Korea* (Bukhaneui Whankyung). Seoul: Ajou Institute of Korean Unification and Health Care.

Ku, Do-Wan. 1996. "The Structural Change of the Korean Environmental Movement." *Korea Journal of Population and Development* 25(1): 155–180.

Lee, Minbu, et al. 2006. *Analysis on the Environmental Change and Natural Hazard in North Korea* (Bukhaneui Whankyung-byunwha-wa Jayun-jaehae). Seoul: Hanwool Academy.

Meadows, Donella H., Jorgen Randers, Dennis L. Meadows, and William W. Behrens. 1972. *The Limits to Growth: A Report for the Club of Rome's Project on the Predicament of Mankind.* New York: Universe Books.

Sung-Jin Kim

L

LAND ETHIC

Aldo Leopold's land ethic is perhaps the most prominent American articulation of an environmental ethic. Leopold's "The Land Ethic" was published posthumously as the capstone essay in *A Sand County Almanac, and Sketches Here and There* (1949). In the view of most readers the substance of his ethic was stated when he asserted that an interaction with nature "is right when it tends to preserve the integrity, stability, and beauty of the biotic community" and "is wrong when it tends otherwise" (Leopold 1949, pp. 224–225). Leopold's ethic has drawn wide attention and conflicting responses, particularly after critical study of it gained momentum in the 1980s. According to one exponent, J. Baird Callicott, Leopold's ethic has been the "most popular among professional conservationists and least popular among professional philosophers" (Callicott 1999, p. 59). Whether lauded or challenged, Leopold's land ethic has cast a long shadow; it would not be an exaggeration to claim that it is the central pillar of contemporary environmental philosophy.

Leopold's land ethic rests on an understanding that humans exist within an integrated community of life that also includes other animals, plants, rocks, soils, and waters. Particularly in his later years Leopold referred to this assemblage as the "biotic community" or, more simply, "the land." Humans are "plain members and citizens" of this biotic community, Leopold contended (Leopold 1949, p. 204), and therefore have a moral obligation to act consistently with the long-term welfare of that community. That welfare is linked to the ways a community functions ecologically; to its capacity, under human use, to remain fertile and productive over the long term.

Leopold devoted considerable effort to understanding how the biotic community functions and needs to function if it is to retain its productive capacity. Late in life he synthesized his conclusions into a normative concept of ecological functioning that he termed "land health." Leopold viewed land health as an appropriate and much-needed goal for all conservation efforts. Just before his death he encouraged colleagues in the conservation movement to embrace it as their goal. With his land ethic Leopold transformed land health from a communal goal into an ethical norm to guide individual behavior. As Leopold explained, the land ethic "reflects a conviction of individual responsibility for the health of the land" (Leopold 1949, p. 221). He restated that point a few paragraphs later by naming the elements of land health: Human actions are morally right when they uphold the "integrity, stability, and beauty" of the biotic community (pp. 224–225).

Leopold's ethic has given respectability to ethical stances that extend moral value beyond human communities (tribe, nation, global village) to cover collections and assemblies of living things such as species and ecosystems. It also has encouraged ethicists to take science seriously as they frame their ethical norms. Leopold summarized the complexity of nature in terms of the ways landscapes function ecologically rather than, for instance, in terms of their physical or biological composition. By making normative use of the modes of functioning of nature, Leopold invited others to embrace ethical stances that respect nature in functional terms. Leopold's ethic emerged out of a lifelong effort to motivate people, particularly private landowners, to live on land in ways that are sustainable. His land ethic, he hoped, would yield practical

Aldo Leopold, Seated Near a Shack, circa 1940. *"The Land Ethic" is possibly Aldo Leopold's most famous writing, and is one of the most prominent works of U.S. environmental ethical literature. Leopold's background as a wildlife scientist and manager lent itself to the development of his environmental ethic.* COURTESY OF THE ALDO LEOPOLD FOUNDATION ARCHIVES.

conservation benefits in terms of improved land-use practices. He implicitly encouraged later conservation advocates to integrate philosophical ideas with on-the-ground conservation labors and evaluate alternative perspectives on the basis of practical consequences.

During his lifetime Leopold (1887–1948) was best known as a wildlife scientist and manager. His professional knowledge extended to multiple-use forestry, grassland management, and the challenges of soil erosion. He disclaimed expertise in agricultural sciences even though farmers and farm organizations regularly sought his professional advice and his academic appointment at the University of Wisconsin was in the Department of Agricultural Economics.

Leopold's early writings on wilderness preservation drew considerable attention, as did his accumulated evidence that wildlife conservation was in many settings best promoted by improving wildlife habitats rather than by imposing tighter limits on hunting, creating additional game preserves, and employing artificial propagation. The frequently quoted essay "Thinking Like a Mountain" in *A Sand County Almanac* has prompted many readers to assume that Leopold was comparatively late in recognizing the ecological roles of predators. However, a broader study of his writings suggests that by the mid-1920s Leopold was aware of the functional benefits of predators. A working forester at the time, however, Leopold also recognized that livestock grazers had a practical need to control predators. Further, he understood that predator control could help private landowners enhance crops of wild game on their lands and that game cropping, even with predator control, could improve alternative land uses. Nonetheless, by the mid-1920s Leopold criticized excessive predator control and pushed for measures to protect predators.

CRUCIAL ISSUES IN THE LAND ETHIC

Critical secondary writing on Leopold's land ethic has tended to dwell chiefly on five overlapping issues:

1. What is the origin of the moral norm in Leopold's ethic?
2. How well does Leopold's concern for the land community as such fit with moral concerns for its human and nonhuman parts?
3. What is the substantive force or content of the land ethic; that is, what did Leopold mean by preserving the biotic community's "integrity, stability, and beauty"?
4. How did Leopold imagine that his ethic might gain traction over time, building on earlier extensions of communal norms?
5. Is Leopold's ethic now largely irrelevant or in need of material revision because of changes in the ways ecologists understand the functioning of nature?

Leopold understood ethics as a body of normative ideals that constrain individuals in daily life as they pursue their self-interest. Individuals are prone to do what is expedient for them personally, he asserted. Applicable ethical schemes require them to broaden their selfish concerns to take account of the welfare of other community members and the "community as such" (Leopold 1949, p. 204). Rarely did Leopold discuss the ethical implications of direct interactions between two or more individuals. Instead, he concentrated on what he saw as the clash between the welfare (expediency) of the individual and the welfare of the surrounding community. He discussed this clash in many other writings, particularly those dealing with conservation economics. In light of this framing of the conservation predicament, the size, composition, and functioning of the relevant moral community were of vital importance for Leopold.

When discussing ethics, Leopold showed special interest in how relevant moral communities expand in size and composition over time, beginning with early peoples. He perceived a slow, long-term expansion in such communities. In several writings he wondered whether this long-term evolution could and would continue, expanding the moral community to include the land as an integrated whole.

Also influential in Leopold's ethical thinking was his conviction that humans are limited beings. As he expressed it, there is much that humans do not know and probably never will know. Humans also have limits on their ability to reason and the expressive capacities of their languages. Even trained scientists, Leopold asserted, know little about the functioning of the land, a reality that exacerbates the challenge of defining land health. Because of these limits, Leopold concluded, it makes sense for people to rely on sentiments and intuitions as well as known facts and reason. It makes sense for them to act humbly and draw lessons from the behaviors of species that have thrived far longer than has *Homo sapiens.*

Leopold viewed his ethic as more than a rule of conduct governing behavior. It was a proposal for a wide-ranging shift in the ways people conceive of and interact with nature, from being separate and apart to being full-fledged community members. His land ethic, Leopold explained, requires "an internal change in our intellectual emphasis, loyalties, affections, and convictions" (Leopold 1949, p. 210). It requires changes in what people want, what they value, and what they deem beautiful.

ORIGIN OF LEOPOLD'S ETHIC

Philosophers studying Leopold's ethic have paid close attention to its moral grounding and to the way it compares with leading approaches to interpersonal ethics. Did Leopold's emphatic concern for the welfare of the biotic community arise out of a belief in the intrinsic moral value of that community apart from the instrumental values that attach to its parts? Is his land ethic, that is, both non-anthropocentric and holistic? Alternatively, did Leopold focus his ethic on the functioning of the community not because the community itself is morally considerable but because the lasting health of the community is instrumentally essential to long-term human welfare? Also, is Leopold's ethic best understood, as it is formally expressed, as a deontological claim, admonishing humans to fulfill their moral duties, or is it better interpreted chiefly as the distillation of pragmatic thinking and experimentation aimed at long-term conservation?

Callicott has asserted that Leopold's ethic is best understood as a claim that the land community as such has intrinsic value and that humans are duty-bound to respect it by fostering its ecological functioning. Thus, Callicott sees Leopold's land ethic as nonanthropocentric and both holistic, extending moral value beyond humans to the land community, and deontological. In supporting his interpretation Callicott has emphasized that Leopold distinguished between conduct that is expedient and conduct that is ethical.

A contrary interpretation was staked out by the philosopher Bryan Norton, also beginning in the 1980s. Norton has viewed Leopold's ethic chiefly as an outgrowth of pragmatic efforts to promote land conservation. Leopold's aim, Norton contends, was to craft an ethic that would prompt people to live in ways that foster long-term human welfare. Leopold's ethic thus remains within the standard anthropocentric ethical paradigm and is best analogized not to Kantian or other deontological perspectives but to the pragmatic moral reasoning of William James, Charles Sanders Peirce, and John Dewey. In Norton's interpretation the land community for Leopold does not possess moral value as such; his ethic thus is neither deontological nor holistic. Instead, the community is simply the soundest practical focus for conservation efforts designed to promote long-term human flourishing. Norton interprets Leopold's distinctions between expediency and ethics as referring only to conflicts about satisfying short-term human preferences at the expense of long-term human well-being. In the long term and taking all humankind into account, expediency and ethics come together.

Both Callicott and Norton have endeavored to promote nature conservation. As they did so, their conflict over interpretations of Leopold quickly turned into a broader dispute about the best way to frame a contemporary environmental ethic. As Leopold scholars, however, they seemed largely to agree on central elements of Leopold's thought. Both read Leopold to include future generations within the relevant moral community. Both conclude that Leopold equated long-term human welfare with the long-term welfare of the biotic community as a whole, thus eliminating conflict between humans and nature. Both believe that Leopold was influenced strongly by Darwinian theories of ethical evolution and, behind them, theories of moral sentiments and virtue-based reasoning.

Few scholars have challenged Leopold's readiness to extend moral value to future generations, a problematic step for many philosophers. In addition, little has been said about his unusual perspective on human welfare. Leopold described human welfare holistically, as a unified totality, not in terms of the summed welfare of individual humans. In addition, Leopold spoke of human welfare after humans embraced the kind of ecological awakening that he deemed essential. Such humans would forgo superfluous wants and would be satisfied by the aesthetic and spiritual benefits of living in a whole and healthy biotic community.

MORAL VALUE OF THE PARTS OF THE COMMUNITY

In a controversial 1980 essay, "Animal Liberation: A Triangular Affair" (Callicott 1989), Callicott asserted that Leopold's concern for community welfare trumped the moral duties that humans owe to other community members as individuals, duties that Leopold also recognized. His essay drew strong responses, leading to an important clarification of Leopold's ethical vision.

For several decades academic philosophers ignored Leopold's ethic chiefly on the ground that moral considerability could never attach, as Leopold seemed to propose, to an intangible aggregate such as the land community. Even ethicists who could take that step found Callicott's interpretation of Leopold troubling. To elevate the whole over the parts was to make the parts morally subordinate. By implication, it seemed morally permissible, if not obligatory, to reduce species populations, humans included, when lower numbers are needed to uphold the healthy functioning of the community. This was the charge of ecofascism: In the name of protecting the land community, humans could and indeed should be killed.

Callicott responded with a refined interpretation of Leopold's land ethic. That ethic, Callicott contended in *In Defense of the Land Ethic,* was not intended to displace existing standards of interpersonal ethics. Instead, it was to be added to them. Leopold's holistic stance, in fact, did justify lethal methods to reduce populations of nonhuman species but certainly not the killing of people. With this reasoning Callicott undercut the charge that Leopold's holism is a form of fascism. Yet as he recognized, he left Leopold's ethic unable to provide clear guidance when the claims of morally worthy individuals call for actions that clash with moral respect for the land community. Thus, to implement Leopold's ethic it became necessary to generate a second-order rule that would integrate the land ethic with existing interpersonal standards.

In several of his writings Leopold commented on the "biotic rights" of parts of nature to continued existence. He applied the term, however, only to species and rare natural communities, not to individual organisms. In the practical work of conservation, efforts to preserve such collective wholes are unlikely to collide with efforts to sustain land health. Accordingly, Leopold's respect for the biotic rights of species and rare communities seems not to conflict with the moral primacy of community health.

SUBSTANCE OF THE LAND ETHIC

Nearly all discussions of the land ethic quote Leopold's famous summation of it. Few commentators, though, have attempted to figure out what Leopold meant when speaking of the "integrity, stability, and beauty" of the land. The word *stability* in particular has been problematic. The term seems to endorse a view of healthy nature as static or unchanging that contemporary ecologists reject. The word *integrity* also has created difficulties. The term typically refers to the totality of species and perhaps also to the full range of biological assemblies that existed in a particular landscape before humans arrived or, more arbitrarily, before humans of European descent arrived. When integrity is defined in this way, people can preserve it only by leaving land untouched and undoing prior human changes, a practical impossibility in places where people live and work. Commentators rarely have taken Leopold's reference to beauty seriously. Beauty, it is assumed, is too vague and subjective a concept to provide objective guidance.

In his writings on Leopold, Callicott examined late twentieth-century writings in ecology to figure out which of several contemporary meanings of stability (and, to a lesser extent, integrity) make the most sense when applied to Leopold's ethic. In a more recent study, *Aldo Leopold's Odyssey* (2006), Julianne Newton (now Warren) attempted to identify with precision the meanings Leopold personally attached to those key words rather than beginning with definitions composed by others. Newton's conclusion is that the substance of Leopold's land ethic cannot be understood without grasping what Leopold meant by land health. His key words, she concludes, were intended to incorporate into his ethic the full body of his writings about land sickness and health. Leopold's writing on land health in turn built on his lengthy effort to construct, from studies of differing landscapes, a "common concept" of how land in general works.

The key attribute of healthy land for Leopold, according to Newton, was apparently stability, by which Leopold meant not a community with an unchanging biological membership but a biotic system that possesses and retains an ability to cycle nutrients over and over at high levels of efficiency without significant loss. "Land is stable," Leopold explained in "Biotic Land-Use" (circa 1942), "when its food chains are so organized as to be able to circulate the same food an indefinite number of times" (Leopold 1999, p. 205). Similarly, integrity for Leopold apparently referred to the suite of species required in a landscape for that landscape to retain its stability. Leopold preferred that positions in nutrient chains be filled by native species but defined health in a way that allows exotics to substitute for native species and permits significant simplifications in community membership as long as nutrient flows remain lengthy and efficient. Beauty for Leopold apparently was an attribute of lands that retained their health, particularly lands that displayed their native integrity.

In this interpretation—that Leopold's key words linked his land ethic to his ideal of land health—insights on Leopold's ethic and its practical implications are best obtained by studying the symptoms of land sickness as he explained them, Leopold's more extended comments on land health, and the illustrations he used in "The Land Ethic" and elsewhere of lands that retain their health despite long human use. Also useful for this purpose are

Leopold's observations that natural communities over evolutionary time tend to increase in biological composition and in their ability to cycle nutrients efficiently without loss. The best land uses, Leopold suggested, are ones that do not impede these apparent evolutionary trends.

EVOLUTION OF A LAND ETHIC

Writing on Leopold's ethic also has paid attention to Leopold's many comments on how ethics emerge over time within moral communities. Most visibly in "The Land Ethic," Leopold drew on the writings of Charles Darwin describing the evolutionary forces that seemingly push communities, over time, to expand the reach of their ethical ideals. Leopold defined an ethic as "the tendency of interdependent individuals or groups to evolve modes of co-operation" (Leopold 1949, p. 202). With little conscious thought, he believed, members of a moral community gradually expand the size and composition of their community, pushed by invisible evolutionary forces. The key to the continued expansion of ethics, then, is to broaden the moral community beyond humans to include the parts of nature with which humans are interdependent. Such an expansion, Leopold believed, is an evolutionary possibility as well as an ecological necessity.

Drawing on the conclusions of anthropologists and historians, several commentators have cast serious doubt on Leopold's evolutionary theories and the conclusions he drew from them. Early peoples often included nonhuman life within their moral calculations. The expansion of ethics to include more people—different tribes, races, ethnic and religious groups, and the like—apparently was accompanied by a contraction in the moral value attributed to wild species. Leopold's errors as anthropologist, however, have not prompted commentators to call his ethic into question given that its moral force rests on different grounds. History aside, Leopold's speculations are vulnerable to claims that evolutionary processes at work in the past have been weakened or displaced by social welfare programs. People who degrade their home landscape may have the ability to remain in place and thrive—to survive in evolutionary terms—as a result of benefits supplied by distant governments; they are not forced to leave or to change their inappropriate land uses.

IS LEOPOLD'S ETHIC OUT OF DATE?

Callicott has been the most prominent commentator who has questioned the validity of Leopold's ethic because of changes in ecological thought since his day. Callicott's concerns are shared by many others, who view Leopold's ethic as a throwback to an earlier, discredited era of ecological understanding.

In his criticism Callicott assumed that Leopold shared the ecological ideas that dominated his day. Like his contemporaries, Leopold viewed natural systems as organized into coherent, persistent communities that retain their composition and functioning until disturbed by external, often anthropogenic, forces. Callicott has undertaken to revise Leopold's ethic to make it more dynamic: that is, to incorporate into it overt reference to the ongoing changes in nature and to the ability of people, without violating the ethics, to alter land at relatively slow spatial and temporal scales.

In her detailed study of Leopold's science and scientific influences Newton implicitly challenged the view that Leopold's science is out of date. She presents Leopold's scientific understandings as peculiar to him, not the same as those of his contemporaries. In her view Leopold and his contemporaries were far more aware of the dynamism of nature than standard historical accounts suggest. Further, Leopold's ethic did not focus chiefly on the biological composition of a community: the part of nature that ecologists see as most dynamic. Instead it focused on the ecological functioning of the community, particularly its ability to cycle nutrients efficiently, an aspect of community functioning that is less prone to change. Thus, contemporary emphases on dynamic changes in community composition may not pose the problems for the land ethic that Callicott discusses.

SEE ALSO *Adaptive Management; Callicott, J. Baird; Leopold, Aldo; Norton, Bryan.*

BIBLIOGRAPHY

Callicott, J. Baird. 1987. *Companion to A Sand County Almanac: Interpretive & Critical Essays.* Madison: University of Wisconsin Press.

Callicott, J. Baird. 1989. *In Defense of the Land Ethic: Essays in Environmental Philosophy.* 1989. Albany: State University of New York Press.

Callicott, J. Baird. 1999. *Beyond the Land Ethic: More Essays in Environmental Philosophy.* Albany: State University of New York Press.

Knight, Richard L., and Suzanne Riedel, eds. 2002. *Aldo Leopold and the Ecological Conscience.* New York: Oxford University Press.

Leopold, Aldo. 1949. *A Sand County Almanac, and Sketches Here and There.* New York: Oxford University Press.

Leopold, Aldo. 1991. *The River of the Mother of God and Other Essays by Aldo Leopold,* ed. Susan L. Flader and J. Baird Callicott. Madison: University of Wisconsin Press.

Leopold, Aldo. 1999. *For the Health of the Land: Previously Unpublished Essays and Other Writings,* ed. J. Baird Callicott and Eric T. Freyfogle. Washington, DC: Island Press for Shearwater Books.

Meine, Curt. 1988. *Aldo Leopold: His Life and Work.* Madison: University of Wisconsin Press.

Meine, Curt. 2004. *Correction Lines: Essays on Land, Leopold, and Conservation.* Washington, DC: Island Press.

Newton, Julianne Lutz. 2006. *Aldo Leopold's Odyssey.* Washington, DC: Island Press/Shearwater Books.

Norton, Bryan G. 2003. *Searching for Sustainability: Interdisciplinary Essays in the Philosophy of Conservation Biology.* Cambridge, UK, and New York: Cambridge University Press.

Eric T. Freyfogle

LANDSCAPE ARCHITECTURE, DESIGN, AND PRESERVATION

Landscape architecture is a diverse but holistic discipline that has emerged since the 1850s as a leading profession in advocating sustainable land-use practices. Landscape architects focus their work on the intersection of human need, environmental sustainability, and aesthetics. Human need is broadly conceived to include practical matters of function, economic need, safety, mental and physical health, and less easily quantified needs such as comfort, engagement with social groups, and self-worth. Care for the natural environment considers how to integrate these human needs into natural systems so as to reduce environmental damage and allow systems to operate more closely to their natural pattern. All of this is done within the framework of creating pleasing landscapes, since those that are appreciated are more often cared for and protected.

As practitioners of a diverse discipline, landscape architects work at different scales and types. Projects can be as small as a walled courtyard or as large as a regional plan. The purposes of these projects may be as diverse as recreation, resource conservation, education, and housing. The field includes such subfields as planting design, natural-resource planning, site planning, heritage-landscape preservation, garden design, urban design, land-use planning, disturbed-land reclamation, wildlife planning, and environmental art. All of this work, no matter its size or scope, is linked by the underlying philosophies of stewardship for land, care for the natural systems that form it, and concern for the people who inhabit it.

EARLY EXAMPLES

These attitudes have evolved over a long period of time because landscape architecture, as an activity rather than as a specialized profession, has been part of human experience since prehistoric times. In the distant past, people learned about natural systems from intimate day-to-day contact. The simple technologies of prehistoric times limited people's ability to modify their environments. Gradually they began to make substantial changes to landscapes. They dammed rivers, channeled water to places it did not go naturally, leveled hills, and planted trees in areas distant from where they naturally occurred. Such activities created environmental problems. In ancient Greece (ca. 1200–323 BCE) and the Cambodian kingdom of Angkor (ninth to fifteenth centuries), for example, the harvesting of wood for agricultural land clearing, construction, and fuel led to extensive deforestation, which intensified local soil erosion.

With the emergence of the Renaissance in Europe, advances in science led to new knowledge about natural systems, as well as to changes in attitudes toward the environment. Medieval fear and awe of nature was replaced by an ever increasing sense that humans could, and should, alter it. Leonardo da Vinci, often cited as the epitome of the Renaissance mind, engineered a large-scale scheme to drain natural wetlands of the Chiana Valley in central Italy. He also advocated rerouting the Arno River to improve the geophysical setting for Florence. In the Renaissance such dreams at times outpaced technology—the Arno was not rerouted, but the marshy Chiana Valley was drained, and thereby transformed into one of the most productive agricultural regions of Tuscany.

Much post-Renaissance environmental thinking was framed within the philosophical perspective referred to today as the Enlightenment. This system emphasized rational thought and science as the basis for human decision making, as well as the importance of individual creativity in solving human problems. The Enlightenment, which influenced Europeanized parts of the world into the twentieth century, served as a critical philosophical system during formative periods of both the profession of landscape architecture and the environmental movement. Both fields came to appreciate natural complexity through Enlightenment philosophy. In the nineteenth century another philosophy, Romanticism, also began to influence human perceptions of nature. Romanticism emphasized emotional, sensory experience of the environment. Taken together, the two philosophies were behind many of the developments in landscape architecture and environmentalism that occurred in the eighteenth and nineteenth centuries.

EIGHTEENTH CENTURY

Developments in the eighteenth century illustrate the influences of both these philosophies. Rationality in economic thought led to an increasing focus on land as private capital and a resource for private, rather than communal, benefit. One resulting process in Great Britain was enclosure, a process that consolidated small feudal land holdings into monolithic estates and thus drove many tenant farmers from the land. Even more extreme were the large plantations of the American South, in which both

The Medieval Town of Cortona, in Tuscany, Italy. *Small towns such as Cortona evidence the delightful way that natural landscape and human-made architecture can complement one another. The town overlooks the Chiana Valley, which was drained during the Renaissance and is now one of the most agriculturally productive regions of Tuscany.* © GILLIAN PRICE/ALAMY.

land and people were exploited as the right of the landowning class. The Industrial Revolution, with its emphasis on science and engineering innovation, also reflected Enlightenment thought. In part as a reaction to industrialization, Romanticism countered with a view that idealized nature. A design approach variously known as the Landscape Gardening School, the Natural Style, the English Style, or the Picturesque evolved from Romantic philosophy and the arts that it influenced. These styles emphasized naturalistic appearance, but not natural processes or preservation of environmental resources.

NINETEENTH CENTURY

With the advent of the nineteenth century, the threads of thought and action began to emerge that eventually led to environmental problems or more recent decades, as well as approaches to their solutions. It is important to note that people were often fully aware of the environmental changes that their activities caused, but continued them out of economic need while experimenting with remedies. For example, owners of southern plantations were conscious, at least by the first decades of the 1800s, that their land clearing and tilling activities caused soil erosion. Planters often reported in agricultural journals both successful and unsuccessful experiments that they carried out to control erosion on their land. In other instances, the public expressed concern about potential environmental impacts from proposed projects. As a case in point, when construction of the Erie Canal was announced in the 1830s, a New York citizens' group raised the issue of possible pollution of Lake Erie with salt water from the Atlantic Ocean. In the middle of the century, thinking about the environment was influenced by Romanticism and other philosophies, especially Transcendentalism, as found in Ralph Waldo Emerson's *Nature* (1836).

Since the early part of the nineteenth century, observers, especially those in America, had noted the rapid pace of environmental change. Before 1820 the botanical explorer and horticulturist François André Michaux observed extensive deforestation in the American East. At around the same time, the painter George Catlin, surveying the drastic

Bridalveil Fall in Yosemite National Park. *Bridalveil Fall is an outstanding example of a waterfall issuing from a hanging valley far above the Yosemite Valley floor. The natural beauty of Yosemite served as a convincing argument for landscape preservation in the mid-1800s, when it became the first large-scale landscape set aside for public use.* © J.A. THOMAS/U.S. GEOLOGICAL SURVEY.

changes in the American West, especially the decimation of buffalo herds, made the farsighted proposal of establishing what he called a "nation's park" to preserve "Nature's works" (Catlin 1926 [1840], vol. 1, pp. 292–293). In fact, the first officially created park, though not a true national park, came less than ten years after he wrote, with the establishment of Hot Springs Reservation (now Hot Springs National Park) in Arkansas.

A different type of appreciation of the environment, focused on the negative effects of human activity, appeared in George Perkins Marsh's *Man and Nature, or Physical Geography as Modified by Human Action* (1864). Marsh's book gave those who had long expressed concern for the natural world, including many early landscape architects, ammunition to support their arguments.

Loss of natural features and systems was often framed in terms of beautiful scenery—a convincing argument in an age when Romantic thought dominated. Horace W. S. Cleveland and Frederick Law Olmsted Sr., both of whom emerged after the Civil War as leaders in landscape architecture, were early advocates of the preservation of beautiful and environmentally significant areas. In the 1850s Cleveland argued for preserving the Middlesex Fells landscape outside of Boston, and later for preserving Minnehaha Falls in Minnesota. In the 1860s Olmsted proposed placing Yosemite Valley under public control for its protection. Yosemite became the first large-scale landscape set aside for public use, but not by the federal government. Instead, it was the state of California, presaging its late-twentieth-century role in progressive environmental legislation, that made Yosemite Valley the first state park in the nation in 1864 (it later became a national park). Within a decade the federal government followed the lead of California. Based in part on the glowing descriptions of Ferdinand V. Hayden and accompanying painter Thomas Moran, over two million acres in the Yellowstone area became the first named national park. The awe inspired by western mountain landscapes led Canada to establish Banff as its first national park in 1885. Landscapes in the East were also deemed worthy of preservation. Niagara Falls became the first cross-border park—an effort that also involved Olmsted.

In the decades after the American Civil War (1861-1865), a number of national organizations formed to advocate active restoration and protection of environments or specific natural features. Among the first was the Arbor Day Foundation, founded by the scientific agriculturist J. Sterling Morton. Alarmed by deforestation in his home state of Nebraska and elsewhere, Morton promoted planting seedlings across the nation. The Audubon Society made protecting birds and their habitats its mission. Founded in 1886 by the appropriately named George Bird Grinnell (and incorporated as a national organization in 1905), the society first focused on reducing slaughter of plumage species. Later it employed conservation wardens, set up sanctuaries, and educated the public. John Muir and others formed the Sierra Club in 1892 to further preservation of California landscapes, especially Yosemite. The group quickly evolved into an advocate for more national parks. Its objectives, as well as differences in philosophies among founders of the group, became clarified when San Francisco revealed plans to dam and flood the scenic Hetch Hetchy Valley near Yosemite Valley. In this struggle, the old emphasis on scenic beauty worked against Hetch Hetchy, which would have been considered spectacular if it were not outshone by Yosemite Valley. O'Shaughnessy Dam was completed in 1923, and the valley was flooded, and remains so today. In the 1980s the U.S. Department of the Interior began considering removing the dam and restoring the valley—a cause that the Sierra Club continues to promote more than eighty years after it lost its first major environmental battle. The arguments that became part of the public debate over Hetch Hetchy, arguments between the value of natural systems in their own right and their value only as resources for human use, continue today.

As influential as these private groups have been, it was through federal action that the largest tracks of land have been preserved, although not for purely environmental purposes. National forests were first established in 1891. Termed reserves, the forests were created to manage consumption with planned harvesting and replanting, rather than to keep trees inviolate. Under Gifford Pinchot, the first chief of the U.S. Forest Service, the service and the forests themselves became institutions professionally run on the basis of current scientific principles. Pinchot advocated wise use for a sustained yield, but did not believe that complete protection, even of outstanding natural areas, was required. This attitude led to much disagreement with Muir and others in the preservation movement.

EARLY TWENTIETH CENTURY

Federal action intensified in the twentieth century. The National Park Service, set up in 1916 with Stephen Mather as its first director, managed existing and future parks. The fledgling service hired its first landscape architect, Charles Punchard, in 1919. In language composed by Frederick Law Olmsted Jr., son of the famous nineteenth-century landscape architect and an inheritor of his firm, legislation establishing the parks set out their rationale as being "to conserve the scenery and natural ... objects and the wildlife therein ... by such means as will leave them unimpaired for the enjoyment of future generations" (as quoted in Albright 1985, p. 36). Unfortunately, just what constituted leaving parks unimpaired meant different things to different groups. Hence, the varied stakeholders in park activities have rarely agreed on the proper direction of park management.

Many landscape architects of the early twentieth century were just as active in environmental causes and issues as had been their nineteenth-century predecessors. Their activities affected all areas of practice, but had special influence on planting design, urban planning, and regional planning. Use of native plants had been a part of landscape architecture since Ossian Simonds proposed use of prairie plants in his projects, such as Graceland Cemetery in Chicago. Others expanded on native plants in designs by suggesting natural groupings for disturbed lands. Frank Waugh, of the University of Massachusetts and later the U.S. Forest Service, advocated study of native plant arrangements and their use in roadside plantings—ideas that he articulated in his classic work *The Natural Style in Landscape Gardening* (1917). Others as well focused on use of native plants, including Beatrix Jones Farrand, who devoted her estate home at Reef Point in Bar Harbor, Maine, to the cultivation of New England natives. Jens Jensen not only valued plants native to the prairie states, but also proposed their use in similar settings in the natural landscapes of Wisconsin, Illinois, Iowa, and Indiana. His designs for parks of the West Side of Chicago typically included miniature versions of regional landscape features: wide, slowly moving streams, which he referred to as prairie lagoons; rockscapes that mimicked the limestone outcroppings found in glaciated regions of Illinois; and large open prairielike spaces fringed with tree species typically found at the fringes of real prairies.

Even in their practice of urban planning, landscape architects of the early twentieth century achieved impressive examples of environmental preservation. This was done largely in the planning of park systems, but also occurred in the design of individual subdivisions. For example, Frederick Law Olmsted Jr. protected the Rock Creek riparian area in Washington, D.C., as part of the park system proposed by the Senate Park Commission of 1900. Walter Burley Griffin designed the Rock Crest subdivision in Mason City, Iowa, to preserve the creek that ran through it by placing house sites on high ground surrounding it and making the stream corridor a wooded backyard for each house.

Aerial View of Greenbelt, Maryland, 1937. *As the economy of the United States and the world declined in the 1930s, many landscape architects took jobs in government programs to plan, design, and oversee work as diverse as park building, community design, water-control construction, and soil stabilization. The city of Greenbelt was a planned community, modeled after English garden cities of the 19th century.* THE LIBRARY OF CONGRESS.

Landscape architects can rightly be considered full participants in the environmental progressive movement of the early twentieth century. This is especially true of those who focused on regional and resource planning. Warren Manning, Clarence Stein, John Nolen, and Frederick Law Olmsted Jr. provided leadership in this effort. Manning's plan for Harrisburg, Pennsylvania, proposed that a minimum of 5 percent of the natural landscape within the city be set aside for protection, a high figure for the early twentieth century. Stein, a cofounder of the Regional Planning Association of America, worked to establish the Appalachian Trail as the first modern multistate footpath. Nolen, although today best known as an inspiration to new urbanists, believed that plans should conform to natural topography, and that land use should fit natural capacities, as he outlined in his 1908 plan for San Diego. Olmsted balanced his more common work in city planning with interest in national parks. Through his firm he worked on a number of environmentally sensitive sites, including Acadia National Park in Maine, Everglades National Park in Florida, and Redwoods National Park in California.

MID-TWENTIETH CENTURY

As much as had been achieved in the early 1900s, it took the twin disasters of the crash of 1929 and the Dust Bowl of the early 1930s to really bring environmental concerns and sustainable-landscape treatments to the forefront of public thought and professional practice. As the economy of the United States and the world declined in the 1930s, many landscape architects, unable to continue private employment, took jobs in government programs to plan, design, and oversee work as diverse as park building, community design, water-control construction, and soil stabilization. For example, landscape architects designed many of the state-park sites and facilities constructed by Civilian Conservation Corps workers. They were also active in the Green

Cities program of the Resettlement Administration, which planned and built model low-cost communities, like Greenbelt, Maryland.

The Dust Bowl, a massive wind-erosion event that crippled the Great Plains for several years, made the general public aware of the effects of environmental damage. A number of concerns voiced earlier, such as that concerning soil erosion, became the basis of new governmental programs. Hugh Hammond Bennett had attempted to point out the potentially drastic consequences of soil erosion, but it took the Dust Bowl to make his arguments graphic. As a result, the Soil Conservation Service, now the National Resources Conservation Service, formed during the 1930s to educate the public on proper soil-preservation strategies. Although staffed largely by soil scientists, the service employed some landscape architects in the planning of some soil-control demonstration sites, particularly near cities.

Environmental concerns took a backseat to military and political survival during the 1940s. World War II and its aftermath caused serious environmental damage in both Europe and Asia. After the war, resource degradation continued as people struggled merely to live. In the United States, which escaped direct war damage, environmental problems resulted from industrial production in munitions production, mining, and other support activities. In some cases, efforts that had started to support the war continued for peacetime use with little consideration about the air, soil, and water pollution that they caused.

As prosperity returned in the 1950s, people began to examine conditions in the environment and take initial steps to lessen damage. The Nature Conservancy was one of the first new organizations to form after the war as a result of these concerns. In 1955 it began the mission for which it is best known, land acquisition, with the purchase of sixty acres (now over seven hundred acres) for the Mianus River Gorge Preserve in Bedford, New York. Other organizations quickly emerged or expanded. This was also a period in which books extolled the value of the environment or warned about the damage being done to it. Marjory Stoneman Douglas championed preserving the Everglades in her classic volume *The Everglades: River of Grass* (1947). She went on to battle wetland destruction by developers and the U.S. Army Corps of Engineers, eventually founding Friends of the Everglades. Aldo Leopold, a noted wildlife expert, scientifically studied nature and emoted about it. He ultimately presented his observations in *A Sand County Almanac* (1949). Although published after death, he was able through its words to communicate what he referred to as a "land ethic," a social and political conscience intended to influence any human intervention in natural landscapes. In *Silent Spring* (1962), Rachel Carson chronicled the damage done through modern pollution, especially from pesticides, using affecting imagery to make her point about the causes of declining bird populations.

LATE TWENTIETH AND EARLY TWENTY-FIRST CENTURIES

Several pivotal events occurred in the 1960s. In mid-decade Congress passed the Wilderness Act, which permanently protected over 9 million acres of significant land from development. Among other pieces of environmental legislation, in 1966 the Endangered Species Preservation Act passed, and then two years later came the Wild and Scenic Rivers Act. But a milestone of U.S. and world environmental policy came in 1969 with passage of the National Environmental Policy Act. This law prescribed a formal process that considers the environmental consequences of any activity receiving federal funding, including public review. This process was codified in required environmental impact statements, many of which were written by landscape architects in consultation with scientists.

The landscape architects Philip Lewis and Ian McHarg brought scientific analysis of landscape capabilities to the discipline by developing the overlay method to study natural and human systems—an approach today done digitally through Geographical Information Systems (GIS). In his seminal work *Design with Nature* (1969), McHarg articulated both a study method and an application of that process to sound land planning, as his firm had done in the Plan for the Valleys in Baltimore County, Maryland. The methods that he outlined in the book and taught at the University of Pennsylvania became the standard by which landscape architects evaluated sites.

As the century progressed, new issues emerged to challenge the profession. Worldwide growth of tourism required consideration of the environmental impacts of both large numbers of visitors and massive tourist facilities on delicate ecosystems. In this discussion, questions of landscape management were central: How could the carrying capacity of an area be determined? Which means of access would protect the ecology? How could sites be used to educate the public on the value of natural systems? Out of these issues there were developed important design guidelines, such as the use of boardwalks for access in sensitive areas.

Such significant philosophical and administrative questions were at one level of the practice of landscape architecture. At other levels, more direct, practical concerns were addressed, such as how to house people in environmentally sensitive and sustainable communities. Village Homes in Davis, California, begun in 1974, demonstrated that sound land-use planning, including retention of on-site runoff and planting of productive species, could be coupled with socially sensitive community building. Village

A Prairie Crossing Home in Grayslake, Illinois. *The Prairie Crossing is a nationally recognized conservation community, planned by Peter Walker and William Johnson. The community contains 359 single family homes, and was built on such ecofriendly principles as environmental protection, healthy living (including all organic farming), and energy conservation.*

Homes became a model of innovative sustainable community design. Other developments took different approaches. Planning at the Woodlands, Texas, aimed at preserving a vital water recharge area for an aquifer that served Houston. As planned by Wallace, McHarg, Roberts, and Todd, a Philadelphia firm of landscape architects and planners, housing occupied less sensitive areas, ponds were created to let water soak into the soil above the aquifer, and numerous pine woods were preserved to retard runoff. Prairie Crossing in Grayslake, Illinois, planned by Peter Walker and William Johnson, represented a different approach to community planning. There buildable sites were clustered tighter than is typical to preserve more land for open space, which includes protected farm land, river corridors, and wildlife habitat.

On a smaller scale, an extensive array of design strategies are now being used to make sites more sustainable and thereby reduce environmental damage. Techniques include xeriscaping (the use of native and adaptable plants to reduce water and other resource consumption), construction of retention and detention ponds to prevent or limit the amount of runoff that leaves a site (as an aid in flood control downstream), energy conservation through site design (including proper building orientation and shading of structures), use of recycled or renewable materials (such as shredded tires to make playground surfaces), and construction of wetlands to purify contaminated water. Landscape architects take seriously the charge to act locally, making sound decisions for a sustainable environment at every scale of work.

Landscape architecture continues in the forefront of efforts in the early twenty-first century to create a more sustainable future for the world. Its practice affects every aspect of sustainability and goes beyond theory and general principles to apply both philosophy and science to real world situations used by people daily. Every action in the landscape—from the most mundane, such as shrub clearing, to the large-scale, such as the construction of a new town—has significant consequences for the local environment. When these local outcomes become aggregated, serious global environmental change can occur. The overriding goal of landscape architecture is, in the words of Robert L. Thayer, to create places "where human communities, resource uses, and the carrying capacity of surrounding ecosystems can all be perpetually maintained" (1994, p. 235).

SEE ALSO *Audubon Society; Environmental Law; Hetch Hetchy; Land Ethic; Landscape Painters and Environmental Photography; Leopold, Aldo; Marsh, George Perkins; Preservation; Sierra Club; Space/Place; Sustainability; Urban Environments; U.S. Forest Service; U.S. National Park Service.*

BIBLIOGRAPHY

Albright, Horace M. 1985. *The Birth of the National Park Service: The Founding Years, 1913–33.* Salt Lake City, UT: Howe Brothers.

Carson, Rachel. 1962. *Silent Spring.* New York: Fawcett Crest.

Catlin, George. 1926 (1840). *North American Indians: Being Letters and Notes on Their Manners, Customs, and Conditions. . . .* Edinburgh, UK: John Grant.

Douglas, Marjory Stoneman. 1947. *The Everglades: River of Grass.* New York, Rinehart.

Hussey, Christopher. 1967. *The Picturesque: Studies in a Point of View.* London: Cass.

Leopold, Aldo. 1949. *A Sand County Almanac, and Sketches Here and There.* Oxford: Oxford University Press.

Marsh, George P. 1864. *Man and Nature, or Physical Geography as Modified by Human Action.* New York: C. Scribner.

McHarg, Ian. 1969. *Design with Nature.* Garden City, NY: American Museum of Natural History.

Nash, Roderick Frazier, ed. 1968. *The American Environment: Readings in the History of Conservation.* Reading, MA: Addison-Wesley.

Nash, Roderick Frazier. 1989. *The Rights of Nature: A History of Environmental Ethics.* Madison: University of Wisconsin Press.

Petulla, Joseph M. 1977. *American Environmental History: The Exploitation and Conservation of Natural Resources.* San Francisco: Boyd and Fraser.

Pregill, Philip, and Nancy Volkman. 1999. *Landscapes in History: Design and Planning in the Eastern and Western Traditions.* New York: John Wiley.

Simmons, I. G. 1993. *Environmental History: A Concise Introduction.* Oxford: Blackwell.

Thayer, Robert L., Jr. 1994. *Gray World, Green Heart: Technology, Nature, and Sustainable Landscape.* New York: John Wiley.

Thompson, Paul B. 1995. *The Spirit of the Soil: Agriculture and Environmental Ethics*. London: Routledge.

Waugh, Frank A. 1917. *The Natural Style in Landscape Gardening*. Boston: R. G. Badger.

Worster, Donald. 1993. *The Wealth of Nature: Environmental History and the Ecological Imagination*. New York: Oxford University Press.

Nancy Volkman

LANDSCAPE PAINTERS AND ENVIRONMENTAL PHOTOGRAPHY

Landscape painting, which began in the late Middle Ages, has played an important role in environmental thought and environmental philosophy, especially in the first eight decades of the nineteenth century in the United States. Although the type of painting that promoted nature preservation ceased to be mainstream art by the end of the nineteenth century, landscape photography, after 1860, began to play a comparable role and has continued to do so to this day.

VIEWS OF NATURE IN THE MIDDLE AGES

In the Middle Ages educated people were taught that nature was not beautiful, for the theology that pervaded the culture of the time saw nature as competing with and detracting from the love of God. For example, Petrarch (1304–1374), while climbing a moutain in 1336 and taking pleasure in the view, reprimanded himself for "admiring earthly things" (Clark 1976, p. 10). Realistic depictions of nature were rare. Images of earthly or natural objects were intended to be symbolic rather than representational. This focus was a characteristic form of thought of the time—"symbolic thinking"—according to which images were associated with biblical stories (Huizinga 1924). The landscape painting of the Middle Ages was essentially a "landscape of symbols" (Clark 1976, chap. 1, pp. 1–31). Mountains, for example, symbolized the wrath of God at the time of the Flood of Noah.

The darkness in depictions of forests reflected a general fear of nature and its estrangement from God, who was otherworldly. In the centuries that followed, the understory of forests gradually brightened with light, signifying the presence of God in nature. By the nineteenth century, the conflict between God and nature was broken. It was possible to love God while loving nature in what otherwise appeared to be realistic, secular painting.

EARLY MODERN VIEWS OF NATURE

As the Middle Ages waned, symbolic thinking was displaced by representational thinking as the primary mode of thought. In the early modern period representation was especially important in the rise of science, but its role was no less important in painting. A central feature of modern painting was the development of perspective, which led to the realistic depiction of spatial relations. In addition, artists began to try to accurately represent the details of nature, which gradually led them to incorporate the science of natural history into their work—especially biology, botany, and geology.

Painters developed close ties with the science of natural history because they were attending to the same properties of natural objects: secondary properties (color, taste, smell, texture, and sound), in terms of the distinction developed by Galileo Galilei and René-Descartes between primary and secondary qualities. Biologists, botanists, and geologists used these properties to classify the objects of their scientific interest. Painters rendered these properties in their artistic work. Writers followed suit, making use of additional secondary properties in their poetry and prose.

Although the transition from the "landscape of symbols" to the "landscape of fact" (Clark 1976) can be presented as a general development among nearly all European artists, American painters of the nineteenth century identified three artists as especially important: Titian (c. 1485–1576), Salvator Rosa (1615–1673), and Claude Lorrain (1600–1682). Titian was considered the first painter to use landscape as background. The human figures, however, were large and the central feature of the paintings. The landscapes themselves were simply generic background decoration. Rosa introduced the sublime into painting in terms of jagged, broken geological features and stormy, tempestuous skies. A weathered stump of a tree in the lower left corner of painters became a tradition among the artists that followed after him. The sublime was an aesthetic appreciation of nature involving the transformation of properties of God into properties of nature. Mountains symbolizing the wrath of God came to be regarded as terrifying but breathtaking natural phenomena (Nicolson 1963). Claude pioneered picturesque beauty in painting.

Initially, the sublime was viewed as the opposite of the beautiful and the primary aesthetic attribute of nature. The sublime was big, the beautiful small; the sublime rough, the beautiful smooth; the sublime terrifying, the beautiful pleasant. The dichotomy between the sublime and the beautiful was established by the philosopher Edmund Burke (1729–1797) in his book *A Philosophical Enquiry into the Origin of Our Ideas of the Sublime and the Beautiful* (1757). William Gilpin (1724–1804), a writer,

Italian Landscape, 1640, by Claude Lorrain. *Claude Lorrain was an early landscape painter whose techniques had a significant influence on nineteenth-century American painters. He is considered the pioneer of the "picturesque," an aesthetic category that seemed to combine both the sublime and the beautiful in landscape painting. Prior to this historical period, landscapes alone were not considered to be an appropriate subject matter for serious painting.* THE ART ARCHIVE/TATE GALLERY LONDON.

artist, and clergyman, attacked Burke's distinction in a series of books analyzing the aesthetic elements of natural scenery, particularly river valleys, arguing that nature also could be beautiful as well as sublime—that is, picturesque. Claude's paintings were the model of picturesque beauty, which combines elements of both the sublime and the beautiful (Hussey 1927): some roughness without being frightening; some harmony without being formal.

The scenic elements of Claude paintings are harmonious rather than chaotic. The human figures are small, but the paintings include ruins or castles to provide a humanizing element. In the nineteenth century the paintings of Rosa (epitomizing the sublime) and Claude (epitomizing the picturesque) served as alternative formulaic styles for American landscape painters.

The admiration for the painting of natural scenery gradually led to an aesthetic appreciation of landscapes similar to those depicted in picturesque painting. People became fascinated with particular landscapes during travel, and it slowly dawned on them that the scenery reminded them of Claude paintings. The transition proved to be a difficult one. In order to facilitate the development of a taste for natural scenery, people used "Claude glasses," a mirror surrounded by a small picture frame, which, when standing with one's back to the natural scene, permitted the traveler to rearview the natural scene much like a framed painting. The glass of the mirror was stained to match the golden brown of the varnish Claude used on his paintings so as to enhance the picturesque aesthetic effect.

VIEWS OF NATURE DURING THE NINETEENTH CENTURY

Although he was not a landscape painter, the writer Henry David Thoreau (1817–1862) articulated a view of nature that emerged out of two traditions that were central to landscape painting: the science of natural history and picturesque travel. Thoreau was considered an important amateur naturalist. In addition, his early writing attempts were heavily influenced by Gilpin. Thoreau began writing *A Week on the Concord and Merrimack Rivers* (1849) shortly after reading twelve of Gilpin's picturesque travel books.

American painting started with a strong focus on portrait painting because it was the most reliable way for artists to receive payment for their work. The first major American landscape painter was Thomas Cole (1801–1848), who is considered the founder of the Hudson River School, a group of artists in New York and New England who became known for the painting of natural scenes, especially in the Adirondacks and the Catskills. Cole focused on landscape painting in large part because he had little ability as a portrait painter. Cole was troubled throughout his career about the type of paintings that would sell. He preferred to paint ideal or composed paintings in which the scene depicted arose from the artist's imagination. Cole believed that the depiction of a real landscape was merely imitation and therefore could not be appreciated for the creativity of the artist, only for the artist's technical skills in painting. Unfortunately, from Cole's perspective, Americans preferred paintings of real places and often wanted to know where they could go to compare his paintings with actual locations in nature. He lamented that, because of the poor tastes of the American people, he was never able to become the great artist he should have been (Novak 1969). On occasion he did paint real places: For example, *The Oxbow (The Connecticut River near Northampton)* (1836) is the kind of painting for which he is best remembered and which served as the inspiration for the rest of the Hudson River School. *The Oxbow* was admired especially because it replaced the formulaic tumultuous sky of Salvator Rosa with a representation of a recognizable meteorological front. Cole was most interested in his religious paintings, but his place in history is based primarily on his work as a landscape painter.

Most members of the Hudson River School did not share Cole's preference for "composed" paintings. After Cole's death, Asher Durand (1796–1886), as the editor of *The Crayon,* a newsletter for artists, became the leader of the Hudson River School. In an editorial in *The Crayon,* he argued for the accurate depiction of real places in landscape painting:

> Let it be remembered that the subject of the picture, the material object or objects from which it is constructed, are the essential parts of it. If you have no love for them, you can have no genuine feeling for the picture which represents them. You may have a kind of admiration for the masterly treatment, and remarkable qualities, but that does not constitute love of Art—it is only an intellectual perception of power.... We *love* Nature and Beauty—we admire the artist who renders them in his works. (Durand 1855)

Durand also emphasized the relationship of landscape painting and nature poetry with one of his most famous paintings, *Kindred Spirits* (1849), which depicts Thomas Cole and William Cullen Bryant (1794–1878), an American Romantic poet. Bryant himself called attention to this connection with a two-volume illustrated book, *Picturesque America* (1872), filled with pen-and-ink drawings documenting the beauty of the United States.

Frederick Church, Cole's only student, became interested in the role of painting in the science of natural history after reading *The Cosmos* (1850) by Alexander von Humboldt (1769–1859), in which Humboldt lamented that he had not taken an artist with him to visually document his scientific studies in South America. A year later Church traveled to South America and retraced Humboldt's steps in order to paint the natural history of the region. Most of Church's paintings were "composed"—that is, not actual places—but he presented them as scientific summaries of the South American landscape. One of Church's currently most famous paintings, *Heart of the Andes* (1859), did not depict a real place but rather was a summary of the typical features of the landscape, with special attention to scientific detail. Art critics recommended that viewers consider portions of the painting separately, much as travelers do in viewing natural scenes. For example, one part of the painting shows a peasant praying before a cross by a road. Another part of the painting depicts a tree precariously growing on a bank that is being undercut by a stream below. The first says something about the religiosity of the region. The second presents a freeze-frame in dynamic natural history: The bank will continue to erode, and the tree will eventually fall into the river.

Church encouraged a fellow artist, Martin Heade, to travel to South America, where he painted studies of hummingbirds. These paintings—for example, *Brazilian Hummingbirds* (1865) and *Magnolia Grandiflora* (ca. 1880)—show that major artists were also influenced by the scientific illustration of eighteenth- and nineteenth-century biologists, botanists, and geologists. Because photography was very primitive until the second half of the nineteenth century, and even then produced only black-and-white images, scientists took art lessons as part of their scientific training and, as a result, developed the same aesthetic sensibilities as landscape painters. Even ideal or "composed" paintings by mainstream artists often

Heart of the Andes, *1859, Frederick Edwin Church.* *One of Church's most famous paintings, this scene does not represent an actual geographical location, but is rather a summary of various features the artist encountered while traveling through South America. Church was the only student of American painter Thomas Cole, the founder of the Hudson River School.* © THE METROPOLITAN MUSEUM OF ART/ART RESOURCE, NY.

produced reliable scientific information because the artist usually used plants, animals, and geological features typical of the region where they painted.

A characteristic feature of nineteenth-century American landscape painting was the presence of intense, bright light. John I. H. Baur (1954) linked all of this painting together, coining the term *Luminism.* The light performs two functions. First, it introduces a religious dimension into seemingly secular paintings. This light might, for example, be a beam coming out of the sky, characteristically accenting a waterfall or a radiant glow brightening the understory of a forest. Second, the light starkly intensifies the materiality of the natural objects represented in the paintings. The result is a presentation of a natural world that is both material and spiritual. (Compare Thomas Cole's *The Pilgrim of the Cross at the End of His Journey* [1844-1845], in which an area of bright light depicts the border between heaven and earth, and Albert Bierstadt's *Sierra Nevada Morning* [1870], in which the same lighting techniques produce a similar effect.)

The Luminists were guided by a belief that all nature is beautiful, as encapsulated in an oft-quoted remark by John Constable (1776–1837): "No, madam, there is nothing ugly; *I never saw an ugly thing in my life:* for let the form of an object be what it may—light, shade, and perspective will always make it beautiful" (Leslie 1951, p. 280). In a similar vein, Ralph Waldo Emerson (1803–1882) wrote, "There is no object so foul that intense light will not make beautiful. . . . Even the corpse has its own beauty" (Emerson 1836), and John Muir (1838–1914) wrote, "none of Nature's landscapes are ugly so long as they are wild" (Muir 1901, p. 4).

This position—that everything natural can be viewed as beautiful—has been named by Allen Carlson (Carlson 1984) as "positive aesthetics." Carlson argues there that "Art is created, while nature is discovered" and that creativity is in the standards created by humans to describe nature but not in nature itself; there is no creativity in nature (Carlson 1984, p. 31). Eugene Hargrove has argued, to the contrary, that creativity is in nature and it is the same as the creativity attributed to God in the late Middle Ages, when the properties of God symbolically manifested in nature were transferred to nature itself in the concept of the sublime. The result is indifferent rather than artistic creativity (Hargrove 1996 [1989]).

ENVIRONMENTAL PHOTOGRAPHY

Luminist painting ceased to be cutting-edge, mainstream art in the 1880s, when it began to be displaced first by impressionism, then postimpressionism and cubism, and ultimately by abstract expressionism. With impressionism artists no longer painted objects lit by strong light and instead painted the light itself in such a way that the objects dropped out of consideration. With postimpressionism and cubism artists gradually abandoned linear perspective and with it the realistic depiction of material objects. With abstract expressionism the quest for accurate representation of nature, begun in the late Middle Ages, was completely abandoned. These new forms of painting severed the link between "art and nature" and instead sought a link between "art and self," making so-called modern art irrelevant environmentally. However, the tradition of representational art continued with the rapid emergence of photography, which quickly elevated itself to an art form in its own right.

Photography began with Louis Daguerre's invention, in France in 1839, of the daguerreotype, which remained the primary form of photography until 1859. Because daguerreotypes required long exposures, did not produce the detail needed for landscape photographs, and were not reproducible through print media, they were mostly limited to portraiture. The glass-plate process developed in the 1850s made landscape photography as we know it today possible and profitable, because multiple copies of each photograph could be easily produced. The technology was fully perfected by 1860, in time to produce chillingly detailed documents of the death tolls on Civil War battlefields in the United States. Shortly thereafter, however, it was put to more pleasant use in nature photography. The poor quality of daguerreotype photography can be strikingly seen in J. D. Hutton's photography of the Great Falls of the Missouri during the W. F. Reynolds's expedition of 1858–1859 (Naef and Wood 1975, p. 29), which is misidentified as a falls on the Yellowstone River. Although Hutton was attempting to reproduce a sketch of the falls made by Meriwether Lewis, the result is so poor that the photograph requires significant explanation to make it intelligible as a nature scene.

In addition to mammoth glass-plate cameras, stereoscopic cameras were also developed at this time, providing smaller three-dimensional views of nature with stunning effect. Nevertheless, the larger photographs, which were more like landscape paintings, proved to be more popular and commercially successful.

The importance of western American nature photography was firmly established internationally with the submission of photographs of Yosemite to the 1867 Paris International Exposition, where they won a bronze medal. Although the photographs were submitted under the name Thomas Houseworth (1828–1915), they are believed to have been made by C. L. Weed (1823–1903) or C. E. Watkins (1829–1916), photographers who worked for Houseworth's company. Weed and Watkins began photographing Yosemite in 1859 and 1861, respectively. Watkins competed with Weed, often photographing the same scenes previously photographed by Weed in order to prove his superiority as a photographer. Eadweard Muybridge (1830–1904) joined the competition, also photographing the same locations. This competition helped call attention to landscape photography as an art form and to Yosemite as a place of spectacular national importance.

Government-sponsored geological surveys also advanced landscape photography in the 1860s. The California State Geological Survey, which came to be a model for other state geological surveys, hired Watkins as a photographer in connection with the survey of Yosemite Valley that was to become a state park as a gift of the federal government. Clarence King, who later became the first head of the U.S. Geological Survey, happened to be in California looking for "the Cotopoxi of the West," a mountain that would be as aesthetically spectacular as an active volcano painted by Church in South America. King joined the California survey and was given the task of surveying Yosemite. There he worked with Watkins and learned the value of landscape photography in geological work. In 1867, when King became the head of the Fortieth Parallel Survey, he immediately hired Timothy O'Sullivan as a photographer. O'Sullivan's work was considered so spectacular and so useful in gaining additional congressional appropriations, that F. V. Hayden's survey of the territories and John Wesley Powell's survey of the Colorado River quickly followed his example. W. H. Jackson (1843–1942) became especially well known for his photographs of Yellowstone and Colorado while working for Hayden.

There was considerable interaction between artists and photographers during this period, especially on the geological surveys. Thomas Moran (1837–1926), most famous for his two paintings of the Great Falls of the Yellowstone, worked with Jackson and sometimes stood in his photographs for scale (see Naef and Wood 1975, p. 222). Bierstadt used a Watkins photograph of the Grizzly Giant in the Mariposa Grove in the creation of his famous painting of the same tree (c. 1864) (Naef and Wood 1975, p. 62). The aesthetic perspectives and goals of the artists and the photographers were the same.

Landscape painting and photography had a tremendous impact on nature preservation in the nineteenth century. Landscape paintings and photographs played a central role in the creation of Yosemite, a California state park, and, a few years later, Yellowstone National Park.

Uinta Mountains, Utah. *This photograph was taken by Timothy O'Sullivan in 1869 while working for the Fortieth Parallel, or King, Survey. The glass-plate process, which replaced daguerreotypes in the 1850s, made landscape photography both possible and profitable.* PHOTO BY T.H. O'SULLIVAN/U.S. GEOLOGICAL SURVEY.

Nearly all of the places painted and photographed by major artists and photographers during that time period are now national parks or national monuments. Frederick Law Olmsted (1822–1903) acknowledged this connection in his report on Yosemite Valley to the California legislature in which he recommended that half of the commissioners for the park be "students of natural science and landscape artists" (Olmsted 1952). A few years later, when landscape photography was better established, he would have most certainly added "landscape photographers."

The creation of these images of then-remote areas permitted the general public to learn about and care about these places, and congressional action to preserve these places quickly followed. In admiring the aesthetic qualities of the paintings and photographs, people were also admiring the aesthetic qualities of the places that the images represented, and their aesthetic enjoyment, as G. E. Moore (1873–1958) put it in another context in *Principia Ethica* (1903), depended on a true belief that the aesthetic qualities depicted continued to exist in the original (Hargrove 1996 [1989]). Just as images of natural objects in the Middle Ages, in terms of symbolic thinking, produced a religious, otherworldly aesthetic experience, the images of natural objects in the modern period, in terms of representational thinking, produced an aesthetic experience of the world as it actually existed a desire to preserve it in a natural state.

The link between images and natural places still plays a critical role in the environmental movement politically, for it makes it possible for people to become committed to protecting places that they may never visit. Without this representational link, the environmental

Yellowstone National Park, Wyoming. *Taken by W.H. Jackson in 1872 while working for the Hayden geological survey, Jackson became especially famous for his photographs of Yellowstone—such as this image of Old Faithful Geyser—and Colorado.* U.S. GEOLOGICAL SURVEY.

movement would have been a local rather than a national and international movement. The original impact of nineteenth-century landscape painting and photography is still appreciated in the work of Ansel Adams (1902–1984), who stands at the end of a tradition started by Weed, Watkins, and Muybridge. The impact, however, is pervasive, for the general public is surrounded by images in photographs and motion pictures that follow the principles and perspectives of nineteenth-century painting and photography. When tourists move back and forth at a scenic view while taking a photograph, they are intuitively employing the compositional techniques developed by those artists and photographers.

SEE ALSO *Environmental Aesthetics; Environmental Art; Hudson River School; Thoreau, Henry David.*

BIBLIOGRAPHY

Baur, John I. H. 1954. "American Luminism." *Perspectives USA* 9: 90–98.

Bryant, William Cullen, ed. 1874. *Picturesque America; Or the Land We Live In.* New York: D. Appleton and Company.

Burke, Edmund. 1757. *A Philosophical Enquiry into the Origin of Our Ideas of the Sublime and the Beautiful.*

Carlson, Allen. 1984. "Nature and Positive Aesthetics." *Environmental Ethics.* 6(1): 5–34.

Clark, Kenneth. 1976 [1949]. *Landscape into Art.* New York: Harper & Row.

Durand, Asher. 1855. "Common Sense in Art." *The Crayon* 1(6): 81.

Emerson, Ralph Waldo. 1836. *Nature.* Boston: J. Munroe and Co.

Hargrove, Eugene C. 1996 [1989]. "Aesthetic and Scientific Attitudes." In *Foundations of Environmental Ethics.* Denton, TX: Environmental Ethics Books.

Huizinga, J. 1924. *The Waning of the Middle Ages: A Study of the Life, Thought, and Art in France and the Netherlands in the XIVth and XVth Centuries.* New York: St. Martin's Press.

Humboldt, Alexander von. 1850. *The Cosmos.* London.

Hussey, Christopher. 1927. *The Picturesque: Studies of a Point of View.* London and New York: G. P. Putnam's Sons.

Leslie, C. R. 1951. *Memoirs of the Life of John Constable, Composed Chiefly of His Letters (1843).* London: Phaidon.

Muir, John. 1901. *Our National Parks.* Boston: Houghton Mifflin.

Naef, Weston J., and James N. Wood. 1975. *Era of Exploration: The Rise of Landscape Photography in the American West, 1860–1885.* Boston: New York Graphic Society.

Nicolson, Marjorie Hope. 1963. *Mountain Gloom and Mountain Glory: The Development of the Aesthetic of the Infinite.* New York: Norton.

Novak, Barbara. 1969. *American Painting of the Nineteenth Century: Realism, Idealism, and the American Experience.* New York: Praeger.

Olmsted, Frederick Law. 1952. "The Yosemite Valley and the Mariposa Big Trees." *Landscape Architecture* 43: 24–25.

Thoreau, Henry D. 1849. *A Week on the Concord and Merrimack Rivers.* Boston: James Munroe and Company.

Wilmerding, John. "Fire and Ice in American Art: Polarities from Luminism to Abstract Expressionism." In *The Natural Paradise: Painting in America 1800–1950,* ed. Kynaston McShine. New York: Museum of Modern Art.

Eugene C. Hargrove

LAST MAN ARGUMENTS

The *last man argument* was devised by Richard Sylvan (before 1983 Sylvan went by the name Richard Routley) and was first published in Routley's article "Is There a Need for a New, an Environmental, Ethic?" (1973). It is a thought experiment designed to show that the prevailing principles of the dominant European and North American tradition are unable to provide a satisfactory basis for an environmental ethic. An adequate ethic of concern for the nonhuman world must therefore have very different foundations.

The shared core assumptions of European and North American ethics (which Sylvan calls a "super ethic") include a *freedom principle,* according to which agents are permitted to act as they please provided that they do not (1) harm others (understood usually, though not always, as other *persons*) or (2) harm themselves irreparably. Sylvan labeled this anthropocentric principle *basic human chauvinism,* because it affirms that moral deliberation and choice involve only human interests and concerns. The last man thought experiment was devised to refute this core principle and thus expose the inadequacy of traditional European and North American ethics as a foundation for an environmental ethic.

Most scholars argue that ethical principles must be universal and therefore applicable not just to actual situations but to all possible situations. It is for this reason that thought experiments are important intuition pumps and play a central role in testing ethical principles. The last man thought experiment is as follows: "The last man (or person) surviving the collapse of the world system lays about him, eliminating, as far as he can, every living thing, animal or plant (but painlessly if you like, as at the best abattoirs). What he does is quite permissible according to basic [human] chauvinism, but on environmental grounds what he does is wrong" (Routley 1973, pp. 207–208).

Because he is the last human survivor, there are no other human interests to be considered, and the chauvinistic liberty principle therefore provides no grounds for moral condemnation of his actions. Nevertheless, it is clear to an environmentally enlightened conscience that the actions of the last man are morally dreadful. If you share Sylvan's intuition that these terminal actions are morally reprehensible, and if there are no anthropocentric considerations to support this intuition, then there must be some nonanthropocentric considerations or values that explain the evil of the acts. (The last man argument is formulated at greater length in Routley and Routley (1980), which Sylvan (then Routley) authored jointly with his then-partner, who later took the name Val Plumwood.)

Holmes Rolston (1975) has set out a parallel "African butterfly" argument in support of nonanthropocentric environmental values. Gratuitous acts of "speciescide" carried out by unscrupulous butterfly collectors, like the actions of Sylvan's last man, are reprehensible to the environmentally enlightened conscience. Robin Attfield (1981) and Mary Anne Warren (1983) have also presented variants of the argument.

A thought experiment that anticipates some aspects of Sylvan's argument was proposed by G. E. Moore in a famous attempt to establish the objective value of beauty.

> Let us imagine one world exceedingly beautiful. Imagine it as beautiful as you can; put into it whatever on this earth you most admire—mountains, rivers, the sea; trees, and sunsets, stars and moon. Imagine these all combined in the most exquisite proportions, so that no one thing jars against another, but each contributes to increase the beauty of the whole. And then imagine the ugliest world you can possibly conceive. Imagine it simply one heap of filth, containing everything that is most disgusting to us, for whatever reason, and the whole, as far as may be, without one redeeming feature.... The only thing we are not entitled to imagine is that any human being ever has or ever, by any possibility, *can*, live in either, can ever see and enjoy the beauty of the one or hate the foulness of the other. Well, even so, supposing them quite apart from any possible contemplation by human beings; still, is it irrational to hold that it is better that the beautiful world should exist, than the one which is ugly? (1903, pp. 83–84)

Moore's intuition is that the beautiful world is objectively better than the heap of filth—and he suggests that this is so quite independently of whether any evaluators exist to contemplate the worlds in question. Nevertheless, according to David Hume (1740), we must be cautious about inferences, however appealing, that draw evaluative conclusions from descriptive premises.

Value intuitions depend crucially on the nature of evaluators. It is not difficult—given our aversion to dung—to share Moore's intuition. This preference, however, is shaped by our biological constitution. It is far from clear that our preference would be shared by, say, a dung beetle or a blowfly. Rather than establishing the objective value of beauty, Moore has established the existence of deep-seated intersubjective aesthetic intuitions shared by humans—and the difficulty of thinking like a blowfly.

Sylvan's argument similarly fails to establish secure objective grounds for nonanthropocentric values. Like other famous thought experiments, however—such as brains in vats (Brueckner 2004) and "trolley" problems (Thomson 1976)—it helps us to regiment our intuitions. The powerful visceral impact of the last man thought experiment reveals a widespread though perhaps not universal biophilia—an affinity for rich, diverse, complex, and beautiful biological systems. The use of cyanide and explosives for fishing on coral reefs, rather than sustainable practices, and the clear-felling and burning of old-growth forests, generate a similarly powerful visceral repugnance that is also widely shared but is also, alas, not universal. The last man thought experiment helps us to understand that the depletion of biodiversity impoverishes not just the biosphere but also the requirements for the sort of human life to which we are adapted and to which many of us aspire.

SEE ALSO *Environmental Philosophy: V. Contemporary Philosophy; Plumwood, Val; Rolston III, Holmes; Speciesism.*

BIBLIOGRAPHY

Attfield, Robin. 1981. "The Good of Trees." *Journal of Value Inquiry* 15: 35–54.

Brueckner, Tony. 2004. "Brains in a Vat." *The Stanford Encyclopedia of Philosophy.* Available from http://plato.stanford.edu/archives/win2004/entries/brain-vat/.

Hume, David. 1888 (1740). *A Treatise of Human Nature*, Vol. 3, ed. L.A. Selby-Bigge. Oxford, UK: Clarendon Press.

Moore, G. E. 1903. *Principia Ethica.* Cambridge, UK: Cambridge University Press.

Rolston III, Holmes. 1975. "Is There an Ecological Ethic?" *Ethics* 85: 93–109.

Routley, Richard. 1973. "Is There a Need for a New, an Environmental, Ethic?" *Proceedings of the Fifteenth World Congress of Philosophy* 1: 205–210.

Routley, Richard, and Val Routley. 1980. "Human Chauvinism and Environmental Ethics." In *Environmental Philosophy*, eds. D. S. Mannison, M. McRobbie, and R. Routley. Canberra: Australian National University.

Thomson, J. J. 1976. "Killing, Letting Die, and the Trolley Problem." *The Monist* 59: 204–217.

Warren, Mary Anne. 1983. "The Rights of the Nonhuman World." In *Environmental Philosophy: A Collection of Readings*, ed. R. Elliot and A. Gare. St. Lucia, AU: University of Queensland Press.

William Grey

LEOPOLD, ALDO

1887–1948

Aldo Leopold was born on January 11, 1887, in Burlington, Iowa, and became one of the most prominent environmental philosophers of the twentieth century. In his childhood he was an avid naturalist, ornithologist, hiker, and hunter. After finishing a master's degree in forestry at the Yale Forest School in 1909, he joined the U.S. Forest Service. Until 1924 Leopold worked exclusively in the Southwest, and from 1924 to 1928 he worked for the Forest Products Laboratory in Madison, Wisconsin. From 1928 to 1932 he conducted game management research for the Sporting Arms and Ammunitions Manufacturers' Institute, and in 1933 he accepted a position as professor of game management at the University of Wisconsin–Madison. He died in Wisconsin on April 21, 1948.

Leopold was among the founders of the field of wildlife management; his *Game Management,* the first text in that field, was published in 1933. He helped create the first designated wilderness area in the United States in 1924 and cofounded the Wilderness Society in 1935. Although he wrote over three hundred articles, he is best known for his posthumously published book *A Sand County Almanac* (1949).

Aldo Leopold, circa 1943. *Aldo Leopold is one of the most prominent environmental philosophers of the twentieth century. He is well known for his work with the U.S. Forest Service, and his famous essay, "The Land Ethic."* COURTESY OF THE ALDO LEOPOLD FOUNDATION ARCHIVES.

A SAND COUNTY ALMANAC

The last and most philosophical essay in the *Almanac* is "The Land Ethic." In that essay Leopold argued that human ethics can, must, and should evolve and that the sense of community on which human ethics is based must expand to include the land itself (i.e., all living and nonliving components of the environment). He went on to say that humans must transform themselves from ecological conquerors to ecological citizens and that changes in a nation's environmental policies must be preceded by similar philosophical changes in that nation's populace. Leopold held that land is a structured ecological mechanism that is replete with energy circuits, capable of being healthy or unhealthy, and vulnerable to sudden anthropogenic disturbances; purely economic, instrumental approaches to land use are ethically impoverished and unsustainable. The essay culminates with the well-known claim that land usage "is right when it tends to preserve the integrity, stability, and beauty of the biotic community. It is wrong when it tends otherwise" (Leopold 1949, pp. 224–225). The land ethic stands as the first systematically holistic ethic in environmental philosophy. However strong people's concern for individual members of the land community may be, concern for the land community itself should be much stronger.

As a whole the *Almanac* is difficult to characterize, partly because of Leopold's efforts to meet common expectations of what "nature books" should be like: personal and narrative, with analytical and theoretical comments tolerated only when "worked into a framework of actual field experience" (Ribbens 1987, p. 93). The overarching theme is an evolutionary and ecological worldview. Part I conveys it indirectly through personal and experiential narrative, Part II conveys it more directly and didactically, and Part III explores its normative implications. One of the most striking blends of personal narrative and philosophical commentary occurs in "Thinking Like a Mountain," a personal, confessional essay in which Leopold not only acknowledges his predator-eradicating past but credits it as a source of sudden and unexpected moral insight. The essay marks a humble approach to the difficulties of making progress on an unfamiliar moral frontier.

It is difficult to determine the ultimate motivation for Leopold's environmental philosophy. Is the land ethic fundamentally prudential and anthropocentric? If that is not the case, is it possible to commend it, as Leopold often does, on grounds of seemingly antithetical enlightened self-interest?

INTERPRETATIONS OF LEOPOLD'S LAND ETHIC

Answering these questions has been a challenge for Leopold scholars. J. Baird Callicott characterized Leopold's land ethic as subjectively deontological but objectively prudential (Callicott 1987, p. 214). Bryan Norton interpreted Leopold as a pragmatist and a value pluralist who embraced "a standard of long-term survival as a noneconomic basis for evaluating a culture's practices and institutions" (Norton 2005, p. 75). Roderick Nash described Leopold as "a battle-scarred veteran of conservation policy wars" who appealed to enlightened self-interest only because it "was the best way to sell his philosophy in the 1930s and 1940s" (Nash 1987, p. 81). Julianne Lutz Newton and Curt Meine denied the presence of an ultimate motivation. Leopold simply encouraged people to evaluate their treatment of land "in both moral and prudential terms" (Newton 2006, p. 320) and "tried to strike a balance between the reality of economic necessity and the truth that standard economic definitions of 'necessity' did not suffice in conservation" (Meine 1988, p. 503). When it came to conservation, Leopold believed that "a profit motive was not enough. Nor, for that matter, was a sentimental motive" (Meine 1988, p. 503).

The scientific foundations of the Leopold land ethic are a source of both strength and vulnerability. Science is regarded as objective and credible, but it is also fallible and dynamic. Ecology no longer conceives biotic communities as stable over time or as exhibiting robust integrity. The Leopold land ethic may be undermined by the so-called shifting paradigm in contemporary ecology and the replacement of the notion of the balance of nature with that of the flux of nature. In that case contemporary environmental ethicists face the question: Can the land ethic be updated in light of the contemporary flux paradigm in ecology?

SEE ALSO *Environmental Philosophy: V. Contemporary Philosophy; Land Ethic; Wilderness.*

BIBLIOGRAPHY

Callicott, J. Baird. 1987. "The Conceptual Foundations of the Land Ethic." In *Companion to A Sand County Almanac: Interpretive & Critical Essays,* ed. J Baird Callicott. Madison: University of Wisconsin Press.

Callicott, J. Baird. 1989. *In Defense of the Land Ethic.* Albany: State University of New York Press.

Callicott, J. Baird. 1999. *Beyond the Land Ethic: More Essays in Environmental Philosophy.* Albany: State University of New York Press.

Leopold, Aldo. 1949. *A Sand County Almanac, and Sketches Here and There.* New York: Oxford University Press.

Meine, Curt. 1988. *Aldo Leopold: His Life and Work.* Madison: University of Wisconsin Press.

Nash, Roderick. 1982. *Wilderness and the American Mind.* New Haven, CT: Yale University Press.

Nash, Roderick. 1987. "Aldo Leopold's Intellectual Heritage." In *Companion to A Sand County Almanac: Interpretive & Critical Essays,* ed. J Baird Callicott. Madison: University of Wisconsin Press.

Newton, Julianne Lutz. 2006. *Aldo Leopold's Odyssey.* Washington, DC: Island Press/Shearwater Books.

Norton, Bryan G. 1991. *Toward Unity among Environmentalists.* New York: Oxford University Press, 1991.

Norton, Bryan.G. 2005. *Sustainability: A Philosophy of Adaptive Ecosystem Management.* Chicago: University of Chicago Press.

Ribbens, Dennis. 1987. "The Making of *A Sand County Almanac.*" In *Companion to A Sand County Almanac: Interpretive & Critical Essays,* ed. J Baird Callicott. Madison: University of Wisconsin Press.

Darren Domsky

LIFE: RESPECT/ REVERENCE

Albert Schweitzer (1875–1965) attempted to combine a systematic nature philosophy with a sense of mystical devotion to the phenomenon of life. Through his reverence for life ethic he sought to create a biocentric worldview that was informed by biological individualism. However, his argument was not developed formally in his written works.

Distracted by his fight against nuclear weapon research and testing after 1945 as well as his medical mission in French Equatorial Africa, Schweitzer made partial drafts but never completed the last two volumes of his *Philosophy of Civilization.* He hoped that his life's work would provide the needed clarity and the definitive example for his philosophy. However, unanswered questions have prevented his work from having a wider influence in environmental thought. It has been concluded by many that Schweitzer could not overcome the inherent problems in trying to reconcile a focus on biological individualism with ecocentric aims. Reverence for life has not been seen in academic circles as a practical ethic or a consistent philosophy.

SCHWEITZER'S ETHICAL MYSTICISM

More recent scholarship has revealed greater sophistication in Schweitzer's thought. Schweitzer intended his concept of ethical mysticism to serve as a balance between his argument for the respect due to all living beings and the need to set this against the reality to take other life to survive. This tension is also at the heart of the conflict between animal rights activism and holistic environmental concerns. Resource managers, for example, often must destroy individuals within a managed population or eliminate invasive exotics entirely for the greater good.

Yet Schweitzer indicated that it is necessary to protect all life that comes within the reach of a moral agent. However, moral agents cannot live up to this standard fully because they must eat and alter their environment to live. The key for Schweitzer was to atone actively for the taking of life through devotional service to all "Life" by promoting a greater good. This creates a deepening and lifelong devotion to ethical duty. Schweitzer became an exemplar of this ethic through his fifty-year medical mission in Africa.

Schweitzer's inspiration came from the Lutheran theology of the Cross. Schweitzer was an ordained Lutheran minister and believed that all would-be disciples of Christ are called upon to follow Jesus's example of self-sacrificing love (*kenosis*) through compassionate missions to eliminate suffering wherever it is found. However, Schweitzer did not see reverence for life as an exclusively Christian ethic. He was a student of world religions and believed that all faiths contain the seeds of ethical mysticism, though he did not find all faiths equally satisfactory in terms of the ethical traditions they promoted.

In a article titled "The Ethics of Reverence for Life" Schweitzer presented the admonition "I see that evil is what annihilates, hampers, or hinders life ... goodness, by the same token, is the saving or helping of life, the enabling of whatever life I can to attain its highest development" (Schweitzer 2002 [1936], p. 129). This argument was derived from German philosopher Arthur Schopenhauer's concept of the will-to-live. For Schopenhauer the will is not intention or desire but refers to the etiological changes in bodies through time. An acorn, for example, contains all the potential manifestations of an oak tree, yet these forms become actualized only over time. The will is the cause for the movement from potentiality into actuality. All bodies possess a will-to-live that gives an organism the instincts and dispositions necessary for the expression of its potential. For Schweitzer, helping all life reach its highest possible development of its will-to-live in light of the needs of others is the basic good in the reverence for life ethic.

APPLICATIONS AND REFINEMENTS

Schweitzer refused to lay down rules for every possible situation in which conflicts between saving life and taking life arise because that would externalize morality in a codified object, leaving it vulnerable to egotistical rationalization, depreciating negotiations, and superficial conviction. However, his life and work show that he recognized that it is necessary to take life to save other life. His practice as a physician, for example, required him to kill countless pathogenic bacteria and parasites to save the life of human and nonhuman patients. In this way he sought to widen and deepen public acceptance of reverence for life and recruit additional moral agents in the promotion of a greater good.

This position is not inconsistent or contradictory. An analogy with pacifism illustrates the underlying dynamics. A pacifist is confronted by a skeptic with the following dilemma: either kill the men who are threatening his or her neighbors or allow them to be killed. The pacifist replies, "I am not concerned primarily with what to do about these dilemmas when they arise; I am concerned with preventing them from arising to begin with through disarmament, institutions to secure world peace, and distributive justice to redress social wrongs." Similarly, the reverence for life ethic seeks ends that secure the best and highest good for all Life.

Schweitzer set out to offer a "path in the thicket" for a civilization that had lost its way (Schweitzer 1949, p. 185). He had come to see the ethical traditions of Kantianism and utilitarianism as morally bankrupt. In their place he sought to weave together several complementary strands of thought at the core of his reverence for life ethic. From Charles Darwin he took the shared evolutionary heritage with all life and the evidence for a social instinct; from David Hume, Arthur Schopenhauer, and Adam Smith, a philosophical basis for sympathy; and from the religious traditions of mysticism, a substantive and personal connection to ethical duty directed to the here and now. In the decades that followed, others moved to establish a life-centered philosophy with greater prescriptive clarity, with or without direct reference to Schweitzer. They include Kenneth E. Goodpaster, Paul W. Taylor (1986), Peter G. Brown, and David K. Goodin.

SEE ALSO *Deep Ecology; Invasive Species; Schweitzer, Albert.*

BIBLIOGRAPHY

Schweitzer, Albert. 1949. *Out of My Life and Thought: An Autobiography,* trans. C. T. Campion. New York: Holt.

Schweitzer, Albert. 2002 (1936). "The Ethics of Reverence for Life." In *Reverence for Life: The Ethics of Albert Schweitzer for the Twenty-First Century,* ed. Marvin Meyer and Kurt Bergel. Syracuse, NY: Syracuse University Press.

Taylor, Paul W. 1986. *Respect for Nature: A Theory of Environmental Ethics.* Princeton, NJ: Princeton University Press.

Peter G. Brown
David K. Goodin

LIMITS TO GROWTH

The 1950s and 1960s were a time of growing awareness of environmental problems, with lethal smog in major cities, warnings that growing air pollution in New York would make the city uninhabitable, the Cuyahoga River catching fire, and predictions of imminent mass starvation and environmental catastrophe from unchecked population growth. Recognizing a crisis situation, the developed nations began to take important steps to address these issues; the United States, for example, created the Environmental Protection Agency, extended the Clean Air Act, and passed the National Environmental Policy Act, Clean Water Act, and Endangered Species Act, all between 1970 and 1973.

EARLY WARNINGS ABOUT GROWTH

In this atmosphere Donella Meadows and her coauthors published *Limits to Growth* (1972), a report on the results of a computer-systems model of the interactions between the human economy and the planetary ecosystem that sustains and contains it. Focusing on population growth, resource depletion, industrialization, pollution, and food production, and assuming exponential growth but finite resources, the model showed that existing trends would lead to resource exhaustion in the coming decades, with potentially catastrophic consequences by the middle of the coming century and perhaps sooner. The report advised that avoidance of this outcome required rapid stabilization of human populations, resource, use and waste emissions. The oil crisis of 1973 initially seemed to confirm some of these conclusions.

The Meadows report was certainly not alone it its conclusions. In *The Entropy Law and the Economic Process* (1971), Nicholas Georgescu-Roegen argued that our economic system, like all physical systems, is subject to the laws of thermodynamics. The laws of thermodynamics (matter-energy cannot be created or destroyed, useful energy dissipates, and disorder increases) limit the physical size of the economy on a finite planet. The increase in entropy entailed by economic production requires compensating inputs of low-entropy energy, and the economy systematically converts low-entropy resources into high-entropy waste, providing human welfare in the process. The economy has access to two sources of low entropy to maintain this process: finite stocks of fossil fuels and other natural resources that we can use as fast as we choose and the finite flow of solar energy, which arrives at a rate beyond our control. As finite stocks become exhausted, the size of the economy will be limited by the flow of solar energy. Eventually, all the material building blocks of our economy must become so dispersed that the fixed flow of solar energy will be inadequate to recycle them, and the economy must collapse. Economic growth can only speed our doom.

Herman Daly also recognized the limits to growth, but he more optimistically argued that a no-growth, steady state economy could be sustained indefinitely. However, the raw materials that serve as the inputs into economic production also serve as the structural building blocks of ecosystems. Structure generates function. Along with other valuable services, ecosystems provide irreplaceable life-support functions that are essential to human survival. Both the extraction of ecosystem structure for economic production and the emissions of waste (known together as throughput) degrade ecosystem services. Long before the threshold of collapse, the diminishing marginal benefits of increasing economic output are likely to fall below the rising marginal costs of ecological degradation, at which point continued economic growth—defined as a "quantitative increase in ... the rate of flow of matter and energy through the economy ... and the stock of human bodies and artifacts"—becomes uneconomic. Economic development—a "qualitative improvement in non-physical characteristics" (Daly 1987, p. 323)—remains possible, however. A sustainable economy cannot extract renewable natural resources faster than they can regenerate, extract nonrenewable resources faster than we can develop renewable substitutes, or spew waste emissions into the atmosphere faster than they can be absorbed. This position became known as strong sustainability. In Daly's view, the sustainable use and just distribution of resources takes precedence over their efficient (wealth-maximizing) allocation.

CORNUCOPIANS AND DOOMSDAYERS

Such claims of limits to growth were met with great skepticism, particularly by conventional economists, who branded their proponents as doomsdayers. Those skeptical of limits became known as cornucopians. From the perspective of the cornucopians, the doomsdayers ignored the capacity of human innovation in general and the free market in particular to adapt to scarcity. As resources become scarce, their prices increase, providing a market incentive to develop substitutes. Relative scarcity might exist, but not absolute scarcity. Almost two hundred years earlier, Thomas Robert Malthus predicted that geometric increases in human population must eventually outstrip linear increases in food production, leading to widespread starvation. Instead, by the time of the Meadows report, humanity was producing more food per capita than at any other time in history, and productivity was continuing to rise. William Stanley Jevons, a nineteenth-century English economist, warned that society must inevitably run out of coal and that "we must not dwell in such a fool's paradise as to imagine we can do without coal what we do with it" (Jevons 1865, p. 145). Both Malthus and Jevons had been proved wrong. In the words of Nobel laureate Robert

Solow, "If it is very easy to substitute other factors for natural resources, then there is, in principle, no problem. The world can, in effect, get along without natural resources" (Solow 1974, p. 11). We can adapt to scarcity simply by dematerializing the economy. Human ingenuity, the ultimate resource, would solve all problems (Simon 1996). The position that human-made capital could substitute for natural capital became known as *weak sustainability.*

The optimism of the cornucopians notwithstanding, wealthier societies made major investments in environmental protection; as a result, many measures of environmental quality began to improve, some dramatically. The doomsdayers claimed that society had acted appropriately on their recommendations. The cornucopians, in contrast, claimed that these outcomes were to be expected anyway. Economic growth, in fact, was not the cause of the problems the doomsdayers presented, but rather the solution. As societies became richer, birth rates declined, so economic growth was the solution to the population problem. A rising tide lifts all boats, in the view of the cornucopians, and in the presence of widespread poverty, growth is a moral imperative. In addition, pollution emissions formed an upside-down parabola when plotted against per capita gross national product (GNP)—poor countries have clean environments, but as they begin industrializing, pollution levels increase. However, as countries get rich enough, their citizens began to demand clean environments, a luxury good. A number of recent books by cornucopians argue that almost all environmental indicators are improving, that resource abundance continues to grow, and that limits-to-growth theorists have again been proved wrong (e.g., Simon 1996; Lomborg 2001).

The concerns of the doomsdayers have increased significantly: They argue that improvements are illusory, that wealthy nations have simply exported their most environmentally damaging industries to poorer nations, and that even if pollution levels taper off with growth, pollution will increase for decades to come. Population growth is slowing but is expected to stabilize at much higher levels than the planet can sustain. Increasing consumption to slow population growth is counterproductive.

CLIMATE CHANGE, RESOURCE DEPLETION, AND DISTRIBUTION

Climate change has convinced many that planetary waste-absorption capacity is the greatest constraint on growth, and some believe that we are already doomed to a catastrophically warmer planet. Diminishing fossil-fuel stocks and rising prices have convinced others that we have reached peak oil, the point at which increasing rates of use overwhelm declining rates of discovery and then begin a steady decline, threatening chaos in a system built on cheap energy. A third group is concerned with renewable-resource depletion and biodiversity loss; the concern is that these trends will result in the collapse of ecological life-support functions. Others believe that water shortages are the greatest threat, or population growth, or toxic wastes. Many believe that we are exceeding all of Daly's tenets for sustainability, overwhelming sources, sinks, and services simultaneously. Such systems thinkers believe we are well into overshoot, currently living off natural capital rather than its yield. They also recognize that we are dealing with complex, dynamically adaptive systems, in which precise prediction is essentially impossible. This is the position taken by the report *Limits to Growth: The 30-Year Update* (Meadows et al., 2004).

Topping off the doomsdayers' woes, distribution has become a major concern. *Limits to Growth* appeared at the end of a forty-five-year decline in both poverty rates and income inequality, which, with the exception of a brief interval in the 1990s, have climbed steadily since. Similar trends are occurring at the global level. In absolute terms the wealthiest nations have amassed far more additional wealth than the poorest—1 percent of gross domestic product (GDP) in the United States equals the GDP of the world's twenty-four poorest nations (Daly and Farley 2004). In relative terms, China, India, and a few other outliers appear to have grown faster than the wealthy nations but have simultaneously experienced unprecedented increases in domestic income inequality. Experiments with green accounting in China have shown that ecological damage and resource depletion accounts for up to one-third of GNP in some regions (Qiu 2007). Real incomes in sub-Saharan Africa, the region with the world's poorest countries, and in the poorest areas of Latin America have actually declined. To make matters worse, the environmental costs of growth, such as climate change, are likely to fall disproportionately on the poor. Growth has done little if anything to alleviate absolute poverty while exacerbating relative poverty.

Conventional economists nonetheless continue to play the role of optimists, their models frequently suggesting that the potential future costs of climate change, heavily discounted, are less than the current costs of mitigation, and that little should be done. The 2006 *Stern Review* on the economics of climate change is considered quite pessimistic and a call for action, but in reality it estimates that even if we do nothing about climate change, the economy will double or triple in size by 2050 (Stern et al., 2006). The report concludes that we should act to mitigate climate change, but under the report's assumptions, this would ironically require supposedly poorer generations to sacrifice for richer ones. The extreme of this cornucopian vision is presented by Thomas C. Schelling, a 2005 Nobel laureate in economics, and other economists who suggest that climate change will matter little because it primarily affects agriculture, which is only a small percentage of GNP (Schelling 2007). Measured in dollars, any consumer good is apparently a perfect substitute for food.

Some argue not only that human wants and needs are satiable but also that, beyond a threshold already passed in the wealthy countries, increasing consumption makes us worse off. With our basic needs satisfied, it is primarily relative wealth that makes us better off, and increasing wealth for all helps no one. In addition to physical limits to growth, we have reached psychological limits, so continued economic growth is undesirable as well as impossible. The horn of plenty is overflowing. As scientists have reached near consensus on the danger of climate change, even some cornucopians and conventional economists are recognizing the need to act, and those who do not risk fading into irrelevance.

CONCLUSION

Arguments in the growth debate range from predictions of inevitable doom no matter what we do—through strong and then weak sustainability—to faith in inevitable sustainability as long as we trust in the market. In our complex ecological-economic system, with uncertain facts and a sample size of one, absolute proof for any position is impossible; uncertainty cannot be eliminated, and the past is a poor guide to the future. The decision whether to act on predictions of the doomsdayers or cornucopians must therefore be weighted by ethical considerations about future generations and risk. The worst-case scenario is acting on the beliefs of the cornucopians if the doomsdayers are correct. In contrast, suppose that we act on the beliefs of the doomsdayers and limit throughput; in that case, even if the cornucopians are correct, market forces will improve efficiency and provide substitutes, and no harm will have been done. If the neocornucopians are correct, the costs of addressing environmental problems are negligible or negative, so even if the chances of catastrophe are vanishingly small, we should act.

SEE ALSO *Consumption; Economics, Ecological; Economics, Environmental; Endangered Species Act; Energy; Environmental Law; Future Generations; Global Climate Change; Pollution; Population; Sustainability; U.S. Environmental Protection Agency.*

BIBLIOGRAPHY

Daly, Herman. 1987. "The Economic Growth Debate: What Some Economists Have Learned but Others Have Not." *Journal of Environmental Economics and Management* 14: 323–336.

Daly, Herman, and Joshua Farley. 2004. *Ecological Economics: Principles and Applications.* Washington, DC: Island Press.

Georgescu-Roegen, Nicolas. 1971. *The Entropy Law and the Economic Process.* Cambridge, MA: Harvard University Press.

Jevons, William Stanley. 1865. *The Coal Question.* London and Cambridge, UK: Macmillan.

Lomborg, Bjorn. 2001. *The Skeptical Environmentalist: Measuring the Real State of the World.* Cambridge, UK: Cambridge University Press.

Meadows, Donella H., Dennis L. Meadows, et al. 1972. *The Limits to Growth: A Report for the Club of Rome's Project on the Predicament of Mankind.* New York: Universe Books.

Meadows, Donella H.; Jørgen Randers; and Dennis Meadows. 2004. *Limits to Growth: The 30-Year Update.* White River Junction, VT: Chelsea Green.

Qiu, Jane 2007. "China's Green Accounting System on Shaky Ground. *Nature* 448: 518–519.

Schelling, Thomas C. 2007. "Greenhouse Effect." In *The Concise Encyclopedia of Economics*, ed. David R. Henderson. Indianapolis: Indianapolis Liberty Fund.

Simon, Julian. 1996. *The Ultimate Resource 2.* Princeton, NJ: Princeton University Press.

Solow, Robert. 1974. "The Economics of Resources or the Resources of Economics." *American Economic Review* 2: 1–14.

Stern, Nicholas, Siobhan Peters, et al. 2006. *Stern Review: The Economics of Climate Change.* London: HM Treasury.

Joshua Farley

M

MARSH, GEORGE PERKINS
1801–1882

George Perkins Marsh, diplomat, public servant, polymath scholar, and pioneer conservationist, was born in Woodstock, Vermont, on March 15, 1801, and died in Vallombrosa, Italy. Lawyer, congressman (1843–1849), U.S. envoy to the Ottoman Empire (1850–1854) and to Italy for an unparalleled twenty-one years (1861–1882), he was an intimate of the best-known scholars of his age on both sides of the Atlantic. A Dartmouth College graduate, Marsh was largely self-taught, notably in mastering twenty languages. Reputed in his own day as an authority on English language and literature, Marsh's enduring fame stems from his insightful study of human impacts on the environment. His classic *Man and Nature* revealed the menace of environmental misuse, explained its causes, and prescribed essential reforms. It linked the collapse of the Roman Empire with the soil exhaustion and erosion attendant on deforestation and exploitative agriculture, warning that similar pressures threatened the new world with the fate of the old. "The Earth is fast becoming an unfit home for its noblest inhabitant, and another era of equal human crime and human improvidence ... would reduce it to such a condition of impoverished productiveness, of shattered surface, of climatic excess, as to threaten the depravation, barbarism, and perhaps even extinction of the species." Continuously in print since its publication in 1864, *Man and Nature* inaugurated much of today's ethical stance toward nature.

In linking culture with nature, science with history, *Man and Nature* was the most influential text of its time next to Charles Darwin's *On the Origin of Species*. With Darwin, Marsh challenged traditional belief in preordained harmony between humanity and the rest of creation. Many before Marsh had noted various specific facets of environmental impact; none had seen or traced the combined effects of human action. Moreover, such influence had previously been assumed largely benign; damage was thought trivial or short-lived. Marsh was the first to see that human impacts were not only enormous and fearsome, but potentially cataclysmic and irreversible.

The long-held general belief was that mankind's God-given mission was to fructify the earth, and the bounty that followed forest clearing, swamp draining, and cultivation attested divine approval. Adverse side effects were easily dismissed, especially in the United States. Soils eroded or exhausted were simply vacated for new lands farther west; forests logged and burned seemed trifling by comparison with the wealth of timber beyond the horizon. Meanwhile, nature left alone would heal itself.

But this recuperative scenario accorded less and less with witnessed facts. A single lifetime saw vast tracts cleared, cultivated—and despoiled. Occupied nature did not heal itself; land once exploited and then abandoned remained for ages, if not forever, depleted. Greed was only partly to blame; much damage was unintended, often unseen. Men did not mean to derange nature; they were blind to the ruin they wrought. But myopia was not incurable. The harm done did not preclude dominion over nature, in Marsh's view; to the contrary, it mandated more alertly intensive governance. Awareness could prompt reform: technology deployed to break nature might also mend it. Ecological processes that revivified

George Perkins Marsh. *A scholar, diplomat, and author, Marsh became a world-renowned champion of environmental conservation and the first to study the potentially damaging impact of humans on the environment.* NPS PHOTO.

the environment could be protected and emulated by man as a coworker with nature. For all its dire portents, a pragmatic optimism pervaded *Man and Nature*. Marsh believed that men could learn to manage the environment for their own sake and for nature's good.

Appearing at the peak of Western resource optimism, *Man and Nature* refuted the myth of limitless plenty and spelled out needs for conservation. In forestry, hydrography, and pedology, Marsh's ecological insights and warnings became virtual gospel, animating watershed management and resource stewardship internationally. Floods and soil erosion during the Dust Bowl and other 1930s disasters rekindled its salience. A 1965 reprint of *Man and Nature* quickened the Earth Day crusade launched by followers of Rachel Carson and Aldo Leopold.

What is known—and feared—about anthropogenic impact has vastly expanded since Marsh's day; anxiety extends to realms and terrors undreamed of by him. Marsh's vision of a self-regulating equilibrium in nature, if left alone, became Frederic Clements's ecological paradigm of the early twentieth century. Although no longer scientifically credible, it continues to pervade popular conceptions of nature among the general public, including many environmentalists. Yet Marsh's appraisal of forest cover and erosion remains largely valid, his watershed control cautions still cogent, his call for stewardship ever more pertinent.

Following *Man and Nature*, Marsh continued to champion ecologically sustainable and socially desirable practices in forestry, irrigation (notably in the American west), and the establishment of parks and nature reserves. Traduced by some as a hard-nosed, resource-focused utilitarian pragmatist, against the aesthetic and ethical idealism of Henry David Thoreau and John Muir, Marsh, in fact, shared their concern for the preservation of wild nature.

Marsh believed social and environmental reform must go hand in hand. His deistic Calvinist-Enlightenment-transcendentalist faith stressed commitment to durable community, aligning respect for past with concern for future generations. To understand and care for nature required rooted residential attachments woefully lacking in pioneer America. Nevertheless, Marsh at first believed that enlightened self-interest would suffice to effect needed reforms. But the overwhelming might of amoral corporate capitalism persuaded him that governmental regulation was essential, lest America suffer the fate of ancient Mediterranean societies ruined by heedless environmental abuse.

SEE ALSO *Darwin, Charles; Future Generations; Leopold, Aldo; Muir, John; Pragmatism; Stewardship; Thoreau, Henry David.*

BIBLIOGRAPHY

Elder, John. 2006. *Pilgrimage to Vallombrosa: From Vermont to Italy in the Footsteps of George Perkins Marsh.* Charlottesville: University of Virginia Press.

Hall, Marcus. 2005. *Earth Repair: A Transatlantic History of Environmental Restoration.* Charlottesville: University of Virginia Press.

Lowenthal, David. 2000. *George Perkins Marsh, Prophet of Conservation.* Seattle: University of Washington Press.

Marsh, George Perkins. 2003 (1864). *Man and Nature; or, Physical Geography as Modified by Human Action*, ed. David Lowenthal. Seattle: University of Washington Press.

David Lowenthal

MARSHES

SEE *Wetlands.*

MENDES, CHICO
1944–1988

Francisco Alves Mendes Filho, better known as Chico Mendes, was born on December 15, 1944, on the Porto Rico rubber estate in the town of Xapuri, state of Acre, in the southwestern part of the Brazilian Amazon rain forest. The son of a rubber tapper, he began tapping rubber at age nine, accompanying his father through the forest as was customary for children at that time. Unlike most of his contemporaries, when he was fifteen, Chico was taught to read.

RUBBER PRODUCTION IN BRAZIL

The exploitation of rubber (*Hevea brasiliensis*), which is native to the Amazon, peaked at the turn of the nineteenth to the twentieth century and was characterized by the contrast between large fortunes and unjust social relations. Poor farmers from the northeast were recruited to work in the Amazon and, once there, were imprisoned by a system of indentured servitude for generations. The high price of rubber; the absence of state oversight in the areas of health, education, transportation services, and labor regulations; and isolation contributed to the social injustice. British rubber cultivation in Asia and the later development of synthetic rubber excluded the Amazon rain forest from the international rubber market.

During World War II a blockade of Asian rubber plantations led the U.S. government to sign agreements with Brazil to reactivate the old rubber estates (*seringais*). These were not plantations but native rubber areas. Brazil maintained that protective policy until the end of the 1960s, allowing thousands of rubber-tapper families to remain in the forest after the price of the raw material fell. Discouraged by low prices and unable to prove title to the land, the native rubber areas were left in the hands of the rubber tappers, who developed a way of life that was based on the use of forest-related products, including rubber, Brazil nuts, fish, game animals, and fruits.

However, that system did not last long. At the outset of the 1970s Brazilian government policy encouraged major landowners from the south to buy land in the Amazon basin for farming and cattle raising. To set up the new farms, forests had to be cleared, and conflict with the rubber tappers became inevitable.

ACTIVISM AND CHANGE

At the age of twenty Chico Mendes taught rubber tappers to read and write. He supported initiatives to help families become economically independent, an action that risked repression by the military regime. In 1973 he gained support from the Christian Base Communities, a Catholic initiative that supported the defense of human rights. In conjunction with the church, starting in 1975, the first rural worker unions began to appear in Acre. Chico soon became their leader.

In 1976, in Brasiléia, a neighboring town to Xapuri, rubber tappers decided to resist deforestation, organizing the first publicly recognized movement to defend the Amazon rain forest. Known as "*empate*" (nonviolent obstruction of tree felling), the tactic soon spread, giving rise to strong resistance to new farms. Although it involved defending the forest, the struggle actually was for social rights and conservation of the forest as a way of life for thousands of families.

After attempting to prevent deforestation, the rubber tappers sought to secure acknowledgment of their rights to the land and their need to make a living from forest-based products. With the support of anthropologists and researchers, the First National Meeting of Rubber Tappers was organized in Brasília in 1985. That meeting produced both the National Rubber Tappers Council, which continues to represent those groups, and the concept of extractive reserves (territorial areas protected by public authority, oriented toward conservation and sustainable use of extractive resources, regulated by concession contracts in accordance with a management plan approved by the environmental agency).

The confrontation between these two ways of using the land worsened in Xapuri, and on December 22, 1988, Mendes was murdered on orders from Darli Alves, a fugitive from justice for crimes committed in the southern part of Brazil. International support stimulated by prizes for and documentaries on Mendes's work from 1986 to 1988 led to the trial and imprisonment of the murderers and forced changes in Brazilian Amazon rain forest policy. In 1990 the first extractive reserves were created in the region, including one named for Mendes, with over 800,000 hectares, putting an end to land disputes in Xapuri.

From 1990 to 2007, 81 conservation units were established for traditional communities, with over 21 million hectares, accounting for 4.3 percent of the Brazilian Amazon and benefiting more than 400,000 people. Mendes's main legacy is the state's recognition of the right of such communities to a type of agrarian reform that respects traditional use of natural resources and ensures access to public education, health care, and development.

The legacy of Chico Mendes extended beyond the social groups to which he was linked. As a result of conflicts related to access to and use of natural resources, the concept of agrarian reform and development has changed, leading to the implementation of an innovative public policy throughout the Amazon basin. A new paradigm has emerged in which local communities that depend on the use of natural resources become conservation allies. As a result of rubber-tapper movements, public policies began prioritizing a new form of development that reconciles environmental protection with social justice. In the years after Mendes's murder that initiative came to be known as sustainable development.

SEE ALSO *Sustainable Development.*

BIBLIOGRAPHY

Keck, Margaret E. 1995. "Social Equity and Environmental Politics in Brazil: Lessons from the Rubber Tappers of Acre." *Comparative Politics* 27(4): 407–424.

Mendes, Chico. 1992. *Fight for the Forest: Chico Mendes in His Own Words.* London: Latin America Bureau.

Revkin, Andrew. 1990. *The Burning Season: The Murder of Chico Mendes and the Fight for the Amazon Rain Forest.* Boston: Houghton Mifflin.

Mary Helena Allegretti

MEXICO AND CENTRAL AMERICA

Environmentalism in Mexico and Central America must be understood in the context of the environmental problems facing that region. Among those problems is a population explosion. Since 1940 Mexico has gone from 28 million to over 100 million inhabitants. Demographic pressures and the model of development (mostly state capitalism in Mexico) have led to a drop in support for conservancy programs and the deterioration of soils, waters, forests, and wildlife. Additionally, the expansion of urban centers has precluded the pursuit of a path to sustainable development. Between 1950 and 1990 Mexico City went from 2 million to 20 million inhabitants. Because of high population densities, unemployment, and a high concentration of industries, many cities in Mexico and Central America are overwhelmed by problems of transportation, air and water quality, garbage disposal, and human health.

ECONOMIC FACTORS

Environmental problems across the region have been accentuated since the transition to a free-market global economy in the late 1980s and early 1990s. The implementation of the North American Free Trade Agreement (NAFTA) in 1994 and the Central American Free Trade Agreement (CAFTA), which includes El Salvador, Guatemala, Nicaragua, Honduras, and Costa Rica, in 2006 has been particularly controversial. Balancing environmental responsibilities with the demands of economic growth has been a difficult challenge for national governments, regional and local nongovernmental organizations (NGOs), and the corporate community.

The emblematic illustration of this problem is the multinational factories constructed in cities along the Mexican-U.S. border that are called *maquiladoras.* Multinational corporations escaped environmental and occupational constraints elsewhere by moving to Mexico and other countries in Central America that have less stringent environmental laws and low labor costs. At the same time relatively attractive wages in combination with drought and unemployment in the central highlands of Mexico emptied villages as young people migrated to work in the *maquiladoras.* As a consequence of that migration, villages and towns grew around factories without water-treatment plants or sewage systems while the *maquiladoras* were dumping toxic waste into local rivers.

Maquiladoras ***in Tijuana, Mexico.*** *Multinational factories called* maquiladoras *can be found along the Mexican-U.S. border. Towns and factories on the Mexican side of the border grew up around factories without water-treatment plants or sewage systems while the* maquiladoras *dumped toxic waste into local rivers.* LUIS ACOSTA/AFP/GETTY IMAGES.

POLLUTION AND CLIMATE CHANGE

Other environmental problems in the region that grew more serious in the last decade of the twentieth century and first decade of the twenty-first include chemical pollution of soils, water, and air in large plantations devoted to export-oriented agriculture; desertification and increased demand for irrigation and drinking water; deforestation, which adds to the problem of water depletion (in spite of policies and regulations, indiscriminate felling and illegal timber trafficking are rampant); illegal trade in fauna and flora; and scarcity of agricultural products for domestic human consumption.

The peoples and environments of Mexico and Central America are affected by the impacts of global climate change. The region has been hit particularly hard by hurricanes since 1988. Seven of the ten most intense Atlantic hurricanes ever recorded have occurred since that time, including Mitch in 1998, the second deadliest on record.

LEGISLATIVE ACTIONS AND NONGOVERNMENTAL ORGANIZATIONS

What is an equitable burden for Latin Americans to assume in the international effort to remedy and limit global change? The current politics of hydrocarbon fuels illustrates the complexity of the problem. The projected economic growth in Latin America and the scarcity of energy in international and regional markets have created a demand for increased production and consumption of hydrocarbon fuels. However, regional and local environmental problems have created a demand for a smaller carbon footprint and a need to address the human and environmental dimensions of the use of hydrocarbon fuels.

Mexicans and Central Americans are aware of current and past ecological problems. The relatively new discipline of environmental history is rescuing from oblivion the contribution of past environmentalists and historical examples of balancing economic development and environmental protection. In this respect Miguel Angel de Quevedo stands out as the most prominent environmentalist of the late nineteenth and early twentieth centuries.

Latin American countries have passed many laws to protect the environment. For instance, in 1988 the Mexican legislature passed the General Law of Ecological Balance and Environmental Protection, an integrated response to the environmental problems of the country and those of Mexico City in particular. About that time the government created the Secretariat of Urban Development and Ecology (SEDUE). Environmental initiatives have driven the regional integration of Central American countries, as evidenced by the creation of institutions such as the Central American Commission on Environment and Development (CCAD) in 1989 and the Central American Alliance for Sustainable Development in 1994.

The philosophy of sustainable development began to emerge in the 1980s. At that time Mexico became an academic leader in Latin America on the theory and application of sustainable development. The Mexican philosopher Enrique Leff, who became the coordinator of the United Nations Environment Program office in Mexico, made many contributions in this area.

The 1990s witnessed the formation of more than a thousand environmental NGOs in Mexico. The unification and alliance of several of them gave birth to the *Partido Verde Ecologista de México* (Green Ecologist Party of Mexico). A milestone in the history of environmentalism in Mexico was the formation of the *Grupo de los Cien* (Group of 100) in 1985, led by the writer Homero Aridjis and founded when a hundred writers and artists published a declaration against pollution. Over the years the *Grupo de los Cien* has called attention to the widespread and unremitting destruction of the Mexican natural environment.

Among the various approaches to environmental concerns in Mexico and Central America the environmental-justice approach has called attention to inequity in the distribution of environmental burdens and risks. For instance, the relationship between public health and ecology has long been understood and its neglect denounced, particularly in regard to urban pollution and the hazards of the *maquiladora* sector near the border.

As North American transnational companies have razed forests and mountains in Central America, a growing number of Catholic clergy (cardinals, bishops, and priests) have come to see protection of the land and water as godly work. Church and community groups have pointed to studies that have found high levels of heavy metals in the soil and water near mines across the region. In spite of the jobs created by those companies, mining and logging are depleting the water table, drying up wells, and forcing formerly productive farming regions to import food. Catholic environmentalism in the region, with its challenges to the powerful and solidarity with the poor, resonates with liberation theology in the minds of many. However, Catholic clergypersons seem to be increasingly in tune with the Vatican's position on the environment. Official Catholic teaching on the environment is based on the belief that creation is a gift of God that must be protected, used responsibly, and shared equitably.

THE ENVIRONMENT AND SOCIAL JUSTICE

The integration of environmental problems and social justice issues is an achievement of environmentalist thinking and activism in Mexico and Central America. In fact, it could be argued that the intellectual tradition in Mexico since the nineteenth century has resisted the typically North American erasure of the human from environmental thinking. The inclusion of history and people in the representation of landscapes in different media and arts to a great extent has been a reaction to a perceived obsession among foreign travelers for representing Mexico and Central America as a wilderness deprived of human history and large urban centers. This vindication of the city is problematic as it also reflects a history of imperial domination by the urban over the rural.

The environmentalist positions taken by the Zapatista rebellion in 1994 in the state of Chiapas in southeastern

Mexico illustrates the complexity of the human-nature/urban-rural environmental thought characteristic of the region. Zapatista documents connect the call for environmental protection with the rights of rural citizens against the centralized authority of the government in Mexico City, political ecology (access to and control of natural resources), and environmental justice. The environmentalism of the Zapatistas also includes the spirituality and biocentrism of traditional indigenous values, which nurture not only Zapatista environmental philosophy but also its commitment to a political struggle without violence.

One response to the environmental crises produced by modernity has been the unearthing and reawakening of forgotten or repressed traditions. This cultural trend can be seen in the recovery of traditional environmental knowledge such as traditional agricultural techniques in Mesoamerica. Environmental and indigenous organizations have established alliances to promote sustainable agriculture and community autonomy in rural Mexico. However, this can lead, as in the case of the Nobel laureate poet and essayist Octavio Paz, to the deployment of poetry for purposes of exposing the legacy of political and ecological destruction over the course of the twentieth century. The connection between embracing environmental thinking and rejecting the legacy of violence associated with utopianism and millenarianism in the twentieth century should not be overlooked, particularly in the context of the civil wars that bled Central America in the 1980s. Mexican and Central American literature is an ideal site for observing this connection, as in the work of José Emilio Pacheco, Pablo Antonio Cuadra, and Ernesto Cardenal.

If the awareness of ecological problems and their many human dimensions among the public and government is not the issue, what is needed to address those problems in Mexico and Central America? The consensus is that in spite of the popularity of environmentalism, the region lacks a powerful environmental movement that would offer unified and strong political support for environmental justice and the protection of wildlife. The challenge has not been environmental awareness and progressive environmental policy and legislation; it has been the inability of governments to implement environmental policy and enforce environmental laws to protect ecosystems and the people who inhabit them.

SEE ALSO *Agriculture; Deserts and Desertification; Environmental Activism; Global Climate Change; Nongovernmental Organizations; North American Free Trade Agreement; Pollution; Sustainable Agriculture; United Nations Environment Programme.*

BIBLIOGRAPHY

Binns, Niall. 2004. *¿Callejón sin Salida? La Crisis Ecológica en la Poesía Hispanoamericana.* Zaragoza, Spain: Prensas Universitarias de Zaragoza.

Howard, Philip. 1998. "The History of Ecological Marginalization in Chiapas." *Environmental History* 3(3): 357–377.

Miller, Shawn William. 2007. *An Environmental History of Latin America.* New York: Cambridge University Press.

Roberts, J. Timm, and Nikki Demetria Thanos. 2003. *Trouble in Paradise: Globalization and Environmental Crises in Latin America.* New York: Routledge.

Simon, Joel. 1997. *Endangered Mexico: An Environment on the Edge.* San Francisco: Sierra Club Books.

Simonian, Lane. 1995. *Defending the Land of the Jaguar: A History of Conservation in Mexico.* Austin: University of Texas Press.

Wright, Angus. 2005. *The Death of Ramon Gonzalez: The Modern Agricultural Dilemma,* rev. edition. Austin: University of Texas Press.

Jorge Marcone

MICROBES

Microbes carry out the functions of life. These entities, too small to see with the naked eye, have the wherewithal to reproduce with sufficient variation to enable evolutionary processes to proceed continually by natural selection. Humans, as with all other large animals and plants, are made up of tens of trillions of cells, whose dimensions are generally in the range of 10 to 30 microns (thousandths of a millimeter). Indeed, each such cell is actually a colony of some 1 to 500 smaller cells, with a 1-micron diameter, called mitochondria and chloroplasts—cells the size of free-living bacteria. Each such cell collective may be regarded as a single ecosystem. Thus ecosystems may range in size from volumes of less than a microliter (a 0.1-millimeter-sided cube) to a system the size of the planet Earth. In addition to cellular life forms, there are also viruses. These entities, having diameters of 20 to 200 nanometers (thousandths of a micron), may also be included in the category of microbes because they carry within them the genes that enable their reproduction, although they are incapable of reproducing in the absence of other living microbial cells.

While life at the microbial level began some 3.5 to 4 billion years ago, ethics is a relatively new invention of humans, who began to use words with the intention of affecting the behavior of other humans within the last 0.1 to 0.2 million years or so. Biologically, ethics are used to enhance the survival and reproductive chances of individual humans and/or groups, societies, or nation-states. Microbes play a role in determining the qualities of the environments that we humans inhabit. The composition of the air we breathe is largely determined by free-living microbes and by chloroplasts, microbes that have become entrapped within the cells of higher plants. Thus if

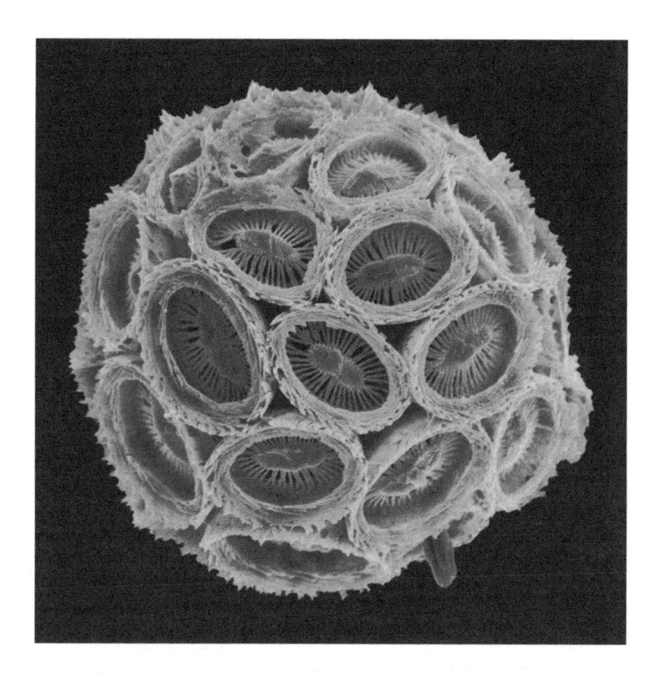

Coccolithopore, a Microbial Plant. *Microbes, too small to see with the naked eye, play a relatively large role in the maintenance of ecosystems. The coccolithopore, an algal organism seen here with scanning electron micrograph takes carbon dioxide from the air and converts it into calcium carbonate plates.* STEVE GSCHMEISSNER/SCIENCE PHOTO LIBRARY/PHOTO RESEARCHERS, INC.

environmental ethics seeks to enhance the survival and reproductive chances of human beings and human civilization, microbes and the ecosystems in which they vitally function should be included within the purview of environmental ethics.

In 1772 Daniel Rutherford discovered the nitrogen cycle driven by bacteria. Bacteria transform the nitrogen in dead organisms to gaseous ammonia and elemental nitrogen as a byproduct of their own metabolic processes. Other bacteria convert these two gases to nitrogenous salts, such as ammonium nitrate, as a byproduct of their metabolic processes. These salts eventually become incorporated, initially, into the proteins of other microbes, and then, via an extensive food chain, into the proteins of higher organisms, upon whose deaths the nitrogen becomes gaseous once more. A similar cycle processes carbon. In the case of carbon, the fossil fuels we burn, the food we eat, and our bodies when we die are in part transformed into carbon dioxide, which is liberated into the atmosphere to become a greenhouse gas. Carbon dioxide in turn becomes food for algae (microbial plants) and plants, which convert this gaseous carbon dioxide into sugars and carbohydrates. The coccolithophore, one such alga, lives in the oceans and is instrumental in acquiring carbon dioxide from the air and converting it into calcium carbonate plates of considerable beauty. This chalk then sinks to the bottom of shallow seas, where it forms the limestone of many coastal cliffs and hills.

Such cycles normally operate in the absence of human intervention. Yet humans modify them for their own purposes. For example, Fritz Haber in 1908 developed a process for artificially producing ammonia from nitrogen. Ammonia later became used to generate fertilizers for increasing crop yields. As a result, when nitrogen-rich salts run off fertilized land, river waters become enriched, and wild plants flourish in them. Here our human ethics requires us to produce food to feed people at the expense of polluted rivers. Similarly, humans interfere in the carbon cycle by producing some 26 billion tons of carbon dioxide per year, in addition to the more than 700 billion tons produced by the respiration of other living organisms both on land and in the oceans. While this provides more carbon dioxide for plants and coccolithophores to thrive on, it is also held to be responsible for part or all of the 0.6-degree-centigrade increase in the temperature of the planet that has occurred over the last century. How to reconcile the requirements of human ethics with those of environmental ethics is a complex and thorny problem.

Many diseases of humans, animals, and plants have microbial causes. The human response to this is to generate cures based on the products of other microbes grown in fermenters. For example, antibiotics such as penicillin are produced by the fungus *Penicillium chrysogenum*. This approach to diseases, mostly caused by bacteria, was received with open arms in the 1940s, when penicillin became widely available. Unfortunately, the overuse of antibiotics has led to the evolution of microbes resistant to such treatments. An alternative approach to diseases caused by microbes (bacteria, viruses, protozoa, and helminths) is to prevent their occurrence with vaccines. This approach led to the elimination of smallpox in 1977 and is close to eliminating polio. As we are beset with new diseases (HIV/AIDS, SARS, the avian flu), new approaches to the generation of vaccines are yielding effective agents to combat such diseases.

At the small end of the scale of living organisms, microbes have a greater influence on the environment than any other type of organism, humans included. They may hold many of the keys to solving the environmental crises that face us.

SEE ALSO *Disease; Energy; Evolution; Global Climate Change; Nanotechnology; Pollution.*

BIBLIOGRAPHY

Elliot, Robert, ed. 1995. *Environmental Ethics*. Oxford: Oxford University Press.

Light, Andrew, and Holmes Rolston, eds. 2002. *Environmental Ethics: An Anthology*. Malden, MA: Blackwell Publishing.

Spier, Raymond E. 2001. *Ethics, Tools, and the Engineer.* Boca Raton, FL: CRC Press.
Vesilind, P. Aarne, and Alastair S. Gunn. 1998. *Engineering, Ethics, and the Environment.* Cambridge, UK: Cambridge University Press.

Raymond E. Spier

MIDGLEY, MARY
1919–

The British philosopher Mary Midgley was born in Dulwich, East London, on September 13, 1919. She graduated from Oxford University in philosophy and ancient history in 1942 and worked as a civil servant and teacher during World War II. In 1950 she married and moved north to Newcastle, where she concentrated on reviewing books and bringing up a family. It was not until 1963 that she took up a lecturing position in the philosophy department at Newcastle University. After "retiring" in 1980, Midgley stayed in Newcastle, where she has continued to work as one of a rare breed: a freelance philosopher. She has published extensively in ethics and environmental philosophy and is an active social commentator on animal ethics, environmental issues, Gaia theory, and the role of science in society.

Animals and Why They Matter was published in 1983, at a time when ethical theory was still firmly entrenched behind human lines. She argued, alongside Peter Singer and a handful of other ethicists, that ethical obligations extend beyond our own species, carefully unraveling and exposing the intellectual confusions that lead to a denial of interspecies ethics and setting out the implications of cross-species ethical obligations for European and North American ethical thought. Unlike Singer's orthodox utilitarian approach to animal ethics, Midgley's is based on the more nuanced concept of human-animal "mixed communities." Her communitarian approach to animal ethics facilitates an otherwise elusive integration of animal ethics with the popular "land ethic" of Aldo Leopold, which is based on the concept of the "biotic community." Midgley's early focus on animals extended into a seminal discussion of our ethical relationships to the wider environment in the essay, "Duties concerning Islands."

A constant theme in Midgley's work is the attempt to uncover the often unnoticed intellectual frameworks or worldviews against which we live our lives, to show how these patterns of thinking affect how we live and act, and, above all, to open them up for critical scrutiny. In the context of environmental and social justice issues, the need for new mental models is, she argues, as crucial as the need for cleaner and more efficient technology. European and North American societies still struggle with corrosive forms of social atomism and individualism that militate against sustained effort to deal with collective human problems, such as hunger, on a global scale. And the dominant view of the environment as a set of resources for people, and of ourselves as detached managers of these resources, has arguably contributed to problems such as climate change and species extinction that now loom so large.

Drawing on Gaia theory as a key source, Midgley attempts to reconstruct our background worldview in a way that is not only more felicitous but that will afford profound understanding of our environmental problems—and inspire us to act. Earth, in her view, is not a lifeless jumble of commodities but an immense, living complexity of interrelated systems. She argues that we are not independent of this living whole but are deeply immersed in it. Once this simple truth is understood—emotionally as well as intellectually—then our reasons for taking care of the environment as an overriding priority become utterly and compellingly clear.

Midgley's work combines careful, detailed analysis with a constant return to wider questions about the place of humans and human activities—especially science—in the bigger scheme of things. This outlook has enabled her to show that apparently conflicting perspectives are, in fact, complementary aspects of a wider whole and that these positions can and should be reconciled. She has argued in this way against the polarization of animal welfare and environmental concerns. She reveals the close connections between reason and emotion through her exploration of the human mind in its evolutionary context, and she emphasizes the importance of imagination in reasoning processes. She argues, in works such as *Science and Poetry* and *The Myths We Live By,* against the supposed opposition between science and myth, science and spirituality. In all these cases she eschews the "intellectual tribalism" whereby one perspective or dimension of an issue is taken to be the whole story and is then ranged in battle against other perspectives. This mistake, she argues, can be noticed only when the wider context is brought into focus. Once this is done, it is possible to achieve bridge building and reconciliation, which in turn have tremendously constructive implications, both practical and theoretical.

The work of "bending thought around to look at itself" or "thinking about thinking," although difficult, is also intensely practical. Midgley's position reveals the potential of philosophy in general—and environmental philosophy in particular—as a practical activity that is both radical and necessary, affording us the much needed ability to rethink our economic social and political institutions when they become problematic—for example, by leading to ecological collapse. Her chief contribution to environmental philosophy is thus an approach concerned not with cleverness and abstraction but with wisdom, with insight, and with change.

SEE ALSO *Animal Ethics; Environmental Philosophy: V. Contemporary Philosophy; Gaia Hypothesis.*

BIBLIOGRAPHY

WORKS BY MARY MIDGLEY

1978. *Beast and Man: The Roots of Human Nature.* Ithaca, NY: Cornell University Press.

1981. *Heart and Mind: The Varieties of Moral Experience.* London: Methuen.

1983. *Animals and Why They Matter: A Journey around the Species Barrier.* Harmondsworth, UK: Penguin.

1983. "Duties concerning Islands." In *Environmental Philosophy,* ed. R. Elliot and A. Gare. Milton Keynes, UK: Open University Press.

1984. *Wickedness.* London: Routledge.

1989. *Wisdom, Information, and Wonder: What Is Knowledge For?* London: Routledge

1991. *Can't We Make Moral Judgments?* Bristol, UK: Bristol Press.

1992. *Science as Salvation.* London: Routledge.

1994. *The Ethical Primate.* London: Routledge.

1996. *Utopias, Dolphins, and Computers: Problems of Philosophical Plumbing.* London: Routledge.

2001. *Gaia: The Next Big Idea.* London: Demos.

2001. *Science and Poetry.* London: Routledge.

2004. *The Myths We Live By.* London: Routledge.

2005. *The Owl of Minerva: A Memoir.* London: Routledge.

Kate Rawles

MILL, JOHN STUART

SEE *Environmental Philosophy: IV. Nineteenth-Century Philosophy; Utilitarianism.*

MILLENNIUM ECOSYSTEM ASSESSMENT

The Millennium Ecosystem Assessment (MA) documents how people are transforming their environment and how environmental transformation is affecting human well-being. As a scientific assessment the report is supposed to be free of ethical judgments. At the same time the intention of the MA is to inform the public and policy makers so that ethical judgments embedded in behavioral and policy decisions are scientifically informed.

The MA was initiated largely by biological scientists who, at the end of the last millennium, were already concerned about the effects of ecological transformations on humanity's future. The four-year study was carried out by some 1,300 environmental scientists, economists, and other social scientists from developed and developing countries who came into the assessment with a wide range of perspectives on issues such as the prospects for new sustainable technologies or the possibility of market solutions to environmental problems. The study frames people's relation to nature in mostly economic terms. Ecosystems are portrayed as natural capital from which ecosystem services flow in support of the human economy, whereas human activities modify and typically deplete nature's capital and thereby also affect the flow of services and future well-being.

KEY FINDINGS

The four main findings of the 2,500-page study, published in four volumes and titled *Ecosystems and Human Well-Being* (2005a), are as follows:

1. Over the past 50 years, humans have changed ecosystems more rapidly and extensively than in any comparable period of time in human history, largely to meet rapidly growing demands for food, fresh water, timber, fiber, and fuel. This has resulted in a substantial and largely irreversible loss in the diversity of life on Earth.
2. The changes that have been made to ecosystems have contributed to substantial net gains in human well-being and economic development, but these gains have been achieved at growing costs in the form of the degradation of many ecosystem services, increased risks of nonlinear changes, and the exacerbation of poverty for some groups of people. These problems, unless addressed, will substantially diminish the benefits that future generations obtain from ecosystems.
3. The degradation of ecosystem services could grow significantly worse during the first half of this century and is a barrier to achieving the Millennium Development Goals (adopted by the United Nations in 2000).
4. The challenge of reversing the degradation of ecosystems while meeting increasing demands for their services can be partially met under some scenarios that the MA has considered, but these involve significant changes in policies, institutions, and practices, that are not currently under way. Many options exist to conserve or enhance specific ecosystem services in ways that reduce negative trade-offs or that provide positive synergies with other ecosystem services (Millennium Ecosystem Assessment 2005b, Synthesis Volume, p. 1).

The assessment investigates five key stressors on global ecosystems—habitat change, climate change, invasive species, overexploitation, and pollution (including both toxics and nutrification)—across thirteen broad ecosystem types: forest (boreal, temperate, and tropical), dryland (temperate grassland, Mediterranean, tropical grassland/savanna, and desert), inland water, coastal, marine, island, mountain, and polar. For the five drivers across thirteen ecosystem types, there was only one combination, habitat change in temperate forests, where the impact of the driver was lessening. Thus, for sixty-four of sixty-five possibilities, stresses on ecosystems are increasing.

The MA investigated alternative policy options to decrease, or even reverse, stresses on ecosystems. The elimination of subsidies that encourage destructive practices such as land conversion and greenhouse gas release would be effective in reducing many kinds of ecosystem transformations. Reducing ecosystem stressors further, however, requires combinations of new economic incentives, behavioral changes, and technological innovations, the specific mechanisms and combinations of which will vary across ecosystems; as well as social, cultural, and political contexts. Integrating complex, context-specific policy responses across stressors, ecosystems, geographic scales, and political boundaries is difficult to even contemplate, let alone describe coherently for policy makers.

IMPACT ON ENVIRONMENTAL UNDERSTANDING, VALUES, AND POLICY DECISIONS

The findings of the MA have been widely cited in academic and popular literature. Material from the MA can be found in the readings of many college courses and textbooks. Numerous regional assessments around the world are amassing knowledge for subsequent global analysis. Nevertheless, the impact of the MA on environmental understanding, values, and policy decisions remains unclear. While the climate assessments of the Intergovernmental Panel on Climate Change (IPCC) have been crucial in shaping climate knowledge, eliciting fresh concerns for the future and focusing policy debates, the MA is much less focused, and so its impacts on people's values, corporate decisions, and public policy are harder to pinpoint.

ETHICAL AND PHILOSOPHICAL ISSUES WITHIN THE ASSESSMENT PROCESS

The choice of a scientific framework for the MA has ethical implications simply because different frameworks highlight, and hence differentially value, different aspects of a problem. The assessment's characterization of ecosystems as capital and their benefits as services evinces a utilitarian ethics and an economic worldview. Perhaps this economic framework accounts for the MA's favorable portrayal of the economic approach known as "payments for ecosystem services." Selecting scientists from developing countries concerned with improving the material well-being of the poor balances the more ecocentric views of scientists from the developed countries. Conventional economists avoid the term *overconsumption*, but the inclusion of ecological economists and sociologists in the work of the assessment has led to an extensive exploration of this issue. Hence even scientific assessments reflect underlying, implicit ethical predispositions and decisions, even if they are never explicitly formulated.

The MA is an example of a new approach to applying science to the complex interactions between social and natural systems. This approach, used most famously in the climate assessments of the IPCC, entails the use of several thousand scientists from a wide range of disciplines to address key policy questions to explore and summarize the findings of the latest scientific literature. Although the focus of the assessment is on the interaction between social and natural systems, natural scientists' portrayals of ecosystems rarely include people, and social scientists rarely include nature in their descriptions of social systems. Numerous other methodological problems had to be overcome, including recognizing and transcending disciplinary assumptions and language in achieving insights into the issues of natural capital and ecosystem services. One of the most important byproducts of the MA was the training of numerous scientists to think and communicate across disciplinary boundaries and to identify critical questions for future research.

SEE ALSO *Ecology: III. Ecosystems; Ecosystem Health; Environmental Policy; Future Generations; Global Climate Change; Intergovernmental Panel on Climate Change.*

BIBLIOGRAPHY

Millennium Ecosystem Assessment. 2005a. *Ecosystems and Human Well-Being.* 4 vols. Washington, DC: Island Press. Available from http://www.millenniumassessment.org/en/Condition.aspx.

Millennium Ecosystem Assessment. 2005b. *Ecosystems and Human Well-Being: Synthesis Reports.* Washington, DC: Island Press. Available from http://www.millenniumassessment.org/en/Synthesis.aspx.

Mitchell, Ronald B.; William C. Clark; David W. Cash; and Nancy M. Dickson, eds. 2006. *Global Environmental Assessments: Information and Influence.* Cambridge, MA: MIT Press.

Norgaard, Richard B. 2008. "Finding Hope in the Millennium Ecosystem Assessment." *Conservation Biology* (April 10).

Richard B. Norgaard

MINING

This entry contains the following:

I. OVERVIEW

Mining is one of the most controversial uses of land. The extraction of materials from the earth is an invasive, nonreversible process that often causes extensive environmental damage, but it is the only known method of acquiring sufficient supplies of many of the raw materials needed to support human well-being. According to U.S. Geological Survey statistics, 2004 world mining production included 14.6 million metric tons of copper, 2,430 metric tons of gold, and approximately 1,340 million metric tons of iron ore.

Mining can be done only in places where there is an economically viable deposit of a desired material. Mineral deposits are the result of complex earth processes that concentrate certain elements in specific locations. Nature determines the distribution of mineral deposits and thus the possible locations for mining; people can choose whether to mine deposits that have been discovered but cannot dictate where the deposits are situated. Mines are often a long distance from the markets that consume their products. From the earliest times flint, salt, gold, and other commodities were traded extensively, making mining the first global industry.

Three sectors dominate the mining industry: metals such as copper, gold, lead, zinc, iron, and nickel; energy minerals, including coal and uranium; and industrial minerals such as sand and gravel for construction, clays for ceramics, and salt for many industrial uses. Each sector uses a range of extraction techniques and faces environmental issues related to the nature of the ore deposits, their chemical composition, and the environmental context of any specific deposit.

HISTORY

The history of mining is linked closely to developments in metallurgy and mineral processing and influenced by political and economic circumstances. Copper, gold, and lead were used in the Middle East by 3500 BCE. Early mining focused on collecting nuggets of pure metal. The recognition of veins of metal-bearing minerals and improved smelting methods to separate metals from rock were the foundations for the Bronze Age and Iron Age. The Romans were the most noteworthy ancient miners; in an advance that allowed miners to reach depths of over 650 feet, they replaced some slave labor with mechanical devices to drain water from mines. Technical innovations were few until the fifteenth and sixteenth centuries, when two methods of removing silver from copper ore—liquation and mercury amalgamation—led to the exportation of silver from the New World to the Old World.

Widespread industrial use of coal started in the eighteenth century in Britain and the nineteenth century elsewhere. Coal often occurs in aerially extensive flat-lying beds, unlike the narrow, often steeply dipping metal-bearing veins that were the main source of most metals except iron until the twentieth century. Coal at the surface of the earth is amenable to large-scale open-pit mining. It is difficult to support the roof of underground coal seams when the coal has been removed; this, combined with the combustible nature of coal and the associated gases, makes underground coal mining particularly dangerous.

TECHNOLOGY AND SOCIAL ASPECTS

Many advances in mining have been tied to improved sources of power. People wielding picks, wedges, and crowbars were the main source of energy until recent times. Explosives were not employed in Europe until the seventeenth century; their use increased greatly after Alfred Nobel's 1863 invention of dynamite. The introduction of steam power in the 1700s was significant, particularly as it powered the Cornish pump, which could remove large volumes of water from deep mines. Steam power was replaced by compressed air in the late 1800s and by electricity in the early 1900s. Much of the industry has been mechanized by combining these forms of power with modern, efficient rock-cutting materials and the introduction of robotics. Mechanized mining has enabled the economic development of large-scale open-pit mines, particularly for coal, iron, and low-grade deposits of base metals.

Important advances in processing include the 1886 discovery of the cyanidation process, in which cyanide is used to dissolve gold from crushed rock; the development in the early 1900s of the flotation process to separate copper, lead, and zinc from finely ground sulfide ore; the adaptation in the 1980s of solvent extraction-electrowinning to process copper and zinc oxide ores by chemically leaching the metal out of crushed rock and then extracting it from solution by electrical methods instead of smelting the ore; and the introduction of bioleaching, which uses bacteria to extract metal from crushed rock. In situ leach mining is an approach that was developed in the 1980s. Instead of removing and crushing the rock and then extracting the metals, a dilute chemical solution

Mining for Coal, West Virginia. *Draglines, such as this one at a mine in Boone County, West Virginia, can weigh millions of pounds and are tall enough to loom over a 20-story building. They are used to expose mountaintop coal seams. Mining practices can cause severe environmental damage in terms of physical ground impact and possible contamination of water, air, and soil.* PHOTO COURTESY VIVIAN STOCKMAN/WWW.OHVEC.ORG.

is circulated through the ore deposit in the ground to leach out the metals. The metal-bearing solution then is collected through extraction wells and processed to obtain the metals. Modern processing methods are generally less environmentally hazardous than older techniques.

The societal aspects of mining have been as important as the technological advances. The Roman Empire structured mining operations throughout much of Europe from about 100 BCE to 500 CE; independent miners returned to the fore when the empire collapsed. Spanish monarchs controlled mining in Central and South America in the sixteenth century. The middle to late nineteenth century saw a great territorial expansion in mining when independent miners opened up the goldfields of the western United States, western Australia, and southern Africa. In the first decade of the twenty-first century it was estimated that thirteen million people were operating similar small-scale independent, or artisanal, mining enterprises, particularly in developing nations. Small-scale informal mining poses special environmental and social problems because it operates largely outside the rule of law.

Modern mining companies emerged around the start of the twentieth century, when capital costs rose to levels far beyond the resources of individual miners. Major multinational mining companies based in Australia, Brazil, Britain, Canada, Chile, China, South Africa, and the United States came to dominate the industry. The balance of power between states, corporations, and local communities is constantly in flux, with alternating periods of nationalization and privatization of resources and growing awareness in the formal mining sector of the importance of sustainable development, social license, and reliable governance.

ENVIRONMENTAL IMPACTS

The environmental impacts of traditional mining can be severe. The physical impacts include large surface pits, noise and vibration during operations, subsidence caused by underground workings, piles of unconsolidated waste material leading to unstable ground, dust pollution, and the influx of sediment into watercourses. Chemical contamination of surface and ground water, air, and soil, which may be long lasting, can be caused by acid mine drainage and the release of metals or the chemicals used to process ore. Little attention was paid to the environmental consequences of mining until the middle to late twentieth century, when social agitation and government regulation forced mining companies to address the issue.

Almost all the environmental impacts of mining can be mitigated, although at a price. Most modern mining companies budget for and implement environmental mitigation programs in all their projects; it is more challenging both legally and technically to remediate the effects of past mining activities and ongoing small-scale mining. It is also difficult to quantify the social impacts of mining and develop appropriate mitigation strategies. Though the benefits of mining are distributed throughout society, the social and environmental costs are concentrated in areas where mines are situated. Defining appropriate distributions of the costs and benefits of mining and determining how much society is willing to pay to mitigate the environmental and social impacts in return for a reliable supply of raw materials are major challenges facing governments, civil society, and the mining industry.

SEE ALSO *Alternative Technology; Energy; Habitat Loss; Land Ethic; Mining: II. Acid Mine Drainage; Pollution; Sustainable Development; Technology.*

BIBLIOGRAPHY

International Institute for Environment and Development and World Business Council for Sustainable Development. 2002. *Breaking New Ground: Mining, Minerals, and Sustainable Development: The Report of the MMSD Project.* London and Sterling, VA: Earthscan Publications.

Lynch, Martin. 2001. *Mining in World History*. London: Reaktion.

Plumlee, Geoffrey S., and M. J. Logsdon, eds. 1999. *The Environmental Geochemistry of Mineral Deposits Part A: Processes, Techniques, and Health Issues. Reviews in Economic Geology*, vol. 6A. Littleton, CO: Society of Economic Geologists.

Temple, John. 1972. *Mining: An International History.* New York: Praeger.

U.S. Geological Survey. 2006. "Mineral Commodity Statistics." In *Historical Statistics for Mineral and Material Commodities in the United States,* comp. Thomas D. Kelly and Grecia R. Matos. U.S. Geological Survey Data Series 140. Available from http://www.minerals.usgs.gov/ds/2005/140.

Maeve A. Boland

II. ACID MINE DRAINAGE

Mineral resource production is vital to modern industrialized societies. The environmental consequences of mineral production, however, have resulted in degradation and destruction of air, water, soil, land, and biological resources. Quantitative estimates are difficult to make, but tens to hundreds of thousands of mine sites in the United States and hundreds of thousands more worldwide probably have caused environmental damage (Custer 2003; Young and Ayres 1992). Environmental restoration is expensive, time-consuming, and a substantial challenge to scientists and engineers. The total costs to clean up mine sites in the United States are thought to run to a few hundred billion dollars (Mining, Minerals, and Sustainable Development 2002). One of the most injurious consequences of metal mining is the formation of acid mine drainage (AMD), a major contaminant to surface and ground waters.

AMD is acidic effluent water discharged from metal mines or mineral-processing wastes containing high concentrations of acidity, sulfate, and metals that are toxic to most forms of life except certain species of microbes (Nordstrom and Alpers 1999). When fresh rock containing pyrite, FeS_2, is exposed to air and water, a rapid reaction ensues, catalyzed by acidophilic bacteria. The general reaction is represented by

$$FeS_2 + \frac{15}{4} O_2 + \frac{7}{2} H_2O \rightarrow Fe(OH)_3 + 2H_2SO_4$$

in which pyrite is oxidized and dissolved, sulfuric acid is produced, and some form of ferric hydroxide is precipitated. The precipitated iron can be seen as orange-red to brown staining or coating on rocks and sediments in mined areas. Other coexisting minerals dissolve rapidly in sulfuric acid and contribute other metals (copper, zinc, cadmium, lead, cobalt, nickel, chromium, aluminum, and manganese) to the water. Such effluent from mine portals, adits, waste rock, and tailings piles typically has pH values of 2 to 4, although in extreme cases the pH can decrease to below 0 (Nordstrom et al., 1999). The high metal concentrations are a greater source of toxicity than is the low pH. In contrast, drinking water has pH values of 6 to 8, and most metals are insoluble in those conditions.

Acid Mine Drainage. *A journalist with Danish Public Broadcasting and Owen Stout of Cabin Creek, West Virginia, examine acid mine drainage at the base of Kayford Mountain, about an hour south of West Virginia. The contaminated water flows constantly from an abandoned deep mine. The West Virginia Department of Environmental Protection has known about the acid mine drainage for years. The tainted water flows into Cabin Creek, which makes its way to the Kanawha River, then flows into the Ohio River, and on to the Mississippi River.* PHOTO COURTESY VIVIAN STOCKMAN/WWW.OHVEC.ORG.

Accidental releases of impounded AMD have caused major devastation to some rivers; for example, the Aznalcollar impoundment in southern Spain broke and released about 7 million tons of acid slurry into the Guadiamar River, contaminating agricultural land, destroying aquatic biota, and threatening the largest European bird refuge, Doñana National Park (Grimalt and Macpherson 1999).

Acid rock drainage (ARD) is a more general term that refers to both AMD and acidic drainage that occurs naturally in mineralized areas where there has been no mining. Mineralized and unmined areas often produce acid waters with elevated concentrations of metals (Runnells et al. 1992). The geochemical process is the same, but mining greatly enhances the rate by exposing much more pyrite-bearing rock to air and water. One estimate from studies at Iron Mountain, California, indicates that mining increased the sulfide-weathering rate by at least two orders of magnitude (Alpers et al. 2007).

The large-scale and hydrogeochemical complexities of the contamination have hindered remediation efforts, although remediation has succeeded at several mine sites worldwide (Fernandez-Rubio 2004). The most common treatment for AMD is lime/limestone neutralization, an effective short-term solution. Because AMD can continue for hundreds to thousands of years, long-term, passive treatment is recommended. Research on wetlands treatment, water management, disposal practices, land reclamation, and phytoremediation holds promise for the long term. New mines can greatly reduce the environmental consequences of mining by incorporating environmental planning and monitoring into the mine plans before mining begins. This approach also saves considerable amounts of money compared to the costs of environmental restoration after mining is finished and compared to legal costs when mining companies are sued for the damages they caused.

Foreign mining companies have taken advantage of the lack of regulatory authority in some underdeveloped poor countries and caused large-scale pollution of resources that destroy or harm the livelihood of native peoples. Mining companies have an ethical and economic obligation to minimize or prevent damage to air, water, soil, land, and biological resources and to prevent injuries to local communities as they exploit mineral resources for the needs of modern society.

SEE ALSO *Ecological Restoration; Habitat Loss; Pollution; Rivers; Sustainability; Water.*

BIBLIOGRAPHY

Alpers, Charles N.; D. Kirk Nordstrom; Ken L.Verosub; and Catherine Helm-Clark. 2007. "Paleomagnetic Determination of Pre-Mining Metal-Flux Rates at the Iron Mountain Superfund Site, Northern California." *Eos Transactions of the American Geophysical Union* 88(23) Jt. Assem. Suppl. Abstract GP41B-04.

Custer, Kelly. 2003. *Cleaning Up Western Watersheds: A Report for the Mineral Policy Center.* Available from http://www.earthworksaction.org/publications.

Fernandez-Rubio, Rafael. 2004. "Sustainable Mining: Environmental Assets." In *Mine Water 2004, Process, Policy, and Progress,* ed. A. P. Jarvis, B. A. Dudgeon, and P. L. Younger, vol. 2. Newcastle upon Tyne, UK: International Mine Water Association.

Grimalt, Joan O., and Enrique Macpherson. 1999. "The Environmental Impact of the Mine Tailing Accident in Aznalcollar (Southwest Spain)." *Science of the Total Environment* 242(1–3): 1–332.

Mining, Minerals and Sustainable Development. 2002. "Mining for the Future. Appendix C: Abandoned Mines Working Paper." *Mining, Minerals and Sustainable Development* 28: 1–20.

Nordstrom, D. Kirk, and Charles N. Alpers. 1999. "Geochemistry of Acid Mine Waters." In *The Environmental Geochemistry of Mineral Deposits Part A: Processes, Techniques and Health Issues,* ed. Geoffrey S. Plumlee and M. J. Logsdon. Littleton, CO: Society of Economic Geologists.

Nordstrom, Darrell Kirk; Charles N. Alpers; Carol J. Ptacek; and David W. Blowes. 1999. "Negative pH and Extremely Acidic Mine Waters from Iron Mountain, California." *Environmental Science & Technology* 34(2): 254–258.

Runnells, Donald D.; Tom A. Shepard; and Enrique E. Angino. 1992. "Metals in Water: Determining Natural Background Concentrations in Mineralized Areas." *Environmental Science & Technology* 26(12): 2316–2322.

Young, John E., and Ed Ayres, eds. 1992. *Mining the Earth.* Washington, DC: Worldwatch Institute.

D. Kirk Nordstrom

III. MOUNTAINTOP REMOVAL

Mountaintop removal is a method of strip mining for coal in mountain regions. Its violent effect on nature and society has generated public controversy in the United States, where it has been practiced since the mid-1980s. Though mountaintop removal mining is done worldwide, its oldest and largest footprint appears in central Appalachia in an area predicted to be the size of the state of Delaware by 2012. The massive scale of mountaintop removal, its irreversible destruction of biologically diverse ecosystems, and its legacy of postbiotic landform complexes distinguish mountaintop removal from earlier forms of strip mining.

THE METHOD

Mountaintop removal involves stripping all vegetative cover and then blasting open a mountain to loosen the overburden, the name the coal industry uses for the communities of soil and resident life forms, defined as "material of any nature ... that lies on top of a deposit of useful materials, ores or coal" (Squillace 1990, p. 138). Mountain ranges have been reduced by as much as 1,000 feet in Kentucky and West Virginia, where up to 2,000 metric tons of explosives are used daily. In the wake of blasting, a twenty-story-high excavator known as a dragline then scoops up the exploded materials and loads them into rock dump trucks, which deposit them in an adjacent hollow to form what is known as a valley fill. The U.S. Environmental Protection Agency (EPA) has predicted that by 2012 more than 2 million acres of the central Appalachian coalfields will have been exploded, mined, regraded, and revegetated in this manner. By the first decade of the twenty-first century more than 2,000 miles of headwaters, including intermittent and perennial mountain streams, had been damaged severely or buried beneath mine spoil (U.S. Environmental Protection Agency 2005). The coal industry defends mountaintop removal mining as the safest and most cost-effective way to meet the energy needs of a nation that produces more than half of its electricity in coal-fired utility plants.

HISTORICAL BACKDROP: LAND ACQUISITION, TECHNOLOGY, AND SOCIAL COMPACTS

The relationship between the coal industry and coalfield communities is characterized by what the anthropologist Marshall Sahlins has called negative reciprocity. In the extractive, nonrenewable coal economy the industry must take out more than it puts back to make a profit. In the twentieth century that negative reciprocity was codified and protected through "instruments of writing" (legal documents) that resolved disputes over land, labor, and resources in favor of the coal industry (Williams 2001).

Those instruments, which form de facto social compacts, include the Broad Form Deeds executed between the 1880s and the 1930s, the 1950 accord between the United Mineworkers and the Bituminous Coal Operators, and the 1977 Surface Mine Control and Reclamation Act (PL 95–87), also known as SMCRA.

The Broad Form Deeds transferred mineral rights to absentee landowners through negotiations that allowed farmers to continue traditional patterns of land use. Between the 1880s and the 1920s thousands of transactions were drawn up between land companies and coalfield residents, deeding mineral rights to the companies in exchange for clear title to the surface. By means of the Broad Form Deed residents gave land companies unfettered access to minerals and timber in exchange for the right to continue farming the land. Ninety percent of the land in that region is owned outright by absentee companies.

In the 1960s and 1970s strip mining was used increasingly throughout the region, with the externalities of acid runoff, deadened streams, and deforested and scarred landscapes. Despite opposition from local communities, strip mining was justified as the most efficient way to get at seams of coal that were too thin to retrieve through underground mining, and residents were held to unfettered access for absentee owners that had been guaranteed by the Broad Form Deeds decades earlier. The giant earthmoving machines used during that period were diminutive precursors of the contemporary draglines and rock dump trucks. The strip mining of the 1960s and 1970s, referred to as highwall, contour, or augur mining (and now as prelaw, that is, before SMCRA, mining), operated by cutting a wedge into the side of a mountain, simultaneously creating a wall perpendicular to a level bench. The bench supported a machine that drilled or cut away the coal from the wall. Refuse, or gob, consisting of mine dust, shale, clay, and poor-quality coal was disposed of in nearby hollows and impounded behind slag dams built of larger chunks of shale and rocks. Water that accumulated behind the impoundments increased the likelihood of severe flooding during heavy rains.

In the 1960s the Pittston Coal Company built a series of slag dams in the Middle Branch at the head of Buffalo Creek in Logan County, West Virginia. In 1972 the uppermost dam burst after a heavy rainfall. What witnesses described as a "wall of blackwater" (Erikson 1976)—132 million gallons—roared through the narrow seventeen-mile-long valley of Buffalo Creek, killing 125 people, injuring 1,100, and leaving 4,000 homeless in a matter of minutes.

That disaster precipitated the 1977 passage of the SMCRA. In addition to prescribing safety standards for surface mining and reclamation operations, SMCRA integrated provisions of the National Environmental Policy Act, the Clean Air Act, the Clean Water Act, and the Historic Preservation Act to facilitate citizen input regarding the propriety and legality of surface mining activities. Some citizen groups, such as Save Our Cumberland Mountains, opposed SMCRA, arguing that environmentally responsible strip mining is not possible (Stacks 1972). In a key provision of SMCRA, Section 515c(1), a coal company is exempt from the requirement to restore the land to its approximate original contour (AOC) if it can demonstrate that flat land will be of greater economic value for postmining development. Claiming that flat land can serve postmining developments such as industrial parks, prisons, shopping malls, airports, and golf courses, coal companies have petitioned for and been granted this exemption in nearly all mountaintop removal mining permits filed since the mid-1980s. In the aggregate the fine print in such agreements has provided the loopholes through which the coal industry has internalized its profits while externalizing its costs.

SOCIAL AND ECOLOGICAL COSTS

The offsite impacts, also referred to as negative externalities, of strip mining and coal-fired utilities plants include the pollution of air, water, and soils; forest decline as a result of acid rain (Ayers et al. 1998); and boom and bust economies at the sites of extraction, resulting in government subsidies during recurrent periods of unemployment and disasters such as mine explosions and flooding. Mountaintop removal mining adds several unprecedented negative externalities to this list.

Impacts on Coalfield Communities Beginning in the mid-1980s, preparations for mountaintop removal mining in the Appalachian Mountains took place out of public view and beyond the awareness of nearby communities. Local citizens first learned that something was changing when customary routes through the mountains were closed. Eventually the use of explosives began shaking their homes, cracking foundations, and destroying water sources, including spring-fed streams and wells. The blasting also precipitated blowouts of water that had accumulated in underground mines, sometimes with enough force to flatten buildings. Rivers grew shallow with siltation and the runoff from denuded slopes and during heavy rains made creeks overflow their banks and roar through the hollows, twisting road signs, ripping apart bridges, and carrying away vehicles, animals, and sometimes people, eleven of whom drowned in the floods of 2001 and 2002. Millions of gallons of blackwater are stored in hollows near coal preparation plants that tower above schools and homes. Black plumes, indicating discharges, in local streams and rivers are reported frequently. As a storekeeper in Stickney, West Virginia,

Aerial View of Mountaintop Removal, at Marthatown, West Virginia. *The mountaintop removal method of strip mining for coal, practiced in the U.S. since the mid-1980s, is largely controversial for the devastating effects the practice causes to an ecosystem. The states of West Virginia and Kentucky have declared states of emergency several times since 2000, due to the severe environmental pollution caused in their Appalachian regions by this form of mining.* FLYOVER COURTESY SOUTHWINGS.ORG/PHOTO COURTESY OF VIVIAN STOCKMAN/WWW.OHVEC.ORG.

put it: "We fear the river above more than the river below" (Hufford 1995, p. 543).

Since 2000 the governors of Kentucky and West Virginia have declared states of emergency in mountaintop removal counties almost annually. On October 11, 2000, a Massey Coal Company impoundment broke in Martin County, Kentucky, sending 300 million gallons of slurry into the Tug Fork River, thirty times more pollution than was released by the *Exxon Valdez*. The EPA called it the worst environmental disaster in the history of the southeastern United States.

Impacts on Biodiversity and Ecology The social and ecological impacts of mountaintop removal mining become clear when one considers that the coves and ridges it destroys support the world's most biologically diverse temperate-zone hardwood forest system. Although this forest is threatened by acid rain from fossil fuel combustion, mountaintop removal adds a further dimension to the ecological assault. In traditional strip mining the location of undisturbed forest above the disturbed mine site allowed for the revegetation of sites downslope with native species. In mountaintop removal disturbed ground upstream of everything else must be reclaimed with exotic species that will grow in nutrient-poor conditions (Edmunds and Loucks 1998).

Postbiotic landform complexes have been replacing an ecosystem that evolved over 300 million years into the world's most biologically diverse temperate zone hardwood system. This system of cove, ridge, and valley associations includes 40 canopy species, with another 40 in the understory, in addition to 1,200 species of botanicals.

Never glaciated, the climate-ameliorating coves sheltered the seed stock that reforested the eastern United States after the Ice Age. The ecologist E. Lucy Braun named this system the mixed mesophytic. Ecologists call it the mother forest. Human interactions with this forest over many generations have resulted in a rich store of knowledge about how to engage its biodiversity. The effects of destroying this region and replacing it with desert species that can survive in the nutrient-poor soils left by strip mining are incalculable. Some ecologists say that it could take 500 years or more to grow hardwoods in such a place; in that span of time the animating bond between collective memory and ecology that formed thousands of years before the European conquest of North America will have been broken.

ETHICS, PHILOSOPHY, POLITICS AND THE LAW

The controversy over mountaintop removal points out a number of contradictions in the system that is supposed to protect private property rights while safeguarding public goods such as air and water quality, soils and biodiversity, and the cultural heritage of a community's surroundings.

Lawsuits and Legislative Battles It was not until the mid-1990s that mountaintop removal mining drew national attention and citizens began organizing against it. After it made the headlines in major journals and newspapers and was featured on television networks, coalitions of local, state, regional, and national citizen groups began campaigning around this issue. In 1998 a West Virginia lawyer named Joe Lovett represented a group that included residents of Pigeon Roost Hollow, the West Virginia Highlands Conservancy, and the Ohio Valley Environmental Coalition. In a landmark case, *Bragg v. Robertson,* the group sued state and federal agencies for breaching a provision of the Clean Water Act that prohibit mining activities within 100 feet of a stream. In allowing mining companies to create valley fills by dumping mine spoil in streams, the plaintiffs argued, the government had failed to protect the stream buffer zone. In 1999 a federal judge ruled in favor of the plaintiffs, noting that valley fills violate the Clean Water Act. Two other judges later issued concurring rulings. The industry has appealed those rulings, and its lobbyists have been working to persuade federal legislators to revise the definition of fill material so that the U.S. Army Corps of Engineers may continue authorizing the dumping of mine waste into headwaters. A protracted battle has continued on the legislative and judicial fronts over the rights of coal companies to violate the stream buffer zone.

Politics: Technology's Subversion of Democracy The effort to overturn the stream buffer zone provision of the Clean Water Act ignores the function that political philosophers ascribe to substantive goods in a democratic polity. Hannah Arendt ascribes twin functions to the physical commons, which serves both to unite stakeholders while forming a buffer that preserves distinct positions (1958). The stream buffer zone functions in both ways. The coal industry, supported by the international banking system, invested its profits in a technology that displaced human workers and communities, and then argued that the laws protecting the commons of clean air and water had to be changed to accommodate the technology of giant earth-moving machinery. Shifting from citizens to engineers the authority to make environmental decisions, the coal industry and its supporters in state and national governments privatize the public goods that give citizens a voice in government. Destroying the headwaters, the coal industry destroys the political grounding of local, regional, and national publics. For local communities, the streams and their names are landmarks for navigating contemporary and historical space. The names for streams also form a set of mnemonic cues to the histories of mountain community life going back many generations. The streams are therefore vital resources for cultural identity and reproduction. Regions may be defined as the ecological limits of large watersheds, in which citizens share a biological stake in the quality of waters upstream. Protected by federal law, the headwaters also provide an anchorage in public life for every citizen in the United States, for whom the capacity to protect the health of headwaters is an indicator of viable democracy. The changing of laws to accommodate technology shifts decision-making authority to corporate engineers, diminishing the substantive commons, and dangerously weakening local, regional, and national democratic capacity.

Epistemological Errors Mountaintop removal mining epitomizes a fundamental error of Western epistemology—the Cartesian mind/body split. This is manifested as the separability of humans from their environments, and the location of human communities outside of ecological systems. Phenomenologists such as Martin Heidegger and Maurice Merleau Ponty, and pragmatists such as John Dewey have identified conceptual trailheads to the way out of this problem. Heidegger, in his *The Question Concerning Technology,* warns against an orientation to technology as a means of development that "challenges forth" a desired outcome rather than "bringing forth" what is already underway in existing relationships. In this regard, he mentions strip-mining and hydro-electric dams, which extract energy as something that can be stockpiled and which threaten to freeze the world (including human beings) into a set of "standing reserves" (Heidegger 1977). Sundering what Dewey calls "the bond that binds the living creature to his environment," (Dewey 1934,

p. 252), the fixed identity of the standing reserve prevents the "bringing forth" of that which is implicit in existing relationships between human and non-human nature. Such relationships would form the object of development in democracy as Dewey envisions it: "the creation of a freer and more human experience in which all share, to which all contribute." (1993, pp. 244–245).

Linking Social and Environmental Justice The federal policies that benefit the coal industry and maintain the status of coal as a source of more than half the electricity produced in the United States have turned the central Appalachian Mountains into a standing reserve of coal. This fixed identity blocks the development of potential alternatives. One term commonly used for the region, *the national sacrifice zone,* highlights the disadvantage at which residents are placed because they experience a disproportionate percentage of the negative externalities generated to meet national energy demands. The groups placed at such a disadvantage are often ethnic minorities who share the conditions of poverty and unemployment that are used to rationalize the concentration of undesirable forms of production and employment such as mineral extraction, petrochemical production, and low-paying factory and service industry jobs.

Some activist groups have used the concept of environmental justice to combat mountaintop removal. That concept highlights the interconnectedness of social and environmental issues in a region that has been neglected by mainstream environmental groups as well as government agencies. In 2003, for example, the Sierra Club mounted an environmental justice campaign against mountaintop removal, and a coalition of groups based in West Virginia, Tennessee, Kentucky, and southeastern Virginia has sponsored Mountain Justice Summer, an annual training camp for citizens interested in working on alternatives to mountaintop removal. Building relationships around the commons of headwaters and biodiversity among citizens from disjunct regions and sectors around the country, such efforts promise to engender and sustain the "genuine care for common experience" called for by Herbert Reid and Betsy Taylor (citing McDermott 1987, p. 91; see also Reid and Taylor, 2003) foundational to democratic capacity and mountaintop retention.

SEE ALSO *Biodiversity; Dams; Energy; Environmental Activism; Environmental Justice; Environmental Law; Environmental Policy; Habitat Loss; Land Ethic; Mountains; Pollution.*

BIBLIOGRAPHY

Appalachian Center for the Economy & the Environment. Available from http://www.appalachian-center.org/index.html.

Appalachian Land Ownership Task Force. 1983. *Who Owns Appalachia? Land Ownership and Its Impact.* Lexington: University Press of Kentucky.

Arendt, Hannah. 1958. *The Human Condition.* Chicago: University of Chicago Press.

Ayers, Harvard; Jenny Hager; and Charles E. Little. 1998. *An Appalachian Tragedy: Air Pollution and Tree Death in the Eastern Forests of North America.* San Francisco: Sierra Club Books.

Braun, E. Lucy. 1950. *Deciduous Forests of Eastern North America.* Philadelphia: Blakiston Co.

Coal River Mountain Watch. Available from http://www.crmw.net

Dewey, John. 1934. *Art as Experience.* New York: Minton, Balch, and Co.

Dewey, John. 1993. *The Political Writings,* ed. Debra Morris and Ian Shapiro. Indianapolis: Hackett Publishing Co.

Edmunds, Stacy, with Orie Loucks. 1998. "A Landscape View of Mountaintop Removal Mining." MA Thesis Oxford, OH: Miami University.

End Mountaintop Removal Action and Resource Center: http://www.ilovemountains.org.

Erikson, Kai. 1976. *Everything in Its Path: Destruction of Community in the Buffalo Creek Flood.* New York: Simon and Schuster.

Heidegger, Martin. 1977. *The Question Concerning Technology, and Other Essays,* ed. William Lovitt. New York: Harper & Row.

Hufford, Mary. 1995. "Context." *Journal of America Folklore* 108 (430): 528–548.

Hufford, Mary. 2005. *Waging Democracy in the Kingdom of Coal: OVEC and the Struggle for Social and Environmental Justice in Central Appalachia 2002–2003.* Philadelphia: Center for Folklore and Ethnography. Available from http://www.sas.upenn.edu/folklore/center/waging_democracy2.pdf.

McDermott, John J. 1987. *The Culture of Experience: Philosophical Essays in the American Grain.* Prospect Heights, IL: Waveland Press.

Montrie, Chad. 2002. *To Save the Land and People: A History of Opposition to Surface Coal Mining in Appalachia.* Chapel Hill: University of North Carolina Press.

Ohio Valley Environmental Coalition. Available from http://www.ohvec.org.

Reid, Herbert, and Betsy Taylor. 2003. "John Dewey's Aesthetic Ecology of Public Intelligence and the Grounding of Civic Environmentalism." *Ethics and the Environment* 8 (1): 74-92.

Sahlins, Marshall. 1972. *Stone Age Economics.* Chicago: Aldine-Atherton.

Squillace, Mark. 1990. *The Strip Mining Handbook: A Coalfield Citizens' Guide to Using the Law to Fight Back Against the Ravages of Strip Mining and Underground Mining.* Washington, DC: Environmental Policy Institute and Friends of the Earth.

Stacks, John F. 1972. *Stripping.* San Francisco: Sierra Club.

U.S. Environmental Protection Agency. 2005. *Mountaintop Mining/Valley Fills in Appalachia: Final Programmatic Environmental Impact Statement (Final PEIS).* Available from http://www.epa.gov/region03/mtntop/index.htm.

West Virginia Division of Culture and History. Available from http://www.wvculture.org.

West Virginia Highlands Conservancy. Available from http://www.wvhighlands.org.

Williams, John Alexander. 2001. *West Virginia: A History.* Morgantown: West Virginia University Press.

Mary T. Hufford

MOUNTAINS

Mountains are dominant features of our physical environment. Covering more than one-fourth of the earth's land surface, mountains occur on all continents and at all latitudes. They are found in 75 percent of the world's nations and provide a home for 10 percent of its human population. Mountains provide abundant natural resources and ecosystem services that support and sustain both mountain dwellers and lowland residents.

Mountains also are rich in cultural significance. Around the globe and throughout history, humans have responded to the evocative presence of mountains in diverse—and sometimes contradictory—ways. These responses have been shaped by the cultural conditions out of which they emerge and find expression in a dizzying array of beliefs and practices, myths and stories, philosophical ideas, and works of art. These socially constructed meanings frame the ways in which human beings perceive, understand, and relate to mountains and mountain environments.

PHYSICAL CHARACTERISTICS

The chief characteristics of mountains are verticality and slope, which together produce wide climatic variability and biological diversity over relatively short distances. For this reason, mountains often serve as ecological islands and refugia (areas where isolated populations of formerly widespread species still exist) that provide critical habitats for many species of rare, often endemic, flora and fauna.

Along with great diversity, mountains are notable for their high geomorphic energy and fragility. The same tectonic processes that build mountains (faulting, folding, thrusting, subduction, earthquakes, and volcanoes) also contribute to their fragility and instability. The presence of steep slopes covered with young, erosive soils further adds to the fragility of mountain environments and increases the potential for severe, long-lasting ecological disturbance.

The diversity and fragility of mountain environments is mirrored in the rich—and threatened—cultural diversity found among mountain communities throughout the world. Relative isolation and inaccessibility, as well as diverse ecosystems, have contributed to a great variety among mountain cultures; at the same time, remoteness and isolation have marginalized these communities and contributed to extreme poverty in some mountain regions. Twentieth-century advances in communications and transportation technology have greatly increased accessibility to mountain regions, bringing both opportunities and challenges to traditional mountain cultures.

MOUNTAIN RESOURCES AND ECOSYSTEM SERVICES

Mountains provide a number of valuable natural resources and ecosystem services for both mountain communities and lowland populations. Many of the plants used as foods and medicines around the world come from mountain regions. The Andes, for example, are the source of more than 200 indigenous varieties of potatoes, and the Himalayas account for approximately 2,000 varieties of rice. The world's population uses medicines developed from plants that grow in mountain environments.

Mountain regions contain 28 percent of the world's forested areas. Besides providing a number of important resources, including timber, fiber, fuels, foods, and fodder for livestock, these forests contribute countless ecosystem services. They provide stability for steep mountain slopes, intercept precipitation, reduce soil erosion, protect water quality, moderate surface temperatures, and provide habitats for numerous forest-dwelling organisms. For many years their relative inaccessibility protected mountain forests and their subsistence-based communities. In recent years, however, growing demand for forest products and increased access to mountain regions have greatly accelerated the rate of deforestation in these areas.

Mountains shape hydrological processes in numerous ways. They influence precipitation patterns, store fresh water in snowfields and glaciers, and regulate the direction and flow of streams and rivers. Most of the major river systems of the world have their headwaters in mountain regions, and mountains provide 80 percent of the earth's surface water. More than one-half of the world's population depends directly upon mountains for the water they use to drink, grow crops, generate electricity, and support industrial processes.

MOUNTAINS AND MINING

Mountains also contain rich deposits of mineral resources, and the history of mining is intimately tied to the history of humanity's relations to mountains, because the tectonic forces responsible for creating most of the world's mountain ranges also provide the intense heat and pressure needed to produce ore deposits of economically valuable metals such as gold, silver, and copper. The discovery and extraction of mountain minerals, as well as key developments in metallurgical processes, have had a dramatic influence on the rise—and fall—of many great civilizations around the world. Gold from the highlands of Nubia and the Sinai Peninsula supplied the great wealth of ancient Egypt. The mountains of southern Spain provided a rich source of tin, lead, silver, copper, iron, and gold for Phoenicians, Carthaginians and other early Mediterranean civilizations. Silver, lead, and gold from the Laurium mines in the mountains of Greece supported the rise of ancient Athens, while the rich gold and silver mines of Mount Pangaeus helped finance the rise of Macedonia under King Philip and Alexander the Great. For more than 400 years, the hills and mountains of Spain, Greece, Britain, Anatolia, Austria, and Transylvania supplied the gold, silver, lead, iron, and tin (a key ingredient in the production of bronze) that contributed to the great material wealth and military might of the Roman Empire.

During the Middle Ages, mountainous regions in Saxony and Bohemia became centers of metal mining in Europe. Between the tenth and twelfth centuries, major copper, silver, and lead mines were established in the Harz Mountains of Lower Saxony and the Erzgebirge ("Ore Mountains") region near the headwaters of the Elbe River. In the centuries that followed, a variety of social, political, and economic factors contributed to the continued expansion of mining activities in Europe, as well as to significant advances in mining techniques and metallurgical processes. These developments culminated in a mining boom in Europe that began in approximately 1450; within the next hundred years, the production of silver in Germany increased fivefold, with much of this production coming from the Harz Mountains.

Rumors of golden cities of El Dorado and "mountains that gushed silver" fueled the imagination of fortune-seeking Europeans arriving in the Americas in the sixteenth century. Less than five years after Hernán Cortés conquered the Aztecs in 1521, Spanish mines were producing silver in several locations across Mexico. Within the next fifty years, major silver strikes in Guanajuato and Zacatecas resulted in the establishment of silver mines throughout the Sierra Madre that yielded tremendous wealth for the new colonial power.

With Francisco Pizarro's defeat of the Inca Atahuallpa in 1533, this pattern repeated itself in the Andes Mountains of Peru, Ecuador, and Bolivia. In 1545 Spanish colonists learned of the rich silver deposits lying beneath Cerro de Potosi in Bolivia, and within a year the Potosi mine was extracting silver from the 15,827-foot mountain. Soon, Cerro de Potosi became known as Cerro Rico ("rich mountain"), and the Spanish idiomatic expression, *valer un potosi* ("worth a potosi"—a fortune) appeared in Miguel de Cervantes' *Don Quixote.* Between 1556 and 1783, the mines of Potosi produced more than 45,000 tons of pure silver. It also is estimated that more than 8 million Indian miners died while working in the Potosi mines.

The discovery of precious metals and other minerals helped drive the westward expansion of the United States and the settlement of the country's western mountains. In 1848 gold was discovered at John Sutter's sawmill, and the California gold rush began in the foothills of the Sierra Nevada. This was followed by major gold strikes during the 1850s and 1860s in places such as Carson City, Nevada; Leadville, Silverton, and Telluride, Colorado; and Helena, Virginia City, and Butte, Montana. Although gold mining in Butte proved to be short-lived, the presence of vast silver and copper reserves earned the city its reputation as "the richest hill on earth." In 1876 European American settlers discovered gold in the Black Hills of South Dakota, which resulted in the establishment of the Homestake Mine in 1877 and the removal of the Sioux Indians from their traditional homelands.

Since the 1990s mountaintop-removal mining has become an increasingly common method of extracting coal from the Appalachian Mountains of the eastern United States. In this process, forests, top soil, and overburden rock are removed from the mountaintops to expose seams of bituminous coal. The coal is then extracted and processed, while the overburden and waste rock are left in nearby valleys and hollows as valley fill. Waste water from processing methods is known as slurry and is impounded in storage pools on site. At current rates, the United States Environmental Protection Agency estimates that approximately 1.4 million acres in the Appalachian Mountains (primarily in West Virginia and eastern Kentucky) will have been mined using mountaintop removal methods by 2010. Although this process is lucrative for mining companies, critics cite a host of environmental and social problems associated with mountaintop-removal mining, including deforestation, habitat loss, surface and groundwater contamination, air pollution, mudslides, and flooding.

People have always looked to mountains as storehouses of valuable natural resources and potential wealth. Humans have used mountain minerals to make tools and weapons; establish currencies; decorate their bodies, artwork and buildings; supply energy; and drive industrial processes. However, the extraction, processing, and use of this mineral bounty have been accompanied by significant environmental and social costs. The tension between the benefits and costs of mineral development in mountain regions has profoundly influenced how humans understand their relationship with mountain environments and their role as agents of environmental change profoundly influenced how people relate to mountain environments and how they understand humanity's role as an agent of environmental change.

Mountaintop Removal. *The towering dragline, center, is dwarfed by the size of this mountaintop removal operation in Boone County, West Virginia, May 2003.* FLYOVER COURTESY SOUTHWINGS.ORG/PHOTO COUTESY OF VIVIAN STOCKMAN/WWW.OHVEC.ORG.

MOUNTAINS AND THE CULTURAL IMAGINATION

In addition to providing an abundant supply of the natural resources and ecosystem services needed to sustain human communities, mountains are extremely rich in cultural significance. Throughout history and across the globe, mountains have spoken to the human sense of the sacred, and the human response to this powerful address has manifested itself in countless forms of religious expression.

Sacred Mountains In many cultures throughout the world, mountain peaks are considered dwelling places of the gods. For the ancient Greeks, Mount Olympus was the home of Zeus and palace of the twelve major gods. For many Hindus, the god Shiva and his consort Parvati (daughter of the god Himalaya) reside on Mount Kailas in Tibet. Other imposing Himalayan peaks also are associated with the divine couple, including Shivling (India), Nanda Devi (India), and Annapurna One (Nepal). Among Tibetan Buddhists, Mount Kailas is the home of Demchog, the Buddha of Supreme Bliss. In Japan, Mount Fuji is the abode of the goddess Konohana Sakuya Hime; in Hawaii, the goddess Pele dwells atop the volcano Kilauea. All major Andean peaks are inhabited by mountain gods known as *apus,* with the supreme *apu* residing atop towering Mount Ausangate (Peru). Many of these sacred mountains serve as temples, shrines, altars, and pilgrimage sites, where devotees go to pray, do penance, and perform sacrifices. For the faithful these mountains serve as fountains of physical and spiritual blessings, bestowing—or withholding—a wide range of divine gifts, including rain, fertility, healing, wisdom, happiness, and success.

In other cultures the gods and their messengers may not live atop mountains, but the mountain heights serve as important meeting places where the divine and human realms intersect. In these cases mountains are sites of

Mount Kailas, Tibet. *Mount Kailas is a sacred site to both Buddhists and Hindus in western Tibet. In addition to their often symbolic importance, mountains provide many valuable natural resources and ecosystem services.* © ISTOCKPHOTO/TCP.

divine revelation. For example, Moses went up Mount Sinai to receive the law and establish the covenant between Yahweh and the children of Israel. In a cave on Mount Hira, the Archangel Gabriel visited Muhammad and began the series of revelations that are recorded in the Koran. For the Plains Indians of North America, the tops of mountains and buttes serve as vision quest sites, where people seek out divine power, protection, and guidance through fasting and prayer.

Mountains also establish sacred geographies. Numerous mountains throughout the world act as the *axis mundi*, or world axis, that stands at the center of the universe, organizing space and uniting the various levels of existence. For Hindus and Buddhists the mythic mountains of Meru and Sumeru, respectively, are located at the center of the world, embodied in the physical form of Mount Kailas. For the Hopi Indians of the southwestern United States, the San Francisco Peaks are the location of the *sipapuni*, or center hole, from which the people emerged and through which they maintain contact with ancestors dwelling below. For the ancient Greeks a similar center hole or world navel (*omphalos*) was located on the slopes of Mount Parnassus, near the site of the famous temple of the oracle at Delphi. Other mountains stand not at the center but at the periphery of the world, holding up the heavens, ordering the landscape, and establishing boundaries. For the Navajo, Apache, and Pueblo Indians of the American Southwest, identifiable peaks mark the four cardinal directions that define and enclose their respective worlds.

The sacred power of mountains does not always manifest itself as divine; it also may take the form of the demonic. Mountains have often been experienced as terrifying and desolate places that inspire fear. For centuries the European Alps were considered a forbidding and dangerous place populated by witches, dragons, ogres, and demons. Mount Hekla, an active volcano in Iceland, was considered the entrance to hell. In many cultures throughout the world, mountains are places from which one's ancestors came and to which, upon death, they return.

Contemporary Mountain Meanings in Europe and North America As these examples and countless others suggest, mountains have held tremendous religious significance

MOUNTAIN TOURISM IN NEPAL

The environmental and social impacts of mountain tourism in the Himalayas are most evident in Nepal. Although closed to foreign visitors as recently as 1950, over the past several decades Nepal has promoted mass mountain tourism through private-sector initiatives with minimal regulation, monitoring, and control. As a result, nearly one million overseas tourists now visit Nepal each year. Most of these visitors participate in some form of mountain tourism, with approximately 25 percent engaging in mountaineering or trekking. The great majority of Nepal's mountain tourism activity is concentrated in the regions surrounding the Annapurna Conservation Area, Langtang National Park, and Sagarmatha National Park (Mount Everest), areas made famous through the writings of foreign mountaineers exploring Nepal in the 1950s and 1960s.

The tremendous influx of foreign tourists since the 1970s has had significant environmental impacts. Increased demand for fuelwood has resulted in dramatic losses in tree and shrub cover. This problem is especially acute in higher-elevation alpine zones, where the harvesting of slow-growing juniper shrubs for fuelwood has resulted in increased soil erosion and denuded landscapes. Other environmental problems associated with high-volume mountain tourism include degraded trails, contaminated water supplies, and large amounts of garbage. The severity of the latter problem has led some to dub Mount Everest the "highest dump on earth."

Mass tourism also has had tremendous social and economic effects on Nepal. On the positive side it generates revenues for tourism-related enterprises and creates employment opportunities. For the Sherpas of the Everest region and for the Gurung, Thakali, and Magar living near Annapurna, tourism revenues have invigorated local economies, raised the standard of living, and supported numerous infrastructure improvements and community-development projects. On the negative side, the influx of tourist money has inflated the costs of basic consumer goods and services and increased economic inequalities and social stratification in some local communities. Much of the revenue generated by tourism does not remain in local mountain communities; instead, it flows back to tourist-generating countries and Nepalese urban centers like Kathmandu. Most tourism jobs pay low wages and are subject to seasonal fluctuations. Negative social consequences associated with mass tourism include the deterioration of traditional values, loss of cultural practices and lifeways, and alienation caused by overwhelming numbers of foreign visitors. During peak seasons the number of nonresidents in the Everest region may be four times greater than the number of native Sherpa residents.

For the people of Nepal, mountain tourism has been, at best, a mixed blessing. Overcoming the host of environmental and social problems associated with mass tourism will require the development of sustainable and equitable policies and effective regulatory practices that conserve fragile mountain environments and benefit local communities.

throughout human history. In contemporary European and North American society, many traditional religious meanings of mountains have taken on secular form. No less powerful than their religious counterparts, these secularized mountain meanings contribute to a rich constellation of ideas and images that continue to inform contemporary understandings of, and relationships to, mountains. For example, the hulking and immovable presence of mountains continues to suggest stability and permanence in a rapidly changing world. Their lofty, snow-covered peaks bespeak purity, wildness, beauty, and transcendence, and they are looked upon by many with awe, wonder, and reverence. They may symbolize one's greatest struggles, trials, and challenges, as well as one's highest goals, vision, and aspirations. They are sought out as sources of inspiration, transformation, and recreation, where many go to test and renew themselves—physically, spiritually, and emotionally.

In Europe and North America, these various mountain meanings have become so well-established that it is tempting to take them as universal. The idea that mountains might evoke images of the demonic rather than divine, that they might represent negative qualities of hubris and unbridled ambition, or that they might be viewed as hideous and ugly deformities seems implausible to many contemporary Europeans and North Americans. Nevertheless, these negative mountain meanings were precisely those that prevailed among many Europeans as recently as the seventeenth century. In *Mountain Gloom and Mountain Glory: The Development of the Aesthetics of the Infinite* (1997 [1959]), Marjorie Hope Nicolson

ALPS

The Alps are a domain of great biological diversity, because of widely varying climate, exposure to sunlight, geology, morphology, hydrology; and also of cultural diversity, with widely varying language, religion, ethnic groups, history and politics. Yet, in any part of the Alps, patterns of life, economy, social and spatial organization are also similar in many respects.

As the main mountain range in Europe, the Alps were for centuries an obstacle to human communication and travel. They were viewed as a hostile terrain due to the imagined presence of negative supernatural forces, and the (often actual) presence of threatening natural or human forces. By the eighteenth century, however, the Alps had begun to attract artists and scientists, and became an example of an inhospitable environment victoriously challenged by humans. Its "sublime" character fascinated visitors, and this "delightful horror" contributed to the growing popularity of Alpinism and tourism. Humans, insignificant and weak, face alpine Nature, infinite and powerful, but they win the confrontation with it by means of mind and spirit (e.g., see Percy B. Shelley's ode *Mont Blanc* and George Gordon Byron's poetic drama *Manfred*, both written in 1816) (Marjorie Hope Nicolson 1959 [1997]).

Following industrialization, and the agricultural, industrial, and commercial development of cities and lowlands, there was a crisis in the Alpine economy and problems of migration and depopulation.

Nonetheless, transit across the Alps increased: Roads crossed the Alps in the early nineteenth century, then railways and tunnels, and finally highways. The traffic was concentrated along some valleys, promoting their economies, but also causing noise and air pollution. Since the late nineteenth century, alpine watercourses have been exploited for hydroelectric plants.

So the Alps were subject to opposing processes: on the one hand, intensive exploitation of transit corridors and tourist areas (with a seasonal overcoming of capacity limits), and on the other hand the depopulation of large areas which reverted to a wild state, where wild animals—predators included—reappeared, and ecological cycles broken for some time were reestablished. This "re-wilding" process is appreciated by ecologists because it favors biodiversity, but farmers, worried about cattle and crops, fight it: Therefore the authorities often allow the killing of wild animals, even members of protected species (wolves, bears).

Some alpine regions maintained a vital culture and economy and a well-balanced relationship with the natural environment without depopulation. This happened where, for instance, local population reached a critical mass, and alpine economy enjoyed comparative advantages in agriculture and manufacture of typically alpine products (e.g., cheese, wine, sausages, delicatessen, in Gruyère, Emmenthal, Val d'Aoste, Valtellina, Trentine, Tirol, and so on), in comparison with the standardized ones of the lowlands; or where demographic and social resistance flows from ideological and political factors (e.g., where ethnic minorities are located).

Around the turn of the twenty-first century, however, climate changes (temperature rise, snow precipitation decrease) hindered winter tourism, which faced growing difficulties. Winter sports development favors bigger and higher tourist resorts, which leads to investing in ski-plants at higher altitude and snowmaking, and penalizes the smaller, lower resorts. This causes a further concentration of development in fewer sites, but in these places there is overbuilding and pollution.

The bigger problems in the Alps thus derive from a paradox: To survive, economically and demographically, either societies open up to the outside, losing their cultural identities and destroying their environments, their two greatest assets; or they close in on themselves, separating from the mainstream of modernity, without suffering its hazards but also without enjoying its opportunities.

The possible solution is the protection of fundamental cultural and natural values through the rule of modernity's local processes. This is only possible if communities do not retire into themselves for an impossible self-defense, but produce a real identification with the land and control the present modifications and opportunities, which are ambivalent because they can have, according to local circumstances, positive or negative issues to communities themselves.

BIBLIOGRAPHY

Bätzing, Werner. 1991. *Die Alpen. Geschichte und Zukunft einer europäischen Kulturlandschaft.* Munich, Germany: C. H. Beck. 2nd edition 2003.

Jouty, Silvain, ed. 2006. *Dictionnaire des Alpes.* Grenoble, France: Glénat.

Kober, Pascal, and Dominique Vulliamy, eds. 2006. *Encyclopédie des Alpes.* Grenoble, France: Glénat.

Nicolson, Marjorie Hope. 1959. *Mountain Gloom and Mountain Glory: The Development of the Aesthetics of the Infinite.* New York: Cornell University Press. 2nd edition, Seattle: University of Washington Press, 1997.

Guglielmo Scaramellini

notes that when seventeenth-century English poets wrote about mountains, they usually portrayed them as ugly and repellent deformities, dangerous to both body and soul. Poets such as John Donne, Andrew Marvell, and their peers drew their images of mountains less from personal experience than from a received literary tradition influenced by several factors: a biblical morality in which mountains symbolized sinful pride; Roman poetry that portrayed mountains as aloof, inhospitable, and hostile; classical notions of beauty emphasizing symmetry, regularity, and proportion; and allegorical and abstract treatments of mountains found in the works of authors such as Augustine and Dante Alighieri. All of these sources contributed to the negative view of mountains that pervades much of seventeenth-century English poetry.

At the close of the century, however, important developments in theology, philosophy, astronomy, and geology were beginning to alter how Europeans conceived of their place in the cosmos and their relationship to the natural world. Although Thomas Burnet's *Sacred Theory of the Earth* (1965 [1684]) still described a world in decline, where the perfectly smooth and rounded orb of God's original creation was now defiled by mountains and other irregularities in the aftermath of the Great Flood, other theologians and philosophers offered a more positive, progressive view of the world, in which nature reflected the orderly design and benevolent purposes of its creator. At the same time, a growing awareness of the vastness of space and time contributed to an emerging aesthetics of the infinite.

Almost immediately, these ideas infused the poets' traditional mountain gloom with an element of ecstatic enthusiasm, typified by the "delightful Horrour and terrible Joy" that John Dennis experienced while traveling through the Alps in 1688. This conflicted emotional response provided the foundation for a new mountain aesthetic of the sublime, with English poets glorifying the awe-inspiring heights of mountains and extolling their irregularities and wildness as glorious expressions of God's plenitude. This new attitude toward mountains reached its highest expression in the works of great Romantic poets such as William Wordsworth, Lord George Gordon Byron, and Percy Bysshe Shelley, all of whom sought out mountains as sources of inspiration and sung praises to their sublime beauty, glory, and divinity.

Nineteenth-century Romanticism exerted a tremendous influence on American transcendentalists such as Ralph Waldo Emerson and Henry David Thoreau, as well as John Muir and other pioneering figures in the American environmental movement. Understood as sources of inspiration, objects of veneration, and places of spiritual renewal, mountains have been a major focus of conservation efforts in the United States. Muir, for example, founded the Sierra Club primarily to protect his beloved Sierra Nevada in California. National parks such as Yosemite, Yellowstone, Grand Teton, Glacier, Denali, and Great Smoky Mountains were established to protect the country's most treasured mountain environments, and the preponderance of mountain regions included in the country's national parks system and wilderness preservation system testifies to the great cultural value Americans attribute to mountains.

Changing attitudes toward mountains among Europeans of the late eighteenth and early nineteenth centuries also are reflected in the rise of mountaineering. No longer forbidding, repulsive, or hostile, the Alps quickly became a tourist destination as Europeans began climbing mountains for sport and adventure. Beginning with Mont Blanc in 1786, climbers reached the summits of all major Alpine peaks in the next hundred years. In 1857 the Alpine Club, the first of many alpine organizations, was established in London, with mountaineering journals and guidebooks appearing shortly thereafter. By the end of the nineteenth century, European mountaineers were looking for new challenges beyond the Alps and exploring the distant peaks of the Caucasus, Karakoram, and Himalayas.

As Europeans came to regard mountains and mountain regions in an increasingly positive light, these changing perceptions found expression in the literature and art of the nineteenth and twentieth centuries. In many cases these new mountain meanings were blended with traditional religious imagery. In Friedrich Nietzsche's *Thus Spoke Zarathustra* (1969 [1883–1885]), for example, mountains retain their traditional role as sites of purification, training, and revelation. The book's prologue begins with Zarathustra, upon attaining wisdom, descending the mountain, like Moses from Sinai, to proclaim his message to humankind below. Hans Castorp, the protagonist of Thomas Mann's novel *The Magic Mountain* (1924), undergoes an extended period of spiritual gestation in a mountain sanatorium that culminates in a revelatory vision of spiritual awakening and rebirth.

In the paintings of the Renaissance, baroque, and neoclassical periods, if mountains appeared at all, they usually provided a highly stylized backdrop for human action. By the nineteenth century, however, mountains began to take center stage in the works of Romantic landscape painters like Albert Bierstadt (1830–1902) and fellow members of the Hudson River School. Bierstadt's mountain scenes are filled with jagged and precipitous peaks, often surrounded with ethereal clouds and illuminated by rays of heavenly light. In paintings such as *Among the Sierra Nevada Mountains*, Bierstadt presents mountains as places of divine splendor and sublime beauty, drawing upon traditional religious imagery and Romantic sensibilities.

Albert Bierstadt,* Among the Sierra Nevada Mountains, California, *1868. *Bierstadt's oil on canvas painting is an example of the numerous depictions of mountains throughout the centuries in the arts. The representation of mountains has assumed various meanings throughout the centuries. They may connote stability and permanence, beauty and transcendence, or even the struggles and trials of contemporary life.* SMITHSONIAN AMERICAN ART MUSEUM, WASHINGTON DC, ART RESOURCE, NY.

Mountains and Conflicts of Meaning Many environmental conflicts of the twentieth and twenty-first centuries have involved mountains and their sometimes conflicting cultural meanings. The heated public debate over the damming of the Hetch Hetchy Valley in Yosemite National Park that occurred from 1900 to 1913 offers a prime example. Opponents of dam construction, led by John Muir, compared the flooding of the mountain valley to the destruction of a sacred temple. Using arguments loaded with religious and Romantic imagery, Muir celebrated the sublime beauty of the mountain landscape. Dam proponents like Gifford Pinchot, meanwhile, saw the Hetch Hetchy Valley and Tuolumne River as a valuable source of freshwater for lowland residents of San Francisco and surrounding communities. Arguing from utilitarian and democratic principles, Pinchot believed the "greatest good of the greatest number" could best be served by flooding the mountain valley to create a reservoir that would provide a reliable supply of freshwater for the growing metropolitan area.

In the 1970s the Hopi Indians of the southwestern United States argued, unsuccessfully, that the expansion of a ski area in the San Francisco Peaks would desecrate their holy mountains and interfere with their religious practices. In this case traditional religious meanings of mountains conflicted with European-American interests in mountain recreation. Similar arguments have been put forward by Native Hawaiians seeking to block geothermal development on the slopes of Kilauea and by Blackfeet Indians opposed to oil and gas drilling in the Badger-Two Medicine area of Montana's Rocky Mountains. Across the globe, such intercultural conflicts are likely to increase as mountain regions become more accessible and the global demand for, and consumption of, mountain goods and services increases. In Nepal and other Himalayan countries the relative ease with which international tourists can now access this once-remote area has resulted in the explosion of a mountain-tourism industry that is rapidly transforming the region's natural and cultural landscape (see sidebar on mountain tourism in Nepal).

MOUNTAIN CONSERVATION AND SUSTAINABLE DEVELOPMENT

In many parts of the world, rapid and unsustainable development is threatening the rich biological and cultural diversity of fragile mountain environments and diminishing the quality and value of the important ecosystem goods and services they provide. In response to these threats, members of the global environmental community have recognized the need to develop policies and practices that promote conservation and sustainable, equitable development in mountain regions. At the Rio Earth Summit in 1992, delegates adopted a global action plan for sustainable development (*Agenda 21*) that includes an entire chapter devoted exclusively to mountains, titled "Managing Fragile Ecosystems: Sustainable Mountain Development."

Since 1992 mountain advocates have been working to implement the strategies outlined in *Agenda 21* (United Nations). The United Nations highlighted the importance of mountain issues by declaring 2002 the International Year of the Mountains, an event that culminated at the Bishkek Global Mountain Summit held in Kyrgyzstan. Recognizing the tremendous diversity of mountain regions and mountain communities throughout the world, the Bishkek Mountain Platform advocates regionally specific conservation and development processes that address the concerns of mountain regions and the lowland communities that depend on them. Through such efforts, mountain advocates are working to protect fragile mountain ecosystems and ensure that the world's mountains will continue to provide for the sustenance—and inspiration—of human communities far into the future.

SEE ALSO *Biodiversity; Earth Summit; Forests; Hetch Hetchy; Mining: I. Overview; Muir, John; Regionalism; U.S. Environmental Protection Agency; U.S. National Park Service.*

BIBLIOGRAPHY

Bandyopadhyay, Jayanta and Shama Perveen. 2004. *Moving the Mountains Up in the Global Environmental Agenda.* Kolkata: Indian Institute of Management Calcutta, Centre for Development and Environmental Policy.

Bernbaum, Edwin. 1997. *Sacred Mountains of the World.* Berkeley: University of California Press.

Bonington, Chris. 1992. *The Climbers: A History of Mountaineering.* London: BBC Books.

Burnet, Thomas. 1965 (1684). *Sacred Theory of the Earth.* Carbondale: Southern Illinois University Press.

Galleano, Eduardo. 1997. *Open Veins of Latin America: Five Centuries of the Pillage of a Continent.* New York: Monthly Review Press.

Gregory, Cedric E. 1980. *A Concise History of Mining.* New York: Pergamon Press.

Mann, Thomas. 1969 (1924). *The Magic Mountain,* trans. H. T. Lowe-Porter. New York: Vintage.

Messerli, Bruno, and Jack D. Ives, eds. 1997. *Mountains of the World: A Global Priority.* New York: Parthenon.

Mountain Forum, The. Available from http://www.mtnforum.org.

Mountain Institute, The. Available from http://www.mountain.org.

Mountain Partnership, The. Available from http://www.mountain partnership.org.

Muir, John. 1961. *The Mountains of California.* New York: Doubleday.

Nepal, Sanjay K. 2002. *Tourism as a Key to Sustainable Mountain Development: The Nepalese Himalayas in Retrospect.* United Nations Food and Agricultural Organisation (FAO) Corporate Document Repository. Available from http:// www.fao.org/docrep/004/y3549e/y3549e10.htm.

Nicolson, Marjorie Hope. 1997 (1959). *Mountain Gloom and Mountain Glory: The Development of the Aesthetics of the Infinite.* 2nd edition. Seattle: University of Washington Press.

Nietzsche, Friedrich. 1969 (1883–1885). *Thus Spoke Zarathustra,* Trans. R. J. Hollingdale. New York: Penguin.

Novak, Barbara. 2007. *Nature and Culture: American Landscape and Painting, 1825–1875.* 3rd edition. Oxford, UK: Oxford University Press.

Pallis, Marco. 1949. *Peaks and Lamas.* Washington, DC: Shoemaker & Hoard.

Reece, Erik. 2006. "Moving Mountains: The Struggle for Justice in the Coal Fields of Appalachia." *Orion* 25(1): 54–67.

Sharma, Pitamber. 1998. *Environment, Culture, Economy, and Tourism: Dilemmas in the Hindu Kush-Himalayas.* Kathmandu: International Centre for Integrated Mountain Development.

United Nations. 1992. *Agenda 21.* United Nations Conference on Environment and Development. Available from http://www. un.org/esa/sustdev/documents/docs_key_conferences.htm.

Scott Friskics

MUIR, JOHN

1838–1914

John Muir was born in Dunbar, Scotland, on April 21, 1838. At age eleven he moved with his family to Wisconsin. Muir is best known for helping shape the modern notion of national parks and helping found the American conservation movement. He was the first to attribute rights to all creatures from snakes to microbes, and his concept of balance and harmony in nature, together with a sense of interdependence among all aspects of the environment, gave his earliest writings an underlying notion of ecology before the term, which was coined in 1866 by Ernst Haeckel, came to America. Muir died in Los Angeles, California, on December 24, 1914.

Muir developed these notions by using a metaphorical and impassioned style of writing rather than making explicit philosophical statements. A strong sense of the human capacity for hubris led him from a position of religious awe to that of a canny conservationist. His democratic belief in the influence of the voting public in recognizing the intrinsic value of the exploited forests and mineral-bearing American landscapes led to his

John Muir, circa 1902. *Muir was a writer and conservationist best known for establishing the modern notion of national parks and as the founder of the Sierra Club. In his writings, Muir espoused a truly nonanthropocentric appreciation of nature in all its forms.* THE LIBRARY OF CONGRESS.

campaign for the establishment of forest reserves and national parks and the formation of the Sierra Club.

In the journal (1867–1869) later published as *A Thousand-Mile Walk to the Gulf* (1992a [1916]) Muir developed a style that not only challenged common assumptions about notions of good and evil in nature but adopted a holistic mode of understanding natural processes. Reacting against the narrow and oppressive Calvinism of his upbringing, Muir took Emersonian transcendentalism into new realms of ecological inclusiveness. For example, the snakes and alligators in the swamps of Florida not only were "unfallen" in religious terms but were "cared for [by God] with the same species of tenderness and love as is bestowed on angels in heaven or saints on earth" (1992 [1916], p. 148).

Combining a holistic sense of the divine with ecological insight, he articulated a nonanthropocentric environmental ethic: "The antipathies existing in the Lord's great animal family must be wisely planned, like balanced repulsion and attraction in the mineral kingdom. How narrow we selfish, conceited creatures are in our sympathies! How blind to the rights of the rest of all creation!" (1992b, p. 148). The fundamental balance in predator-prey dynamics has to be observed as dispassionately as are the analogous dynamics in minerals if the community of animals, plants, minerals, landscapes, and weather is to be appreciated in all its interrelationships: "When we try to pick out anything by itself, we find it hitched to everything else in the universe" (1992b, p. 248). The natural processes by which things are hitched to one another must be viewed dispassionately, including destructive processes.

For Muir, the traditional judgments of culture that separated death and destruction from creation and growth had prevented ecological insights into the processes of nature. For example, when people can see that "everything is flowing—going somewhere, animals and so-called lifeless rocks as well as water," they may see that in Alaska "out of all the cold darkness and glacial crushing and grinding comes this warm abounding beauty and life to teach us that what we in our faithless ignorance and fear call destruction is creation finer and finer" (1992b, p. 841).

Muir deployed different modes of writing for different purposes and different audiences. He was an empirical scientist, celebratory nature writer, campaigning conservationist, authoritative naturalist, and wilderness polemicist. The keys to his concept of a series of "nation's parks" were long-term protection for future generations and short-term renewal through direct contact with nature by urban populations that then would vote for the protection of those iconic landscapes and their ecologies. Muir thus embraced tourism as an inspirational and educational conservation strategy.

Early in his writing life he came to see that the urban and Eastern audiences for his writing could become a constituency for the preservation of forests, wilderness, and iconic landscapes. On successive trips to Alaska he developed a growing respect for indigenous lifestyles and skills, together with concern about the future of those communities. For Muir culture, in the form of his writings, was a mode of communication with nature that could be used to mediate a sustainable relationship with it. If this required a preaching tone in his writing, he could rise to the challenge: "The outcries we hear against forest reservations come mostly from thieves who are wealthy and steal timber by wholesale. ... Any fool can destroy trees. ... But [God] cannot save them from fools—only Uncle Sam can do that" (1992b, pp. 604–605).

Muir had the ability to conduct an environmental campaign and employ the power of pressure-group environmental politics. He worked for a bill to preserve Yosemite National Park, teaming with Robert Underwood Johnson: Johnson lobbied in Washington, and Muir wrote up the evidence and arguments in a series of articles for Johnson's *Century Magazine.* Muir also worked with Charles Sargent's Forestry Commission on a tour of American forests to establish thirteen forest reserves in the face of opposition from lumber and mineral interests. As the founding president of the Sierra Club, Muir set up a pressure group of influential Bay Area businesspersons to defend the new Yosemite National Park and, through outings and meetings, expand the club's sphere of influence. The strain of the club's campaign against the building of Hetch Hetchy Dam, and the campaign's eventual failure, hastened the end of Muir's life but became a crucial lesson in his legacy.

SEE ALSO *Conservation; Ecology: I. Overview; Environmental Politics; Hetch Hetchy; Holism; Sierra Club; Wilderness.*

BIBLIOGRAPHY

Fox, Stephen. 1985. *The American Conservation Movement: John Muir and His Legacy.* Madison: University of Wisconsin Press.

Gifford, Terry. 1992. "Introduction." In *John Muir: The Eight Wilderness-Discovery Books.* Seattle: Mountaineers.

Gifford, Terry. 2006. *Reconnecting with John Muir: Essays in Post-Pastoral Practice.* Athens: University of Georgia Press.

Muir, John. 1992a *The Eight Wilderness Discovery Books.* Seattle: Mountaineers.

Muir, John. 1992b (1916). *A Thousand-Mile Walk to the Gulf.* New York: Penguin.

Nash, Roderick. 1989. *The Rights of Nature: A History of Environmental Ethics.* Madison: University of Wisconsin Press.

Terry Gifford

N

NAESS, ARNE
1912–

The philosopher Arne Naess was born in Oslo, Norway, on January 27, 1912, and received a doctorate in philosophy from the University of Oslo in 1938. He is best known for his invention of the term *deep ecology* to indicate that environmental issues are questions of ethics and philosophy rather than science and politics. Through a combination of his ideas and his persona, Naess was probably the most influential living environmental philosopher in the first decade of the twenty-first century.

EDUCATION AND EARLY WORKS

In the 1930s Naess traveled to Vienna as a young student to join the Vienna Circle, working closely with Moritz Schlick and Rudolf Carnap in the field of analytic philosophy. In 1937 Naess became the youngest full professor in Norwegian history, and over subsequent decades he wrote a series of textbooks on introductory logic and the history of philosophy that became the foundation for reform of his nation's university system, which required for many years that all students take a semester of philosophy before working in their chosen disciplines. His first book, *Truth as Conceived by Those Who Are Not Themselves Professional Philosophers* (1938), used a survey approach to demonstrate that ordinary people hold a range of views on truth similar to those voiced by philosophers.

During World War II Naess was active in the clandestine resistance against the Nazi occupiers, and after the war he led a reconciliation project to bring war criminals together with the parents of the Norwegian soldiers they tortured and killed. In the Cold War period, Naess was asked by the United Nations to lead a philosophical effort to study the worldwide uses of the term *democracy*. The resulting book, *Democracy in a World of Tensions* (1951), revealed that the word could mean almost anything; because of that disturbing conclusion, the book never was reprinted.

LATER WORKS AND INFLUENCE ON ENVIRONMENTAL ETHICS

In mainstream philosophy Naess is best known for his work in the philosophy of language in *Interpretation and Preciseness* (1953) and *Communication and Argument* (1966). Other major theoretical works in English include *Scepticism* (1968), *Gandhi and Group Conflict* (1974), and *The Pluralist and Possibilist Aspect of the Scientific Enterprise* (1969).

Naess is a mountaineer and for a few years in the early 1950s, with his ascent of Tirich Mir, held the record for the highest mountain ever climbed. A decade later, inspired by Rachel Carson, he resigned from his professorship to devote himself to environmental issues. *Ecology, Community, and Lifestyle* (1989 [in Norwegian 1976]) is his main theoretical work in environmental philosophy and the work in which the theory of Deep Ecology is articulated in depth. It is an environmental philosophy, not an ethic, that encourages each individual to think of nature as the ground of his or her own interest so that the greatest sense of self-realization will encompass a "Self" of the environment and become "Self-realization." People should situate their identities and interests in nature uniquely, developing their own "ecosophies" that build on a personal sense of place and a duty of care for the earth and fit into their immediate surroundings with greater attention and dignity.

Together with George Sessions, Naess politicized Deep Ecology by putting forth a platform of eight points that turned his concept into an ethical manifesto:

1. The flourishing of human and nonhuman life on the earth has intrinsic value. The value of nonhuman life-forms is independent of the usefulness they have for narrow human purposes.
2. The richness and diversity of life-forms are values in themselves.
3. Humans have no right to reduce that richness and diversity except to satisfy vital needs.
4. Present human interference with the nonhuman world is excessive, and the situation is worsening rapidly.
5. The flourishing of human life and cultures is compatible with a substantial decrease in the human population.
6. Significant change of life conditions for the better requires change in economic and technological policies.
7. Quality of life should be given more primacy than a high standard of living.
8. Those who subscribe to these points have an obligation to implement the necessary changes.

This platform was adopted by radical environmental groups such as Earth First! as their guiding philosophy, but Deep Ecology might have reached its greatest popular prominence when Senator Al Gore wrote in his 1992 book *Earth in the Balance* that "we must change the fundamental values at the heart of our civilization" to solve global environmental problems (p. 144). By the dawn of the twenty-first century many people embraced this view even if they did not realize where the idea came from.

In 2002, at age eighty-eight, Naess published *Life's Philosophy*, a personal account of his history through ideas. It became the number one best-seller in Norway and catapulted its author to a new level of fame. In 2005 the *Selected Works of Arne Naess* was published in ten volumes. It is perhaps the most comprehensive publication of the works of any living philosopher.

At age ninety-five Naess continued to speak out in the name of nature and conservation and remained optimistic that humanity will be able to improve its relationship to the world around it "by the twenty-second century" (Sessions 1995, p. 501).

SEE ALSO *Deep Ecology; Earth First!; Environmental Philosophy: V. Contemporary Philosophy.*

BIBLIOGRAPHY

WORKS BY ARNE NAESS

Naess, Arne. 1989. *Ecology, Community, and Lifestyle: Outline of an Ecosophy*, trans. David Rothenberg. New York: Cambridge University Press.

Naess, Arne, with Per Ingvar Haukeland. 2002. *Life's Philosophy: Reason and Feeling in a Deeper World*, trans. Roland Huntford. Athens: University of Georgia Press.

WORKS ABOUT ARNE NAESS

Brennan, Andrew, and Nina Witoszek, eds. 1999. *Philosophical Dialogues: Arne Naess and the Progress of Ecophilosophy*. Lanham, MD: Rowman & Littlefield.

Gore, Al. 1992. *Earth in the Balance: Ecology and the Human Spirit*. Boston: Houghton Mifflin.

Sessions, George, ed. 1995. *Deep Ecology for the Twenty-first Century*. Boston: Shambhala.

David Rothenberg

NANOTECHNOLOGY

Nanotechnology is defined by the U.S. National Nanotechnology Initiative as the "understanding and control of matter at dimensions of roughly 1 to 100 nanometers, where unique phenomena enable novel applications." At this small scale (1 billionth of a meter), matter takes on novel properties, such as greater reactivity, electrical conductivity, and penetrability. Nanotechnology was used as long ago as 10 C.E. in the first century to develop paints of different colors that varied only by the size of the particles. Yet it was not until 1959 that the idea of nanotechnology was introduced by Richard Feynman, a physicist at the California Institute of Technology, in a talk titled "There's Plenty of Room at the Bottom." Although Feynman did not use the word *nanotechnology*, he suggested the possibility of precisely manipulating atoms and molecules. In size, nanoparticles range from the size of several atoms to the size of biological molecules, such as proteins. Although nanoparticles exist in nature, only recently developed tools such as atomic-force microscopy and better scientific understanding of materials have allowed humans to manipulate matter at the nanoscale to achieve certain goals. As an example, in 1989 IBM performed a famous experiment using atomic-force microscopy to move individual xenon atoms to form the letters "IBM."

Nanotechnology is a broad term that encompasses a diverse set of applications, tools, and methods linked together primarily by scale. The technology can be applied to medicine, food, agriculture, manufacturing, health, consumer products, and environmental science and remediation. In medicine, gold and dendrimers (complex organic molecules) are used to specifically target and destroy cancer cells without the horrible side effects of chemotherapy. In renewable energy, barriers to greater use of solar energy include the high cost of materials for the panels. Some nanoparticles convert light to energy with greater efficiency, and so less material is needed. Hence, nanomaterials can be used in solar panels to lower their cost. In the environment, iron oxide nanoparticles are used to remove arsenic, a potent toxin, from drinking water, and DNA-based nanoparticles are being developed for tracking pollution in natural environments. Finally, in agriculture, nanosensors are being considered for detecting agricultural growing conditions and responding only when needed for the timed release of fertilizer, water, or pesticides.

Environmental issues and ethics intersect with nanotechnology in several ways. Ensuring the environmental safety of nanomaterials, making decisions about funding priorities or their environmental release, and developing systems for broader oversight all involve value judgments. Ethical issues associated with nanotechnology generally fall into categories that also apply to other technologies, such as utilitarian accounting of risks, costs, and benefits and the examination of fundamental moral principles such as autonomy, justice, nonmalfeasance, or beneficence (Beauchamp and Walters 1999). Environmental ethics includes questions related to how nanotechnology affects the preservation of natural, nonhuman living systems; what obligation nanomaterial producers have to protect the environment; and how much certainty about environmental safety is needed prior to environmental release.

FUNDING

Several nations have significant research programs in nanotechnology, including Japan, the United States, and nations of the European Union. The United States was one of the first nations to fund nanotechnology research and development in a coordinated fashion at the federal level. The U.S. National Nanotechnology Initiative (NNI) formally began in 2000 with a research program funded at $400 billion. Key figures in the formation of the initiative were Mihail Roco of the National Science Foundation (NSF) and Neal Lane, President Bill Clinton's science adviser. The budget for the initiative grew to $1.4 billion in 2008. Over half of this funding goes to the U.S. Department of Defense, and a significant proportion goes to the NSF, the Department of Energy, and the National Institutes of Health. Smaller amounts are allocated to the Department of Commerce, Department of Agriculture, U.S. Environmental Protection Agency (EPA), Department of Transportation, and other agencies. Nearly every public agency has some responsibility for nanotechnology on account of the diverse methods, products, and applications of the technology.

In 2006, the NNI channeled about 8 percent of its funds into the societal dimensions of nanotechnology, which includes educational programs, research, and programs focused on ethical, legal, and social implications (ELSI); and environmental health and safety (EHS) research (National Research Council 2006). This designation of funding is an important statement on the part of the U.S. government: an acknowledgment that emerging technologies are embedded within social and cultural systems and that it is important to study societal dimensions alongside technological development. Yet some scholars argue that ELSI programs and research have not received as much support as they should. In the first four years of the NNI, Ira Bennet and Daniel Sarewitz (2006) estimate that less than 0.5 percent has specifically gone to ELSI work. Regardless, several interdisciplinary research teams across the United States are studying a range of ELSI issues from proper oversight of nanomaterials to public attitudes about nanotechnology and ethical issues arising from future nanotechnology applications. In 2005, the NSF funded the Center for Nanotechnology and Society at Arizona State University and the University of California, Santa Barbara.

Arguably, there has also been little funding for the study of environmental health and safety issues associated with nanotechnology. Funding for risk-related research is about 1 percent of total funding for the NI (Maynard 2006), and only a portion of this goes specifically to studies of environmental risk. In 2008 legislative hearings in the United States focused on the need to increase that percentage. The lack of significant levels of funding for EHS issues has created a situation in which over five hundred nanotechnology products are on the market while little is known about their environmental transport, fate, and toxicity. Applications are growing and will continue to progress from chemical nanomaterials in consumer products (passive nanostructures) to active biologically based nanostructures and nanosystems that can respond, change, and move through human and natural systems (Roco and Bainbridge 2005).

Toxicology studies suggest that nanoparticles are more toxic at lower concentrations than their larger counterparts, owing to their higher surface-area-to-mass ratio and greater reactivity. Little is known, however, on the amounts and nature of nanomaterials released into the environment (e.g., from manufacturing plants), and there are few comprehensive, field-based studies on nanoparticles and their risks to the environment, although frameworks for risk analysis have been developed (Morgan 2005). Buckyballs (made out of 60 carbon atoms and shaped like a soccer ball) and carbon nanotubes (thin-walled tubes made of carbon atoms) are two types of nanoparticles that have been shown to be harmful to fish and microorganisms. In 2008 government organizations in Japan, the European Union, and the United States proposed increasing resources for environmental health and safety studies of nanotechnology applications.

OVERSIGHT

A subset of ethical questions focuses on oversight of nanotechnology products. Typically, formal government regulation is limited to utilitarian considerations such as the risks, costs, and benefits of commercial approval. Yet oversight also encompasses ethical principles of fairness and justice, for example, through the distribution of risks and benefits among human communities and ecosystems. Another oversight question involves who or what entities should bear primary responsibility for overseeing nanotechnology products. Still another oversight issue is whether consumers and patients should have rights to know and choose products generated from nanotechnology (the ethical principle of autonomy). Finally, should oversight consider harm to natural systems and human responsibilities to protect the environment for its intrinsic value or for its use by future generations?

U.S. oversight of nanotechnology currently relies on agencies and regulations that have provided oversight for related technologies or products. There is no specific statute or mechanism for oversight of nanotechnology products. Yet existing laws and systems may not adequately address the novel properties and unique challenges of nanoproducts (Kuzma 2006; Davies 2007). National and international stakeholders and government organizations are currently considering how best to oversee the applications and products of nanotechnology.

Most chemical and consumer nanotechnology products do not require premarket testing and safety data. The EPA (2007) administers a voluntary stewardship program in which data generated by industry are collected by the agency, but there are no formal requirements for regulatory approval. The stated policy of the U.S. Food and Drug Administration (FDA; 2007) is to treat nanoscale drugs, devices, food products, and biologics as substantially equivalent to their larger counterparts. The nonprofit group Environmental Defense Fund partnered with DuPont to develop a risk-analysis framework for nanomaterials, and they are encouraging companies to use it. On the international level, the Organization for Economic Cooperation and Development is helping to form partnerships among nations to generate and share data about the safety of nanomaterials. The International Standards Organization and other standard-setting bodies are designing safety standards for nanotechnology products.

CONTROVERSY

Some stakeholders, citizen groups, and organizations, citing negative past experiences with other emerging technologies, such as biotechnology, are skeptical of the promise of nanotechnology. Although there has not been massive public outrage about nanotechnology, several interest groups have objected to the voluntary approaches to overseeing nanotechnology in place in 2008. The Erosion, Technology, and Concentration Group called for a moratorium on nanotechnology products in the marketplace until more information is obtained. The British Royal Society recommended a precautionary approach to the environmental release of freestanding nanoparticles until more is known about their transport and effects. The International Center for Technology Assessment (2007), in partnership with several other nonprofit organizations, has developed principles for regulation that include public transparency and mandatory premarket testing by independent experts. The center has also filed a legal petition against the FDA for its lack of premarket safety testing of cosmetics and sunscreens with nanoparticles.

One prominent example of debate about nanotechnology in the environment involves silver-coated nanomaterials. There are on the market in nanosilver washing machines, refrigerators, socks, and food-packaging materials. Silver ions kill microorganisms, and products coated with nanosilver stay fresh longer. Initially, these products were entering the marketplace with no premarket testing. However, several environmental and other interest groups brought attention to the potential release of silver ions into the environment (e.g., from the wash cycle of nanosilver-coated washing machines) and petitioned the EPA to regulate them. After about a year of consideration, the EPA decided in 2007 to regulate nanosilver under the U.S. pesticide law, the Federal Insecticide, Rodenticide, and Fungicide Act. This relatively strong statute requires premarket tests to be submitted to the EPA and products to be registered. However, the manufacturer must claim that its product is designed to kill pests or germs. As a result, some companies that produce nanosilver materials are removing this pest-killing claim from their products to avoid regulation.

As of 2008, producers of most nanomaterials are primarily responsible for ensuring the safety of their products through voluntary programs. This approach seems fair in that the financial burden of safety testing is placed on the entity that stands to profit. However, a body of literature suggests that citizens are less confident when oversight is placed in the hands of those who have vested interests. Public confidence is greater when independent bodies are responsible for oversight and regulation is mandated by government (Macoubrie 2006).

People also prefer that information about nanotechnology products be made available to them (Macoubrie 2006). As of 2008, there is a lack of transparency associated with what nanoproducts are in the marketplace and what environmental health and safety studies have been done on them. Manufacturers hold the information, since protecting their confidential business information and intellectual property is important for their survival.

The dilemma is that this lack of transparency, which seems necessary for business, may violate consumer rights to know and choose nanoproducts (autonomy).

To explore and increase dialogue about these and other policy issues, scholars have called for "upstream public engagement" as a mechanism for integrating the public's views about nanotechnology into decision making early in the process (Wilsdon and Willis 2004). Some upstream-public-engagement exercises have been conducted for nanotechnology (Toumey 2006), although there is not yet a formal and systematic way to get the public's input on nanotechnology oversight.

Ethical arguments about nanotechnology and the natural environment include viewpoints about the sanctity of nature. Nanotechnology involves the control and manipulation of matter at the atomic scale, and some people believe that humans do not have the right to tamper with nature at this fundamental level. Similar debates occurred in the context of genetically engineered organisms. Several groups are considering appropriate oversight for nanotechnology and the best way to learn from past technologies (David and Thompson 2008; Paradise et al. 2008).

In the context of nanotechnology, broad questions about the role of humans in the environment, human obligations to protect ecosystems at all or some costs, and the intrinsic value of nonhuman entities and Earth need more consideration. Because there are only a handful of environmental applications of nanotechnology on the market, there is an opportunity to study the ethical and societal implications of nanotechnology, create upstream-public-engagement activities, and proactively develop oversight-assessment strategies prior to full-scale development, use, and release. Also calling for quick yet careful consideration are the environmental impacts of nanomaterials released into the environment as by-products or waste from manufacturing, since there are hundreds of consumer products with nanomaterials already on the market.

SEE ALSO *Future Generations; Industrial Ecology; U.S. Environmental Protection Agency; U.S. Food and Drug Administration.*

BIBLIOGRAPHY

Beauchamp, Thomas, and Leroy Walters. 1999. *Contemporary Issues in Bioethics*. 5th edition. Belmont, CA: Wadsworth.

Bennet, Ira, and Daniel Sarewitz. 2006. "Too Little, Too Late? Research Policies on the Societal Implications of Nanotechnology in the United States." *Science as Culture* 15(4): 309–325.

David, Kenneth, and Paul B. Thompson. 2008. *What Can Nanotechnology Learn from Biotechnology? Social and Ethical Lessons for Nanoscience from the Debate over Agrifood Biotechnology and GMOs.* Oxford, UK: Elsevier.

Davies, Clarence. 2007. *EPA and Nanotechnology: Oversight for the 21st Century.* Washington, DC: Project on Emerging Nanotechnologies.

International Center for Technology Assessment. 2007. "Principles for the Oversight of Nanotechnologies and Nanomaterials." Available from http://www.icta.org/doc/Principles%20for%20the%20Oversight%20of%20Nanotechnologies%20and%20Nanomaterials_final.pdf.

Kuzma, Jennifer. 2006. "Nanotechnology Oversight: Just Do It." *Environmental Law Reporter* 36: 10913–10923.

Macoubrie, Jane. 2006. "Nanotechnology: Public Concerns, Reasoning, and Trust in Government." *Public Understanding of Science* 15: 221–241.

Maynard, Andrew. 2006. *Nanotechnology: A Research Strategy for Addressing Risk.* Washington, DC: Project on Emerging Nanotechnologies.

Morgan, Kara. 2005. "Development of a Preliminary Framework for Informing the Risk Analysis and Risk Management of Nanoparticles." *Risk Analysis* 25(6): 1621–1635.

National Research Council. 2006. *A Matter of Size: Triennial Review of the National Nanotechnology Initiative.* Washington, DC: National Academies Press.

Paradise, Jordan; Susan M. Wolf, Gurumurthy Ramachandran; et al. 2008. "Developing Oversight Frameworks for Nanobiotechnology." *Minnesota Journal of Law, Science, and Technology* 9(1): 399–416.

Roco, Mihail C., and William S. Bainbridge. 2005. "Societal Implications of Nanoscience and Nanotechnology: Maximizing Human Benefit." *Journal of Nanoparticle Research* 7: 1–13.

Toumey, Chris. 2006. "Science and Democracy." *Nature Nanotechnology* 1: 6–7.

U.S. Environmental Protection Agency. 2007. "Nanoscale Materials Stewardship Program and Inventory Status of Nanoscale Substances under the Toxic Substances Control Act: Notice of Availability." *Federal Register* 72(133): 38083–38085.

U.S. Food and Drug Administration. 2007. "Nanotechnology: A Report of the U.S. Food and Drug Administration Nanotechnology Task Force, July 25, 2007." Available from http://www.fda.gov/nanotechnology/taskforce/report2007.pdf.

Wilsdon, James, and Rebecca Willis. 2004. "See-Through Science." London: Demos. Available from http://www.demos.co.uk/files/Seethroughsciencefinal.pdf.

Jennifer Kuzma

NATIONAL PARK SERVICE

SEE *U.S. National Park Service.*

NATIONAL SCIENCE FOUNDATION

The statute creating the U.S. National Science Foundation (NSF), which President Harry Truman signed in 1950, specified that the agency support the mathematical, physical, biological, and engineering sciences. The emphasis on hard science, on big science, and on training engineers, meant to counter scientific advances in the Soviet Union, became entrenched with the 1957 launch of the Soviet satellite *Sputnik*. The governing National Science Board, raising concerns about "objectivity, verifiability, and generality," opposed the use of funds to support the social sciences. One member likened the social sciences to Pandora's box, except in limited areas, such as demography, where the science could be closely tied to empirical, mathematical, and experimental methods. When President John F. Kennedy sought to move the NSF toward applied and socially relevant research, the NSF began to support some projects in the social sciences. Pressure from Congress, culminating in 1968 with a radical amendment to its underlying statute, compelled the NSF to apply research to national needs, and as the agency funded projects dealing with politically salient and controversial issues, such as the environment, it created programs for the social sciences, such as economics. The requirement to fund applied, rather than just basic, research not only brought more support to the social sciences but also opened what had been a closed question: whether the NSF could support projects examining the social, legal, and ethical implications of science and technology. This entry describes how the NSF has responded to that question.

By the early 1970s, many NSF officials, especially those in the biological sciences, had become aware of public controversies over new technologies and recognized the need to investigate the normative conditions and consequences of the increasing power, and hence responsibility, that science gave society. Skeptics questioned whether research into values was suitably scientific, how it would be organized, and whether it would unduly emphasize risks over benefits. In 1976, following the recommendation of the Ethical and Human Values in Science and Technology program, an advisory committee established with the National Endowment for the Humanities, the NSF made its first grants in this area, some cofunded by the National Endowment for the Humanities until the latter withdrew in 1980.

The budget cuts of the Ronald Reagan administration in the early 1980s, which were aimed especially at the social sciences and at science education under the NSF, would have eliminated the Ethical and Human Values in Science and Technology program had not it moved to the Directorate for Scientific, Technological, and International Affairs. Although the program suffered substantial budget cuts, support from the scientific community and from NSF Director John Slaughter helped preserve the program. In 1986 a new NSF director posed an even greater threat by proposing to zero-out the roughly $1 million budget of the Ethics and Values in Science and Technology program in favor of funding large-scale engineering research centers. Dr. Rachelle Hollander, who had become the director of the program in 1980, helped bring the scientific community, including such groups the American Association for the Advancement of Science, into a lobbying effort that persuaded the Congressional oversight committees to restore program funding. Ethics and Values in Science and Technology, now housed in the Directorate for Biological, Behavioral, and Social Sciences, saw its funds distributed through an "ethics across the foundation" approach, which required a proposal-by-proposal management and negotiation effort. In this effort Hollander, by persuasion and collaboration, worked with a growing number of NSF programs, obtaining from them contributions that in effect increased by one-third the overall budget of the Ethics and Values in Science and Technology program.

Between 1977 and 1985 the Ethics and Values in Science and Technology program made more than 150 awards, many in the areas of the environment, hazards, and agriculture, including animal welfare. The timing of this support (along with the intellectual oversight of managers of the Ethics and Values in Science and Technology program) accounts in large part for the successful development of the field of environmental ethics, which drew little enthusiasm from mainstream departments of philosophy. Between 1986 and 1994, a period of bureaucratic turmoil and reorganization at the NSF, the Ethics and Values in Science and Technology program (renamed the Ethics and Values Studies program), together with the Science and Technology Studies program, moved into the new Directorate for Biological, Behavioral, and Social Sciences, under the general rubric of the Societal Dimensions of Engineering, Science, and Technology.

As public concern over the social, ethical, and cultural implications of emerging technologies grew—one need only think of nuclear power, recombinant DNA technology, pesticides, and the assessment of environmental risk—the NSF, from 1994, included in its budget specific funding for research in ethics and values studies and in science and technology studies. The Ethics and Values Studies program also continued to collaborate with, and thus garner additional support from, larger NSF projects. An ethics component is included in several big-ticket NSF initiatives, for example, the foundation-wide Integrative Graduate Education and Research Training program, which makes individual grants in excess of $2.5 million.

Congressional and public concern continued to direct the NSF toward applied research that would lead to measurable results in solving national problems. In 1996 a task force recommended that the competitive review of proposals center on two questions: "(1) What is the intellectual merit of the proposed activity? (2) What are the broader impacts of the proposed activity?" These criteria, which were implemented in 1997, required applicants to integrate the intellectual merits of their proposed research with its broader impacts (the social merits). The consideration of the impacts and thus the social value of proposed research brought a normative dimension to every project. At the same time, the NSF gave greater weight to the relation of science and society and to research, including normative and conceptual analysis, concerning the formation of science policy.

In 1976 only two major academic centers for ethics and public policy existed: the Hastings Center in New York, which specializes in medical ethics, and the Center (now Institute) for Philosophy and Public Policy at the University of Maryland. With NSF support—and later funding from the program on the Ethical, Legal, and Social Implications of the Human Genome Project at the National Institutes of Health—ethics and policy programs and centers have now become established in universities across the country. Professional organizations, for example, the Association for Practical and Professional Ethics (1991), formed for philosophers, social scientists, and scientists who see the study of the normative dimensions of science and technology as a legitimate field of scholarship. Several journals, such as *Science, Technology, and Human Values* and *Science and Engineering Ethics*, were founded to publish papers exploring the "value," or social and cultural, dimensions of engineering, information technology, the environmental and social sciences, and technology.

Grants made under the Societal Dimensions of Engineering, Science, and Technology rubric have included normative and conceptual analyses of issues in the social sciences, but far more often address the social implications of environmental, engineering, and biological sciences and technology. Thus principal investigators in Ethics and Values Studies projects are as likely to be humanists (for example, philosophers) and natural scientists as social scientists. By ensuring that the projects it funded directly related to the social consequences and ethical implications of science and technology, the Ethics and Values Studies program to some extent avoided the skepticism often directed toward the social sciences within the NSF and by oversight committees. The Ethics and Values Studies program now appears integral to the attempt to understand and evaluate the broader impacts of scientific research.

As of 2007 scholars seeking support for research on the normative dimensions and social implications of science and technology may apply not only to the now integrated Science and Society Program, which includes the Societal Dimensions of Engineering, Science, and Technology program and awards grants from an annual budget of approximately $3 million, but also to other NSF programs, such as those involving science policy, ethics in engineering, and nanotechnology, all of which have been sensitized to the need for normative analysis. Much of this progress can be attributed to Hollander's nearly thirty years of leadership at the Ethics and Values Studies program. She retired in 2006.

SEE ALSO *Nanotechnology; Risk Assessment; Technology.*

BIBLIOGRAPHY

Hollander, Rachelle. 1987. "In a New Mode: Ethics and Values Studies at the NSF." *Science, Technology, and Human Values* 12(2): 59–61.

Hollander, Rachelle, and Nicholas H. Steneck. 1990. "Science- and Engineering-Related Ethics and Values Studies: Characteristics of an Emerging Field of Research." *Science, Technology, and Human Values* 15(1): 84–104.

National Science Foundation. "The National Science Foundation: A Brief History." July 15, 1994. Available from http://www.nsf.gov/about/history/nsf50/nsf8816.jsp.

Mark Sagoff

NATIVE AMERICANS

The cosmologies, ways of knowing, and value systems of North American indigenous peoples (NAIPs) have attracted the attention of environmental philosophers and ethicists worldwide. A central reason for this attraction is the inspiration NAIP worldviews have provided for critics of the dominant Western European worldview and practices. NAIP worldviews articulate alternatives to the exploitative and ecologically destructive attitudes of the dominant society by directing thought, perception, and action in ways that promote a more harmonious, sustainable, and respectful human-environment relationship.

NAIP AND WESTERN WORLDVIEWS

In its *Agenda 21* of 1992 the United Nations has drawn attention to the environmental practices of indigenous people around the world as models of holistic thinking, ecological sustainability, and environmental protection. Although it is difficult to characterize this relationship precisely, the ideas of being connected to the earth, coming from or being intimately connected to the land, belonging to the land, and being keepers of the land express how most NAIP worldviews frame the relationship between people and the land.

These ideas are indicative of a holistic way of understanding and engaging the world. Accordingly, they are used to criticize the fragmenting tendencies of Western thought, such as the idea of a radical separation between rational human beings and the land, which has characterized Western thought from Plato through Thomas Aquinas, René Descartes, and Immanuel Kant. The attraction to the worldviews of indigenous peoples has resulted from what critics view as fundamental failures and inadequacies of Western European metaphysical, epistemological, and normative systems. Metaphysical systems are ultimate explanations of the fundamental properties of nature and why things happen. Epistemological systems articulate ways of knowing or principles of knowledge acquisition. Normative systems tell people what is of value and what people ought to value.

DEFINITIONS AND NOMENCLATURE

The use of the term *North American indigenous peoples* in this entry represents an attempt to avoid some of the problems that labels create. The term *Indian* is acceptable to some people even though it is not based in NAIP languages. Others find it unacceptable because it helps propagate colonial attitudes. Some accept *Indigenous* (or *indigenous*) *peoples*, but this term generally is used to refer to the wider global community of what have been called native peoples. Some people insist on being called by their national names (e.g., Haudenosaunee, Anishnaabek). For this entry the more neutral term has been chosen

NAIP worldviews typically do not assume a radical detachment of body from mind, nature from culture, and rational beings from nonrational beings. They emphasize the importance of acting in accordance with Earth and its balances and respect for a creator or a greater power. The lowercase term *creator* is used in this entry to refer to both concepts. Many NAIP cultures today use expressions such as *all my relations* to indicate how they view their connection to the earth. People are fully relational beings; they are what they are by virtue of their relations to all others, including trees and rocks. People are first members of a community of beings and second distinct and private individuals. Origin stories describe how humans emerged from the land, not from an intellectual, noncorporeal realm.

The NAIPs who have been at the center of this attention belong to two linguistic groups: the Algonquian and Iroquoian. The reason for this focus is partly historical. British tradition has dominated the writing of North American history in regard to NAIP worldviews, and the British were engaged primarily with Algonquian and Iroquoian peoples; therefore, these three traditions have come to dominate the way in which NAIP worldviews have been conceived in the environmental arena. Other nations and cultures have become foci of study, partly because more NAIPs are participating in the dominant scholarly and publishing systems. The appeal of the environmental movement to Plains (especially Sioux) and West Coast NAIPs also has received more attention. Only a few of the roughly five hundred nations and two hundred linguistic communities that lived in the Americas before contact with Europeans, not to mention other continents, are mentioned in this entry. This is unfortunate, because the NAIP presence was far from insignificant. William Denevan (2007) figures the total NAIP population at contact was 40 million to 100 million (the U.S. census of 2000 puts that number between 2,475,956 and 2,663,818, or fewer than 1 percent of the U.S. total population).

The environmental practices and principles of NAIPs, such as taking only what is needed, avoiding the burning of woodlands and prairies, and using everything one takes, have been cited in an attempt to change the behavior of the so-called developed world and its conception of its relationship to ecosystems. Because NAIP worldviews would have people perceive the world as alive and occupied by other persons (nonhuman animals, trees, and rocks) who are conceived as relatives, they encourage Westerners to come into more intimate contact with and develop an awareness of the world as a community and in so doing become more respectful of Earth.

Many of the Algonquian and Iroquoian people have articulated this view in the concept of Mother Earth, the living giver and sustainer of life, who is owed the deepest respect. Accordingly, the practice of thanksgiving for members of the Earth community who give themselves so that others can live is guided by the principle of reciprocity (e.g., giving tobacco in exchange for the life they take to sustain themselves). This conception of Earth grounds what has come to be known as traditional ecological/environmental knowledge (TEK). Such knowledge involves different modes of becoming aware of the life force that is Earth. The drumming circle, for instance, produces the sound that represents the heartbeat of Mother Earth to which drummers, singers, and dancers connect.

CARICATURES OF NAIP WORLDVIEWS

NAIP cultures often have been romanticized, and the concept of the NAIP personality has been constructed as the noble savage (or ignoble savage). These caricatures have deep roots in North American and European history. From the seventeenth century through the eighteenth misrepresentations by, among others, Thomas Hobbes, John Locke, and Jean-Jacques Rousseau have cast NAIPs as politically and legally naive and/or innocent. Archie Belaney (Grey Owl) and Thomas Seton (the Boy Scout movement), with a different intent, viewed

Makah Indians at a Whaling, 1999. *The environmental practices of Native American indigenous peoples (NAIPs) are often seen as more ecologically friendly than Western practices. Certain principles, such as taking only what is needed and using everything one takes, are often cited as criticism of the "developed" world. However, this comparison does not always hold, as many environmentalists disapprove of other NAIP practices, such as the Makah whale hunt in Washington State.* AP IMAGES.

NAIPs as incapable of harming the land because they live in an honorable and constant harmony with it.

This caricature implies that NAIPs lack the capacity for reasoning and decision making and thus operate at the level of animal intuition. Where Rousseau saw this state as a golden age of humanity, Locke saw it as a ground for justifying the appropriation of NAIP land. A number of influential colonists used this conception to declare the lands of North America *vacuum domicilium* ("empty place"; in Australia the term *terra nullius* was used) because, as nonrational people, NAIPs did not understand what it meant to own the land and thus could not lay proper claim to it. The term *noble savage* was used by Rousseau and others to criticize what they saw as the degeneration of their own societies. Baron Lahontan, for example, recalled or perhaps fabricated accounts of his discussions with Adario (the Huron sachem [leader] Kondarionk) as a series of syllogistic arguments. He constructed those accounts to contrast the uncorrupted Indian mind with the degenerate European mind. To some extent later writers (e.g., Henry David Thoreau, J. Fenimore Cooper) and more contemporary proponents of wilderness awareness (e.g., Tom Brown Jr.) continue to propagate such ideas. Tom Brown was educated in the art of tracking by Stalking Wolf, one of the last Apaches to be trained in the old ways. He, along with others, continues the commentary on the degeneration of Western civilization.

REFUTATIONS OF THE CARICATURES

These one-sided and often distorted descriptions are challenged not only by NAIP teachers and scholars but by non-NAIP scholars as well. Careful perusal of early colonial records such as the Records of Indian Affairs of the 1600 and 1700s supports stories of shrewd Indian businessmen trading at Albany and Montreal and with the Hudson Bay Company. They describe treaty-making processes that involved articles of agreement and arguments over the foundations of an intercultural law. Oral tradition, as John Borrows (2002) (Anishnaabe) and J. Baird Callicott (1989) show, corroborates the view that the NAIP relationship to the land was law-governed or normative. Many stories in these traditions are not about naive, innocent Indians living in perfect harmony with Mother Earth but about a people who could violate the creator's law and face consequences for doing so (e.g., stories involving Nanabush). Engagement with the land is a struggle to find balances between good and evil, individuality and community, peace and war. Humans must establish a balance of opposites, to use a phrase borrowed from the 1999 work of Taiaiake (Gerald) Alfred (Kanien'kehaka, Mohawk).

Stories of failing to achieve balance are corroborated by archival and archaeological records. The Haudenosaunee, aided by guns supplied by the Dutch, for example, used those weapons in what were known as the Beaver Wars to conquer Huronia (southern Ontario) and control the beaver trade. Long before those wars the origin story of the Iroquois Confederacy told of the creator's displeasure with the warring ways of the Haudenosaunee. Deganawidah, the peacemaker, was sent to bring the Great Law of Peace to the Haudenosaunee precisely because they were not living in peace and harmony even before contact. Poor irrigation methods resulted in salinization of soil, forcing communities such as the Hohokam to move. After contact, overtrapping in colonized areas resulted in a shift of the fur trade toward the west. Mistakes were made by NAIPs because of wrong intent (e.g., greed), inadequate planning (poor engineering), and bad judgment about the limits of what could be taken from Earth; hence, there was a need for normative direction.

It is wrong to suggest that seeking harmony implies an absence of human intervention and alteration of the land from a NAIP point of view. From the Aztecs, Incas, and Maya to the Anasazi and Cahokia, Native Americans were builders of towns and cities, devisers of political institutions, and planters of vast fields of crops long before contact. William Cronon (1983) describes how controlled burns were used to engineer environments. According to many practitioners of TEK, that practice extended all the way to the West Coast. Controlled burns fostered the maintenance of grazing lands for ungulates that the people would hunt. Increasing work in ethnobotany in collaboration with NAIPs is beginning to show how NAIPs used various techniques to shape the landscape for the production of edible plants, particularly berry-producing plants. Others used systems of weirs to trap fish. Still others developed manufacturing facilities for the production of trade goods such as wampum beads, turquoise, and copper.

With the historical understanding of NAIPs becoming more complex, conceptions of their worldviews have become more nuanced. As increasing numbers of NAIP writers, such as Vine Deloria Jr., Paula Gunn Allen, and N. Scott Momaday, have found it necessary to enter the dominant scholarly fray to correct and better represent NAIP perspectives and cultures, this complexity has become more evident. Until approximately the 1970s almost all that was written about NAIP environmental thought was by non-NAIP writers. However, since that time a growing number of NAIPs have helped enable the philosophical complexities of NAIP environmental thought to be brought to the attention of scholars, artists, and those in the general social/political arena. Others, such as Thurman Lee Hester, Sandy Grande, Viola Cordova, Jace Weaver, and Dennis McPherson, have brought those complexities more specifically into the environmental thought/ethics arena. Publications by NAIPs in the journal of environmental philosophy *Environmental Ethics* are also on the rise.

NAIP ENVIRONMENTAL PHILOSOPHY

As interpretive horizons have opened, awareness of NAIP environmental philosophy has deepened. The more people understand about dodemic (totemic) identification, for instance, the more they understand how complex the connection to the land is. Iroquoian and Algonquian peoples were and still are organized into clans according to their dodemic identity (e.g., wolf, turtle, bear). In various ways, dodems differentiate people's and clans' roles and responsibilities. Dodemic identity is connected to normative function. If one considers other cultural practices, such as the pipe ceremony and wampum belts and strings, and the way they were used in condolence ceremonies, acts of reciprocity, and the giving of thanks—all of which prepared minds and hearts to face the creator—it becomes clear that the connection to Earth was and still is not only of a normative but of a spiritual order. Prayers of thanksgiving for the waters, the plants, the creatures of Earth and air, and the wind are constant reminders of what it takes to have a good mind and lead a good life. This kind of awareness engages a sense of the sacred in the world, because the creator is in some way immanent, infused in all creation. This idea sometimes is captured in the idea of a sacred circle of life, a living mystery. Giving thanks, then, is more than an obligation or expression of gratitude for what is provided; it represents a recognition of the responsibility to maintain clarity and sincerity.

The connection to Earth also involves evil. Haudenosaunee at-the-woods-edge ceremonies took place when travelers arrived at a community. They would be greeted just outside the community where they would be fed and allowed to rest. Wampum belts and strings would be used to "wipe away" the sweat from their bodies and the blood that the thorns and briars of the forest had inflicted on them. The forest was not just a kind benefactor of humanity but also a place where evil lurked. Moreover, in Algonquian traditions, manitous (some of whom are evil) sought to deceive and kill humans. To be connected to the land, therefore, was also to be attuned to its destructive forces.

To further complicate matters, the role of the trickster brings ambiguity to NAIP epistemologies. The trickster (e.g., Coyote, Raven) can take many human and nonhuman shapes. What appears to be a human being can in fact be a coyote and vice versa. It is not clear, therefore, what could verify a perception. Epistemic sensibilities thus must become attuned to a range of possibilities that go beyond what people might perceive to be the truth. Knowledge acquisition is not a straightforward task but involves reflection at various levels. It is not difficult to understand, then, why NAIPs avoid making confident assumptions about, for instance, rights to exploit and own the land.

THE NAIP CONCEPT OF LAND AND TERRITORY

North American history is replete with instances of colonials systematically employing misrepresentations of NAIP thought in order to exploit the NAIPs and their lands. Governor John Winthrop (known for his appeal to *vacuum domicilium*), for example, argued that because Indians had taken payment for a piece of land, that land belonged to the English. Roger Williams, who better understood his NAIP neighbors, countered Winthrop's claim by arguing that when sachems such as Miantonomo agreed to allow colonists onto their land, they were not transferring ownership but were giving rights to use the land.

Thus, the concept of territory is much more appropriate than that of property. Although others might have had a right to use a territory, they first had to establish an agreement with its occupiers. If they wanted to hunt in or travel through a territory, emissaries carrying wampum belts or strings would be sent to the villages of that territory to request permission. Failure to follow this protocol or an equivalent (e.g., carrying a calumet or peace pipe) while traveling or hunting on another group's land could be construed as an act of aggression and justification for war.

Part of the reason land ownership could not be transferred as it could be in European practices was the fact that territory was a communal responsibility. Communities, not individuals, laid claim to territory. They were keepers of the land. Thus, to impose the concept of property in the sense of having the right to alienate or exploit what is owned was a means for undermining NAIP systems of territorial governance and the underlying conceptual framework of those systems.

NAIPS AND THE ENVIRONMENTAL MOVEMENT

Similarly, NAIP environmental thought has been misrepresented and manipulated in service of the environmental movement. Despite good intentions, that movement has exploited certain caricatures, such as the image of the ecologically sensitive "Crying Indian" (Iron Eyes Cody, who was not a NAIP) in the "Keep America Beautiful" campaign of 1971. Some environmental historians trace the use of the idea of the ecological Indian to an 1810 meeting between the Shawnee war chief Tecumseh (Tekamthi), who led the western nations against the United States in the War of 1812, and General William Henry Harrison, who led the U.S. Army. Tecumseh is alleged to have said to Harrison after being invited to sit on a chair, "The sun is my father and the earth is my mother; she gives me nourishment and I repose upon her bosom."

By 1877, the connection had become popularized and the idea of Mother Earth began to be used by NAIPs (e.g., the Wanapum chief Smoholla) to gain popular support in their fight to retain traditional territories and resist colonial expansion. Sam Gill (1987) suggests that the emergence of the concept of Mother Earth in NAIP cultures is a political fiction based on Western ideas and images; it serves as a critique of U.S. expansionism by establishing an unassailable moral authority for the NAIP position. However, there are periodic references to an earth mother in earlier records, and previous descriptions of NAIP cosmologies suggest that Mother Earth is not incompatible with NAIP traditions. Origin stories, including those recorded by the Jesuits in the seventeenth and eighteenth centuries, form a pattern that implies an emergence of humanity from the earth. It is possible, then, that the popularization of the concept in the 1800s was part of a cultural and political evolution in which the Mother Earth cosmology was given more explicit expression after having been precipitated by the NAIP struggle for political recognition.

Since the 1800s various forms of activism and resistance have been employed by NAIPs in their struggle to establish land management practices (fairer resource allocations), protect sacred sites, and safeguard health. The ongoing Yucca Mountain case in Nevada, involving the storing of nuclear energy wastes, has been resisted by Shoshone and Paiute people partly to protect Mother Earth and partly to assert treaty rights. Tulalip efforts to gain resource management rights in Washington (a struggle related to the 1974 Bolt decision recognizing comanagement rights for the Tulalip) are similarly motivated. In other areas, NAIPs have used road blockades and armed resistance (as in the Oka crisis in Canada) to reclaim land, protect sacred sites against development, and assert their treaty and aboriginal rights. The Oka crisis in Canada in 1990 began with Kahnasatake Mohawk efforts to reclaim a sacred site that was being developed for a golf course, and terminated in an armed standoff against Quebec police and the Canadian military. The Grassy Narrows Ojibway have been victims of mercury poisoning from pulp plants on the English River in northern Ontario, and the James Bay Cree have resisted initiatives to remove them from their land so that hydroelectric developments could proceed. The long list of environmental and legal disputes over disproportionate distribution of harms and benefits from waste management and resource extraction continues to grow as NAIPs struggle for a voice in the areas of policy making and decision making. Justice issues that concern the land can be divided roughly into two categories: (1) giving voice to NAIPs and recognizing treaty and aboriginal rights; and (2) balancing the harms and benefits (Vecsey and Venables 1980; Weaver 1996).

Conflicts between environmentalists and NAIPs over various environmental practices such as hunting suggest that the attempted marriage between environmental and NAIP philosophy and ethics may not be an entirely happy one. The Makah whale hunt (Washington State), hunting in national parks, and the harvesting of whales for ceremonial muktuk in the north are examples of conflict between conservationists and NAIP principles. These conflicts warn against making simplistic connections between NAIP philosophy and the aims of environmental protection.

SEE ALSO *Animal Ethics; Biocentrism; Black Elk; Callicott, J. Baird; Environmental Justice; Holism; Land Ethic; North America; Pantheism; Sustainability; Traditional Ecological Knowledge.*

BIBLIOGRAPHY

Alfred, Gerald R. [Taiaiake]. 1999. *Peace, Power, Righteousness: An Indigenous Manifesto*. Don Mills, Ontario, Canada: Oxford University Press.

Booth, Annie, and Harvey M. Jacobs. 1990. "Ties That Bind: Native American Beliefs as a Foundation for Environmental Consciousness." *Environmental Ethics* 12(1): 27–43.

Borrows, John. 2002. "Nanabush Goes West: Title, Treaties, and the Trickster in British Columbia." In *Recovering Canada: The Resurgence of Indigenous Law*. Toronto and Buffalo, NY: University of Toronto Press.

Caduto, Michael J., and Joseph Bruchac. 1991. *The Native Stories from Keepers of the Earth: Told by Joseph Bruchac*. Golden, CO: Fulcrum.

Callicott, J. Baird. 1989. "Traditional American Indian and Western European Attitudes toward Nature: An Overview." In *In Defense of the Land Ethic: Essays in Environmental Philosophy*. Albany: State University of New York Press.

Callicott, J. Baird, and Michael P. Nelson. 2004. *American Indian Environmental Ethics: An Ojibwa Case Study*. Upper Saddle River, NJ: Prentice Hall.

Cronon, William. 1983. *Changes in the Land: Indians, Colonists, and the Ecology of New England*. New York: Hill and Wang.

Deloria, Vine, Jr., and Clifford M. Lytle. 1983. *American Indians, American Justice*. Austin: University of Texas Press.

Denevan, William M. 2007. "The Pristine Myth: The Landscape of the Americas in 1492." Available from http://jan.ucc.nau.edu/alcoze/for398/class/pristinemyth.html.

Gill, Sam D. 1987. *Mother Earth: An American Story*. Chicago: University of Chicago Press.

Goulet, Jean-Guy A. 1998. *Ways of Knowing: Experience, Knowledge and Power among the Dene Tha*. Vancouver: University of British Columbia Press.

Grande, Sandy Marie Anglas. 1999. "Beyond the Ecologically Noble Savage: Deconstructing the White Man's Noble Savage." *Environmental Ethics* 21(3): 307–320.

Johnston, Basil. 1976. *Ojibway Heritage*. Toronto: McClelland and Stewart.

Krech, Shepard, III. 1999. *The Ecological Indian: Myth and History*. New York: Norton.

Martin, Calvin. 1978. *Keepers of the Game: Indian Animal Relationships and the Fur Trade*. Berkeley: University of California Press.

Martin, Paul S.; George I. Quimby; and Donald Collier. 1975 (1947). *Indians before Columbus: Twenty Thousand Years of North American History Revealed by Archeology*. Chicago: University of Chicago Press.

Neihardt, John G. 1932. *Black Elk Speaks: Being the Lifestory of a Holy Man of the Oglala Sioux*. New York: Morrow.

Ryan, Allan J. 1999. *The Trickster Shift: Humour and Irony in Contemporary Native Art*. Seattle: University of Washington Press.

United Nations. 1992. *Agenda 21*, Vol. 3; *Indigenous People*. Report of the United Nations Conference on Environment and Development, Rio de Janeiro, June 1992. Available from http://www.un.org/.

Vecsey, Christopher, and Robert W. Venables, eds. 1980. *American Indian Environments: Ecological Issues in Native American History*. Syracuse, NY: Syracuse University Press.

Waters, Anne S., ed. 2003. *American Indian Thought: Philosophical Essays*. Malden, MA: Blackwell.

Weaver, Jace. 1996. *Defending Mother Earth: Native American Perspectives on Environmental Justice*. Maryknoll, NY: Orbis Books.

Bruce Morito

NATURAL LAW THEORY

Natural law theory—the theory that the laws of human morality can be derived from an ordered, rational, and purposive universe—has existed in various forms since ancient Greek times. Central to natural law theory is the idea that the laws guiding human conduct are greater than or predate personal self-interest, the needs of the community, and the good of the state. Those laws are natural in that they are the same as the laws that govern the natural world. They can be identified through the use of rational thinking combined with right-minded observation of the physical universe. Many advocates of natural law theory are theists and subscribe to the notion of a rational divine will as the supreme ordering force, but the concept does not depend on the existence of a divinity.

In general, natural law theories of morality assert three things:

1. The principles for human conduct exist in a universe governed by rational laws.
2. Those laws can be discerned by human reason to determine the moral good.
3. Moral principles are universal and unchanging.

HISTORICAL BACKGROUND

Ancient Greek philosophy emphasizes the concept of virtue (*arête* in Greek), which requires a constant striving toward perfection in accordance with one's true nature. Inherent in this principle is one of the kernels of natural law theory: Nature has provided fundamental principles that guide the development of all living things. Aristotle (384–322 B.C.E.) often is recognized as the father of natural law. In the *Rhetoric* he noted that, aside from the "particular" laws that each people has set up for itself, there is a "common" law that is in accordance with nature.

Natural law theory was developed further in Hellenistic and Roman stoicism. The stoics believed that there are empirical standards of truth and justice that are revealed to people by right reason according to nature. Theirs was a philosophy of cosmic determinism, or a world governed by inviolable natural laws. Stoicism includes an account of human freedom that holds that human reason is free to choose either to follow or to turn away from the laws of nature. Virtue consists of a will that is in accordance with nature. The most influential natural law stoics were Cicero (106–43 B.C.E.), Seneca (4 B.C.E.–65 C.E.), Epictetus (55–135 C.E.), and Marcus Aurelius (121–180).

Thomas Aquinas (1225–1274), especially in the *Summa Theologica,* established the concepts of natural law morality still used by the Catholic Church. The fundamental principle of Catholic natural law is that good is to be done and evil is to be avoided. The source of good is God, who is perfect actuality: "The only perfectly good thing in existence is God, who is pure actuality. But anything else is good to the extent that its potentialities are brought to actuality. A fully developed oak tree is better than a warped or stunted one" (Aquinas quoted in O'Conner 1968, p. 19).

The modern period in natural law theory began with Hugo Grotius (1583–1645), who argued that natural law does not depend on the existence of God. That argument went directly against Aquinas's doctrine of natural law.

Natural law is associated with, though not identical to, the concept of natural rights. In the seventeenth century natural law was linked to a liberal political theory. On the eve of the French Revolution, John Locke (1632–1704) envisioned humans as being born into a state of nature in which they are rational, tolerant, and happy. In that state humans enjoy the natural rights of life, liberty, and property. Society, according to his theory, exists to protect those individual natural rights.

CONTEMPORARY USE

Natural law in the ancient and theistic sense requires belief in a teleological universe, one that is purposive and in accordance with some higher principle, a concept that is challenged by a contemporary post-Darwinian view of the universe as a cosmic accident and/or belief in personal divinity. Recently, though, there has been a resurgence of natural law theorists. This movement sometimes is called the new natural law, or neo-natural law. The most prominent contemporary natural law theorists are the Australian John Finnis (1979), the Americans Germain Grisez and Robert P. George, and the Canadian Joseph Boyle. New natural law is usually nontheistic and focuses on "basic human goods," such as human life, knowledge, and aesthetic experience, which are self-evidently and intrinsically worthwhile.

Natural law is invoked in current public policy debates about biotechnology. Theorists affirm the need for biotechnological research to promote the vital natural goods of human health and human knowledge but emphasize the need to focus on technologies that produce those goods in nondefective ways, or ways that respect the integrity of the human person. For this reason they condemn the use of human embryos for research or new technologies. They also condemn assisted suicide, human cloning, and genetic selection of offspring.

APPLICATION TO ENVIRONMENTAL ETHICS

At its most basic, natural law requires that everything in the universe act in accordance with its nature. This principle applies not only to humans but also to nonrational natural beings. Because of its appeal to nature to guide human action, natural law theory is compelling to some in the field of environmental ethics. It is seen as restoring the proper relationship of humankind to the natural world. Traditionally, however, natural law theory is applicable only to humans. As rational beings, humans are the only entities who can participate in the rational laws of the universe. Also, it was used most often by those who related the concept of an ordered universe to the idea that there is a state of equilibrium in nature. As that idea has fallen out of favor in ecology, so have appeals to natural law in environmental ethics.

There has been a connection between Catholic natural law and environmental ethics. Pope Benedict XVI saw in the environmental movement a promising route for the recovery of the natural law tradition and an antidote to the cultural situation in the West, in which notions of spirituality and morality have been excluded from the idea of rationality. For Benedict moral norms are a priori and absolute, not matters of individual taste and subjectivity, as they commonly are thought to be in contemporary Western culture. Benedict argued that contemporary ecological knowledge supports the idea of limits in nature that humans disregard at their own peril.

SEE ALSO *Environmental Law; Environmental Philosophy: III. Early Modern Philosophy; Private Property; Utilitarianism.*

BIBLIOGRAPHY

Donnelly, Bebhinn. 2007. *A Natural Law Approach to Normativity.* Burlington, VT: Ashgate.

Finnis, John. 1979. *Natural Law and Natural Rights.* New York: Oxford University Press.

Forte, David F., ed. 1998. *Natural Law and Contemporary Public Policy.* Washington, DC: Georgetown University Press.

O'Connor, D. J. 1968. *Aquinas and Natural Law.* London: Macmillan.

Simmons, A. John. 1992. *The Lockean Theory of Rights.* Princeton, NJ: Princeton University Press.

Erin Moore

NATURE CONSERVANCY

The Nature Conservancy (TNC) is the world's largest environmental organization by amount of revenue. Created in 1950 and incorporated in Washington, D.C., in 1951, it attempted initially to protect land valued by its founders, largely university ecologists who wanted to save properties useful for studying unique biological features undisturbed by humans. Those scientists were pioneers in establishing the preservation of natural conditions as a primary environmental issue.

EARLY MISSION

In its first twenty years TNC succeeded in the narrow mission of buying properties in their natural state to stop development. It earned a reputation as the real estate arm of the conservation movement. Whereas other groups lobbied politicians and regulators or educated the public, TNC simply bought property. Most of its early efforts were confined to the eastern United States.

In the early years its scientific goals became blurred by opportunism. In a rush to amass acreage, it bought not just undisturbed ecological treasures but purchased green space of all kinds, including woodlots and part of a golf course. In 1971 alarmed ecologists on the board of governors hired a science director, Robert Jenkins, who had earned a doctorate at Harvard, to restore a science-based land-acquisition process.

Jenkins refocused the mission on protecting natural diversity, as biodiversity was called at that time. He launched TNC's most enduring legacy in 1974: the Natural Heritage Program. By 1989 he had opened offices for the program in every state. The state offices, staffed by scientists, assembled a computerized catalog of unique species and ecosystems. Those data enabled TNC to identify sites with true diversity, enabling the group to fulfill its mission: "to preserve plants, animals and natural communities that represent the diversity of life on Earth by protecting the lands and waters they need to survive."

BIODIVERSITY AND LAND PRESERVATION

In 2000 TNC spun off the Natural Heritage Program into a separate organization: NatureServe. That move accompanied a marked shift in the way the organization identified lands for protection. TNC began to move away from its single-minded focus of a quarter century of saving lands and waters rich in biodiversity. It moved toward saving lands and waters of every kind, no matter how rich. In that effort it began working with the World Wildlife Fund, Conservation International, and other groups.

Although not abandoning the Natural Heritage data, TNC divided the planet into thirteen "habitat types," such as boreal forests, savannas, and tropical grasslands. It then set a target, in concert with other conservation groups, to save 10 percent of all those habitat types globally by 2015. Using that new perspective, TNC research showed that many habitats that had been taken for granted had lost far more acreage than had "treasures" such as tropical rain forests. A far greater percentage of temperate grasslands had been lost, for example, than tropical rain forests. Biodiversity associated with less-vaunted habitat types thus was at equal risk.

As TNC's methods to identify lands evolved, so did its methods for protecting them. TNC protected land in its early years mainly through the use of fee-simple purchase, but it soon embraced conservation easements. In an easement TNC would buy development rights, leaving other rights with the landowner. The use of easements exploded late in the twentieth century, and many groups began to use them to conserve land nationwide.

In the 1990s TNC staff in far-flung field offices adopted a flurry of other tools to save land. They engaged in community activism, regulatory influence, public land management, lobbying, and even for-profit enterprise to foster business development compatible with conservation. Throughout that period TNC continuously expanded its reach overseas, with operations in Canada, Latin America, the Pacific Islands, and Asia.

Overseas, TNC's flexible approach to saving land became especially prominent. In South America, for example, it partnered with a U.S. corporation, American Electric Power (AEP), to save a vast undeveloped forest in Bolivia. AEP, a Columbus, Ohio–based coal-burning utility, sought land for carbon credits to manage global climate change. In China it signed a deal to develop a conservation plan for 67,000 square kilometers in Yunnan. It negotiated in 2008 to manage floodplain risks downstream from the Three Gorges Dam.

As TNC began to protect biodiversity by all means possible, it promoted new thinking about the value of nature. Like no other organization, it put boots on the ground to back the conviction of a small band of scientists in the wake of World War II who believed that ecological preservation matters to everyone. TNC's entrepreneurial conservationists continued to press their efforts to stop irreversible loss of biodiversity as renowned ecologists such as E. O. Wilson, one of Jenkins's professors at Harvard, continued to trumpet the cause later in the century. TNC thus helped make a concern that had been the concern only of scientists a worry of the public at large.

At the same time TNC promoted the use of the capitalist system to protect natural resources. The leaders of the organization did not let the philosophical debate between environmental idealism and capitalism hinder their actions. Decades before other conservation groups they engaged business professionals in finance, real estate, law, and marketing to work with leaders from the nonprofit and for-profit sectors to advance conservation.

MEANS VERSUS ENDS

TNC insiders periodically wrestled with the obvious contradictions in their work between means and ends. However, pragmatism—"following the money" and a willingness to work with anyone—remained the guiding force. In the 1970s the president, Patrick Noonan, a champion of that notion, was accused of taking tainted money to fund conservation. His trademark response typifies the approach taken by TNC and, increasingly, other groups: "The problem with tainted money is there taint enough."

SEE ALSO *Biodiversity; Environmental Activism; Land Ethic; Nongovernmental Organizations; Preservation.*

BIBLIOGRAPHY

Birchard, Bill. 2005. *Nature's Keepers: The Remarkable Story of How the Nature Conservancy Became the Largest Environmental Organization in the World.* San Francisco: Jossey-Bass.

Ginn, William J. 2005. *Investing in Nature: Case Studies of Land Conservation in Collaboration with Business.* Washington, DC: Island Press.

Nature Conservancy. Available from http://www.nature.org.

Weeks, W. William. 1997. *Beyond the Ark: Tools for an Ecosystem Approach to Conservation.* Washington, DC: Island Press.

Bill Birchard

NEW ENVIRONMENTAL PARADIGM

In the 1970s the rapid growth of environmental awareness in the United States stimulated the development of ecological perspectives across a range of academic disciplines in addition to those which already existed in ecology and were developing in environmental science, earth science, and other natural sciences. As philosophers were elaborating versions of an environmental ethic, particularly those acknowledging the intrinsic value of the natural world, social scientists began to flesh out an environmental paradigm. Whereas philosophical efforts to develop an environmental ethic had a strong normative element, social scientists tended to adopt a Kuhnian perspective (see Kuhn 1996) by positing the need for an environmental paradigm to account for the anomalies of ecological problems and limits. The efforts of the social scientists tended to be less normative but shared with the philosophical work a desire to break with the strong anthropocentrism that had dominated Western thinking among scholars and laypersons.

DISCIPLINARY PARADIGMS AND SOCIETAL WORLDVIEWS

The concept of a new environmental paradigm (NEP), which was stimulated by the work of natural scientists such as Rachel Carson, Paul Ehrlich, Barry Commoner, and Garrett Hardin, had two general thrusts: At the disciplinary level social scientists critiqued the dominant anthropocentric assumptions (paradigms) of their disciplines and urged the adoption of a new environmental paradigm (later termed an ecological paradigm) to enable social scientists to appreciate the significance of ecological problems and their implications for modern industrial societies (Dunlap 1980).

At the societal level social scientists attempted to clarify the dominant social paradigm (DSP) through which Western societies viewed the world and demonstrate how its anthropocentrism contributed to environmentally destructive practices. They also explicated an alternative social paradigm or worldview that could help foster a more ecologically sustainable society and began to assess empirically the degree to which one was emerging. Efforts at the disciplinary and societal levels had the shared goal of overcoming an anthropocentric outlook and were designed to encourage both scholars and laypeople to "view the world ecologically" (Olsen et al. 1992).

Within academia efforts to establish an environmental/ecological paradigm have met with considerable success not only in the natural sciences, where fields such as ecology and environmental science are firmly grounded in such a paradigm, but also in the social sciences. Until the 1970s mainstream social science operated from a strongly anthropocentric human exemptionalism paradigm built on the largely implicit assumption that the exceptional characteristics of modern societies—heavy reliance on science, technology, and sophisticated social organization—exempted humans from the ecological constraints faced by other species and primitive societies. The prominence of global problems such as ozone depletion, loss of biodiversity, and climate change has made an exemptionalist stance increasingly difficult to defend, and the emergence of fields such as ecological economics, conservation psychology, and environmental sociology represents the institutionalization of a new environmental/ecological paradigm in the social sciences.

A Kuhnian paradigm shift has occurred or is in progress across numerous academic disciplines. An environmental/ecological perspective acknowledging that modern industrial societies remain ecosystem-dependent despite their technological sophistication and may be undermining their ecological viability because of unbridled technological development has taken root in many natural and social science disciplines. Unlike the Copernican, Darwinian, and Freudian paradigmatic revolutions that preceded it, the shift from human exemptionalism to an environmental/ecological paradigm does not have a singular instigator, although figures as far back as George Perkins Marsh helped lay the groundwork for Carson, Commoner, Ehrlich, Hardin, and the many other popularizers of the new paradigm.

EMPIRICAL RESEARCH

As scientific evidence supporting an environmental/ecological paradigm emerged in the 1970s, environmentalists used it to claim that contemporary societies are ecologically destructive and need to shift toward more sustainable practices. In the process many environmental activists and environmentally concerned citizens saw the necessity of replacing the current DSP—based on the assumption that limitless resources, promethean science and technology, and the ingenuity of humans ensure inevitable progress and a future of abundance—with a more ecologically realistic worldview. This potential shift in societal worldviews, which is analogous to the shift in disciplinary paradigms, attracted the attention of social scientists.

An early effort to measure public acceptance of a new environmental paradigm through the use of an NEP scale consisting of items that tap beliefs about the balance of nature, limits to growth, and human domination of nature soon was supplemented by more methodologically complex approaches. Stephen Cotgrove (1982); Lester Milbrath (1984); and Marvin Olsen, Dora Lodwick, and Riley Dunlap (1992) developed sets of bipolar items posing NEP and DSP choices as opposite ends of a continuum and asked survey respondents to choose between them. Their studies also drew on increasingly elaborated depictions of the two contrasting paradigms offered by numerous social analysts that included a wider range of components such as centralization versus decentralization and participatory democracy versus top-down decision making.

Those studies provided insights into the competing perspectives of groups such as environmental activists and industrialists, finding that environmentalists strongly endorsed the NEP and industrialists endorsed the DSP. However, what first appeared to be a rapid growth in acceptance of the NEP began to ebb with the sociopolitical shifts in the 1980s exemplified by the election of Ronald Reagan in the United States and Margaret Thatcher in the United Kingdom. Both personified the DSP, and the conservative trend they led slowed the growth of the NEP even though the discovery of new ecological problems continued to pose anomalies for the DSP. Societal worldviews are less affected by disconfirming evidence than are scientific paradigms, and particularly in the United States the growing power of the conservative movement and its success in demonizing "environmental extremism" hindered growth of the NEP.

Also, the emergence of sustainable development as a compromise between the traditional growth-is-good perspective and the 1970s limits-to-growth perspective created a more complicated situation. The endorsement of sustainable development (and implicit acknowledgment that past growth generated negative ecological impacts) by progressive industrialists, government officials, and laypersons as well as by most of the environmental community has made it more difficult to capture alternative worldviews by using a simple DSP/NEP dichotomy.

MENTAL MODELS AND THE NEP SCALE

As a result, efforts to measure the DSP and the NEP as polar opposites have faded, and a revision of the original NEP Scale has become the preferred measure of ecological worldviews (Dunlap, Van Liere, Mertig, and Jones 2000). Its validity was strengthened when anthropologists found that the mental models laypersons employ to understand ecological problems consist of beliefs akin to those tapped by the NEP Scale (Kempton, Boster, and Hartley 1995). The NEP Scale continues to be used in a growing number of nations to assess the degree to which various publics and more specific groups endorse key components of an environmental/ecological paradigm or worldview. It should be helpful in tracking the evolution of the new environmental paradigm throughout societies worldwide.

SEE ALSO *Carson, Rachel; Ehrlich, Paul; Environmental Philosophy: V. Contemporary Philosophy; Future Generations; Limits to Growth; Sustainable Development.*

BIBLIOGRAPHY

Cotgrove, Stephen. 1982. *Catastrophe or Cornucopia: The Environment, Politics and the Future.* New York: Wiley.

Dunlap, Riley E., ed. 1980. "Ecology and the Social Sciences: An Emerging Paradigm." *American Behavioral Scientist* 24(1): 1–151.

Dunlap, Riley E.; Kent D. Van Liere; Angela G. Mertig; and Robert Emmet Jones. 2000. "Measuring Endorsement of the New Ecological Paradigm: A Revised NEP Scale." *Journal of Social Issues* 56: 425–442.

Kempton, Willett; James S. Boster; and Jennifer A. Hartley. 1995. *Environmental Values in American Culture.* Cambridge, MA: MIT Press.

Kuhn, Thomas S. 1996. *The Structure of Scientific Revolutions.* 3rd edition. Chicago: University of Chicago Press.

Milbrath, Lester W. 1984. *Environmentalists: Vanguard for a New Society.* Albany: State University of New York Press.

Olsen, Marvin E.; Dora G. Lodwick; and Riley E. Dunlap. 1992. *Viewing the World Ecologically.* Boulder, CO: Westview Press.

Riley E. Dunlap

NIETZSCHE, FRIEDRICH

SEE *Environmental Philosophy: IV. Nineteenth-Century Philosophy.*

NONGOVERNMENTAL ORGANIZATIONS

By virtue of their numbers alone, which approached forty thousand worldwide in 2008 according to the Union of International Organizations, nongovernmental organizations (NGOs) have become significant players on the international stage. Much of their growth has been a product of heightened globalization spurred by the emergence of the Internet in the 1990s as NGO numbers increased from six thousand to twenty-six thousand (Nye 2004). That dramatic expansion tells only part of the story, however, for it is in their qualitative contributions that the role and growing importance of NGOs in the twenty-first century are best seen.

STRUCTURE

NGOs pluralize world politics by offering multiple channels of access across traditional nation-state borders. They act as both allies and adversaries to states, forming networks that advocate policy changes and define ethical standards.

Their activities affect almost every economic, political, or social facet, from health care to history, language to law, theology to ethics, and culture to security and defense. However, it is in the environmental arena where NGOs have had the greatest impact, from the climate change initiatives of the Sierra Club to the tropical forest protection programs of Rainforest Action Network to the campaign to protect the ozone layer by Environmental Defense Fund.

The environmental NGO community is not homogeneous, ranging from NGOs that work directly within the system and hire platoons of attorneys and MBAs to those which espouse monkeywrenching and embrace the label *ecoterrorists.* Just as there are many shades of green in the environmental NGO community, there is an array of advocacy methods among NGOs. Some groups focus on lobbying or litigation, others emphasize scientific research or property acquisition, and still others concentrate on monitoring international law. Beyond these mainstream approaches, participatory strategies such as grassroots networking and community education have gained more emphasis.

DEFINITION

However, despite agreement on their growing importance, there is no generally accepted definition of NGOs; the term has different connotations for different people. In one fundamental sense NGOs can be defined by what they are not: They are not governmental but nongovernmental. This is perhaps their greatest strength and greatest weakness. NGOs are nonparochial and much more flexible than states. They often are touted as the ideal antidote to traditional governmental bureaucracy, but this lack of statehood can be a handicap. NGOs are not subjects of international law and remain second-class citizens in international organization settings. Despite their frequent contributions to global governance, their only rights are those of observers, not voting entities.

The World Bank defines NGOs as "private organizations that pursue activities to relieve suffering, promote the interests of the poor, protect the environment, provide basic social services, or undertake community development." The United Nations conceptualization offers a few more particulars, identifying four generally accepted characteristics. The United Nations, in fact, is largely responsible for the popular use of the term *NGO* as a result of its categorization in 1945 in Article 71 of Chapter X of the UN Charter granting consultative status through the Economic and Social Council (ECOSOC). The four UN characteristics start with the fact that an NGO is a nonprofit organization. Second, it operates independently of government; this is a complicated stipulation in that some NGOs receive governmental funding. Third, these groups must be noncriminal in nature. Fourth, NGOs are not constituted as political parties. Although they may be active in an election process, they do not run for office or serve.

Over the years, in this context, the socially constructed image of NGOs has come to be highly positive. Scholars such as Joseph Nye (2004) point out that NGOs utilize a vast reservoir of soft power to shape policy through attraction rather than compulsion of economic and military might. Others, such as Margaret Keck and Kathryn Sikkink (1998), point out that NGOs are a considerable democratic force, empowering disenfranchised areas of the world by giving voice to those who had none before. However, it is important to offer a few qualifiers. NGOs often have a narrow membership, and at times they exhibit little democratic accountability. Like their domestic counterparts, interest groups, NGOs have been accused of being elitist and operating on behalf of special interests. A key difference is that the interests NGOs pursue, unlike those of most interest groups, are not exclusionary benefits; people benefit from clean air or clean drinking water whether or not they support the Natural Resources Defense Council.

Because NGOs are not homogeneous, they are difficult to classify. To complicate matters, NGOs are constantly changing. Some smaller NGOs serve a specific purpose and die out. Others, such as Conservation International, are born out of political differences within a parent NGO, in this case The Nature Conservancy. More typically, the larger, richer NGOs of the global North actively foster the birth of smaller grassroots NGOs, hoping to continue a beneficial relationship for both parties in the future. Others, such as Birdlife International, offer an umbrella structure to link up with domestic-based organization such as the National Audubon Society in the United States.

ACTIVITIES

One way to categorize NGOs is by thematic scope. That is, what do NGOs seek to do? How is their mission oriented? Are they single-issue or multi-issue organizations? Do they target the protection of a particular set of species as in the case of the Royal Society for the Protection of Birds, or are they interested in broad biodiversity protection as in the case of the World Conservation Society?

Another approach is to categorize them by geographic scope. Where does the NGO operate? Is it a grassroots organization targeting specific local initiatives, such as the Friends of the Wekiva in central Florida or the Nobel Prize–winning Wangari Maathai's Greenbelt Movement in rural Kenya? Is it a nationally based group such as the Wilderness Society in the United States or perhaps more broadly but still regionally focused such as the Defenders of Wildlife in North America? Or is the NGO a truly global organization with both a global agenda and a global impact, such as the World Wildlife Fund and the World Wide Fund for Nature?

NGOs also vary widely in size, hierarchical structure, and financial budgets. The Sierra Club has over 1.3 million members in the United States alone, whereas Earthwatch has 20,000 globally. Some are highly centralized, such as Earthjustice Legal Defense Fund, whereas others have a loose federal structure, such as Greenpeace International. By and large, the wealthy environmental NGOs are found in the northern hemisphere, and the poorer groups in the southern hemisphere.

Environmental NGOs are an alternative power source to nation-states that enhance environmental justice around the world by creating new transnational political coalitions through the creation and maintenance of civil society. At times NGO activity targets business interests, particularly with "name and shame" tactics that hurt corporations such as Home Depot and Royal Dutch Shell in the pocketbook with the threat of consumer boycotts of their products: timber and petroleum, respectively. At other times this may entail direct clashes with states, as occurred when Greenpeace's *Rainbow Warrior* chased the French throughout the South Pacific to bear witness to that country's nuclear testing. Another example is Earth Island Institute filing a suit against the U.S. government for failing to enforce the Marine Mammal Protection Act and calling attention to Mexico's incidental takings of dolphin in fishing for tuna in the eastern tropical Pacific.

Sometimes the impact is much more conciliatory, such as Conservation International enhancing capacity building in Ghana with a shade coffee growing cooperative. Sometimes the activity of an NGO takes the form of participation in global dialogue, such as the 1992 United Nations Conference on Environment and Development (UNCED), popularly known as the Earth Summit, in Rio de Janeiro, Brazil, and the World Summit on Sustainable Development in Johannesburg, South Africa, in 2002. NGOs clearly are a set of actors with increasing influence in global politics, from the establishment of fundamental environmental ethical standards to specific policy formation.

SEE ALSO *Biodiversity; Earth Summit; Ecotage and Ecoterrorism; Forests; Global Climate Change; Globalization; Greenpeace; Ozone Depletion; Sierra Club.*

Major Nongovernmental Organizations

NGO	Contact Information	Mission	Area of Activity	Country Representation
National Audubon Society	225 Varick Street 7th floor New York, NY 10014 Phone: 212- 979-3000 http://www.audubon.org	Conservation and restoration, with particular focus on birds and their habitat	United States	United States
Birdlife International	Wellbrook Court Girton Road Cambridge CB3 0NA UK Phone: +44 (0)1223 277 318 Fax: +44 (0)1223 277 200 http://www.birdlife.org/	Conservation of bird habitat and biodiversity	Global	Belgium, Ecuador, Fiji, Japan, Jordan, Kenya, UK and partners worldwide
Center for International Environmental Law	1350 Connecticut Avenue, NW Suite #1100 Washington, DC 20036 Phone: 202-785-8700 Fax: 202-785-8701 http://www.ciel.org/	Legal counsel, policy research, advocacy and training to promote international law protecting environment and human health	Global	Switzerland and United States
Conservation International	2011 Crystal Drive, Suite 500 Arlington, VA 22202 Phone: 703-341-2400 http://www.conservation.org	Biodiversity Conservation	Global	Western Hemisphere, Africa, Asia-Pacific
Defenders of Wildlife	1130 17th Street, NW Washington, DC 20036 Phone: 202-682-9400 http://www.defenders.org	Biodiversity conservation, particularly regarding US Endangered Species Act	Canada, Mexico & United States	Canada, Mexico and United States
Earth First!	http://www.earthfirst.org/	Considers itself a movement (instead of an organization) for wilderness conservation through civil disobedience and monkey-wrenching	Global	Australia, Canada, Czech Republic, Netherlands, Philippines, UK, United States
Earth Island Institute	300 Broadway, Suite 28, San Francisco, CA 94133 Phone: 415-788-3666 http://www.earthisland.org/	Biological and cultural diversity protection	Global	United States
Earth Liberation Front	http://www.earthliberationfront.com/	Underground environmental movement with no leadership, membership or official spokesperson	United States	United States
Earthjustice Legal Defense Fund	426 17th Street, 6th Floor Oakland, CA 94612-2820 Phone: 510-550-6700 http://www.earthjustice.org/	Non-profit public interest law firm defending natural resources, wildlife, healthy environment for people	United States focus	United States with alliances in Canada, Latin America, and Russia
Earthwatch Institute	3 Clock Tower Place Suite 100 Box 75 Maynard, MA 01754 Phone: 978-461-0081 http://www.earthwatch.org/	Scientific field research and education to promote a sustainable environment	Global	Australia, Japan, UK, United States
Environmental Defense Fund	257 Park Avenue South New York, NY 10010 Telephone: 212-505-2100 http://www.edf.org	Protecting environmental rights, including access to clean air, water, food, healthy ecosystems	United States focus	United States
Friends of the Earth	1717 Massachusetts Avenue Suite 600 Washington, DC 20036 Phone: 202-783-7400 http://www.foe.org/	Promotes justice and a healthy environment	Global	United States

[Continued]

Table 1, part 1. CENGAGE LEARNING, GALE.

Major Nongovernmental Organizations

NGO	Contact Information	Mission	Area of Activity	Country Representation
Friends of the Earth International	Secretariat PO Box 19199, 1000 gd Amsterdam, The Netherlands Phone: +31 20 622 1369 http://www.foei.org/	Conservation and environmental rights	Global	Global
Greenpeace - International	Ottho Heldringstraat 5 1066 AZ Amsterdam, The Netherlands Phone: +31 20 7182000 http://www.greenpeace.org/international/	Began in opposition to whaling and nuclear testing while protecting Antarctica but now targets climate change and other global environmental issues	Global	Global
Greenpeace -USA	702 H Street, NW Washington, DC 20001 Phone: 202-462-1177 http://www.greenpeace.org	Began in opposition to whaling and nuclear testing while protecting Antarctica but now targets climate change and other global environmental issues	Global	United States
Izaak Walton League of America	707 Conservation Lane Gaithersburg, MD 20878 Phone: 301-548-0150 http://www.iwla.org/	Conservation	United States	United States
League of Conservation Voters	1920 L Street, NW Suite 800 Washington, DC 20036 Phone: 202-785-8683 http://www.lcv.org/	Advocates election pro-environmental candidates who will adopt and implement sound environmental policies	United States	United States
Natural Resources Defense Council	40 West 20th Street New York, NY 10011 Phone: 212-727-2700 http://www.nrdc.org/	Conservation	United States and Global	China and United States
National Wildlife Federation	11100 Wildlife Center Drive Reston, VA 20190-5362 Phone: 1-800-822-9919 http://www.nwf.org/	To inspire Americans to protect wildlife	United States	United States
The Nature Conservancy	4245 North Fairfax Drive, Suite 100 Arlington, VA 22203-1606 Phone: 703-841-4850 http://www.nature.org/	Conservation	Global	Global
The Ocean Conservancy	1300 19th Street NW 8th Floor Washington, DC 20036 Phone: 202-429-5609 http://www.oceanconservancy.org	Marine conservation	Global- Marine	United States
Rainforest Action Network	221 Pine Street, 5th Floor San Francisco, CA 94104 Phone: 415-398-4404 http://www.ran.org/	Conservation of rainforests and people near those forests	Rainforests across the globe	Japan and United States
Royal Society for the Protection of Birds	The Lodge Potton Road, Sandy Bedfordshire SG19 2DL Phone: +44 01767 680 551 http://www.rspb.org.uk/	Biodiversity conservation, especially wild birds	United Kingdom	United Kingdom
The Sea Shepherd Conservation Society	PO Box 2616 Friday Harbor WA 98250 Phone: 360-370-5650 http://www.seashepherd.org/	Marine conservation	Global - Marine	Australia, Brazil, Canada, Belgium, Ecuador, France, Netherlands, UK, United States
The Sierra Club	85 Second Street, 2nd Floor San Francisco, CA 94105 Phone: 415-977-5500 http://www.sierraclub.org/	Conservation	United States	United States

[Continued]

Table 1, part 2. CENGAGE LEARNING, GALE.

Major Nongovernmental Organizations

NGO	Contact Information	Mission	Area of Activity	Country Representation
The Wilderness Society	1615 M St, NW Washington, DC 20036 Phone: 1-800-THE-WILD http://www.wilderness.org/	Protect wilderness and inspire Americans to care for wild places	United States	United States
Union of Concerned Scientists	2 Brattle Square Cambridge, MA 02238-9105 Phone: 617-547-5552 Fax: 617-864-9405 http://www.ucsusa.org/	Researches practical solutions to global warming, nuclear weapons, vehicle pollution, GMOs, etc.	Global	United States
World Conservation Society	2300 Southern Boulevard Bronx, New York 10460 Phone: 718-220-5100 http://www.wcs.org/	Biodiversity conservation and management of urban wildlife parks such as the Bronx Zoo	Global	United States
World Resources Institute	10 G Street, NE - Suite 800 Washington, DC 20002 Phone: 202-729-7600 Fax: 202-729-7610 http://www.wri.org/	Think tank researching practical ways to protect the earth and improve people's lives	Global	United States
Worldwatch Institute	1776 Mass. Ave. N.W. Washington, DC 20036-1904 Phone: 202-452-1999 http://www.worldwatch.org/	Think tank focusing on sustainable development issues such as climate change, resource degradation, population growth, and poverty	Global	United States
World Wildlife Fund – WWF US	1250 24th Street NW Washington, DC Phone: 202-293-4800 Fax: 202-293-9211 http://www.worldwildlife.org/	Biodiversity Conservation	Global	United States
World Wide Fund for Nature – WWF International	Av. du Mont-Blanc 1196 Gland, Switzerland Phone: +41 22 364 91 11 Fax: +41 22 364 48 92 http://www.panda.org/	Biodiversity Conservation	Global	Global

Table 1, part 3. CENGAGE LEARNING, GALE.

BIBLIOGRAPHY

Duke University Libraries. "NGO Research Guide." Available from http://library.duke.edu/research/subject/guides/ngo_guide/igo_ngo_coop/ngo_wb.html.

Global Policy Forum. 2006. "NGOs." Available from http://www.globalpolicy.org/ngos/index.htm.

Keck, Margaret E., and Kathryn Sikkink, eds. 1998. *Activists beyond Borders: Advocacy Networks in International Politics.* Ithaca, NY: Cornell University Press.

Mathews, Jessica T. 1997. "Power Shift." *Foreign Affairs* 76(1): 50–66.

Nye, Joseph. 2004. "The Rising Power of NGO's." *Daily Times* (Pakistan), June 27. Available from http://www.dailytimes.com.pk/default.asp?page=story_27-6-2004_pg3_7.

Princen, Thomas, and Matthias Finger. 1994. *Environmental NGOs in World Politics: Linking the Local and the Global.* London and New York: Routledge.

Simmons, P. J. 1998. "Learning to Live with NGOs." *Foreign Policy* 112: 82–96.

Wapner, Paul. 1996. *Environmental Activism and World Civic Politics.* Albany: State University of New York Press.

Weiss, Thomas G., and Leon Gordenker, eds., 1996. *NGOs, the UN, and Global Governance.* Boulder, CO: Lynne Rienner.

Michael M. Gunter Jr.

NORTH AMERICA

Distinct, influential, and sometimes conflicting philosophical images and values about what humans are, what nature is, and what the relationship between humans and nature is and should be have been at the root of environmental controversies in North America over the last 500 years. Before European contact and the subsequent settlement of North America very different philosophical images and values reigned; those perspectives persisted, along with modifications of them. The confluence of those distinct philosophical ideas shaped both the

George Catlin, **Buffalo Chase Over Prairie Bluffs,** ***1844.*** *Scholars are undecided as to the environmental impact of the first populants of Native America (beginning roughly 14,000 years ago). Although it is commonly assumed that North American Indians had an environmental ethic similar to that of contemporary philosophers, others claim that their population of the continent caused mass destruction of native flora and fauna.* THE ART ARCHIVE.

environmental attitudes of North Americans and the landscape of the continent.

NORTH AMERICA BEFORE 1492

Both the practices and the environmental ethics of pre-Columbian North Americans are contested. Roughly 14,000 years ago Asian big game hunters crossed the exposed land bridge between what is now Russia and Alaska and also sailed along the western coast of North America to populate or greatly increase the population of North America. According to one theory (Martin 1967), those skilled hunters quickly swept across North America, Central America, and part of South America, slaughtering the unfamiliar megafauna; indigenous populations were drawn by the easy pickings of unwary animals, leaving extinguished species in their wake. Other scholars (Krech 1999) have suggested that although the arrival of big game hunters and the disappearance of many North American species corresponded, the demise of those species was attributable as much to changes in climate and perhaps other factors as it was to the newly arrived human hunters.

In environmental circles it commonly is assumed that before contact North American Indians (First Nations in Canadian parlance) had an inclusive environmental ethic on a par with the most inclusive contemporary environmental ethics, such as Deep Ecology and Aldo Leopold's land ethic. Although details varied from tribe to tribe, North American Indians considered nonhuman animals, rocks, rivers, mountains, and traditional myths and stories to have direct moral standing (Callicott and Nelson 2004). Contemporary ideas about the behavior of native North American Indians range from the assumption that they were skilled and intensive managers of the land to the belief that they were passive denizens of North America, a perspective that seems inconsistent with prevailing ethnographic and archaeological evidence. How is it possible to reconcile an inclusive ethic with the Pleistocene extinction hypothesis or the overshoot of ecological carrying capacity, such as overuse of water resources in the Southwest six to eight centuries ago? N. Scott Momaday (1976) suggested that a profound ethical change occurred when the big game hunters gradually came to see the land of North America as home, then as beautiful, and then as intrinsically valuable.

Although the argument about an indigenous North American land ethic is an ongoing debate, knowledge about the extent of Indian impact is becoming more extensive. Many assumptions about Indian environmental ethics are premised on what is known about what pre-Columbian Indians did. In the last two decades of the twentieth century and the first decade of the twenty-first, however, much of the dogma surrounding those practices was challenged. For example, until recently the dominant figure for the North American human population was approximately 1 million. Currently, it is known that there were at least 4 million and up to 16 million inhabitants in 1491. Researchers also have come to appreciate that those humans affected the North American continent for over 14,000 years, sometimes in a very intensive fashion. For example, in the midwestern United States it is known that there were large agricultural complexes, each of which covered up to 200 to 300 acres, that supported thousands of people. It also is known that humans, through the intensive use of fire, actively and continually maintained much of what is considered the original prairie of North America and determined the composition and shape of forest ecosystems throughout the continent. Additionally, vast complexes of ceremonial and burial mounds, complex systems of trading, and cities housing tens of thousands of people (e.g., Cahokia in Illinois, which eventually overshot its carrying capacity and scattered its residents) existed in North America.

However, the myths, stories, and legends of precontact Indians seem to indicate very inclusive systems of ethics that guided and continue to govern subsistence practices of hunting, fishing, and gathering; those practices cumulatively shaped species distribution, diversity, and productivity (Blackburn and Anderson 1993, Frey 2001). Arguably, it was the animism—the belief that nature and/or natural entities are imbued with an indwelling spirit (Nelson 2004) and therefore possess what might be considered a kind of intrinsic value—of native North Americans that undergirded that inclusive ethic.

1492 TO 1776

Although Europeans had made forays into North America for perhaps 500 years, European settlement (or invasion) began to affect North American inhabitants and landscapes more seriously in the early part of the sixteenth century. Most immediately, waves of European-introduced diseases, generally introduced accidentally, began to decimate the populations of native North Americans. Because that anthropogenic disease regime effectively reduced the native population by as much as 90 percent and because the disease spread ahead of the settlers, Europeans felt justified in concluding that North America was a wilderness of continental proportions that was theirs for the taking. That assumption was affirmed by a tendency to perceive native peoples as having more the status of nonhuman wildlife than that of full-fledged humans worthy of moral consideration and respect. In light of the Puritan religious background of the early New England colonists, it may be safe to say that the conquest of the North American inhabitants and landscape was more than a consistent proposition; it was an expected and highly moral vocation. The Puritan leader and witch hunter Cotton Mather summarized this mentality when he asserted that "what is not useful is vicious."

At the same time that that internal pressure created a willingness to affect the North American landscape with little regard for the land or its native inhabitants, a more global market created a lucrative outlet for the products of the relatively unexploited North American continent, and that confluence had a major impact on the fish and wildlife populations. North Atlantic cod, which when dried provided a rich and easily transportable source of protein, and beaver pelts from English and French colonies, which when felted provided sought-after hats, are two notable examples. Exploitation also occurred in Spanish colonies, but it existed alongside traditions of caring for the common good; for example, water resources were shared as a community responsibility in places such as California and New Mexico. Although all Europeans participated in global market arrangements, the British did so with the greatest enthusiasm and the steepest ecological costs.

The meeting between native North Americans and the newly arrived Europeans was as much a clash of ideologies and ethics as it was a clash of technologies. Europeans brought with them not only the ability to alter the landscape but also the willingness to do that. Native North Americans, in contrast, had the ability to alter their landscape more seriously than they did but lacked the willingness and need to do so. Inclusive native ethics were inconsistent with the narrowly anthropocentric ethic of the European settlers. Although tempered, that Euro-American ideology is still in operation.

1777 TO 1899

In 1893 the historian Frederick Jackson Turner pronounced that the American frontier had closed three years earlier. Although historians have challenged that interpretation, the idea of a closed frontier resonated with Americans who saw the transition from the completion of Manifest Destiny (the belief that westward expansion and territorial acquisition all the way to the western coast of the United States was inevitable) to the beginnings of a new and gentler ethic of relationship with nature. Analogously, Canadians linked the development of their nation to the exploitation of raw natural resources—staples such as furs and wheat—but did not develop a strong conservation ethic from that economic precariousness. Temporally and effectively corresponding to the beginning and the height of the Industrial Revolution, the late eighteenth and early nineteenth centuries saw the most brazen exploitive environmental practices that the technology of the time would allow. In the early part of that period North Americans of all types seemed limited only by their technical ability to affect nature and not at all by their willingness to do so. However, arising with these exploitive customs were challenges to that prevailing environmental ethic.

After the American Revolution the United States began purchasing and conquering what is now the Lower 48, a process that was complete by 1853 and that established the geography needed for accelerated economic exploitation. Canada, remaining until the present under the sovereignty of the British Crown, effectively emerged from British colonialism by 1873. Farmers moved quickly to fill the temperate North American continent and displaced native peoples and their environmental practices, except in the subarctic and arctic northern territories of Canada and Russian-owned Alaska. Laws such as the U.S. Homestead Act of 1862 and the Canadian Dominion Lands Act of 1872 served as powerful examples of a philosophical and ethical predisposition that manifested itself first legally and then on the land. These and similar laws privatized the public domain with an emphasis on small farmers, but often corporations such as railroads and mining companies fraudulently took advantage of the laws and wreaked ecological havoc for short-term economic gain.

This era also saw the boom and eventually the bust of many of the extreme extractive practices in North America. For example, the end of Great North Woods lumbering and massive commercial hunting occurred during that period. That era also witnessed the removal of American Indians from some would-be park areas and the relegation of most American Indians in the United States to reservations. The abuse of the environment in the form of the intentional slaughter of the herds of buffalo that fed certain Indian tribes (arguably a kind of biological warfare) is an example of the indifference and even contempt that the inhabitants of European descent in North America had for both the rights of nature and the rights of the native peoples. Similarly, mining rushes in the West extracted immense mineral wealth from the earth without considering the attendant environmental damage, such as deforestation, erosion, pollution, and habitat destruction (Isenberg 2005). The end of those practices, however, was due primarily to the end of the readily accessible resources that were the focus of the cut-and-run practices of the extractive industries. However, two separate and at times competing natural resource philosophies—resourcism and preservationism—were emerging at that time, neither of which allowed for the types of abuses seen in the past. The era of unthoughtful and uncontested resource exploitation in North America came to an end at about the turn of the twentieth century.

Largely in response to the profligacy of nineteenth-century capitalist development, many Americans began to question practices and reform policies, especially those concerning public lands. Figures such as George Perkins Marsh and John Wesley Powell warned of the social and economic harm that was likely to result from continued environmental degradation. Fearing that privatizing public lands was proceeding without a plan and in wasteful ways, reformers moved the federal government toward protecting land in addition to selling or giving it away. In 1872 the U.S. Congress reserved the first national park at Yellowstone; Canada followed in 1887 by reserving Rocky Mountain Park, later renamed Banff National Park. By 1891 the U.S. president had the power to reserve lands from settlement in what were known as forest reserves and now are called national forests. By the early twentieth century the Canadian Forest Service began recognizing problems with traditional harvest practices, but its regulatory presence remained weak for decades. Still, government-protected areas such as the national parks and forests were limited in terms of the types of economic activities that could be pursued there. Thus, the reforms initiated government involvement in checking economic exploitation and effectively announced that some environments were better left alone or managed with a longer-term perspective.

Besides government reform, private individuals and organizations instituted change in environmental practices. Early in the nineteenth century a few reform-minded farmers recognized the problem of soil erosion and attempted to initiate new practices on their farms that would yield long-term economic benefits without the ecological costs associated with most farming techniques (Stoll 2002). Organizations were even more active. In 1887 the Boone and Crockett Club was established by Theodore Roosevelt, a complex environmental figure known as much for setting aside vast tracks of land for national forests as he was for slaughtering vast numbers of African wildlife for American museums. Equally complex was the mission of the Boone and Crockett Club. Although its focus was the conservation of dwindling game populations and the habitats in which they lived, the foundational value of the club seems anthropocentric. That changed, however, in 1892 when John Muir founded the Sierra Club. Although Muir's Sierra Club did not shun activities that humans enjoyed, it had an ultimate goal of preserving wild nature. The Sierra Club operated under the philosophy that exposure to wild nature would increase people's knowledge of a place and thus prompt in them a love for wild areas and a willingness to act on their behalf.

In an important early paper in environmental ethics the historian Lynn White, Jr. (1969), argued that the environmental abuses visited upon the North American continent by its denizens of European descent were simply the manifestation of a certain interpretation of their imported religious tradition, Christianity. Although White blamed the despotic interpretation of the human-nature relationship in the Christian tradition (the interpretation that informs humans that the earth is there for their use and abuse, that it is God's desire that people "dominate and subdue" His creation), he did not blame Christianity itself, a point that often is missed in reactions to his argument. Hence, for White and for environmental philosophers after him it was the philosophical and ethical predisposition, coupled with the emerging technological power, of the Old World arrivals that facilitated the radical alteration of the North American landscape in that period. By the end of the nineteenth century North America thus had experienced both massive ecological devastation and the roots of philosophical shifts that would thwart or at least complicate continued the pursuit of profit at the expense of nature.

1900 TO 1955

The early part of the twentieth century was marked by perhaps the most dramatic environmental battle in North American history and certainly the most continuously recognized one. The battle over the Hetch Hetchy Valley in Yosemite National Park pitted two contrasting environmental philosophies against each other. As early as 1864 George Perkins Marsh, the U.S. ambassador to both Turkey and Italy, had challenged the narrow and ultimately paradoxical anthropocentric justification of resource exploitation that had reigned on the North America continent. Employing the notion of the usufruct (use without destruction), Marsh suggested that it was inappropriate to believe and act as if the earth had been given to humans for "consumption" or "profligate waste" (Marsh 1864, p. 34). In 1905 the equally broad-minded and European-trained Gifford Pinchot was appointed the first chief of the U.S. Forest Service. Both Marsh and Pinchot developed their environmental philosophies in reaction to what they viewed as an overly exclusive and shortsighted human use of natural resources that they encountered in Europe. Although Pinchot argued for a more democratic and far-reaching form of conservation than the one he had encountered in Europe—he suggested that people should strive for "the greatest good of the greatest number for the longest time" and defined conservation as "the wise use of the earth and its resources for the lasting good of men" (Pinchot 1998, pp. 326–327)—he remained an anthropocentrist, stating that "there are just two things on this material earth—people and natural resources" (Pinchot 1998, p. 326).

In the Hetch Hetchy battle Pinchot argued from the point of view that the valley should be dammed to provide publicly owned water and electricity for San Francisco because those were the higher human benefits of resource use (Righter 2005). That stance pitted Pinchot against his old friend John Muir, who argued that the valley should be left as it was, a position that came to be known as preservation. Muir's position was a manifestation of the nineteenth-century transcendental philosophy of Ralph Waldo Emerson and Henry David Thoreau, who believed that only in nature could one witness the handiwork of God and transcend ordinary existence to find higher truths. Hence, for Muir, to destroy Hetch Hetchy was to display "a perfect contempt for Nature" (Muir 1992, p. 716) akin to destroying temples and churches. Here, for the first time since the European conquest, one can see a North American environmental philosophy suggesting that nature has a kind of value that transcends instrumental and anthropocentric ends, although Muir and his allies believed that tourists should benefit from such preservation (Righter 2005).

In the mid-twentieth century Aldo Leopold attempted to meld those two environmental philosophies. Although he worked to improve farming techniques and secure other human ends, Leopold also argued that people should judge the morality of actions, policies, and laws by their tendency to "preserve the integrity, stability, and beauty of the biotic community" (Leopold 1949, pp. 224–225), a community inclusive of human beings. Perhaps most important, Leopold's environmental philosophy turned away from a Judeo-Christian worldview and toward an evolutionary-ecological one premised on an assumed continuity between humans and the nonhuman world. A good example of Leopold's melded philosophy can be found in his ideas about wilderness preservation. Leopold and other preeminent ecologists of that time suggested that certain places in the United States should be set aside as designated wilderness areas as early as the late 1910s. However, Leopold's early rationale for wilderness preservation was dominated by arguments for human recreation that were common among other early wilderness thinkers. Later in his thinking about wilderness Leopold began to supplement such arguments with nonanthropocentric viewpoints that suggested that wilderness should serve as a place to house otherwise unwelcome wildlife and ultimately as a base datum of healthy land.

Although the moral will to enact a more inclusive environmental ethic was present by the mid-twentieth century, certain distractions prevented its full blossoming. The financial and psychic cost of two world wars that came on either side of the Great Depression, the Dust Bowl, and the New Deal legacy of U.S. President Franklin Delano Roosevelt (employing economic stimulus practices such as the Civilian Conservation Corps that transformed nature in a dramatic and narrowly anthropocentric manner), in combination with the advent of urban sprawl and subsequent forms of pollution, had an impact on the land and served as distractions from the development of an environmental ethic distinct from anthropocentrism. Moreover, the institutionalization of conservation in bureaucracies such as the U.S. Forest Service/Canadian Forest Service, the U.S. National Park Service/Canadian Dominion Parks Branch, the U.S. Soil Conservation Service, and the U.S. Bureau of Reclamation all meant that the North American political economy proved most influential in shaping national environmental policies and practices in both the United States and Canada. The emphasis on resourcism in Canada tended to go unchallenged even longer than was the case in the United States. At the same time, however, the science of ecology was emerging and beginning to influence and alter environmental discourse. This fusion of science and ethics soon would manifest itself in powerful and far-reaching ways.

1955 TO 1970

In the early 1950s a near replica of the battle over Hetch Hetchy Valley occurred at Echo Park within Dinosaur National Monument in Colorado. This time, however, nature won and the dam was not built. The political compromise arising from the debate, though, allowed the construction of Glen Canyon Dam, effectively creating a conceptual cleavage between sacred lands (those within the national park system) and profane lands (those outside the system). This episode, which was fought in Congress and in the national media, marked a revival of wilderness activism and paved the way for an ascendant environmentalism with a focus on particular, local issues and problems.

Environmentalism was becoming a popular, as opposed to an exclusive, concern, at least among many middle-class white Americans. In fact, many people attribute the emergence of their personal environmental concern as well as the dawn of the environmental movement to Rachel Carson's *Silent Spring* (1962). Even though the DDT that the book warned of was not banned until 1972 in the United States and 1985 in Canada, Carson's warning about unrestrained alteration of and impact on the natural world triggered the popular perception that the environment was endangered and worth worrying about. Other debates, such as Paul Ehrlich and Barry Commoner's debate over whether human population or technological invention was the fundamental environmental issue, also helped popularize environmentalism. More than ever North Americans debated environmental questions publicly and began to challenge narrowly anthropocentric treatments of nature forcefully.

A number of legislative successes for environmental causes emerged from that popular concern. In 1963 (2006 in Canada) the United States passed the Clean Air Act. In 1964 the U.S. Wilderness Act was passed, which ultimately would preserve nearly 5 percent of the country (one-half of that in Alaska) as designated wilderness. The National Environmental Policy Act (NEPA) that was signed into law on January 1, 1970, required federal projects to prepare an environmental impact statement (EIS) cataloging expected effects along with various alternatives (the Canadian equivalent went into effect in 1995 and is known as the Canadian Environmental Assessment Act). NEPA also opened the EIS to a period of public comment, giving interested citizens an opportunity to voice their concerns and furnishing the opportunity for lawsuits to compel more environmentally ethical planning. In 1970 Senator Gaylord Nelson of Wisconsin formalized environmentalism as a popular and urgent matter by sponsoring legislation creating an annual Earth Day. Originally established as a day for "teach-ins" focused on environmental problems, it evolved into a celebration of ecological values.

Meanwhile, academia was witnessing a great change. Departments of ecology were springing up at universities all over the European and North American world, supporting research that could be and was being employed in defense of nature. With the dominance of the ecosystem concept, ecology also appeared to have arrived as a full-fledged and quantifiable science. After that period of explosive growth in North Americans' concern for and willingness to act on behalf of the environment, a variety of philosophically astute and environmentally conscientious philosophers and other academics began to get into the game.

1971 TO THE PRESENT

Environmental philosophy and ethics, along with a number of other environmental disciplines (from history to sociology, economics to literary criticism), emerged in the early 1970s. It can be said that North America, along with England, Australia, and Norway, has been a point of origin for the field of environmental philosophy and perhaps its epicenter. Impelled by the first Earth Day, J. Baird Callicott taught the first course in environmental ethics in the world in 1971 at the University of Wisconsin–Stevens Point. In 1979 Eugene Hargrove launched the discipline's first and still preeminent journal, *Environmental Ethics.* Although the University of Georgia took the early

Dinosaur National Monument, U.S. *Dinosaur National Monument is a part of the Uinta Mountains, sitting on the border between Colorado and Utah. In the 1950s, a plan was proposed to build a dam in Echo Park, in the middle of the monument. A nationwide campaign of protest prevented construction of the dam, and many consider this event as the start of a successful conservationist era.*

institutional lead as the seat of environmental philosophy, eventually the University of North Texas became the continent's and eventually the world's leading philosophy department focused on environmental philosophy. Pioneering Canadian environmental philosophers included Allen Carlson, Peter Miller, and Bob Jickling. Nearly every

North American university now teaches at least one course in environmental ethics (though not always in the department of philosophy), and many have at least one scholar who focuses on the field.

Between 1971 and 1979 philosophers interested in environmental ethics and philosophy worked in relative isolation from one another. By the time they began to discuss their work more publicly, various positions already had emerged. Those positions centered on the matter of who or what deserved direct moral standing and who or what merited only indirect moral standing. The seemingly dominant way to discuss this topic was in reference to who or what had intrinsic value and why. There are dozens of textbooks, five or six journals, thousands of articles, and hundreds of books in the field. Moreover, there are a few graduate degree programs in environmental philosophy, undergraduate majors, and related university programs and courses featuring the works of environmental philosophers. Environmental philosophers also have begun to infiltrate conservation science in various ways.

This period also marked the beginning of what might be called the second wave of the environmental movement, beginning in the mid-1980s. That wave had a much more global and systemic focus than its predecessor, concentrating, for example, on issues such as the precipitous loss of biodiversity and the impending sixth great extinction, stratospheric ozone depletion, acid rain, social justice and human rights, and rapid global climate change as much as it did on more local forms of environmental harms. Old distinctions such as that between conservation and preservation no longer seemed to fit or make sense, although some people still employed them.

At the same time, clearly prompted by environmental philosophies such as Deep Ecology, environmental activism became much more radicalized with activist groups such Earth First!, whose motto was "No compromise in defense of Mother Earth," and Greenpeace, a Vancouver-based organization that grew to have more than 2.5 million members. Environmental politics also became more polarized as the result of some profound shifts in values. The U.S. (1973) and Canadian (1996) endangered species acts, for example, represent a significant moral shift suggesting that species and other categories of animal populations merit direct moral standing and deserve to exist for their own sake, apart from the impact they may or may not have on more narrowly conceived human economic interests. Dramatic and ongoing battles over owls, wolves, grizzly bears, and salmon have been the direct result of, among other things, this philosophical and ethical change.

A number of current and near future topics promise to occupy North American environmental philosophy in the coming years. Although historically an exceptionally relevant and respected discipline, philosophy took a turn toward specialization in the twentieth century and, in the opinion of some people, lost some of its relevance and influence. Many environmental philosophers, however, seek to make their work relevant to science and policy. Although it holds great promise, this renewed commitment to relevance and interdisciplinarity will continue to present a challenge to environmental philosophy. Making a commitment to relevance and thinking of ways to account conceptually and ethically for the moral relevance of human and nonhuman individuals as well as the environmental collectives that serve as the focal point of much contemporary environmental concern and navigate between the good of each when they are in conflict are issues that present another challenge to environmental philosophers. Finally, working to craft philosophical and ethical systems that account for the dominant ecological paradigm focused on flux and change presents a particularly difficult problem for environmental philosophy and for much of environmental discourse.

SEE ALSO *Anthropocentrism; Christianity; Conservation; Deep Ecology; Ecology: V. Disequilibrium Ecology; Emerson, Ralph Waldo; Forests; Hetch Hetchy; Land Ethic; Leopold, Aldo; Marsh, George Perkins; Mexico and Central America; Muir, John; Pinchot, Gifford; Pollution; Preservation; Sierra Club; Species; Thoreau, Henry David; U.S. Forest Service; U.S. National Park Service; White, Lynn, Jr.; Wilderness.*

BIBLIOGRAPHY

Blackburn, Thomas C., and Kat Anderson, eds. 1993. *Before the Wilderness: Environmental Management by Native Californians.* Menlo Park, CA: Ballena Press.

Callicott, J. Baird. 2003. "The Implication of the 'Shifting Paradigm' in Ecology for Paradigm Shifts in the Philosophy of Conservation." In *Reconstructing Conservation: Finding Common Ground,* ed. Ben A. Minteer and Robert E. Manning. Washington, DC: Island Press.

Callicott, J. Baird, and Michael P. Nelson. 2004. *American Indian Environmental Ethics: An Ojibwa Case Study.* Upper Saddle River, NJ: Prentice-Hall.

Carson, Rachel. 1962. *Silent Spring.* Boston: Houghton Mifflin.

Frey, Rodney, in collaboration with the Schitsu'umsh. 2001. *Landscape Traveled by Coyote and Crane: The Worlds of the Schitsu'umsh: Coeur d'Alene Indians.* Seattle: University of Washington Press.

Hargrove, Eugene C. 1989. *Foundations of Environmental Ethics.* Englewood Cliffs, NJ: Prentice Hall.

Isenberg, Andrew. 2005. *Mining California: An Ecological History.* New York: Hill and Wang.

Krech, Shepard III. 1999. *The Ecological Indian: Myth and History.* New York: W.W. Norton & Company.

Leopold, Aldo. 1949. *A Sand County Almanac, and Sketches Here and There.* New York: Oxford University Press.

Marsh, George Perkins. 1864. *Man and Nature; or, Physical Geography as Modified by Human Action.* New York: C. Scribner.

Martin, Paul S. 1967. "Pleistocene Overkill." *Natural History* 76: 32–38.

Momaday, N. Scott. 1976. "A First American Views His Land." *National Geographic* (July): 13–18.

Muir, John. 1992. *John Muir: The Eight Wilderness Discovery Books.* London: Diadem Books.

Nash, Roderick Frazier. 1989. *The Rights of Nature: A History of Environmental Ethics.* Madison: University of Wisconsin Press.

Nelson, Michael P. 2004. Animism, in *Encyclopedia of World Environmental History,* Shepard Krech, J.R. McNeill and Carolyn Merchant, eds. New York: Routledge, pp. 55–56.

Pinchot, Gifford. 1998. *Breaking New Ground* (Commemorative edition). Washington, DC: Island Press.

Righter, Robert W. 2005. *The Battle over Hetch Hetchy: America's Most Controversial Dam and the Birth of Modern Environmentalism.* New York: Oxford University Press.

Stoll, Steven. 2002. *Larding the Lean Earth: Soil and Society in Nineteenth-Century America.* New York: Hill and Wang.

Turner, Frederick Jackson. 1920. *The Frontier in American History.* New York: H. Holt and Company.

White, Lynn, Jr. 1969. "The Historic Roots of Our Ecologic Crisis." *Science* 155: 1203–1207.

Michael P. Nelson
Adam M. Sowards

NORTH AMERICAN FREE TRADE AGREEMENT

The North American Free Trade Agreement (NAFTA) between Canada, Mexico, and the United States entered into force in January 1994 after three years of negotiation and extensive public debate. Canada and the United States had established the Canada-U.S. Trade Agreement (CUSTA) in 1989, so the new arrangements involved the addition of bilateral agreements between Canada and Mexico and between the United States and Mexico. The primary purpose of any trade agreement is to reduce barriers to trade with the expectation that trade liberalization will contribute to greater economic growth and prosperity. Trade is one element of the general process of globalization along with capital flows (foreign investment), migration, cultural and religious exchanges, and others. Because trade and other forms of economic integration contribute to economic growth, they have an impact on human well-being and the use of natural and environmental resources. This impact in turn means that trade and globalization have ethical dimensions related to distributive justice, the rights of future generations, and the integrity of ecosystems. The implications of trade liberalization and globalization for the environment have been the subject of intense debate since the 1980s; the political controversy surrounding NAFTA has helped to crystallize various positions on these issues.

There are two general approaches to trade liberalization. Multilateral trade liberalization is carried out by the 152 countries that belong to the World Trade Organization (WTO) through negotiation and private judicial proceedings aimed at resolving trade disputes. Multilateral agreements in the context of the WTO are based on the requirement that member countries treat all other members in the same fashion. In contrast, regional and bilateral agreements, the other approach to trade liberalization, are discriminatory. NAFTA is a regional trade agreement in which lower trade barriers are extended by each member only to the other members, each of which is free to pursue whatever trade policies it wishes with nonmembers. More ambitious regional agreements such as the European Union (EU) require harmonization of trade policies by the members as well as other measures to free up the movement of capital and labor.

David Vogel (2000) suggests that increasing environmental regulation in the 1970s and 1980s, along with rapid growth in trade and a rising number of trade agreements, led to tensions between environmentalists and advocates of free trade. Those who favor trade liberalization worry that environmental regulations will become de facto trade barriers, whereas environmentalists fear that progress in protecting the environment will be undermined as firms either relocate to countries with lax environmental standards (sometimes referred to as "pollution havens") or use the threat of job losses to push the government to lower standards in the name of competitiveness. Environmentalists also note that trade itself is polluting as energy and other resources are used to carry out international shipping, which may also spread invasive species and other forms of pollution. Trade advocates point out that concern for the environment tends to increase as a country's standard of living rises; they further argue that, because trade contributes to rising living standards, trade agreements actually encourage the development of better environmental policies.

NAFTA went further than most free trade agreements in its inclusion of provisions on capital flows (investment) and the environment. Mary Tiemann (2000) notes that national technical standards related to the environment were expressly allowed by the agreement, which also subordinates its provisions on trade liberalization to international environmental agreements on the ozone layer, hazardous waste, and trade in endangered species. Despite these provisions the debate leading up to the ratification vote in the United States was contentious, and President Bill Clinton elected to introduce side agreements on the environment and labor to increase the likelihood that Congress would approve the agreement. The North American Agreement on Environmental Cooperation (NAAEC) and the North American Agreement on Labor Cooperation (NAALC) were ratified along with NAFTA in late 1993.

The principal goal of the NAAEC is to ensure that the three governments actually enforce their environmental regulations. It also aims to promote cooperation on environmental issues and increased environmental safeguards. The organization created to oversee the agreement is the Commission for Environmental Cooperation (CEC), based in Montreal. The CEC is an international organization with several special features. First, in addition to a council made up of representatives from national governments, nongovernmental organizations (NGOs) can participate in its operations through membership in the joint public advisory committee. Further, individual citizens have the right to submit complaints to the CEC secretariat that a country's environmental laws are not being enforced, which reviews the complaint and provides technical information to the council, which adjudicates the case. The focus of the CEC is on air and atmosphere, biodoversity, pollutants, and water. The CEC has also been working on devising methods to measure the environmental impact of NAFTA in particular and, beginning in 2003, of trade in general.

Assessing the environmental impact of NAFTA is complicated because so many other events occurred as NAFTA was being implemented. Some observed changes that have taken place since NAFTA went into effect may have been caused or influenced by NAFTA, but many more would have happened even if NAFTA had not been approved. For example, based on data available from the Energy Information Administration (2008), North American greenhouse-gas emissions increased by more than 17 percent between 1993 and 2005. NAFTA probably played little if any role in this change, which was driven primarily by the energy and environmental policies of the United States, the producer of some 85 percent of North American greenhouse-gas emissions.

In addition to NAAEC, there are several agreements between the United States and Mexico directed at monitoring and controlling pollution along the border between the two countries. Chantal Carpentier (2004) and Vogel (2000) both conclude that cross-border trade under NAFTA has had a minor impact on the environment. Unfortunately, the various environmental agreements adopted in conjunction with NAFTA do not appear to have had positive impacts on the environments that were expected, perhaps because they have been underfunded by the three governments. But these arrangements do lay the groundwork for increased environmental cooperation in the future, and the innovative structure of the CEC holds out the possibility for extensive citizen participation in attaining important environmental goals.

SEE ALSO *Atmosphere; Biodiversity; Future Generations; Nongovernmental Organizations; Ozone Depletion; Pollution; Species; Water; World Trade Organization.*

BIBLIOGRAPHY

Carpentier, Chantal Line. 2004. "Trade and Environment in North America." IRRP Working Paper Series No. 2004-09f. Montreal, Canada: Institute for Research on Public Policy.

Commission for Environmental Cooperation (CEC). 2008. Available from http://www.cec.org/home/index.cfm?varlan=english

Energy Information Administration. 2008. U.S. Department of Energy. Available from http://www.eia.doe.gov

Tiemann, Mary. 2000. "NAFTA: Related Environmental Issues and Initiatives." CRS Report for Congress 97-291 ENR. Washington, DC: Congressional Research Service. Available from http://www.ncseonline.org/NLE/CRSreports/04Mar/97-291.pdf.

Vogel, David. 2000. "International Trade and Environmental Regulation." In *Environmental Policy*, eds. Norman J. Vig and Michael E. Kraft. 4th edition. Washington, DC: CQ Press.

E. Wesley F. Peterson

NORTON, BRYAN
1944–

Bryan Norton was born on July 19, 1944, in Marshall, Michigan. He is best known in the field of environmental ethics for his rejection of nonanthropocentrism—a philosophical worldview in which nonhuman nature is treated as an object of independent moral concern—and challenging the view that an adequate environmental ethic must acknowledge intrinsic natural value and direct moral duties to nature. Norton has been the leading figure in the "pragmatist wave" that began in the mid-1990s, and his work is also significant for its practical, policy orientation and interdisciplinary character. He is among a handful of environmental philosophers who have expanded the intellectual universe of environmental ethics to incorporate disciplines in the natural and social sciences, including conservation biology, environmental economics, and environmental policy and management.

EARLY WORKS

As with most philosophers whose writing shaped the field during its first two decades, Norton's interest in environmental philosophy developed well after his graduate studies in the philosophy of language, science, and epistemology at the University of Michigan, where he received a doctorate in 1970. At Ann Arbor he wrote a dissertation on the metaphilosophy of Vienna Circle philosopher Rudolf Carnap; that work later resulted in Norton's first book, *Linguistic Frameworks and Ontology* (1977).

Norton's career in environmental ethics and philosophy can be traced from its beginnings in environmental value theory in the 1980s, to a growing emphasis on epistemological issues (especially justification) in the late 1980s and early 1990s, to its later focus on language and the pragmatics of environmental communication. His first publications in the journal *Environmental Ethics* (Norton 1982, 1984) focused on the rights of nonhumans and future generations; the philosophical problems afflicting both positions led him to conclude that rights-based approaches in environmental ethics are not viable.

Norton was skeptical about environmental philosophers' emerging focus on the intrinsic value of nature, which he felt was both philosophically flawed and politically unnecessary. Accordingly, in the mid-1980s he developed "weak anthropocentrism," an approach that did not require the recognition of the intrinsic value of nature (Norton 1984). Norton described weak anthropocentrism as accommodating a broadly instrumentalist and pluralistic theory of environmental value within a humanistic worldview. His argument emphasized the role of nonhuman nature as a teacher of human ideals and a good shared between present and future generations. Weak anthropocentrism thus was positioned as a moderate alternative to both narrow forms of economic valuation ("strong anthropocentrism") and nonanthropocentric, intrinsic-value-of-nature arguments.

Norton continued to develop this line in his 1987 book *Why Preserve Natural Variety?*, which surveyed the range of economic and ethical options for valuing species and ecosystems (biodiversity). In that book Norton presented the concept of "transformative value" as the normative core of weak anthropocentrism. Contact with wild species and ecosystems could, he argued, prompt individuals to evaluate critically and transform their exploitative, consumer-centered preferences into more environmentally benign ideals compatible with an ecologically enlightened worldview (Norton 1987). The significance of his argument was that one could justify the protection of endangered species and ecosystems on anthropocentric grounds while steering clear of both economistic and nonanthropocentric arguments.

LATER WORKS

In the 1990s Norton's weak anthropocentric stance evolved into a more explicitly pragmatic approach to environmental philosophy. As early as 1988 he tapped into pragmatism in his 1988 contrarian reading of Aldo Leopold, author of *A Sand County Almanac*. Norton argued that Leopold subscribed to an informal pragmatist epistemology and theory of environmental value, an interpretation that went against the dominant reading of Leopold as perhaps the major nonanthropocentrist in the history of environmental ethics (Callicott 1989, 1999).

Norton's writing assumed an increasingly strong policy focus in the 1990s. Among other things, that orientation led him to reverse the traditional logic of applying philosophical principles to practical problems in favor of a greater emphasis on the problems themselves. That approach appeared in *Toward Unity among Environmentalists* (1991), in which he described an emerging policy consensus among environmental activists—in short, support for multivalue, sustainable ecosystem management—that existed despite disagreements over deeper environmental values and motives. In *Toward Unity,* Norton concluded that the anthropocentric-nonanthropocentric debate in environmental ethics was not as important as previously thought because it did not thwart political agreement on common policy goals. Termed the convergence hypothesis, Norton's argument proved controversial, attracting rebuttals from a number of nonanthropocentrists (Minteer 2009).

Later in the 1990s Norton's work continued to explore a range of philosophical and policy questions, including the relationship between economic and pragmatic approaches to environmental valuation, the role of ecological science in the policy process, and analyses of ecosystem health, biodiversity, and sustainability (Norton 2003). His 2005 book *Sustainability: A Philosophy of Adaptive Ecosystem Management* marked a return to the analytical pragmatism of Carnap that Norton had explored early in his career. In *Sustainability,* Norton proposed an integrated scientific and value discourse for discussions of sustainability across the environmental science, policy, and management domains. In doing so, he reframed environmental problems as linguistic failures rather than ethical ones, in the process setting a new agenda for environmental philosophy in the pragmatics of language and communication.

SEE ALSO *Biodiversity; Callicott, J. Baird; Carson, Rachel; Commoner, Barry; Conservation Biology; Convergence Hypothesis; Earth First!; Economics, Environmental; Ehrlich, Paul; Endangered Species Act; Environmental Philosophy: V. Contemporary Philosophy; Environmental Policy; Global Climate Change; Greenpeace; Hargrove, Eugene; Intrinsic and Instrumental Value; Leopold, Aldo; Pragmatism; Sustainability.*

BIBLIOGRAPHY

WORKS ABOUT BRYAN NORTON

Callicott, J. Baird. 1989. *In Defense of the Land Ethic: Essays in Environmental Philosophy*. Albany: State University of New York Press.

Callicott, J. Baird. 1999. *Beyond the Land Ethic: More Essays in Environmental Philosophy*. Albany: State University of New York Press.

Minteer, Ben A. 2009. *Nature in Common? Environmental Ethics and the Contested Foundations of Environmental Policy*. Philadelphia: Temple University Press.

WORKS BY BRYAN NORTON

Norton, Bryan G. 1977. *Linguistic Frameworks and Ontology. A Re-Examination of Carnap's Metaphilosophy*. The Hague, Netherlands: Mouton.

Norton, Bryan G. 1982. "Environmental Ethics and the Rights of Future Generations." *Environmental Ethics* 4(4): 319–337.

Norton, Bryan G. 1984. "Environmental Ethics and Weak Anthropocentrism." *Environmental Ethics* 6(2): 131–148.

Norton, Bryan G. 1987. *Why Preserve Natural Variety?* Princeton, NJ: Princeton University Press.

Norton, Bryan G. 1988. "The Constancy of Leopold's Land Ethic." *Conservation Biology* 2(1): 93–102.

Norton, Bryan G. 1991. *Toward Unity among Environmentalists*. New York: Oxford University Press.

Norton, Bryan G. 2003. *Searching for Sustainability. Interdisciplinary Essays in the Philosophy of Conservation Biology*. Cambridge, UK, and New York: Cambridge University Press.

Norton, Bryan G. 2005. *Sustainability. A Philosophy of Adaptive Ecosystem Management*. Chicago: University of Chicago Press.

Ben A. Minteer

NUCLEAR POWER

When the nucleus of an atom is divided or joined to another nucleus, an enormous amount of energy is produced. When these processes of fission (division of the nucleus of an atom) or fusion (joining the nucleus of an atom with another) take place under controlled conditions, this energy can be used to drive turbines that can do work: for example, to propel a large vehicle such as a ship or to generate electricity. These same processes, however, can be adapted to produce weapons with massive destructive power. Nonmilitary uses of *nuclear power* refer to the generation of energy through a controlled process of fission in a nuclear reactor.

NUCLEAR POWER THROUGHOUT THE WORLD

According to figures of the Power Reactor Information System (PRIS) of the International Atomic Energy Agency (IAEA), an independent international organization that was established in 1957 to promote the safety and security of the peaceful use of nuclear science and technology, there were at the beginning of 2008 a total of 439 nuclear power plants in operation in the world in 30 countries, with a total net installed capacity of 372,202 Gigawatts. At this time, there were also five nuclear power plants in long term shutdown, whilst 34 nuclear power plants were under construction (IAEA 2008). At the end of 2006, there were 103 licensed nuclear power plants in the United States located at 65 sites, predominantly situated in the eastern half of the country. As of 2008, about 20 percent of the electricity used in the United States is generated by nuclear reactors, placing nuclear power second to electricity generated by coal (IAEA 2003a: 7-9; IAEA 2007a). In comparison, over 78 percent of the domestic energy production in France comes from its 59 nuclear reactors (IAEA 2007a). Other countries with relatively large numbers of nuclear reactors include Japan (with 55 operational nuclear reactors that generate 30 percent of its electricity), Russia (with 31 reactors that generate about 16 percent of its electricity), South Korea (with 20 reactors that generate about 39 percent of its electricity), the United Kingdom (with 19 reactors that generate just over 18 percent of its electricity), Canada (with 18 reactors that generate about 16 percent of its electricity), Germany (with 17 reactors that generate about 32 percent of its electricity), and Ukraine (with 15 reactors that generate 47.5 percent of its electricity) (IAEA 2007a).

In its *Annual Report 2006*, the IAEA (2007b: 3) points out that about three-quarters of the world's operating nuclear reactors are 20 or more years old. Since the life span of a nuclear reactor is between 20 and 40 years, depending on the type of technology and design used, it can be expected that decisions about decommissioning a large number of nuclear plants, as well as investing large sums of money to replace these reactors will have to be made over the next two decades. The British government already announced in January 2008 that up to 10 new nuclear reactors will have to be built in the United Kingdom by 2020 in order to replace those that need to be decommissioned. Similarly, Russia announced in March 2008 that it will put four new nuclear reactors into operation by 2020.

In the United States, no new nuclear power plants have been ordered since the late 1970s, since the anticipated growth in electricity demand slowed, and nuclear construction costs soared. The nuclear accident at Three Mile Island in 1979 strengthened the reluctance of utilities to further invest in nuclear reactors. Instead of ordering new reactors, the current trend in the U.S. is rather to increase the

Nuclear Plant near Berwick, Pennsylvania. *The use of nuclear power is becoming increasingly popular as a replacement for fossil fuels becomes more and more necessary. Proponents also cite the cheap cost of production and lack of carbon dioxide by-products as other reasons in support of nuclear power. However, the drawbacks of nuclear power in terms of safety and health can be seen in such instances as the Chernobyl disaster.* AP IMAGES.

efficiency of existing nuclear plants. The last new reactor (Watts Bar 1) was completed in 1996 (IAEA 2003a: 6-9). Most of the reactors that are currently under construction, or recently completed, are located in developing countries such as China, Russia, South Korea, and India.

THE BENEFITS AND DRAWBACKS OF NUCLEAR ENERGY

All commercial nuclear power plants generating electricity make use of fission technology, of which there are many varieties; the commercial use of fusion reactors is projected to become a reality only in forty to fifty years. Because nuclear fuel releases exponentially greater amounts of energy than that produced by chemical fuels, and because this fuel (usually uranium^{235}) is abundant and therefore relatively cheap, many policy makers view nuclear power as an attractive option for the provision of a steady supply of energy for the world's economy as the price of fossil fuels continues to rise. These proponents of nuclear power point to two other key advantages of nuclear power: It emits almost no carbon dioxide, one of the major greenhouse gases that have contributed so heavily to global climate change; and renewable energy sources like wind, the Sun, and waves are, in their present state of development, unable to replace fossil fuels.

The main concern of critics of nuclear power is that the by-products of fission are radioactive, and some of them remain so for hundreds of thousands of years. High doses of exposure to radioactive by-products can cause a variety of cancers and genetic deformities. Hence the generation of nuclear energy poses two major safety issues: the permanent disposal of nuclear waste and the operation and decommissioning of nuclear reactors. These concerns are not merely hypothetical: Widespread radioactive contamination resulted from the partial reactor meltdown in 1979 at Three Mile Island in Pennsylvania and from the 1986 disaster in Chernobyl, in Ukraine, which led to the evacuation and resettlement of more than 200,000 people. The risk of exposing present and future generations to unsafe levels of radioactivity is at the core of the political and ethical controversies that surround the use of nuclear energy.

THE MILITARY CONTEXT OF NUCLEAR ENERGY

The first fission nuclear reactors were built in the United States in the early 1940s. They were the product of research conducted by the Manhattan Project (1942–1946), a top-secret program (funded mainly by the United States with help from the United Kingdom) that met its goal of creating the world's first nuclear weapon in 1945. The military origins of nuclear energy are never far from the minds of its critics. Although touted by its champions as a purely peaceful, nonmilitary application of atomic energy, the use of nuclear power poses the risk that plutonium—one of the by-products of nuclear fission—can find its way back from civilian nuclear plants to military applications, thus contributing to the problem of the proliferation of nuclear weapons. With the emergence of so-called rogue states and increasing concerns about global terrorism since the attacks of September 11, 2001, the issue of nuclear proliferation has acquired renewed urgency, as reflected in the search for weapons of mass destruction in Iraq and the monitoring of nuclear programs in Iran and North Korea. Concerns have also arisen about the possibility that terrorist groups might acquire and use a dirty bomb, a "radiological dispersal device" (RDD) that spreads radioactive material with a conventional explosive, such as dynamite.

DEBATES OVER NUCLEAR ENERGY

Given the controversies that swirl around the use of nuclear energy, its future remains cloudy. Since the initial use of nuclear power for civilian purposes in the 1950s, a sociopolitical and ethical debate has raged, centering on the issues of operational safety, costs, permanent disposal of waste, and the risk of proliferation. These debates are dominated by two central questions: (1) Do the benefits of nuclear power outweigh its risks and costs? (2) What principles should guide decision making about nuclear power?

The first question usually elicits utilitarian answers on both sides: Proponents contend that even with all risks and costs considered, the widespread implementation of nuclear power would be a net gain for society; opponents argue that the long-term health and ecological risks are so dire that renewable energy sources are preferable. Both sides of this utilitarian cost-benefit analysis quote facts and figures, often citing sympathetic scientific authorities. Supporters of nuclear power such as the World Nuclear Association (WNA) typically claim that it will provide us with an endless supply of energy, that it is cheap, that it is safe if managed properly, that the problem of long-term storage of high-level nuclear waste will be solved in due course, that we do not have any viable alternatives to provide the energy needs of both industrialized (developed) and developing countries, and that it could help us to reduce emissions of greenhouse gases in the fight against global climate change (World Nuclear Association 2008). It is not just major lobbying groups such as the WNA who are pushing for this form of power—even some notable environmentalists, such as Greenpeace cofounder Patrick Moore, have done an about-face on the question of nuclear energy, viewing it as the only technologically feasible alternative to fossil fuels. He writes that "nuclear energy may just be the energy source that can save our planet from another possible disaster: catastrophic climate change" (Moore 2006).

Quoting different facts and figures, opponents of nuclear power (of which the various reports of Greenpeace [2008] or the Heinrich Böll Green Political Foundation [2008] are examples) typically claim that uranium mining is environmentally destructive and makes use of carbon-intensive technologies; that nuclear power is not safe; that it generates high-level waste that will need long-term management, even with underground burial facilities (geological disposal); that nuclear power is much more expensive than it is claimed to be if the costs of decommissioning of plants and the management of long-term waste storage are included; that nuclear power is economically viable only through direct or indirect state subsidies; that nuclear power exposes the world to the risks of radioactive materials and proliferation; and that nuclear power is not the solution to the challenge of climate change.

The irony is, however, that although this emphasis on quantifiable costs, risks, and benefits is highly important and revealing, the utilitarian debate about nuclear power has fallen into gridlock for a very long time now, with neither side able to convince the other of the merits of its arguments.

NUCLEAR ENERGY AND DECISION THEORY

A similar standoff has resulted from debates about what kind of principles should inform decision making about the development, use, maintenance, and management of nuclear power. Kristin Shrader-Frechette, in her critical analysis of the nuclear industry and its historical evolution (in which she has also contributed substantively to a critical analysis of quantitative risk assessment), argues that typical forms of governmental decision making about nuclear power run against the grain of the principle of prima facie political equality (1983a, 1983b, 1993, 2002). Instead of following the democratic principles of procedural justice and informed consent, nuclear power decisions usually exemplify a heavy-handed, top-down, authoritarian approach, dominated by secrecy and even deception. She seeks to demonstrate that such procedures engender environmental injustices that single out already marginalized members of society, disproportionately exposing them to the risks of nuclear power while insulating those who benefit from it. A hypothetical example of such injustice would be to offer a very poor, drought-stricken country in sub-Saharan Africa the option of establishing several nuclear waste storage facilities within its borders with a view to stimulate the local economy.

Shrader-Frechette's approach clashes head-on with that of nuclear-power advocates who argue that the confidentiality of energy deliberations is a nonnegotiable principle because national security is at stake; strategic decisions about the provision of energy cannot, therefore, be exposed to public questioning and scrutiny. Such considerations, they believe, justify their rejection of appeals to incorporate measures of open discussion and deliberative democracy in decisions about nuclear energy.

This deadlock shows that further analyses are needed to gain an understanding of channels through which ethical issues related to nuclear power can be better understood and resolved. There have been promising contributions in these areas from the philosophy of science and sociology. Philosophers of science have focused on the relationship between science and technology, particularly on how conceptions about the nature of science can influence—indeed, have influenced—policy decisions about technology development, management, and use. Prominent examples of this approach are found in the work of Robert Frodeman, Carl Mitcham, Christine Turner, and Roger Pielke who show that a simplistic, positivist account has fostered the notion that science works with nothing but unequivocal empirical data and the logical conclusions that follow from them. From this view it follows that deliberations about the significance, use, and applications of science are left to politicians and managers, who eventually decide which technology-development paths should be followed or avoided.

Pielke (2003) points out the irony that, within this decisionistic model, so-called value free science is called upon by decision makers to provide objective facts on the basis of which policy choices can be made. Science is thus invoked to justify policy choices, but only by precluding public discussion of these choices on the grounds that the policies are portrayed as indubitable (scientific) truths. Within the nuclear debate this appeal to the inherent objectivity of science has been used in the past to settle policy issues (for instance, the very choice of adopting nuclear power as a source of energy) without a thorough discussion of the manner in which nuclear technology affects the values of society.

Some scholars propose a different conceptualization of science, one that views it as a social institution, not merely as an insular domain of objective truth. They argue that science affects society in ways that cannot be fully portrayed and assessed by science itself. Accordingly, they argue for a reconceptualization of the relationship between science and policy making, one that takes cognizance of the uses to which science can be put by various societal interest groups. Appeals to science itself often cannot settle disputes between these groups (Sarewitz 2000). Hence there is a need for a constant critical assessment of all claims about the significance and policy implications of science.

Pielke (2003) argues that scientists should actively participate in this process of assessment, whereas Frodeman and Mitcham (2007) introduce the notion of a broad, deep, and critical interdisciplinary approach that fosters collaboration among scholars from the sciences, social sciences, and humanities. They hope to encourage such cooperation not only by asking the "fundamentally humanistic question of what counts as *pertinent* knowledge: Knowledge for what?" but also by describing and assessing the values of society "as it struggles to address questions of social and environmental justice, human freedom and responsibility, and the proper roles of the public and private sectors." An example of such an interdisciplinary approach to nuclear power can be found in the work of Turner and Frodeman (1995), which combines the resources of philosophy and geology in asking how the proposed Yucca Mountain site for the permanent disposal of nuclear waste can be evaluated by the criteria of both science and social values (Frodeman 1995).

Frodeman and Mitcham (2000) have proposed the integration of science and the common good through a process of critical public conversation. Critics of their work have argued that it neglects the power dynamics at play in deep ethical differences about issues such as nuclear power or the permanent disposal of nuclear waste (Rouse 2000). Frodeman and Mitcham's later work on interdisciplinary approaches to these issues does critically address the power relations underlying such differences (2007).

Sociological perspectives on the "social, institutional and technical processes through which technologies become constructed, enacted and maintained" (Irwin et al. 2000, p. 81; Jasanoff et al. 1995) indicate a pathway that can be further explored to engage with some of the power relations that are implicit in any public conversation. Alan Irwin focuses on the discourse used by authoritative speakers to establish and maintain confidence in civilian nuclear power—even in the face of severe obstacles such as major nuclear accidents. He examines the justification of nuclear power (a) in a discourse of modernity in which nuclear power is portrayed as one of the first manifestations of Big Science and thus inevitably constitutes social progress; (b) in technical arguments based on the assumption that any technical problems will surely be ironed out by advancing scientific understanding (which means that nuclear "accidents can be dismissed as part of the technological learning curve or else viewed as a consequence of non-scientific design choices"); (c) in a view of the environment "as resilient and robust in the face of radioactive contaminants provided emissions are kept at a 'reasonable level'"; and (d) in the opinion that the acceptability of risks can be determined objectively and rationally on quantitative scales that function independently from institutional processes and particular contexts (Irwin et al. 2000, pp. 82–83).

Perhaps these critical perspectives from the philosophy of science and sociology will fail to find a receptive audience among real-world decision makers. But faced with the specters of global climate change and rising prices for ever scarcer fossil fuels, these officials will be searching for grounds on which to decide whether to replace aging nuclear plants with newer ones or to turn instead toward heavy investments in research into renewable energy technologies. The contributions of philosophers and sociologists can help to shift the debate about nuclear power from utilitarian arguments about its costs and benefits to dialogues about the form of rationality at work in science and policy making—perhaps even to public conversations between decision makers and the citizenry that can self-consciously and critically consider the impact of policy choices on the values, aspirations, and ideals of society.

SEE ALSO *Chernobyl; Energy.*

BIBLIOGRAPHY

Frodeman, Robert. 1995. "Geological Reasoning: Geology as an Interpretive and Historical Science. *GSA Bulletin* 107(8): 960–968.

Frodeman, Robert, and Carl Mitcham. 2000. "Beyond the Social Contract Myth: Integrating Science and the Common Good." *Issues in Science and Technology* 16(4): 341–352.

Frodeman, Robert, and Carl Mitcham. 2007. "New Directions in Interdisciplinarity: Broad, Deep, and Critical." *Bulletin of Science, Technology & Society* 27(6): 506–514.

Greenpeace International. 2008. "End the Nuclear Age." Available from http://www.greenpeace.org/international/campaigns/nuclear.

Heinrich Böll Stiftung: The Green Political Foundation. 2008. *Nuclear Issues Papers: Nuclear Power—Myth and Reality.* Available from http://www.boell.de/ecology/ecology-1232.html.

IAEA. 2003a. *Country Nuclear Power Profiles: USA.* Available from http://www-pub.iaea.org/MTCD/publications/PDF/cnpp 2003/CNPP_Webpage/countryprofiles/USA/USA2003.htm.

IAEA. 2003b. *Country Nuclear Power Profiles: France.* Available from http://www-pub.iaea.org/MTCD/publications/PDF/cnpp2003/CNPP_Webpage/countryprofiles/France/France2003.htm.

IAEA. 2007a. *Nuclear Power Reactors in the World.* Vienna: IAEA. Available from http://www-pub.iaea.org/MTCD/publications/PDF/RDS2-26_web.pdf.

IAEA. 2007b. *Annual Report 2006.* Report GC(51)/5. Vienna, IAEA. Available from http://www.iaea.org/Publications/Reports.

IAEA. 2008. *Power Reactor Information System (PRIS).* Available from http://www.iaea.org/cgi-bin/db.page.pl/pris.main.htm.

Irwin, Alan, Stuart Allen, and I. Welsh. 2000. "Nuclear Risks: Three Problematics. In *The Risk Society and Beyond: Critical Issues for Social Theory*, ed. Barbara Adam, Ulrich Beck, and Joost van Loon. London: Sage.

Jasanoff, S.; G. E. Markle; J. C. Peterson; and T. Pinch. 1995. *Handbook of Science and Technology Studies.* London: Sage.

Moore, Patrick. 2006. "Going Nuclear: A Green Makes the Case." Washingtonpost.com, April 16, 2006. Available from http://www.washingtonpost.com/wp-dyn/content/article/2006/04/14/AR2006041401209_pf.html.

Pielke, Roger A. 2003. "The Significance of Science." Center for Science and Technology Policy Research, University of Colorado. Available from http://sciencepolicy.colorado.edu/admin/publication_files/2003.15.pdf.

Rouse, J. 2000. "Commentary on 'Science's Social role.'" *Issues in Science and Technology: Forum.* Available from http://findarticles.com/p/articles/mi_qa3622/is_200001/ai_n8882817/pg_16.

Sarewitz, D. 2000. "Science and Environmental Policy: An Excess of Objectivity." In *Earth Matters: The Earth Sciences, Philosophy, and the Claims of Community*, ed. Robert Frodeman. Upper Saddle River, NJ: Prentice Hall.

Shrader-Frechette, Kristin. 1983a. *Nuclear Power and Public Policy: Social and Ethical Problems with Fission Technology.* Boston: Kluwer.

Shrader-Frechette, Kristin. 1983b. *Four Methodological Assumptions in Risk-Cost-Benefit Analysis.* Springfield, VA: National Technical Information Service.

Shrader-Frechette, Kristin. 1993. *Burying Uncertainty: Risk and the Case against Geological Disposal of Nuclear Waste.* Berkeley: University of California Press.

Shrader-Frechette, Kristin. 2002. *Environmental Justice: Creating Equality, Reclaiming Democracy.* New York: Oxford University Press.

Turner, Christine and Robert Frodeman. 1995. "The Need to Bring Philosophy Back into Geology: Efforts at the USGS." Invited paper, Geological Society of America Annual Meeting, Seattle, October 25, 1994.

World Nuclear Association. 2008. "The Need for Nuclear." Available from http://www.world-nuclear.org/why/why.html.

Johan Hattingh

O

OCEANS

The oceans, which cover about 71 percent of the surface of Earth, are composed of a great variety of life-filled habitats that range from the dark and cold of the deep ocean-floor trenches to the sunlight and warmth of tropical lagoons. The oceans support a tremendous degree of biological species diversity and account for just less than half of annual carbon fixation via photosynthesis. The great volume of ocean water absorbs carbon dioxide, oxygen, and other gases as well as heat from the Sun. Major currents such as the Gulf Stream and the Humboldt Current circulate both warm water and cold water across the face of the planet. Marine environments are thus critical to the composition of the atmosphere, the production of weather, and the long-term regulation of climate. Humans always have utilized marine resources ranging from edible algae, to cleansing sponges, to whale oil, to petroleum pumped from offshore platforms. Travel across the ocean surface is integral to the international shipping of goods. Coasts are prime locales for recreation, such as surfing and snorkeling. Humans have evolved unique maritime cultures, with art, myth, technology, and even religion focused on the sea.

RELIGIOUS AND MYTHIC TRADITIONS

In Middle Eastern and European myths the oceans are critical to primordial creation events. The opening chapter of Genesis gives aquatic and marine environments priority in the narrative of creation as God's spirit "moves over the face of the waters" (Genesis 1:2). God closes the separation of the land from the waters by declaring the physical structure of the emerging planet to be "good" and, by using the Hebrew word *tob* ("beautiful"), granting the oceans inherent worth. The ancient world associated the oceans with chaos, mystery, and uncontrolled physical force. The story of Noah's ark is one of many variants of the flood myth in cultures worldwide that describes the oceans exceeding their boundaries and serving as agents of divine displeasure or universal social reorganization.

In Greco-Roman cosmology water is one of the four basic elements, along with earth, air, and fire. The capricious and temperamental deity of the oceans, Poseidon or Neptune, although rewarding sailors who respect him with fair winds, often invokes his physical power, causing storms and other disasters. In ancient Roman art, sea monsters were associated with chaos, whereas dolphins symbolized rescue and therefore religious salvation or the afterlife. In the religions of eastern and southern Asia the oceans also convey universal values, such as the vastness and depth of the Buddha. Ancient Greco-Roman and biblical conceptualization of the oceans strongly influenced Western philosophical and literary perceptions until the modern era.

Regional religions, especially those of cultures dependent on oceanic resources, often personify sea creatures and emphasize their organic productivity. In the *Kumulipo,* a chant recounting the Hawaiian creation myth, the coral polyp, a keystone species of tropical reefs, is the first creature born of the divine pair of primary deities, Kumulipo and Po'ele. The *Kumulipo* describes the faunal diversity of the reef, including fishes, turtles, lobsters, and other invertebrates, such as starfish, sea cucumbers, limpets, and mollusks. The impressive inventory classifies marine life by similarity of form and by habitat.

Tribes of the Pacific Northwest treated salmon as a separate animal society, with their villages situated beneath the waves. Regional religions often deify marine phenomena such as tidal waves and marine organisms such as sharks and whales, which also may serve as family or clan totems. Cultures dependent on oceanic resources utilize laws and rituals to guide the management of marine harvest. The cultures of the Pacific Northwest have a First Fish ceremony, which precludes the harvesting of salmon for several days at the beginning of an upstream run, ensuring the escape of an adequate number of spawning adults to replenish the stocks.

THE HISTORY OF MARITIME MANAGEMENT

The ancient Greeks were the first western philosophers to consider the oceans, and Aristotle (384–322 BCE) wrote treatises on the biology and diversity of marine organisms. Before the twentieth century, however, philosophical ethicists usually did not treat the oceans as a separate topic and thus wrote little that specifically concerns marine issues. Much of the intellectual tradition concerning the oceans is the product of legal thinkers such as Hugo Grotius, who in 1625 published *Mare Librum*, which argued for international freedom of the high seas for the shipping trade and other commercial uses. During the twentieth century the implementation of treaties and conventions concerning harvesting and environmental care of the oceans greatly expanded. Some early efforts in multinational cooperation include the North Pacific Fur Seal Treaty (1911), which restricted harvesting of both fur seals and sea otters; the Migratory Bird Treaty (1918), which regulated the capture, transport, and sale of migratory birds such as waterfowl; and the International Convention on the Regulation of Whaling (1946), which restricted the harvesting of whales and established the International Whaling Commission.

The regulatory trend since the 1970s has been toward negotiating conventions and agreements for managing specific regions such as the Mediterranean, protecting the high seas from destructive activities such as the dumping of radioactive waste, and expanding national jurisdictions farther offshore. A particularly influential instrument is the United Nations Convention on the Law of the Sea (UNCLOS III) of 1982, which has tied the right to manage the seas to responsibility to protect their biotic and environmental resources and encouraged individual countries to manage the adjoining continental shelves. Before World War II most nations claimed exclusive jurisdiction over three to six nautical miles of ocean contiguous to their shorelines. Since that time many nations, including the United States, have claimed Exclusive Economic Zones of two hundred nautical miles.

ENVIRONMENTAL ETHICS AND THE SEA

One of the first ethical models applied to fisheries in response to the environmental movement of the 1960s was Garrett Hardin's tragedy of the commons, which describes the human tendency to overutilize commonly held resources. Before the modern period the high seas were treated as common property, open to any nation able to send ships to harvest from them. That resulted in the depletion of a number of species of marine mammals, including the Atlantic gray whale and Steller's sea cow, both now extinct, and all three species of right whales, which are endangered. Hardin's model suggests not just regulating resource extraction but requiring those utilizing ocean resources to contribute to their care and management, a value reflected in UNCLOS III.

More recently the philosopher J. Baird Callicott (1992) proposed extending the principles of Aldo Leopold's land ethic to the oceans. This model encourages humans to view themselves as participants in the ocean community and recognize the inherent value of the living organisms that inhabit the seas. In 2004 Susan Bratton suggested using Rachel Carson's writings on the oceans to develop an ecotonal ethic of the oceans that compensates for human perceptual limitations in understanding ocean processes and ecosystems. The oceans present a suite of environmental issues that differ from those on land because of the prevalence of public domain, the limited exploration of the deep seas, and the international nature of ocean conservation. Recent volumes dedicated to ocean values include L. Anthea Brooks and Stacy VanDeveer's *Saving the Seas* (1997) and Dorinda Dallmeyer's *Values at Sea: Ethics for the Marine Environment* (2003). Since the 1970s academic dialogue and publication concerning ocean ethics have been oriented increasingly toward specific cases, four of which are discussed below.

Shoreline Development and Sea-Level Rise A major source of damage to inshore ecosystems such as those of barrier islands, tidal marshes, inlets, and coral reefs is the human attempt to stabilize the inherently dynamic boundary between the oceans and the land. Protection of docks and channels is of course necessary to shipping and international commerce. When they are not interfered with, natural processes such as major storms and long-term changes in climate restructure and relocate shorelines as sea levels rise and fall. Contemporary archaeologists find the remains of ancient cities and settlements below the current tide line.

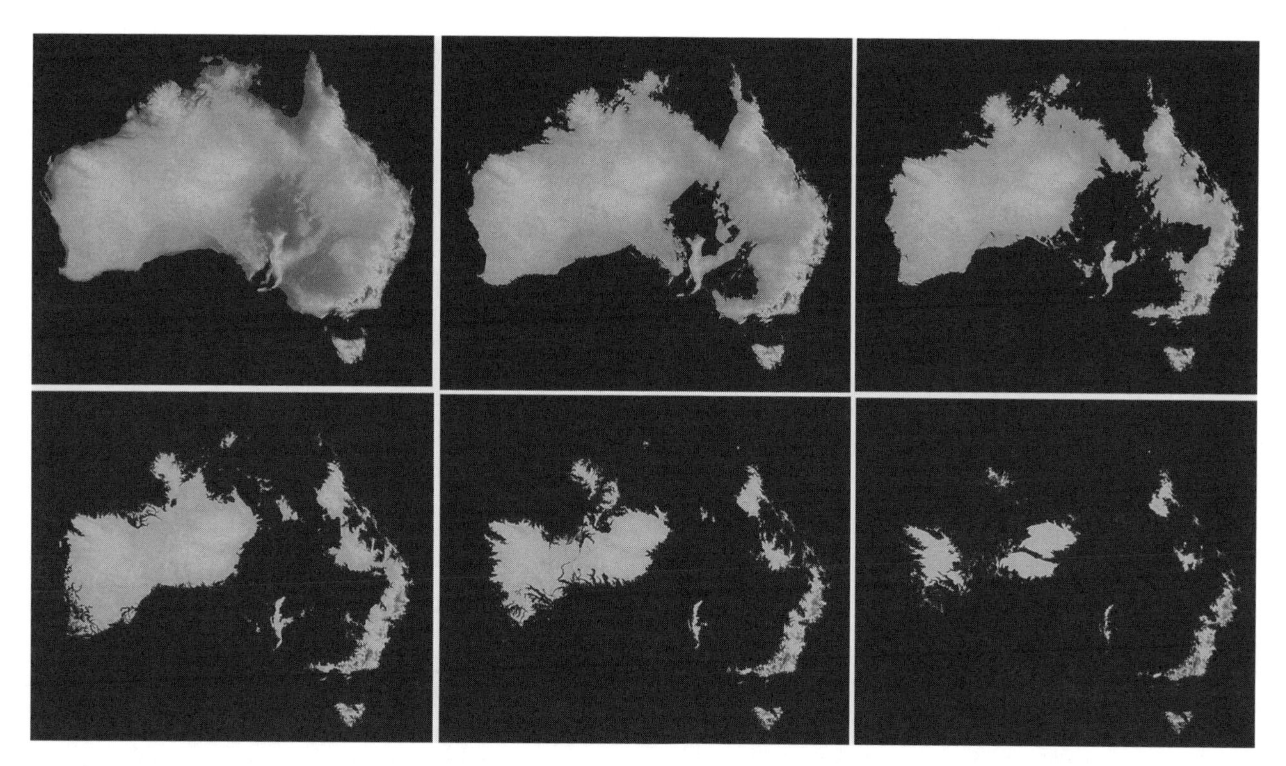

Projection of Australia as Sea Levels Rise. *In this series of images, U.S. geographer Stephen Young has calculated how the Australian continent would appear if the seas were to rise by increments of 100 meters (a total of 500 meters in the final image). Such a rise in sea levels would transform Australia from a massive continent into an archipelago of smaller islands.* **STEPHEN YOUNG/SALEM STATE COLLEGE VIA GETTY IMAGES.**

At the height of the continental glaciations, ending about 18,000 years ago, shallow-water regions currently below the low tide line, such as part of the Irish Sea, were above sea level. This former terrestrial zone was flooded slowly as the glaciers melted. In the contemporary period accelerated changes in global climate and atmospheric change resulting from the use of fossil fuels and land clearing are speeding the recession of the ice caps; thus, sea levels are continuing to rise, threatening to flood many low-lying coastal regions, such as the coasts of Florida, the Polynesian islands, and much of Bangladesh. Aside from climate change, storms, tsunamis, and other disturbance events continually re-form coastlines through the processes of erosion and deposition. Hurricanes, typhoons, and other major weather events can move great volumes of material, overwashing barrier islands, cutting shorelines back many meters, and flattening entire dune systems. On Cape Hatteras, North Carolina, for example, one major storm can fill an inlet across the islands, leaving a bridge spanning sand flats while cutting through roads and forming new inlets completely across the islands at other locations.

Preindustrial cultures usually avoided permanent construction on unstable shorelines, elevated buildings above flood levels, or occupied beaches by using temporary camps and portable infrastructure. The demand for permanent facilities for industry served by paved roads and the development of recreational or seasonal residences within an easy walk of the ocean have made modern coastal communities vulnerable to severe damage and financial loss from both sea-level rise and storm-induced flooding. The sprawl and inflated land values of modern coastal cities also encourage construction on high-risk properties.

When Hurricane Katrina flooded New Orleans in 2005, the older portion of the city, established on higher ground, avoided complete inundation because of its elevation. Wards at lower elevations were under several meters of water, causing major loss of life and billions of dollars of property damage. New Orleans, with its hurricane-breached levees, is also a lesson in the difficulty of protecting coastal municipalities with artificial barriers. Immobile sea walls such as the one in Galveston, Texas, offer little protection from the most severe storm surge while disrupting the natural process of sand deposition, which provides the buffer zone between land and sea.

The current engineering solution to eroding shorelines is often to dredge millions of cubic meters of sand from the ocean bottom and "nourish" the beaches. The dredging damages ecosystems on the ocean bottom.

Many of the major sand deposits are of Pleistocene origin and are not being replenished, and so beach nourishment may not be a long-term solution, at least in terms of current geologic resources. Global climate change, especially if the primary trajectory is toward warmer temperatures and the melting of the glacial ice sheets, will increase the impacts of storm surge and coastal erosion caused by sea-level rise. Some scientific models also conclude that cyclonic storms of oceanic origin will have greater average intensity.

The issues presented by shoreline change include whether some natural environments should be exempt from human development, at-risk coastal environments should remain in the public domain, and governments should prevent development in coastal areas with a high risk of flooding. Ethicists have invoked the precautionary principle in the case of beaches known to be dynamic. Coastal regulations may require buffer zones or exclude highly unstable sites from construction. Barrier islands, for example, have high value for recreation and protect mainland shorelines from wave damage. The designation of Cape Hatteras and Cape Lookout as U.S. National Seashores has made those areas accessible to the public while reducing storm-inflicted property damage. The concept of sustainable energy production based on renewable energy is a concrete approach to the issue of accelerated sea-level rise because a reduction in emissions of greenhouse gases would slow global climate change.

Fisheries and Harvest of Marine Life Humans long ago discovered that inshore marine resources such as coral reefs are easily overexploited. Polynesian cultures have chiefs who regulate the taking of reef resources and enforce the seasons for the harvesting of different species as well as the numbers and sizes of the organisms removed. Tribes and nations have fought wars over access to lucrative fisheries, such as those for salmon and cod. Industrialization has increased the human capacity to deplete marine fisheries. When engine-powered craft and winches began to displace fishing with hand lines and nets hauled by muscle power in the nineteenth century, fishers who long had fished by hand protested the potential damage to spawning beds and to the invertebrates and bottom-dwelling creatures on which many fisheries depend. The tonnage harvested increased, as did the amount of by-catch, or marine creatures caught and killed by accident. Scientists of the nineteenth century, including Thomas Huxley, rejected the concerns of experienced commercial fishers, whom they considered uneducated. Huxley believed the oceans were so vast and productive that their fisheries could not decline significantly. The industrialized methods, however, damaged the fertile shallow banks off Britain, and fishers had to travel farther and farther out to sea to find fish.

Current ecological research indicates that many fisheries management programs that are based on obtaining maximum sustainable yield have resulted in declines not just in fish population numbers but also in the size and quality of the fish harvested. The mathematical formulas used to predict yield did not incorporate adequate information about fish demography, population geography, ocean food webs, or environmental variations in ocean conditions, such as fluctuations in ocean temperature caused by El Niño. Oceanographers using underwater cameras have surveyed ocean bottom sediments plowed by trawling and discovered that invertebrate populations critical to the food chains that support commercial fisheries have been disturbed. Another example is the Chesapeake Bay of the eastern United States, where the use of dredges has destroyed the oyster beds necessary for oyster larvae to establish and grow on the solid substrate of old shells.

The degradation of some ocean fisheries has been so great that many, such as the ground fisheries of Georges Bank off New England, have been closed temporarily by national governments or international agreements. The larger fishing vessels of heavily capitalized fleets and the wealthier countries often have driven less well financed or indigenous fisheries out of business. International fishing fleets are reducing fish availability for regional fisheries from the coasts of Africa to the shores of the Bering Sea. The human populations of forty countries worldwide are dependent on fish or marine food sources for a majority of their protein intake. All but one of those nations would be classified as less developed.

Ethical models applied to fisheries include the precautionary principle, sustainability, the inherent value of all species, and ecojustice or environmental justice. The precautionary principle holds that new capture and processing technologies should not be deployed to harvest an entire fishery or used under the supposition of extracting the maximum sustainable yield until the impacts of the technology have been studied fully and understood. Unfortunately, regulation of new fisheries often lags behind the development of the technology to exploit them, and so the international norm is still overharvesting of species such as orange roughy and Patagonian toothfish (Chilean sea bass).

The concept of sustainability requires leaving resources available for harvest in the immediate and distant future. This requires closing or limiting extraction from fisheries in decline and may mandate less destructive methods of capture, such as the use of hand lines. The concept of the inherent worth of all species requires reduction of incidental catch (by-catch) of nontarget organisms such as sea turtles and seabirds. This may be accomplished by banning technology prone to accumulating incidental harvest, such as drift nets; improving technology, for example, by installing turtle excluder devices on shrimp nets; or simply closing offending fisheries by, for example, banning the use of trawl nets near seabird rookeries. Ecojustice calls for defense of the livelihood of indigenous and inshore fishing communities and fair access to ocean resources. These principles also encourage the preservation of cultural and linguistic diversity.

An additional issue for fisheries is the proliferation of fish and aquatic species culture. Salmon farming has displaced the harvesting of wild stocks through much of northern Europe. Salmon farms harbor diseases, produce waste, and disturb coastal ecosystems. Shrimp culture for export has spread ponds through coastal southeastern Asia, displacing mangrove forests and natural lagoons. Scientists are skeptical about the use of hatchery-reared salmon to replenish oceanic stocks. Interbreeding between hatchery fish and wild fish may decrease the genetic viability and ecological fitness of the salmon populations. Environmental ethical responses include ecosystemic rights in which there would be a requirement for the preservation of substantial and representative areas of natural ecosystems and the concept of maintaining wilderness areas or zones free of human interference, which may be applied to marine as well as terrestrial habitats. The global trend toward industrializing and privatizing fisheries is also an environmental justice issue in that it replaces family-owned fishing boats with large corporate fleets or fish farms.

Ocean Pollution Human pollution of ocean environments is of two geographic origins. The first is direct dumping or release of materials and toxins into the oceans either from ships or by discharge in ports or on the coast. Examples include ships spilling or leaking oil, sewage being pumped into estuaries by towns, oceanic dumping of sludge, freighters emptying bilges, and a ship's crew tossing garbage overboard. Lisa Newton, Catherine Dillingham, and Joanne Choly (2006) proposed three ethical imperatives that would prevent disasters such as the *Exxon Valdez* grounding in Prince William Sound, Alaska: holding those accountable legally responsible and enforcing the law, protecting the last remaining stretches of wild coast by restraining economic development, and pursuing sustainable energy polices that would reduce dependence on oil. Ocean drilling for oil and gas may also cause spills, and adds to greenhouse gas emissions. Offshore deposits of gas hydrates, such as methane hydrate, represent a major untapped energy source on the continental shelves, and may contain twice the carbon of all other fossil fuel reserves. Methane hydrate is trapped in ice crystals in relatively deep water; thus new technologies are necessary to exploit it. As methane is among the most powerful of greenhouse gases, the extraction of gas hydrates could add significantly to global climate change and sea-level rise.

The second source of ocean pollution is items and substances, often originating in the middle of continental landmasses, that are carried long distances by rivers or the atmosphere and eventually reach the ocean. Fertilizer spread in the midwestern United States reaches the Gulf of Mexico via the Mississippi River, and pesticides spread on marshes or farm fields make their way into aquatic food chains and eventually end up in coastal sediments. Atmospheric transport of nitrogen and sulfur compounds released by factories, electrical power generation, and automobiles produces acid rain over estuaries such as Chesapeake Bay. The additional nitrogen may act as a fertilizer, stimulating algal blooms. As the algae die and bacteria draw oxygen from the water during the process of decomposition, even marine environments can become anoxic (without oxygen) and thus fatal to fish and other animals.

Regulation of ocean pollution was slow to develop because much of the offending material was diluted or washed into international waters, where the impacts were not visible to the public. Although offshore disposal is regulated by the Convention on the Prevention of Marine Pollution by the Dumping of Wastes and Other Matter of 1972, also known as the London Dumping Convention, people still have difficulty comprehending the damage their activities do to the oceans. Among the environmental problems people have difficulty seeing—at least without the assistance of marine science—is the spread of dead or toxic zones where cities have deposited their sludge and other wastes or where major rivers have carried high loads of pollutants out to sea. A dead zone extending for many kilometers has appeared offshore from the Mississippi River Delta. Although the causes of this environmental disaster are not known completely, marine ecologists suspect pollutants such as fertilizers and industrial effluents. If this happened in a forested or agricultural region, the U.S. public would be outraged.

A potential environmental ethical response to both the damage caused by dead zones and the plowing of the ocean bottom with drag nets is to invoke ecosystemic equity and integrity: Humans should use the same standards for protecting ocean habitats that they use for managing habitats on land and should not protect some marine ecosystems while degrading others. Further, people should maintain the basic ecoystemic function and components of the ecosystems they purposefully harvest (bottom fisheries) or affect coincidentally (the dead zones). In the case of bottom disturbance, regulating more destructive fishing methods is appropriate and helps conserve the reproductive potential of the fish. In the case of dead zones, humans should reduce pollution inputs until the zones can recover their natural species diversity and food webs.

The public is more likely to observe trees dying from acid rain than to notice the impacts of additional nitrogen on an estuary. Under the concept of ecosystemic equity, both a mountain spruce forest and submerged flats of eel or turtle grass merit human care and protection. Regulation of ocean pollution has made increasing use of the precautionary principle. The 1996 Protocol of the Convention on the Prevention of Marine Pollution by the Dumping of Wastes and Other Matter allows oceanic disposal only of listed materials; all other forms of waste are excluded until proved safe and officially added to the inventory of permitted wastes.

Marine Parks There has been a global expansion of networks of marine parks and reserves. The twentieth-century strategy was primarily to protect the aesthetic and the tourism-worthy. A majority of marine parks currently steward coral-studded lagoons, kelp forests, rocky reefs, and other ecosystems popular with swimmers, scuba divers, and recreational fishers. Application of the concept of ecosystem equity, however, implies that the representative areas of open continental shelves, deep trenches, and even shallow banks such as Georges Bank, which is critical to North Atlantic fisheries, should be protected from human disturbance. Recent scientific studies have found that setting aside protected areas in more ordinary habitats preserves fish spawning and feeding areas and maintains populations of a wide variety of marine organisms, including invertebrates taken as by-catch.

SEE ALSO *Coral Bleaching; Environmental Justice; Global Climate Change; Greenpeace; Hunting and Fishing: I. Overview; Hunting and Fishing: IV. Angling; Hunting and Fishing: V. Commercial Fishing; Hurricane Katrina; Land Ethic; Leopold, Aldo; Pollution; Polynesia; Precautionary Principle; Species; Sustainability; Tragedy of the Commons.*

BIBLIOGRAPHY

Bratton, Susan. 2004. "Thinking Like a Mackerel: Rachel Carson's *Under the Sea-Wind* as a Source for a Trans-Ecotonal Sea Ethics." *Ethics and the Environment* 9(1): 1–22.

Brooks, L. Anthea, and Stacy D. VanDeveer. 1997. *Saving the Seas: Values, Scientists, and International Governance.* College Park: Maryland Sea Grant College.

Callicott, J. Baird. 1992. "Principal Traditions in American Environmental Ethics: A Survey of Moral Values for Framing an American Ocean Policy." *Ocean and Coastal Management* 17: 299–308.

Dallmeyer, Dorinda G., ed. 2003. *Values at Sea: Ethics for the Marine Environment.* Athens: University of Georgia Press.

Dean, Cornelia. 1999. *Against the Tide: The Battle for America's Beaches.* New York: Columbia University Press.

Hall, Stephen J. 1999. *The Effects of Fishing on Marine Ecosystems and Communities.* Oxford, UK, and Malden, MA: Blackwell Science.

Hardin, Garrett. 1968. "The Tragedy of the Commons." *Science* 162: 1243–1248.

Koslow, Tony. 2007. *The Silent Deep: The Discovery, Ecology, and Conservation of the Deep Sea.* Chicago: University of Chicago Press.

Newton, Lisa H.; Catherine K. Dillingham; and Joanne Choly. 2006. *Watersheds 4: Ten Cases in Environmental Ethics.* Belmont, CA: Thomson Wadsworth.

Norse, Eliot A., and Larry B. Crowder. 2005. *Marine Conservation Biology: The Science of Maintaining the Sea's Biodiversity.* Washington, DC: Island Press.

Ray, G. Carelton, and Jerry McCormick-Ray. 2004. *Coastal-Marine Conservation: Science and Policy.* Malden, MA: Blackwell.

Sobel, Jack A., and Craig Dalgren. 2004. *Marine Reserves: A Guide to Science, Design, and Use.* Washington, DC: Island Press.

Taylor, Joseph. E. 1999. *Making Salmon: An Environmental History of the Northwest Fisheries Crisis.* Seattle: University of Washington Press.

Susan Power Bratton

ODUM, EUGENE

1913–2002

Eugene Odum, born on September 17, 1913 in Newport, New Hampshire, was one of the most influential figures in twentieth-century ecology. Best known for his advocacy of the concept of ecosystems and holistic ecosystem-based perspectives in ecology and environmental problem solving, he has been called the father of modern ecology.

Gene Odum received his Ph.D. in zoology from the University of Illinois in 1939, writing a dissertation on environmental influences on bird physiology. While at the University of Illinois he came under the influence of the animal ecologist Victor Shelford. Odum's holistic conception of ecological systems was strongly influenced by Shelford's and Frederic E. Clements's conception of the plant-animal community as a complex superorganism. According to this conception, the plant-animal community responds as a dynamic unit to changes in abiotic conditions, progressing through a sequence of developmental stages until it reaches a self-perpetuating stable equilibrium. Odum defended a modified version of Clements's succession theory throughout his career, but instead of treating the plant-animal community as the fundamental unit of analysis, Odum adopted Arthur Tansley's concept of an ecosystem. Odum's mature theory of ecosystem growth and development draws on thermodynamic and cybernetic concepts found in the work of Alfred J. Lotka, Raymond Lindeman, G. Evelyn

Hutchinson, and his own younger brother Howard T. Odum.

In the 1940s at the University of Georgia, Gene Odum began writing *Fundamentals of Ecology*, which was to become the most successful textbook in the history of ecology. *Fundamentals of Ecology* was first published in 1953, a second edition (coauthored with his brother Howard) appearing in 1959, and a third edition in 1971. *Fundamentals of Ecology* is notable for its emphasis on the ecosystem as the fundamental unit of analysis for ecology and its systematic integration of general ecological principles, applied conservation science, and environmental advocacy. Odum believed that ecological science could and should function as a foundation for environmental problem solving and ecologically sustainable economic and social development. *Fundamentals of Ecology* is credited with attracting a generation of students that shared Odum's optimistic vision of ecology as an integrated-systems science with the potential to transform humanity's relationship with the natural environment in positive ways.

The popularity of Odum's style of ecosystem ecology reached its peak within academic ecology by the end of the 1960s. In plant-community ecology in the 1970s, Clements's succession theory was effectively replaced by more reductionistic models of plant communities that emphasized their lack of internal coherence and the radical contingency of successional trajectories. Evolutionary ecology also emerged as a powerful new integrative perspective that challenged the group-selectionist assumptions in Odum's model of ecosystem development. By the end of the 1980s, holistic, systems-oriented ecosystem ecology was effectively marginalized in academic ecology in the United States. Nevertheless, ecosystem ecology in the tradition of Gene and Howard Odum continues to be developed and practiced by ecologists around the world.

In the last two decades of his life, Gene Odum devoted more of his time to promoting and presenting his vision of ecology and sustainable living to a broader audience. In 1989 he published *Ecology and Our Endangered Life-Support Systems*, and in 1998, *Ecological Vignettes: Ecological Approaches to Dealing with Human Predicaments*. In these works Odum argued that principles of ecosystem growth and development apply equally to human socioeconomic systems, and he developed a prescriptive model of social organization based on these principles. For example, Odum believed that ecosystems with more extensive mutually beneficial interactions between system components are more stable than ecosystems with less such interactions, and that diversity of components promoted system stability by enabling redundancy in critical ecosystem functions. Consequently, he recommended modes of social organization that increased cooperative relationships and decreased competition, and argued for diversification of energy and food sources to ensure stable supplies of these resources.

Odum argued that this transformation of human social systems to ecological maturity requires a new set of ethical and political values. The values of free-market industrial capitalism may be adaptive in the early stages of economic and social growth, but become liabilities in later stages when limiting constraints on growth become evident. The transition to a more harmonious relationship between humanity and the natural environment requires a shift in values that promotes efficient use of resources, recycling and reduction of waste, higher degrees of cooperation and diversity among social groups, and greater appreciation of the dependence of human welfare on the quality of environmental resources. Such values promote the conditions for stability of mature social systems, just as they do the stability of mature ecosystems.

Odum's ethical writings have been compared to Aldo Leopold's land ethic, though Odum's rhetoric is more consistently anthropocentric than Leopold's. Nevertheless, Odum's style of ecosystem ecology has often been used to underwrite the views of nonanthropocentric environmental ethicists and deep ecologists.

SEE ALSO *Conservation Biology; Ecology: III. Ecosystems; Holism; Leopold, Aldo; Sustainability.*

BIBLIOGRAPHY

Craige, Betty Jean. 2001. *Eugene Odum: Ecosystem Ecologist and Environmentalist.* Athens: University of Georgia Press.

Odum, Eugene P. 1971. *Fundamentals of Ecology*, 3rd edition. Philadelphia: W. P. Saunders.

Odum, Eugene P. 1993. *Ecology and Our Endangered Life Support Systems*, 2nd edition. Sunderland, MA: Sinauer Associates.

Odum, Eugene P. 1998. *Ecological Vignettes: Ecological Approaches to Dealing with Human Predicaments.* Amsterdam: Harwood Academic Publishers.

Kevin deLaplante

ORGANIC FARMING

The term *organic farming* was first used in 1940 by Lord Walter Northbourne in *Look to the Land* (p. 81), to describe an alternative to *chemical farming.* Organic farming is a method of agricultural production that eschews chemical fertilizers and pesticides. But organic farming is not simply a way of growing food; it is also a social movement. From its inception, strong ideological notions relating to soil, health, nature, politics, science, and spirituality have offered both a critique of and alternative to modern orthodox farming practices. At times the subject of passionate and polarizing debates, organic farming has often been considered marginal, antiquated, or even subversive. Around the turn of the twenty-first century, it became increasingly important economically and culturally, but new questions have begun to emerge regarding the effect of this success on its ideological and philosophical underpinnings.

ORIGINS

Organic farming is based on traditional agricultural practices, but its emergence as both a method and a movement was mainly a reaction to the increasing industrialization of agriculture in the early twentieth century. Philip Conford describes the history of the early movement, centered in England, in *The Origins of the Organic Movement* (2001). At this time, organic farming was mainly promoted by members of the social and intellectual elite. Their political motives ranged widely, but organic thought was consistent in a number of aspects. Belief in the importance of soil and the necessity of humus (organic matter) for sustainable production of healthy crops has always been central. More than merely a scientific argument, this idea is based on a holistic worldview that emphasizes interconnectedness and an agriculture that mimics natural systems. Early organic advocates believed that healthy soil was connected to healthy crops, healthy crops to healthy people, and healthy people to healthy societies. This holistic worldview is fundamentally at odds with the scientific reductionism of modern agriculture, and the organic movement has long been based on opposition to this emerging orthodoxy.

Other concerns of the early movement included preserving rural populations, culture, and livelihoods. Organic proponents opposed agricultural mechanization and argued that agriculture rather than industry should form the basis for a healthy nation. In this way, the early organic movement embodied philosophical and political beliefs similar to the agrarianism of Thomas Jefferson and others. Early organic thought also parallels agrarian philosophy in its promotion of small, independent, diversified farms, and its rejection of agriculture singularly focused on economic efficiency.

Perhaps the most prominent and influential of the early organic proponents was Sir Albert Howard. Howard developed and promoted a systemic vision of agricultural production and land stewardship based on returning composted organic waste to farm fields, and rejected the economic materialism of industrial agriculture. Howard emphasized the role of organic farming in promoting individual and social health, and maintaining soil fertility for future generations. A prolific lecturer, Howard also presented his ideas in a number of published works, including *An Agricultural Testament* (1940) and *The Soil and Health: A Study of Organic Agriculture* (1947).

Though a trained scientist himself, Howard states in *An Agricultural Testament* that "Instead of breaking up the subject into fragments and studying agriculture in piecemeal fashion by the analytical methods of science, appropriate only to the discovery of new facts, we must adopt a synthetic approach, and look at the wheel of life as one great subject and not as if it were a patchwork of unrelated things" (p. 22). Like many in the early movement, his holistic vision of farming was connected to a Christian spirituality that included a concept of nature as divine. Like many others, he was also highly influenced by Asian thought.

Another important figure in the development of organic farming was the Austrian philosopher Rudolph Steiner. Like Howard, he rejected the materialism and reductionism of modern agricultural practices and emphasized the living soil as the basis for health, vitality, and spiritual connection. Steiner's philosophy was influenced by Franz Brentano, Friedrich Nietzsche, Hinduism, and Theosophy, but was based most heavily on the works of Johann Wolfgang von Goethe and Christian spirituality. Steiner published several works on Goethe's epistemology, spirituality, and concepts of nature and science. Though his writings span a variety of subjects, among his most important and lasting influences was his development of *Biodynamic Agriculture*, an intensive, integrated production method that conceives of the farm as a self-contained system balanced by the interconnected life-forces of both the earth and cosmos.

Both Steiner and Howard were important influences for Jerome Irving Rodale, a magazine publisher and health crusader who would become the foremost advocate for organic farming in the United States. Unlike most of his European counterparts, Rodale promoted organic in a way that was primarily pragmatic, secular, and apolitical. Among his most important contributions were his promotion of organic gardening, which introduced its methods to non-farmers, and his establishment of long-term research trials comparing organic and conventional production methods.

Rodale's publications were highly influential for the American counterculture, which embraced organic farming and gardening passionately during the late 1960s and early 1970s. During this time, young people participated in a vast rural migration to conduct utopian experiments in homesteading and communal living. Warren Belasco details this phenomenon in *Appetite for Change: How the Counterculture Took on the Food Industry* (1989). This "back to the land" movement was based on a neo-agrarian

Examining Organic Produce, Springfield, Virginia. *Ogranic food—produced without chemical pesticides —has become more and more popular, and is found in grocery stores and markets throughout the world. Proponents of organic food cite that it is more environmentally friendly than chemical dependent and/or land-intensive agriculture.* **AP IMAGES.**

worldview that rejected consumerism in favor of self-sufficiency, simplicity, and a closer connection to the natural world. Like the early organic movement, it emphasized small farms, soil conservation, and crop diversity, as well as environmentalism and energy conservation. In contrast, its social goals were promoted more through lifestyle decisions than political advocacy.

This movement formed the basis for a community of like-minded individuals who worked to develop, define, and promote organic farming and production standards. By the late 1980s organic food had gained significant popular status, and was beginning to command premium prices. At the same time, production standards had become more complex, and certification schemes were increasingly sophisticated and numerous. Recognition of the emerging economic importance of organic products and a perception that their commercial development was limited by the lack of consistent standards led to the legislative institutionalization of organic farming through the development of national certification programs.

A MOVEMENT IN CRISIS

In the United States, a National Organic Program (NOP) administered through the Department of Agriculture was authorized through the 1990 farm bill to establish rules defining appropriate practices and a process for certifying organic farms and production facilities. Initially organic growers and advocates were encouraged by this legitimization, but the implementation and repercussions of the NOP have caused considerable frustration and disillusionment over time. Appropriate materials and practices have always been controversial, and remain contested.

Enacted in 2002, the NOP Final Rule describes organic farming as "A production system that is managed in accordance with the [Organic Foods Production] Act and regulations in this part to respond to site-specific conditions by integrating cultural, biological, and mechanical practices that foster cycling of resources, promote ecological balance, and conserve biodiversity" (Subpart A, Section 205.2). As such, the NOP ignores social movement issues related to health, labor standards, farm size, energy conservation, and rural development. Critics have claimed that the NOP has robbed organic farming of its philosophical basis, and facilitated its commercialization and industrialization. Since its implementation there have been substantial increases in the size of organic farms, the number of farms growing both organic and conventional produce, the prevalence of organic products originating outside of the United States, and the concentration of ownership in the organic food processing and retail sectors.

In response, many traditional organic producers have sought to differentiate themselves and their products based on their adherence to traditional elements of the organic philosophy. One result has been the emergence of the terms *beyond organic* and *moreganic*, the *local food* movement, and the reframing of organic principles using terms such as *civic agriculture*, coined by Thomas Lyson in his 2004 book *Civic Agriculture: Reconnecting Farm, Food, and Community*. The rapid development of civic agriculture and the local food movement are evident in the proliferation of farmers markets and *community supported agriculture*, which invites consumers to participate directly in the labors, risks, and rewards of food production. The growth of such direct marketing practices is consistent with organic movement support for small, diverse, independent farm enterprises, and is being led by farmers committed to traditional organic production methods.

For almost a century, organic farming has been an evolving, negotiated, values-based activity related to ideological notions regarding the relationship between nature, society, and food production. Small, diverse, ecologically conscious growers continue to maintain farm operations based on the kind of holistic, postmaterialist, agrarian values that have been the hallmark of organic philosophy. Organic farming practice is thriving today, but the organic farming movement is very much in crisis. Caught between two equally uncertain and divergent scenarios, its adherents are struggling simultaneously to maintain as many of its principles as possible within the current rules, and to redefine a movement that has lost its name, and is unable to divest itself from the system that has usurped it.

SEE ALSO *Agriculture; Berry, Wendell; Factory Farms; Farms; Genetically Modified Organisms and Biotechnology; Jackson, Wes; Shiva, Vandana; Sustainable Agriculture; U.S. Department of Agriculture.*

BIBLIOGRAPHY

Belasco, Warren J. 1989. *Appetite for Change: How the Counterculture Took On the Food Industry.* New York: Pantheon Books.

Conford, Philip. 2001. *The Origins of the Organic Movement.* Edinburgh, Scotland: Floris.

DeLind, Laura B. 2000. "Transforming Organic Agriculture into Industrial Organic Products: Reconsidering National Organic Standards. *Human Organization* 59(2): 198–208.

Guthman, Julie. 2004. *Agrarian Dreams: The Paradox of Organic Farming in California.* Berkeley: University of California Press.

Howard, Sir Albert. 1940. *An Agricultural Testament.* Oxford, UK: Oxford University Press.

Howard, Sir Albert. 1947. *The Soil and Health: A Study of Organic Agriculture.* New York: Devin-Adair.

Lyson, Thomas A. 2004. *Civic Agriculture: Reconnecting Farm, Food, and Community.* Medford, MA: Tufts University Press.

Northbourne, Lord Walter. 1940. *Look to the Land.* London: Dent.

Steiner, Rudolph. 1958. *Agriculture: A Course of Eight Lectures: Given at Koberwitz, Silesia, 7 to 16 June 1924.* London: Bio-Dynamic Agricultural Association.

U.S. Department of Agriculture. 2002. *National Organic Program Final Rule.* Washington, DC: Federal Register (7 CFR, Part 205).

Taylor Reid

ORTEGA Y GASSET, JOSÉ

1883–1955

José Ortega y Gasset, the most important Spanish philosopher of the twentieth century, was born on May 9, 1883, in Madrid. He held the chair of metaphysics at the Central University of Madrid. He was elected to the constituent assembly during the Second Republic. After the outbreak of the Spanish Civil War, he lived in exile, first in Argentina and various parts of Europe, settling in Portugal 1945 and returning permanently to Spain in 1948.

THE FOUNDATIONS OF ORTEGA Y GASSET'S ENVIRONMENTAL PHILOSOPHY

Ortega y Gasset's philosophy, beginning with his programmatic statement (1914), "I am I and my circumstance" (Ortega y Gasset 1963, p. 45) has been called "the first expression of an ecological approach in philosophy" (Rogers 1994, p. 505), Ortega y Gasset's concept of "circumstance" having been influenced by Jakob von Uexküll's (1864–1944) concept of the *Umwelt* (surrounding world, environment) of the organism and by Edmund Husserl's (1859–1938) phenomenological concept of the *Umwelt* (Marías 1970). Uexküll's concept is biologically oriented, focusing on the structure of the organism. Husserl's concept is phenomenological, focusing on the practical world (later the "life-world") as it is given to consciousness. Rejecting Husserl's transcendental turn, Ortega y Gasset was able to draw on both Uexküll's naturalistic concept and Husserl's phenomenological concept.

Although Ortega y Gasset did not concentrate on environmental issues in the contemporary sense, he did emphasize the impact of the environment on culture and individuals. His theory of "vital reason" began to take shape in his account of a forest in El Escorial, the reality of which he views as a function of vital experiences (Ortega y Gasset 1963, pp. 59–69; Marías 1970). Noting the sad state of ethical theory, he wrote, "There are people who believe in good faith that we have no obligations toward the rocks and therefore have tolerated advertisers' smearing with pitch … the venerable rocks of the mountain ranges" (2007, p. 99). The rocks of the mountain can only be what they are, "venerable," as a function of individual perspectives (Marías 1970).

ORTEGA Y GASSET ON HUNTING

Ortega y Gasset's most important work on broadly environmental issues is his *Meditations on Hunting* (1942), originally written as a preface to a memoir on big game hunting by his friend Edward, Count Yebes. It has become one of the most influential philosophical works on this subject, especially among hunters. His approach is explicitly existential: "The life that we are given has its minutes numbered, and in addition it is given to us empty. … Thus, the essence of each life lies in its occupations" (2007, p. 35). Ortega y Gasset notes that most human work is unsatisfying, draining life rather than filling it up. Therefore, the human being "finds it essential to divert" himself or herself (2007, p. 29). Hunting—along with dancing, racing, and conversation—is one of the diversions traditionally practiced by human beings for their own sake. Hunting is an activity in which, rather than losing time, we are "gaining it, filling it satisfactorily and as it should be filled" (2007, p. 37).

Distinguishing hunting from fighting, Ortega y Gasset defines hunting as "what an animal does to take possession, dead or alive, of some other being that belongs to a species basically inferior to its own" (2007, p. 62). But Ortega y Gasset insists that it is essential to hunting that the hunted animal has its chance to avoid capture. "There is, then, in the hunt as a sport a supremely free renunciation by man of the supremacy of his humanity. Instead of doing all that he could do as man, he restrains his excessive endowments and begins to imitate Nature—that is, for pleasure he returns to Nature and re-enters it" (2007, p. 63).

In addition to this hunter's ethos, Ortega y Gasset recognizes ethical issues pertaining to the activity of hunting itself. "Every good hunter is uneasy in the depths of his conscience when faced with the death he is about to inflict on the enchanting animal. He does not have the final and firm conviction that his conduct is correct. But neither, it should be understood, is he certain of the opposite" (2007, p. 98). But Ortega y Gasset insists that "the greatest and most moral homage we can pay to certain animals on certain occasions is to kill them with certain means and rituals" (2007, p. 101). He attacks "photographic hunting" as an ethical "mannerism" (2007, p. 103), because "In the preoccupation with doing things as they should be done—which is morality—there is a line past which we begin to think that what is purely our whim or mania is necessary. We fall … into a new immorality …, which is a matter of not knowing those very conditions without which things cannot be" (the goal of killing the animal is essential to hunting). "The hunter seeks this death because it is no less than the sign of reality for the whole hunting process. To sum up, one does not hunt in order to kill; on the contrary, one kills in order to have hunted" (2007, p. 105). Only in hunting is one truly an active participant in the countryside, intensely attuned to every aspect of it, perceiving it from both the hunter's and the prey's perspective. Hunting does not objectify the prey; it is the natural and appropriate response to the animal itself. "The only adequate response to a being that lives obsessed with avoiding capture is to try to catch it" (2007, p. 129).

Describing the hunter as "the alert man" (2007, p. 138), Ortega y Gasset views sport hunting as "a vacation from his human condition" (2007, p. 121), a reprieve from the draining tedium of everyday life. "Life is a terrible conflict.… Hunting submerges man deliberately in that formidable mystery and therefore contains something of religious rite and emotion in which homage is paid to what is divine, transcendent, in the laws of Nature" (2007, p. 106).

CRITICISMS OF ORTEGA Y GASSET'S VIEWS ON HUNTING

Questions about Ortega y Gasset's views on hunting can be raised from a number of perspectives. Ortega y Gasset's anthropocentric position is based in part on his hierarchical view of the natural world, but this view is itself difficult to sustain. There is no defensible sense in which the hunted animal is necessarily "inferior" to the hunter. The fact that Ortega y Gasset approaches sport hunting purely from the point of view of "diversion" limits the meaning that can be found in it. He makes it clear that eating the hunted animal is not part of hunting and has nothing to do with the meaning of the hunt. A truly ecological approach to hunting might replace Ortega y Gasset's hierarchy of superior hunter and inferior hunted with the notion of "webs" to which both hunter and hunted belong. One who hunts as an end in itself can do so as a member of an ecosystem, taking from it, by hunting, a portion of nourishment and giving back to it by participating in and protecting the very wildness that makes hunting possible in the first place. From this perspective hunting can be more than a mere diversion.

SEE ALSO *Animal Ethics; Hunting and Fishing: I. Overview; Hunting and Fishing: II. Recreational Hunting.*

BIBLIOGRAPHY

Jones, Alan Morris. 1997. *A Quiet Place of Violence: Hunting and Ethics in the Missouri River Breaks.* Bozeman, MT: Spring Creek Publishing.

Marías, Julián. 1970. *José Ortega y Gasset: Circumstance and Vocation.* Norman: University of Oklahoma Press.

Ortega y Gasset, José. 1963. *Meditations on Quixote,* trans. Evelyn Rugg and Diego Marin. New York: Norton.

Ortega y Gasset, José. 2007. *Meditations on Hunting,* trans. Howard B. Wescott. Belgrade, MT: Wilderness Adventures Press.

Rogers, W. Kim. 1994. "Ortega and Ecological Philosophy." *Journal of the History of Ideas* 55(3): 503–522.

J. Claude Evans

OUTER SPACE

Humankind lives in a gravity cocoon that contains all the ingredients—rocks, gases, water—that allow life and consciousness to flourish. This cocoon is probably one of many similar tiny dots floating in the vastness of space, but whether there is life elsewhere is unknown. What is remarkable about Earth is that it is a distillation of a tiny fraction of the elements of the universe: the ones that provide rocks, water, and atmosphere. The universe consists mostly of hydrogen gas, but a small number of heavy rock-forming elements condense at certain distances from nascent stars.

Many terrestrial, or Earthlike, planets will have abundant water as long as they reside in their stellar habitable zone (Kasting et al. 1993). Water is one of the most abundant molecules in the universe, and typically clouds of ice-rich comets form at the outer reaches of the domain of a new star. Scattering of comets and asteroids into the inner or habitable zone planets ensures that water and volatile gases collect on the newly formed extrasolar terrestrial planets (Morbidelli et al. 2000).

Martian North Polar Ice Cap. *This image of the north polar region of Mars, taken from a high resolution camera on the European* Mars Express *spacecraft, shows layers of water, dust and ice in perspective view. The cliffs are nearly 2 kilometers high, and the dark material could be volcanic ash. The possibility of life on other planets would likely dramatically alter our conceptions of society and the environment. It seems clear that the tourism industry will be focusing on outer space sometime in the near future.* **PHOTO COURTESY OF ESA/DLR/FU BERLIN (G. NEUKUM).**

EXTRATERRESTRIAL SOCIETY

Earth arguably is a closed system in that human society is bound to it and to the limits of its resources. There may be other closed systems: terrestrial planets around nearby stars that have sentience and society as well. The detection of life in any location other than Earth will change what people think of as society. The global society, with its quirks and conveniences and homogeneity, has emerged in recent years. How will human views of society and environment change when people learn that versions of them exist on other worlds?

It is somewhat likely though not probable that scientists will find microbes in the Mars polar ices or along the edges of the steam vents of Saturn's moon Enceladus (Spencer and Grinspoon 2007). It is more likely that in the middle of the twenty-first century it will be possible to detect the composition of Earthlike atmospheres and surfaces around other stars. These observations could be of high enough fidelity to indicate whether a global biosphere altered the nature of that atmosphere or surface.

Humankind's view of life certainly will change if microbes are found on other bodies in the solar system.

Will they be related genetically to Earth organisms, or will they have worked out an evolutionary method all their own? During the first 600 million years after Earth formed, the inner planets still were being bombarded with leftover planetary building blocks. Calculations based on shock-wave physics from atomic explosions show that a certain percentage of the planetary surface, dislodged by those collisions and blasted into space, never experiences temperatures greater than 1,000 degrees Celsius. Rocky blocks would get transported throughout the solar system and fall on other planets as they do today (Gladman et al. 1996). During reentry into a new atmosphere, many of those blocks never experience temperatures greater than 1,000 degrees Celsius (Melosh 1988). Thus, from the perspective of life the early solar system is best viewed as a vast sea with constant transport of material across the voids and its delivery to other worlds. Life would have opportunities to take such rides. People would be forced to face the fact that the environment, if defined by life, would include fragile enclaves that might be found on Mars or the moons of Saturn. Some of the ancestors of human society may live elsewhere, but they will deserve respect if they are discovered.

Alternatively, the solutions that life worked out to use its environment to replicate and evolve may be entirely different on Mars or Enceladus. In this case people's sense of environment must be reexamined because this finding would show that life is able to evolve under conditions that people cannot now imagine.

THE NEW SPACE

The days of Earth as the cradle and unique home of human civilization are drawing to a close. In rapid succession there have been successful private trips to space, and a large number of investors and players in the new space business are preparing to take tourists, investigators, honeymooners, artists, poets, engineers, and doctors into space. Some go on suborbital flights with just minutes of microgravity. Others circle Earth and see their home planet for what it is: a globe

The Human Outpost in Space, 2002. *The International Space Station (ISS) is a massive scientific laboratory in low Earth orbit, allowing it to be seen with the naked eye. NASA's goal is to have a similar outpost on the Moon by 2026. One of the research areas aboard the ISS is biology; scientists hope to improve their understanding of the effects on the human body of long-term exposure in outer space. The results of this and similar research are expected to contribute to the possibility of human colonization in space.* NASA.

overflowing with life, floating in emptiness. In the early decades of the twenty-first century space hotels may accommodate visitors in an environment with relaxing zero gravity and spectacular vistas where sport, art, and culture may evolve in new ways. Eventually commercial trips may take passengers to the moon, and one day adventurers and tourists may explore lunar highlands and fissures never before seen by humans.

In civil space efforts the National Aeronautics and Space Administration (NASA) is spending billions of dollars to develop a new and safe transportation system that can shuttle astronauts to the International Space Station. That system also may be part of the means of transportation to the Moon and may play a prominent role in NASA's stated goal of building a human outpost on the nearest solar system by 2026. Investment in the future lunar infrastructure is strong, with NASA studies of habitats, vehicles, and in situ resource utilization ongoing. It is envisioned that some versions of these systems will be used to explore nearby asteroids, ancient rocks that can bring and have brought global disaster to Earth in an instant.

Perhaps most ominously or excitingly, it appears that another space race is shaping up. The United States will have to ground the space shuttles in 2010 and contract for rocket rides with Russia to visit the International Space Station (ISS). The new NASA manned space transportation architecture will not be ready until 2014, at which time it will be able to take over transport between Earth and the ISS. The Chinese have an aggressive program of lunar robotic and manned exploration of the moon. China has leveraged its soft power by partnering with countries that the United States does not want to deal with or regards as an investment risk. The influence of the Chinese space program has increased as it has helped develop space system capabilities for Third World countries that cannot afford them on their own.

A conservative 2006 estimate of the value of the potential market for space tourism by 2021 is $1 billion (Taylor 2006). Space tourists already have gone to the ISS despite objections from NASA. Economies of scale and rapidly developing technologies make it likely that space tourism will grow along with the international push to do more and invest more to put more humans into space.

ENVIRONMENTAL ETHICS IN SPACE

Society will be different in space. Some see a utopian or utopian-libertarian universe of infinite space and infinite possibilities. However, as far as can be foreseen, people will be transported in small, highly dependent groups. They will bring their sense of ethics and environment with them, but how will they adapt to the realities of living in space? Dropping a tool with a mass of a kilogram outside the space station during repairs is not a big deal because it can be tracked and the risks of collision with the ISS can be mitigated. However, those people will be living on palpably finite and severely limited resources for the first time, and colonies off Earth are likely to develop a set of environmental or conservation values and ethics very different from those of their forebears. Moreover, the distinctive experience of the minicultures that evolve when humans are isolated is well documented at scientific stations such as the one at the South Pole.

The Moon is vast: Is leaving a burned-out rocket engine behind on it a significant environmental concern? Can people explore the Moon scientifically without regard for what is happening to its environment now or, as a consequence of what they do now, in the future? There is already a lot of human-made debris on the Moon, including Apollo landers with their buggies and equipment for experiments and a dozen or so crashed probes that provided movies of approach and immolation. Are these relics garbage littering the otherwise pristine surface of the Moon, or are they proud monuments to the achievements of the human race? Would people be better off if their motto was "leave nothing but footprints"? The footprints left on the Moon during the Apollo visits will last many times longer than a plastic bag thrown on the ground in a park on Earth, yet it is difficult not to see the extraordinariness of this first lone human boot print on a world other than Earth.

The exploration of space already has led to the extraterrestrial exportation of a familiar environmental ethic: the unexamined, consumer-driven sense of infinite vastness to absorb humankind's effluent that people implicitly employ in their everyday lives on Earth. Leaks of atmospheric gases, chunks of metal, tools, gloves, waste, garbage, burned-out rocket stages, dead satellites, and abandoned landers on the Moon, Mars, and Venus are among the environmental legacies of humankind to space. The debate about space debris is in full swing, largely because low-Earth-orbit pollution, even in the vastness of space, is bad enough that it is likely to cause a catastrophe on the ISS, along with the threat that some of the larger objects pose to humans, other life-forms, and ecosystems on Earth when they fall back down through the atmosphere onto the surface of the planet.

A turning point in the issue of space debris occurred in October 2006, when the Chinese chose to test an antisatellite weapon on one of their dead communications satellites. The debris from the explosion instantly increased the total amount of space debris (10,000 tracked objects larger than 4 centimeters) by 10 percent. Worse, the orbit of the target was high enough that debris scattered into orbits covering a wide range of altitudes, polluting some of the orbital ranges used most heavily by commercial satellites. Because of the high orbit of the target, much of the debris will take centuries to degrade into lower orbits as a result of friction with the upper reaches of the atmosphere. Most of the debris will burn up, but it will be raining pieces of Chinese space hardware for a century.

The United States performed a similar feat in the interest of keeping the world safe from hydrazine. A large military satellite had gone astray, and it was in a low enough orbit that blasting it to pieces would result in a debris cloud that would reenter the atmosphere in months. The U.S. test in February 2008 was equally successful, although because its stated goal was to destroy a hydrazine tank that if it survived reentry would be hazardous to people and the fact that it created a more responsible debris cloud constitute a weak argument that the United States holds the moral high ground on ground-based antisatellite tests.

Nevertheless, the fundamental ethical underpinnings of the lost gloves and wrenches floating in near-Earth space, the abandoned equipment on other planets, and the spraying of near-Earth space with debris from military exercises are the same. Space is an infinitely large sink; if one drops an empty bag of potato chips, nobody will ever see it and it will never be a part of the phenomenological landscape. It has been learned painfully on Earth that with increasing population density that type of thinking is flawed. If space, like land or ocean or sky, is to be exploited for human purposes, must it also become the ultimate wastebasket, as has been the history of the human use of all other media?

The expansion of people and then society into outer space inevitably will carry with it an environmental ethic. If the safety and survival of crews are at stake, it is certain that environmental protection or sustainability will be on the bottom of the list of priorities, and it can be assumed that space will be considered a convenient dumping ground. The occasional radioactive power source or debris cloud of heavy metals will be ejected randomly into space if it is expedient for the safety of the astronauts, whether they are governmental or private space voyagers.

Outer space is currently in the realm of science and engineering, yet when one considers that people soon may be exporting Earthly genes to other planets, one must conclude that they also will be exporting human culture. The conditions of a human mission to Mars are likely to be highly stressful; situations in which the moral and ethical fortitude of the astronauts will be tested are sure to arise. It is therefore important that a nascent colony or even an expedition be well informed about both ethics and environmental ethics. Stress reduction and time to adapt psychologically must have a high priority. Efficient and effective mechanisms for conflict resolution must be in place. Time and energy should be put into fostering a culture that places a premium on respect and boundaries, one in which each astronaut is motivated first and foremost by compassion for his or her colleagues. The possibility that romantic relationships may develop should be anticipated and analyzed, and a policy regarding such relationships must be formulated.

To some researchers the idea of living off new worlds and building colonies on other planets is absurd. The expense would be enormous, and the benefits and risks both to the colonists and to Earth-bound humanity are unknown. To others, however, it appears to be inevitable. It is human destiny to explore and then colonize. The frontier now is space, they think, and it has no limits.

Those who think that exploring space and colonizing other planets is absurd believe that the required spatial and temporal scaling of the concepts of exploration and colonization stretches those concepts to the breaking point. Christopher Columbus's voyage of discovery and colonization covered only 5,000 kilometers, and he made the pelagic crossing from the Canary Islands to Cuba in just over a month. Moreover, when he arrived, he could breathe the air and drink the water. He was back in Spain in less than a year. Mars, the nearest planet that can be explored by humans, is 23 million kilometers distant at its closest approach to Earth. By current means of propulsion it could be reached in at best 150 days. Once explorers arrive there, water must be mined from the ice, and the atmosphere is oxygenless. The nearest star, Alpha Centauri, is more than four light-years distant. Even if advanced propulsion could be developed for that journey, for the astronauts who attempt the trip it would mean forever leaving the society they know. If nearly unlimited energy could be harvested, it would be possible for humans to travel at speeds close to the speed of light. Albert Einstein's special relativity has been used to calculate both the energy necessary to get to Alpha Centauri and the effects of time dilation. For the astronauts on such a mission it might take ten years to get to the nearest star. During those years, moving at near-light speed relative to Earth, approximately ten thousand years would have gone by on Earth. Near-light speed is a time machine, and astronauts on a trip to Alpha Centauri would return twenty years older to an Earth that was twenty thousand years in the future. A nearly infinite supply of energy would be required to reach those velocities, although such a supply is not ruled out by the laws of physics.

Nevertheless, Carl Sagan (1994) and others have argued that there is a moral imperative to colonize other planets. By spreading its genes beyond one small world the human race would survive even in the case of a

Boot print of Buzz Aldrin. *This image commemorates one of the first steps ever taken on the Moon, by American astronaut Buzz Aldrin on the* Apollo 11 *mission in July 1969. In the pristine environment of the Moon, this boot print will last for millions of years, until it is slowly eroded away by atoms from the solar wind.* NASA.

planetary catastrophe. However, that thought may lead to the idea that there is little reason to try to avoid an anthropogenic planetary catastrophe. Having colonized other planets, the human race would survive even after a human-made catastrophe on Earth, so why should the prospect of such a catastrophe concern people from the point of view of environmental ethics? After humans use up Earth to the point where it is no longer habitable, they will emigrate to other planets; this is precisely the environmental ethic of the frontier.

THE OUTER SPACE WILDERNESS

For most of human civilization the world was infinite, with more land, trees, and resources than could be used. Now humankind lives on a small world where the fuels that have driven human technology are being depleted. Forests are disappearing. The limits of freshwater, ocean resources, and the atmosphere are being felt, but the ultimate consequences are unknown.

Habitat loss for wild species is one of the most profound changes that the planet has experienced as a result of human civilization. The interconnected nature of plants, animals, seasons, and water is disrupted over much of the planet, and as a result species extinction is occurring at a rate faster than it has after nonanthropogenic global catastrophes such as asteroid impacts.

National park and wilderness systems have been somewhat effective at preserving natural habitat in perpetuity. Wolves have returned to Yellowstone, and the entire ecosystem is adjusting. People treasure wild spaces not just for what Earth was before they arrived and flourished but also for their natural beauty and the diversity of animals and birds and insects, with which people have an evolutionary and genetic bond.

Space, the new frontier, is in most respects effectively infinite. The immediately accessible extraterrestrial surfaces, such as those of the Moon and Mars, are huge but not unlimited. However, beyond them are Ganymede and Titan and Pluto and eventually Earthlike worlds around other stars, most of them more light-years distant than the number of years in a human lifetime. Mars has about the same solid surface area as Earth because although it is smaller in circumference, it has no oceans.

Planetary park systems that would preserve unique and diverse elements of planetary environments have been proposed (Cockell and Horneck 2004, 2006). Through human exploration of Mars and other places with robotic spacecraft, it has been discovered that there is natural beauty on other worlds as well. A planetary park system would designate large swaths of planetary surfaces to be wilderness if that becomes necessary, although it would appear to be difficult to preserve a portion of the surface of any planet from the corrosive effects of oxygen if that planet is to have a breathable atmosphere. An anticolonial extraterrestrial environmental ethic, in contrast, might require that other planets not be touched by robotic or human spacecraft or be used as a dumping ground for hardware falling from orbit. Such an ethic carries with it more than the practical issues of preservation of the unique and beautiful; it extends the concept of wilderness and people's duty to consider their impacts on the environment into space.

On Mars there is the possibility that colonies of organisms live within the ice or in lava tubes. Current NASA planetary protection policies require primarily that a landed spacecraft on Mars not disrupt any life to the extent that it is not analyzable. Some consideration is given to avoiding contaminating a possible biosphere, and even greater consideration is given to avoiding contaminating Earth with alien microorganisms from returned samples.

THE CONDUCT OF HUMANS IN SPACE

These considerations suggest that the injection of environmental ethics into the use, exploration, and eventual colonization of space is necessary. Philosophical practices such as contemplation and meditation may play a vital role in establishing a self-sustaining and self-correcting astronaut minisociety on which future colonists can build to create an extraterrestrial society that is based on mutual respect and sustainability. Exploring methods of stress reduction and coping mechanisms to adapt psychologically to extended periods away from Earth and its familiar environments must have a high priority. Efficient mechanisms for conflict resolution must be in place. Time and energy should be put into developing a culture that retains mutual respect and boundaries among the astronauts and the extraterrestrial environment, including space stations and spaceships and is motivated by compassion for colleagues. Whether this will result in a code of conduct that includes explicit respect for the space environment, a set of international laws, or both requires discussion. It is important to develop methods that introduce a conscious awareness of the values that spacefarers bring with them and to examine how they change with the realities of working and living in space.

SEE ALSO *China; Environmental Philosophy: V. Contemporary Philosophy; Intergenerational Justice; Land Ethic; Ozone Depletion; Preservation; Water; Wilderness.*

BIBLIOGRAPHY

Cockell, C. S., and G. Horneck. 2004. "A Planetary Park System for Mars." *Space Policy* 20: 291–295.

Cockell, C. S., and G. Horneck. 2006. "Planetary Parks—Formulating a Wilderness Policy for Planetary Bodies." *Space Policy* 22: 256–261.

Gladman, Brett. J.; Joseph A. Burns; Martin Duncan; et al. 1996. "The Exchange of Impact Ejecta between Terrestrial Planets." *Science* 271(5254): 1387–1392.

Kasting, James F.; Daniel P. Whitmire; and Ray T. Reynolds. 1993. "Habitable Zones around Main Sequence Stars." *Icarus* 101(1): 108–128.

Melosh, H. J. 1988. "The Rocky Road to Panspermia." *Nature* 332: 687–688.

Morbidelli, A.; J. Chambers; J. I. Lunine; et al. 2000. "Source Regions and Time Scales for the Delivery of Water to Earth." *Meteoritics & Planetary Science* 35(6): 1309–1320.

Sagan, Carl. 1994. *Pale Blue Dot: A Vision of the Human Future in Space.* New York: Random House.

Spencer, John, and David Grinspoon. 2007. "Planetary Science: Inside Enceladus." *Nature* 445: 376–377.

Taylor, Chris. 2006. "Hurtling into the Space Tourism Industry." CNNMoney.com. Available from http://money.cnn.com/2006/03/30/technology/business2_futureboy0330/index.htm.

Mark A. Bullock

OZONE DEPLETION

Ozone depletion, the thinning of ozone in the upper atmosphere, has been observed since the late twentieth century and is due to anthropogenic emissions of ozone-depleting chemicals. Stratospheric ozone shields Earth's surface from damaging solar radiation, which can harm humans, other animals, plants, and ecosystems. Steps to control ozone-depleting chemicals began in the late 1970s in the United States and were followed by the institution of the international Montreal Protocol in 1987. Ozone depletion raises ethical issues associated with indirect and diffuse harm, collective action, and obligations under conditions of risk and uncertainty.

THE SCIENCE OF OZONE DEPLETION

Depletion of stratospheric ozone is caused primarily by anthropogenic emissions of halogen source gases—gases containing chlorine and bromine that break down in the upper atmosphere. Halogen source gases include chlorofluorocarbons, historically used in refrigeration, aerosol propellants, and air-conditioning systems; bromine-containing chemicals called halons, used as fire retardants; and methyl bromide, used as an agricultural fumigant. Although some halogen source gases arise from natural sources, these are not the primary cause of ozone depletion.

Ozone (O_3) is a molecule that occurs in the troposphere (lower atmosphere) and stratosphere (upper atmosphere). In the troposphere, excess ozone is considered a pollutant. It damages plant and animal tissues and is a major component of smog. Ozone in the upper atmosphere, however, has a protective effect. It shields Earth's surface from ultraviolet (UV) radiation—particularly UV-B radiation, with

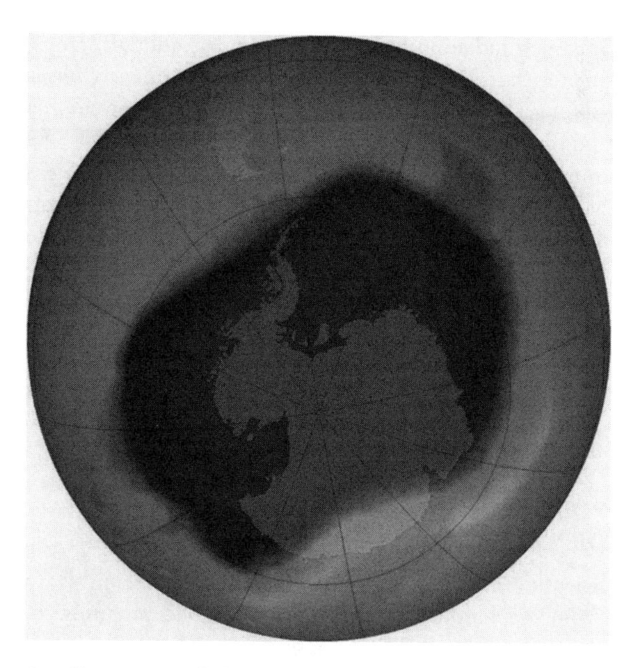

Satellite Image of the Ozone Hole. *This image provided by NASA in 2006 shows the atmosphere's ozone "hole," a region where there is severe depletion in the layer of ozone (a form of oxygen) that protects Earth from the Sun's ultraviolet rays. At the time this image was recorded, on September 24, 2000, the Antarctic ozone hole was approximately 11.4 million square miles.* AP IMAGES.

wavelengths of 280 to 315 nanometers—which can cause health problems in humans and other animals as well as damage in wild plants and agricultural crops.

Stratospheric ozone breaks down through a set of chemical reactions involving reactive chlorine and bromine molecules. Although many ozone-depleting chemicals do not react directly with ozone, when these gases move from the troposphere to the stratosphere, they are converted into reactive halogen gases that catalyze the breakdown of ozone molecules.

The chemical breakdown of ozone by reactive halogen gases has had stronger effects on some parts of the globe than on others. Ozone depletion has been most significant at the poles, particularly over Antarctica, where seasonally severe ozone depletion, known as the ozone hole, has been observed since the 1980s. Winter weather conditions in the southern polar region favor rapid breakdown of ozone, and this leads to the formation of an ozone hole in austral late winter and early spring. Thus, although global ozone has been reduced on average by 4 percent, seasonal declines of up to 37 percent have been observed over Antarctica.

EFFECTS OF OZONE DEPLETION ON BIOLOGICAL SYSTEMS

Ozone depletion leads to increased ultraviolet radiation, and such radiation harms humans, other animals, crops, and ecosystems. Humans' increased exposure to ultraviolet radiation can cause skin cancer (both carcinoma and melanoma), cataracts, and suppression of the immune system. In addition, UV-B can damage the eggs of fish and amphibians; alter the DNA, lipids, and proteins of plants; modify plant-growth patterns; and reduce the yields of some crops. UV-B can also harm phytoplankton and has the potential to disrupt marine food webs. Though there are experimental studies showing effects of UV-B on plants and animals, the ecological consequences of increased ultraviolet radiation are not well understood, since they involve complex interactions.

POLICY RESPONSES TO OZONE DEPLETION

Scientific research linking anthropogenic chemicals to ozone depletion gained significant public attention in 1974 after the chemists Mario Molina and Sherwood Rowland published a paper in *Nature* outlining the mechanisms by which chlorofluorocarbons could destroy ozone. Later that year Molina and Rowland presented their work at a scientific meeting and publicized their findings to the press. Media coverage of their hypothesis led to significant debate and to reduced consumer use of chlorofluorocarbons. Then in 1978 the U.S. Environmental Protection Agency banned nonessential use of chlorofluorocarbons in aerosol sprays.

A comprehensive international response to ozone depletion came later. Shortly after a 1985 international meeting in Vienna to consider restrictions on ozone-depleting chemicals, new research revealed the Antarctic ozone hole, a severe thinning of stratospheric ozone over the South Pole. This renewed concern and led DuPont, a leading chlorofluorocarbon producer, to increase investment in the development of substitute chemicals. These events, in conjunction with pressure from U.S. environmental groups, laid the groundwork for the Montreal Protocol.

The Montreal Protocol on Substances that Deplete the Ozone Layer, an international treaty, was instituted in September 1987. The protocol, which went into effect in 1989, outlined a timeline for the reduction of certain chlorofluorocarbons and halons. The treaty required signatories from developed nations to cap chlorofluorocarbon emissions at 1986 levels by the middle of 1989, then to cut emissions to 50 percent of 1986 levels by 1998. Halons regulated under the treaty were to be reduced to 1986 levels by 1992. Less developed countries were allowed a ten-year grace period before restrictions went into force.

The initial cuts sought by the Montreal Protocol were modest and by themselves would not solve the problem of ozone depletion. However, the protocol included ongoing assessment provisions, which provided a process for collecting improved data on emissions of ozone-depleting chemicals, tracking ozone depletion, and synthesizing scientific research on its causes. These provisions allowed the protocol to be adapted as new data emerged showing ozone depletion to be worse than originally believed and identifying additional chemicals contributing to the problem.

In light of new data, the Montreal Protocol was strengthened at subsequent international meetings. A 1990 agreement entered into in London required complete phaseout of chlorofluorocarbons and halons by 2000; regulated two additional chemicals, carbon tetrachloride and methyl chloroform; and established a fund to assist developing nations in reducing ozone-depleting chemicals. Subsequent modifications to the treaty include the Copenhagen Amendment of 1992, which added controls on hydrochlorofluorocarbons and methyl bromide and accelerated timelines for reducing ozone-depleting chemicals; the Montreal Amendment of 1997, which sped reductions of methyl bromide; and the Beijing Amendment of 1999, which mandated earlier control of hydrochlorofluorocarbons.

The regulation of ozone-depleting chemicals kick-started industry investments in research on and development of substitutes, including less-damaging chlorofluorocarbons, hydrochlorofluorocarbons, and hydrofluorocarbons. These alternative chemicals were much less potent than early chlorofluorocarbons. In addition, it was found that many aerosols could be produced without ozone-depleting chemicals at all. Mechanical pumps and chemical propellants such as carbon dioxide could be used instead. Effective substitutes for the strongest ozone-depleting chemicals also have been found for refrigerants, foams, solvents, and many fire retardants.

PROGNOSIS FOR OZONE RECOVERY

Although some ozone-depleting chemicals remain in use, overall production has declined substantially, and the total concentration of ozone-depleting chemicals has decreased since the 1990s. Because atmospheric lifetimes of some ozone-depleting chemicals are long (up to 100 years), a substantial lag time can occur between cuts in ozone-depleting chemicals and effects on ozone recovery. Nevertheless, according to the 2006 *Scientific Assessment on Ozone Depletion* (World Meteorological Organization 2007), stratospheric ozone levels outside the polar regions have stabilized. Severe ozone depletion continues to be observed at the poles, but variability in ozone levels in the polar regions is attributable to variable weather conditions rather than to increased concentrations of ozone-depleting chemicals. The 2006 assessment predicts a return to pre-1980 ozone levels by midcentury for most latitudes, with recovery over Antarctica occurring between 2060 and 2075.

ETHICS OF OZONE DEPLETION

Stratospheric ozone depletion raises a number of ethical issues. As mentioned above, ozone depletion can harm humans, nonhuman animals, plants, and ecosystems. Yet unlike paradigmatic cases of ethically significant harm, the harm caused by ozone-depleting chemicals was initially unknown, unintentional, indirect, and diffuse. Producers of ozone-depleting chemicals were at first unaware that their products could harm the environment and/or people, and they had no intention to cause harm. The harm was indirect because it was not the chemicals themselves, but rather the effects of the chemicals on stratospheric ozone, that caused damage. Lastly, the harm was diffuse in that it was not the action of a single individual, but rather the cumulative effects of many corporations and individuals producing and using ozone-depleting chemicals, that caused the problem.

The philosopher Dale Jamieson (2007) argues that nonparadigmatic ethical problems are more challenging than paradigmatic ones. Paradigmatic cases involve direct and immediate harm, with an obvious perpetrator and recipient, as when a single individual knowingly strikes another. In such cases, assignment of moral responsibility is straightforward. Nonparadigmatic cases are more complex. For example, should people be held responsible for harm that they did not know they were causing? Do corporations and nations have the same moral responsibilities as individuals? What obligations exist when the nature and magnitude of the harm are unknown? When harm is caused by the joint actions of many parties, how should responsibility be allocated?

Two types of issues are of central concern in the ozone case: issues of collective action, and issues of ignorance, risk, and precaution. Collective-action issues arise when the realization of a particular goal—in this case, prevention and mitigation of damage to stratospheric ozone—requires cooperation of multiple parties. Collective-action problems can be difficult to resolve when incentives for individual actors do not support collectively beneficial outcomes. In the ozone case, for example, in the absence of regulation, individual chemical companies had little incentive to invest in alternatives to ozone-depleting chemicals, because ozone-depleting chemicals were profitable and the costs of developing and producing alternatives were unknown. This collective-action problem was largely overcome by regulation. Once the Montreal Protocol went into effect, all companies were required to cut production of ozone-depleting chemicals and to seek alternatives.

It is controversial whether individual actors have moral obligations to refrain from contributing to collective-action problems in the absence of regulation or other forms of collective agreement. Some argue that unilateral restraint under such circumstances will accomplish little. In the ozone case, however, the unilateral move by the United States in the late 1970s to ban ozone-depleting chemicals in aerosol sprays may have induced later collective action to resolve the ozone problem.

A second issue of concern in the ozone case involves the appropriate ethical response to uncertainty and risk. Ozone depletion was suspected well before discovery of the Antarctic ozone hole. The initial science on ozone depletion was based on chemical theory and was borne out by empirical data only later. When the problem was first being discussed, the nature and extent of the problem and associated risks were unknown.

There is no philosophical consensus on the appropriate ethical response to risk, though it is widely agreed that the magnitude of the risks and the reversibility or irreversibility of potential harms are important factors to be considered. Some moral philosophers and environmental advocates favor a precautionary approach, where action is taken to avert harm associated with an activity even when the nature and magnitude of the potential harm are not fully known. Others hold that risks should be considered in the context of cost-benefit analyses, in which potential risks are weighed against potential benefits. In the ozone case, this would involve asking whether the benefits of continuing to use ozone-depleting chemicals outweigh the risks of predicted ozone depletion. The cost-benefit approach is intuitively appealing and widely employed, but can be difficult to operationalize. Cost-benefit analysis requires quantification of risks and benefits to enable their comparison on a single scale (often monetary). Some object to cost-benefit analysis on the basis that not all goods are commensurable and that some goods should not be given a price. On the opposite side of the ledger, those who support cost-benefit analysis and oppose the precautionary principle emphasize that it is sometimes worth taking risks when significant benefits can be gleaned by doing so. In the ozone case, the debate over precaution was not fully resolved, yet by incorporating provisions for ongoing assessment, the Montreal Protocol stimulated research and innovation that clarified both the importance of restricting ozone-depleting chemicals and the feasibility of doing so.

The regulation of ozone-depleting chemicals through the Montreal Protocol and its successor agreements is often highlighted as an environmental success story from which we can learn. Indeed, international cooperation to address the ozone problem has been impressive, with each successive agreement accelerating reductions in ozone-depleting chemicals. Technological innovation has produced viable substitutes in almost all economic sectors, emissions of ozone-depleting chemicals have declined substantially, and full recovery to pre-1980 levels of stratospheric ozone is expected during this century.

Global climate change shares a number of features with ozone depletion: Effective control of greenhouse-gas emissions requires collective action, and while the scientific understanding has improved substantially, the precise magnitude and effects of climate change remain uncertain. Whether lessons from the ozone case can be applied to help address climate change will depend on similarities and differences in the structure of each problem from scientific, economic, political, and social perspectives. Despite parallels, there are important differences. While it may have been relatively easy for individual consumers to stop using aerosol sprays, for example, it may be more difficult for individuals significantly to reduce their use of fossil fuels. In addition, the kinds and sources of greenhouse gases are more diverse and diffuse than was the case with ozone-depleting chemicals, and this makes coordination to solve the problem more complex. Nevertheless, development of alternative technologies was much more successful than anticipated in the ozone case; perhaps the same will hold true for greenhouse gases and climate change.

SEE ALSO *Alternative Technology; Atmosphere; Cost-Benefit Analysis; Global Climate Change; Pollution; Precautionary Principle.*

BIBLIOGRAPHY

Björn, Lars Olaf, and Richard L. McKenzie. 2008. "Ozone Depletion and the Effects of Ultraviolet Radiation." In *Photobiology: The Science of Life and Light*, ed. Lars Olaf Björn. 2nd edition. New York: Springer.

Fahey, D. W., et al. 2007. "Twenty Questions and Answers about the Ozone Layer: 2006 Update." In *Scientific Assessment of Ozone Depletion: 2006*. Geneva: World Meteorological Organization. Available from http://www.esrl.noaa.gov/csd/assessments/2006/chapters/twentyquestions.pdf.

Gillespie, Alexander. 2006. *Climate Change, Ozone Depletion, and Air Pollution: Legal Commentaries with Policy and Science Considerations*. Boston: Martinus Nijhoff.

Jamieson, Dale. 2007. "The Moral and Political Challenges of Climate Change." In *Creating a Climate for Change: Communicating Climate Change and Facilitating Social Change*, ed. Susanne C. Moser and Lisa Dilling. New York: Cambridge University Press.

Kauffman, Joanne M. 1997. "Domestic and International Linkages in Global Environmental Politics: A Case-Study of the Montreal Protocol." In *The Internationalization of Environmental Protection*, ed. Miranda A. Schreurs and Elizabeth C. Economy. New York: Cambridge University Press.

Makhijani, Arjun, and Kevin R. Gurney. 1995. *Mending the Ozone Hole: Science, Technology, and Policy*. Cambridge, MA: MIT Press.

Molina, M. J., and F. S. Rowland. 1974. "Stratospheric Sink for Chlorofluoromethanes: Chlorine Atom-Catalysed Destruction of Ozone." *Nature* 249: 810–812.

Parson, Edward A. 2003. *Protecting the Ozone Layer: Science and Strategy*. New York: Oxford University Press.

Raffensberger, Carolyn, and Joel Tickner, eds. 1999. *Protecting Public Health and the Environment: Implementing the Precautionary Principle*. Washington, DC: Island Press.

Sunstein, Cass R. 2005. *Laws of Fear: Beyond the Precautionary Principle*. New York: Cambridge University Press.

UN Environment Programme. 2006. *Environmental Effects of Ozone Depletion and Its Interactions with Climate Change: 2006 Assessment*. Nairobi, Kenya: Author. Available from http://ozone.unep.org/pdfs/Final%20UNEP%202006%20Report.pdf.

World Meteorological Organization. 2007. *Scientific Assessment of Ozone Depletion: 2006*. Geneva: World Meteorological Organization.

Marion Hourdequin

P

PAGANISM

Phenomena labeled *Pagan* vary considerably but have, or allegedly have, in common a stress on physical being and belonging. In the mid-twentieth century, *Paganism* was chosen as the name of a religious movement that claimed to be reviving ancient nature veneration and polytheism. In North America, it is common to call this religion *Neo-Paganism* to distinguish it from pre-Christian religions (such as those of classical Rome). *Pagan* is also used in a derogatory sense to allege that some people, religions, activities, attachments, and passions are worldly, bodily, and materialistic, and fail to value the transcendence privileged by Christianity and some other monotheistic religions.

ORIGINS AND APPLICATIONS

Pagan derives from a Latin term for an administrative district. A Roman *pagus* was something like a parish, a township, or a neighborhood perhaps. In the early twentieth century, however, it was thought that *pagus* originally referred to rural rather than urban places. The notion that people in the countryside live in close harmony with the seasons, and with their environment more generally, was linked to the theory that Christianity was initially more successful in the cities than in the conservative backwaters of rural agricultural communities. This understanding had obvious appeal to people who claimed to be reviving pre-Christian religions (initially drawing on classical Egyptian, Greek, and Roman sources), contested the denigration of nature, and were dissatisfied with the alienation from nature that they experienced in urban and suburban modernity. The designation of Paganism as a "nature religion," in contrast with religions that have either transcendent deities or self-realization as their primary focus, shows recognition of the central place that nature has in Paganism.

Paganism has also been used to label adherents of nonmonotheistic religions and cultures. In particular, indigenous people have been called *Pagan* when they continue to practice ancestral, local religions. Some indigenous people, rhetorically admitting to being *Pagan*, challenge others' rights to define them negatively and contest their denigration of place, land, and belonging. All uses of the term *Pagan*, whether as a chosen self-appellation or as a derogatory epithet, carry associations with physicality, materiality, belonging, and also require careful consideration of the categories "nature" and "natural."

SELF-IDENTIFIED PAGANS

As documented by Ronald Hutton in *Triumph of the Moon* (1999), Paganism was created as a new religion in the early to middle twentieth century by people who blended a number of elements together. In addition to stressing the positive value of natural (nonurban) places, Pagans popularized forms of European esotericism, such as the practice of magic, claimed to be reviving various forms of pre-Christian (Egyptian, Greek, Roman, Celtic, and Germanic) polytheism, and integrated these with folklore. Pagans developed attractive rituals and new calendar customs for natural living and earth-based spirituality. The continuing international spread and diversification of Paganism, discussed in Graham Harvey's *Listening People, Speaking Earth: Contemporary Paganism* (2006), is evidence of the popularity of this effort and its evolution.

Paganism has no single central authority or hierarchy, and no sacred scriptures. Pagans are inspired in their quest to understand and celebrate the natural world by a wide range of literature, including botanical treatises, ancient epics, archaeological reports, and fantasy fiction (see Clifton and Harvey 2004). Some Pagans belong to subdivisions or denominations, which they prefer to call "traditions" or "paths," such as Druidry, Heathenry, Witchcraft (or "the Craft"), and Goddess Spirituality. There are also many ethnic Paganisms (e.g., Hellenic, Lithuanian, Norse), drawing on regional and ancestral traditions, archaeology, history, and folklore. In some respects these efforts parallel indigenous revitalization movements that emphasize locally meaningful knowledge and lifestyles. The many variations in ways of being Pagan are usually celebrated by Pagans, some insisting that "diversity is natural, many different types of trees make up a forest."

Common ground among Pagan movements is found in seasonal and calendar festivals. Most Pagans mark the solstices and equinoxes with rituals, and many add other festivals to their annual cycles. These can be honored as key points in the changing seasons (the beginning of summer or the end of winter) or as significant moments in the relationship between Earth and Sun (the longest day or the longest night). Communal events can draw humans into seasonal relations and encourage them to see their personal life cycles as paralleled by changes in larger, cosmic events. Human birth, growth, aging, death, and possibly rebirth can be celebrated as matching the growing and diminishing strength of the Sun in the annual cycle of the seasons. Creativity, intentionality, and productivity are not only human attributes but are also recognizable in the natural world. Springtime festivals may celebrate the growth of plants, the birth of young animals, the increase of light and warmth, and an intention to live creatively in the following seasons.

POLYTHEISM AND MAGIC

In addition to the relationship between the Sun and Earth, full and new moons are of considerable importance to many Pagans. They are valued as particularly good times to venerate divine beings and perform magic. The moon is important to Pagan polytheism and magic because Pagans inherited the idea that the Moon is a favored manifestation of a divine being, a goddess, willing and eager to help people perform magic for the benefit of life.

Paganism is generally polytheistic. While this last term draws attention to a contrast with the monotheistic assertion that there is only one deity, the important issue for polytheists is not how many deities exist, but what results from engaging with a particular *kind* of deity. Polytheism generally encourages a celebration of the material, physical world and embodied life. Rather than an interest in a possible afterlife, it promotes locally meaningful attempts to live as good members of communities conceived of as including not only humans but all other living beings, including deities. Common elements in many Pagan rituals are songs and speeches addressed to the Goddess, where "the Goddess" can be an individual divine being or an entire community of deities. The "Goddess" can also serve as an all-embracing category equivalent to *life* or *nature*; and in some cases it refers to the powerful sense of unity experienced in ritual gatherings. Lack of dogma allows Pagans to celebrate together without requiring conformity to particular beliefs or manifestos. The essence of Paganism is the celebration and enhancement of life for oneself and others.

Pagans explain magic as the art of causing change according to will or the art of changing consciousness according to will. In these formulations, *will* refers to the best sense one has of oneself. Thus, magic requires effort to live up to one's highest standards and to seek to improve oneself. Practitioners are encouraged to think that in rituals they are empowered to change their consciousness about the way things are, improving their understanding and appreciation of the world (realizing that what is bad for an individual may be good for a community, or vice versa). Practitioners are also encouraged to seek to cause change to improve life. The precise means by which magic works, the natural or metaphysical energies that might be drawn on, and the tools for manipulating either consciousness or the world vary according to particular traditions and teachers. In all cases, magic promotes a participative and positive engagement with the world, rooted in the notion that all acts (however small they may seem) have effects that may be of considerable scope.

ANIMISM AND SHAMANISM

If polytheism has been labeled *Pagan* (whether in the positive or negative sense), so too have animism and shamanism. *Animism* has borne two meanings. Initially, it summarized a theory that religion is, by definition, founded in a "belief in spirits." Here, *spirits* meant "non-empirical entities" and the theory asserted the falsity of religions. However, when Pagans and indigenous people talk about spirits, they rarely focus on questions of belief or matters of definition. They might talk about spirits of ancestors, the elements (earth, air, fire, and water), or place, adding the term *spirits* to indicate that these are somehow living, communicative beings who participate in ceremonies. In a second, more useful use, *animism* refers to the view that the world is a community of persons, most of whom are not human, but all of whom deserve respect. *Persons* here points to the importance of

Pagan Celebration in Rakov, Belarus. *Neo-Pagans dance around a fire to celebrate Ivan Kupala, with festivities similar to those of Mardi Gras. Paganism is actually a newer religion, blending different forms of pre-Christian polytheism. Although Paganism is strongly associated with nature, it does not necessarily follow that all pagans are environmental activists.* AP IMAGES.

personal relationships with other beings. Like polytheistic Paganism, this variety of animism roots people in the everyday, physical world and even more radically rejects the transcendentalizing project of monotheism.

Along with the understanding of animism, the understanding of shamanism has also changed. *Shamanism* was initially attractive as a label for ecstatic techniques that seemed to result in powerful visions of alternative realities. But shamanism is beginning to be seen as the element of animism in which experts mediate between human and other-than-human communities. The individual experience of transcendence has been replaced by a desire to learn skills for communicating with neighboring beings. Animism and shamanism have become increasingly important as Paganism has moved from its Romantic roots and negated the dualism that separates culture from nature. Pagans may insist either that humans are part of nature, or that the whole world is cultural, in both cases pointing to the inadequacy of language influenced by dualism. Pagans have also insisted that a dualism of good and evil cannot be mapped onto the more messy relationships that make shamanism (as mediation) necessary. All of this may be seen as part of the process of indigenizing Paganism.

INDIGENIZING NATURE

Pagan understandings of nature have evolved quickly since the mid-twentieth century. The Romantic notion that authenticity and meaning are best found in wild places has been replaced by a more indigenous stress on the importance of belonging and participation. Wilderness may remain important, but now as a domain of richer diversity and communicative possibility, rather than in opposition to domains of human culture. Celebration of biodiversity may even lead to *more* (sub-)urban ritual as gardens are recognized as richer habitats than most agricultural land. Along with the shift in academic understandings of the origins of the term *Pagan* (from country dweller to local citizen), Pagans have increasingly insisted that all places are sacred, or at least part of nature. Just as Roman shrines at key natural places enshrined or animated the concept of a limit to human utilization and commodification of the world, at least nodding toward the rights of other-than-human life, as Ken Dowden (2000) suggests, so contemporary Pagans have contested globalization on the grounds of a more participatory cosmology. In opposing anthropocentricism, Pagans seek to realign human culture away from consumerism toward respectful, local engagement. Experience of belonging in place and of valuing the rumors of ancient practices aid in this indigenizing process.

PAGAN ECOLOGY AND ENVIRONMENTALISM

Paganism is an umbrella term for a religion focused on nature following a variety of practices. Pagan festivals celebrate natural seasons and alignments between Earth, Sun, and Moon. They express the positive value of human participation in natural cycles, but may equally seek to inculcate the notion that plants, animals, and planets act intentionally and relationally, that is, culturally. Leitmotifs of Pagan discourse point to the centrality of materiality, physicality, performance, of secular belonging, of place and emplacement. This is not to assert, however, that as celebrants of a religion of nature, Pagans must be or inevitably are environmental activists. Nature is certainly central to Paganism, but Pagans may prefer to work out their Earth ethic in the quietist domains of their homes (eating organically and recycling) rather than to engage in confrontational activism. They may also insist that Paganism consists more in being at home in the world (one sense of *ecology*), as achieved during seasonal festivals that reconnect people to abiding patterns of cosmic life, than in engaging in environmental activism focused on human activities.

SEE ALSO *Biodiversity; Christianity; Environmental Activism; Pantheism; Romanticism; Urban Environments.*

BIBLIOGRAPHY

Clifton, Chas S., and Graham Harvey, eds. 2004. *The Paganism Reader*. London: Routledge.

Dowden, Ken. 2000. *European Paganism: The Realities of Cult from Antiquity to the Middle Ages*. London: Routledge.

Harvey, Graham. 2006. *Listening People, Speaking Earth: Contemporary Paganism*. London: C. Hurst.

Hutton, Ronald. 1999. *The Triumph of the Moon: A History of Modern Pagan Witchcraft*. Oxford, UK: Oxford University Press.

Graham Harvey

PANTHEISM

Western religious and philosophical traditions do not obviously ground an ecological and environmental ethic. Even those who favor such a program for grounding an environmental ethic might agree that any attempt to ground an environmental ethic in such traditions requires interpretive dexterity and a good deal of charity. Those who think that the route from such traditions to sound environmental ethics and attitudes is too tortuous, if it exists at all, may well look to other traditions for a philosophical or spiritual basis.

Pantheism, with roots in both Eastern and Western traditions, is an obvious source for grounding environmental ethics, because, from a pantheistic point of view, nature itself is divine, and what is in any sense divine is not something to be trifled with or exploited. As John Toland, author of the *Pantheisticon*, wrote, "In a Word, every Thing in the Earth is organic.... This justifies my Answer to a German Inn-Keeper, who impertinently importuned me to tell him, what Countryman I was? The Sun is my Father, the Earth my Mother, the World's my country, and all Men are my relations" (1976 [1751], pp. 32–33). Similarly, Bertrand Bronson sees the pantheistic ethos of John "Walking" Stewart as grounding not just an environmental ethic but all morality—or perhaps an environmental ethic is an ontological grounding of all other morality. "The universe is altogether composed of eternal and indestructible matter. All matter is one infinite whole... , Nature.... It follows from this community of matter that the interests of the whole material universe are intimately the interests of every individual man. This is the basic truth of morality" (1943, pp. 146–147).

WHAT IS PANTHEISM?

Pantheism is a metaphysical and religious position. It is broadly defined as the view that "God is everything and everything is God.... The world is either identical with God or in some way a self-expression of his nature" (Owen 1971, p. 74). Similarly, it is the view that everything that exists constitutes a unity, and that this all-inclusive unity is in some sense divine (MacIntyre 1967, p. 34). Baruch Spinoza's *Ethics* (1675) is regarded as the most thoroughgoing account of a pantheistic position. Aside from Spinoza, other possible philosopher pantheists include some of the pre-Socratics, Plato, Laozi, Plotinus, Friedrich Schelling, Georg Wilhelm Friedrich Hegel, Giordano Bruno, John Scotus Erigena, and Paul Tillich. Possible pantheists among literary figures include Ralph Waldo Emerson, Walt Whitman, D. H. Lawrence, and Robinson Jeffers. For a notion of just what pantheism is, the Force in *Star Wars*, while very different from Spinoza's "singular substance," is a good popularization. Pantheism is the traditional religious alternative to theism, and many profess pantheistic beliefs, though obscurely. The central claims of pantheism are prima facie no more fantastic than those of theism, and possibly less so.

UNITY AND DIVINITY

Different versions of pantheism offer different accounts of unity and divinity. The central problem of pantheism is to determine how to understand these central terms. For example, philosophical Daoism is one of the best articulated and thoroughly pantheistic positions there is. The Way (*dao*) is the central unifying feature, but understanding just what is meant by the Way and how it operates and what its implications are is a hermeneutic and philosophical task in its own right. What kind of unity is, or should be, claimed by pantheists, and which, if any, is plausible? There may be acceptable alternative criteria, and each will have implications for a pantheistic understanding of ethics, including environmental ethics. Like theism, pantheism is by no means a univocal view.

Attributing unity to the world simply on the basis of all-inclusiveness is irrelevant to pantheism. To understand the world as everything is to attribute a sense of unity to the world, but there is no reason to suppose that this sense of all-inclusiveness is the pantheistically relevant unity. Similarly, mere numerical, class, or categorical unity is irrelevant, since just about anything (and everything) can be one or a unity in these senses. Unity may also be explained in terms of divinity. The all-inclusive whole may be a unity because it is divine, either in itself (Spinoza's substance) or because of a divine power informing it. The latter is the case in the thought of some pre-Socratics, for whom the unifying principle is divine because it is immortal and indestructible. Neither of these positions satisfactorily explains the relation between unity and divinity, or why divinity might be seen as a basis of unity. Less naturally, the question arises as to whether the all-inclusive whole is divine because it is a unity.

PANTHEISM, THEISM, ATHEISM

When pantheism is considered as an alternative to theism, it involves a denial of at least one, and usually both, central theistic claims. Theism is the belief in a personal God, who in some sense transcends the world. Pantheists usually deny the existence of a personal God—a minded being having the properties of a person, such as having intentional states. As an alternative to, and denial of, theism and atheism, pantheism denies that God is a completely transcendent being distinct from the natural world. The dichotomy between immanence and transcendence (that is, between a god that is part of this world and a god that is outside of it) has been a principal source of philosophical and religious concern in Western and non-Western traditions, and all major traditions have at times turned to pantheism as a way of resolving difficulties with the theistic notion of a transcendent deity.

A. H. Armstrong says that the term *pantheistic* is a "large, vague term of theological abuse" (1976, p. 187), primarily, it seems, because it is deemed as an expression of atheism. With some exceptions, pantheism is nontheistic, but it is not atheistic. It is a form of nontheistic monotheism, or even nonpersonal theism. The primary reason for equating pantheism with atheism is the assumption that belief in God must be belief in a personal God. In his nonpantheistic phase, Samuel Taylor Coleridge claimed, "Every thing God, and no God, are identical positions" (McFarland 1969, p. 228). H. P. Owen wrote, "If God (theos) is identical with the Universe (to pan) it is merely another name for the Universe. It is therefore bereft of any distinctive meaning; so that pantheism is equivalent to atheism" (1971, pp. 69–70). Similarly, Arthur Schopenhauer (1951, p. 40) said that "to call the world God is not to explain it; it is only to enrich our language with a superfluous synonym for the word 'world.'" If what Schopenhauer, Coleridge, and Owen want to show is that believing in a pantheistic God is a confused way of believing in something that can adequately be described apart from any notion of deity, they are mistaken.

PANTHEISM, ETHICS, ECOLOGY

Pantheists, like theists, tend to be moral realists. They believe that some kinds of actions are objectively right and others wrong, and that what is right or wrong is independent of what any person thinks is right or wrong. With the exception of religious ethics, moral realism has not been a widely accepted philosophical position since about the 1970s.

Not accidentally, pantheism is often taken to be a view inherently sympathetic to ecological concerns. There is a tendency to picture pantheists outdoors and in pastoral settings. This tendency has roots in the Stoics' veneration of nature, and in the much later nature mysticism and perhaps the pantheism of some nineteenth-century poets, such as William Wordsworth and Whitman. This notion was fostered in the twentieth century by possible pantheists such as John Muir, Jeffers, Lawrence, and Gary Snyder, who explicitly identify with and extol nature, and claim that people's close association and identification with nature and the natural is necessary to well-being. Belief in a divine unity and some kind of identification with that unity are seen as the basis for an ethical framework and way of life that extends beyond the human to nonhuman and nonliving things. The divine unity is, after all, all-inclusive.

Consider some examples of alleged connections between pantheism, ethics, and ecology. Armstrong wrote,

> Plotinus may give us a lead to a better understanding of the world and may help us to adjust our attitudes and evaluations in a way which may help us to deal with some of the most pressing problems of our time, and especially to do something towards closing the gap between man and non-human nature which has been widening through the Christian and rationalist centuries with, as we are now beginning to see, disastrous results. (1976, p. 188)

Armstrong denies that Plotinus was a pantheist. But he does claim that there is in Plotinus a metaphysical basis for an environmental ethic in which there are significant pantheistic elements. He suggests how aspects of Plotinus's thought can change our attitudes and evaluations concerning nonhuman nature. Grace Jantzen (1984, pp. 156–157) makes a claim similar to that of Armstrong's in her proposal of a model of the world as God's body. Armstrong's view concerning Plotinus and Jantzen's view concerning the implications of her model for ethics and ecology are often taken to be true of pantheism in general. For religiously inclined nontheists, pantheism is supposed to have the resources capable of (in Armstrong's words about Plotinus) "closing the gap between man and nonhuman nature which has been widening."

Whatever critics allege the shortfalls of pantheism to be, there is a prominent, if not prevalent, view that its implications would be a good thing for ecology and for aspects of ethics dealing with the nonhuman, as well as the human. The work of Spinoza is most prominently cited in this context (Naess 1980; Mathews 1990). Genevieve Lloyd (1980), for example, attempts to extract from Spinoza a ground for environmental ethics, arguing that this can be done without assigning rights to nonhuman natural entities and nature as a whole on the basis of his system.

Although Spinoza is the best-known pantheist, looking in his metaphysics for a foundation for environmental ethics is, as Lloyd points out, not without its difficulties. After all, Spinoza rejected animal rights, and despite his view that man is part of nature, environmental ethics is in "apparent tension" with "his treatment of morality as circumscribed by what is good for human beings ... [and his view] that other species can be ruthlessly exploited for human ends" (Lloyd 1980, p. 294).

Whether or not Spinoza provides a suitable metaphysical basis for an environmental ethic depends, in part, on whether his metaphysics and ethics are plausible. For that reason alone, one might be suspicious of grounding an environmental ethic on Spinoza's philosophy. But Spinoza's particular system aside, it is often supposed that pantheism can offer a more suitable basis for an environmental ethic, and perhaps for ethics generally, than the Judeo-Christian tradition and some nonreligious alternatives, such as utilitarianism, contractarianism, and Kantian views. It is unlikely, however, that the committed theist or utilitarian would, or can, agree that pantheism, if true, offers a better basis for an environmental ethic than their own ethical theories—for metaethical reasons. The meaning of key ethical terms and the conditions governing their use in normative ethical theories are described in terms of normative principles characteristic of a particular system. Utilitarians, for example, would not allow that a pantheist's ethical reasoning provides a sound basis for moral deliberation unless utility, defined in terms of happiness or some other greatest good, were the pantheist's supreme normative principle, which it is not.

Harold W. Wood Jr., a founder of the Universal Pantheist Society, claims that pantheism provides the foundation for an environmental ethic not offered by the Judeo-Christian tradition. He wrote, "Instead of a 'conquer the Earth' mentality, pantheism teaches that respect and reverence for the Earth demands continuing attempts to understand ecosystems. Therefore, among religious viewpoints, pantheism is uniquely qualified to support a foundation for environmental ethics" (1985, p. 157). He talks about interacting with "God-as-nature." With the important exception of Spinoza, pantheists generally do not equate God with nature, but rather believe that nature is a divine unity.

The idea of unifying principles is also present in nature mysticism, as manifest in the work of Wordsworth and some other Romantics. Nature mysticism (as found in the poetry of Wordsworth, Jeffers, and Snyder) has in common with more philosophically robust versions of pantheism the idea that unity is rooted in nature, and this common feature explains why nature mysticism and philosophical pantheism are often conflated with one another. They are distinguishable, however, for nature mysticism need not attribute divinity to nature.

As noted, a pantheistic environmental ethic cannot self-consistently be anthropocentric. Arne Naess (1980) pointed out that Spinoza's anthropocentrism was inconsistent with his pantheism. The pantheist's ethics—environmental ethic and ethics more generally—is metaphysically grounded in the divine unity, the unifying principle that accounts for our commonality with other living and nonliving things and the grounds for extending our notion of moral community to those other living and nonliving things. Everything that is part of the divine unity (as everything is) can be thought of as also part of

Caspar David Friedrich,* Wanderer above a Sea of Fog, *Ca. 1817. *The pantheistic tradition views nature as divine, and asserts that anything divine should not be trifled with or exploited. Moreover, pantheists believe in an all-inclusive God who is "everything and everywhere;" this allows the religion to be an obvious source of environmental ethic. Friedrich's painting of the Wanderer depicts the artist's interest in nature, and alludes to the idea that morality involves oneness with nature.* **BILDARCHIV PREUSSISCHER KULTURBESITZ / ART RESOURCE, NY.**

the moral community. Aldo Leopold wrote, "The land ethic simply enlarges the boundaries of the community to include soils, waters, plants, and animals, or collectively, the land.... A thing is right when it tends to preserve the integrity, stability, and beauty of the biotic community. It is wrong when it tends otherwise" (1949, pp. 219, 240). Looking to pantheism as a metaphysical justification of, for example, Leopold's "land ethic" is not unreasonable—or at least no more unreasonable than pantheism itself is.

Philosophically minded ecologists and others who argue, on the basis of pantheistic metaphysics, that our notion of moral community must be enlarged to include nonhuman and even nonliving entities, also claim that practical consequences are involved. The issue is not merely one of providing a rational basis for an environmental ethic. The metaphysically minded ecologist or pantheist claims that the desired results can be obtained only by changing our concept of what constitutes the moral community.

The pantheist, like the theist or atheist, takes reality as determining ethical requirements. Since unity is predicated for some evaluative consideration, value is a focal point and principal concern for pantheism. The pantheist's ethical discourse is related in complex ways to the nature of the divine unity. The pantheist tries to discern and live in accordance with the unity and the kind of values intrinsically associated with it. This is clearly seen in Daoism. To act correctly, one acts in accordance with the Way (*dao*), which is the unifying principle. The *Daodejing*, like most other primary scriptural sources, is at one and the same time an ethical treatise on how to live and a metaphysical treatise about the nature of reality. The Indian doctrine of karma can also be interpreted pantheistically: Action that accords or defies the all-pervasive principle of retribution is by its very nature associated with value and promotes the good.

ETHICS, THE ENVIRONMENT, AND IMMORTALITY

Living an ethical life for the pantheist means, in part, living a life in accordance with ultimate reality. In attempting to conform to ultimate reality, the pantheist is no different from the theist, Daoist, Confucian, Buddhist, or atheistic existentialist. Pantheists strive to live in accord with reality as the divine unity of which they are a part. The nature of the unity is the metaphysical basis for a regulative ideal of how one should live. One achieves well-being and happiness only to the extent that one pursues and achieves the ideal. Living in accordance with the unity is to live in accord with one's nature, with the nature of other things, and with conditions in the world generally.

The notion of an ethical and valuable life, of pursuing an ideal and realizing the good life, is linked to the idea that living involves a *telos* or goal. Thus, pantheistic ethics can be seen as partly teleological, even Aristotelian. Pantheistic ethics may also be related to the theistic notion of salvation. Jeffers, for example, writes,

> I believe that the universe is one being, all its parts are different expressions of the same energy, and they are all in communication with each other, influencing each other, therefore parts of one organic whole.... The parts change and pass, or die, people and races and rocks and stars; none of them seems to me important in itself, but only the whole. This whole is in all its parts so beautiful, and is felt by me to be so intensely in earnest, that I am compelled to love it, and to think of it as divine.... This whole alone is worthy of the deeper sort of love; and... there is peace, freedom, I might say a kind of salvation, in turning one's affections outward toward this

> one God, rather than inwards on one's self, or on humanity. (Quoted in Sessions 1977, p. 512)

Consider further this celebrated passage from Emerson's *Nature*:

> In the woods is perpetual youth. Within these plantations a decorum and sanctity reign, a perennial festival is dressed, and the guest sees not how he should tire of them in a thousand years. In the woods we return to reason and faith. There I feel that nothing can befall me in life—no disgrace, no calamity (leaving me my eyes) which Nature cannot repair. Standing on the bare ground—my head bathed by the blithe air, and uplifted into infinite space—all mean egoism vanishes. I become a transparent eye-ball; I am nothing. I see all. The currents of the Universal Being circulate through me; I am part or particle of God. (1971 [1836], p. 10)

Pantheism, like Aristotelianism and theism, has its own notion of the *telos* of life and its own concepts of the good life.

SEE ALSO *Daoism; Emerson, Ralph Waldo; Jeffers, Robinson; Land Ethic; Leopold, Aldo; Muir, John; Naess, Arne; Paganism; Process Philosophy; Snyder, Gary; Spinoza, Baruch; Teleology.*

BIBLIOGRAPHY

Armstrong, A. H. 1976. "The Apprehension of Divinity in the Self and Cosmos in Plotinus." In *The Significance of Neoplatonism*, ed. R. Baine Harris, 187–198. Norfolk, VA: International Society for Neoplatonic Studies, Old Dominion University.

Attfield, Robin. 1983. *The Ethics of Environmental Concern.* Oxford, UK: Blackwell.

Bronson, Bertrand. 1943. "Walking Stewart." In *University of California Publications in English*, no. 14. Berkeley: University of California Press.

Emerson, Ralph Waldo. 1971 [1836]. "Nature." In *The Collected Works of Ralph Waldo Emerson*, Vol. 1: *Nature, Addresses, and Lectures.* 6 vols. Cambridge, MA: Belknap Press of Harvard University Press.

Grula, John W. 2008. "Pantheism Restructured: Ecotheology as a Successor to the Judeo-Christian, Enlightenment, and Postmodern Paradigms." *Zygon* 43(1): 159–180.

Jantzen, Grace. 1984. *God's World, God's Body.* Philadelphia: Westminster.

Leopold, Aldo. 1949. "The Land Ethic." In his *Sand County Almanac.* New York: Oxford University Press.

Levine, Michael P. 1994. *Pantheism: A Non-theistic Concept of Deity.* London: Routledge.

Lipner, J. J. 1984. "The World as God's 'Body': In Pursuit of Dialogue with Ramanuja." *Religious Studies* 20: 145–161.

Lloyd, Genevieve. 1980. "Spinoza's Environmental Ethics." *Inquiry* 23: 293–311.

MacIntyre, Alasdair. 1967. "Pantheism." In *Encyclopedia of Philosophy*, Vol. 6, ed. Paul Edwards. New York: Macmillan and Free Press.

Mathews, Freya. 1990. *The Ecological Self.* London: Routledge.

McFarland, Thomas. 1969. *Coleridge and the Pantheist Tradition.* Oxford, UK: Oxford University Press.

Naess, Arne. 1977. "Spinoza and Ecology." In *Speculum Spinozanum, 1677–1977*, ed. Siegfried Hessing. London: Routledge and Kegan Paul.

Naess, Arne. 1980. "Environmental Ethics and Spinoza's Ethics." *Inquiry* 23: 331–325.

Owen, H. P. 1971. *Concepts of Deity.* London: Macmillan.

Passmore, John. 1974. *Man's Responsibility for Nature: Ecological Problems and Western Traditions.* New York: Scribners.

Sayre-McCord, Geoffrey. 1988. "Introduction: The Many Faces of Moral Realism." In *Moral Realism*, ed. Geoffrey Sayre-McCord. Ithaca, NY: Cornell University Press.

Schopenhauer, Arthur. 1951. "A Few Words on Pantheism." In *Essays from the Parerga and Paralipomena*, trans. T. Bailey Saunders. London: George Allen and Unwin.

Sessions, George. 1977. "Spinoza and Jeffers on Man in Nature." *Inquiry* 20: 481–528.

Swinburne, Richard. 1977. *The Coherence of Theism.* Oxford, UK: Oxford University Press.

Taylor, Paul. 1975. *Principles of Ethics.* Encino, CA: Dickenson Publishers.

Toland, John. 1976 (1751). *Pantheisticon.* New York: Garland.

Wood, Harold W., Jr. 1985. "Modern Pantheism as an Approach to Environmental Ethics." *Environmental Ethics* 7: 151–163.

Michael P. Levine

PASSMORE, JOHN ARTHUR
1914–2000

John Passmore, an Australian philosopher, was the author of the first philosophical monograph on environmental ethics, *Man's Responsibility for Nature* (1980; first edition, 1974). Although Passmore's anthropocentric approach was widely rejected in favor of nonanthropocentric alternatives in the decades immediately following the appearance of his book, Passmore was enormously influential in persuading philosophers that environmental issues merited serious attention.

Unlike his professional colleagues in the 1960s and 1970s, who ignored environmentalists' charges that the anthropocentrism of Western traditions of metaphysics, morality, and science made these traditions incapable of criticizing or condemning large-scale environmental destruction, Passmore took these charges seriously. In *Man's Responsibility for Nature*, he evaluates arguments for these charges and for the solutions critics were proposing: a radical revision of Western metaphysical, moral, and scientific theories. Passmore argued that while the dominant metaphysical traditions of Western thought are guilty as charged, Western moral and scientific traditions are not. Divested of the metaphysical baggage that has distorted them, these latter two branches of the Western tradition, despite their anthropocentrism, provide sufficient resources to condemn our environmentally destructive practices. He presented his argument in three parts.

In part 1 of *Man's Responsibility for Nature* Passmore considered and largely endorsed the charge that the Christianizing of Western metaphysics encouraged Westerners to see human beings as possessing a value and destiny fundamentally divorced from that of the natural world and to view the latter's value and purpose as limited to its utility in helping us to fulfill divine injunctions to multiply on and subdue Earth. This conception of our status in relation to nature, Passmore agreed, licenses our adoption of environmentally destructive practices. But although this view of our relation to nature has long been dominant, he argued, it has always had rivals. Even within Christianity, we find minority traditions holding that our role is to act as nature's steward, preserving its fruitfulness for the future, and/or to cooperate with nature in the ongoing creation of the world.

In part 2 he considered and rejected the charges that Western moral and scientific traditions lack the means to condemn environmentally destructive practices and thus should be replaced by nonanthropocentric alternatives. Passmore's strategy was to adopt the anthropocentric outlook typical of Western moral and liberal political theories and then apply it to current controversies about human exploitation of nature to show that it neither blinds us to the problematic character of human exploitation of nature nor denies us grounds for criticizing destructive exploitation as harmful, unjust, selfish, and/or wrong. From an anthropocentric perspective, he argued, any environmental practice is problematic whose consequences are undesirable in themselves without also being unavoidable byproducts of those human practices genuinely essential to desirable social life. Using this anthropocentric definition, Passmore argued that he is no less able than nonanthropocentrists to recognize that industrial pollution, depletion of nonrenewable resources, extermination of wild species and wilderness areas, and unrestricted human reproduction are all problematic. In each case, Passmore argued, the responsible practices have consequences that are undesirable while being largely nonessential for desirable social life. Hence, our standing moral and social traditions can and rationally must condemn the greed, insensitivity, and shortsightedness that motivates these practices, as well as the callousness and injustice inherent in imposing their avoidable consequences upon their unwilling victims, present and future.

Unless the dominant metaphysical conception of humanity's relation to nature can be displaced by conceptions more in keeping with the minority traditions of stewardship and/or cooperative partnership with nature, the forces of greed, shortsightedness, and insensitivity cannot readily be overcome by appeals to morality and justice alone, Passmore felt. He argued that Western scientific reasoning is playing an important role in helping to bring about this change in perspective. Science has challenged the old belief that Earth's resources exceed our ability to consume them, forcing upon us the realization that they may not sustain future generations if loving stewardship is not practiced. Furthermore, as scientific understanding of ecological complexity increases, it becomes increasingly evident that bending nature to our will is an impossibility. Ongoing creativity in art, science, morality, and culture will become impossible unless we learn to partner with nature to develop more sustainable practices. Passmore concluded that enlightened anthropocentrism, coupled with conceptions of humanity as stewards of nature and cooperative partners with nature, possesses the necessary resources to condemn our current destructive practices and demand their reform—if only we can find the will to act accordingly.

SEE ALSO *Christianity; Stewardship.*

BIBLIOGRAPHY

Hargrove, Eugene. 1979. "The Historical Foundations of American Environmental Attitudes." *Environmental Ethics* 1: 209–240.

Passmore, John. 1977. "Ecological Problems and Persuasion." In *Equality and Freedom: International and Comparative Jurisprudence*, ed. Gray Dorsey, Vol. 2, 431–442. Dobbs Ferry, NY: Oceana Publications.

Passmore, John. 1980 (1974). *Man's Responsibility for Nature: Ecological Problems and Western Traditions*, 2nd edition. London: Duckworth. 1st edition, 1974.

Passmore, John. 1999. "Philosophy and Ecology." In *Proceedings of the Twentieth World Congress of Philosophy*, Vol. 1: *Ethics*, ed. Klaus Brinkmann, 141–151. Bowling Green, OH: Philosophy Documentation Center.

Routley, Richard [later Richard Sylvan]. 1973. "Is There a Need for a New, an Environmental, Ethic?" In *Proceedings of the Fifteenth World Congress of Philosophy*, vol. 1, 205–210. Sofia, Bulgaria: Sophia Press.

White, Lynn, Jr. 1967. "The Historical Roots of Our Ecologic Crisis." *Science* 155: 1203–1207.

Jennifer Welchman

PATENTING LIFE

The extension of patents to living matter has raised a host of political, economic, ethical, and practical concerns. New science and new institutions offer expanded opportunities for individuals and organizations to claim ownership and control of components of life or entire living organisms. This entry examines the historical context for ownership of living matter, the modern structure of ownership, the current status of ownership and control, and an array of issues triggered by expanded private ownership of life.

THE HISTORICAL CONTEXT FOR OWNERSHIP OF LIFE

The modern debate about ownership of life is embedded in the history of the science and the institutions designed to advance and use that science. Before recorded time, most people lived as hunters and gatherers, gathering plants and hunting animals for their immediate use but not really controlling or nurturing those organisms. As peoples settled down and began to identify and improve, adopt and adapt, a range of plants and animals as foods, they exerted ownership by asserting individual or communal possession of breeding stock or by controlling knowledge about its characteristics (Diamond 1997).

The concept of intellectual property rights—the right to exclusive ownership and exploitation of useful inventions—originally came in the form of trade secrets. In some cases these secrets involved know-how that could be passed along to others and repeatedly used to make or improve goods or services (e.g., bread making or brewing). In other cases, innovations were embedded in a living reproducible organism, but only the products of those organisms (e.g., grafted fruit trees) were widely exploited. This de facto form of intellectual-property protection was relatively inefficient, however, because the owner had to make often expensive and wasteful efforts to keep the intellectual property secret.

MODERN PROPERTY-PROTECTION MECHANISMS

Explicit, legally sanctioned property rights for technologies, products, processes, or business practices are relatively new. Our modern patent system is firmly rooted in Enlightenment England. The evolving common-law system in England first accepted and confirmed that individuals have the right to the "fruits of their labor" (Locke 1950 [1689]) and then developed a patent system that ultimately protected "inventions" of new and useful products, processes, and technologies.

Patents are essentially a bargain between society and inventors. Inventors get an exclusive period of twenty years to exploit their inventions in exchange for full disclosure of their methods. To patent an invention, the technology, process, product, or business practice must be "novel, useful and non-obvious." This precludes patents on mere discoveries (e.g., an unimproved plant from nature), concepts not reduced to practice, and previously known ideas and products. Patent rights are granted only in the country of application, and rights must be enforced by the inventor. There are no automatic exemptions for others to use patented material for further advances, so researchers and those seeking to commercialize follow-on inventions need to negotiate with related patent holders. Once the patent has expired, the protected knowledge enters the public domain.

Living matter was excluded from early patent systems because the state of science and technology was at a level where new plant varieties or animal breeds could not meet the tests of being novel, useful, and nonobvious. Modern science began to break down the divide between living and nonliving inventions. While modern biology is usually dated from the 1860s, with the emergence of Charles Darwin's theory of evolution (1859) and Gregor Mendel's laws of inherence (1866), it took another generation before these concepts were reduced to practice and systematically applied to breeding plants and animals.

The legal system began to adapt in 1930, when the United States allowed patents on asexually propagated plants, such as fruit trees produced by grafting. Then in the 1940s a number of European countries created property protection, called plant breeders' rights, for sexually reproducing plant varieties. Plant breeders' rights provide developers of new, stable, and uniform plant varieties an eighteen-year right to control the sale of propagating material and to levy royalties for use by others. As with patents, these rules are nationally based; there are no automatic international rights. Plant breeders' rights are cost-effective but offer less protection than patents, as farmers may save seed for replanting and researchers are allowed to use existing varieties to develop new varieties.

The biotechnology era may be dated from 1953, with the discovery by James Watson and Francis Crick of the double-helix structure of deoxyribonucleic acid (DNA), but it practically began in 1973 when Herbert Boyer and Stanley Cohen invented a method of cloning genetically engineered molecules in foreign cells—what has become known as the "cut and paste" method.

The legal system converged with the new science in 1980, when the U.S. Supreme Court granted a utility patent for claims related to a human-engineered microorganism, on the basis of a definition of the organism as a new composition of matter (*Diamond v. Chakrabarty*, 447 U.S. 303, 206 *USPQ* 193). This decision was followed in 1985 by a utility patent for a genetically modified plant, and in 1988 for a genetically modified animal (*Ex parte Hibberd*, 227 *USPQ* 443 [Bd. Pat. App. & Inter.]).

Since the 1980s these developments in the United States have been replicated around the world. In some cases governments legislated or courts ruled in ways that adjusted national rules to conform to the U.S. system. Meanwhile, international negotiations knit together groups of national systems through treaties. Finally, in 1995 the World Trade Organization Agreement and its associated Trade Related to Intellectual Property (TRIPs) Agreement extended intellectual-property rights internationally, requiring all 150 member countries to provide inventors the option of either patent protection or, in the case of living organisms, some form of protection, such as plant breeders' rights. As of 2006, all member states of the World Trade Organization, including developing countries, are required to provide protection.

CONTEMPORARY OWNERSHIP OF LIFE

As a result of the evolution of science and the complementary development of property-rights systems, there are now thousands of ownership claims to living matter and the tools and structures needed to make use of that living matter. One way to examine the scope of ownership is to take a slice through the information, knowledge, tools, and processes used to manipulate living matter.

In 2008 more than 1,500 organisms, ranging from unicellular yeast to the human genome, are being or have been sequenced. While there was a great flurry of patenting activity of some gene sequences in the 1990s, as of 2008 most genomic information is freely and publicly available in databases. About the only people with effective property protection in this domain are the owners of the machines and inputs used in the sequencing process.

There are also a variety of public and private gene banks, where living plants (mostly seeds), microbes, and animal and human tissues (e.g., blood and cell lines) are preserved. While much of this genetic material is in the public domain (e.g., the more than 600,000 seed accessions of the eleven gene banks of the Consultative Group on International Agricultural Research), there is increasing private interest in genetic material, especially human tissues (e.g., in Iceland and Estonia, where a mix of public and private interests have been developing population gene banks) (Sheremeta and Knoppers 2007).

While the conventional tools of plant and animal breeding have now all entered the public domain, the tools of biotechnology are so recently patented that most remain closely held by firms, universities, hospitals, or public research labs. Some large multinational companies have a bundle of complementary technologies that enable them to operate freely, but most firms and public research programs need to negotiate licenses from other patent holders to assemble the rights to undertake research or to commercialize the resulting products of that research.

Downstream in the products of biotechnology, two somewhat different systems have emerged. Plants and animals can be protected by plant breeders' rights or animal-pedigree rules. While there are no conclusive data, one can probably safely say that virtually all commercially developed plant varieties and animal pedigrees are protected by one or more property mechanisms. Meanwhile, patents have been used in a variety of ways to protect the products of biotechnology. In the plant and animal world, most countries of the Organization for Economic Cooperation and Development offer protection to inventors who insert proprietary genes in plants, and in the United States and the European Union, inventors can claim patents on an entire multicellular organism that is genetically modified. Although Canada, among others, has declined to extend patents to multicellular organisms, the Supreme Court of Canada, in its 2004 review of *Monsanto v. Schmeiser*, ruled that while patents are not permitted on whole plants, the presence of the Roundup Ready gene in every cell of the plant provides the right to exert control over an entire plant (Phillips 2007a). So the rule of thumb for farmers wanting to use commercially developed genetically modified plants and animals is that they should conclude licenses with the owners.

Patenting human genetics is different. Although most countries allow patents on genes and in some cases organs, no country will allow patents on entire humans. Such an extension of rights is deemed by all to offend the *ordre public*. Nevertheless, patenting has proceeded apace. A recent study by Kyle Jensen and Fiona Murray (2005) concluded that 4,270 human genes (in 4,382 claims), nearly 20 percent of all known human genes, have been claimed in U.S. patents. Nearly two-thirds (63%) of these patents are assigned to private firms, with the rest assigned to individual researchers, hospitals, research labs, and universities.

CONTROVERSIES

There has been vociferous debate about the most appropriate systems to deal with intellectual advances—those ideas, recipes, formulas, and processes that generate increases in economic productivity and social well-being. Some controversies involve the ethics of what we are doing, while others relate to the ethics of the outcomes we are generating (for a survey of the economic issues, see Phillips and Stovin 2000).

The Morality of Ownership of Life Many are unhappy with the scope of what is patentable. There is virtually unanimous agreement that it is immoral to patent whole humans, and no systems allow such patents, but there is a wide range of views about what other potential living matter should be excluded or have restricted protection.

Many of the cases related to living matter (e.g., *Moore v. Regents of the University of California* 793 P.2d 479, Cal. 1990) have touched on this issue, but frequently the issue has been related more to who should own the invention rather than whether a patent should be issued.

Nevertheless, many environmental groups, developmental nongovernmental organizations, church groups, and farmer advocates have called for patents to be removed from all higher life forms, including animals and plants, arguing that patenting living organisms is either immoral, inequitable, or inefficient. Also, developmental nongovernmental organizations and indigenous communities have called for much tighter control on patenting of indigenous knowledge and genetic resources and seek broad exemptions from patent enforcement for indigenous farmers and for humanitarian purposes.

Because moral issues are hard to legislate, governments have generally set up systems to allow individuals or groups to raise specific concerns. The TRIPS Agreement permits an *ordre public* provision to address noneconomic values in the patent system. On a case-by-case basis, patents can be refused should the commercial exploitation of the invention violate public order. European Directive 98/44 on the Legal Protection of Biotechnological Inventions (European Commission 1998), for example, states that processes that use human embryos for commercial purposes and processes that clone human beings violate the *ordre public*. In practice, this provision is usually invoked by a third party in an opposition procedure after the patent has been granted. Some systems provide an opportunity to challenge patents while they are being evaluated. Australia, the European Patent Office, France, Germany, India, and Japan currently have opposition processes, and a recent U.S. report has recommended establishing an opposition procedure in the U.S. patent system.

Intellectual-Property Rights and the Freedom to Operate The creation of private intellectual-property rights for biotechnological innovations in the past thirty years has opened the system to substantial private involvement, and this has raised the concern that intellectual property rights make research and commercialization more complex, costly, and inequitable.

One of the most pressing issues for many scientists and companies is the freedom to operate in a world of overlapping and interwoven claims to intellectual property. One often cited example is golden rice, which was developed by public-sector scientists in Switzerland for public-good purposes. When they realized that seventy patented technologies owned by thirty-two different entities had been used, they decided that they did not have the capacity to acquire freedom to operate without assistance, and ultimately decided to assign the product to a large multinational company to commercialize it (Kryder et al. 2000). A big part of the problem is that markets for intellectual property are just beginning to emerge. It is still expensive and time-consuming to search for patent owners or their agents; negotiations for licensing are often protracted; and enforcing rights through contracts or the courts can be prohibitively expensive.

There is rising concern that patents constrain research activity. There is clear evidence in the plant biotechnology industry that since the extension of property rights, many multinational firms have entered and as of 2008 a few dominate some crops. Even though smaller ventures may be able to do research, no start-up firms have successfully commercialized any new genetically modified crops. Meanwhile, in the field of human health, Mildred K. Cho et al. (2003) surveyed clinical laboratory directors in the United States and found that 53 percent had decided not to develop some new clinical genetic tests because of patent concerns.

Patents are also having an effect in product markets. Percy Schmeiser's unsuccessful challenge of Monsanto's patent on Roundup Ready canola was at one level a question of freedom to operate—in this case Schmeiser attempted to assert a farmer's privilege to seed. The Canadian courts ruled that Monsanto has both the right to patent their invention and the authority to commercialize it under their own rules (a contemporary case in the United States, *Monsanto v. McFarling*, delivered similar results). A slightly different argument was made by a group of organic farmers in Saskatchewan, Canada, in an unsuccessful class-action suit (*Hoffman and Beaudoin v. Monsanto and Bayer*). The organic producers argued that the presence of openly pollinated genetically modified canola released by the defendants made it impossible for organic-canola producers to assure foreign buyers of their organic status, and this led to losses in the European Union. The trial courts, upheld at the appellate level, ruled that there was no basis for a class-action suit because there was too much diversity in the farming population. Any claims would have to proceed through other means. This case left unclear whether firms commercializing genetically modified crops are liable for any economic damage they cause other producers (e.g., by commingling in fields or the supply chain).

There is no doubt, however, that firms producing genetically modified crops are liable for health and safety risks. In 2000 Aventis CropScience ended up recalling StarLink hybrid feed corn, with the insecticidal Bt gene, which was not approved by the U.S. Food and Drug Administration for human consumption. In 2000 the feed corn, after two years of use as a commercial feed, was found

Altered Tomatoes. *A German biologist examines a genetically transformed tomato plant. The transgenic plant has a safety lock, which means that the release of its altered genes, for example, through pollen, is not possible.* AP IMAGES.

commingled with food corn in the U.S. and other food systems. The company spent an estimated $100 million to remove the variety from the market, paid penalties to the regulators, and settled a class-action suit on behalf of farmers with a $110 million settlement (Phillips 2007b).

Patents and Traditional Knowledge Although only a small part of the agricultural revolution in Europe in eighteenth and nineteenth centuries can be traced to species and varieties of plants from other ecosystems, the major advances in the twentieth century are directly attributable to the collections of landraces and germplasm assembled from around the world. With the availability of the modern tools of biotechnology, there has been a renewed interest in examining genetic resources—regardless of whether they are located in situ in traditional settings or ex situ in modern seed banks—to identify and adapt any useful genes.

The debate about the source and use of traditional varieties of crops (and even human genetic traits) has shown up in two venues. First, a number of groups have disputed and formally challenged patents issued in the United States and the European Union for purportedly new varieties that are actually traditional cultivars from developing countries. High-profile cases include the Mexican yellow bean (a U.S. patent by Pod-Ners for Enola was challenged in U.S. courts by the International Centre for Tropical Agriculture), Indian turmeric (a patent for its wound-healing properties issued in the United States was challenged and subsequently canceled), the Indian neem tree (a series of patents for industrial and pharmacological properties assigned to W. R. Grace and Company issued in the United States and the European Union were challenged and struck down in Europe), and basmati rice (a patent in 1997 by RiceTec Inc. was challenged in the United States and subsequently amended). These and other proven or alleged acts of biopiracy have increased debate about the concept of invention, the definition of prior art, and the ethics of using traditional knowledge and genetic resources in patents.

Second, there is extensive international debate about, and effort directed toward, protecting traditional knowledge and indigenous genetic resources through legal agreements. Efforts are underway at the International Labor Organization, the United Nations (via various declarations on rights of indigenous peoples), the development banks, and the Convention on Biological Diversity. The International Undertaking on Plant Genetic Resources for Food and Agriculture (1983) and the International Treaty on Plant Genetic Resources (2001), which establish genetic materials as the common heritage of humankind, have led to further negotiations under the aegis of the TRIPS Agreement (article 27.3b), the Doha Declaration (article 17), the Convention on Biological Diversity (article 8j), and the World Intellectual Property Organization. Nevertheless, the issue remains unresolved and a point of contention among indigenous communities and governments around the world (Phillips and Onwuekwe 2007).

CONCLUSION

At one level, the debate about ownership of living matter is not new. A variety of strategies for protecting inventions related to living organisms—including trade secrets, private contracts, and commercial strategies—predate patents. What is new is that patents have created a new mind-set about the opportunities and threats of global development using living matter.

Some believe that patents on living matter are absolutely essential for scientific and commercial development. Without patents, they argue, development would slow or cease on many projects related to food, health, and the environment. They argue that patents offer an open, transparent, and relatively economically efficient means of encouraging investment and commercialization. Moreover, patents facilitate our broader governing system by creating clearly defined assets and sets of owners, which provide the basis for financing, marketing, regulatory assessment, safety monitoring, and enforcement. All key regulatory processes formally or informally rely on the ownership assigned through patents. Without any owner, there might not be any concerted effort at development.

Others see patents on life as a dangerous development. They see patents as contributing to winner-take-all races among large multinational corporations and developed countries that seek to dominate nature. The fear is that by and by patents on living matter will distort our economic system, deliver inequitable economic effects, generate undesirable social outcomes, and ultimately exacerbate many of the problems we wish to resolve.

At root, the issue of ownership and control is a longstanding and fundamental division in cultures and societies. The debate over their role in biotechnology and nature is just one in a long line of ownership controversies.

SEE ALSO *Convention on Biodiversity; Darwin, Charles; Evolution; Genetically Modified Organisms and Biotechnology; Seed Banks; Shiva, Vandana; Transgenic Animals.*

BIBLIOGRAPHY

Cho, Mildred K.; Samantha Illangasekare; and Meredith A. Weaver; et al. 2003. "Effects of Patents and Licenses on the Provision of Clinical Genetic Testing Services." *Journal of Molecular Diagnostics* 5(1): 3–8.

Diamond, Jared. 1997. *Guns, Germs, and Steel: The Fates of Human Societies.* New York: Norton.

European Commission. 1998. "Directive 98/44/EC of the European Parliament and of the Council of 6 July 1998 on the Legal Protection of Biotechnological Inventions." *Official Journal of the European Community,* L series, no. L213. Available from http://eur-lex.europa.eu/.

Jensen, Kyle, and Fiona Murray. 2005. "Intellectual Property Landscape of the Human Genome." *Science* 310: 239–240.

Kryder, R. David; Stanley Kowalski; and Anatole Krattiger. 2000. "The Intellectual and Technical Property Components of Pro–Vitamin A Rice (GoldenRice^{TM}): A Preliminary Freedom-to-Operate Review." *ISAAA Briefs,* no. 20. Ithaca, NY: International Service for the Acquisition of Agri-biotech Applications. Available from http://www.isaaa.org.

Locke, John. 1950 (1689). *A Letter Concerning Toleration.* New York: Macmillan.

Phillips, Peter. 2007a. "Farmers' Privilege and Patented Seeds." In *Accessing and Sharing the Benefits of the Genomics Revolution,* ed. Peter Phillips and Chika Onwuekwe, 49–64. Dordrecht, Netherlands: Springer.

Phillips, Peter. 2007b. *Governing Transformative Technological Innovation: Who's in Charge?* Oxford, UK: Edward Elgar.

Phillips, Peter, and Chika Onwuekwe, eds. 2007. *Accessing and Sharing the Benefits of the Genomics Revolution.* Dordrecht, Netherlands: Springer.

Phillips, Peter, and Derek Stovin. 2000. "The Economics of Intellectual Property Rights in the Agricultural Biotechnology Sector." In *Agricultural Biotechnology in Developing Countries: Towards Optimizing the Benefits for the Poor,* ed. Matin Qaim, Anatole F. Krattiger, and Joachim von Braun, 259–280. London: Kluwer.

Sheremeta, Lorraine, and Bartha Knoppers. 2007. "Beyond the Rhetoric: Population Genetics and Benefit-Sharing." In *Accessing and Sharing the Benefits of the Genomics Revolution,* ed. Peter Phillips and Chika Onwuekwe. Dordrecht, Netherlands: Springer.

Peter W. B. Phillips

PESTICIDES

Although it has no biological validity, the term *pest* is assigned to any organism that is doing something that someone finds unpleasant. The word *pesticide* comes from the Latin *pest,* "a plague," and *caedere,* "to kill." Pesticides are chemicals, cultured organisms, or their products that are used to disrupt an organism's physiology long enough to kill it or severely reduce its growth. There are at least twenty-one groups of pesticides, each defined by the organism or organisms they control. The most familiar are insecticides (insects), herbicides (plants), fungicides (diseases), and rodenticides (rodents). Other groups include algicides (algae), avicides (birds) nematicides (nematodes), piscicides (fish), and silvicides (trees and woody plants).

Agricultural applications account for 83 percent of all the pesticides used in the United States. Industry, commerce, and government use 13 percent, and home and garden users are responsible for only 3 percent. Herbicides account for 58 percent of all the pesticides used, insecticides 28 percent, fungicides 8 percent, and all others 7 percent (U.S. Environmental Protection Agency 2001). Agricultural uses generally are regarded as essential to continued production of abundant food. Many believe that improved pesticide technology will eliminate the problems pesticides cause and enhance the sustainability of agriculture, but that claim is debatable. Pests and pesticides have been scientific issues for many years. Only recently have they become subjects of philosophical inquiry.

PESTICIDE REGISTRATION

Most nation-states have some sort of pesticide registration. In some developing countries procedural and data requirements are few to nonexistent, primarily because of fiscal constraints but also because of lack of awareness of the importance of registration. The laws are not well implemented in some countries. Pesticide regulators have the goals of providing protection from adverse effects and gaining the benefits of pesticides. Those objectives are achieved through registration and control of the pesticide label, which allows control over use, performance claims, use directions and precautions, packaging, and advertising. Registration protects the interest of the public and the rights of the manufacturer by ensuring proper use and environmental and human protection.

In many developed countries pesticides have been subject to some kind of governmental regulation for decades. The public is aware of potential problems with pesticides and fearful because of the mistakes that have occurred. Nearly everyone knows something negative about DDT (banned in 1971 in the United States; all use ended in 1973), and many are aware of problems caused by the use of Agent Orange during the Vietnam

Crop Duster, Hebron, Maryland. *A farmer sprays chemicals on a field of snap beans. The beans had been showing signs of white mold due to heavy rainfall. The use of pesticides in agriculture has been heavily debated, with the main issue being production (of crops and goods) versus protection (of humans and the environment).* AP IMAGES.

War. The public does not know about the complex and continually reviewed procedures necessary for the registration of a pesticide before it is used. Registration is not a matter simply of recording ownership and paying a nominal fee. It mandates compliance thorough a regulatory process that demands proof of safety but usually not proof of efficacy.

The U.S. system is among the most complex and successful; however, it is not perfect, and there are many complaints from persons who argue that protection is not sufficient and also from manufacturers who find the process slow, expensive, and unnecessarily cautious. The United Kingdom (UK) used to have a voluntary approval process in which a consensus was reached among the manufacturer, the government, and users about appropriate regulation. It was abandoned in the mid-1980s, and the UK and the European Union now regulate advertising, storage, application, and crop use. Many nations follow the standards put forth by the United Nations Food and Agriculture Organization Codex Committee on Pesticide Residues, which establishes maximum residue limits for pesticides in food and guides countries on safety regulations for the use, storage, and analysis of pesticides. Egypt reduced pesticide use 90 percent from 1971 to 2002. No herbicides are permitted on crops grown for export.

FEDERAL INSECTICIDE, FUNGICIDE, AND RODENTICIDE ACT

Pesticide development after World War II created the need for stronger laws. The U.S. Department of Agriculture (USDA), supported by the pesticide industry, developed the 1947 Federal Insecticide, Fungicide, and Rodenticide Act (FIFRA). No other federal law controls pesticides and their labeling. The FIFRA added two new ideas to pesticide regulation. The first was that all pesticides intended for shipment in interstate commerce must be registered with the secretary of agriculture before shipment. The second was that the USDA had control over all precautionary statements on a pesticide label and was empowered to review the public presentation of safety procedures. Manufacturers have the burden of proof.

These provisions stopped the shipment of untested or improperly labeled products in interstate commerce. Withholding registration effectively stopped unapproved uses. The USDA could withhold registration until data were provided to prove that a pesticide could achieve the degree of pest control claimed and did not cause human harm. The act protected users from physical injury and economic loss and protected the public from injury (previously only purchasers were protected). Manufacturers had to prove effectiveness, and pesticides were defined and limited to economic poisons, which the act defines as "any substance or mixture of substances intended for preventing, destroying, repelling, or mitigating any insects, rodents, fungi, weeds, and other forms of plant or animal life or viruses except viruses on or in living man or animals."

THE PESTICIDE USE DEBATE

Public debate about pesticide use began with publication of Rachel Carson's *Silent Spring* (1962), which was followed by *The Pesticide Conspiracy* (Van den Bosch 1978), which was less widely read but had notable effects. There have been many books and papers published since that time, but none has had Carson's influence.

The debate about pesticides is similar to other debates about agricultural technologies (e.g., biotechnology). Thomas DeGregori expresses the enduring scientific perception of the problem when he states that "public discourse is being driven by emotional language" (DeGregori 2002, p. 125). Norman Borlaug argues that scientific progress "must not be hobbled by excessively restrictive regulations" (Borlaug 2001, p. 28). If the needs of the 8.3 billion people who are likely to be alive in 2025 are to be met, Borlaug claims, conventional technology and biotechnology will be essential. He also claims that "extremists in the environmental movement, largely from the rich nations or the privileged strata of society in poor nations, seem to be doing everything they can to stop scientific progress in its tracks" (p. 29). Borlaug acknowledges the debt of agriculture to the environmental movement, whose efforts have led to "legislation to improve air and water quality, protect wildlife, control the disposal of toxic wastes, protect soil, and reduce loss of biodiversity" (p. 29). However, antitechnology extremists have gone too far and their policies will have "grievous consequences for the environment and humanity." Borlaug's claim is ironic in that those in the environmental movement continue to pursue improvement of the things for which he thanks them.

The 1947 FIFRA was a truth-in-packaging law that emphasized the value of production over protection. The ensuing DDT debate and other pesticide problems (e.g., the dependence of production on pesticides, pesticide resistance, the lack of sustainability of the agricultural production system) have shifted regulation and public discussion away from the initial emphasis on production toward the protection concerns of the environmental movement.

According to the scientistic view of science, "science can determine a fact, that these facts represent objective reality, and that values or beliefs play no role in determination of facts" (Barker and Peters 1993, p. 5). Scientism sees science as objective and value-free and states that it is not the scientist's task to create or change social, economic, or political policy. Scientism is a caricature of the actual nature of science, which is socially embedded, value-driven, theory-laden, and dynamic, so that today's scientific "truth" is tomorrow's falsehood. Scientists attempt to understand and explain the natural world, and technology applies scientific findings to the world. In general, science has been regarded by the public as good and technology has been judged to be good or bad depending on how it is used (Boulter 1997). However, science and its underlying values have moved from being viewed as an unalloyed public benefit to being regarded with suspicion, if not distrust. Scientists used to be seen as being guided by wholesome curiosity and a search for the elusive truth. Now the public is wary. Science, like all human activities, is influenced by social, economic, and political pressures. It is equally well known that some bad consequences that scientists said were unlikely actually have occurred. For example, there are pesticides in some drinking water supplies and food, and nitrates pollute water. Yet there is little debate in the agricultural community about how such things come to be. Within the agricultural community these situations are regarded as unfortunate technological accidents, not moral failures.

Public and scientific debates about pesticides frequently are based on scientific facts—often selected facts—but the disputants nearly always disagree about the story (Charles 2001). All stories are in some sense true, especially when one knows the preconceptions of the storyteller. The dispute is over the goodness of the characters (their virtue or lack of virtue), the plot (why is this happening?), the editing (what facts count?), and how it all will end (Charles 2001). What one hears or reads in these conflicting stories is often not a reasoned debate of the issues. It is a presentation and defense of one of the polar views: Pesticides are good and are required to feed the expanding human population versus pesticides are bad (for a variety of reasons, especially the unpredictability of future effects) and will not help feed people. Pesticides, many argue, are unpredictable, threatening, and a product of the type of human folly that leads people to believe that they can and should control the environment.

PERCEPTION OF RISK

Science can measure risk and determine the probability of the occurrence of a defined risk. Safety, in contrast, is a normative personal or political judgment. Judgment of safety is not and should not become a scientific decision. Science creates the data on which many judgments and decisions are based, but scientists, through the scientific process, cannot determine individual actions or public policy on the basis of their data. Something may be described as unsafe because it is found through observation to increase the risk of undesirable consequences. For example, a pesticide can have high human toxicity or be an environmental hazard. Scientists can measure the risk (the likelihood) of human or environmental effects. Farmers may decide not to use a pesticide, insurance companies may charge high premiums if it is used, and legislative bodies may pass laws requiring special use precautions. Scientists may agree with those actions, but science does not create them.

People perceive risk in different ways depending on where they live, how rich or poor they are, their level of education, their friends, the scientific evidence they are aware of, what they read, and so on. Perception of risk may differ from the facts as determined by scientific study. But what degree of risk is acceptable and to whom? The answer may be determined legislatively, or it may be determined by one's perception of the risk. Many people are much more likely to accept a very risky activity (e.g., mountain climbing, hang gliding) if they assume the risk voluntarily, the likely effects are perceived to be delayed, the risk is a known common hazard, there are no alternatives available, and the consequences are thought to be reversible. When the opposite situation prevails, risks are accepted less readily. There are 2,000 to 3,000 cases of pesticide poisoning in the United States each year but only about 30 deaths. There are several thousand cases of pesticide poisoning and many more deaths in developing countries each year. In the United States pesticides are regarded as more risky and dangerous than the data show they really are. This is the case because they are seen as uncontrolled, involuntary risks with irreversible, severe, rapid consequences. There is uncertainty about long-term negative effects on human health or the environment. Pesticides are perceived as things that are likely to be misused and are regarded as dreaded, uncommon hazards.

There should be debate about whether pesticides can be hazardous to humans or the environment. They are toxic to people and will poison and may kill if they are not used properly. Many prescription pharmaceuticals, household cleaning agents, aspirin, automotive fuel, and other common products also are dangerous if they are not used properly. Their inherent toxicity does not change with use, but the possibility of danger increases with improper use. Neither stupidity nor ignorance increases the inherent toxicity of anything, but both increase risk.

SEE ALSO *Agricultural Ethics; Agriculture; Carson, Rachel; Environmental Law; Environmental Policy; Food Safety; Genetically Modified Organisms and Biotechnology; Risk Assessment; U.S. Environmental Protection Agency; U.S. Food and Drug Administration.*

BIBLIOGRAPHY

Barker, Anthony, and B. Guy Peters. 1993. "Science Policy and Governments." In *The Politics of Expert Advice: Creating, Using, and Manipulating Scientific Knowledge for Public Policy*, ed. Anthony Barker and B. Guy Peters. Pittsburgh: University of Pittsburgh Press.

Borlaug, Norman E. 2001. "Ending World Hunger: The Promise of Biotechnology and the Threat of Antiscience Zealotry." In *Of Frankenfoods and Golden Rice: Risks, Rewards, and Realities of Genetically Modified Foods*, ed. Frederick H. Buttel and Robert M. Goodman, 25–33. Madisons of the Wisconsin Academy of Sciences, Arts and Letters.

Boulter, D. 1997. "Scientific and Public Perception of Plant Genetic Manipulation—A Critical Review." *Critical Reviews in Plant Science* 16: 231–251.

Carson, Rachel. 1962. *Silent Spring*. Boston: Houghton Mifflin.

Charles, Daniel. 2001. "Telling the Story." *Transactions of the Wisconsin Academy of Sciences, Arts and Letters* 89: 15–23.

DeGregori, Thomas R. 2001. *Agriculture and Modern Technology: A Defense*. Ames: Iowa State University Press.

U.S. Environmental Protection Agency. 2001. *Pesticides Industry Sales and Usage—2000 and 2001 Market Estimates*. Available from http://www.epa.gov.oppbead1/pestsales.

Van den Bosch, Robert. 1978. *The Pesticide Conspiracy*. Garden City, NY: Doubleday.

Robert L. Zimdahl

PHENOMENOLOGY

Feelings have a central place in ethics. Although their importance or centrality varies across major schools of thoughts, how we feel about various kinds of activities and behaviors forms at least part of what an ethical theory will seek to explain. Environmental ethics focuses special attention on the activities and behaviors that affect the natural environment. And so it focuses attention as well on the feelings we have regarding these behaviors, and on the constituent parts of that environment, including, among other things, plants, animals, species, ecosystems, and human beings. Phenomenology studies the nature of our first person experiences, which include our moral feelings toward the environment and our behaviors with respect to it. And so a discussion of phenomenology is appropriate in any comprehensive examination of environmental ethics.

THE PHENOMENOLOGICAL METHOD

Phenomenology is the study of conscious experience, *as* experienced. As such, it is intrinsically a first-person endeavor, an examination of the nature of conscious experience, rather than a third person characterization of the objective features of experience.

Etymologically, *phenomenology* is the study of phenomena, or of how things appear. Implicit in this is the distinction between how things appear and how they are. Consequently, phenomenology is concerned with how things appear to subjects of experience rather than with how they are independently of such experiences. This focus upon our experiences of objects in abstraction from questions concerning their existence apart from experience is what is sometimes referred to as phenomenological reduction.

From the standpoint of the natural sciences, we can, for example, study the neurobiology of the perceptual states of organisms, and the causal interactions between such organisms and their environments. We can study the behavior, linguistic and otherwise, of sentient animals, or examine the influence of social and cultural practices on human beliefs and desires. In each case, we are examining experience from a third person point of view: We are concerned with the objective properties necessary and/or sufficient for something to be a conscious experience, or for some entity to undergo a conscious experience. Phenomenology, on the other hand, studies what it is like to have a conscious experience. Its concern is the nature of experience as it is lived, that is, as it is experienced by a subject of experience, described from the first-person point of view.

It might seem that phenomenology is an entirely personal, indeed, individualistic undertaking. As I write this sentence, I can feel the firmness of the chair beneath my body, I have a visual sensation of the computer screen in front of me, I feel a throbbing pain in one of my toes, and I am fighting the urge to get up and smoke a cigarette. No doubt, the current experiences of the reader are considerably different. If phenomenology were nothing but the cataloging of such individual experiences, it would amount to little more than what Immanuel Kant (1925 [1781], p. 21) called a "random groping," that is, a mere sampling of empirical descriptions without attention to any underlying theoretical framework.

However, phenomenology seeks more than such a cataloging of individual experiences. It is an attempt to describe and explain immanent structures within all conscious experiences as such. This is at once it most controversial feature, and, if successful, its greatest theoretical strength. It is controversial in that it apparently involves an inductive generalization from one case to all (typical) cases. That is, the phenomenologist looks at his or her own individual conscious experiences and proceeds, on this basis, to make general statements about the nature of all experiences as such.

The phenomenologist thinks that if others engage in serious and unprejudiced reflection, they will discover their own experience to be similar.This is essential to phenomenological method: Claims to have uncovered or described some universal aspect of or immanent structure within experience are always subject to empirical disconfirmation by any other conscious subject of experience. It is as though the claims of the phenomenologist are always followed by an implicit "Right?" or "Don't you agree?" Their purported universality presupposes the tacit agreement of the reader or listener. Conversely, as purportedly universal claims, they are open to disconfirmation by the existence of even a single individual who, "upon serious and unprejudiced reflection," observes his or her own experience to be otherwise. (Although Martin Heidegger, a leading phenomenologist, prefaced his own claims by the caveat "approximately and for the most part.")

THE PHENOMENOLOGICAL TRADITION

The phenomenological movement in philosophy has its most explicit roots in the early twentieth century works of Edmund Husserl. This tradition continues through the works of Martin Heidegger (1962 [1926]), Maurice Merleau-Ponty (1962 [1945]), Jean-Paul Sartre (1956 [1943]) and others, and remains prevalent today in the work of contemporary Continental philosophers. It is typical for those working within this tradition to view phenomenology as a foundational discipline within philosophy. Phenomenology, that is, is understood as the proper starting point for dealing with fundamental philosophical problems, rather than beginning from within epistemological, metaphysical, or ethical frameworks.

Husserl's first major work dealing with phenomenological themes, *Logical Investigations* (1970 [1900–1901]) shows the influence of both Franz Brentano (1995 [1874]) and Bernard Bolzano (1973 [1837]), among others. From Brentano, Husserl inherited an interest in descriptive psychology. Rather than focusing on the causal relations of mental states, descriptive psychology seeks to describe and classify them. Bolzano's work on logic emphasized the distinction between subjective and objective ideas or mental representations. While subjective ideas are, in Kantian terms, "mere modifications" of the subject (Kant 1925 [1781], p. 182), objective representations point to something that has an existence apart from individual subjects of experience. The combination of these two influences can be seen in Husserl's interest in the structural or immanent features found in all conscious states (or, more narrowly, within certain kinds of conscious experiences) as such.

Following Brentano, Husserl believed that the most fundamental structural feature of conscious states was their intentionality, that is, the fact that they are of or about something, that they are directed toward an object (or state of affairs). That is, if I am seeing, there is something that I see. If I remember, there is something I remember. If I hope, there is something for which I hope. For Brentano (1995 [1874], p. 88), intentionality is in fact the mark of the mental, involving an "intentionally in-existing" object. For Husserl, the ontological status of this intended object apparently varies between his earlier and later works. In *Logical Investigations*, it seems that the intentionality of experience is best viewed as a structural feature of consciousness, and that the intended object, if there is one, is typically an existent object in the natural world. In his later works, beginning with *Ideas* (1969 [1913]), intentionality consists in a relation to a kind of mental object, leading to the charge that Husserl moved from a realist position in his earlier works to an idealist one later on. The ontological status of such phenomena or objects *as* intended remains a controversial one both within and apart from the phenomenological tradition.

THE ROLE OF PHENOMENOLOGY IN ENVIRONMENTAL ETHICS

Phenomenology is concerned with the universal or immanent structural features of our consciousness of objects, while environmental ethics is concerned, among other things, with the moral status of (human and) nonhuman aspects of the natural world. A typical claim of environmental ethicists is that at least some nonhuman parts of the environment have intrinsic or inherent value, that is, that they have a kind of value that is in some important way independent of their relations to human beings and their various desires and needs. But this claim appears to run counter to the prevalent view that all values are relative to valuers, and that the only valuers are conscious beings, typically human beings or other sentient animals. And so the claim that natural entities have intrinsic or inherent value apparently runs afoul of the belief that values, of necessity, can be values only for human (or other sentient) beings. While environmental ethicists such as Holmes Rolston (*Philosophy Gone Wild*, 1986) seek to deny that the value of natural entities need be dependent upon human valuers, philosophers such as J. Baird Callicott (*In Defense of the Land Ethic*, 1989), while stressing that values need not be human-centered, concede that the source of all value must lie "in the breasts" of human or other valuing subjects.

This constitutes a core issue for environmental ethics, as it probes to the very foundations of such an ethics. But it is here that phenomenological methodology may have something to contribute. Phenomenology looks at the nature of conscious experience of objects, including, of course, our experiences of objects in the natural environment. The claim that such objects have some kind of value (whether intrinsic or otherwise) often stems from our experiences of those objects. For many it seems clear that we experience natural objects as having value, and so the role of phenomenology is in examining those experiences.

A key question is whether we experience the values of natural entities as residing in them, independently of our individual wants and needs, or whether we experience those values as in some way projected by us upon those objects. Does the value of these entities "lie in the eye of the beholder" as does (it is often said) the beauty of works of art, or does it lie in the objects themselves, as an objective or natural property, such as its size, shape, and weight?

A phenomenological approach to this question requires investigating the nature of our subjective conscious awareness of such objects. Is the value that we associate with such objects a contingent, idiosyncratic, feature of our individual experiences, or is it a necessary and universal feature of conscious experiences as such? Answering such a question goes beyond the scope of this discussion, but the role of phenomenology in such an inquiry is evident. We must look to our first-person experiences of such entities. Do we experience natural value as something we contribute to natural objects, or as something we discover in them? And if the latter, what is the necessary structure of consciousness in virtue of which such attributions of value can be understood as objectively valid and not merely as subjective fancies? These are phenomenological questions regarding the very nature of our first person subjective awareness of natural objects, and our answers to them have profound and pervasive implications for environmental ethics.

Debates in ethics, environmental and otherwise, are often settled by an appeal to our moral intuitions, that is, to our shared individual feelings about certain kinds of entities and behaviors. As the study of the immanent structure of such subjective experiences, phenomenological method has a critical role to play in our understanding of any ethics of the environment.

SEE ALSO *Deep Ecology; Environmental Philosophy: V. Contemporary Philosophy; Environmental Philosophy: VI. Postmodern Philosophy.*

BIBLIOGRAPHY

Bolzano, Bernard. 1973 (1837). *Theory of Science,* trans. Burnham Terrell. Dordrecht, Netherlands, and Boston: Reidel.

Brentano, Franz. 1995 (1874). *Psychology from an Empirical Standpoint,* ed. Oskar Kraus, trans. Antos C. Rancurello, D. B. Terrell, and Linda L. McAlister. London and New York: Routledge.
Callicott, J. Baird. 1989. *In Defense of the Land Ethic.* Albany: State University of New York Press.
Heidegger, Martin. 1962 (1926). *Being and Time,* trans. John Macquarrie and Edward Robinson. New York: Harper.
Husserl, Edmund. 1969 (1913). *Ideas: General Introduction to Pure Phenomenology,* trans. W. R. Boyce Gibson. London: Allen & Unwin.
Husserl, Edmund. 1970 (1900–1901). *Logical Investigations,* trans. J. N. Findlay. London: Routledge and K. Paul.
Kant, Immanuel. 1925 (1781). *Immanuel Kant's Critique of Pure Reason,* trans. Norman Kemp Smith. London: Macmillan.
Merleau-Ponty, Maurice. 1962 (1945). *Phenomenology of Perception,* trans. Colin Smith. London and New York: Routledge.
Rolston, Holmes, III. 1986. *Philosophy Gone Wild: Essays in Environmental Ethics.* Buffalo, NY: Prometheus Books.
Sartre, Jean-Paul. 1956 (1943). *Being and Nothingness: An Essay on Phenomenological Ontology,* trans. Hazel E. Barnes. New York: Philosophical Library.

Kent Baldner

PINCHOT, GIFFORD
1865–1946

Gifford Pinchot was born in Simsbury, Connecticut, on August 11, 1865. A major figure in the history of conservation in the United States, Pinchot played a key role in shaping environmental consciousness during the late nineteenth and early twentieth centuries. As the first chief of the U.S. Forest Service (1905–1910) he helped define the mission and policies as well as the organizational structure and institutional culture of that agency. He was the driving force behind the Progressive conservation agenda of President Theodore Roosevelt and the leading spokesperson for the Progressives' utilitarian philosophy of conservation. Pinchot died on October 4, 1946, in New York City.

The oldest son in a wealthy and influential family, Pinchot decided at an early age to become a professional forester. After his graduation from Yale University in 1889, he studied forestry with the leading foresters of France, Germany, and Switzerland. In Europe he learned forest management methods that emphasized long-term planning, efficiency, and profitability; those principles later would be the hallmarks of his conservation philosophy.

Upon returning home, Pinchot proclaimed himself the first scientifically trained American forester and embarked on a long public career that blended passion for forestry with political ambitions and ideals. As a member of the National Forest Commission (1896)

Gifford Pinchot. *Pinchot is often referred to as the "father of the Forest Service," being named chief of the U.S. Forest Service and contributing to the development of the agency's mission, policies, and structure.* NPS PHOTO.

and a special forest agent with the Department of Interior (1897), Pinchot played an important role in the expansion of the federal forest reserve system under President Grover Cleveland. In 1898 he became chief of the U.S. Division of Forestry. Seven years later, when the Division of Forestry became the U.S. Forest Service, he was named its first chief. Between 1901 and 1908 Pinchot was a key member of Roosevelt's "Tennis Cabinet" and helped launch many of the president's conservation initiatives.

Throughout his career Pinchot was at the center of political controversies. The most famous was his long battle with his former friend and hiking companion John Muir over the flooding of the Hetch Hetchy Valley in Yosemite National Park. In 1909 Pinchot and Secretary of Interior Richard Ballinger engaged in a highly publicized battle over allegedly fraudulent Alaskan land claims. That controversy led to Pinchot's dismissal as chief of the Forest Service in 1910 and ultimately contributed to the splintering of the Republican Party in 1912.

Among his many accomplishments Pinchot helped establish the Society of American Foresters and, with the financial support of his father, played an instrumental role in founding the Yale Forest School. He was a leading figure in the rise of the Progressive Party in the 1910s and eventually served two terms as the Republican governor of Pennsylvania (1922–1926 and 1930–1934).

Throughout his career Pinchot advocated conservation policies that promoted "the use of natural resources for the greatest good of the greatest number for the longest time." Although Pinchot credited his fellow Progressive W. J. McGee with formulating that phrase, Pinchot was its most visible and zealous proponent. Echoing Jeremy Bentham's (1748–1832) utilitarian maxim ("greatest happiness of the greatest number"), Pinchot and his

Progressive colleagues added a temporal element ("for the longest time") to emphasize the role of conservation in ensuring the continued availability of natural resources for the benefit and use of future generations.

In addition to its debt to utilitarianism, Pinchot's conservation philosophy was influenced by George Perkins Marsh's 1864 book *Man and Nature; or Physical Geography as Modified by Human Action.* For Pinchot, Marsh's account of past civilizations whose declines could be linked directly to the destruction of their forests and watersheds served as a warning to a rapidly growing nation bent on the unrestrained exploitation of its rich but limited supply of natural resources. Of equal concern to Pinchot was the inequitable distribution of the social and economic benefits derived from that unregulated assault on the public domain.

Pinchot summarized his philosophy of conservation in terms of three guiding principles: (1) the wise use of natural resources for the benefit of people currently alive; (2) the prevention of waste and the preservation of resources for the benefit of future generations; and (3) the development and preservation of natural resources for the benefit of the many, not the profit of the few. For Pinchot, conservation was an all-encompassing idea that could address national environmental and social ills while delivering equality, freedom, and lasting peace. To be effective, however, he believed that conservation policies needed to be based on sound science and required strong regulatory efforts by the federal government and in particular a powerful executive branch.

Occasionally Pinchot's ideas about conservation put him at odds with preservationists such as John Muir, who called for the protection of certain areas from all forms of development. More often, however, Pinchot's conservation ethic found its strongest opposition among those who demanded unregulated access to natural resources for private gain, often at the expense of land health and public welfare.

SEE ALSO *Conservation; Hetch Hetchy; Muir, John; Roosevelt, Theodore; U.S. Forest Service; Utilitarianism.*

BIBLIOGRAPHY

Hays, Samuel P. 1959. *Conservation and the Gospel of Efficiency: The Progressive Conservation Movement, 1890–1920.* Cambridge, MA: Harvard University Press.

Marsh, George Perkins. 1864. *Man and Nature; or, Physical Geography as Modified by Human Action.* New York: Charles Scribner.

McGeary, M. Nelson. 1960. *Gifford Pinchot: Forester-Politician.* Princeton, NJ: Princeton University Press.

Miller, Char. 2001. *Gifford Pinchot and the Making of Modern Environmentalism.* Washington, DC: Island Press/Shearwater Books.

Pinchot, Gifford. 1947. *Breaking New Ground.* New York: Harcourt, Brace.

Scott Friskics

PLUMWOOD, VAL

1939–2008

Val Plumwood created and articulated a critical ecofeminist version of ethics and political philosophy and importantly lived as a philosopher sage with and within the natural world about which she wrote. She was born Val Morrell on August 11, 1939, into a poor family that ran a poultry farm near Sydney, Australia. She studied philosophy at the University of Sydney in the 1960s. In the 1970s she was a prominent member of a group of philosophers at the Australian National University that set in motion the first wave of Australian environmental philosophy. This group argued that environmental problems stemmed not merely from faulty policies, practices, and technologies but also from underlying human attitudes toward the natural world that were built into European and North American thought. These thinkers were especially critical of the anthropocentric idea that only humans matter morally and that people have no obligation to protect nonhuman nature for its own sake.

With her second husband, the philosopher Richard Routley, Plumwood coauthored a number of important early treatises, including several articles arguing that human chauvinism (exceptionalism) was detrimental to environmental ethics (Routley and Routley 1979). The Routleys divorced in 1981, and Val became the sole inhabitant of a stone house she had built with Richard in a temperate rain forest in southern Australia. Through her experiences in living in this rural environment she acquired a deep knowledge of nature that became legendary. She changed her name to Val Plumwood, after Plumwood Mountain—the location of her home—which in turn was named after the plumwood tree.

Plumwood was an independent scholar and took irregular teaching positions at a number of places, including Macquarie University, the University of Sydney, Murdoch University, the University of Tasmania, North Carolina State University, the University of California at Berkeley, and the University of Montana. The Australian National University awarded Plumwood a Ph.D. in 1991. She was also an important environmental activist and in the 1970s and 1980s played a key role in an environmental campaign to save rain forests in eastern Australia. She was attacked by a crocodile while she was canoeing alone through Kakuda National Park (Australia) in 1985. After three "death rolls" in the water, she escaped the crocodile's jaws with horrific injuries and crawled for hours through tropical swamps before she was rescued. She wrote about the experience of being prey, which ironically reaffirmed her vegetarianism (Plumwood 1999).

Much of Plumwood's environmental philosophy focused on analyzing, critiquing, and providing alternatives to the dualisms that she believed lie at the heart of the domination of women, nature, and others. In her view the division between mind and matter that supposedly sets humans apart from nature became codified into an opposition between reason and nature in the European and North American tradition. This polarity, in turn, informed many categories of thought and created an ideology of dualisms that deemed "nature" to be inferior to "reason." Plumwood contended that this dualistic ideology was used to legitimize the subjugation of many social groups, including women, people of color, the working class, the poor, colonized peoples, indigenous peoples, and nonhuman nature. This led to the central ecofeminist insight that struggles for social justice and environmentalism are inseparable.

In 1990 Janna Thompson ventured a critique of Plumwood's outlook in her article "A Refutation of Environmental Ethics." She argued that that there were no foundations beyond sentience for environmental ethics; in response Plumwood (1991) argued that many entities beyond sentient animals—such as rocks, mountains, and ecosystems—possess teleological properties that are worthy of respect and that any adequate environmental philosophy must incorporate an ethic that challenges the purely instrumental human treatment of nature.

In her first book, *Feminism and the Mastery of Nature* (1993), Plumwood develops a feminist critique of dualisms and instrumentalism to argue that the master form of European culture's rationality has been unable to acknowledge its dependence on nature, women, and other dominated groups of people that it constructed as inferior. This "rational" distortion, she argues, has shaped the basic categories of European and North American thought and has threatened the survival of people and nonhuman nature. In "Wilderness Skepticism and Wilderness Dualism" Plumwood (1998) teases out the androcentrism, anthropocentrism, and Eurocentism that she believed to be embedded in the idea of wilderness; she further analyzes the roles dualisms have played in constructing the traditional concept of wilderness as a nonhuman nature that is opposed to human culture.

Plumwood steered her ecofeminism through both environmental ethics and political philosophy, developing what she called a critical-feminist-socialist ecology. At the time of her death, her work was directed toward death as a philosophical theme, especially in her unpublished article "Tasteless: Towards a Food-Based Approach to Death" (2007).

SEE ALSO *Animal Ethics; Anthropocentrism; Ecological Feminism; Environmental Activism; Sylvan, Richard; Vegetarianism.*

BIBLIOGRAPHY

WORKS BY VAL PLUMWOOD

With Routley, Richard. 1979. "Against the Inevitability of Human Chauvinism." *Ethics and Problems of the 21st Century*, eds. Kenneth E. Goodpaster and Kenneth M. Sayre. Notre Dame, IN: University of Notre Dame Press.

1986. "Ecofeminism: An Overview and Discussion of Positions and Arguments." *Australasian Journal of Philosophy* 64: 120–138.

1991. "Ethics and Instrumentalism: A Reply to Janna Thompson." *Environmental Ethics* 13: 139–149

1993. *Feminism and the Mastery of Nature*. London: Routledge.

1998. "Wilderness Skepticism and Wilderness Dualism." In *The Great New Wilderness Debate*, eds. J. Baird Callicott and Michael P. Nelson. Athens: University of Georgia Press.

1999. "Being Prey." In *The New Earth Reader: The Best of Terra Nova*, eds. David Rothenberg and Marta Ulvaeus. Cambridge, MA: MIT Press.

2000. "Integrating Ethical Frameworks for Animals, Humans, and Nature: A Critical Feminist Eco-Socialist Analysis." *Ethics and the Environment* 5: 285–322.

2002. *Environmental Culture: The Ecological Crisis of Reason*. London: Routledge.

2007. "Tasteless: Towards a Food-Based Approach to Death." Unpublished manuscript from the Forum on Religion and Ecology, Harvard University Center for the Environment.

WORKS ABOUT VAL PLUMWOOD

Thompson, Janna. 1990. "A Refutation of Environmental Ethics." *Environmental Ethics* 12: 147–160.

Mark Woods

POLLEN FLOW

The possibility of the movement of genes from engineered crops to wild relatives has been one of the primary concerns associated with the release of genetically modified crops. Such gene flow could result in the evolution of increased competitive ability in wild relatives, making them more noxious weeds (Ellstrand 2003a, Hancock 2003, Snow et al. 2003).

FREQUENCY OF CROP/WILD HYBRIDIZATIONS

Although the early consensus was that hybridization between crops and their wild relatives occurred infrequently, later research showed that crop/wild hybridizations are relatively common. The reproductive barriers between wild and domesticated taxa are so minimal that they are often considered subspecies, and breeders can readily transfer genes between them. A number of parameters, such as breeding system, flowering time, hybrid viability, and isolation distance can affect the rate of gene transfer, but if compatible relatives are within the area where pollen is dispersed, genes will eventually escape. Evidence for crop introgressions into wild populations of native relatives has been provided for many crops (Ellstrand et al. 1999).

Concerns about the deployment of genetically engineered crops have now shifted to the issue of whether transgenes will persist in native environments and have negative consequences. It has been generally assumed that crop/weed hybrids would be poorly adapted in nature and that transgenes would not, therefore, spread and persist; a few experiments have now shown, however, that the initial hybrids between crops and their native relatives do occasionally have fitness equal or superior to the wild antecedents and that genes from crops often persist for long periods in natural populations (Ellstrand 2001).

THE EFFECT OF TRANSGENES ON NATIVE POPULATIONS

The impact of transgene escape into wild populations will be strongly associated with the plant characteristic that the gene affects and the invasiveness of its wild progenitors (Hancock 2003). Transgenes that have a neutral effect on fitness, such as the marker genes used to recognize transgenic plants during experimental development, might spread randomly in natural populations but would have no subsequent impact on native fitness. Genes with detrimental effects on growth and development, such as male sterility or reduced woodiness in trees, would most likely be selected against in the natural environment and would not spread beyond a narrow area adjacent to commercial plantings. The transgenes associated with pest resistance would have variable effects, depending on the invasiveness of the recipient species and the level of natural control. If a wild species is an agronomic weed, the escape of a herbicide-resistance gene could make it a more noxious pest. Virus- fungal- and pest-resistant genes could increase the fitness of wild populations and make them better competitors if damage from the pest is controlling the size of natural populations. Those transgenes with direct positive effects on fitness, such as those broadening environmental tolerances, could result in dramatic adaptive shifts and have a major impact on the fitness of native populations, depending on the invasiveness of the species.

It has been suggested that the escape of transgenes into native populations could have a negative impact on levels of genetic diversity (Rissler and Mellon 1996). The addition of the transgene itself would, however, actually increase genetic diversity slightly, and any subsequent loss in genetic diversity would occur only at those loci tightly linked to a selectively beneficial transgene. The genes that are adjacent to such a transgene would be "dragged" along, possibly replacing any native diversity at these loci. The relative impact on native diversity would still be small, because the loci tightly linked to the transgene would make up only a small fraction of the species genome.

Another concern that has been expressed is that transgenes will have unexpected secondary genetic effects in natural populations. There could be "epistatic effects," where the transgenes interact uniquely with those of the native species, or "pleiotropic effects," where the transgenes influence more than the target trait. Although these possibilities can never be completely excluded, it is unlikely that the transgenes that have been selected for deployment will have dramatically different effects in the wild than they would in a cultivated background. The crops were originally derived from the wild species, and the transgenes have faced numerous evaluations, from the initial transformations to the final field screens, before release. Genes have been moved from native species to crop species by conventional breeders without any unexpected ramifications.

ENVIRONMENTAL IMPACT OF TRANSGENIC CROPS

Concerns have been raised about the environmental impact of crops engineered to produce pharmaceutical compounds, the so-called "pharma" or "industrial" crops (Ellstrand 2003b). The transgenes producing these products might escape into native populations via pollen flow and present a risk to humans and other animals that might consume them if the compounds are toxic. They also could reduce the fitness of native populations if they had a significant negative effect of competitive ability. It would seem prudent to regulate the release of these engineered industrial crops, as with other transgenic crops, taking into account the likelihood that someone or something could be harmed by them. If they do produce compounds that are potentially toxic or have a significant effect on reproductive fitness, their ability to reproduce needs to be severely restricted through mechanical means, or they need to be engineered into a crop that does not have any proximal native relatives. Although several methods of engineering sterility have been proposed or developed, none is ready for deployment (Chapman and Burke 2006).

One additional concern that is commonly expressed is that the introduction of transgenic crops could contaminate organically grown crops, in which purity from transgenes is a requirement. Relatively short isolation

A Bumble Bee Foraging for Pollen. *One of the major concerns with using genetically modified crops is the possible effects on native plants through hybridization. Some believe that transgenes could have a negative effect on genetic diversity. Other concerns include the contamination of native populations from "pharma crops," via pollen flow.* ELIZABETH SELLERS/NBII.GOV.

distances can be used to prevent all but a small percentage of transgene contamination into nontransgenic crops, but at present there are no restrictions on where genetically engineered (GE) crops are grown in relation to non-GE ones.

CONCLUSION

In summary, transgenes will escape into natural populations through pollen flow if compatible relatives are in proximity. Decisions on the risk of transgenic crops to native relatives should be based on three questions about risks: (1) Is a compatible relative present in the areas of deployment? (2) Is the native relative highly invasive? (3) Will the engineered trait significantly affect the invasiveness of the native relative? The degree of risk associated with the unrestricted release of potential GE crops depends on the answers to these three questions.

SEE ALSO *Agriculture; Genetically Modified Organisms and Biotechnology.*

BIBLIOGRAPHY

Chapman, Mark A. and John M. Burke. 2006. "Letting the Gene out of the Bottle: The Population Genetics of Genetically Modified Crops." *New Phytologist* 170: 429-443.

Ellstrand, Norman. C. 2001. "When Transgenes Wander, Should We Worry?" *Plant Physiology* 125: 1543–1545.

Ellstrand, Norman. C. 2003a. *Dangerous Liaisons? When Cultivated Plants Mate with Their Wild Relatives.* Baltimore, MD: Johns Hopkins University Press.

Ellstrand, Norman. C. 2003b. "Going to 'Great Lengths' to Prevent the Escape of Genes that Produce Specialty Chemicals." *Plant Physiology* 132: 1770–1774.

Ellstrand, Norman. C.; H. C. Prentice; and J. F. Hancock. 1999. "Gene Flow and Introgression from Domesticated Plants into Their Wild Relatives." *Annual Review of Ecology and Systematics* 30: 539–563.

Hancock, J. F. 2003. "A Framework for Assessing the Risk of Transgenic Crops." *BioScience* 53: 512–519.

Rissler J., and M. Mellon. 1996. *The Ecological Risks of Engineered Crops.* Cambridge, MA: MIT Press.

Snow A. D.; D. A. Andow; P. Gepts; et al. 2003. "Genetically Engineered Organisms and the Environment: Current Status and Recommendations." *Ecological Applications* 15: 377–404.

Jim Hancock

POLLUTION

Pollution refers to energy or substances that contaminate the environment, causing harm to humans or other living organisms. It can also refer to the act or process of releasing these pollutants. Major examples include air pollution, water pollution, and soil contamination caused by the release of toxic chemicals, particulates, or radioactive substances. Less well-known examples include noise pollution (noise that is loud enough to be physically harmful or annoying), thermal pollution (changes in water temperature that affect aquatic life), and light pollution (light from cities that interferes with animal life or astronomical observations). These examples illustrate that a substance can be harmless or even beneficial when present in small quantities while becoming harmful at higher concentrations. Because the status of these substances as pollutants depends on their quantity, there is room for debate about whether to label them as pollutants whenever they rise above "natural" levels or only if they cause harm to particular organisms or biological systems. Efforts to alleviate pollution played a major role in the growth of the environmental movement during the middle of the twentieth century, and the contribution of pollution to global climate change is one of the most important contemporary environmental concerns.

HISTORY

Humans have released pollution since prehistoric times. For example, early efforts at metal grinding and mining resulted in a variety of harmful emissions, culminating in widespread lead pollution during the Roman Empire. Nevertheless, the Industrial Revolution of the 1700s and 1800s produced air and water contamination on an unprecedented scale. These concerns became especially prominent within the environmental movement during the mid-twentieth century. London, which had suffered from smog for centuries, experienced a severe episode of air pollution in 1952 that killed thousands of people in a few days. Rachel Carson's classic book *Silent Spring,* published in 1962, questioned the wisdom of releasing large quantities of pesticides into the environment and especially challenged the widespread spraying of DDT. A famous 1969 fire on the Cuyahoga River in northeastern Ohio stimulated concerns about water pollution. Public outcry regarding these environmental problems led to the creation of the U.S. Environmental Protection Agency (EPA) in 1970 and the passage of several important laws, including clean air acts in a number of countries, the U.S. Clean Water Act of 1972, and the U.S. Toxic Substances Control Act of 1976.

A number of the most famous environmental and public-health disasters of recent years were connected to pollution. In Japan the discovery of Minamata disease in the 1950s (a severe human neurological syndrome caused by mercury contamination in seafood) highlighted the hazards of heavy metal pollution in wastewater. One of the most catastrophic industrial accidents of modern times occurred in Bhopal, India, in 1984. Toxic methyl isocyanate (MIC) gas was accidentally released from a Union Carbide pesticide plant, killing up to 5,000 people within days and causing serious health problems in some 100,000 people in the ensuing decades. During the 1970s the Love Canal chemical waste landfill (near Niagara Falls, New York) became a symbol of the dangers of old toxic-disposal sites. The many serious illnesses suffered by residents of a neighborhood built over the site prompted the EPA to remove the citizens from their homes and reimburse them. The case galvanized support for the U.S. Comprehensive Environmental Response, Compensation, and Liability Act (CERCLA) of 1980, often called Superfund. It established fees, paid by the petroleum and chemical industries, to help pay for the cleanup of heavily contaminated waste sites for which responsible parties could not easily be identified.

EXXON VALDEZ

The oil tanker *Exxon Valdez* ran aground on March 24, 1989, in Prince William Sound, Alaska. This incident illustrates the wide range of impacts on humans and the environment from pollution-related environmental disasters. The tanker spilled more than 10 million gallons of crude oil, damaging 1,000 miles of Alaskan coastline and killing thousands of seabirds, sea otters, fish eggs, and other organisms. Millions of dollars were lost because of decreased fishing and tourism, contributing to the bankruptcy of the Chugach Native American group. The spill spurred economists to improve their techniques of contingent-valuation analysis, a set of approaches used to measure how much the public values particular environmental areas and how much of that value is lost by contamination. In the case of Prince William Sound, the public assessed the lost environmental value at several billion dollars.

ETHICAL AND PHILOSOPHICAL SIGNIFICANCE

Pollution raises a number of important ethical and philosophical questions. Although it was central to the environmental movements of the mid-twentieth century, many philosophers have questioned whether it provides the most appropriate motivation for environmental concerns. For example, Deep Ecologists such as Arne Naess have criticized those who focus solely on pollution and resource degradation as "shallow" environmentalists, concerned primarily about the needs of wealthy humans in the developed world. Nevertheless, a strong case can be made that those concerned about nonhuman organisms, developing countries, and underprivileged groups should also be worried about pollution. In the 1996 book *Our Stolen Future*, which Vice President Al Gore likened to *Silent Spring*, Theo Colborn and her coauthors argued that low-dose exposure to a variety of industrial chemicals was much more harmful to wildlife and humans than previously thought. They claimed that many of these toxins are endocrine disruptors, interfering with the delicate hormonal system of animals and causing species losses, damaged eggs, skewed sex ratios, disturbed mating practices, and abnormal development.

Kristin Shrader-Frechette argued in her book *Environmental Justice* (2002) that pollution is also taking a disproportionate toll on vulnerable groups: children, minorities, and poor citizens of developing countries. Many particularly hazardous pesticides that have been banned in wealthy countries are still produced there and then exported abroad, where they are often used improperly. Another concern is that, as environmental regulations have increased in Europe and North America, polluting industries (along with old, high-emission factory equipment) are moving to developing countries such as China, where they are producing serious threats to human health. Harmful electronic wastes (from computers, cell phones, and other electronic equipment) have also been exported to developing countries, where they are often recycled by poor workers under extremely toxic conditions. The Basel Convention, which entered into force in 1992, is an international treaty designed to reduce and regulate the flow of hazardous waste between countries. In response to critics who argued that the original treaty was inadequate, a Basel Ban Amendment (which prohibits the transfer of hazardous waste from a number of developed countries to developing countries)

Water Pollution in China, 2005. *A woman collects plastic bottles near a river in China's Jiangxi Province. The riverwater is polluted with a reddish dye discharged from a small paper factory nearby.* AP IMAGES.

was proposed in 1995 and has been ratified by a number of countries, as well as the European Union.

An important ethical question is how to determine acceptable levels of pollution. Many analysts argue that market forces should play an important role in these decisions. In a particularly striking example, Lawrence Summers, the former chief economist of the World Bank (and former president of Harvard University) signed an internal World Bank memo in 1991 that was later leaked to the *Economist* magazine ("Let Them Eat Pollution" 1992). The memo argued that developed countries ought to export more pollution to developing countries. The rationale was that the economic costs of pollution-related deaths or injuries are lower in developing countries because wages are so much lower. A more moderate and perhaps more justifiable appeal to market forces for regulating pollution involves setting a "cap" on the allowable emission levels of a particular pollutant and then allowing a group of industries to "trade" rights to emit the substance under the cap.

A common objection to many of these economic approaches is that they focus primarily on overall efficiency and not on maintaining a fair distribution of health risks among members of the population. Numerous ethicists argue that people have rights not to suffer significant risks of harm or death from pollution, no matter what the economic benefits to others might be. This position raises difficult problems of its own, however, because of the need to decide what level of risk is acceptable; critics frequently argue that it is impossible to eliminate all pollution-related risks. One solution is to claim that there is some threshold of risk that is so small (sometimes called *de minimis*) that it can be ethically ignored. Others worry that the aggregation of many *de minimis* risks could still pose significant ethical concerns.

Another important philosophical issue is how to balance the perspectives of experts and citizens when making societal decisions about pollution. The psychologist Paul Slovic has found, for example, that ordinary citizens appear to incorporate a wide variety of considerations (e.g., voluntariness, dread, and fairness) into their risk assessments, whereas experts focus more narrowly on the probability of a particular harm. Some

ACID RAIN

Sulfur and nitrogen compounds released from power plants, factories, and motor vehicles can react in the atmosphere to form acidic precipitation. This acid rain harms insects, aquatic organisms, forests, soils, and possibly human health. Acid rain has the potential to produce adverse effects that are very distant from the source of the pollution, much like the effects of chlorofluorocarbons on the ozone layer, the damage caused by fertilizer and pesticide runoff into waterways, and the global warming caused by greenhouse gases. These long-distance forms of pollution raise a number of difficult ethical and social issues:

- They frequently require regulatory cooperation across state and national boundaries.
- They create significant challenges for scientists who seek to identify precise cause-effect relationships between pollutants and harms.
- They make it more difficult to assign ethical responsibility to polluters, because small releases by many individuals can aggregate into large-scale effects.
- The agents who gain economic benefits by causing the pollution are frequently not the same individuals who suffer ill effects from it.

commentators argue that these differences justify respect for the risk perception of nonexperts, but critics such as Cass Sunstein argue that citizens' perceptions usually involve poor assessments of risk probabilities. Another concern about deferring to experts is that, in situations of scientific uncertainty, ordinary citizens sometimes have local expertise that rivals the understanding of scientists. The sociologist Brian Wynne argues, for example, that analysts made serious mistakes when evaluating risks from radioactive contamination of British sheep following the Chernobyl nuclear accident of 1986. Notably, they failed to account for details of sheep behavior and soil type that were well known to farmers. In her book *Taking Action, Saving Lives* (2008), Shrader-Frechette emphasizes another reason to doubt many expert analyses: Polluting industries fund questionable studies designed to obtain results that further their interests. In order to mitigate the effects of financial conflicts of interest on scientific research, she calls for independent sources to fund more studies on the environmental and public health effects of pollution.

PERSISTING ISSUES

The contribution of pollution to climate change is perhaps the most profound environmental issue of the twenty-first century. Emissions of greenhouse gases such as carbon dioxide are warming Earth's atmosphere, producing a wide range of deleterious effects that could include increased flooding in some areas, drought in other places, increased spread of some diseases, and major species extinctions. Some less certain but particularly serious potential consequences include a massive rise in sea levels (especially if ice sheets on Greenland or West Antarctica melt) and alteration of the Gulf Stream that warms Europe. Developing international cooperation on policies to mitigate climate change is one of the major challenges facing world leaders.

As nations attempt to lower their emissions of carbon dioxide in response to climate change, they are debating another crucial issue: pollution risks from nuclear power plants. Although these plants do not directly emit greenhouse gases, they have the potential to release large amounts of radioactive material, either through reactor accidents or through improper waste disposal. The possibility of accidents has become more worrisome as countries with poor industrial safety records begin to pursue nuclear power. In the United States, the plan to locate a repository for high-level radioactive waste at Yucca Mountain, Nevada, has become a focal point for nuclear concerns. Experts from the U.S. Department of Energy and the EPA have claimed that the site would be safe for thousands of years, but critics have questioned the validity of these estimates. The case has stirred renewed citizen and activist scrutiny of "official," "expert" analysis: Critics again are questioning claims to objectivity and seeking to uncover any hidden conflicts of interests. Such issues illustrate the importance of an informed and vigilant citizenry that can deliberate intelligently about the difficult ethical issues raised by pollution.

SEE ALSO *Deep Ecology; Energy; Environmental Law; Environmental Philosophy: V. Contemporary Philosophy; Global Climate Change; Pesticides; U.S. Environmental Protection Agency.*

BIBLIOGRAPHY

Carson, Rachel. 1962. *Silent Spring*. Boston: Houghton Mifflin Company.

Colborn, Theo; Dianne Dumanoski; and John Peterson Myers. 1996. *Our Stolen Future*. New York: Dutton.

Davis, Devra. 2002. *When Smoke Ran Like Water*. New York: Basic Books.

International Atomic Energy Agency. 2002. *An International Review of the Yucca Mountain Project TSPA-SR.* Vienna: IAEA.

Irwin, Alan. 1995. *Citizen Science.* London: Routledge.

Lapierre, Dominique, and Javier Moro. 2003. *Five Past Midnight in Bhopal: The Epic Story of the World's Worst Industrial Disaster.* New York: Warner Books.

"Let Them Eat Pollution." 1992. *The Economist* 322 (8): 66.

Markowitz, Gerald, and David Rosner. 2002. *Deceit and Denial: The Deadly Politics of Industrial Pollution.* Berkeley: University of California Press.

Naess, Arne. 1973. "The Shallow and the Deep, Long-Range Ecological Movement." *Inquiry* 16: 95–100.

Selinger, Evan, and Robert Crease. 2006. *The Philosophy of Expertise.* New York: Columbia University Press.

Shrader-Frechette, Kristin. 1991. *Risk and Rationality: Philosophical Foundations for Populist Reforms.* Berkeley: University of California Press.

Shrader-Frechette, Kristin. 1993. *Burying Uncertainty: Risk and the Case against Geological Disposal of Nuclear Waste.* Berkeley: University of California Press.

Shrader-Frechette, Kristin. 2002. *Environmental Justice: Creating Equality, Reclaiming Democracy.* New York: Oxford University Press.

Shrader-Frechette, Kristin. 2008. *Taking Action, Saving Lives: Protecting Environmental and Public Health.* New York: Oxford University Press.

Slovic, Paul. 1987. "Perception of Risk." *Science* 236: 280–285.

Sunstein, Cass. 2002. *Risk and Reason: Safety, Law, and the Environment.* Cambridge, UK: Cambridge University Press.

Wynne, Brian. 1989. "Sheep Farming after Chernobyl: A Case Study in Communicating Scientific Information." *Environment* 31(2): 10–15.

Kevin C. Elliott

POLYNESIA

The Polynesian islands form a triangle in the Pacific Ocean from Aotearoa (New Zealand) at the southwestern corner, to Rapa Nui (Easter Island) at the southeastern corner, to the Hawai'ian archipelago at the northern apex. Lying within the triangle are Samoa, Tonga, the Cook Islands, the Society Islands (including Tahiti), the Marquesas, and many more islands. Genetic analysis indicates that the Polynesians are descended from both the indigenous Taiwanese and the peoples of Melanesia, north and east of Australia. Those peoples began expanding farther eastward into the pelagic Pacific around 3,000 years ago. By the fourth century CE, the Polynesians had reached Rapa Nui, and by the fifth, Hawai'i.

HUMAN SETTLEMENT IN POLYNESIA

Aotearoa was peopled by the Polynesians last, only about 1,000 years before the present time. Genetic analysis and radiocarbon dating of chicken bones found in Chile indicate that they reached South America (3,600 miles from Rapa Nui) a century before the Spanish arrived. Sweet potatoes, native to South America, appeared on mid–South Pacific islands as early as the eleventh century. These facts suggest that the Polynesians were trading with South American Indians for half a millennium before Europeans crossed the Atlantic. The navigational skills required for locating, settling, and traveling between tiny bits of land in the largest ocean indicate great sensitivity to the environmental interface of land, sea, air, sky, and the behavior of other animals—in regard to winds, currents, stars, and birds—by means of all five senses.

COSMOLOGY AND CULTURE

Polynesia is united by dialects of a common language and variations of a common material, social, and cognitive culture. Its cosmology is among the most distinctive features of Polynesian cognitive culture and is especially germane to environmental ethics. Although there are many popular and local Polynesian origin myths, a more abstract and esoteric evolutionary epic is extant in both Hawai'i and Aotearoa—the northernmost and southernmost enclaves of Polynesia, respectively—suggesting a common ancestor in the Society Islands, which can be thought of as the cultural motherlands of eastern Polynesia.

An especially detailed expression of that epic is given in the *Kumulipo*, a Hawai'ian genealogical chant composed for Kalani-nui-ia-mamao, a Big Island chief, in the seventeenth century (Johnson 1981). This chief's genealogy is traced all the way back to the coral polyp, and the progression forward from there is quite similar to that of modern biology, moving from coelenterates (corals), to annelids and nematodes (segmented and unsegmented worms), to echinoderms (starfishes and sea urchins), to mollusks (shellfish), and on to marine and terrestrial vertebrates. The biological knowledge recorded in the *Kumulipo* was greater than that which existed in Europe when James Cook made contact with the Polynesians in the eighteenth century.

Of course, missing from the *Kumulipo* is Charles Darwin's major nineteenth-century scientific achievement: the concept of natural selection. From a scientific point of view, the explanation of the proliferation and increased complexity of species by natural selection is the crux of the modern theory of evolution. However, from an environmental-ethical point of view, whether evolution is blindly driven by natural selection or bootstrapped by reproductive *mana* (spirit power) is irrelevant. What is ethically important is the sense of relationship, of kinship, between human and nonhuman life implied in an evolutionary understanding of origins.

SOCIETIES AND SOCIAL ORDER

In most of Polynesia—Aotearoa being the exception—the islands are relatively small and the resource base is limited. Except for birds, some of them flightless and very vulnerable, there were virtually no native wild animals to hunt, although the Polynesians did introduce pigs. In addition to pork, they subsisted mainly by fishing the surrounding waters and cultivating enclosed fishponds and growing taro, the staple of their diet, and a few other domesticated food plants that they imported.

As island populations grew, highly organized and stratified societies developed to maximize the yield of the limited resource base. The volcanic island homes of the Polynesians were divided into wedge-shaped watersheds—bounded on two sides by lava ridges and on the third by the sea—steeply descending from the mountainous interior to the coastal lowlands. Commoners worked the land and sea. Their activities were overseen by stewards (*konohiki*) coordinated by aristocratic administrators (*ali'i*) who were governed by high chiefs (*ali'i nui*), all ruled by a paramount chief (*moi*) who might reign over a whole island. War between neighboring paramount chiefs was common as they struggled to expand their territories to aggrandize themselves but also to relieve population pressure and perhaps keep growing populations in check.

Polynesian social order was maintained by an elaborate system of taboos (the word is of Polynesian origin) or *kapus,* especially the *'aikapu,* or sacred eating, according to which men and women eat apart from one another and certain foods are forbidden to women. Food flowed up that social food chain from the bottom to the top. Polynesian social organization thus mirrored the natural ecological organization of Polynesian lands, for in ecology, energy flows through ecosystems from the bottom of the trophic pyramid to the top, from autotrophic plants, to herbivores, to omnivores, to carnivores. The analogue in Polynesian thought to energy in ecological thought is *mana*, of which the paramount chief, at the apex of the social food chain, has the most. In their conception of their social order, as in the scientific conception of the ecological order, Polynesians included the land as the base of their society. In the Hawai'ian dialect they call themselves *kama-aina*, "children of the land." Love of the land, *aloha aina*, is a pervasive sentiment, and *malama aina*, serving or caring for the land, is a cardinal duty. When the social order and the ecological order are in perfect harmony, everything is *pono* in paradise.

Aotearoa is different from the rest of Polynesia, consisting of two relatively large islands and lying in temperate, not tropical, latitudes, with richer and more diverse terrestrial and marine natural resources. The Maori social order is less vertically organized into gentry and common people and more laterally organized into tribes and clans, with each of the *hapu* established by its own voyaging canoe that came to Aotearoa from a legendary Polynesian motherland. Thus, there was less pressure on the Maori to cultivate the land intensively and efficiently or to develop a highly organized and stratified society. Hence, individuals experienced a greater sense of identity with their tribe, clan, and extended family.

LAND OWNERSHIP

In the modern West, land usually is owned by individuals. This concept of individual ownership is based on a theory crafted and defended by the English philosopher John Locke (1632–1704), who argued that when an individual "mixes" labor with an uncultivated piece of land—clearing it of rank vegetation and planting it with crops—its natural productivity greatly increases. Thus, the industrious individual has a right to "enclose" it and exclude others, to make private property of it. When land becomes private property, it may be "alienated": transferred, bequeathed, divested, bought, and sold. One may husband the land—manage it sustainably—and pass it on to one's offspring in its full productive capacity. Alternatively, one may overwork, mine, or strip the land, thus drawing from it not a living but a windfall profit to invest in another enterprise: One may sell or abandon it, move away, and leave the land in a ruined condition.

In the Maori concept of land ownership, by contrast, the tribe, not the individual, owns the land. Unlike an individual, whose lifetime is finite, a tribal community endures if not forever, at least indefinitely. Moreover, Maori tribal identity is merged with a particular landscape, and an individual's identity is merged with that of the tribe. Thus, for the Maori dislocation from ancestral lands constitutes a loss of tribal and therefore personal identity. Correlatively, from a traditional Maori point of view, sustainable development is the only kind of development that makes sense. The Treaty of Waitangi, signed by a number of Maori tribal chiefs in 1840, effectively established New Zealand as a British colony and guaranteed the Maori certain rights in perpetuity, including land rights. It was largely ignored until 1975, when the Waitangi Tribunal was established to adjudicate Maori claims under the treaty, including those regarding Maori cultural associations with various sites.

ENVIRONMENTAL ETHICS

Ethics often is rendered paradoxical by disparities between attitudes, values, and beliefs, on the one hand, and actions, practices, and behavior, on the other hand. For example, how is it conceivable that militant crusades, brutal inquisitions, and genocidal pogroms went on in the name of Christianity, which professes ethical precepts such as turning the other cheek, walking the extra mile, and loving one's enemy? Environmental ethics is no exception. For all the environmental sensitivity of the Polynesians evident in their navigational skills and for all their evolutionary kinship with and knowledge of the local biota, and despite the fact that many of their social organizations mirrored the ecological organizations of their island habitats and the fact that in some Polynesian societies personal and tribal identity were melded and vested in their homelands, Polynesian environmental behavior has not been beyond reproach.

The Hawaiian royalty wore robes made from the red and yellow feathers of two species of honeycreeper; it took thousands of those small birds to make one cloak. The flightless birds that had enjoyed freedom from mammalian predators were devastated by Polynesian hunters, their dogs and pigs, and the rats that had stowed away on their voyaging canoes. The moa of Aotearoa is only the most infamous example. Hundreds of species of birds went extinct during the interval between the arrival of the Polynesians and the arrival of Europeans from Aotearoa to Hawai'i to Rapa Nui. Lowland forests were cleared for the cultivation of crops, and the composition of upland forests changed radically as a result of anthropogenic causes that are still under investigation.

If the extinction of the Moa and the other flightless birds on Aotearoa is the most notorious instance of the hundreds of avifauna extinctions wrought by the Polynesians, the deforestation of Rapa Nui is the most notorious instance of the ecological devastation they perpetrated. The major reason its Polynesian inhabitants cut down the trees of Rapa Nui was to move their distinctive giant stone busts from the place where they were quarried and carved to the places where they were erected. Ironically, although these were the images of the beings that the islanders propitiated to sustain them, they turned out to be the instruments of their destruction. Without forests to hold the soil, it rapidly eroded. Birds had no place to roost and nest. The people had no wood to make boats, and so they could not go to sea to fish. They literally destroyed the Rapa Nui ecosystem and thus destroyed themselves.

Global climate change will continue to pose a particular challenge to Pacific island peoples, including the Polynesians, during the remainder of the twenty-first century and beyond. The predicted rise in sea level will swamp some low-lying atolls. On all the islands, even those with high mountains, coastal areas will be affected. Island peoples are more vulnerable to the increased frequency and intensity of cyclones, fueled by higher ocean temperatures, than are those who live on large land masses where they can seek refuge in the interior. Rising ocean temperatures may have significant impacts on the marine biota and thus on marine resources. Thus, a discussion of Polynesia and environmental ethics would be incomplete without mention of the concept of environmental justice and the duties owed by those most responsible for global climate change to those affected most adversely by it.

Endangered Hawaiian Monk Seal. *Lifeguards kept beachgoers away from this Monk Seal, on the endangered animals list, who spent two days lounging on the North Shore of Oahu in 2007. The species has shown a rapid decline in population since the rapid spread and development of humans throughout the Hawaiian islands.* PHIL MISLINSKI/GETTY IMAGES.

SEE ALSO *Asian Philosophy; Environmental Justice; Extinction; Global Climate Change; Land Ethic; Sustainability.*

BIBLIOGRAPHY

Diamond, Jared. 2005. *Collapse: How Societies Choose to Fail or Succeed.* New York: Viking.

Johnson, Rubellite Kawena. 1981. *Kumulipo: The Hawaiian Hymn of Creation.* Honolulu: Topgallant Pub. Co.

Kirch, Patrick Vinton, and Roger C. Green. 2001. *Hawaiki, Ancestral Polynesia: An Essay in Historical Anthropology.* Cambridge, UK, and New York: Cambridge University Press.

Orbell, Margaret. 1985. *Hawaiki: A New Approach to Maori Tradition.* Christchurch, New Zealand: Canterbury University Press.

Worthy, Trevor, and Richard Holdaway. 2002. *The Lost World of the Moa: Prehistoric Life of New Zealand.* Bloomington: Indiana University Press.

J. Baird Callicott

POPULATION

In the year 1 C.E. there were approximately 200 million people on Earth—less than the population of the contemporary United States alone. It took nearly two thousand years for the global population to increase sixfold: In 1850 the world population was an estimated 1.26 billion. The next increase of nearly sixfold has taken only 150 years, less than one-tenth the time of the previous sixfold jump: As of 2008 the world human population was 6.5 billion. This rapidly increasing growth rate of population, with attendant steep increases in consumption of natural resources, threatens the well-being of Earth's current and future inhabitants.

Environmental philosophers hold differing views on how humans affect the environment. Many questions and controversies have arisen in relation to the issue of population: Are resources disappearing? How do consumption patterns of a rising population change the human impact on the planet? Is there an optimal size of the human population? To what extent do humans have duties to other humans, other species, and future generations? Do ever-increasing rates of population growth augur more hunger, environmental degradation, and poverty? How do cultural and religious attitudes about gender, birth control, reproduction, and the institution of motherhood affect the size of families? How do gender, race, and class affect reproductive choices? How can population growth be restrained?

MALTHUS'S GLOOMY ARITHMETIC

Until the early nineteenth century little thought was given to human population growth except as evidence of the success of the human enterprise on Earth. That view changed with the initial publication, in 1796, of *An Essay on the Principle of Population* by the English political economist Thomas Malthus (1766–1834). Malthus argued that population increases geometrically or exponentially (1, 2, 4, 8, 16, 32, and so on), whereas agricultural productivity can increase only arithmetically (1, 2, 3, 4, 5, 6, and so on), leading to an inevitable strain on resources. For example, a farm couple might own and cultivate a hectare of land and from its yield feed themselves and four children; if those four children each had four children of their own, then in the next generation the same hectare of land must feed sixteen people; then, at the same rate of fertility, there would be sixty-four mouths to feed in the following generation and then 256 in the one following that.

On Malthus's calculation the ingenuity of the farm family might allow it to double the productivity of its hectare of land during the lifetime of the first generation, but any further doubling in productivity would be difficult. For example, suppose that the first-generation farm family produces 100 bushels of corn on its hectare of land, which would be adequate for six people. The next generation manages to produce 200 bushels on the same hectare, which must be divided not among twelve but among sixteen people. With great effort the subsequent generation might eke out an additional 100 bushels of corn from the same hectare for a total of 300, but that would have to be divided among sixty-four people. The share per person of food resources would therefore have gone, in the course of only three generations, from 6:100 to 16:200 to 64:300. At the same rate of fertility, in yet the next generation 256 people must share 400 bushels of corn if, by the most ingenious means, the original hectare can be made to produce as much—in which case the person to food ratio will be 256:400. At this point the hectare of land has reached its maximum possible productivity; hence the person-to-food ratio will shrink in the following generation to 1,024:400.

At this point several scenarios are imaginable: (1) starvation reduces family numbers to the "carrying capacity" of its hectare of land; (2) uncultivated wild land is appropriated and made to produce corn; (3) cultivated land belonging to another family is taken by force. Of course, an enlightened farm family might have foreseen the consequences of its own fertility and consciously limited its fertility rate to two—the replacement rate—by one means or another. But there are only two means of achieving a steady-state population: reduced fertility or increased mortality.

The "population problem," first articulated by Malthus, is captured only iconographically in the parable of the farm family and their single hectare of land. In the real world the human population consumes and depends on resources other than corn—indeed, on resources other than food. The fertility rate of the actual global human population varies from decade to decade, having peaked in the 1980s; as of 2008 it was running only a few tenths of 1 percent above the replacement rate of approximately 2.1. As the fertility rate exceeds the replacement rate, the population grows by an annual percentage. Even a seemingly small rate of increase of only 1.3 percent would result in a doubling of the population in just fifty-three years. The fertility rate and the rate of population increase is not uniform throughout the world. In some regions fertility rates and rates of population increase are negative and in others positive; a region's population growth or decline is also affected by emigration and immigration.

JULIAN SIMON VS. THE EHRLICHS ON POPULATION GROWTH

Malthus's concerns echo loudly in the debate between the late Julian Simon and Paul and Anne Ehrlich on natural resources and the size of the human population. According to Julian Simon (1981), natural resources are not limited; he argues that when one resource—such as petroleum—is depleted, its price rises, stimulating research and the development of substitutes such as biofuels. Simon thus encourages unbridled consumption of current natural resources, which generates wealth, which in turn may be invested in the development of new technologies to meet the increasing demands of a growing human population. He rejects research calling into question the patterns of consumption and trading practices of the wealthy nations of the global north. He thus contends that economic incentives working in free markets will result in less pollution and a better environment. In sum Simon believes resources will expand as a result of human ingenuity, and the environment will be shaped to fit human needs. He encourages the creation of artificial substitutes for things in nature and argues that our survival capacities will increase from generation to generation, despite (or because of) changes such as reduction in the number of species in the world.

Simon questions the reasoning behind negative views about population growth. For Simon human intelligence is the ultimate resource. He rejects policies that pressure people to have fewer children and instead argues that population growth offers positive benefits, despite short-term costs. According to Simon human talents and capabilities offer endless possibilities that can translate into innovative solutions to challenges such as pollution abatement and resources availability.

In Simon's view population growth will not lead to more famine and desertification. The world eats better now than it ever has before, even in poor countries, according to Simon. He argues that when more food is needed, as both more land is brought under cultivation and advancing agricultural technology increases production per hectare, more food will be available. In addition, even as population increases, the number of farmers decreases while the amount of land per farmer is rising, and he views this as economically more efficient. In Simon's view environmental degradation, habitat loss, and species loss are problems only if economic losses also occur.

In contrast to Simon's view, the Ehrlichs (1998) contend that humans pose a dangerous threat to the environment. According to the Ehrlichs the growth of human population and consumption is responsible for the earth's increasingly degraded environment and global insecurity. They believe that effective remediation is possible, but only if there is a halt and then a reversal of human population growth. They question the optimistic representations of the future of economic growth because such projections, in their view, do not include the significant environmental costs of pollution, environmental health risks, and faltering ecosystems. They question the ability of current market mechanisms to allocate resources properly and doubt that technological advances will be able to address the problems of depleted natural resources and environmental degradation. They do not believe that curbing the use of natural resources should be left to the market. Although some economic studies have focused almost exclusively on the negative economic effects of reducing the use of fossil fuels such as oil, the Ehrlichs argue that there are many possible benefits to reducing per capita energy consumption—especially improving the health of both human and nonhuman beings.

The Ehrlichs also distinguish between biological wealth, or natural capital, and economic wealth and capital. They argue that human consumption and pollution deplete biological wealth and threaten entire ecosystems; they note that there are no artificial substitutes for some natural resources such as air, land, and water, which are subject to unprecedented deterioration. Acid rain, water pollution, toxic waste, climate change, deforestation, and loss of biodiversity are a few of the many environmental problems that the Ehrlichs attribute to the unsustainable scale of human population and patterns of consumption. They point to anthropogenic climate change as a potential cause of biodiversity loss and argue that extinctions of species will in turn cause more serious disruptions of ecosystems such as forest destruction in Eastern Europe by acidic air pollution, the desiccation of the Aral Sea in the former Soviet Union, and desertification in the Sahel region of Africa.

The Ehrlichs believe that limiting the human population size and reducing consumption are preconditions of a sustainable future. In their view eating is one of humanity's most ecologically destructive activities. They call attention to the problems associated with increasing the production of food, including use of synthetic agricultural fertilizers, irrigation, and chemical pesticides in green-revolution technology, which has been touted as the key to boosting food production to keep pace with population growth. They cite substitutions of synthetic pesticides for natural pest control, inorganic fertilizers for natural ones, and chlorination for natural water purification as examples of unsatisfactory attempts to create artificial alternatives to ecosystem services. They argue that humanity's struggle both to feed the poor and to overfeed the rich is one of the principal causes of environmental degradation. They also note that human population growth and proportionately increasing pressures on food production result in the urban sprawl that devours agricultural land, which in turn spurs the conversion of forests and other natural plant communities into cropland for food production. Thus the task of saving the remaining forests is made harder because of the demands of a growing human population and its need for more food and wood products.

"TRAGEDY OF THE COMMONS," FAMINE, AND FUTURE GENERATIONS

Garrett Hardin (1974) critically examines the once-popular metaphor of the earth as a spaceship that we all live on and must share equally. Hardin prefers the metaphor of a lifeboat. Each wealthy nation can be conceived of as a lifeboat, full of rich people with low fertility rates, whereas the rapidly reproducing people of poor countries are swimming in water, begging for admission to a lifeboat. He argues that we need to recognize the limited carrying capacity of any lifeboat. Sharing in accordance with the spaceship ethic will soon swamp lifeboats and everyone will drown.

According to Hardin spaceship ethics is problematic because it leads to the "tragedy of the commons." Using the example of pastureland, Hardin argues that an owner adequately cares for his privately owned pasture because overshooting its carrying capacity will lead to a deterioration of the health of his herd and, therefore, economic losses. A farmer would recognize and restrain himself within the carrying capacity of his privately owned pasture. If, however, a pasture becomes a commons open to all, it is less likely, Hardin argues, that each individual will refrain from overshooting the carrying capacity. If one person increases his herd by one, the health of all the animals grazing on the commons will suffer, including his own; the value of adding an additional animal will benefit him alone, whereas the cost of doing so will be shared by all.

In Hardin's view common ruin is inevitable if there is no "mutual coercion mutually agreed upon"—regulations, in a word—and he points to air and water as examples of resources that are polluted because they are treated as an unregulated commons. The economic benefits of a polluting industry flow to its owners, whereas the costs of pollution are borne by all. Hardin argues that a free good is likely to become an overused or wrongly used good. He calls into question relief for nations in need by a world food bank, which he regards as a commons in disguise that will eventually bring ruin upon all who share in the commons, according to Hardin. Despite its good intentions, he believes such a system of sharing would encourage the population-growth differential between rich and poor countries. Because poor countries have faster rates of population growth than wealthy countries, this trend, he argues, would only increase with a global system of sharing. He contends that with increasing population growth some nations threaten to exceed—or have already exceeded—their carrying capacity. Hardin argues that when assistance is offered from abroad in order to save poor people plagued by famine, this well-intentioned humanitarianism backfires by diminishing the quality of life for those who remain and for future generations. He calls for those in the lifeboat to consider their commitment to future generations as well as to themselves.

William Aiken (1980) calls into question Hardin's judgment that it is a nation that has a carrying capacity. Aiken argues that the biological notion of carrying capacity is not applicable to the concept of a nation. Nor, in his view, is the natural environment, artificially carved up into nations, a boat that will necessarily "sink" when extra people are added. Aiken suggests that the concept of carrying capacity is ambiguous. Because technology leads to continual increases in the human carrying capacity of the environment, there is no way to determine that capacity precisely. Surely, he argues, there is an absolute limit to the number of people Earth can accommodate, but what that limit is or whether we are beyond it, closely approaching it, or still far from it is not known. Aiken also notes that Hardin focuses on mortality and ignores alternative fertility-focused methods of reducing population such as birth control.

CONSUMPTION PATTERNS AND POPULATION

Humans do not all consume the same amount of resources and generate the same amount of waste. Resource consumption and waste generation vary significantly between developed and developing nations. Only one-fifth of the planet's population lives in industrialized north, but it consumes more than two-thirds of the world's resources. The rest of the world shares what remains. The United States has less than 5 percent of the world's population and uses approximately 25 percent of the world's resources. In *Women and the Environment* (1993) Annabel Rodda notes, for example, that in the industrialized nations the average person is likely to consume more than 200 pounds of paper and 900 pounds of steel per year, compared to approximately 17 and 94 pounds, respectively, consumed by the average Third World resident. The industrialized nations use significantly more energy than the rest of the world; Julie Sze (1997) notes that the average citizen in the United

Total, Urban and Rural Populations by Major Area, Selected Periods, 1950–2050

Major Area	Population (millions)					Average Annual Rate of Change (percentage)			
	1950	1975	2007	2025	2050	1950–1975	1975–2007	2007–2025	2025–2050
Total population									
Africa	224	416	965	1394	1998	2.48	2.63	2.04	1.44
Asia	1411	2394	4030	4779	5266	2.12	1.63	0.95	0.39
Europe	548	676	731	715	664	0.84	0.24	–0.12	–0.30
Latin America and the Caribbean	168	325	572	688	769	2.65	1.77	1.02	0.45
Northern America	172	243	339	393	445	1.40	1.03	0.82	0.50
Oceania	13	21	34	41	49	2.03	1.49	1.05	0.65
Urban population									
Africa	33	107	373	658	1234	4 .76	3.90	3.15	2.52
Asia	237	574	1645	2440	3486	3.54	3.29	2.19	1.43
Europe	281	444	528	545	557	1.84	0.54	0.18	0.08
Latin America and the Caribbean	69	198	448	575	683	4.21	2.55	1.38	0.69
Northern America	110	180	275	337	401	1.98	1.33	1.11	0.70
Oceania	8	15	24	30	37	2.60	1.44	1.17	0.89
Rural population									
Africa	192	309	592	736	764	1.92	2.03	1.21	0.15
Asia	1174	1820	2384	2339	1780	1.75	0.84	–0.11	–1.09
Europe	267	232	204	170	107	–0.57	–0.41	–1.00	–1.84
Latin America and the Caribbean	98	126	124	113	87	1.01	–0.06	–0.50	–1.08
Northern America	62	64	63	56	44	0.11	–0.02	–0.65	–1.00
Oceania	5	6	10	12	11	0.88	1.60	0.78	–0.04

SOURCE: Table 1.3 from United Nations Department of Economic and Social Affairs/Population Division. *World Urbanization Prospects: The 2007 Revision,* February 2008.

Table 1. CENGAGE LEARNING, GALE.

States uses energy at the rate of 3 Japanese, 6 Mexicans, 12 Chinese, 33 Indians, 147 Bangladeshis, or 422 Ethiopians. The environmental footprint of First World residents is much deeper than that of the typical citizen of the Third World.

WASTE

Residents of the industrialized nations generate more waste than people living in the rest of the world. For example, the average person in the United States produces almost 2,000 pounds of solid waste per year. Americans and Europeans are consumers of high-tech consumer electronics such as computers, cell phones, and televisions, which now constitutes the fastest-growing part of municipal waste in the United States and Europe. According to a 2001 Environmental Protection Agency report, this discarded electronics waste generated approximately 70 percent of the heavy metals and 40 percent of the lead now found in landfills in the United States.

Citizens of wealthy nations are largely responsible for toxic dumping, the destruction of biodiversity, and soil and water depletion. Often race is the main factor in the location of hazardous-waste disposal sites in the United States. This inequitable burden also occurs on a global scale. Developed countries produce large amounts of waste that are often transported to poor nations and can cause environmental degradation. Despite international regulations, approximately 80 percent of the electronic waste generated in a year in the United States is being exported to poorer countries such as China, Pakistan, and India, and to countries in West Africa. Large containers of computer parts are shipped to less-wealthy nations and workers in these countries crack open and melt computer parts over open flames to retrieve metals. The toxic chemicals, vapors, and particles released in this process include lead, polyvinyl chloride (PVC), and polycyclic aromatic hydrocarbons (PAHs), all of which are persistent toxics. According to Elizabeth Grossman (2006), samples taken in Guiyu, China, in 2005 found levels of copper, lead, tin, and cadmium 400 to 600 times higher than what would be considered normal and safe. These samples also found polychlorinated biphenils (PCBs), PAHs, brominated flame retardants, nonphenols, phthalates, and triphenyl phosphates. As a result of this exported e-waste, the air and water in this region have been severely polluted.

POVERTY ECONOMICS AND POPULATION

Affluent, developed countries have lower fertility rates and thus lower rates of population growth than poor, developing counties. Jack Hollander (2003) believes that poverty is the root of population growth—indeed, of all environmental problems; he asserts that only a free and affluent society can achieve zero population growth and environmental protection. Hollander therefore argues, in contrast to the Ehrlichs, that population growth is not a serious long-term global problem. He points out that the Ehrlichs' predictions of massive famines have not been borne out. Instead of focusing directly on limiting population growth—by, for example, withholding food aid or promoting contraception—he recommends economic development, technological progress, and unregulated capitalism to eliminate poverty and thus to slow and ultimately halt population growth.

Hollander also believes that eliminating poverty will improve environmental quality. He thinks that for the 80 percent of the world's people who are not affluent, life's basic necessities take on a higher priority than environmental quality. He maintains that when people have economic and educational opportunities, the human population will stabilize and efficient agriculture will reduce or eliminate hunger and the demand for more land for food production. He contends that economically secure people demand environmental protection.

The Ehrlichs insist, in opposition to Hollander, that a rich person contributes much more to the damaging of Earth's life-support systems than does one living in poverty. Restricting the size of human population, not increasing affluence, should be the first priority for healing the planet, in their view. The Ehrlichs call attention to the energy trap. More energy is required to give all human beings an affluent lifestyle; although developing and deploying these energy sources would be very difficult and costly; using that much energy would create an even larger threat to ecosystems. The Ehrlichs therefore emphasize two key imperatives: reducing consumption and waste generation by the rich and limiting the population growth of all humans, rich or poor.

GENDER, RACE, CLASS, AND POPULATION

Gender, race, and class have been prominent issues in debates about human population. Some argue that gender, race, and class are often not adequately addressed in Ehrlich-style population-restraint perspectives. Recommendations of a reduction of population growth can be seen as racist and classist in condemning the rapid growth that occurs mostly in areas outside of the ambit of the developed countries. Such critics argue that the Ehrlichs' arguments fail to address the reasons for rapid growth in the third world.

Vandana Shiva (1989) argues that there is a link between the destruction of nature and the oppression of women. She argues that agriculture has shifted into two sectors: the cash-mediated masculine sector and the subsistence-oriented feminine sector. As a result the cash economy draws men away from the land, increases women's workload in producing subsistence, and disrupts ecosystems because of the green revolution's focus on growing irrigation-dependent cash crops through the use of synthetic chemicals. Shiva asserts that, as more land is diverted to cash crops and degraded by green-revolution technologies, women have less access to land and other resources but increased burdens in food production for family subsistence. As a result of the environmental degradation caused by industrial agriculture, Shiva notes that women must walk longer distances for water, fodder, and fuel.

Val Plumwood (1991) argues that numerous studies have shown that, in the Third World, ecologically insensitive, high-technology agriculture and forestry strengthen the control of the elites over natural resources and aggravate social inequalities, including men's control over women. Ecofeminists such as Shiva and Plumwood argue that the key to stanching population growth in developing countries is not withholding food aid and promoting contraception or increasing affluence and consumption: They believe that the solution lies in ensuring women's economic and reproductive autonomy. When women are empowered with education, economic means, and reproductive choices, they are more likely to be able to choose to have fewer children because their status may not be as dependent on bearing many children. In addition, they may not feel the need to increase the economic workforce of children as a means of making ends meet. Fewer births may benefit these women in several ways including their health and their efficiency in managing natural resources.

The ecological feminist Chris Cuomo stresses the importance of considering categories such as gender, race, class, and sexuality in an analysis of the complex issues involved in population growth (1994). She points out that some approaches to population control lack a critical analysis of the many social factors underlying gender oppression. She notes that these social factors—including the institution of motherhood and attitudes about sexuality and women's bodies—which contribute to the growth of human population, are ignored in many biology-based theories about "carrying capacity" and "standard of living." She calls attention to the inadequacy of viewing humans as a homogenous species and thus failing to recognize the impact of gender, race, and class on population growth. Cuomo asks, "Why do women bear many children, even in areas or communities where high population density impacts on individual lives very directly, through overcrowding, shortages of food and other necessities, poor health and hygiene, and the obvious destruction of local land and species?" (1994, p. 95).

Cuomo argues that sexism, the institution of motherhood, racism, classism, cultural factors, sexuality, and health issues fuel population growth. The ecofeminist approach acknowledges and examines in detail the need for the disempowerment of women in various ways: in terms of control over their own bodies, their roles in culture, and their sexuality, and their identities as they relate to the environment in which they live. According to this view, women's systematic oppression in patriarchal societies directly relates to the degradation of the natural environment.

Ecofeminist writers have linked women's oppression and the feminization of poverty to human population growth. Ronnie Zoe Hawkins (1992) notes that, although women often seek to limit family sizes, they are sometimes denied access to the means for doing so. Cultural beliefs and values in a patriarchal society often pressure women into bearing many children, even at the expense of their own health. Forms of birth control and abortion may be prohibited by religious or cultural views or both. In addition, women may be alienated from their own bodily functions and processes.

The roots of population growth are thus seen to lie in the poverty and patriarchy that form institutionalized barriers to women's freedom of reproductive choice. For example, many Third World cultures discourage the open discussion of birth control, and contraceptive devices are not readily accessible. In many cultures male babies are seen as more valuable than females; the overwhelming majority of abortions in countries such as China and India are performed to prevent the birth of females, resulting in de facto gendercide. Although this practice may limit population growth because there will be fewer mothers to bear children, feminist critics have pointed with alarm to its grave ethical implications.

Of all the roles traditionally assigned to women, motherhood is the one that is most common across cultures. The ideal of a good mother as a woman constantly bound to her children, physically and emotionally, willing to sacrifice herself and put the children's welfare before anyone else's, including her own, is a demanding ideal, but its perceived nobility offers insight into the reasons that some women continue to reproduce in circumstances where high population lowers their standard of living. Another explanation turns on the economic value of children as part of a family workforce in poor agrarian communities. Third World women often participate more than men in the food system. Women in Africa produce more than 70 percent of Africa's food. Andy Smith (1997) argues that it is often in the economic interest of Third World women to have more children in order to raise more export crops and earn more money.

Smith argues that population-control measures are needed most urgently among the prosperous citizens of the United States because of their rapid rate of consumption of resources. Sze (1997) notes that Third World immigrants and refugees of color with high fertility rates threaten to outbreed the low-fertility white populations of industrialized countries. Sze argues that white fears of "Third World-ification" by Latinos and Asians assume that the world's people of color are to blame for environmental degradation caused by overpopulation.

Some further argue that methods of population control have been threats to the reproductive health among women of color. For example, Smith and Lori Gruen call attention to the history of forced sterilization of women of color and the history of U.S. contraceptive companies marketing dangerous drugs such as Depo-Provera to other countries.

CONCLUSION

While gender, race, and class issues are crucial for consideration in analyses of human population size and growth, it is important that such studies do not use women, particularly women of color, as scapegoats. Reproductive choice is a human rights issue as well as an environmental issue. Policies geared toward the empowerment of all women in their reproductive choices are necessary in order to move toward a more sustainable human population and a healthy planet for all.

SEE ALSO *Anthropocentrism; Biodiversity; Consumption; Deserts and Desertification; Ecological Feminism; Economics, Ecological; Environmental Citizenship; Environmental Policy; Forests; Future Generations; Habitat Loss; Hunger; Intergenerational Justice; Plumwood, Val; Shiva, Vandana; Species; Tragedy of* *the Commons; U.S. Environmental Protection Agency; Waste Management.*

BIBLIOGRAPHY

Aiken, William. 1980. "The 'Carrying Capacity' Equivocation." *Social Theory and Practice* 6(1): 1–11.

American Association for the Advancement of Science. "AAAS Atlas of Population and Environment." Available from http://atlas.aaas.org.

Cuomo, Chris. 1994. "Ecofeminism, Deep Ecology, and Human Population." In *Ecological Feminism*, ed. Karen Warren. New York: Routledge.

Davis, Angela. 1981. *Women, Race, and Class*. New York: Vintage.

Ehrenreich, Nancy, ed. 2008. *The Reproductive Rights Reader: Law, Medicine, and the Construction of Motherhood.* New York: New York University Press.

Ehrlich, Paul R., and Anne H. Ehrlich. 1998. *Betrayal of Science and Reason: How Anti-Environmental Rhetoric Threatens Our Future.* Washington, DC: Island Press.

Grossman, Elizabeth. 2006. *High Tech Trash: Digital Devices, Hidden Toxics, and Human Health.* Washington, DC: Island Press.

Gruen, Lori. 1993. "Dismantling Oppression: An Analysis of the Connection between Women and Animals." In *Ecofeminism: Women, Animals, Nature*, ed. Greta Gaard. Philadelphia: Temple University Press.

Hardin, Garrett. 1974. "Lifeboat Ethics." *Psychology Today* (September): 38–43. Available from http://www.garretthardinsociety.org/articles/art_lifeboat_ethics_case_against_helping_poor.html.

Hawkins, Ronnie Zoe. 1992. "Reproductive Choices: The Ecological Dimension." *APA Newsletters* 91(1): 66–73.

Hollander, Jack. 2003. *The Real Environmental Crisis: Why Poverty, Not Affluence, Is the Environment's Number One Enemy.* Berkeley: University of California Press.

Mies, Maria. 1986. *Patriarchy and Accumulation on a World Scale: Women in the International Division of Labour.* London: Zed Books.

Plumwood, Val. 1991. "Nature, Self, and Gender: Feminism, Environmental Philosophy, and the Critique of Rationalism." *Hypatia* 6(1): 3–27.

Rodda, Annabel. 1993. *Women and the Environment.* London: Zed Books.

Shiva, Vandana. 1989. *Staying Alive: Women, Ecology, and Development.* London: Zed Books.

Simon, Julian. 1981 *The Ultimate Resource.* Princeton, NJ: Princeton University Press.

Simon, Julian. 1990. *Population Matters: People, Resources, Environment, and Immigration.* New Brunswick, NJ: Transaction.

Smith, Andrea. 2005. *Conquest: Sexual Violence and American Indian Genocide.* Cambridge, MA: South End Press.

Smith, Andy. 1997. "Ecofeminism through an Anticolonial Framework." In *Ecofeminism: Women, Culture, Nature*, ed. Karen Warren. Bloomington: Indiana University Press.

Sze, Julie. 1997. "Expanding Environmental Justice." In *Dragon Ladies*, ed. Sonia Shah. Cambridge, MA: South End Press.

Van DeVeer, Donald, and Christine Pierce, eds. 1994. *The Environmental Ethics and Policy Book: Philosophy, Ecology, Economics.* Belmont, CA: Wadsworth.

Warren, Karen, ed. 1996. *Ecological Feminist Philosophies.* Bloomington: Indiana University Press.

Warren, Karen. 2000. *Ecofeminist Philosophy: A Western Perspective on What It Is and Why It Matters.* Lanham, MD: Rowman & Littlefield.

Cecilia Herles

POSTCOLONIAL ENVIRONMENTAL ETHICS

The term *postcolonial environmental ethics* refers to the view that any globally relevant environmental ethic must recognize the legacy of Euro-American resource extraction from subordinate states over the last five centuries. The term *postcolonial* draws attention to the fact that the contemporary period has followed the demise of Euro-American colonial empires that lasted from the late 1400s through the 1960s and that people in the West still are affected by neocolonial attitudes toward nature and the human categories (indigenous/women) that are conceptually connected with nature. At the core of these attitudes is Eurocentric diffusionism (Blaut 1993), the idea that Euro-American cultures have dominated world resources over the last five centuries because of their innate cultural economic superiority over dominated cultures in Africa, Asia, and the Americas; that is, colonial resource extraction is ethically justified. The utilitarian philosopher John Stuart Mill embodied this attitude when he described British colonies as "hardly to be looked upon as countries,… but more properly as outlying agricultural or manufacturing establishments belonging to a larger community.… The West Indies…are the place where England finds it convenient to carry on the production of sugar, coffee, and a few other tropical commodities" (Mill 1965 [1848], p. 693).

Postcolonial environmental philosophers typically critique contemporary economic and cultural globalization for its neocolonial policies. Postcolonial environmental ethics today works constructively toward an environmental ethic that will replace the Euro-American hierarchy with a more diverse, horizontal, historically infused view of environmental problems that integrates environmental and social justice. It therefore brings forward sources of understanding that often are marginalized, such as understanding based on gender, race, caste, and class. Instead of separating people from nature—imagining nature as a pristine sanctuary separate from human beings—it works to undercut conceptual and practical dualisms such as nature versus culture, feminine versus masculine, emotion versus reason, and the Orient versus the West.

HISTORICAL BACKGROUND

In the wake of Christopher Columbus's claim to have discovered the Western Hemisphere, Pope Alexander VI issued the papal bull *Inter Caeteras,* granting the world east of the Canary Islands to Portugal. Ferdinand and Isabella of Spain were granted the lands to the west so "that in our times especially the Catholic faith and the Christian religion be exalted and be everywhere increased and spread, that the health of souls be cared for and that barbarous nations be overthrown and brought to the faith itself" (Alexander VI 1917 [1493]). In 1498 the Portuguese sailor Vasco da Gama established the eastern water route to the same territory that Columbus coveted by rounding the Cape of Good Hope and landing at Calicut in southwestern India. Thus, Europe, whether looking east or west, had staked its claim to the non-European world as its colony. The reason was primarily environmental, for as Madhav Gadgil and Ramachandra Guha wrote, "Colonialism's most tangible outcome (one whose effects persist to this day) related to its global control of resources" (1992, p. 116). The 24 acres of land available to support each European in 1491 increased to 120 "ghost acres" per European inhabitant with the advent of colonialism (Gadgil and Guha 1992).

As the historian J. M. Blaut has recounted, in the years from 1561 to 1580, 85 percent of the silver in the world came from the Americas. Potosí, a silver-mining city in the Andes, had a population in the 1570s of 120,000 people, more than Paris, Rome, or Madrid. Sugar exports from Brazil in 1600 were double the value of all English exports to the entire world in that period. Blaut estimated that more than a million people were working for the European economy in the Western Hemisphere at the close of the sixteenth century (Blaut 1993).

European economic, cultural, and military dominance over the last five centuries has been due largely to smallpox, not innate European superiority. Although the numbers are disputed, most historians believe that the Western Hemisphere was more populous than Europe in 1491, with perhaps 112 million people, roughly 95 percent of whom died within 130 years of European contact (Mann 2006). The American wilderness that explorers such as René-Robert de La Salle encountered in his voyage up the Mississippi in 1681, "a solitude unrelieved by the faintest trace of man" (Mann 2006, p. 360), was in fact an artificial wilderness created by genocide and disease. John Winthrop, the first governor of the Massachusetts Bay Colony, observed in 1634, "For the natives, they are neere all dead of small Poxe, so as the Lord hathe cleared our title to what we possess" (quoted in Crosby 1986, p. 208).

POSTCOLONIAL ENVIRONMENTAL THEORY

Postcolonialism has influenced many disciplines in the humanities, but it often has remained human-centered. Postcolonial environmental ethics therefore issues a challenge to understand colonialism and its contemporary legacy in terms of the ways in which colonial exploitation of resources is intertwined with the exploitation of people who are understood as being connected to nature. Thus, women, indigenous peoples, and people of color occupy ambiguous spaces when they are constructed in colonial narratives as intermediaries between nature and culture.

Postcolonial environmental ethics is pluralist because its proponents recognize that resistance and solutions grow out of local conditions. A model is the Self Employed Women's Association (SEWA), a trade union founded in 1972 by Elaben Bhatt because in India most women (94 percent) work outside the wage labor economy, finding employment in the informal sector. Their contributions to the economy and to environmental stewardship usually go unrecognized. SEWA embraces Gandhian principles of *satya* (truth), *ahimsā* (nonviolence), *sarvadharma* (integrating all faiths, all people), and *khadi* (propagation of local employment and self-reliance). The association includes a Forest Worker's Campaign because many poor women sustain themselves and their families through the collection of forest produce. The workers raise and sell saplings as well as produce from their nurseries. SEWA contends that the government forest department undercuts women's activities and has launched a campaign to "Feminise Our Forests."

Informed by SEWA's emphasis on local self-reliance (*khadi*), postcolonial environmental ethics often opposes attempts by the so-called First World to develop the Third World as another chapter in what historian Ranajit Guha (1989) termed the "idiom of improvement." The economist William Easterly (2006) compared the West's post–World War II attempt at Third World development with Rudyard Kipling's urgent plea to the United States to colonize the Philippines:

> Take up the White Man's burden—
> Send forth the best ye breed—
> Go bind your sons to exile
> To serve your captives' need

Authors such as Vandana Shiva (2001) and Deane Curtin (1999, 2005) have questioned the attempts of U.S. Agency for International Development (USAID), the World Bank, the International Monetary Fund, and the World Trade Organization to extend the ownership society to the genetics of seeds, medicines, and the human genome. These efforts threaten the security of dispossessed peoples, who need dependable access to nature for health, safe food and water, and fundamental human dignity.

SEE ALSO *Christianity; Ecological Feminism; Environmental Pluralism; Globalization; India and South Asia; Shiva, Vandana.*

BIBLIOGRAPHY

Alexander VI. 1917 (1493). "Papal Bull *Inter Caeteras.*" In *European Treaties Bearing of the History of the United States and Its Dependencies*, ed. Frances Gardner Davenport, Vol. 1, 71–78. Washington, DC: Carnegie Institution of Washington. Available from http://www.wadsworth.com/history_d/templates/student_resources/0155082620_murrin/sources/ch01-01.html.

Blaut, J. M. 1993. *The Colonizer's Model of the World: Geographic Diffusionism and Eurocentric History*. New York: Guilford Press.

Crosby, Alfred W. 1986. *Ecological Imperialism: The Biological Expansion of Europe, 900–1900*. Cambridge, UK, and New York: Cambridge University Press.

Curtin, Deane. 1999. *Chinnagounder's Challenge: The Question of Ecological Citizenship*. Bloomington: Indiana University Press.

Curtin, Deane. 2005. *Environmental Ethics for a Postcolonial World*. Lanham, MD: Rowman and Littlefield.

Easterly, William. 2006. *The White Man's Burden: Why the West's Efforts to Aid the Rest Have Done So Much Ill and So Little Good*. New York: Penguin Press.

Gadgil, Madhav, and Ramachandra Guha. 1992. *This Fissured Land: An Ecological History of India*. Delhi, India, and New York: Oxford University Press.

Guha, Ranajit. 1989. "Dominance without Hegemony and Its Historiography." In *Subaltern Studies 6: Writings on South Asian History and Society*. Delhi, India: Oxford University Press.

Mann, Charles C. 2006. *1491: New Revelations of the Americas before Columbus*. New York: Vintage Books.

Mill, John Stuart. 1965 (1848). "Of the Competition of Different Countries in the Same Market," Book 3, Chapter 25, *Principles of Political Economy with Some of Their Applications to Social Philosophy*, ed. John M. Robson. Toronto: University of Toronto Press (with Routledge and Kegan Paul).

Shiva, Vandana. 2001. *Protect or Plunder? Understanding Intellectual Property Rights*. London and New York: Zed Books.

Deane Curtin

POVERTY

SEE *Environmental Justice.*

POWELL, JOHN WESLEY

1834–1902

John Wesley Powell is best known as the one-armed explorer of the Colorado River, but his later work as an advocate of land reform earned him consideration as North America's first great bioregional thinker. As a leading intellectual of his time, he also made lasting contributions to geology, linguistics, and anthropology. Born on March 24, 1834, in Mount Morris, New York, Powell grew up on a succession of family farms in Ohio, Illinois, and Wisconsin.

An ardent abolitionist, he enlisted in the U.S. Army at the outbreak of the Civil War, ultimately attaining the rank of major. Wounded at the Battle of Shiloh in 1862, he suffered amputation of his right arm but eventually returned to active duty and served in the Vicksburg and Nashville campaigns.

Soon after the war he conceived the idea of exploring the largest remaining blank spot on the map of the United States south of Alaska: the canyons of the legendary Colorado River and its tributaries. Powell aimed to map the region, ascertain its character, and assess its usefulness.

On May 24, 1869, at Green River Station, Wyoming Territory, Powell and nine others shoved four wooden dories into fast water and floated out of sight, bound either for history or oblivion. The loss of a boat and its provisions to an early rapid strained group cohesion. Later, in the Grand Canyon, weeks of scant food, brutal heat, arduous labor, and terrifying rapids prompted three men to leave the expedition and attempt to walk to civilization. They died in the attempt. Powell and the remaining five (one other had departed in June) safely reached the Virgin River in Nevada Territory on August 30, after a wilderness journey of some 900 miles.

The success of the expedition led Congress to create the Geographical and Geological Survey of the Rocky Mountain Region, with Powell in charge. In that capacity, through the 1870s, Powell explored and mapped the Colorado Plateau. He also pursued his long-standing interest in Native American ethnography, producing *Introduction to the Study of Indian Language* in 1877. Two years later he became head of the newly created Bureau of Ethnology in the Smithsonian Institution.

Powell produced an even more groundbreaking work in 1878. His *Report on the Lands of the Arid Region of the United States* proposed a thorough overhaul of laws and policies governing settlement of western lands. He based his argument on two ideas, which, he said, the government and its people would ignore at their peril.

The first was that the land had limits. Most of the United States west of the hundredth meridian received fewer than twenty inches of precipitation annually, which

John Wesley Powell, 1873. *Best known for his explorations of the Colorado River, Powell is also remembered for his contributions to geology, linguistics, and anthropology. He is seen here talking to a Paiute Indian during a survey of northern Arizona.* U.S. GEOLOGICAL SURVEY.

was inadequate to sustain unirrigated agriculture. The nation's homestead and preemption laws, which were predicated on development of 160-acre farms, were therefore a prescription for failure. Powell's second idea was that the character of institutions governing settlement—systems for water rights and land tenure, for instance—would directly shape society's prospects for wealth, justice, and democracy.

Powell offered an alternative system, which he continued to revise for another dozen years, culminating in three articles published in *Century Magazine* in 1890. The cornerstone of Powell's plan was to redraw the political landscape of the West as a system of what he called "watershed commonwealths," achieving "local self-government by hydrographic basins" (Powell 2001, pp. 306, 308). Citizens in each watershed would use the land as they collectively wished, while the interlocking interests of irrigators, stock raisers, loggers, and others kept abuses to a minimum.

Powell published the articles because he was under attack. Since 1881 he had directed the U.S. Geological Survey, which Congress formed by combining Powell's survey with two others. In 1888, at Powell's urging, Congress authorized the Irrigation Survey within the Geological Survey to identify reservoir sites and other lands necessary for development of irrigation agriculture throughout the West. As a result of events that Powell neither controlled nor advocated, the work of the Irrigation Survey led to a temporary ban on settlement within the public domain in 1889. The resulting furor destroyed much of Powell's public influence, and he resigned from the Geological Survey in 1894.

He continued as director of the now-renamed Bureau of American Ethnology until his death in 1902; among his important contributions to anthropology is *Indian Linguistic Families of North America*. Nevertheless, Powell's legacy is richest as a philosopher of land use. He held a largely utilitarian view of nature and was no environmental saint, but like few others in his time or since, he fathomed the interrelation of society, land, and water and spoke its truth relentlessly.

SEE ALSO *Land Ethic; Utilitarianism.*

BIBLIOGRAPHY

Powell, John Wesley. 2001. *Seeing Things Whole: The Essential John Wesley Powell*, ed. William deBuys. Washington, DC: Island Press.

Stegner, Wallace. 1954. *Beyond the Hundredth Meridian: John Wesley Powell and the Second Opening of the West.* New York: Houghton Mifflin.

Worster, Donald. 2001. *A River Running West: The Life of John Wesley Powell.* New York: Oxford University Press.

William deBuys

PRAGMATISM

Pragmatism as a school of philosophy is concerned with practical consequences of ideas. Because environmental issues are practical by their very nature, the central doctrines of pragmatism have been of interest to philosophers and others concerned with environmental ethics.

ORIGINS AND BASIC PRECEPTS OF PRAGMATISM

The formalization of pragmatism began in the United States during the 1870s as a reaction to the epistemological foundationalism and mind-body dualism of René Descartes, the doctrine of the transcendental ego and the "thing in itself" advanced by Immanuel Kant, and the sensory atomism of British empiricists such as John Locke. Although the movement had many contributors, including Jane Addams, George Herbert Mead, and F. C. S. Schiller, the most influential among the founding pragmatists were Charles S. Peirce (1839–1914), William James (1842–1910), and John Dewey (1859–1952).

Reduced to its most basic form, pragmatism is a family of theories of meaning, truth, and inquiry. Pragmatists hold that the meaning of a concept lies in its conceivable practical consequences. Peirce treated truth as a limit toward which disciplined scientific inquiry would eventually converge. For James true ideas are those that we can assimilate, validate, corroborate, and verify. Dewey identified truth with warranted assertibility.

What these treatments of truth have in common is a rejection of key elements of traditional correspondence and coherence theories of truth, as well as an emphasis on the active, selective, reconstructive activity of human inquiry. Peirce, James, and Dewey agreed that inquiry begins with a state of doubt or disequilibrium that is genuinely felt (and not just theoretical or feigned); that it proceeds by way of problem formation, hypotheses, and tests; and that, when successful, it results in a new state of belief or equilibrium in which the subjective and objective conditions that occasioned the original doubtful state have both been satisfied. Dewey characterized this as a process of accommodation of an organism to relevant external conditions and alteration by the organism of relevant external conditions. These two processes together he termed "adjustment" (Dewey 1934, p. 12).

PRAGMATISM AND NATURE

In addition to these basic tenets, there are pragmatism's commitment to the continuity between human life and the rest of nature; its rejection of control or mastery over nature in favor of management of undesirable conditions; its doctrine of fallibilism (or the view that absolute certainty regarding existential affairs is unattainable); and its treatment of ethics as contextual and constructive. Although the roots of pragmatism are in the sciences, it is not scientistic: It rejects the idea that the methods and the conclusions of the natural sciences are applicable to all areas of experience, as well as the notion of the natural sciences as value-free. Each of the founding pragmatists accepted some variety of evolution, and each held some variety of naturalism (as opposed to supernaturalism).

Although Peirce, James, and Dewey shared these and other basic ideas, they differed regarding other matters, including issues that are now debated by environmental philosophers. Peirce and James, for example, accepted various forms of panpsychism, or the view that preference or feeling at the very least (Peirce), or perhaps even consciousness (James), pervades the universe. It was Dewey, however, whose published work most directly interfaces with contemporary debates among environmental philosophers.

Dewey, for example, responded to Thomas Huxley's argument that the world comprises two orders: a "cosmic" order of struggle and strife and an "ethical" order of sympathy and cooperation. He rejected the underlying dualism of Huxley's argument, suggesting instead that there is only one nature, that human beings are a part of nature, and that the relation of human beings to the rest of nature is like that of a gardener who artfully utilizes one part of nature in order to manage another part (Dewey 1898). Dewey's concept of the relation of human beings to their environment thus anticipates the work of the pioneering environmentalist Aldo Leopold, for whom management was a key concept.

PRAGMATISM AND THE ENVIRONMENT

Early in the twentieth century, Dewey's critique of idealist and mystic approaches to environmental issues anticipated some issues currently under debate. He understood idealism, for example, to posit ultimate, inherent environmental value or values that transcend mere human valuation. Contemporary versions of this view have been advanced by those such as Holmes Rolston III, who has held that there are inherent values in nature that transcend human experience and that should be the basis of human values.

Dewey understood the mystic position to hold that rational debates about the environment are pointless—that environmental value is romantically or aesthetically felt and thus requires an attitude of direct vision. A contemporary version of this view has been held by those such as Michael Zimmerman, who has argued for a spiritual appreciation of nature that can only be precipitated by a new type of nondualistic awareness.

Dewey criticized both of these positions. Against the idealist he argued that it is not some ultimate or transcendent value beyond our experience that properly informs our decisions, but instead careful deliberation about real alternative courses of action that is based on the best available evidence. Against the mystic he argued that a romantic or aesthetic appreciation of nature might be a good place to begin deliberation, and that aesthetic qualities pervade inquiry about alternative courses of actions, but that inquiry into wider environmental concerns would be blocked if human relations to the rest of nature were to remain purely romantic or aesthetic.

ENVIRONMENTAL PRAGMATISM

The term *pragmatism* as used in the phrase *environmental pragmatism* has been used by Andrew Light and others to denote an approach to environmental philosophy that incorporates some of the central insights of the founding pragmatists in a general sense but that is at the same time broader and more open-ended. In 1996 Light and Eric Katz published a watershed collection of essays titled *Environmental Pragmatism.* Their aim was to vitalize what they regarded as a sluggish and deadlocked discipline by transcending received methodological dogmas. They argued that environmental ethics up to that point had been too dependent on theoretical considerations and that in order to establish its relevance as a discipline it must begin to address areas of practical decision making in which experimental results trump what Dewey termed the vice of "intellectualism"—that is, neglect of concrete experiences. They issued a call for metatheoretical pluralism that would be open to the "plausibility of divergent ethical theories working together in a single moral enterprise" (Light and Katz 1996, p. 4). Since the publication of that volume, Light has been actively engaged with issues such as environmental justice and ecological restoration.

The work of Bryan G. Norton provides an excellent example of how this new, broader pragmatism can be married to the insights of the classical pragmatists. Norton's distinction between felt and considered preferences, for example, recalls Dewey's distinction between what is merely valued and what has proved to be valuable. Both Norton and Dewey reject what Norton has termed *strong anthropocentrism*, or the view that felt preferences are sufficient guides for action with respect to the environment. Both prefer versions of "weak anthropocentrism," a view that rejects the underlying subject-object dualism of strong anthropocentrism, with its tendency toward the domination of nature. Weak anthropocentrism advances a transactional view of nature according to which the source of environmental value lies within considered preferences that are the result of experimental inquiry, although the locus of such value may be in nonhuman nature.

Although Norton's work is built on strong theoretical foundations, it is also manifestly practical. His experience at the Environmental Protection Agency led him to conclude that the agency suffers from a structural defect that he terms *towering*: He argues that the two towers of scientific analysis and policy decisions fail to connect (Norton 2005, p. 27). His alternative, a pragmatic model that he terms "adaptive management," attempts to build bridges between the two towers by emphasizing the experimental nature of successful inquiry of all types. He argues for a methodological naturalism that "advocates developing self-corrective processes in public discourse, whether scientific or evaluative" (p. 203). Norton's work is thus grounded in experimentalism.

Norton attempts to supplant metaphysical and ideological debates about environmental issues with deliberation based on the development of sets of "indicators." Indicators express but do not represent values. They are "certain processes and changes in the system [that are] important enough to monitor. Once these processes are chosen for monitoring, it will be possible to identify goals that can be stated as desired levels to be achieved and maintained with respect to the chosen indicators" (Norton, p. 453). Norton describes his method as a kind of "disciplinary stew" in which ordinary language and commonsense policy discourse make up the broth, whereas the meat and potatoes is the work of the empirical sciences (p. 461).

Paul B. Thompson's work stands out as an excellent example of the ways in which environmental pragmatism can be applied to the ethics of food biotechnology. Like Norton, Thompson has been strongly influenced by the work of the founding pragmatists; like Dewey, for example, he just brackets traditional metaphysical debates such as those that concern the existence of inherent or intrinsic values in nature. A key feature of his work is his rejection of "rational-choice" decision models in favor of more pragmatic "context" models that take into account both the social institutions that inform and condition decisions about values, and the needs and interests of specific communities at specific times and places. In his view, "people do not frame their lives as a series of objectives for which they are seeking the most effective means," (Thompson 2007, p. 286) but rather, "people apply their own values in selectively adopting or rejecting scientific knowledge claims" (p. 287). For Thompson the task of the environmental philosopher is to function as a kind of liaison between scientific communities and the individuals and communities that are attempting to find ways of applying their values amid changing circumstances.

Rejection of rational-choice models is also a key element in the work of the pragmatist economist Daniel W. Bromley. Bromley rejects the positivist dogma that "correct decisions necessarily follow from the correct methods of discovering the 'truth' about monetary values of nature and nature's many services" (Bromley 2004, p. 85). Applying his alternative pragmatist model, he treats preferences as constructed rather than fixed, stable, and known. He follows Dewey in arguing that warranted assertions about environmental issues are not the result of the identification of "causes" that are external to deliberation. They are, instead, the result of "the incessant working out of... what seems the better thing to do in the current setting and circumstances" (p. 86).

Environmental pragmatism in the Netherlands exhibits a rich blend of classical pragmatism at the same time it builds on the type of discourse ethics developed by Jürgen Habermas. In their 2002 proposal for a pragmatic environmental ethics, Jozef Keulartz, Michiel Korthals, Maartje Schermer, and Tsjalling Swierstra embraced the founding pragmatists' antifoundationalism, antidualism, and antiskepticism. They argued that work toward Dewey's goal of a creative democracy will be more oriented to process than to product and that it must develop procedures to ensure that all involved parties have their say in ethical debates. But they also recognized that procedures can become deadlocked and that substantive interventions are sometimes necessary. They therefore embraced activities such as "studying problem translations, sketching possible future scenarios, and developing new moral vocabularies" in order to facilitate participation across communities of interest (Keulartz et al., 2002 p. 16). They are concerned to demonstrate methods by which environmental ethics can begin to cope with the dynamic character of our technological culture.

SEE ALSO *Environmental Philosophy: V. Contemporary Philosophy; Norton, Bryan; Rolston III, Holmes.*

BIBLIOGRAPHY

Bromley, Daniel W. 2004. "Reconsidering Environmental Policy: Prescriptive Consequentialism and Volitional Pragmatism." *Environmental and Resource Economics* 28: 73–99.

Dewey, John. 1898. "Evolution and Ethics." In *The Collected Works of John Dewey: The Early Works, 1882-1898*, Vol. 5, ed. Jo Ann Boydston. Carbondale and Edwardsville: Southern Illinois University Press, 1972.

Dewey, John. 1909. "Nature and its Good: A Conversation." In *The Collected Works of John Dewey: The Middle Works, 1899–1924*, Vol. 4, ed. Jo Ann Boydston. Carbondale and Edwardsville: Southern Illinois University Press, 1977.

Dewey, John. 1934. *A Common Faith*. In *The Collected Works of John Dewey: The Later Works, 1925-1953*, Vol. 9, ed. Jo Ann Boydston. Carbondale and Edwardsville: Southern Illinois University Press, 1986.

Keulartz, Jozef; Michiel Korthals; Maartje Schermer; and Tsjalling Swierstra. 2002. *Pragmatist Ethics for a Technological Culture*. Dordrecht, Germany: Kluwers.

Light, Andrew and Eric Katz, eds. 1996. *Environmental Pragmatism*. New York: Routledge.

Norton, Bryan G. 2005. *Sustainability: A Philosophy of Adaptive Ecosystem Management*. Chicago: University of Chicago Press.

Thompson, Paul B. 2007. *Food Biotechnology in Ethical Perspective*. 2nd edition. New York: Springer.

Larry A. Hickman

PRECAUTIONARY PRINCIPLE

The basic message of the precautionary principle (PP) is that on some occasions, measures against a possible hazard should be taken even if the available evidence is not sufficient to consider the existence of that hazard as a scientifically established fact. Thus, PP is about acting to prevent harm in situations in which it is uncertain whether there is a threat or, if there is a threat, how serious it is.

PP has been used primarily in environmental contexts but also in regard to risks to human health. The areas in which it has been thought applicable are highly diverse and range from the regulation of chemicals and genetically modified organisms to research into life-extending medical therapies. It generally has been discussed in the context of threats of serious or irreversible harm, that is, threats that cannot be compensated for easily, such as species extinction.

PP refers both to principles that can be applied by decision makers and policy makers in general (PP in the broad sense) and to principles of national or international law (PP in the legal sense). In the broad sense, some commentators have interpreted it as one or another familiar principle from decision theory, such as maximin, a decision rule that states that an agent should choose the action for which the worst possible outcome is the least bad. They typically take the assumedly risk-neutral strategy of maximizing expected utility (MEU) to be the default rule in risk management and regard PP as more risk averse than MEU.

PP in the broad sense is regarded sometimes as a moral principle and sometimes as a principle for decision making that can be justified on moral or prudential grounds. Those who attempt to make a moral argument for PP have, among other things, appealed to some principle of responsibility emanating from the huge scope of human action, as Hans Jonas did in *The Imperative of Responsibility* (1984) before the term *precautionary principle* was coined. Another suggestion refers to a Rawlsian thought experiment, in which agents behind a veil of ignorance choose a PP. John Rawls (1921–2002) was a philosopher who argued for the maximin principle by using a thought experiment in the form of a hypothetical bargaining situation in which agents agree on the basic principles for a just society. To ensure fair bargaining in the thought experiment, the participants are supposed to be behind a "veil of ignorance." This means that each person is ignorant about what his or her own chances in society will be. He or she does not know anything about his or her sex, ethnic origin, skills, personal characteristics, intelligence, and so on. Rawls argued that these bargainers would choose a maximin principle that maximizes the position of the least well-off in society.

Several commentators also treat PP as a principle of common sense, citing proverbs such as "better safe than sorry" and "an ounce of prevention is worth a pound of cure."

It is fruitful to distinguish between precaution and precautionary principles. Someone might take precautions in a particular case without subscribing to any principle of precaution. For a principle to be present, it may be necessary to demand that the agent at a minimum subscribe to a claim such as "precaution should be taken in situations of type T." The fact that the agent believes that precaution should be taken in the particular situation S is not sufficient. In addition, there must at least be a claim such as "precaution should be taken in situation S and in relevantly similar situations," with the meaning of "relevantly similar" somehow specified.

Various versions of PP have been included in several international legal and policy documents. PP has had proponents primarily among environmentalists, and it has been subject to a heated debate, in particular since the mid-1990s.

HISTORY OF THE PRECAUTIONARY PRINCIPLE

The origin of the term *precautionary principle* is not clear. Obviously, examples of precautionary regulatory and policy measures in regard to health and environmental issues appeared long before the term was coined. One example is the British Alkali Act of 1874. A more recent example is the so-called Delaney Clause, a 1958 amendment to the U.S. Federal Food, Drug, and Cosmetic Act, which banned the use of potentially carcinogenic food additives. A common claim in the literature is that PP first emerged in West German environmental law in the 1970s under the name of *Vorsorgeprinzip*. The United Nations *World Charter for Nature* (1982) contained PP-like wording without using the phrase *precautionary principle*. The term was used explicitly in the Ministerial Declaration of the Second International Conference on the Protection of the North Sea (1987).

IMPORTANT USES OF THE PRECAUTIONARY PRINCIPLE

The most prominent and frequently cited version of PP is probably the one found in Principle 15 of the 1992 Rio Declaration (United Nations Conference on Environment and Development 1993):

> In order to protect the environment, the precautionary approach shall be widely applied by States according to their capabilities. Where there are threats of serious or irreversible damage, lack of full scientific certainty shall not be used as a reason for postponing cost-effective measures to prevent environmental degradation.

This version of PP has been included nearly verbatim in many other documents, although the English text of the declaration does not use the term *principle*. However, translations into several other languages use an expression that corresponds to it directly, and the passage cited here commonly is referred to as expressing PP.

PP has had a prominent place in the European Union (EU), at least since its inclusion (without a definition) in the Maastricht Treaty (1992). An important document is the *Communication from the Commission on the Precautionary Principle* of the Commission of the European Communities issued in 2000 and endorsed at the European Council's meeting in Nice in December 2000. The text and reasoning in these documents are reflected in several other EU documents, such as the so-called General Food Law.

Though not in any way binding, the Wingspread Statement on the Precautionary Principle that was adopted by a conference of environmentalists in 1998 has had considerable influence in the PP debate:

> When an activity raises threats of harm to human health or the environment, precautionary measures should be taken even if some cause-and-effect relationships are not fully established scientifically (cited in Raffensperger and Tickner 1999, pp. 354f).

The Wingspread version has been cited by both advocates and critics of PP. Its advocates regard it as an example of a powerful version of PP that would have the ability to change policy in a more precautionary direction, and its critics see it as an example of an extreme, rigid, and stifling principle.

VERSIONS OF THE PRECAUTIONARY PRINCIPLE

Whether one regards PP in the broad sense or the legal sense, there are numerous versions of it. They can be classified in at least three perhaps overlapping groups that regard PPs as (1) rules of choice; (2) epistemic rules or principles; or (3) procedural requirements.

When interpreted as rules of choice, many existing versions of PP have four common elements and a common structure: (1) the threat dimension; (2) the uncertainty dimension; (3) the action dimension; and (4) the prescription dimension. The common structure can be expressed in the following "if" clause containing the four dimensions:

> *If* there is (1) a threat, which is (2) uncertain, *then* (3) some kind of action (4) is mandatory.

The actual phrasing varies among different versions of PP, and the action can refer to abstaining from action as well. Some versions of PP are very strict, whereas others are significantly more permissive. The Wingspread Statement is an example of PP as a rule of choice.

Epistemic versions of PP are rules not for choosing courses of action but for determining what to believe. One example would be to demand that evidence suggesting a causal link between an activity and possible harm should be given greater weight than it would in other circumstances. In this way, some commentators have wanted to apply PP in the appraisal or assessment of risks rather than in the management of risks. One possible version of PP along these lines is the use of precautionary defaults, that is, a cautious or pessimistic assumption that is used in the absence of adequate information in the assessment of risks; that assumption can be replaced when such information is obtained. Epistemic versions of PP have been discussed primarily in academic contexts.

Procedural versions of PP are not algorithms for choosing particular courses of action but requirements for how such decisions are to be made. For instance, what arguments are to be considered legitimate? On whom does the burden of proof fall? The version of PP from the Rio Declaration belongs to this broad group of PPs. In this category, burden-of-proof requirements have been discussed extensively. Commentators who propose burden-of-proof versions of PP claim that those who wish to carry out a potentially harmful activity should be required to demonstrate that that activity will be safe before it is allowed to proceed rather than demanding that regulatory authorities provide evidence of harm before banning the activity. Thus, for example, genetically modified (GM) foods and the use of hormones as growth promoters in meat production should be presumed to be harmful until the producers can provide evidence that the products are safe. Another procedural version of PP concerns the level of scientific evidence required to trigger precautionary action. Some PP proponents have suggested that comparatively scant scientific evidence should suffice to warrant treatment of something as harmful (pending further information), whereas a lot more evidence should be required for it to be accepted as safe.

THE PRECAUTIONARY PRINCIPLE AND ENVIRONMENTAL RISK MANAGEMENT

There are two primary ways in which PP has been thought to be able to contribute to improved management of environmental risks. The first is the possibility of increased safety. Proponents of PP often contrast their approach with that of traditional cost-benefit analysis. Citing historical examples, they argue that cost-benefit analysis has led to overly large risks being taken. They also have been critical of the idea of safe levels, as expressed in concepts such as assimilative capacity and threshold doses for toxic substances and radiation. (A threshold dose is a level of exposure to, for instance, a toxic substance below which no harmful effects occur.) Researchers may be mistaken about the threshold or about which level is safe or whether there is a safe level at all, and it is better to err on the side of caution. In radiation protection, this is expressed in the acronym ALARA, which recommends that exposure be kept *a*s *l*ow *a*s *r*easonably *a*chievable.

The second application concerns the role of science in policy. Proponents of PP argue that too much faith in scientific evidence has led to the neglect of some risks. Science has limitations, and sometimes policy makers have failed to consider those limitations. For example, in pure science, type I errors (false positives: concluding that there is a phenomenon or an effect when there is none) generally have been regarded as much more problematic than type II errors (false negatives: missing an existing phenomenon or effect). However, from a policy point of view, type II errors such as believing that a highly toxic substance is harmless may be more serious. Burden-of-proof and other procedural versions of PP have been seen as offering possibilities for balancing this situation. In addition, it is hoped that PP, in particular its procedural versions, will offer a way of ensuring public influence and legitimacy. This view has been put forward by those who see risk management as an activity run by experts with too little regard for public concern about environmental matters.

OBJECTIONS TO THE PRECAUTIONARY PRINCIPLE

Obviously, not everyone shares these views, and PP has been subject to criticism. There are at least five main types of objections to it. First, it has been criticized for being ill defined and too vague to be useful in regulation. It is true that there are definitional problems with PP and that there are many different versions of it. However, this criticism is not necessarily devastating because vague principles and concepts abound in regulation; they can be made more precise through explication and practice. Furthermore, some argue that to be operational, political principles such as PP need a certain vagueness or flexibility that allows for different interpretations.

Second, PP has been criticized for forcing decision makers to pay unreasonable attention to extremely unlikely scenarios. In the extreme, it thus may be self-refuting. If PP is understood as prohibiting courses of action that may lead to harm, PP will prohibit every action, including the action of taking precautionary measures, because any action may have unforeseen harmful consequences, and it never can be proved that there are no such consequences. A principle prohibiting every course of action, then, is self-refuting. Some decision theorists have produced a version of this argument in which a formalized version of PP is shown to be incoherent or incompatible with reasonable desiderata.

The third objection is a related but weaker claim. According to this objection, PP is not incoherent but counterproductive. That is, the precautionary measures prescribed by PP would lead to more risk taking, not less. For instance, precautionary measures against pesticides or potentially unsafe GM crops might lead to famine. This is known as risk trade-off. Increased costs induced by regulation might lead to greater risks than the ones against which the regulatory measures were supposed to safeguard. This argument has been proposed by a number of U.S. commentators who see PP as a disguised form of trade protectionism.

The fourth objection is that PP is a value judgment, not a scientific judgment. This objection is dismissed by proponents of PP because it is a principle for decision making, and such principles are necessarily normative.

The fifth objection is that PP does not take science seriously and marginalizes the role of science in policy making. This objection is relevant, but it may have more to do with differing views on how to treat scientific uncertainty. Several PP proponents emphasize that decisions should be based on science but that more attention should be paid to the handling of scientific uncertainty.

Proponents of PP have responded to these and similar objections by amending or revising their versions of PP. For instance, to meet the charges of incoherence and counterproductivity, it has been suggested that PP should be supplemented with some sort of de minimis principle, which requires that extremely improbable scenarios be disregarded. An example of an attempt at amending PP can be found in the 2000 communication from the Commission of the European Communities, which states that precautionary measures should, among other things, be proportional to the chosen level of protection, nondiscriminatory, consistent with other similar measures, based on an examination of potential costs and benefits of action (and lack of action), subject to review in the light of new scientific evidence, and capable of assigning responsibility for who should produce the scientific evidence needed for a more thorough risk-benefit assessment.

Such proposals aim at making PP more operative. However, their implementation might mean that in the end very little is left of PP.

SEE ALSO *Cost-Benefit Analysis; Economism; Environmental Law; Environmental Policy; Extinction; Food Safety; Genetically Modified Organisms and Biotechnology; Pesticides; Rio Declaration; Risk Assessment.*

BIBLIOGRAPHY

Commission of the European Communities. 2000. *Communication from the Commission on the Precautionary Principle.* Available from http://www.ec.europa.eu/environment.

Gardiner, Stephen M. 2006. "A Core Precautionary Principle." *Journal of Political Philosophy* 14(1): 33–60.

Jensen, K. K. 2002. "The Moral Foundation of the Precautionary Principle." *Journal of Agricultural and Environmental Ethics* 15(1): 39–55.

Jonas, Hans. 1984. *The Imperative of Responsibility: In Search of an Ethics for the Technological Age.* Chicago: University of Chicago Press.

Peterson, M. 2007. "Should the Precautionary Principle Guide Our Actions or Our Beliefs?" *Journal of Medical Ethics* 33(1): 5–10.

Raffensperger, Carolyn, and Joel Tickner, eds. 1999. *Protecting Public Health and the Environment: Implementing the Precautionary Principle.* Washington, DC: Island Press.

Sandin, P. 2007. "Commonsense Precaution and Varieties of the Precautionary Principle." In *Risk: Philosophical Perspectives,* ed. Tim Lewens. New York: Routledge

Sandin, P.; M. Peterson; O. Hansson; et al. 2002. "Five Charges against the Precautionary Principle." *Journal of Risk Research* 5(4): 287–299.

Sunstein, Cass R. 2005. *Laws of Fear: Beyond the Precautionary Principle.* Cambridge, UK, and New York: Cambridge University Press.

Trouwborst, Arie. 2002. *Evolution and Status of the Precautionary Principle in International Law.* The Hague, Netherlands: Kluwer Law International.

United Nations Conference on Environment and Development. 1993. *The Earth Summit: The United Nations Conference on Environment and Development,* introduction and commentary Stanley P. Johnson. London and Boston: Graham and Trotman/Martinus Nijhoff.

Per Sandin

PRESERVATION

By the late nineteenth century Americans began to think seriously about setting aside areas of land to protect them from commercial development. Proponents of preservation believed that certain places should be shielded from human exploitation and devoted to less intrusive human ends—recreational, aesthetic, and spiritual—or protected simply as a gesture of respect for the landscape itself. Preservation became an early focal point of a set of concerns that later fell under the rubric of environmentalism. Preservation later came to overlap and compete with related philosophies such as conservationism (sometimes called resourcism) and restorationism.

The idea of preservation gives rise to two philosophical questions: First, what does it actually mean to preserve something? Second, what should be preserved? The historical and contemporary debates over preservation center on these questions.

EARLY PRESERVATION: 1800s TO 1960s

Early gestures at environmental preservation focused as much on the preservation of heritage (or the preservation of origins)—whether of the nation or the human species—as they did on preserving particular physical places. The American transcendentalist Henry David Thoreau (1817–1862) worried that the preservation of wildness

Avalanche Peak, at Yellowstone. *Avalanche Peak is a part of the Absaroka Mountain Range, on the eastern border of Yellowstone National Park. The peak is one of the most popular sites for hiking in the park. Although popular for recreational activities, national parks are concerned first and foremost with preservation and conservation. Yellowstone was established in 1872, when the main focus of preservation in the United States was on wilderness areas.* NPS. PHOTO BY BOB GREENBURG.

(often misquoted as "wilderness") was necessary to offset the looming development of America; he wrote, "in Wildness is the preservation of the world" (Callicott and Nelson 1998, p. 37). Concerned mainly with protecting the remaining wild places or wilderness areas in the United States, John Muir (1838–1914) was also a leading nineteenth-century preservationist. Muir grounded his preservation efforts in a variety of arguments: heritage ("going to the mountains is going home"), a wide-range of instrumental values (from watershed protection to mental therapy for "tired, nerve-shaken, over-civilized" urbanites), and even deeper intrinsic value ("This Sierra Reserve . . . is worth the most thoughtful care . . . for its own sake") (Callicott and Nelson 1998, pp. 48–62).

Muir's commitment to a specific place—Yosemite's Hetch Hetchy Valley—pitted him against the utilitarian-motivated U.S. Forest Service chief, Gifford Pinchot (1865–1946). Pinchot proposed damming the Tuolumne River to provide cheap water and electricity to the burgeoning human population of San Francisco (a proposal that was considered conservationist at the time), whereas Muir proposed preserving the valley from this human encroachment. This famous philosophical and political battle sharply and dramatically delineated the distinction between conservation and preservation. This distinction still divides the American environmental movement as well as environmental philosophy and serves as the touchstone of American natural resource education today.

From 1919 until his death, the American ecologist, forester and environmentalist Aldo Leopold (1887–1948) wrote extensively on the importance of wilderness preservation. His early writings focused almost exclusively on the recreational value of such areas, whereas his later writings reflected on the value of preservation to science as a criterion of ecological normality and a measure of "land health."

From the late 1800s to the 1960s, preservation efforts focused largely on setting aside large landscapes and ecosystems such as national parks and wilderness areas in the national forests. The desire for preservation appears to be correlated with our success at fulfilling our

Manifest Destiny (the belief that the United States was destined to expand from the Atlantic to the Pacific seaboards) and a growing sense that we had conquered enough, that it was time to set some areas aside for the preservation of our human and national heritage. The preservation of specific places was codified in the U.S. Wilderness Act of 1964, which sought to establish "a National Wilderness Preservation System" (Callicott and Nelson 1998, pp. 120–130). This characterization is echoed in contemporary discussions of preservation. The philosopher G. Stanley Kane, for example, defines *preservation* as "setting aside areas that still remain undisturbed and protecting them against human encroachment," and he defines *restoration* as "bringing degraded areas back to something resembling an unspoiled condition" (Kane 2000, p. 221). Early preservationist philosophy was manifested in the creation of environmental groups such as the Sierra Club in 1892, the Wilderness Society in 1935, and the Nature Conservancy in 1950.

PRESERVATION SINCE THE 1960S

Although philosopher-scientists such as Leopold and Victor Shelford (1877–1968) had urged the preservation of representative ecosystems, their concerns were not translated into public preservation policy, which was motivated by the aesthetic appreciation of monumental scenery (as served by the national park system) and the desire to provide outdoor recreation (as served by the national wilderness preservation system). Beginning in the late 1960s, however, preservationist concern shifted from scenic landscapes suitable for recreation to the preservation of species of plants and animals. As early as 1920, scientists began noticing with alarm the loss of species—what later became known as the loss of biological diversity or biodiversity. The biologist Francis Sumner, for example, wrote in 1920 of the "importance of saving from destruction the greatest possible number of living species of animals and plants, and saving them, so far as possible, in their natural habitats and in their natural relations to one another" (Nelson and Callicott 2008, p. 32). In the 1970s and 1980s, the biologist Edward O. Wilson became the most prominent proponent of the preservation of biodiversity. In 1985 biologists such as Michael Soulé, Reed Noss, and David Ehrenfeld founded the Society for Conservation Biology as a "mission-driven" effort to preserve Earth's biological diversity.

Environmentalists began to see that biodiversity preservation was a more urgent concern than was the preservation of heritage landscapes. The concern for the preservation of biodiversity was codified in 1973 in the U.S. Endangered Species Act (ESA), which implicitly attributes intrinsic value to, and confers legal rights on, species, subspecies, and distinct population segments, protecting them from the "consequence of economic growth and development untempered by adequate concern and conservation" (Endangered Species Act of 1973, Sec. 2(a)(1), p. 3). The ESA is one of the most powerful conservation laws in the world, shaping much of contemporary discourse about preservation; it has, therefore, become the focus of much antienvironmental critique. In fact, environmentalists themselves sometimes criticize the ESA for its overemphasis on various species and its implicit indifference to the fate of entire ecosystems.

Since the 1960s preservationists have come to focus on four main areas of concern:

- species;
- ecosystems, which include biota and abiota (the nonliving parts of an ecoystem) with an emphasis on the preservation of the functions or processes performed or the services provided by the ecosystem (e.g., nitrogen cycle, carbon budget, water filtration);
- community, which emphasizes the preservation of certain end states of biota (e.g., wilderness, grassland, wetland);
- genetic diversity.

Contemporary preservation efforts have been buoyed by scientific advances such as the ability to readily quantify and understand DNA and the realization that in the face of environmental change, it is genetic diversity (i.e., heterozygosity, allelic diversity, inbreeding coefficient, and population subdivision and structure) that promotes a species' or population's chances for survival. For example, in addition to the preservation of species, the ESA, in later amendments, allows for the preservation of distinct population segments (DPS). Although the ESA does not precisely define a DPS, most scientists use the term to refer to a population representing an important component in the evolutionary legacy of the species (U.S. Fish and Wildlife Service 1996). Conservation geneticists, however, have suggested that DPSs become more definite when defined in terms of genetic diversity and future evolutionary potential, or what are sometimes called evolutionarily significant units. There are, however, limits to scientists' understanding of the relationship between population viability and genetic diversity. Moreover, efforts to champion the preservation of species in more precisely quantifiable terms still entail normative decisions—such as what constitutes "significant" in the evolutionarily significant unit.

CRITIQUE OF PRESERVATION

There are significant disagreements among philosophers about the meaning and goals of preservation. What does it mean to preserve something? Why would you want to preserve something? It might be tempting to think that all these various foci of preservation all really converge on the same thing, that a focus on preserving scenic landscapes would result in preserving biodiversity, and further that the end result of restoration would be the same as that of preservation. There is reason to think, however, that this is simply not true, that different foci would lead to different actions on the ground with different end states. For example, the philosopher Sahotra Sarkar has pointed out that

> biodiversity conservation... cannot be identical with wilderness preservation.... [They] differ not only with respect to their explicit and implicit long-term objectives, but also with respect to their justifications, their immediate targets and obstacles, and the strategies that are likely to achieve these targets.... [Sometimes] the tasks of biodiversity conservation and wilderness preservation converge, but at least as often they do not. (Nelson and Callicott 2008, p. 231)

There is also a growing scientific literature indicating that actions that maximize the conservation of a species are not necessarily those that maximize the preservation of overall biodiversity, much less scenic or recreationally suitable landscapes. For example, scientists have growing doubts about the value of umbrella species—large "charismatic" species with large home ranges, the preservation of which was once assumed to preserve many other smaller, less "popular" species that might also exist in the critical habitat of the umbrella species.

One standard criticism suggests that preservation upholds interests of nature over the interests of humans. This criticism has been pressed most sharply by scholars and activists from the developing world. In 1989 the Indian scholar Ramachandra Guha (Callicott and Nelson 1998) pointed out that certain preservation tendencies (most notably wilderness preservation) have been ethnocentric and therefore not easily transferable to other contexts around the world without grave human consequences. Similarly, the protected-areas scholar David Harmon, echoing the views of the environmentalist Norman Meyers, suggests "that the whole notion of 'setting aside' has in fact done great damage to the conservation movement around the world" given the lack of attention that has been paid to varying ecologies in various parts of the world and the "top-down" fashion in which such environmentalism is perceived (Callicott and Nelson 1998, p. 228).

Defenders of preservation sometimes concede that they are choosing nature over humans. Philip Cafaro and Monish Verma, for example, argue that when human needs "conflict with measures that are necessary to preserve species, we believe they should be met in ways that preserve wild nature" (Rothenberg and Ulvaeus 2001, p. 60). Other preservationists deny the conflict between nature and humans. The wilderness advocate Dave Foreman, for example, suggests that these criticisms of preservation emanate from "Third World jingoism" and "chronic anti-Americanism" and that preservation "need not conflict with the needs and rights of the downtrodden" (Nelson and Callicott 2008, pp. 399–400).

Another more conceptual criticism suggests that preservation either creates or perpetuates a mentality that alienates humans from nature, whereby humans are despoilers of nature, chronic ecological malefactors. In this view the measure of successful preservation, then, is the degree to which human intervention is absent. This conceptual alienation opens up the door for the misanthropy and elitism that we have sometimes seen in the environmental movement The book *Defending the Earth: A Dialogue between Murray Bookchin and Dave Foreman* (Bookchin and Foreman 2001) nicely captures the tension between advocates and opponents of this viewpoint.

Others have taken exception to preservation strategies that attempt to reconcile the dualism between humans and nature; these critics view such strategies as preventing preservation efforts in areas that are moderately or heavily affected by humans. Referring to a concern about the ways in which preservation (in this case, of the Arctic National Wildlife Refuge in Alaska) can divert attention from other, equally important environmental issues (such as the agrarian landscape), the writer Wendell Berry confesses that he "made a sort of vow... [to not] support any more efforts of wilderness preservation that were unrelated to efforts to preserve economic landscapes and their human economies.... We can[not] preserve either wilderness or wilderness areas if we can't preserve the economic landscapes and the people who use them" (Berry 2008, p. 601). William Cronon likewise laments the need to ignore and even erase the rich legacy of erstwhile human settlement in the Apostle Islands of Wisconsin in order to create a "proper" Apostle Islands wilderness area (Nelson and Callicott 2008).

BEYOND PRESERVATION

Instead of a focus on the preservation of either processes (e.g., evolutionary) or end states (e.g., wilderness areas or biodiversity reserves), some have suggested that the goal should be preservation (or conservation or restoration) of an appropriate human relationship with nature. In this approach preservation implies the implementation of virtues such as humility, respect, attentiveness, and care. On this view the problem of preservation is the problem of figuring out how humans ought to relate to nature. Some have suggested that preservation is much more a gesture of respect than it is a desire to preserve a state or process. For example, the philosopher Andrew Light writes that the value of restoration lies "in the revitalization of the human relationship with nature" (Kane 2000, p. 95).

SEE ALSO *Biodiversity; Conservation; Endangered Species Act; Environmental Activism; Environmental Aesthetics; Environmental Law; Guha, Ramachandra; Hetch Hetchy; Leopold, Aldo; Muir, John; Nature Conservancy; Pinchot, Gifford; Sierra Club; Society for Conservation Biology; Thoreau, Henry David; Utilitarianism; Wilderness; Wilderness Act of 1964; Wilson, Edward O.*

BIBLIOGRAPHY

Allendorf, Fred W., and Gordon Luikart. 2006. *Conservation and the Genetics of Populations.* Boston: Wiley-Blackwell.

Berry, Wendell. 2008. "Hell, No. Of Course Not. But..." In *The Wilderness Debate Rages On: Continuing the Great New Wilderness Debate,* ed. Michael P. Nelson and J. Baird Callicott, 601–602. Athens: University of Georgia Press.

Bookchin, Murray, and Dave Foreman. 1991. *Defending the Earth: A Dialogue between Murray Bookchin and Dave Foreman.* Cambridge, MA: South End Press.

Callicott, J. Baird, and Michael P. Nelson, eds. 1998. *The Great New Wilderness Debate.* Athens: University of Georgia Press.

Endangered Species Act of 1973 (61 U.S.C. 1531–1544, 87 Stat. 884).

Kane, G. Stanley. 2000. "Restoration or Preservation? Reflections on a Class of Environmental Philosophies." In *Environmental Restoration: Ethics, Theory, and Practice,* ed. William Throop. Amherst, MA: Humanity Books.

Light, Andrew. 2000. "Restoration or Domination? A Reply to Katz." In *Environmental Restoration: Ethics, Theory, and Practice,* ed. William Throop. Amherst, MA: Humanity Books.

Nelson, Michael P., and J. Baird Callicott, eds. 2008. *The Wilderness Debate Rages On: Continuing the Great New Wilderness Debate.* Athens: University of Georgia Press.

Rothenberg, David, and Marta Ulvaeus, eds. 2001. *The World and the Wild.* Tucson: University of Arizona Press.

U.S. Fish and Wildlife Service. 1996. "Endangered Species Program." *Federal Register* 61: 4722. Available from http://www.fws.gov/endangered/POLICY/Pol005.html.

Michael P. Nelson
John A. Vucetich

PRIVATE PROPERTY

The term *property* is used to refer to entities for which an individual or group has certain rights. Important types of property include real property (land), personal property (other physical possessions), and intellectual property (inventions and intellectual and artistic creations). Private property is defined in contrast to public property: The possessor of private property customarily has exclusive rights to dispense with the property as he or she sees fit, including its use, transfer, and ownership.

NATURAL RIGHTS

To talk about property is to talk about property rights. Theorizing about property rights in the Anglo-American philosophical tradition has its roots in theorizing about natural rights and natural law. To claim a natural right is to assert that some rights are grounded in human nature rather than in legislation or other acts of government. Claims about natural rights and natural law purport to say something about what legal rights should be rather than what they are. Human rights and property rights often are represented as natural rights.

The Dutch thinker Hugo Grotius (2005 [1625]) secularized the idea of natural law. In his treatment natural law theory became a naturalistic inquiry into the question of which social arrangements were most conducive to the betterment of humankind. Grotius argued that there would be laws of nature, as dictated by the requirements of human nature, even if there were no deity. According to the Grotius scholar Stephen Buckle (1993), property rights are a social creation generated and defined by agreement but still are natural rights in a crucial way insofar as they are inevitable in human social life.

John Locke followed in the footsteps of Grotius. That made his work enormously influential among secular philosophers even though Locke was a theist. Locke (1960 [1690]) argued that God gave the world to humankind in common for the betterment of humankind and therefore intended that people should have the right to do what they need to do to put the earth to work. Individual persons own their own selves. Persons are God's property, but in relation to other humans, the individual alone holds the right to decide how his or her body is to be put to work.

This right to choose how to put people's bodies to work would be useless in that original state and God would be leaving humankind to starve unless people also were at liberty to make a living by laboring on otherwise unowned objects in the world. People normally are not at liberty to seize what already belongs to someone else, but when a resource is not owned, a person can come to own it by mixing his or her labor with it in a way that makes it more useful. Thus, people acquire a crop by virtue of being the ones who planted and harvested it and acquire the land underneath the crop by virtue of being the ones who made that land more productive than it had been in its unappropriated wild condition.

Working within the Lockean tradition, William Blackstone characterized property as the "sole and despotic dominion which one man claims and exercises over the external things of the world, in total exclusion of the right of any other individual in the universe" (1765–1769). In practice, though, property rights in the Anglo-American tradition are and always have been hedged with restrictions. The dominion to which Blackstone referred is limited by easements, covenants, nuisance laws, zoning laws, regulatory statutes, and, more generally, the public interest.

THREE KINDS OF RULES

In normal cases property is protected by a property rule that states that no one may use it without the owner's permission. In other circumstances one might say property is protected by a liability rule that states that no one may use it without compensating the owner. In a third kind of case one might say property is protected by an inalienability rule that states that no one may use it even with the owner's permission (Calabresi and Melamed 1972).

The fundamental rationale for liability rules is that sometimes it costs too much or is impossible to get consent to use someone's property and sometimes the contemplated use is compellingly important. For example, people have no right to use their cars to run over a neighbor's fence. Yet every time one pulls out of a driveway, there is some risk that that person's plans will go awry and he or she will accidentally damage someone's property. Whereas a property rule would require people to get advance permission from every property owner against whom they run the risk of committing accidental trespass, a liability rule requires instead that they compensate the owner of a fence after the fact if they accidentally damage it. The analogous rationale for an inalienability rule is that there are forms of property so fundamental that people would cease to be persons in the fullest sense if they were to, for example, sell them. A person may, for example, regard his or her kidney or vote as his or her property yet deny that this confers right to sell such things. In this respect the person would be treating his or her right as inalienable.

Spur Industries, Inc. v. Del E. Webb Development Co. 494 P.2d 701 (Ariz. 1972) is a case that was settled almost simultaneously with the publication of Guido Calabresi and A. Douglas Melamed's article. In this case the judge, who had not read the article, ruled in favor of the housing developer Del Webb, granting an injunction against the feedlot operator Spur Industries while holding that Del Webb had to compensate Spur. The judge reasoned that Spur's property claim was valid but that because the feedlot was a public nuisance, Spur could be forced to move, with compensation paid by Del Webb because Del Webb's housing development brought the public to the nuisance. Thus, Spur's property right in effect was protected by a liability rule rather than a property rule. The takings clause of the Fifth Amendment to the U.S. Constitution can be interpreted as specifying that in cases of compelling public interest property titles are protected by a liability rule rather than a property rule.

PROPERTY RIGHTS AS RIGHTS TO EXCLUDE

In 1913 Wesley Hohfeld distinguished between rights and liberties. One is at liberty to use something just in case one's using that thing is not prohibited. One has a right to a thing just in case one's using it is not prohibited, plus one has the additional liberty of being able to prohibit others from using that thing. That is, the difference between a mere liberty and a full-blooded property right is that with a property right there is an owner who holds a right to exclude other would-be users.

In the contemporary era, though, the term *property rights* generally is understood to refer to a bundle of rights that could include the rights to sell, lend, bequeath, use as collateral, or even destroy. (John Lewis [1888] generally is regarded as the first person to use the "bundle of sticks" metaphor.) The fact remains, though, that at the heart of any property right is a right to say no: a right to exclude nonowners. In other words, a more or less Blackstonian right to exclude is not just one among many sticks in a property bundle. Rather, property is a tree, and whereas other sticks are branches, the right to exclude is the trunk. One conceptual reason for saying this is that without the right to say no, the other rights are reduced to mere liberties rather than genuine rights. For example, one can be the owner of a bicycle in a meaningful sense even if for some reason one has no right to lend it to anyone else. By contrast, if one has no right to deny another permission to lend it to anyone else, one is not the owner of the bicycle in any normal sense. Thus, there is a conceptual reason why, among the various sticks that make up the bundle of rights people call property, the right to exclude is the most essential.

In addition to this conceptual reason, there is a practical reason why the right to exclude is the core of a property right. The evolution of property law is driven by an ongoing search for ways to enable people to avoid commons tragedies. Commonly held resources are subject to indiscriminate overuse. If a productive asset is held in common, there is an additional risk that no one will be willing to bear the cost of investing in making the resource more productive. In practice, communal regimes also lead to indiscriminate dumping of wastes that range from piles of unwashed dishes to substances that lead to ecological disasters that threaten whole

Private Beach in Malibu, California. *Many theorists believe that the right to exclude nonowners is at the core of any property right. While this American homeowner boasts such rights to private property, other nations have an eminent domain clause that stipulates that coastlines are the property of the state, and therefore no property owner can prohibit access to the sea. Private property as it is commonly understood is defined in contrast to public property, and grants its possessor exclusive rights to dispense with the property as he or she sees fit.* AP IMAGES.

continents. When people's activities impose a cost on innocent bystanders, economists describe those costs as external costs or externalities. There also can be external benefits, or positive externalities, as in a case where one person notices that a second person has developed an innovative new technique for plowing land and copies that technique. The first person is reaping some of the benefit of the second person's research without having paid for it. Property in patents is one tool for "internalizing" the benefits of research and development.

Centuries before the analytical tools of game theory, ecology, and contemporary economics were available, Locke noted that the option of fencing the commons and creating private property enables people to make land more productive than land is when left in the wild. The reason for this is that the right to exclude empowers owners to capture the positive benefits of productivity and at the same time internalize the negative impact of overuse. The right to exclude is for practical purposes the essence of property in general, not only private property. A national park service must be able to exercise a right to exclude cattle ranchers to do its job of protecting national parks. An Israeli kibbutz must be able to exercise a right to exclude nonmembers to be able to feed its members.

FORMS OF PROPERTY

Commonly held property is not necessarily an ecological disaster. For example, the open-field agricultural practices of medieval times gave peasants exclusive cropping rights to scattered thin strips of arable land in each of the village fields (Ostrom 1990). The strips were private only during the growing season, after which the land reverted to the commons for the grazing season. The custom of "stinting" allowed the villagers to own livestock only in proportion to the relative size of their land holdings in the growing season. Governance by custom enabled communal owners to avoid commons tragedies. David Schmidtz and Elizabeth Willott (2003) described a modern case of successful communal management.

Private property is not necessarily an ecological success. Privately managed parcels also are subject to indiscriminate dumping of wastes and various other uses that ignore spillover effects on neighbors. One advantage of private property is that owners can buy out one another and reshuffle their holdings in a way that minimizes the extent to which their activities bother one another. However, this does not always work out well, and the reshuffling itself can be wasteful because there are transaction costs. Such costs include all the expenses incurred in concluding a transaction: commissions, time and money spent on transportation to and from the market, equipment and space rentals, time waiting in line, and so on.

A plausible social goal would be to have a system of property that minimizes the sum of transaction costs and the cost of externalities. For example, a public freeway may minimize transaction cost if the alternative is a private toll road, but the toll road may invest revenues in road repairs, minimizing the external cost of damage to the road caused by thousands of other drivers. There often is a trade-off. Harold Demsetz (1967), Robert Ellickson (1993), and Carol Rose (1986) considered which mix of private and public property best meets this goal. Especially when there are far-flung externalities among people who do not know one another, there is no easy way to determine which mix of private and public property is best. The difficulties in detecting such externalities, tracing them to their source, and holding people accountable for them are difficulties in any kind of property regime.

Privatization exists in different degrees and can take different forms. Simply parceling out land or sea is not always sufficient to stabilize possession of resources that make land or sea valuable in the first place. Suppose, for example, that fish are a fugitive resource, that is, a resource known to migrate from one parcel to another. In that case owners have an incentive to grab as many fish as they can whenever a school passes through their territory. Thus, simply dividing fishing grounds into parcels may not help fishers avoid collectively exceeding sustainable yields. It depends on the extent to which the sought-after fish migrate from one parcel to another and on conventions that are evolving continuously to help neighbors deal with the inadequacy of their fences or other ways of marking off territory. Thus, not all forms of privatization are equally good at internalizing externalities. Communal management generally does not work, but privatization per se is not a panacea.

SEE ALSO *Economics, Environmental; Environmental Politics; Land Ethic; Natural Law Theory; Tragedy of the Commons.*

BIBLIOGRAPHY

Blackstone, William. 1765–1769. *Commentaries on the Laws of England.* Oxford, UK: Clarendon Press.

Buckle, Stephen. 1993. *Natural Law and the Theory of Property: Grotius to Hume.* Oxford, UK: Clarendon Press.

Calabresi, Guido, and A. Douglas Melamed. 1972. "Property Rules, Liability Rules and Inalienability: One View of the Cathedral." *Harvard Law Review* 85: 1089–1128.

Demsetz, Harold. 1967. "Toward a Theory of Property Rights." *American Economic Review* 57(2): 347–359.

Ellickson, Robert C. 1993. "Property in Land." *Yale Law Journal* 102: 1315–1400.

Grotius, Hugo. 2005 (1625). *Rights of War and Peace,* ed. Richard Tuck. Indianapolis: Liberty Fund.

Hardin, Garrett. 1968. "The Tragedy of the Commons." *Science* 162: 1243–1248.

Hohfeld, Wesley. 1964. *Fundamental Legal Conceptions, as Applied in Judicial Reasoning,* ed. Walter Wheeler Cook. New Haven, CT: Yale University Press. (Orig. pub. in two parts in 1913 and 1917.)

Lewis, John. 1888. *Treatise on the Law of Eminent Domain in the United States.* Chicago: Callaghan.

Locke, John. 1960 (1690). *Two Treatises of Government,* ed. P. Laslett. Cambridge, UK: Cambridge University Press.

Ostrom, Elinor. 1990. *Governing the Commons: The Evolution of Institutions for Collective Action.* Cambridge, UK, and New York: Cambridge University Press.

Rose, Carol. 1985. "Possession as the Origin of Property." *University of Chicago Law Review* 52(1): 73–88.

Rose, Carol. 1986. "The Comedy of the Commons: Custom, Commerce, and Inherently Public Property." *University of Chicago Law Review* 53: 711–787.

Schmidtz, David. 1994. "The Institution of Property." *Social Philosophy and Policy* 11: 711–781. Revised in Schmidtz 2008.

Schmidtz, David. 2008. *Person, Polis, Planet: Essays in Applied Philosophy.* New York: Oxford University Press.

Schmidtz, David, and Elizabeth Willott. 2003. "Reinventing the Commons: An African Case Study." *University of California, Davis, Law Review* 37(1): 203–232. Revised in Schmidtz 2008.

Spur Industries, Inc. v. Del E. Webb Development Co. 494 P.2d 701 (Ariz. 1972).

David Schmidtz

The author thanks research assistant Jacqueline Figueroa for thoughtful feedback.

PROCESS PHILOSOPHY

Process philosophy names a range of philosophical theories. The term *process* implies rejection of substances as inherently or ontologically primary. It normally identifies philosophies that assert the independent existence of nature and the location of human beings within it. These philosophies encourage people to take seriously the course of natural events, and, indeed, the rise of evolutionary thinking was one key source of process philosophies. However, in the twentieth century, developments in physics took on primary importance, and evolutionary thought fell into the background.

ALFRED NORTH WHITEHEAD

In the twenty-first century, process philosophy refers chiefly to that of Alfred North Whitehead (1861–1947) and those influenced by him, and it is on this broadly Whiteheadian tradition that this entry will focus. Whitehead was a mathematician, logician, and theoretical physicist who devoted extensive attention to relativity theory. He arrived at his philosophical views partly through reflection about physical fields and partly by the analysis of his own experience. He thought that all the indivisible entities that constitute the world are something for themselves as well as for others. They are all momentary experiences consisting largely of unconscious feeling and emotion. This view can be called *panexperientialism.* This term is better than the more common *panpsychism* because the latter suggests consciousness and even high-grade experience, whereas in panexperientialism consciousness is limited to fairly complex animals. Whitehead's point is that to be at all is to be something in and of itself and not simply an object in the experience of others.

Humans as subjects are aware that some feeling states are preferable to others. They can thus form some judgments from their own experience that are applicable to what occurs elsewhere. In Whitehead's magnum opus, *Process and Reality* (1929), intensity serves as the measure of the value of an experience. In *Adventures of Ideas* (1933) he developed a much more complex theory of value in which "strength of beauty" plays the central role.

Whitehead believed that human experience, like processes in the physical world, consists of a sequence of distinct indivisible momentary experiences rather than a continuous flow. The later experiences inherit from the earlier ones by including them, or, alternatively, the earlier experiences flow into the later ones. This relation is called a *feeling* or a *prehension.* The ultimate locus of subjectivity, and therefore of value, is the individual experience.

Examination of an individual, momentary experience clarifies the implications for ethics. A moment of human experience inherits extensively from predecessor moments. For adult human beings, this is the most important relation. However, each momentary experience inherits from other sources as well, most obviously from neuronal events. These mediate both new sense data and new bodily feelings. The new occasion is an integration of what it inherits from its personal past and the new stimuli from the body.

There is not first an occasion of experience that then inherits from the past and the body. The new occasion comes into being through the synthesizing of these relations. Thus relations are prior to the entity that is synthesizing them, although the priority is not temporal, because the relations and their synthesis all come into being at once.

Whereas in substance thought relations are external, in this process vision they are internal. In substance thought, relations among substances are chiefly spatial, and their changes do not affect the nature of the substances. For process thought, relations to past occasions are internal to the one that is becoming. A human experience is largely constituted by its relations, both to other persons and to the rest of the world.

Whitehead himself may have come to this radically relational vision more from field theory than from introspection. A physical field is constituted by a multiplicity of local events. But one can conceive of the events equally well as what the field is at a locus. It does not consist of entities that could exist elsewhere. What exists elsewhere is, by virtue of its location, a different entity. Whitehead concluded that every indivisible event is what it is largely as a fresh integration of features of its given world, one unique to each occasion.

Since, in such a model, the ethical concern is to increase value, the question is how occasions achieve more or less value. The greater value is achieved by the inclusion and integration of more of what the world offers. Diversity in these offerings makes possible greater value in the integration. But since diverse elements taken by themselves are often incompatible, simple occasions are forced to exclude most of them. More complex occasions are able to convert what would otherwise be mutually exclusive into a contrast in which the diversity is retained and the whole is richer for having both the different elements and the new values acquired through contrasting them.

Whitehead thus emphasized the importance of the environment. However, each occasion also transcends what it inherits from its world. It achieves value through some flash of novelty that introduces fresh contrast. This is the basis of the vibratory character of primitive physical entities as well as of the creative elements introduced by life. What novelty is possible is determined by what is given. The novelty itself is an expression of the creativity of the universe and is derived from the eternal order of potentiality, which Whitehead called the primordial nature of God. Thus the individual creatures that make up the environment both call forth respect for themselves by their intrinsic value and also contribute to the value of human experience in two ways: first, by the intensity or richness of their own individual experience and, second, by their diversity, which provides for their perceivers the possibility of rich contrasts.

CHARLES HARTSHORNE

Charles Hartshorne (1897–2000) has been the most influential philosopher in the development and promulgation of Whitehead's philosophy. He combined his philosophical pursuits with an interest in birds, and was himself deeply concerned about such ecological matters as the loss of habitat for other species. Hartshorne was a metaphysician and philosopher of religion. He focused attention on what Whitehead treated rather briefly as the consequent nature of God.

Both Whitehead and Hartshorne believed that what happens fleetingly in the world is cumulatively real in God. Without this retention of value in the divine life, the transitoriness of all things would undercut the importance of what happens in the world. Because what we do to ourselves and to our fellow creatures we do also to God, this process tradition undergirds ethics with religious faith. Hartshorne called it *panentheism.* His ethical goal was contribution to the divine life; so he called his ethics *contributionist.*

RESPONSES TO THE ENVIRONMENTAL CRISIS

Process thinkers were prepared by their philosophical understanding to respond to the environmental crisis. Although few engaged in explicit formulation of environmental ethics, much of their writing has been permeated by environmental concerns. Noteworthy examples are Ian G. Barbour, a physicist, and David Ray Griffin, a philosopher of religion.

John B. Cobb, a theologian, drew together the implications of process philosophy for environmental ethics, and in 1972 he published *Is It Too Late? A Theology of Ecology* (revised in 1955). Meanwhile, Charles Birch, a Whiteheadian biologist specializing in ecology, was trying to get the World Council of Churches (WCC) to respond to the new concern. This effort succeeded in 1975 at the Nairobi meeting. The WCC changed the earlier call for "just and participatory societies" to ones that are "just, participatory and *sustainable.*" Birch gave the plenary address hailing and explaining this move. Subsequently, Birch and Cobb teamed up to write *The Liberation of Life* (1981), in which they argue for an ecological view of living things to replace the materialist one still dominant in the sciences.

Whitehead's philosophy supports concern both for ecosystems and for individual animals. The WCC concerned itself only with the former. The Whiteheadian philosopher of religion Jay McDaniel joined Birch in organizing a WCC meeting at Annecy, France, in 1988, with an emphasis on individual animals. The report produced by this meeting had little influence in the WCC, but it led to the publication of *Liberating Life* (McDaniel 1990). McDaniel also led in the American Academy of Religion in dealing theologically with animals. Additionally, he has written extensively on ecological spirituality.

Since 1972 Cobb has worked with the Whiteheadian economist Herman Daly contributing an essay to his *Toward a Steady-State Economy*, published in 1973.

In 1994 they jointly authored *For the Common Good*, proposing that economics be rethought on different assumptions. Daly has been a leader in ecological economics.

A number of religious thinkers, influenced by process philosophy, have been prominent in ecofeminism. These include Karen Baker-Fletcher, Rita Nakashima Brock, Carol Christ, Nancy R. Howell, Carol Johnston, Catherine Keller, Sandra Lubarsky, Sallie McFague, Mary Elizabeth Moore, Rebecca Parker, and Marjorie Suchocki.

CRITIQUES

Some of these ecofeminists, other ecofeminists, ecologically concerned Buddhists, and Deep Ecologists criticize Whiteheadian philosophy for advocating a "hierarchy of values" that judges some creatures to be of greater intrinsic value than others. They hold that this is too closely connected with anthropocentric and/or hierarchical thinking, locating human beings at the top, and evaluating others according to their proximity to humans. Defenders of Whitehead's gradation of values respond that practical decisions based on such judgments are inescapable and that Whitehead's criteria for judging the amount of value are not crudely anthropocentric. Although whales resemble humans less than do monkeys, the occasions of whale experience are probably richer. There may be very different creatures elsewhere in the universe with higher grades of value. God's experience, as process thinkers imagine it, is incomparably richer.

Clare Palmer offers the most thorough criticism of Whitehead as environmental ethicist. She considers the judgment that other beings resemble humans in having subjectivity to be "dominating" or "colonizing." She argues that the generalization from human experience suppresses the difference of the other. One response is that the usual ways of describing their otherness deprive them of intrinsic value altogether, whereas Whitehead attributes intrinsic value to all.

Palmer shows that a variety of ethical approaches can find a foothold in Whitehead's philosophy, but she concludes that it is finally consequentialist. Consequentialists fail to attend to relations to the past, such as promises that have been made. Whitehead certainly encouraged the increase of value, but occasions aim at value not only in their future but also in the present. Because each occasion is composed largely of relations to the past, such relations as promise keeping are not be ignored. Cobb dealt with technical ethical questions of this sort in his contribution to Daly's *Toward a Steady-State Economy.*

Palmer finds some of Whitehead's language about nonhuman animals demeaning. Although Whitehead's basic emphasis was to reduce the difference between human beings and other species, she is right that he took human superiority for granted. She also notes rightly that Whitehead thought of civilization as pure gain. He did not appreciate the quality of life in hunting and gathering societies or their arguably superior environmental record. Overall it is true that, even though Whitehead does not emphasize evolutionary and historical progress, he did take many of the humanist assumptions of his time for granted. To follow his philosophy does not require repetition of all his judgments.

Lisa Sideris has followed the theologian James Gustavson in formulating norms for evaluating environmental ethical positions. She judges, surprisingly, that process thought fails to emphasize affective judgments. Although Whitehead used reason extensively in his arguments and considered the emergence of rationality as a great gain, his ultimate appeal was to intuition. Emotion is the primary reality and is involved in every judgment.

Sideris argues for a radically theocentric ethics and judges that process ethics is too anthropocentric. On the one hand, she objects to the view that God aims at the increase of value in the world and that faith in God contributes to well-being. On the other, she objects that whereas process thinkers oppose the sacrifice of animals for minor human gains, they would sacrifice other creatures to meet urgent human needs. This is true. Process thinkers oppose the fragmentation of thought and do not want an environmental ethics that threatens deep concern for the well-being of every human, and especially for the poor and oppressed. Their goal is a single ethic that integrates concern for the biosphere, animal suffering, and human flourishing. They give less attention to refining ethical formulations than to promoting broader changes in attitude and in spirituality. In particular, they think the theoretical changes most urgently needed are in such fields as agriculture and economics.

SEE ALSO *Animal Ethics; Ecological Feminism; Economics, Ecological; Ecotheology; Environmental Philosophy: V. Contemporary Philosophy; Pantheism.*

BIBLIOGRAPHY

Barbour, Ian G. 1993. *Ethics in an Age of Technology*. San Francisco: HarperSanFrancisco.

Birch, Charles, and John B. Cobb Jr. 1981. *The Liberation of Life from the Cell to the Community*. Cambridge, UK: Cambridge University Press.

Birch, Charles; William Eakin; and Jay B. McDaniel, eds. 1990. *Liberating Life: Contemporary Approaches to Ecological Theology*. Maryknoll, NY: Orbis Press.

Cobb, John B., Jr. 1995 (1972). *Is It Too Late? A Theology of Ecology*, rev. edition. Denton, TX: Environmental Ethics Books.

Cobb, John B., Jr. 2004. "Thinking about Whitehead and Nature." In *Whitehead's Philosophy: Points of Connection,* ed. Janus A. Polinowski and Donald W. Sherburne. Albany: State University of New York Press.

Daly, Herman E. ed. 1973. *Toward a Steady-State Economy*. San Francisco: Freeman.

Daly, Herman E., and John B. Cobb Jr. 1994. *For the Common Good: Redirecting the Economy toward Community, the Environment, and a Sustainable Future.* 2nd edition. Boston: Beacon Press.

Griffin, David Ray, ed. 1988. *The Reenchantment of Science: Postmodern Proposals*. Albany: State University of New York Press.

Hartshorne, Charles. 1978. "Foundations for a Humane Ethics: What Human Beings Have in Common with Other Higher Animals." In *On the Fifth Day: Animal Rights & Human Ethics,* ed. Richard Knowles Morris and Michael W. Fox. Washington DC: Acropolis Books.

McDaniel, Jay B. 1990. *Earth, Sky, Gods, & Mortals: Developing an Ecological Spirituality*. Mystic, CT: Twenty-Third Publications.

Palmer, Clare. 1998. *Environmental Ethics and Process Thinking*. Oxford, UK: Clarendon Press.

Palmer, Clare. 2000. "Religion in the Making? Animality, Savagery, and Civilization in the Thought of A. N. Whitehead." *Society and Animals* 5: 287–304.

Sideris, Lisa H. 2003. *Environmental Ethics, Ecological Theology, and Natural Selection*. New York: Columbia University Press.

John B. Cobb Jr.

Q

QUEER THEORY

Queer theory, which emerged as an academic area of study in the early 1990s, is an interdisciplinary body of scholarship that attempts to understand the way power works as a productive and normalizing force that shapes personal identity, human experience, knowledge, policy making, and political movements. For example, queer theorists question the norms of gendered embodiment that make a person recognizable as a man or a woman and their relationship to sexual norms such as heterosexuality and monogamy. Though influenced by feminist theory and gay and lesbian studies, queer theory builds on the philosopher Michel Foucault's (1926–1984) insight that power is productive rather than being only restrictive and that identity and resistance are effects of societal practices that normalize certain types of behavior and bodies compared with others. As a force of normalization, power takes on the appearance of being natural and inevitable while inducing conformity in the individual. Queer theorists have used Foucault's insights about power to interrogate understandings of gender identity and desire.

The term *queer* often is used as an umbrella designation to refer to lesbian, gay, bisexual, and transgender people; however, within the context of queer theory the term is best understood as a verb. In other words, queer theory aims to queer—to question and defamiliarize—the identities, concepts, knowledge, and experiences that shape lives, values, societies, policies, and academic inquiry. Central to queer theory is a critique of heteronormativity, which is understood both as the institutionalized assumption that heterosexuality and conventional modes of being in a relationship such as marriage and monogamy are normal and natural and as the pressure and obligation to conform to those modes as a condition for respect, legitimacy, recognition, and citizenship.

Queer environmental philosophy seeks to theorize the role of assumed norms of identity, experience, and embodiment that inform the deployment of terms such as *nature, the natural, sustainability, future generations*, and *the common good.* These are key terms which underlie environmental philosophy, policy, and activism, but queer environmentalism recognizes and critiques their heteronormative bias. Queer environmentalism also makes visible and critiques connections between the disavowal of that which is understood to be queer and environmental destruction. There are many areas of overlap between queer environmental philosophy and ecological feminism. In order to understand the commonalities between queer environmental philosophy and ecological feminism, as well as their differences, one must first understand some of the ways in which feminism has influenced queer theory.

QUEER THEORY AND FEMINISM

Queer theory is influenced by and shares the feminist critique of the sex-gender system of oppression and compulsory heterosexuality, both of which rely on the presumed naturalness and inevitability of the connection between biological sex, gender, and desire/sexuality. Both feminism and queer theory expose the social and cultural nature of the sex-gender system and its long-standing role in rationalizing the oppression of women and sexual minorities.

Despite some feminist disagreement with the critique of identity found in queer theory, that critique has roots in feminist interrogation of the concept of woman as well as Foucault's understanding of power and subjection. Foucault understood subjection both as the subjection of the individual to dominating force and the production of the individual as a subject. Building on that concept, queer theory characterizes all identity categories as effects of power and not as pure sites of resistance to power, and its critique of identity includes a critique of the subjects around which feminism and lesbian and gay studies have been organized: woman and lesbian and gay. In 1990 the philosopher Judith Butler used Simone de Beauvoir's famous question, "What is a woman?" as a point of departure and asked how the subject of feminism (woman) is "produced and restrained" by the power structures through which women's liberation is sought (Butler 1990, p. 2). Gender, Butler argued, is performative, the materialization of historical and cultural norms that make gender and sex visible, real, and meaningful (Butler 1990, 1993, 2004). Butler's performative theory of gender was a major contribution to feminist theory and was foundational for queer theory.

ECOLOGICAL FEMINISM AND QUEER ENVIRONMENTALISM

There also are areas of overlap between ecological feminism and queer environmentalism. For example, both are critical of gender and nature essentialism and value dualisms. Essentialism is the view that all the members of a group (e.g., women) share unchanging characteristics, an essence that is necessary for membership in that group. Queer environmentalists and ecological feminists contend that essentialist definitions do not portray gendered reality or the relationship between humans and nature accurately. Both queer environmentalism and ecological feminism have been influenced by Donna Haraway's (1991) concept of the cyborg, calling into question distinctions between human beings, nature, and technology. Chris Cuomo distinguished between ecofeminism and ecological feminism, arguing that much of ecofeminism is premised on universal, essentialist conceptions of woman, woman's connection to nature, and feminine virtues that ignore how white, wealthy women in the North benefit from globalized exploitation of the earth, poor people, and people of color (Cuomo 1992).

In the 1990 essay "The Power and Promise of Ecological Feminism" Karen J. Warren argued that a critique of value dualisms is an important contribution of ecofeminism. Value dualisms are opposites that categorize dominant conceptions of reality: man/woman, human/animal, culture/nature, mind/body, reason/emotion, and white/black, for example. These opposites are value dualisms because in each pair the second term is subordinated to the first and the meanings attached to the first term define cultural norms. In value dualisms the terms in each pair are "seen as exclusive (rather than inclusive) and oppositional (rather than complementary)," and characteristics associated with the first term are considered to be better than those associated with the second (Warren 2000, p. 46). In an attempt to bridge the gap between ecofeminism and queer theory, Greta Gaard proposed a queer ecofeminism that challenges the oppression of the erotic, an oppression shaped by "reason/erotic" and "heterosexual/queer" dualisms that define Western culture and are integral to the oppression of queers, women, indigenous people, and nature (Gaard 1997, pp. 118–119).

Queer environmentalism differs from ecological feminism in the centrality of Foucault's concept of power and Butler's performative theory of gender to its analysis of environmental problems. For instance, oppression is the predominant concept of power that fuels the ecological feminist critique of interlocking systems of oppression (e.g., racism, sexism, speciesism, classism, oppression of the earth) that support violence against all oppressed humans, nonhuman animals, and nature. Although queer theorists do not deny the existence of oppression, they consider analyses of power centered on oppression to be incomplete. When power is understood as primarily oppressive, the world is divided into those who have power (those who exercise power) and those who do not (those over whom power is exercised), and resistance becomes a matter of making visible power imbalances that characterize oppression, explaining their injustice, and seeking liberation for the oppressed. However, queer theorists in general and queer environmentalists in particular are interested in the ways in which power operates in the myriad coincidences of knowledge and power that generate norms, condition subjectivity, and increase the institutionalized management of individuals.

QUEER CRITIQUES OF ENVIRONMENTALISM

Queer analyses of environmental problems focus on the ways in which the discourse of environmentalism produces subjects that ultimately legitimate rather than undermine globalized capitalism and its role in environmental destruction. Catriona Sandilands argued that environmentalism is a "normalizing discourse" that assumes that there are limits inherent in nature (Sandilands 1999, p. 80). Consequently, both radical and mainstream environmentalism tend to focus on "self denial" as the hallmark of the environmentally responsible citizen who is implicitly juxtaposed against people and governments in the "Third World" that have failed to exercise appropriate self-discipline (Sandilands 1999, pp. 80–81). Even though many, including ecofeminists, critique specific forms of population control, Sandilands contended that most environmentalists have not challenged population/fertility management as a goal (Sandilands 2004).

Éric Darrier (1999) also crtiqued environmental discourse, focusing specifically on the function of the concept of nature in environmental philosophy and policy. Like Sandilands, he was critical of the values and practices people are obliged to adopt in the name of "the common good," "sustainable development," or "the proper functioning of the 'ecosystem'" (Darrier 1999, p. 217). Darrier proposed a "green aesthetics of existence" rather than a green ethics of existence, a practice he defined as critically self-reflective regarding the assumed truths of nature that are used to justify environmental policies and encourage the adoption of particular values or modes of being (Darrier 1999).

Some theorists, such as Alison Kafer and Eli Clare, have examined connections between queerness, disability, and nature. Kafer argued that many environmentalist narratives about experiences with nature, including ecofeminist narratives, assume able-bodiedness (Kafer 2005). That disregard of disability, Kafer contended, contributes to conflicts between environmentalists and advocates of disability rights over issues such as the accessibility of public lands and parks. According to Kafer, many environmentalists assume that making trails and canoe launches accessible, for example, will exceed limits that are inherent in nature, as if making trails more accessible would "insert the all-too-human into the 'wilderness'" (Kafer 2005, p. 139). For Kafer the absence of the disabled body in environmentalism ultimately assumes and naturalizes a human/nature dualism that many environmentalists want to critique. As an alternative, she proposed Clare's 1999 narratives of hiking as important sources of insight into the ways in which ideas of nature can be shaped by queerness and disability. For Kafer and Clare, queering nature involves understanding that nature and disability are not mutually exclusive, that there is nothing "unnatural" about disability.

Queer environmentalism calls into question the normalization of the concepts of the human, nature, future generations, and population that inform even discourses mobilized on behalf of environmental protection. It proposes persistent questioning of the ways even well-intentioned environmental philosophy and policy can promote the exploitative interests of globalized capitalism.

SEE ALSO *Ecological Feminism; Ethics of Care.*

BIBLIOGRAPHY

Butler, Judith. 1990. *Gender Trouble: Feminism and the Subversion of Identity*. New York: Routledge.

Butler, Judith. 1993. *Bodies That Matter: On the Discursive Limits of "Sex."* New York: Routledge.

Butler, Judith. 2004. *Undoing Gender*. New York and London: Routledge.

Clare, Eli. 1999. *Exile and Pride: Disability, Queerness, and Liberation*. Cambridge, MA: SouthEnd Press.

Cuomo, Chris J. 1992. "Unraveling the Problems in Ecofeminism." *Environmental Ethics* 14(4): 351–363.

Cuomo, Chris J. 1998. *Feminism and Ecological Communities: An Ethic of Flourishing*. London and New York: Routledge.

Darrier, Éric. 1999. "Foucault against Environmental Ethics." In *Discourses of the Environment*, ed. Éric Darrier. Oxford and Malden, MA: Blackwell.

Gaard, Greta. 1997. "Toward a Queer Ecofeminism." *Hypatia* 12(1): 114–137.

Haraway, Donna J. 1991. *Simians, Cyborgs, and Women: The Reinvention of Nature*. New York: Routledge.

Kafer, Alison. 2005. "Hiking Boots and Wheelchairs: Ecofeminism, the Body, and Physical Disability." In *Feminist Interventions in Ethics and Politics: Feminist Ethics and Social Theory*, ed. Barbara S. Andrew, Jean Clare Keller, and Lisa H. Schwartzman. Lanham, MD: Rowman & Littlefield.

Sandilands, Catriona. 1999. "Sex at the Limits." In *Discourses of the Environment*, ed. Éric Darier. Oxford and Malden, MA: Blackwell.

Sandilands, Catriona. 2004. "Eco Homo: Queering the Ecological Body Politic." In *Environmental Philosophy as Social Philosophy*, ed. Cheryl Hughes and Andrew Light. Charlottesville, VA: Philosophy Documentation Center.

Warren, Karen J. 1990. "The Power and the Promise of Ecological Feminism." *Environmental Ethics* 12(3): 125–146.

Warren, Karen J. 2000. *Ecofeminist Philosophy: A Western Perspective on What It Is and Why It Matters*. Lanham, MD: Rowman & Littlefield.

Kim Q. Hall

R

REGAN, TOM
1938–

Tom Regan, best known as the philosophical father of the animal-rights movement, was born on November 28, 1938, in Pittsburgh, Pennsylvania. Regan taught for over thirty years at North Carolina State University, receiving emeritus status upon retirement. Major honors include multiple teaching awards, research awards, the William Quarles Holladay Medal, the Gandhi Award, and the Joseph Wood Krutch Medal. His work has focused on animal rights and on the philosophy of G. E. Moore. His major publications include *The Case for Animal Rights* (1983) and *Bloomsbury's Prophet: The Moral Philosophy of G. E. Moore* (1987). Both of these books were nominated for the Pulitzer Prize and the National Book Award. He also authored *All That Dwell Therein: Essays on Animal Rights and Environmental Ethics* (1982), edited *Animal Sacrifices: Religious Perspectives on the Use of Animals in Science* (1986), and authored *Defending Animal Rights* (2001). Of his many articles, two are major contributions to environmental ethics: "The Nature and Possibility of an Environmental Ethic" (1981) and "Does Environmental Philosophy Rest on a Mistake?" (1992). He founded the Tom Regan Animal Rights Archive at North Carolina State University. Tom and Nancy Regan also created the Culture and Animals Foundation to promote animal rights through art.

In his 1981 classic essay "The Nature and Possibility of an Environmental Ethic," Regan argued that a genuine environmental ethic is an ethic *of* the environment, not merely an ethic *for the use of* the environment. The latter sort of ethic regards the environment and the entities inhabiting it as valuable only inasmuch as they benefit humans (are *instrumentally* valuable). An ethic of the environment, by contrast, must hold (1) that nonhuman beings are valuable in and of themselves (are *inherently* or *intrinsically* valuable), and (2) that both conscious and nonconscious beings are inherently valuable. Regan then shows that arguments against the possibility of an environmental ethic rest on unjustified moral claims. Next, Regan exposes the failure of arguments attempting to establish that an environmental ethic is unnecessary for environmental protection. Finally, Regan sketches what it would mean for a being to be inherently valuable/good: Its value is independent of conscious beings, is an objective property arising from its other properties, and appropriately inspires respectful admiration and preservation. Regan concludes his prolegomena to any future environmental ethics by posing two unanswered questions: What property makes a being inherently good? How can we know which beings are inherently good?

In his seminal work *The Case for Animal Rights* (1983), Regan suggests some answers. Here Regan lays out a meticulous case for the inherent value of all "subjects-of-lives." Subjects-of-lives are experiencing, goal-directed beings with welfares. Clear cases of subjects-of-lives, Regan holds, are mammals over one year old. Such beings are rights bearers, having valid claims against moral agents. How can we know this? Regan claims that only rights theory containing the subject-of-a-life moral principle meets the criteria for a satisfactory ethical theory: consistency, scope, precision, conformity with reflective rational intuitions, and simplicity/parsimony.

It would be wrong to conclude that Regan, with the subject-of-a-life proposal, has ruled out an environmental ethic. He explicitly leaves open its possibility by stating that being the subject-of-a-life is sufficient, but not claimed to be necessary, for inherent value. He thus grants that nonconscious beings might be inherently valuable, although he says it is "radically unclear" how this could be so (1983, p. 246). Nevertheless, Regan holds that his rights view is compatible with an environmental ethic according inherent value to individuals. An example is Paul Taylor's individualistic ethic, defended in *Respect for Nature* (1986). Regan's framework is incompatible with a holistic environmental ethic, however; such an ethic accords inherent value solely to wholes such as species and ecosystems, attributing merely instrumental value to individuals.

In 1992 Regan published another key contribution to environmental ethics, "Does Environmental Ethics Rest on a Mistake?" Here Regan argues that the emerging paradigm for an environmental ethic is fatally flawed. According to the paradigm, natural entities are due respect from moral agents because of the noninstrumental value of these entities, with different and greater intrinsic value attributed to wild beings than to domestic beings. Regan proceeds to argue that no theory of intrinsic value can meet all these requirements for an environmental ethic. A theory according intrinsic value to beings will accord them either equal intrinsic value or different intrinsic value. The first type of theory (e.g., Taylor's biocentric egalitarianism) cannot accord more intrinsic value to wild beings than to domestic beings. The second type of theory would establish a hierarchy of intrinsically valuable beings, with lower-ranking entities being sacrificed for the sake of higher-ranking ones. Such a theory cannot ground respect for natural entities, which merely occupy positions in a hierarchy. Moreover, such a theory is indistinguishable in practice from one positing a hierarchy of instrumentally valuable beings, with a top level of intrinsic value (e.g., an ecosystem, for the sake of which individuals can be culled). This second type of theory is more parsimonious than theories that attach intrinsic value hierarchically, although it too provides no grounds for respecting natural entities.

If Regan is correct, an individualistic environmental ethic, already challenged to justify its criterion of intrinsic value, must abandon the claim that wild and domestic beings have different degrees of intrinsic value. Individualists and holists rejecting Regan's approach must show that he is mistaken here.

Regan's animal-rights philosophy has received its share of criticisms from environmental ethicists. Both J. Baird Callicott (1989) and Mark Sagoff (1984) have argued that the animal-rights view would commit moral agents to protecting all rights bearers, even prey. They hold that this would have disastrous implications for the environment. A different objection has been raised by ecofeminists such as Josephine Donovan (1993). According to this criticism, moral-rights theory is based on a patriarchal worldview that enshrines reason and individualism while denigrating emotion and community, lying also at the foundations of Callicott's approach to environmental ethics. Instead of making rights and justice paramount in ethics, ecofeminists urge that care, partiality, and nature should be at the center.

Regan has replied to these and other critics in *Defending Animal Rights* (2001). He argues that moral agents do not have the general obligation to assist prey (p. 19). Nonhuman predators do not violate prey rights, since these predators are not moral agents. Moreover, there is no duty to assist prey animals against their innocent attackers, because in general they can fend for themselves. Against the ecofeminist critique, Regan argues that it is based on misrepresentation, undefended claims, false dilemmas, and the same biases embodied by patriarchal views (pp. 54–64).

Whether one thinks Regan has succeeded or failed in his criticisms of major environmental ethical theories and in his defenses of animal-rights philosophy, he has made ground-breaking contributions to the view that moral worth is not confined to humanity.

SEE ALSO *Animal Ethics; Callicott, J. Baird; Ecological Feminism; Intrinsic and Instrumental Value; Taylor, Paul.*

BIBLIOGRAPHY

Callicott, J. Baird. 1989. *In Defense of the Land Ethic: Essays in Environmental Philosophy.* Albany: State University of New York Press.

Donovan, Josephine. 1993. "Animal Rights and Feminist Theory." In *Ecofeminism: Women, Animals, Nature*, ed. Greta Gaard, 167–194. Philadelphia: Temple University Press.

Regan, Tom. 1981. "The Nature and Possibility of an Environmental Ethic." *Environmental Ethics* 3(1): 19–34.

Regan, Tom. 1982. *All That Dwell Therein: Essays on Animal Rights and Environmental Ethics.* Berkeley: University of California Press.

Regan, Tom. 1983. *The Case for Animal Rights.* Berkeley: University of California Press.

Regan, Tom, ed. 1984. *Earthbound: New Introductory Essays in Environmental Ethics.* New York: Random House.

Regan, Tom, ed. 1986. *Animal Sacrifices: Religious Perspectives on the Use of Animals in Science.* Philadelphia: Temple University Press.

Regan, Tom. 1987. *Bloomsbury's Prophet: The Moral Philosophy of G. E. Moore.* Philadelphia: Temple University Press.

Regan, Tom. 1992. "Does Environmental Ethics Rest on a Mistake?" *The Monist* 75(2): 161–182.

Regan, Tom. 2001. *Defending Animal Rights.* Urbana: University of Illinois Press.

Sagoff, Mark. 1984. "Animal Liberation and Environmental Ethics: Bad Marriage, Quick Divorce." *Osgood Hall Law Journal* 22(2): 297–307.

Taylor, Paul. 1986. *Respect for Nature*. Princeton, NJ: Princeton University Press.

"Tom Regan Animal Rights Archive." Available at http://www.lib.ncsu.edu/animalrights

Evelyn B. Pluhar

REGIONALISM

A *region* is a portion of the earth's surface unified by a high degree of internal functional integration and formal consistency. For instance, the corn belt of the United States, which stretches from Ohio to Iowa, functions much like a vast machine, with components such as cornfields, farmhouses, county roads, feedlots, and meat-packing plants. The recurrence of these components in the landscape gives the region a uniform—some might say monotonous—appearance, smell, and sound. Because there is a perceptible consistency in the region's landscape and in the manners, customs, and culture of its people, regions are usually recognized by ordinary people, and their names are part of vernacular geography.

Regional is an adjective attached to forms (artifacts, institutions, beliefs) that are characteristic of, but not necessarily unique to, a region. We speak of a regional landscape, cuisine, dialect, architecture, or costume. In traditional human societies (folk cultures), almost all cultural traits were regional because interaction with people outside the region was limited and most artifacts and practices were adapted to the needs and opportunities of the local environment. Regional distinctiveness has been greatly reduced by modern communication and transportation technology, which increases long-distance interaction and permits functional integration on the continental and global scales.

Regionalism is self-conscious attachment to, and preservation of, regional forms. The members of a folk culture are not regionalists, and their reproduction of regional forms is not regionalism, because they reproduce regional forms out of habit and necessity. These are the only forms they know, or the forms best suited to their physical environment. Regionalism is, on the other hand, a disposition to prefer regional forms (of speech, dress, food, and architecture), when other and perhaps more cost-effective alternatives are available, simply because they are regional.

For example, in Texas, cowboy boots and cowboy hats were, originally, folk forms of dress that emerged from precedents available in the Anglo-Hispanic borderlands to meet the needs of men who spent much of their lives in the sun, with their feet in stirrups. They wore these articles because almost every other man they had ever seen wore them and because they were tools well adapted to cattle ranching in a hot climate. Today most men and women who wear cowboy boots and hats are, in contrast, self-consciously identifying themselves with the region. They are saying, "I'm from Texas and I'm proud of it!" Regional idioms, accents, foods, and the like may all become expressions of regionalism once they are used, self-consciously, as symbols of identity, pride, or defiance. Regional folk culture becomes vernacular regionalism when, and to the extent that, it is elective and expressive rather than habitual and necessary.

In addition to the vernacular regionalism of ordinary people, there is an intellectual regionalism of artists and scholars who write regional literature and history, or restore, preserve, and develop regional folkways, crafts, arts, and architecture. Intellectual regionalism is sometimes antiquarian and curatorial, an exercise in preserving the memory of a regional culture that is lifeless, however beautifully embalmed. More often, intellectual regionalists aspire to keep regional culture alive with new and creative work within the regional tradition. Their understanding of tradition has much more to do with the ideas of the poet and critic T. S. Eliot or the philosopher Alasdair MacIntyre than with Morris dancing, quilting bees, or Civil War reenactment. The best-known example of American intellectual regionalism is the work the writers, known as the Fugitives or Agrarians; centered at Vanderbilt University in the 1930s; they worked to restore pride in the southern way of life.

Vernacular and intellectual regionalism is frequently explained as a reactionary movement on the part of people who fear spatial homogenization and the universal rationality of enlightened modernity. This interpretation reveals more about the prejudices and plans of the interpreter than it does about the motives of regionalists. Its defects are the assumptions that an international style (of everything) is inevitable, that dislike and fear of modernity is irrational, and that regionalism is, at best, equivalent to throwing the covers over one's head at the sound of the bogeyman coming up the stairs. More portentously, critics of regionalism like to suggest, with very little evidence, that it is the gateway to xenophobia and fascism. Regionalism does, indeed, dispute the universal pretensions of enlightened modernity; at least in its intellectual form, however, it does so in an enlightened manner. Far from indulging in fanciful nostalgia, regionalism criticizes and challenges the core assumptions of modernity. Regionalists are not typically cowards who have thrown the covers over their head; they are, more often than not, men and women with the courage to look modernity in the eye and ask it some very hard questions.

SEE ALSO *Globalization; North America; Space/Place.*

BIBLIOGRAPHY

Davidson, Donald. 1938. *The Attack on Leviathan: Regionalism and Nationalism in the United States.* Chapel Hill: University of North Carolina Press.

Eliot, T. S. 1949. *Notes towards the Definition of Culture.* New York: Harcourt, Brace.

Glassie, Henry. 1968. *Pattern in the Material Folk Culture of the Eastern United States.* Philadelphia: University of Pennsylvania Press.

MacIntyre, Alasdair. 1988. *Whose Justice? Which Rationality?* Notre Dame, IN: Notre Dame University Press.

Scruton, Roger. 2006. *England: An Elegy.* London: Continuum.

Weaver, Richard M. 2000. In *Defense of Tradition: Collected Shorter Writings of Richard M. Weaver, 1929–1963,* ed. Ted J. Smith III. Indianapolis, IN: Liberty Fund.

Jonathan M. Smith

RESOURCE MANAGEMENT

Resource management refers to decision making about how to conserve, allocate and use the goods and services available to satisfy people's value demands. Such decision making is fundamental to people's interactions with one another and with the environment, as depicted in Harold Lasswell's model of the human social process: "participants → seeking to maximize values (gratifying outcomes) → utilize institutions → affecting resources" (Lasswell 1971, p. 18). The term *resource management* often is used to refer specifically to decision making about the goods and services available from the natural world, or natural-resource management. This includes decisions about allocating the benefits and costs of resource use among current members of society and between current and future generations. To some people resource management is simply a matter of good planning: carefully making use of available resources to provide social utility while ensuring that there is enough left to meet future needs. That conception, however, masks important ethical and philosophical debates about which approach to management is most appropriate, what the goals of management should be, and whether it is even acceptable to characterize aspects of nature as resources to be managed.

RESOURCES

A basic utilitarian definition of a resource might be "something that can be used by humans." Anthropocentric thinking of that type dominates Western models of natural resource management, although there have been dissenting voices through the years, such as George Perkins Marsh, Henry David Thoreau, John Muir, and Aldo Leopold. Critics of anthropocentrism object to the implication that the value of the nonhuman world lies solely in its ability to satisfy human desires. They point to the many failures of attempts to control or dominate nature and argue that treating the natural world as a collection of resources that can be manipulated and used to achieve human ends is misguided and arrogant.

Alternatives to anthropocentric resource management draw on a variety of philosophical traditions that posit a more equal relationship between humans and the natural world, including the worldviews of some aboriginal societies, in which humans are considered part of nature and there is an emphasis on respect for the nonhuman world; the land ethic of Aldo Leopold, in which humans are members of a broader community of the land and "a thing is right when it tends to preserve the integrity, stability, and beauty" of that community (Leopold 1949, pp. 224–225); the "democracy of all God's creatures" promoted by Saint Francis of Assisi (White 1967, p. 1206); and the more recent human-nature egalitarianism of deep ecology. In those alternative conceptions treating elements of nature as resources to be managed or assigning value to nature solely on the basis of instrumental considerations makes little sense.

Whether or not nature has intrinsic value, humans do make use of and derive benefits from the natural world. In addition, as the claims of the critics of anthropocentrism demonstrate, human values and attitudes toward nature are multidimensional, ranging from an emphasis on use and dominion to aesthetic appreciation and moral concern about natural rights. When resources are defined more broadly to include anything that provides value, resource management can encompass decisions about providing the full spectrum of those values, including protecting wilderness areas for moral reasons as well as managing other areas for recreation or consumptive use. Furthermore, it is not just elements of the natural world that are at stake; cultural, heritage, spiritual, and other human resources often are included in modern understandings of resource management, and managing human behavior can be as important as managing nature.

The physical characteristics of resources also have implications for management because those characteristics dictate the ways in which particular resources can be used or allocated to provide value. Some resources provide materials or inputs for human activities, whereas others have the capacity to assimilate and process wastes or outputs. Some resources are bought and sold in markets and have readily identifiable prices that provide an indication of their value to humans. Others, known as nonmarket resources, do not have easily determined monetary values, making it more difficult to assess their importance relative to goods and services that are exchanged in the market. Some resources, such as forests and fisheries, are renewable because when they are harvested at an appropriate rate, they can regenerate and provide a continuing supply in the future. In this sense renewable resources form a stock that can be maintained while a flow of benefits is withdrawn. Other resources, such as oil and gas, metals, and most other minerals, are nonrenewable because they cannot be regenerated within human time scales once they are consumed or converted.

Two other key characteristics of resources for management are subtractability, which indicates whether use by one person reduces the capacity of the resource to provide goods or services to others, and excludability, which indicates how difficult it is for users or managers to exclude additional users. When resources are subtractable and nonexcludable (common pool resources), they are likely to be overused unless rules or other institutional structures limit access and exploitation.

APPROACHES TO RESOURCE MANAGEMENT

Approaches to resource management have varied with the scarcity and physical characteristics of resources and have evolved with changes in values, knowledge, and technology. For example, when the American West first was occupied by settlers of European descent, resources seemed highly abundant and the principal management concerns were to allocate those resources in a fair manner and ensure that they were used to encourage settlement and economic development. Accordingly, the Western system of water rights was built on principles of "first in time, first in right," and "use it or lose it," which determine priority among users and help ensure that water is put to use but pay little attention to conservation or the effects of extraction on other users or other components of the system. Similarly, land was allocated for free to those who would homestead and farm, grazing of cattle and sheep was permitted on public lands without charge, and mining and timber rights were granted to encourage the exploitation of those resources.

With the closing of the frontier in the second half of the nineteenth century the supply of free land dwindled, and this, together with evidence of the damage caused by overuse—such as the effects of sheep ranching in the Yosemite area, the impacts of extensive land clearing and logging throughout the West, and the crashes of prominent wildlife populations—fostered the development of two main competing streams of conservation-oriented thinking about resource management. The preservationist philosophy championed by John Muir called for setting aside entire regions of the landscape to protect wilderness values, provide sanctuaries for wildlife species, and preserve examples of wild nature. Preservationists pushed for the protection of Yosemite and other national parks, and this philosophy continues to inform the management of protected areas in both terrestrial and marine environments. In contrast, the progressive conservation of Gifford Pinchot advocated scientifically informed use of resources to maximize the benefits for all: "Conservation means the greatest good to the greatest number for the longest time" (Pinchot 1947, p. 325). This utilitarian philosophy, which often is applied to the management of a single resource, informed sustained yield management policies for forests and fisheries and has become ingrained in many resource management agencies.

Management for a single use does not deal with the interactions of the different demands that humans make on their environment or the effects that utilizing one resource may have on other aspects of the environment and other people's values. In response to increasing pressures and conflicts among demands such as recreation, water supply, ranching, agriculture, and forestry, the concepts of multiple-use and integrated resource management evolved.

Under a multiple-use approach resources are managed to provide multiple benefits for a variety of users, with the overall objective of maximizing human utility. The overlap between multiple use and sustained yield is evident in legislation such as the Multiple Use Sustained Yield Act of 1960, which requires management of U.S. national forests for a sustained yield of products and services, including uses such as range, outdoor recreation, timber, and wildlife. Although multiple-use management involves a degree of integration, the idea of integrated resource management has been extended and interpreted much more broadly to mean coordinated management of social, ecological, and economic systems toward goals such as sustainable development and sustainability.

Approaches to resource management that are explicitly systems-oriented became more prevalent near the end of the twentieth century. For example, in the early 1990s many federal and state land management agencies embraced ecosystem management. The principles of ecosystem management may include managing ecological systems holistically to maintain their integrity, taking account of the disturbance regimes in which those systems evolved, managing adaptively to accommodate and learn from change, and involving the public and incorporating social considerations in management. Ecosystem-based management is a close relative of ecosystem management, which emphasizes management on the basis of ecosystems rather than management of ecosystems. Another important application of systems-oriented thinking to resource management is found in the work of C. S. Holling and his collaborators on panarchy theory, which considers the adaptive cycles and interactions of multiple economic, social, and ecological systems operating at different temporal and spatial scales and encourages management to "maintain ecological resilience as well as the social flexibility needed to cope, innovate, and adapt" (Holling 2001, p. 404). As with other contemporary models of resource management, systems-based approaches often encourage public participation and the use of traditional ecological knowledge in decision making.

GOALS OF MANAGEMENT

Underlying this diversity of management approaches is an equally diverse array of goals toward which resource management can be directed. Among the more prominent goals are the following:

- Maximizing the extraction and use of a single resource in the short term;
- Maximizing the yield of a single resource that is sustainable over the longer term;
- Maximizing the sustainable benefits of multiple resources at the same time;
- Preserving natural landscapes or systems;
- Reducing harmful pollution or achieving an economically optimal level of pollution;
- Internalizing the positive and negative externalities caused by human activities;
- Protecting human health;
- Preserving charismatic or otherwise highly valued species;
- Preserving biodiversity;
- Maintaining ecosystem health or land health;
- Maintaining ecological integrity or ecosystem integrity;
- Maintaining the resilience and adaptive capacity of systems;
- Maintaining ecosystem functions;
- Keeping systems within their historical range of variability;
- Sustainable development;
- Sustainability.

Since the report of the World Commission on Environment and Development in 1987, the goals of resource management often have been tied to the concept of sustainable development—or sustainability—of social, economic, and ecological systems. Definitions of sustainability vary, however, from sustaining the combined stocks of natural and human-created capital, to sustaining the stock of natural capital separately, to managing within safe minimum standards, which may entail precautionary management to avoid crossing thresholds and causing irreversible change.

Throughout history humans have sought to manage resources, in some cases more successfully than others. Different approaches to resource management reflect different understandings of the appropriate relationship between humans and the natural world. Whether people attempt to manage resources actively toward particular goals or manage their own behavior to allow natural systems to operate without extensive human influence, experience suggests that failure to respect the limits on the capacity of natural systems or to recognize and adapt to changes in that capacity over time can be disastrous.

SEE ALSO *Biodiversity; Conservation; Deep Ecology; Economics, Ecological; Economics, Environmental; Environmental Policy; Land Ethic; Limits to Growth; Muir, John; Preservation; Sustainability.*

BIBLIOGRAPHY

Diamond, Jared. 2005. *Collapse: How Societies Choose to Fail or Succeed.* New York: Viking.

Ehrenfeld, David. 1978. *The Arrogance of Humanism.* New York: Oxford University Press.

Grumbine, R. Edward. 1994. "What Is Ecosystem Management?" *Conservation Biology* 8(1): 27–38.

Hardin, Garrett. 1968. "The Tragedy of the Commons." *Science* 162: 1243–1248.

Holling, C. S. 2001. "Understanding the Complexity of Economic, Ecological, and Social Systems." *Ecosystems* 4(5): 390–405.

Hussen, Ahmed M. 2000. "The Economics of Sustainable Development." In *Principles of Environmental Economics: Economics, Ecology, and Public Policy,* 2nd edition. London and New York: Routledge.

Kellert, Stephen R. 1996. *The Value of Life: Biological Diversity and Human Society.* Washington, DC: Island Press for Shearwater Books.

Lasswell, Harold D. 1971. *A Pre-View of Policy Sciences.* New York: American Elsevier.

Leopold, Aldo. 1949. *A Sand County Almanac, and Sketches Here and There.* New York: Oxford University Press.

Ostrom, Eleanor. 1990. *Governing the Commons: The Evolution of Institutions for Collective Action.* Cambridge, UK, and New York: Cambridge University Press.

Pinchot, Gifford. 1947. *Breaking New Ground.* New York: Harcourt, Brace.

Sessions, George, ed. 1995. *Deep Ecology for the Twenty-First Century.* Boston: Shambhala.

Stanley, Thomas R., Jr. 1995. "Ecosystem Management and the Arrogance of Humanism." *Conservation Biology* 9(2): 255–262.

White, Lynn, Jr. 1967. "The Historical Roots of Our Ecologic Crisis." *Science* 155: 1203–1207.

Wilkinson, Charles F. 1992. *Crossing the Next Meridian: Land, Water, and the Future of the West.* Washington, DC: Island Press.

World Commission on Environment and Development. 1987. *Our Common Future: Report of the World Commission on*

Environment and Development. New York: Oxford University Press.

Worster, Donald. 1994. *Nature's Economy: A History of Ecological Ideas,* 2nd edition. Cambridge, UK, and New York: Cambridge University Press.

Murray B. Rutherford

RIO DECLARATION

The United Nations Conference on Environment and Development (UNCED), the Earth Summit, held in Rio do Janeiro in June 1992, produced a short document titled *Rio Declaration on Environment and Development.* This document was to be named the *Earth Charter*, but developing nations objected that such a name focused too much on the earth and nature, and not enough on people and development, so the title was changed. It was renamed the *Rio Declaration* for the lack of any consensus about a title with more explicit reference to its contents. The declaration states twenty-seven principles, most given in a sentence or two, while a few form short paragraphs. Although it was only six pages long, there were lengthy arguments during the proceedings over nuances of phrasing. Toward the close of the conference a document was produced, and this has since been signed by almost every nation. The United States signed, with some protests about possible misinterpretations of the language of the declaration.

Even before the conference started, developing nations had already made it clear that they did not want an earth charter. In the discussion, a First World country delegate suggested an earth charter, a short creed, that "should be framed and put in the room of every child of the world." The retort from a Third World delegate: "Not every child has a room, maybe not even a bed!" Representatives of developing nations argued that direct concern for nature was an elitist luxury of First World nations, an inhumane overlooking of human poverty. "Ecologists care more about plants and animals than about people," complained Gilberto Mestrinho, governor of the Brazilian state of Amazonas. Or concern for saving the Earth was insincere, critics objected, unless accompanied by large donations from the wealthy nations to those in developing nations being asked to preserve nature.

"Human beings are at the centre of concerns for sustainable development," so the *Rio Declaration* begins in principle 1. It goes on to say that people are entitled to "a healthy and productive life in harmony with nature." Alternative language, which was rejected, read, "Human beings are entitled to live in a sound environment, in dignity and in harmony with nature for which they bear the responsibility for protection and enhancement." Principle 4 reads, "In order to achieve sustainable development, environmental protection shall constitute an integral part of the development process and cannot be considered in isolation from it." Development is clearly the dominating motif, with environmental conservation subsidiary to it.

Principle 7 reads, "States shall cooperate in a spirit of global partnership to conserve, protect and restore the health and integrity of the Earth's ecosystem. In view of the different contributions to global environmental degradation, States have common but differentiated responsibilities. The developed countries acknowledge the responsibility that they bear in the international pursuit of sustainable development in view of the pressures their societies place on the global environment and of the technologies and financial resources they command."

Environmental ethics with any direct concern for animals, plants, species, or ecosystems was essentially stripped from the draft language for the declaration. Its ethics was much more subdued than had been anticipated, because the rich-poor controversy became so unexpectedly intense. "Changes in life styles of the rich to those that are less polluting and wasteful is essential to reaching sustainable development." So proposed the developing nations in a draft text that the developed nations rejected. The objection was not so much to eliminating waste and pollution as to the suggestion that poverty in the South (the developing nations) was the fault of overconsumption in the North (the wealthy nations).

There were widespread complaints that world population growth was insufficiently addressed in the *Rio Declaration*, as well as overall at the Earth Summit, due to ideological and religious objections. The motivations suppressing attention to population control were as often implicit as explicit: that population reduction is an effort to reduce the number of non-Western (or non-Northern) people in the world, or that population control is an easier route than sharing unequally distributed resources, or that population control violates human rights or national sovereignty, or that the large populations of the poor really consume less than the limited but extravagantly consuming populations of the wealthy nations. The *Rio Declaration* mildly says, "States should . . . promote appropriate demographic policies" (principle 8). Developing nations were much more anxious to thrust blame on the developed countries for their overconsumption.

The *Rio Declaration* contrasts, tellingly, with a much earlier UN document called the *World Charter for Nature.* This charter begins, "Every form of life is unique, warranting respect regardless of its worth to man" (United Nations General Assembly 1982). A total of 112 nations endorsed this charter, though the United States vigorously opposed it. This statement was largely aspirational; few took it to require any serious changes in policy. In contrast, the *Rio Declaration*, coupled with the massive *Agenda*

21, which accompanied it, was expected to produce changes in behavior. The diplomatic negotiations formulating the document became a kind of morality play of developed nations versus developing nations, North versus South, rich versus poor, development versus conservation.

Disappointment in the *Rio Declaration* led original advocates of an *Earth Charter* to continue their efforts, and such efforts continued during the decade following the United Nations Conference on Environment and Development. A version was completed in March 2000 at The Hague, Netherlands, and efforts to gain subscribers continue. Thousands of organizations have endorsed it, including the United Nations Educational, Scientific, and Cultural Organization (UNESCO), but not yet the United Nations General Assembly. The first principle of the *Earth Charter*, "Respect Earth and life in all its diversity," states, "Recognize that all beings are interdependent and every form of life has value regardless of its worth to human beings." The latter phrasing recalls that of the *World Charter for Nature* and was inserted with an eye to the adoption of the *Earth Charter* by the United Nations General Assembly.

The *Rio Declaration* contains some key themes that are working their way into law: the principle that the polluter pays, responsibility for spillover damage from one country to another, intergenerational equity, public participation, a precautionary approach, environmental-impact assessments, differential responsibilities, healthy environments. Despite its shortcomings, the *Rio Declaration* serves a useful purpose as a negotiated multinational instrument that can serve as an icon for environmentally responsible development.

SEE ALSO *Convention on Biodiversity; Earth Charter; Earth Summit; Ecology: III. Ecosystems; Population; United Nations Educational, Scientific and Cultural Organization.*

BIBLIOGRAPHY

Earth Charter Initiative. *The Earth Charter*. Available from http://www.earthcharter.org/

Foo, Kim Boon. 1992. "The Rio Declaration and Its Influence on International Environmental Law." *Singapore Journal of Legal Studies* 1992: 347–364.

United Nations Conference on Environment and Development. 1992. *Rio Declaration on Environment and Development.* UN Document A/CONF.151/26 (Vol. 1). Available from http://www.unep.org/Documents.Multilingual/Default.asp?DocumentID=78&ArticleID=1163

United Nations General Assembly. 1982. *World Charter for Nature.* UN General Assembly Resolution No. 37/7 of October 28, 1982. Available from http://www.un.org/documents/ga/res/37/a37r007.htm

Holmes Rolston III

RISK ASSESSMENT

In an uncertain world most important choices involve a risk of losing something of value. Risk assessment is a process of discovering the possible adverse effects of decisions and figuring out what to do about them. People try to identify and measure the likelihood of adverse effects, put a value on them, and compare them with the costs and benefits of alternatives. This is an important part of any process of rational decision making.

TYPES OF RISK ASSESSMENT

The relative importance of a decision (and the importance of the risks involved), including the expected costs and benefits of accepting, reducing, or eliminating risk, can determine whether the appropriate way to assess risk is through a quick and intuitive judgment or through a more formal and often more technical and costly process. A rational person might do a quick and intuitive risk assessment to help decide when to buy new tires for a car; the same person is likely to take more care, think more formally, and get more information and advice before deciding whether to undergo a risky medical procedure to diagnose or treat a possible cancer. Businesses assess risk before deciding whether to invest in a new venture, and governments perform risk assessments to decide whether to introduce a new program or regulation. In any risk assessment the aim is to decide whether it is most reasonable to introduce or create a risk in exchange for a benefit, try to eliminate or reduce the cause or mitigate the effects of an identified risk, or accept a risk or adapt to the bad consequences if they occur.

RISK ASSESSMENT IN PUBLIC POLICY

Although risk assessment is something individuals and businesses do more or less explicitly or formally every day, the term has acquired a special sense in the context of its use in justifying and explaining decisions that involve public policy. Risk assessment has been part of government activities for as long as governments have been concerned about threats to public and environmental health from disease, pollution, war, and technologies. However, risk assessment for public policies developed as a formal discipline in the 1960s, a period when there was broad support in the United States and other industrially developed countries for regulations aimed at protecting humans and their environments.

Between 1965 and 1980 the U.S. Congress enacted more than thirty major laws aimed at protecting health and the environment that established or strengthened at least a dozen regulatory agencies. Those laws set the agenda and the process for most environmental regulations for several decades. Explicit attention to the complicated nature of environmental risks led to the growth and influence of research into risk assessment as some advocacy groups lobbied for stricter regulations and others responded to newly promulgated regulations by arguing that they went too far and cost too much. All that activity prompted scientists and engineers to continue to develop formal methods of risk assessment that could improve the quality and acceptability of complex regulatory decisions. In the process the methods and use of formalized risk assessments themselves became subjects of ethical and political concern.

One reason for these controversies in environmental policy making is that different parties to the debates have different political agendas and also may have fundamentally different attitudes toward human responsibilities to nature. Risk assessments tend to disaggregate the problems people face and present them as far as possible in terms of their effects on human health and welfare. The advocates of risk assessment often encourage people to monetize or quantify the components of problems to make the various factors comparable and then aggregate those components into an overall evaluation of the alternative prospects. Some people regard this process as exemplifying the idea of rational decision making, whereas others regard it as being designed to blind people to some of the most important concerns and thus distorting important values.

HOW SAFE IS SAFE ENOUGH?

One of the stimulants to developing risk assessment techniques was the controversy over the safety of nuclear power plants in the United States in the 1970s and 1980s. Nuclear power was turning out to be an expensive and at best marginally profitable source of electricity, and the newly established Nuclear Regulatory Commission, responding to concerns about the safety of large reactors, considered promulgating new regulations that would have added to the costs of producing electricity. Some risk assessors and nuclear engineers began to argue that reactors were already safe enough and that further safety measures were unnecessary and too costly to be justified. Those assessors admitted that any nuclear reactor presents an inevitable small risk of a serious accident but argued that the level of risk was socially acceptable. Their conclusion was based on studies that examined the levels of risk the public accepted in other areas of life and how much the public seemed to be willing to pay to reduce a risk to life, health, or the environment. Critics of the status quo argued that because the consequences of a major nuclear accident were potentially so catastrophic, the probability of an accident must be reduced to very close to zero for that technology to be socially acceptable.

ECOLOGICAL RISK ASSESSMENT

The Preamble to the National Environmental Protection Act establishes a U.S. "national policy which will encourage productive and enjoyable harmony between man and his environment... [and] promote efforts which will prevent or eliminate damage to the environment and biosphere and stimulate the health and welfare of man..." The U.S. Environmental Protection Agency (EPA) was established in 1970 to oversee the enforcement of this and other environmental laws. In the first two decades of its existence, however, the EPA focused almost exclusively on protecting human health and welfare from nature while ignoring the need to protect ecosystems from the impact of human activity. This tendency was exacerbated by the rising influence of human health-based risk assessments in guiding EPA policies. In an attempt to correct this imbalance, the EPA around 1990 developed Ecological Risk Assessment (ERA) guidelines, initially for use in its Superfund program.

The aim of ERA is to evaluate potential adverse effects that human activities have on the living organisms that make up an ecosystem. An ERA is supposed to identify stressors that will produce an adverse effect on an ecosystem. A stressor is any physical, chemical, or biological agent that can induce changes in organisms. Stressors may be physical (e.g., dams, construction, etc.), chemical (e.g., pesticides, effluents, etc.), or biological (e.g., introduced species or pathogens). Adverse effects are those that are considered potentially harmful to the healthy functioning of an ecosystem.

ERAs so far have had only limited impacts on environmental policies. Part of the reason is that ecosystems are complex and often harder to understand than human health effects. There is more disagreement about what is good or bad for an ecosystem and about which ecological entities are most important to protect.

BIBLIOGRAPHY

National Environmental Policy Act of 1969, P.L. 91-190, *U.S. Statues at Large* 83 (1970): 853, codified at *U.S. Code* 42 (1982), § 4332.

As this controversy developed, the question for risk assessment became: How safe is safe enough? The proponents of nuclear power wanted to look at risk in a general and uniform way and see the problem as determining a general value of a probabilistic increment of the risk of death, illness, or environmental loss. The critics insisted that some risks are different from others and that it is necessary to consider the qualities of the different risks involved as well as the measure of a probability-weighted outcome to determine what is socially acceptable.

HOW PEOPLE JUDGE RISK

The question is partly empirical. It involves issues that are economic as well as issues that are psychological and may involve other cultural factors as well. Research that examined how people thought about this question had several important findings. First, most people find it very difficult to think clearly and consistently about risk and probability. People find it especially hard to compare different risks directly in a coherent way. It is easy to be confused, especially when experts disagree, about the significance of low levels of exposure to harmful substances. For these and other reasons, some experts on risk question the value of using economic data on consumer willingness to pay to reduce some increment of risk through the use of optional safety features in, for example, automobiles or preventive medicines, to support conclusions about social preferences for generally acceptable trade-offs among risks, costs, and benefits. To interpret consumer behavior as revealing a preference for trading off risks, costs, and benefits, it is necessary to assume that consumers are aware of the risks involved, are factoring that knowledge into their economic choices, and are behaving consistently. Research on the psychology of choice casts doubt on those assumptions.

Psychologists and anthropologists have shown that people are influenced by features of a risk other than the magnitude of harm and the probability that it will occur when they judge whether particular risks are acceptable. Social scientists have identified a number of factors that may explain, for example, why many people find the risks posed by nuclear power plants unacceptable but show little concern about technologies or products that pose a much greater likelihood of death or harm. These findings and the debates they generate suggest that any answer to the question "How safe is safe enough?" requires answering more basic questions about whether the goal of risk assessment in public policy should be primarily to protect public health and the environment or instead to determine what will satisfy public values. Risk assessment thus becomes entangled with moral and political issues.

RISK ASSESSMENT AND RISK EVALUATION

As the role and methods of risk assessment came to be seen as more complex and controversial, several expert studies sponsored by the independent National Academy of Sciences (NAS) tried to offer guidance. The main advice in those studies was to separate the scientific aspects of risk assessment, which the NAS sometimes referred to as risk estimation, from the more inherently controversial and value-laden activity of risk evaluation. The idea was that risk assessment should be able to provide useful advice that is free from the moral and political considerations that are part of making important decisions about public policy. The hope behind those recommendations was that experts in risk assessment might focus on the objective and measurable components of risk while the subjective and value dimensions could enter at a later stage of evaluation or political deliberation.

As some of the experts who helped write those NAS reports were aware, however, it is not easy to make risk estimation value-free. A risk assessment must begin by identifying which risks to study and what to count as a risk, an exercise that often requires making value judgments. Suppose, for example, that one is interested in reducing occupational risk and learns that exposure to a certain chemical acts synergistically with tobacco smoke to produce cancer rates in workers exposed to both that are higher than the aggregate of the two independent risks. One must decide whether this risk should be included in an assessment of occupational risk or whether it is acceptable to ignore it in a study because it is a risk to smokers who make a personal lifestyle choice and thus is not a risk to workers in general. Any choice one makes here involves an ethical judgment.

A second class of issues that cause problems in separating the scientific from the value-laden determinants of risk assessment has to do with how to treat different kinds of uncertainty. In the absence of direct epidemiological evidence of the risk of low levels of exposure to many carcinogenic substances, for example, risk assessors must rely on studies conducted on animals that involve exposure to high levels of the substances involved. Experts disagree about the proper way to extrapolate from high risk to animals to low risk to humans. The NAS studies recommended dealing with those uncertainties by reporting results as ranges rather than single numbers or point estimates of risk. This solution is not always available, however, and it does not get to the problem of determining how conservative or precautionary one should be in selecting a number or describing a range. This is another way in which simply identifying and estimating risk involves making value judgments.

The problems involved in understanding and communicating about uncertainties are among the most difficult issues that risk assessment must confront, especially in areas involving important environmental risks. This is the case because uncertainty can be of different kinds and can be relevant in different ways. Some of these issues are well illustrated in risk assessments of global climate change. In 1988 the United Nations organized the Intergovernmental Panel on Climate Change (IPCC), a group of thousands of leading scientists around the world, and as stated in the "Principles Governing IPCC Work," assigned it the task of reporting on "the scientific basis of risk of human-induced climate change, its potential impacts and options for adaptation and mitigation" (1998, p.1).

UNCERTAINTY

The IPCC published four reports that reflected the consensus of the world scientific community on those questions. Each report indicated that there is much uncertainty about human-induced climate change, but the nature of that uncertainty has changed dramatically during the two decades of IPCC reporting. In its First Assessment Report, published in 1990, the IPCC reported that global climate change could be observed but that there was no way to determine with certainty whether human activities were contributing significantly to it. By 1995 the IPCC scientists could agree that "the balance of evidence suggests a discernable human influence on global climate," but that language still supports uncertainty about the amount of the human contribution. In its fourth report, published in 2007, this uncertainty was removed. The IPCC concluded that the evidence now showed that "[m]ost of the observed increase in globally averaged temperature is very likely due to the observed increase in anthropogenic greenhouse gas concentrations."

This is as close as scientists ever get to stating that a conclusion is certain in a complex world, but in light of the conservative language, the degree of certainty (which the report explains means greater than 95 percent confidence) can be appreciated only by comparing this statement with the earlier IPCC conclusions. Moreover, many other uncertainties—about more localized changes and about which areas and populations will be hurt the most and which may even benefit from climate change—are left untouched or even increased by research that virtually eliminates uncertainty about what is happening on a global scale. These remaining and shifting uncertainties create many problems for risk evaluation.

One issue illustrated by those uncertainties involves the way the distribution of risk can interact with both the perspective taken in a risk assessment and the sometimes different perspectives on determining what does and should matter most to people. Consider a situation in which an average individual in an exposed population has a $1/n$ chance annually of suffering an adverse consequence. This level of individual risk can be distributed across a population in different ways. To take two possibilities, $1/n$ individuals in the population will with certainty suffer the consequence each year but the identities of those individuals cannot be known in advance; alternatively, there is a $1/n$ chance that the entire population will suffer the consequence in any specific year. Those risks can seem identical from an individual's perspective, but from the perspective of the population as a whole the two ways of realizing the risk can be quite different. The former possibility involves no risk at all, but the latter might involve a catastrophic risk. It is clearly an important ethical consideration to determine whether and how risk to the group is to be balanced against risk to an average individual within the group or to a particular group among a larger population. Attempts to balance the risks as seen from more than one perspective also can require changing the level of risk exposure to the individual or the group.

ETHICAL ISSUES

It is not possible to eliminate the possibility of adverse consequences from activities that are deemed socially desirable. Many of the ethical issues surrounding risk assessment can be seen as asking how people should measure and respond to those risks, but at least one concern raises a different sort of question. Some people are bothered by the assumptions of most risk assessments that all values are comparable and can be measured on a single scale. This concern is especially significant when the chosen scale is money and the determinant of value is an aggregation of individuals' willingness to pay for things or changes in human welfare. In these assessments the value of human life and health and the value of natural treasures and healthy ecosystems are treated as commodities or goods with a price. The worry of some critics of this approach to thinking about complicated questions of policy and social values is not whether ethical judgments must be made in the process of risk assessment but whether ethical issues are involved even in considering some possibilities as appropriate subjects of formal risk assessment.

SEE ALSO *Cost-Benefit Analysis; Economics, Environmental; Environmental Law; Environmental Policy; Global Climate Change; Intergovernmental Panel on Climate Change; Nuclear Power; Pollution; Precautionary Principle.*

BIBLIOGRAPHY

Douglas, Mary, and Aaron Wildavsky. 1982. *Risk and Culture: An Essay on the Selection of Technical and Environmental Dangers*. Berkeley: University of California Press.

Fischhoff, Baruch; Stephen Watson; and Chris Hope. 1984. "Defining Risk." *Policy Sciences* 17: 123–139.

Glickman, Theodore S., and Michael Gough, eds. 1990. *Readings in Risk*. Washington, DC: Resources for the Future.

Intergovernmental Panel on Climate Change. 1995. *IPCC Second Assessment Report.* Summary for Policymakers. Available from http://www.ipcc.ch/pdf/climate-changes-1995/spm-science-of-climate-changes.pdf

Intergovernmental Panel on Climate Change. 2008. *IPCC Fourth Assessment Report: Climate Change 2007.* Summary for Policymakers. Available from http://www.ipcc.ch/pdf/assessment-report/ar4/syr/ar4_syr_spm.pdf

Keeney, Ralph, and R. Winkler. 1985. "Evaluating Decision Strategies for Equity of Public Risks." *Operations Research* 33: 955–970.

MacLean, Douglas, ed. 1986. *Values at Risk.* Totowa, NJ: Rowman and Allanheld.

MacLean, Douglas. 2001. "Risk Analysis." In *Encyclopedia of Ethics*, 2nd ed., ed. L. Becker and C. Becker, 1515–1518. New York: Routledge.

National Academy of Sciences Committee on Risk and Decision Making. 1982. *Risk and Decision Making: Perspectives and Research.* Washington, DC: National Academy Press.

National Research Council. 1983. *Risk Assessment in the Federal Government: Managing the Process.* Washington, DC: National Academy Press.

Schelling, Thomas. 1984. "The Life You Save May Be Your Own." In *Choice and Consequence.* Cambridge, MA: Harvard University Press.

Shrader-Frechette, Kristin. 1991. *Risk and Rationality: Philosophical Foundations for Populist Reforms.* Berkeley: University of California Press.

Slovic, Paul; Baruch Fischhoff; and Sarah Lichtenstein. 1979. "Rating the Risks." *Environment* 21: 14–39.

Starr, Chauncey. 1969. "Social Benefit Versus Technological Risk." *Science* 165: 1232–1238.

Stern, Paul C., and Harvey V. Fineberg, eds. 1996. *Understanding Risk: Informing Decisions in a Democratic Society.* Washington, DC: National Academy Press.

World Meteorological Association and the United Nations Environment Programme, Intergovernmental Panel on Climate Change. 1998. "Principles Governing IPCC Work." Available from http://www.ipcc.ch/pdf/ipcc-principles/ipcc-principles.pdf

Douglas MacLean

RIVERS

Rivers are the blue ribbons of the earth. Over geological time they have shaped the lay of the land through erosion, flooding, and meandering—carving sinuous paths from headwaters in the mountains to estuaries at the mouth. They are the perfect mediators between aquatic and terrestrial ecosystems, providing habitat for land and water species alike. Carrying and depositing sediments, they form intricate complex landscapes with fertile floodplains at the center of their basins. The basin or watershed comprises the entire catchment area that drains surface water into a river, which carries it to the ocean. Connecting land, air, and ocean, rivers play a crucial role in the hydrological cycle: Evaporated from the ocean and transported through the atmosphere, water returns to land as snow or rain and, seeping through soil into streams, it ends up in a river, again on its journey back to the ocean.

Every river has a distinctive flow pattern, determined by the size of its basin and features such as topography, geology, climate, and vegetation. The flow is a crucial driver for the health of the river system, determining the conditions for animal and plant life. Over the course of the year, rivers might have very different flow signatures, depending on snow melt and seasonally variable rain. The Mekong River, for example, carries fifty times more water in the summer monsoon period than in its long dry season. Although it is the seventh-longest river in Asia, the twelfth-longest in the world (4,180 km), and the tenth-largest by *annual average* volume, it has the highest flow variability and, when swollen by summer rains, it turns out to be the third largest by volume, surpassed only by the Amazon and Brahmaputra.

The Amazon is by far the largest river: It delivers more freshwater to the ocean than any other river, 20 percent of the global river discharge. Its drainage area, the Amazon Basin, is enormous: 6.5 million square kilometers, almost 40 percent of South America. Its annual average discharge is 219,000 cubic meters per second. In comparison, the Nile, with its 6,695 kilometers, is the longest river (300 kilometers longer than the Amazon's 6,387 kilometers), but it discharges on average a mere 2,830 cubic meters per second. During certain periods of the year, no water at all reaches the mouth of the Nile at the Mediterranean Sea, revealing the most recent and influential driver of river flows nowadays: humanity. During some seasons all the Nile's water is taken for irrigation.

RIVERS IN CULTURE

Throughout recorded history humans have shaped rivers for irrigation, navigation, and flood protection. The ancient Sumerian culture and, later, the Assyrian civilization (2400 B.C.E.) flourished in the Fertile Crescent, an area whose astounding fecundity was due to its main river systems, the Tigris and Euphrates. These peoples constructed elaborate irrigation and flood-control projects that led to the emergence of cities and sophisticated tax-based economic and legal systems. Like most irrigation-based societies, their civilizations collapsed because of soil depletion, salinization, and a high vulnerability to invasion. By the time the Greek historian Herodotus (c. 484–c.425 B.C.E.) visited the area—parts of present-day Turkey, Syria, Iran, and Iraq—the Persians had established their rule. "Rivers . . . they revere," he wrote about the Persians. And he continued, "They will neither urinate nor spit nor wash their hands in them, nor let anyone else do so" (Herodotus, Book 1, Chapter 138, Section 2).

The only irrigation-based ancient civilization that proved to be sustainable was that of the Egyptians. Their secret lay in the fact that their method of irrigation was basin-based—that is, it made use of the natural rise and fall of the Nile's seasonal fluctuations and hydrology. The river's yearly flooding rejuvenated the land not only with water but also with a new layer of soil—a rich layer of sediments. This annual flood-deposition cycle continued until the Aswan High Dam disrupted the natural flow pattern of the Nile in the twentieth century. A famous bas-relief from 3100 B.C.E.—showing one of the ancient kings ceremonially holding a hoe for cutting an irrigation ditch—testifies to the existence and importance of irrigation waterworks going at least back to that time. For at least 5,000 years the peoples of Egypt maintained an irrigation-based culture.

The flooding of the Nile precipitated the development of mathematics: The world's first numerical system was invented in order to apportion the land after floods and sediment depositions had obliterated boundaries (*geometry* literally means "land measurement") and to determine planting and harvesting schedules. Herodotus rightfully called Egypt "the gift of the Nile." Many Egyptians worshiped Hapi, the god of the Nile; although he was the father of all gods, he was depicted with breasts, a symbol of his capacity to nurture with life-giving fluids.

Rivers are revered in many traditions. Along African rivers one finds small offerings like a bowl of fruit or a bundle of flowers; rivers are of vital importance in Africa, the world's most arid continent. For Hindus in India, rivers are goddesses. The doorways of early Hindu temples were decorated with images of the two main rivers: Ganga rode on a crocodile and Yamuna on a tortoise. "Mother Ganga"—known internationally by its Anglicized name, Ganges—is still the most sacred river in India. Her generative and purifying powers are invoked in various initiation rites and bestow longevity on the one who drinks of her waters. Pilgrims visit her banks for daily ritual bathing, which heals the weak and sickly and washes away sins. People give the bodies of the dead to the river or spread their ashes over her, and she carries them in a last transition to the land of their ancestors.

In many cultures rivers represent the capacity for transformation. The most definitive transition is the one between life and death, symbolized in Greek mythology by the rivers Styx and the Lethe. Indeed, the latter washes away human memories so one can begin anew the afterlife. At the dawn of Greek philosophy, Heraclitus used the image of a river to symbolize the ever-changing flux of nature. In many traditions time is represented as a river. In various literary texts we also find river symbolism. In Dante's *Inferno* ice—not fire—is at the center of hell, and frozen stasis is its most severe punishment. Small rivulets—trickles thawed out of ice—symbolize the narrow path to renewal and hope.

Rivers have been anchors of civilization and bones of contention (the word *rival* is cognate with *rivulet*). They cover great distances and connect a wide diversity of people and places, from high-altitude snowy mountains to lowland deltas. They are rich ecological and cultural corridors. Many cities have arisen at the bank of a river or at a confluence of two rivers. Such rivers have shaped a valley and at a strategic site a city arose: the bottleneck, the valve, of flows of sediments and trade material, in and out of the valley—hence a center of commerce.

In the early 1500s Leonardo da Vinci and Niccolò Machiavelli furtively conspired to mastermind a diversion of the Arno River from the city of Pisa. It would have deprived the city of water, thereby giving victory to Florence after a ten-year war. The plan, if successful, would have turned Florence into a world power with access to a seaport. For various reasons the plan fell apart, inspiring Machiavelli to compare fortune to a river—something that is unpredictable, violent, and irresistible. Even though that scheme to divert the Arno did not come to fruition, the underlying paradigm of the control of fortune through a powerful combination of economic, engineering, and strategic forces was a precursor of modern river management.

RIVERS IN THE MODERN ERA

In early modern times rivers were relegated to the industrial and mercantile back regions of town and turned into functional arteries for transport, for harbors, and for other economic purposes. Or they were exiled underground; or turned into open sewers, carrying diseases; or dried, paved over, and turned into roads to make space for traffic.

Since the Industrial Revolution most of Europe's rivers have been treated as cheap waste transportation to the sea. Heavily polluted, they have harmed human health and degraded the quality of coastal and marine waters. The biodiversity of thousands of kilometers of waterways was affected. The same happened in the United States, culminating in a famous incident in 1969: A thick layer of oily industrial pollutants on the Cuyahoga River in Cleveland, Ohio, caught fire from the sparks of a passing train. Even the revered Ganges and Yamuna rivers turn into open sewers after they pass through Delhi and Varnassi, respectively. Each enters a city alive and leaves nearly dead, soiled by sewage and other waste, anaerobic with a zero dissolved-oxygen level as gaseous sludge rises from the bottom and floats to the surface.

Along the Mississippi River farming and logging on a massive scale had caused immense erosion by the early twentieth century. Vast amounts of topsoil were washed down the river into the Gulf of Mexico, a trend that had a disastrous impact on impoverished farmers. The Tennessee River is a major tributary of the Mississippi. The Tennessee Valley Authority (TVA) was created in 1933 as a federally owned corporation to provide the residents of the Tennessee River watershed, one of the areas most severely affected by the Great Depression, with a regional development plan based on modernizing the area's economy and society. The complex plan for flood control and electricity generation involved the creation of twenty-nine dams and 1,050 kilometers of navigation channels. It is today the largest public power company in the United States.

Dam building has been promoted as the prime technological approach to riverine development. It was at the heart of President Franklin Roosevelt's program of New Deal public works, which, during the Depression years of the 1930s, provided jobs for masses of unemployed Americans. FDR, however, was only carrying on in the tradition begun by his elder distant cousin and predecessor, President Theodore Roosevelt, who had initiated the twentieth-century American modernization paradigm in 1901: "Great storage works are necessary to equalize the flow of streams and to save the flood waters," he proclaimed (Postel and Richter 2003, p.1). In the same spirit of progress, Prime Minister Nehru of India honored dams in the 1940s as "the Temples of Modern India"; India went on to build some 3,600 dams.

The twentieth century saw an unprecedented impact on rivers, mainly through dam building, leaving a legacy of approximately 800,000 dams worldwide, of which 48,000 are 15 meters or higher, with almost half of these large dams built in China. The collective weight of the harnessed water and trapped sediments in the reservoirs has caused a measurable change in the angle of the earth's axis and the speed of its orbital movement. One-fourth of the global sediment ends up in reservoirs behind dams instead of nourishing floodplains and estuaries. Silt's color accounts for the names of the Yellow River and the Colorado River (Colorado is Spanish for "colored"), but dams have robbed them of the significance of their names: As released from the major dams on these great rivers, the water is as clear as glass. Some of the smaller reservoirs behind dams have silted up to form marshy plains.

Dams provide almost 20 percent of the world's electricity supply. The massive dam at Itaipu (built from 1975–1991), for example, on the mighty Parana River between Brazil and Paraguay, provides Sao Paolo (a city of 11 million people) as well as Rio de Janeiro (a city of 6 million) with energy and furnishes 20 percent of Brazil's electricity and 93 percent of Paraguay's. The Itaipu dam shifted the course of the seventh-biggest river in the world. Echoing Nehru in a secular fashion, the American Society of Civil Engineers called the dam one of the seven wonders of the modern world.

Dams come at high costs. They profoundly disrupt long-established ecological, hydrological, and cultural systems. Dams have displaced 40–80 million people, either by forced eviction or because of the loss of their traditional livelihood as a result of dam-induced environmental change. The stagnant waters of the reservoirs breed mosquitoes and are infested with freshwater snails that carry parasites, generating diseases such as Schistosomiasis (also known as *bilharzia*, or snail fever), commonly found in Asia, Africa, and South America. Snail fever, endemic to Egypt, is exacerbated by the Aswan High dam and associated irrigation projects along the Nile.

Dams have triggered protests all around the world. Modern communication technologies—email, blogs, YouTube videos—have facilitated global networks supporting local communities, social movements, and nongovernmental organizations organized to oppose dams. Influential advocacy groups, such as International Rivers, question the conventional or modern "development" model that dams epitomize. Among other things, they research other ways of meeting people's needs for water, energy, and protection from damaging floods. In 1997 the World Commission on Dams (WCD), an independent panel to review large dams, was created, led by the International Union for Conservation of Nature (IUCN), and the World Bank. The Commission initiated a broad-based research process resulting in its 2000 report, *Dams and Development: A New Framework for Decision-Making*, which acknowledges the important benefits of dams for human development but also judges that the price paid—both socially and environmentally—is unacceptable (WCD 2001).

Once widely regarded as a symbol of progressive technology, dam building is now commonly viewed as a threat to the earth's ecosystems. Nevertheless, major dams are still being built and planned. A series of dams are planned for the headwaters of the Ganges River in India. The Tehri Dam, which began to fill in 2004, already affects the flow of the Ganges River from the Himalaya Mountains. Sunderlal Bahuguna, Indian activist and philosopher in the tradition of Mahatma Gandhi, has gone on many hunger strikes to stop the Tehri Dam Project, and was forcibly moved to a government issued home upstream. A leading Indian scientist, G. D. Agarwal, former dean of the Indian Institute of Technology at Kanpur, has decided to go on a fast unto death against the damming of the River Bhagirathi "to oppose the destruction of this ecological marvel and the epitome of Hindu cultural faith" (Agarwal 2008).Where for Nehru the dam was the temple, now the river is the religious center of attention.

At the beginning of the twenty-first century, one in ten of the world's major rivers no longer reached the sea for part of the year because of upstream use of their waters, mainly for irrigation. The Nile, the Yellow, the Indus, the Rio Grande, and the Colorado rivers, along with many others, now end in sand, sometimes hundreds of kilometers before they have a chance to reach their mouths and fertilize their deltas to maintain ecologically rich estuarial ecotones that mix sweet and salt water into one of the most productive kinds of ecosystems on Earth. The number of foreshortened rivers may increase during the twenty-first century. Although irrigation already accounts for 70 to 80 percent of human use of freshwater, more and more countries are pushing their water resources, including rivers, to the limit in order to increase food production.

Many rivers are thus but a shadow of their former selves. The blue lines on the map are often tokens of faded glories. Whereas rivers once symbolized transition, they are now themselves in transition.

RIVERS IN TRANSITION

In German romanticism the Rhine and Danube—with their castles, vineyards, and associated ancient legends—loomed large in the cultural imagination. This popular veneration did not spare the rivers from pollution and other forms of environmental degradation. A renewed environmental consciousness, however, is transforming cultural attitudes toward rivers. Human and ecosystem water needs ought to be balanced, according to the water experts Sandra Postel, director of the Global Water Policy Project, and Brian Richter, director of the Freshwater Initiative of the Nature Conservancy (Postel and Richter 2003). More than 60 percent of the world's 227 largest rivers have been fragmented by infrastructures such as dams and diversions. Rivers are turned on and off instead of flowing by natural rhythms. Their main functions are providing hydroelectric power, irrigation for agriculture, and shipping routes for trade; their traditional ecological functions and services have been systematically undermined. With great efficiency the periodic floods of untamed rivers shaped river channels and redistributed sediment, creating habitats essential to fish and other riverine life. Because floodwaters are no longer getting cleansed by floodplain wetlands, more pollution is reaching inland and coastal seas. Hence in many parts of the world, the harnessing of rivers for economic gain is causing more harm than good.

In 2000 the European Parliament adopted a water-policy framework, the European Community Water Framework Directive (the WFD), an unprecedented attempt to design a regime for managing water quality and instream flow for each watershed, thus creating a basic geographic unit for resource planning. The WFD aims to manage whole watersheds or river basins in a holistic manner (a strategy referred to as integrated water-resources management [IWRM]) at the basin or watershed level in order to integrate land and water—upstream and downstream water, surface water, groundwater, and coastal water. A more transparent and participatory transnational governance alternative has thus replaced a politically fragmented, top-down management approach. In an effort to achieve multiple-stakeholder consensus, hydrological and engineering expertise has been complemented by ecological concerns along with urban, agricultural, industrial, and recreational interests. Because water connects all elements of society, an integrative water ethic becomes essential, one that views local problems of water quality and quantity in transregional and global political contexts.

The major water laws were created in an era of economic control of rivers. The emerging new paradigm is based on the concept of ecological health. This radical shift calls for a dislodging of entrenched laws and policies. Around the globe, in areas as diverse as Australia, South Africa, Europe, and Texas, governments are implementing policies that establish allocation of water for ecosystem support—mainly minimum instream flows to maintain environmental quality and sustainability.

Because rivers are the anchors of cultures, many cities are revitalizing communities and ecosystems by reconnecting them to the rivers that run through them. Rivers are resurfacing in the public imagination as cultural and ecological corridors, creating a cultural rejuvenation. Although some of these projects are driven solely by economic motives, most of them stimulate awareness of the river's ecological, cultural, and economic wealth and inspire public education about local water issues. Some of them explicitly aim for increasing stewardship of the river and engaging communities in river-related projects. New urban river projects—such as river-walk promenades, theaters, cafés, and restaurants along old quays—are beginning to appear in many places. This trend often comes with a gentrification of dilapidated harbor neighborhoods as old warehouses turn into high-priced riverfront housing developments.

For example, the Los Angeles River is about to be unlocked from its concrete ditch and restored by means of a riparian-community-based process meant to create a blue ribbon along housing developments, parks, and walkways, thus revitalizing river and community alike. New Orleans plans "RiverSphere" as a forum for art, science, and technology focused on the Mississippi. In an era of fluvial imagination, river festivals burst forth: there is the Brisbane River Festival in Queensland, Australia; the Hudson River Festival in New York City; London's Thames Festival; and Danube day throughout southeastern Europe. New urban water cultures are emerging that celebrate

The Concrete Channel of the Los Angeles River. *The Los Angeles River was once the primary water source for the Los Angeles basin, but frequent flooding led some to search for a sollution to control the unpredictable river flows. Since the late 1930s, the river has been primarily a flood control channel. Environmentalists and others hope that the river can soon be returned to its more natural state.* PHOTO BY IRENE KLAVER.

rivers as part of a deepening appreciation of waterways and growing involvement in river basins. The "One River Mississippi" festival celebrates the river along the Father of Waters (which is what "Mississippi" means in the Cherokee language) at various places, all at the same time.

Also in rural areas river restoration is underway: Remeandering—done by many of the same engineering firms that straightened them fifty years ago—gives floodplains back to the rivers. And there is a movement to learn to live with the floods—riparian restoration and nurturing wetlands to take out pollutants, fertilizers, and pesticides before the water drains back into the river.

With the combined application of interdisciplinary engineering, hydrological expertise, and ecological sensitivity, the twenty-first century promises a renewal of river cultures, perhaps even a respect approaching the ancients' reverence for these great waterways. New management regimes are seeking to work with, not against, the great rivers to enhance their flow and allow them the freedom of flood in a natural cycle of renewal and rejuvenation for humans and the entire manifold of life that flows into and out of these vital arteries the earth.

Rivers are archives. They record deep and shallow time, revealing what happened a million years in the past as well as a moment ago upstream. Rivers are the sinews of the world; without them life unravels

SEE ALSO *Atmosphere; Biodiversity; Carson, Rachel; Dams; Farms; Food; Hinduism; Nongovernmental Organizations; Roosevelt, Theodore; Salmon Restoration; Sustainability; Waste Management.*

BIBLIOGRAPHY

Agarwal, G. D. 2008. *Letter from GD Agarwal to the Government of India.* April 14. Available from http://tapasya-bhagirathi.blogspot.com/

Alley, Kelley. 2002. *On the Banks of the Ganga: When Wastewater Meets a Sacred River.* Ann Arbor: University of Michigan Press.

Conca, Ken, 2006. *Governing Water: Contentious Transnational Politics and Global Institution Building.* Cambridge, MA: MIT Press.

Herodotus. 1921–1924. *The Histories.* 4 vols. Loeb Classical Library. A. D. Godley, ed. London: W. Heinemann; and New York: G. P. Putnam's Sons.

Kibel, Paul Stanton. 2007. *Rivertown: Rethinking Urban Rivers.* Cambridge, MA: MIT Press.

Lebel, Louis, John Dore, Rajesh Daniel, and Yang Saing Koma, eds. 2007. *Democratizing Water Governance in the Mekong.* Bangkok, Thailand: Mekong Press.

Masters, Roger D. 1999. *Fortune Is a River: Leonardo da Vinci and Niccolò Machiavelli's Magnificent Dream to Change the Course of Florentine History.* New York: Plume Press.

Magris, Claudio. 2001. *Danube.* London: Harvill Press.

Pearce, Fred. 2006. *When the Rivers Run Dry: Water—The Defining Crisis of the Twenty-First Century.* Boston: Beacon Press.

Postel, Sandra, 1999. *Pillar of Sand: Can the Irrigation Miracle Last?* New York: W.W. Norton.

Postel, Sandra, and Brian Richter. 2003. *Rivers for Life: Managing Water for People and Nature.* Washington, DC: Island Press.

Raffles, Hugh. 2002. *In Amazonia: A Natural History.* Princeton, NJ: Princeton University Press.

Reid, Jan. *Rio Grande.* 2004, Austin: University of Texas Press

Reisner, Marc. 1993. *Cadillac Desert: The American West and Its Disappearing Water.* New York: Penguin.

Wohl, Ellen. 2004. *Disconnected Rivers: Linking Rivers to Landscapes.* New Haven, CT: Yale University Press.

Wolf, Aaron T. 2002. *Conflict Prevention and Resolution in Water Systems.* Northampton, MA: Edward Elgar.

World Commission on Dams (WCD). 2001. *Dams and Development: A New Framework for Decision-Making.* London: Earthscan. Available from http://www.dams.org/report/contents.htm

Worster, D. 1985. *Rivers of Empire: Water, Aridity, and the Growth of the American West.* New York: Oxford University Press.

Irene Klaver

ROLSTON, HOLMES, III

1932–

Holmes Rolston III was born in the Shenandoah Valley of Virginia on November 19, 1932. His father was a rural pastor. Rolston grew up with the Blue Ridge Mountains on the horizon; from his earliest years he recalls wandering the woods and swimming the creeks of his native landscape.

Rolston's formal education included a B.A. in physics from Davidson College (1953), a Ph.D. in theology from Edinburgh University (1958), and (after several years' service as a pastor back in Virginia) an M.A. in the philosophy of science from the University of Pittsburgh (1968). At each step Rolston felt dissatisfaction with the reigning concepts of nature and with people's mistreatment of nature itself. He especially disliked the common insistence that nature was value-free. He came to realize the need for an environmental philosophy that could undergird a richer appreciation of life on earth. His forty-year career in the Philosophy Department at Colorado State University, beginning in 1968, was largely devoted to creating such a philosophy.

Rolston's 1975 article, "Is There an Ecological Ethic?" helped to jump-start interest in environmental ethics in academic philosophy. In 1979 he helped to found the first journal in the field, *Environmental Ethics,* and as of 2008 remained an associate editor. During this time he was developing his own nonanthropocentric environmental ethics, first in articles later collected in *Philosophy Gone Wild* (1986) and then in a systematic presentation in the book *Environmental Ethics* (1988).

Rolston is best known for his sustained, ingenious, and uncompromising advocacy of the idea that values inhere objectively in nature. He holds that individual organisms, biological species, and ecosystems may all possess intrinsic values—values based on what they themselves are—in addition to their instrumental values to human beings. These intrinsic values ground duties to treat nature with respect and use it with restraint. Rolston insists that human beings are sometimes morally required to put values discovered in nature above their own preferences or self-interest.

In addition to defending nature's intrinsic values, Rolston attempts to enrich our account of nature's instrumental values to humans (including life-supporting, economic, recreational, aesthetic, scientific, and spiritual values). He also argues for nature's "systemic value": the creative capacities of the earth's ecosystems to generate intrinsic and instrumental values over evolutionary time. Rolston's goal is a comprehensive and accurate account of the way in which nature ought to be valued, one that not only does justice to human beings' uniquely complex and important roles on earth and to the new values brought forth by consciousness, but that also appreciates nonhuman nature for what it is.

Rolston's value arguments are built upon detailed, scientifically informed descriptions and an appreciation of the natural entities in question. They have been so influential that casual observers often define environmental ethics as the position that nature has intrinsic value, or equate environmental ethics with nonanthropocentrism. Strictly speaking, this is false, since one can consistently hold an anthropocentric environmental ethics. Rolston, however, finds such ethical outlooks to be inadequate, either as guides to practice or as complements to a modern scientifically informed worldview.

Rolston has also worked to specify what respect for nature might mean for policy issues such as protected-areas management, endangered species and biodiversity conservation, wilderness preservation, sustainable development, corporate environmental responsibility, and population policy. *Conserving Natural Value* (1994) provides a comprehensive, nuanced account of Rolston's positions on many of these issues. Throughout his writings he places a premium on reining in human consumption in order to preserve wild nature.

In addition to his efforts to formulate an environmental ethics, Rolston has endeavored to overcome the modern split between science and religion. Motivation for this project goes back at least to his time as a pastor to evolution-fearing but nature-loving southern farmers. This work, also widely influential in philosophical circles, led to his invitation to give the Gifford Lectures at the University of Edinburgh in 1997–1998, later published as *Genes, Genesis, and God* (1999). In 2003 he was awarded the John Templeton Prize in Religion, the money from which he used to endow a chair at his alma mater, Davidson College. Rolston has been university distinguished professor at Colorado State University since 1992.

Often referred to as "the father of environmental ethics" and later sometimes as its "grandfather," Rolston lectured on all seven continents. He strongly influenced environmental ethics through his six books, fifty authored chapters, and more than one hundred articles; his many generous efforts to help philosophers, theologians, and scientists further their own work; and his practical conservation efforts at the local, state, federal, and international levels.

SEE ALSO *Biodiversity; Conservation; Consumption; Environmental Philosophy: V. Contemporary Philosophy; Intrinsic and Instrumental Value; Population; Preservation; Sustainable Development; Wilderness.*

BIBLIOGRAPHY

WORKS BY HOLMES ROLSTON III

1986. *Philosophy Gone Wild: Essays in Environmental Ethics.* Buffalo, NY: Prometheus Press.

1988. *Environmental Ethics: Duties to and Values in the Natural World.* Philadelphia: Temple University Press.

1994. *Conserving Natural Value.* New York: Columbia University Press.

1999. *Genes, Genesis and God: Values and Their Origins in Natural and Human History.* Cambridge, UK: Cambridge University Press.

Philip Cafaro

ROMANTIC POETRY, ENGLISH

The Romantic period of the late eighteenth and early nineteenth centuries has historically been seen as a time of a renewed interest in the natural world, and many attitudes, ideas, and landscapes that writers of the period celebrated in their poetry have endured in contemporary environmental thinking. Influenced by such European writers as Jean-Jacques Rousseau, the English poets of the period saw nature as a means of resisting and opposing the artifice and corruption of many aspects of culture. Indeed, one of the signal ideas of the Romantic period is that nature and culture are opposed rather than conforming, as Alexander Pope and other Enlightenment figures had thought. A central motivating idea for the Romantics was that human nature was perfectible in nature and made corrupt only by culture. Thus such early Romantic poets as Charlotte Smith, William Wordsworth, and Samuel Taylor Coleridge wrote lyric poems featuring the poet fleeing various instances of culture (cities, crowds, families) and seeking the isolation of various forms of wilderness.

INDIVIDUAL IMAGINATION AND THE NATURAL WORLD

In this version of the Romantic lyric, the speaker returns to spiritual health by finding in the natural world solace and wholeness. The most influential example of this is Wordsworth's "Lines Composed a Few Miles above Tintern Abbey," a poem that is also a kind of manifesto for how the natural world can make the individual a better person, morally, spiritually, and physically. Many poems by Smith and Coleridge make the same point. Coleridge's pronouncement in "The Eolian Harp" (1796) that the "one Life within us and abroad" is the source of all solace is a succinct encapsulation of a quintessentially Romantic and environmental idea. The key arguments here are that the harmony and beauty of the natural world point to a fundamental reality obscured and even destroyed by much human activity and that nature is imbued with, or is a manifestation of, a powerful spirit, consciousness, or force to which human consciousness may return.

Wordsworth developed these ideas in dozens of poems, including his epic autobiography *The Prelude*, which argues that an early and sustained connection to the natural world is necessary for the development of a healthy and creative imagination. Indeed, literary criticism of the early twenty-first century has heralded Wordsworth as the father of the nascent ecological vision of Romanticism (Bate 1991; Buell 1995), and also as the originator of a key myth of modern environmentalism, or at least one of its deepest and perhaps most unscrutinized ideologies: the idea that the natural world is a source of health and that cities and other forms of human culture produce pollution and disease, both physical and spiritual.

The Romantic poets did not think of themselves as belonging to a school or movement, however, and each had distinctive ideas about the natural world. Indeed, their belief in the sanctity of the individual imagination almost guaranteed much diversity of thought. The emphasis on the individual's encounter with the natural world is also a central feature of much environmental writing by such later figures as Ralph Waldo Emerson and Henry David Thoreau in the mid-nineteenth century, and as Robinson Jeffers, Edward Abbey, Aldo Leopold, and many others in the twentieth. There are two connected ideas here: that nature offers places of solitude in which to discover one's individuality and that the true quality of nature can only be appreciated and understood by the solitary observer.

ANIMAL ETHICS AND THE ROMANTIC PERIOD

Although Coleridge would give up both poetry and his idealization of nature to become a philosophical idealist by the early 1800s, the extraordinary popularity of his *Rime of the Ancient Mariner* has earned him a perhaps outsized influence on environmental thought. This self-consciously mysterious and symbolic poem tells the story of the mariner's apparently random killing of a beautiful and companionable animal and the revenge enacted upon him by both the natural and supernatural worlds. The poem has been read variously as revealing humanity's crimes against nature (and nature's ultimate revenge), as arguing for sympathy and kindness toward animals, and as revealing a deep and hidden moral (or amoral) order in the physical world—all of which ideas resonate with environmental thought.

Concern for the status and well-being of animals is a consistent and important feature of Romantic-period environmental thought. Such early Romantic poets as

Anna Barbauld and Robert Burns wrote popular poems that seem to condemn cruelty to animals, and the young Coleridge's "To a Young Ass," much mocked at the time, asserted solidarity with animals. Though, as a radical and apocalyptic Christian, William Blake generally valued human imagination over the necessarily fallen and corrupt physical world, his poetry brilliantly and consistently sees animals, and human treatment of them, as symbolic of actual sin and potential redemption. The "peasant poet" John Clare is remembered today in part for the careful description of birds and mammals in his poetry, and for his powerful evocation of sympathy for the well-being of animals and the ecosystems that support them. Clare is important too for decrying the privatizing of public grazing lands and for extolling the virtue of careful and sustained observation of animals and landscapes, as opposed to the rare moments of epiphanic insight celebrated by most of the other Romantic poets.

Probably the most committed animal-rights poet and most radical environmental thinker of the Romantic period was Percy Bysshe Shelley, who wrote two tracts on the "necessity" of vegetarianism, in which he developed the radical view of meat eating (and the production and slaughter of livestock) as the "root of all evil." His argument, ultimately about the interconnectedness of things, is that how and what we eat, for instance, affects our physical environment as well as our moral being. Though Shelley's thinking often overreaches, the daring and vigor of his imagining of the relationship between consciousness, culture, and the physical world anticipates much contemporary environmental thinking the most of all the Romantic poets.

Shelley is simultaneously a radical materialist and a radical idealist. Important poems such as "Mont Blanc," "Ode to the West Wind," and *Prometheus Unbound* argue that the laws of the world are ultimately those of nature, which have long been obscured by the corrupting cycles of human development. Shelley seems to have been the only Romantic poet aware of the deep implications of the new science of geology: that the Earth was far older than the Bible stated and also that humankind may be insignificant in the vastness of geologic time. Shelley was invigorated by the thought that if natural forces, rather than human forces, were primary, then human culture (which Shelley saw as producing predominantly violence and pain) was ultimately a veneer and could be replaced, through the forces of nature and healthy human imagining, by the kind of ecological utopia imagined at the end of *Prometheus Unbound*: a place of permanent spring and renewal.

JOHN KEATS AND LORD BYRON

Though John Keats and Lord Byron (George Gordon Byron) have long been seen as important figures of the English Romantic period, their contributions to environmental thinking, relative to the other four central figures, have been less obvious. Both poets were more interested in the nature and value of poetry and in the life of the poet than in the natural world as an end in itself. Though Byron wrote memorably about the natural world in many poems, he did so in the mode of ironic posing as a late Romantic. Yet his descriptions of his love of animals, and the centrality of physical passion to human nature, are central and powerful. He would no doubt be surprised and pleased that his most influential contribution to environmental thinking is his comic nihilistic poem "Darkness" (1816), which imagines a world in which the sun dies, and human morality and then human life are slowly extinguished: "The World was void, / The populous and the powerful was a lump, / Seasonless, herbless, treeless, manless, lifeless— / A lump of death." The poem was probably written in response to the abysmally cold summer of 1816, caused by the massive release of ash by the explosion of the Indonesian volcano Tambora the year before. Now the poem has powerful resonances with nuclear winter and catastrophic climate change.

Keats managed to absorb much of the spirit of the period and to reflect it in some of the most beautiful and powerful verse ever written. His "Ode to Autumn" (1819), one of the most perfect nature poems of the period, revels in the ability of language to reproduce the experience of observing and being content with the natural world. The poem is also tinged with the recognition that the bliss and contentment of a harmonious relationship with nature is threatened by winter and death. Indeed, Keats's awareness of mortality, of the inevitability of change, provides a necessary correction to the Wordsworthian idealization of the natural world as static and harmonious, due primarily to Wordsworth.

Though the Romantic poets produced an astonishing variety of ideas about the natural world, their belief that writing about the natural world could itself be significant and beautiful has had the most profound effect on later writers. Romanticism virtually invented nature writing as we know it today.

SEE ALSO *Abbey, Edward; Animal Ethics; Emerson, Ralph Waldo; Jeffers, Robinson; Landscape Painters and Environmental Photography; Leopold, Aldo; Romanticism; Ruskin, John; Thoreau, Henry David; Wordsworth, William.*

BIBLIOGRAPHY

Bate, Jonathan. 1991. *Romantic Ecology: Wordsworth and the Environmental Tradition.* London: Routledge.

Bate, Jonathan. 2000. *Song of the Earth.* Cambridge, MA: Harvard University Press.

Buell, Lawrence. 1995. *The Environmental Imagination: Thoreau, Nature Writing, and the Formation of American Culture.* Cambridge, MA: Harvard University Press.

Kenyon-Jones, Christine. 2001. *Kindred Brutes: Animals in Romantic-Period Writing.* Aldershot, UK: Ashgate.

Morton, Timothy. 1994. *Shelley and the Revolution in Taste: The Body and the Natural World.* Cambridge, UK: Cambridge University Press.

Oerlemans, Onno. 2002. *Romanticism and the Materiality of Nature.* Toronto: University of Toronto Press.

Onno Oerlemans

ROMANTICISM

There are two main traditions of Romantic literature and philosophy with significant implications for environmental philosophy. The emergence of Romanticism as a dominant school of thought in Germany toward the end of the eighteenth century marked a significant change in conceptions of nature and the natural sciences. British Romanticism developed slightly later, from similar sources, and was predominantly a literary movement. The German Romantics, by contrast, were as concerned with the scientific study and depiction of nature as with its portrayal in art, and indeed considered these two realms of human activity to be inseparable. As a result of the emphasis of German Romanticism on the beauty and freedom of nature, its philosophy and literature played an influential role in the worldwide inception and development of environmental preservation, appreciation, and ethics. British Romanticism too, with its focus on the relationship of humans to nature, fostered new ways of viewing and valuing the natural world.

ORIGIN OF THE CONCEPT OF ROMANTICISM

Although the Romantic period was commonly contrasted with the Neoclassical period that preceded it, the specific sense of the term *Romantic* as it began to be used in the early nineteenth century was conceived in 1800 by Friedrich Schlegel (1772–1829) to evoke a "return to the Romans" (1982 [1800], p. 89) and the Italian medieval romance (1991 [1798–1800], p. 31), both of these as part of a projected urbanity that could embrace a multitude of styles and genres while retaining a universal and progressive significance. The term *Romanticism* also refers to the idea of a *lingua romana*, or a language of (all) the people. Romanticism evolved in reaction to what were conceived to be the negative consequences of Enlightenment thought, especially its association with the industrialization and homogenization of Europe. Despite their reaction against scientific conceptions that envisioned and depicted nature mechanistically, Romantic thinkers were not antiscientific, and indeed they made significant contributions to the philosophy of nature and the natural sciences.

In addition, Romanticism allied itself with progressive political thought, in particular, the ideals of equality and freedom espoused by the French Revolution. The philosophy of Jean-Jacques Rousseau (1712–1778) postulated that humans in the state of nature are neither good nor evil and that it is society that corrupts humans or cultivates good citizens. His educational theory emphasized the role of the study of nature in properly educating citizens. These two aspects of his thought thus provided a link between the study of nature and progressive politics. Rousseau's last published work, *The Reveries of the Solitary Walker* (1782), expressed the need for humans, even in a fully realized civil society, to reconnect with the natural environment to actualize fully their human essence.

PHILOSOPHICAL FOUNDATIONS OF GERMAN ROMANTICISM

Romantic philosophy of nature was deeply influenced by the philosophy of Immanuel Kant (1724–1804) and the literary and scientific studies of Johann Wolfgang von Goethe (1749–1832), both of whom, in different ways, advocated intertwining science and art. In his *Critique of Judgment*, Kant describes a "technic of nature," by which he meant that for scientific explanations to systematize all the discrete laws and regularities arrived at through observation of nature, humans must go beyond empirical cognition and think of nature in an artistic manner. This necessity of conceiving nature as a unified whole is an a priori demand that our faculty of judgment carries within itself, and cannot be justified objectively. Nonetheless, Kant argues, without what he calls the "fiction" of the theoretical construct of the unity of nature, scientists will never rest content with their results. Kant postulated the organism, a self-regulating natural being, as the most satisfying metaphor for describing nature as a totality. This conviction led to the Romantic conception of nature as an organic whole. In addition, Kant's description of the sublime aesthetic judgment contributed to the Romantic fascination with wild landscapes depicted in visual art and poetry.

As a natural scientist, Goethe insisted on a more empirical approach to nature than Kant's articulation of the transcendental conditions for the possibility of making and unifying judgments, although the primal phenomena that Goethe identified in nature attributed spirituality to nature. This attribution inspired the Romantic philosophy of nature, which saw nature as a kind of subject in its own right. Goethe identified what he considered to be the two great driving forces of nature: polarity, on the one hand, and enhancement or intensification, on the other. Polarity, according to Goethe, is a property of nature insofar as it is thought of as natural, and intensification is a property of nature insofar as it is spiritual. He called polarity "a state of constant attraction and repulsion," and intensification "a state of ever-striving ascent." These two forces affect mind and body equally, and are interdependent for their existence. Goethe argued that phenomena like magnetism and metamorphosis are "originary," in the sense of strictly belonging neither to the realm of matter nor to that of spirit.

LITERARY FORMS OF GERMAN ROMANTICISM

These originary phenomena were named "potencies" by Novalis (the pseudonym of Friedrich Leopold, Freiherr von Hardenberg; 1772–1801), the Romantic poet, novelist, scientist, and mathematician. Novalis called Romanticism a "qualitative potentization" (1965–2005, vol. 2, p. 545), or an elevation to a higher level of potency, by which he meant that when one elevates ordinary experience to the mystical level, the finite to the infinite, the known to the unknown, or the material to the spiritual, one is romanticizing it. The idea behind Romantic thought was that by re-enchanting nature through art, a closer affinity might be created between humans and nature—an affinity that in turn might lead to an increase in knowledge without destruction of nature. The natural world was not to be conceived of as a machine to be harnessed and manipulated for practical purposes, nor as an object upon which one could experiment without compunction, but rather as an organism whose needs and endeavors could oppose or complement, but always affect, those of humans.

The word *organic* was first used around 1799 by the Romantic poet Friedrich Hölderlin (1770–1843) to project onto nature the peculiarly human activities of "self-action," art, and reflection; to attribute to the whole of nature systematic form (1988 [1794–1800], p. 54). While Hölderlin's poetry can be read as a continuation of the Romantic project of re-enchanting nature by creating art to reinstill a sense of beauty, magic, and wonder in nature, he viewed the human desire for a unification with nature as ultimately unfulfillable. Organic nature is nature as appropriated and thus inevitably transformed by humans. Standing in opposition to organic nature is what Hölderlin designated as "aorgic" nature, nature prior to human intervention, a realm of being that precedes the subject/object distinction and that is thus not conceptualizable or knowable (1988 [1794–1800], p. 53). The difference between the aorgic and the organic is akin to the difference between an unlimited striving and a series of constraining forms that bring nature into presence. According to Hölderlin, these two forces are continually in tension with each other, and humans will never know nature as it is in itself, however much they may wish to.

FRIEDRICH SCHELLING

Friedrich Schelling (1775–1854), who made nature the central focus of his systematic thought, was perhaps the best-known German Romantic philosopher. In his philosophy of nature, Schelling combined the Romantic quest to integrate the diverse areas of human inquiry, Kant's critical philosophy and insights into how to overcome the gap between nature and human freedom, Goethe's emphasis on grounding all theoretical philosophy of nature in experience, and the idealist perspective of Johann Gottlieb Fichte (1762–1814), as well as the most radical scientific theories of his day.

Schelling not only insisted, like Kant, that theories of nature must reflect a unified conceptualization of nature, but also expressed complete confidence that nature agrees with the maxims of reflective reason (1988 [1797], p. 41). Schelling set out to show that nature in itself is, in fact, systematic, and that its purposiveness is not simply limited to the form of our cognition. Calling his philosophy of nature a "speculative physics," Schelling argued that nature is in fact the realization of an overarching rationality. What Georg Wilhelm Friedrich Hegel accomplished for the moral world, customs, social life, politics, and history, Schelling first outlined for the physical world of nature. In nature, spirit should recognize itself.

After 1801 Schelling turned away from a Fichtean preoccupation with the centrality of the "I" and of freedom as the highest principle of all philosophy, toward the philosophy of nature. This change of direction arose directly out of his reading and discussion of the chemical theories of Antoine-Laurent Lavoisier, his attendance at the lectures of Karl Friedrich Hindenburg on experimental physics in Leipzig, his reading of Carl Kielmeyer on organic powers and Alexander von Humboldt on electrophysiology, and his research into the theories on biology and medicine of John Brown, as well as his enduring allegiance to Kant's speculative scientific theories on force, life, and matter.

Schelling broke with Fichte on the nature of the absolute ego. For Fichte, nature is nothing more than the pure negative, the illusory "not-I," of the absolute ego. Originally, Schelling emphasized the power of mind or spirit (Fichte's absolute ego) to expand outward, to be determined or restricted only by the negative force of consciousness, or the not-I (rather than the in-itself of Kant, which lay outside of the mind). In contrast to Fichte, Schelling argued that the limitation or restriction came from the absolute ego's striving to know itself, and that the natural world arose through the interaction of the creative ego and the constrictive formative force. Schelling stated that his philosophy of nature, or speculative physics, regards active nature (*natura naturans*)—as opposed to passive nature (*natura naturata*), which empirical science takes as its object—as a subject in its own right and the proper focus of all theory (2004 [1799], p. 202). Schelling's assertion of an independent philosophy of nature as active put him at the center of German Romanticism.

Although Schelling accorded active nature a kind of freedom, it was a freedom within the bounds of law, in contrast to the theory of vital force of Johann Friedrich Blumenbach (1752–1840), which Schelling believed illegitimately implied the complete self-determination of nature. Following Kant, Schelling asserted that the behavior of nature can be scientifically accounted for only as law-governed, and that such lawful freedom can hold only if we take the organism to be a fundamental concept, not only in biology, but also in chemistry and physics. He thus moved beyond the logical analysis of the a priori conditions for the lawfulness of nature toward the real experience of the organism as a freely self-motivated natural entity.

There seems to be a fundamental correspondence between the systematizing power of the human (and divine) mind and the organization of nature. This structure can be seen in the simplest of organized forms. Schelling set his philosophy apart from pure idealism by insisting that two distinct types of philosophy are equally necessary: a transcendental philosophy that understands nature as the visible organism of our understanding and a philosophy of nature that explains the ideal as arising from and explainable from the real.

Schelling called nature the "original duplicity" in its character as both subject and object, and claimed that the opposite tendencies that pervade nature through magnetism, electricity, and sexuality manifest this duplicity (both mechanically and vitally) in nature's productivity (1867, p. 201). The process of the conflict of forces in this expanded sense can be experienced through the senses both in the realm of the inorganic, through magnetism, electricity, and chemical processes, and in the realm of the organic, through sensibility, irritability, and metamorphosis. Schelling argued that the heart of natural science was the experiment, since in the experiment "nature [is] compelled to act under certain definite conditions, which either do not exist in it at all or else exist only as modified by others" (2004 [1799], pp. 196–197). Only through such acts is it possible to gaze into the internal structure of nature.

In his essay "On Human Freedom," Schelling, following Jakob Boehme (1575–1624), described nature as effecting the self-revelation of the divine itself. God, according to Schelling in this essay, enters freely into a relation of *love* with nature. It is this relation that allows the self-manifestation of the divine, of human freedom, and of ethics. Here Schelling outlines a modified, dynamic Spinozan pantheism in which nature provides the ground for not only human freedom and divine actualization, but also evil. Schelling insists that nature, as God's ground, remains eternally separate from God, but he also accords a kind of freedom to nature because of its origin in the divine (1987 [1809], pp. 251–252). The separation of nature as ground and nature as existence in God, and the dual beginning that springs forth from this difference, also allows for personal existence, love, and human freedom. Such a loving relationship extends not only to the relationship between God and nature and between God and humans, but also and essentially to the relationship between humans and nature, whose intimate interconnectedness can be seen in the analogous structure of organism and system.

BRITISH ROMANTICISM

British Romanticism emerged out of German Romanticism. The poet Samuel Taylor Coleridge (1772–1834) in particular was strongly influenced by the philosophy of Schelling, as well as by Kant and Fichte. Coleridge's poetry and theoretical writings express the belief that a new conception of self in its relation to nature might emerge through consideration of the relation of words as "living powers" to thought and being. William Wordsworth (1770–1850) and Coleridge together initiated a new way of considering and responding to the natural world, often along the themes of journey and return through self-transformation—themes borrowed from the metaphysics of German Idealism. Wordsworth celebrated the spiritual power of raw natural beauty as a resource for self-renewal and education. John Keats (1795–1821) also used poetry to celebrate the beauty of nature and the eternal solace that nature can offer in the midst of human suffering. In contrast, William Blake (*Jerusalem*), Percy Bysshe Shelley ("Ozymandias"), Mary Shelley (*Frankenstein*), and Lord Byron ("Darkness") offered nightmarish visions of the results of human destruction or false manipulation of nature.

INFLUENCES OF GERMAN ROMANTICISM ON ENVIRONMENTAL PHILOSOPHY

The American Transcendentalists Ralph Waldo Emerson (1803–1882) and Henry David Thoreau (1817–1862) were influenced by German Romanticism in their articulation of the fundamental unity and harmony of humankind and nature, and the essential spirituality of all creation. They shared in the post-Kantian acceptance of a tempered convergence of religion and science.

Emerson, following Kant, strove to see in the beauty of nature a reconciliation of the realms of science and freedom, and like Schelling, he accorded a kind of morality to nature. The organism and the organic principle of form were central concepts for Emerson and Thoreau in their reflections on nature and writing. John Muir (1838–1914), the American naturalist, shared the Romantic conviction that there is a divine presence in nature, and he translated his concern for wilderness and appreciation of the sublime in nature into an initiative for nature preserves and the first U.S. national parks.

SEE ALSO *Emerson, Ralph Waldo; Landscape Painters and Environmental Photography; Muir, John; Pantheism; Romantic Poetry, English; Ruskin, John; Spinoza, Baruch; Thoreau, Henry David; Wordsworth, William.*

BIBLIOGRAPHY

Cunningham, Andrew, and Nicholas Jardine, eds. 1990. *Romanticism and the Sciences.* Cambridge, UK: Cambridge University Press.

Fichte, Johann Gottlieb. 1970 [1795]. *Science of Knowledge (Wissenschaftslehre), with the First and Second Introductions,* ed. and trans. Peter Heath and John Lachs. New York: Appleton-Century-Crofts.

Goethe, Johann Wolfgang von. 1949. *Gedenkenausgabe der Werke, Briefe und Gespräche,* 24 vols., ed. Ernst Beutler. Zürich, Germany: Goethestiftung für Kunst und Wissenschaft.

Goethe, Johann Wolfgang von. 1988. *Scientific Studies,* ed. and trans. Douglas Miller. New York: Suhrkamp Publishers.

Goethe, Johann Wolfgang von. 1992 (1790). *Die Metamorphose der Pflanzen,* ed. Rudolph Steiner. Stuttgart, Germany: Verlag Freies Geistesleben.

Hölderlin, Friedrich. 1988 (1794–1800). *Essays and Letters on Theory,* trans. and ed. Thomas Pfau. Albany: State University of New York Press.

Jähnig, Dieter. 1989. "On Schelling's Philosophy of Nature." *Idealistic Studies* 19(3): 222–230.

Kant, Immanuel. 1987 [1791]. *The Critique of Judgment,* trans. Werner S. Pluhar. Indianapolis, IN: Hackett.

Larson, James L. 1979. "Vital Forces: Regulative Principles or Constitutive Agents? A Strategy in German Physiology, 1786–1802." *Isis* 70(2): 235–249.

Larson, James L. 1994. *Interpreting Nature: The Science of Living Form from Linnaeus to Kant.* Baltimore, MD: Johns Hopkins University Press.

Novalis (pseudonym). 1965–2005. *Schriften: Die Werke Friedrich von Hardenbergs,* 6 vols., ed. Richard Samuel; Hans-Joachim Mähl; and Gerhard Schulz. Stuttgart, Germany: Kohlhammer Verlag.

Novalis (pseudonym). 2007. *Notes for a Romantic Encyclopedia,* trans. and ed. David W. Wood. Albany: State University of New York Press.

Pepper, David. 1984. *The Roots of Modern Environmentalism.* London: Croom Helm.

Richards, Robert J. 2002. *The Romantic Conception of Life.* Chicago: University of Chicago Press.

Roe, Nicholas, ed. 2005. *Romanticism: An Oxford Guide.* Oxford: Oxford University Press.

Schelling, Friedrich Wilhelm Joseph von. 1867. "Introduction to the Outlines of a System of Natural Philosophy or on the Idea of Speculative Physics and the Internal Organization of a System of this Science," trans. Tom Davidson. *Journal of Speculative Philosophy* 1:4: 193–220.

Schlegel, Friedrich. 1982 [1800]. "Dialogue on Poetry," trans. Ernst Behler. In *German Romantic Criticism,* ed. A. Leslie Willson. New York: Continuum.

Schelling, Friedrich Wilhelm Joseph von. 1985 [1802]. "On the Relationship of Philosophy of Nature to Philosophy in General." In *Between Kant and Hegel: Texts in the Development of Post-Kantian Idealism,* trans. George di Giovanni and H. S. Harris. Albany: State University of New York Press.

Schelling, Friedrich Wilhelm Joseph von. 1987 [1809]. "Philosophical Investigations into the Essence of Human Freedom and Related Matters," trans. Priscilla Hayden-Roy. In *Philosophy of German Idealism,* ed. Ernst Behler. New York: Continuum, 217–284.

Schelling, Friedrich Wilhelm Joseph von. 1988 [1797]. *Ideas for a Philosophy of Nature,* trans. Errol E. Harris and Peter Heath. Cambridge, UK: Cambridge University Press.

Schlegel, Friedrich. 1991 [1798–1800]. *Philosophical Fragments,* trans. Peter Firchow. Minneapolis: Minnesota University Press.

Schelling, Friedrich Wilhelm Joseph von. 2004 [1799]. *First Outline of a System of the Philosophy of Nature,* trans. Keith Peterson. Albany: State University of New York Press.

Elaine Miller

ROOSEVELT, THEODORE
1858–1919

Born in New York City on October 27, 1858, Theodore Roosevelt was the twenty-sixth president of the United States, a historian, a naturalist, a big-game hunter, and a cofounder of the Boone and Crockett Club. Roosevelt's presidency is noted for its support for the conservation of natural resources. Roosevelt withdrew from development a total of 234 million acres of public lands and set them aside as national monuments and parks, national forests, and federal wildlife refuges. He created or empowered federal agencies to manage that land on an unprecedented level. Roosevelt died on January 16, 1919, in Oyster Bay, New York.

POLITICAL CAREER

Roosevelt's childhood interest in natural history and ornithology led him to consider becoming a zoologist. He published two ornithology books while in college, the first of his three dozen books on a variety of topics. Instead of becoming a zoologist, he graduated from Harvard and enrolled at Columbia Law School, though he entered politics without getting a law degree. He served in the New York State Assembly from 1882 to 1884. In 1883, he purchased a cattle ranch in what is now North Dakota and worked it on and off for the next five years. The experience taught him about life and the environment in the semi-arid Western frontier region.

In 1887 Roosevelt co-founded the Boone and Crockett Club, a gentleman hunter's club with a strong interest in the preservation of large game animals. It fought for the preservation of Yellowstone Park, contributed to the passage of the Forest Reserve Act in 1891—the basis of the national forest system—and helped found the Bronx Zoo in 1895.

Roosevelt served on the U.S. Civil Service Commission (1889–1895) and the New York City Police Board (1895–1897) and became assistant secretary of the U.S. Navy (1897–1898). When the Spanish-American War began in April 1898, he resigned to form the First United States Volunteer Cavalry regiment, better known as the Rough Riders. He returned a war hero and was elected governor of New York in 1898. In 1900 he was elected vice president of the United States, and in September 1901 he became president after the assassination of William McKinley.

Theodore Rooselvelt with John Muir, 1903. *President Roosevelt and John Muir often camped together at Glacier Point in Yosemite National Park, having lengthy talks about conservation. Although he supported the protection of natural wonders, Roosevelt rejected the impractical idea of of eliminating all development in wild areas.* THE LIBRARY OF CONGRESS.

ACHIEVEMENTS AS A CONSERVATIONIST

Besides trustbusting, Roosevelt remains in the American political consciousness for his strong support of the conservation movement. Shortly after becoming president, Roosevelt asked Gifford Pinchot of the Division of Forestry and Frederick Newell of the U.S. Geological Survey to help craft an integrated program for federal management of natural resources. Like them, Roosevelt feared letting private industry continue to exploit public lands without federal regulation and believed that only federal scientific management could save American natural resources and, by extension, democracy from industrial monopolies and their wastefulness.

During his first term in office Congress passed the Newlands Reclamation Act in 1902, bringing nearly three million acres of semiarid land in a dozen states under irrigation and making settlement there feasible. After the 1904 presidential election Roosevelt made conservation a central domestic issue.

Roosevelt supported both preservation and conservation. He favored the preservation and protection of natural oddities and wonders but rejected as impractical the preservationists' idea of eliminating development in all natural or wild areas. He urged Congress to pass the Antiquities (or National Monuments) Act of 1906, which allowed the president to set aside scientifically or historically important areas such as the Grand Canyon for protection. He also established the first federal wildlife refuges and reservations to protect wildlife and its habitats. At the same time Roosevelt promoted utilitarian conservation and planned natural resource development. In 1905 Congress transferred the sixty-three million acres of federal forest reserves to the Department of Agriculture and created the U.S. Forest Service to manage the national forests for timber and watershed protection.

Roosevelt appointed commissions and held conferences to highlight his conservation agenda. In 1907 he appointed the Inland Waterways Commission to examine multiple-purpose development of river basins. His Governors Conference on Conservation in 1908 inspired individual states to establish state forests, forest services, and conservation boards. Roosevelt also expanded conservation to include the health of children, waste in war, and civic beautification. The Country Life Commission, for example, attempted to stop the disintegration of rural life, which some social critics considered of vital importance for the political survival of the nation. Congress, however, resented the expansion of executive power and cut off funding.

After leaving the presidency in 1909, Roosevelt undertook an African expedition to collect specimens for the Smithsonian Institution. After his defeat in 1912 as a third-party candidate for president, he explored a newly discovered river in Brazil. The trip nearly killed him, but his party mapped the river and collected specimens. In Roosevelt's honor, Brazil renamed the river Rio Roosevelt. His last years were spent writing articles and reviews for various publications. Among his many accomplishments, Roosevelt's conservation work may be his greatest legacy.

SEE ALSO *Conservation; Environmental Law; Pinchot, Gifford; Preservation; U.S. Department of Agriculture; U.S. Forest Service.*

BIBLIOGRAPHY

Collins, Michael L. 1989. *That Damned Cowboy: Theodore Roosevelt and the American West, 1883–1898.* New York: P. Lang.

Cutright, Paul Russell. 1956. *Theodore Roosevelt, the Naturalist.* New York: Harper.

Cutright, Paul Russell. 1985. *Theodore Roosevelt: The Making of a Conservationist.* Urbana: University of Illinois Press.

Gould, Lewis L. 1991. *The Presidency of Theodore Roosevelt.* Lawrence: University Press of Kansas.

Hays, Samuel P. 1999. *Conservation and the Gospel of Efficiency: The Progressive Conservation Movement, 1890–1920.* Pittsburgh, PA: University of Pittsburgh Press.

Morris, Edmund. 1979. *The Rise of Theodore Roosevelt.* New York: Coward, McCann and Geoghegan.

Morris, Edmund. 2001. *Theodore Rex.* New York: Modern Library.

Roosevelt, Theodore. 1913. *An Autobiography.* New York: Macmillan.

James G. Lewis

ROUTLEY, RICHARD

SEE *Sylvan, Richard.*

RUSKIN, JOHN

1819–1900

The Victorian art and social critic John Ruskin was born in London on February 8, 1819. Among his primary concerns was the accurate description of nature. His 1834 essay in *Loudon's Magazine* discussed, as an example, the causes of the color of the Rhine. That commitment to exactness and curiosity about nature, articulated explicitly or implicitly within the terms of natural theology, defined his thought throughout his life. In the first half of his career, Ruskin's principal intellectual preoccupations were with the moral testimony of art, particularly art representing nature. Ruskin understood the best art as art that, first, respected the empirical truths of nature through accurate representation and, second, embodied a great artist's interpretation of the moral truths of nature. To see the natural world properly, in this respect, was an act of love and a way of understanding ethical teaching. The five volumes of Ruskin's *Modern Painters* (1843–1860) offered a detailed account of the English painter J. M. W. Turner as an ideal artist, faithful in the representation, and wise in the interpretation, of nature. They proposed as a general conviction that to "see clearly is poetry, prophecy, and religion,—all in one" (vol. 5, p. 333).

Ruskin's close observation of nature contributed to his exceptional drawings and watercolors. In the early 1850s he supported, with equivocation, the principles of the Pre-Raphaelite Brotherhood to represent the natural world faithfully. His concern with the representations of art gradually expanded into a preoccupation with the condition of the world in which artists worked. *Unto This Last* (1860) marked, at least as Ruskin saw it, his formal turn to economics and politics. Inseparable from his political arguments were new questions about the meaning of nature and how industrial modernity treated it. Although Ruskin had rejected his evangelical Christianity in 1858, only for the briefest moments thereafter did he lose faith in the divinely inspired meaning of the natural world. In the 1860s, looking for new ways to discuss what human beings could learn from the natural world, he explored mythological readings of nature, including Greek myth in *The Queen of the Air* (1869). Ruskin's late scientific textbooks—*Love's Meinie* (1873–1881), *Deucalion* (1875–1883), and *Proserpina* (1875–1886)—endeavored to draw the best wisdom from such mythological approaches to nature.

Politically, Ruskin in the 1870s paid attention to rural communities and the contamination of the natural world, which he regarded as a betrayal of humanity's God-given duty of care. The public letters of *Fors Clavigera* (1871–1884) developed plans for an alternative agrarian community—the Guild of St. George—that privileged hand labor over mechanical labor, and that, in its statement of principles, required a basic commitment to respect life and beauty: "I will not kill nor hurt any living creature needlessly, nor destroy any beautiful thing, but will strive to save and comfort all gentle life, and guard and perfect all natural beauty, upon the earth" (vol. 28, p. 419). In his last years, Ruskin's impatience with industrial pollution prompted frequent statements admonishing the nation and recurrent private lament. In 1878 he entered into an angry argument with the Corporation of Manchester, which was trying to turn the lake of Thirlmere into a reservoir. Elsewhere he mourned the loss of humbler natural sites that he had personally valued (see the elegy for Croxted Lane in *Fiction—Fair and Foul*, pt. 1 [1880]). Most dramatically, in his public lectures *The Storm-Cloud of the Nineteenth Century* (1884), he argued that a "plague wind" was infecting England (vol. 34, p. 31). That wind, he said, suggested an approaching environmental calamity born of divine displeasure. From many years of observation, Ruskin adduced evidence for the palsied behavior of nature: "blanched Sun,—blighted grass,—blinded man" (vol. 34, p. 40). He concluded that natural disorder confirmed moral gloom. His lectures, the most extreme development of Ruskin's reading of nature's lessons for human beings, offered prototypes of early-twenty-first-century rhetoric of environmental apocalypse.

Ruskin's legacy in environmental matters was significant in his own century. Alongside William Morris, he influenced the development of the arts and crafts movement, with its favoring of traditional industries and nonmechanical labor. His initiation of the Guild of St. George lay behind later back-to-the-land movements and socialist utopias (see, for instance, the work of Edward Carpenter). The development of the Garden City initiative, late Victorian campaigns for clean air, and the foundation of the National Trust owe inspiration to Ruskin's writings, which continue to be cited as protoenvironmentalist and prescient in its strong commitment to the organic harmony of living things.

SEE ALSO *Environmental Aesthetics; Environmental Art; Environmental Philosophy: IV. Nineteenth-Century Philosophy; Romanticism; Wordsworth, William.*

BIBLIOGRAPHY

Ruskin, John. 1903–1912. *The Complete Works of John Ruskin*, 39 vols., ed. E. T. Cook and Alexander Wedderburn. London: Allen.

Francis O'Gorman

RUSSIA AND EASTERN EUROPE

Environmental ethics in the Commonwealth of Independent States and East European countries was initially formed under the strong influence of Russian science, Russian culture, and, up to the early 1990s, Marxist ideology. For a variety of geographical, historical, and political reasons, Russia was the dominant intellectual center in Eurasia. Even after the 1990s, East European countries and the Commonwealth of Independent States remained under the influence of Russian high culture, which depends not only on the spread of the Russian language and the popularity of Russian literature and philosophy but also on the unprecedented development of theoretical and scientific thought in the Soviet Union in the twentieth century. Scientific schools, emanating from the Soviet Academy of Sciences or the leading universities of Russia, still function more or less successfully in all of these countries.

Russian scientists—among them Vladimir I. Vernadskii, Nikita N. Moiseev, Vladimir N. Sukachev, Nikolai V. Timofeev-Resovskii, and Alexander L. Yanshin—made fundamental contributions to the development of ecology in the twentieth century. Because of this strong tradition, the Soviet scholarly community was receptive to the ideas of environmental ethics developed by Albert Schweitzer and the Italian industrialist and economist Aurelio Peccei.

PHILOSOPHICAL UNDERSTANDING OF HUMAN INTERACTIONS WITH NATURE

During the Soviet period, ecological literature that analyzed the state of the environment paid special attention to the role of sociopolitical factors, which were considered primary for understanding and resolving emerging ecological problems. The view was that only in the capitalist world do ecological catastrophes arise.

Under socialism, the ecological relation between humans and nature was explored from the vantage point of dialectical materialism. According to this concept, the interaction between humans and nature is a natural historical process, the concrete contents and movement of which depend on the level of development of productive forces and on the class that controls the means of production. In other words, the relation between humans and nature is not constant and natural, but is determined by social factors. Humans are related to nature through practical activity and material production. This relation is one not of abstract observation of nature, but of interaction through labor and practical activity.

Under presocialist socioeconomic institutions, there is spontaneous interaction between society and the natural environment. Under socialism, however, society has transitioned from spontaneous interaction with nature to guiding nature and being responsible for its future course. The view was that real guidance of the interactions between society and nature is possible only with public ownership of the means of production, sociopolitical unity, and humane social relationships.

Initially, on the basis of these philosophical and methodological considerations, the Soviet socioecological literature even attempted to claim that there was no ecological crisis under socialism, since socialism engaged in planned, predictable development of the economy. Yet socialism needed more theoretical development, particularly for working out the practical, legal, socioeconomic mechanisms for harmonizing the interaction of society and nature. Unfortunately, this need was not fully understood and put into practice by the authorities of the Soviet Union and other socialist countries. Thus, under socialism, these counties were unable to resolve the contradictions inhering in interactions between nature and society, contradictions that increased with technological progress.

ENVIRONMENTAL PROBLEMS AND DISASTERS

The high concentration of industry in cities during the Soviet period degraded the environment around them and spawned a variety of health problems. No less than 10 percent of Russian cities have high levels of soil, air, and water pollution. Almost every city with a population of over 1 million people has ecological problems. Many of these and other environmental problems, as cataloged by William Carter and David Turnock (1993), are a legacy of the Soviet era. Areas of environmental degradation include the region around the Aral Sea (in Kazakhstan, Uzbekistan, and Tukmenistan), the former Semipalatinsk nuclear test site in Northern Kazakhstan, and the industrial zones of Donbass in Ukraine and of the Kola Peninsula and the South Urals in Russia.

During the period of glasnost in the 1980s, new nongovernmental organizations began engaging the state on environmental issues, but the collapse of the Soviet Union and other communist states and the economic crises that ensued effectively terminated these discussions. Since the early 1990s, clear improvement in the environmental situation has not occurred. This failure was a result in part of the transition to market economies, in which protection of the environment had low priority. While more progress was made in the relatively richer East European countries, several problems require urgent attention. These include uncontrolled management of dangerous waste, unreliable treatment of waste water, deterioration of air quality in highly populous urban and industrial areas, and contamination and damage to the soil by pesticides and overcultivation. In addition, multinational oil corporations have motivated Kazakhstan, Uzbekistan, and Azerbaijan to allow large-scale drilling in the shallow waters of the Caspian Sea, at the risk of destroying vulnerable ecosystems.

By far the worst environmental damage resulted from the accident at the Chernobyl nuclear power plant on April 26, 1986, when a breakdown of the fourth reactor resulted in the release of a large amount of radioactivity into the environment. The explosion released thirty to forty times the combined fallout from the atomic bombing of Hiroshima and Nagasaki. Belarus (the European part of the Russian Federation) and Ukraine suffered the most, with Belarus receiving about 60 percent of the fallout. A year after the disaster, the radiation across most of the European territory of the Soviet Union primarily involved long-lived cesium isotopes. In areas closer to the disaster, strontium-90, various isotopes of plutonium, and several other highly radioactive elements continue to pose hazards. By 2005, 56 direct deaths had occurred, and the expectation is that there will eventually be as many as 4,000 cancer deaths among the 600,000 most exposed and as many more among the 6 million people living nearby. Ten-year forecasts of the continuing pollution of European territories have been made. Most of the pollution will largely disappear between 2006 and 2016, but in the Bryansk region pollution will remain until 2092.

ADDRESSING ENVIRONMENTAL PROBLEMS

As of 2008, several groups are seeking to address these environmental problems. For example, the Regional Environmental Center for Central and Eastern Europe, which publishes the quarterly magazine *Green Horizon*, seeks to assist in solving environmental problems in this region by providing information on the environment, promoting public participation, and encouraging cooperation among environmental stakeholders.

Western countries eventually overcame their neglect of the environment. Several characteristics of nonsocialist societies played a role in that process. These characteristics include transparency, public availability of environmental information, the emergence of mass environmental movements, and advocacy by ecologically oriented political parties leading to the adoption of environmental laws. As a result, under constitutional states, environmental problems started to be resolved more effectively than under socialism because needed laws were rapidly adopted and enforced.

A NEW ENVIRONMENTAL ETHIC

Socioecological literature of the Soviet period included serious theoretical works arguing that contradictions in the interrelations between society and nature are inevitable quite apart from the socioeconomic structure of society. This argument was based on an analysis of how the system of society and nature functions under the laws of nature, particularly the conservation laws of physics and chemistry, which preclude the destruction of matter and energy. This literature concluded that contradictions are inevitable for such systems because a society cannot develop without using natural resources and producing

Brown Bear, Kamchatka, Russia. *The brown bear (*Ursus arctos*), shown here crossing a stream in the Valley of the Geysers, Kronotsky Zapovednik Reserve, is found in many parts of the world, but the species is most populous in Russia. Over the last century, habitat encroachment and hunting by humans has dramatically reduced the number of brown bears globally; in most populations they are considered endangered or threatened. Conservation efforts are fighting to increase the numbers of brown bears in Russia and elsewhere.* IGOR SHPILENOK/NATUREPL.COM.

waste and this reliance sooner or later leads to specific ecological collisions. In coming to these conclusions, the scientific literature, unlike the ideological literature, did not idealize Eastern socialist economies and did not bow before the received notion that the presocialist societies of the West would inevitably break down.

Environmental ethics in the Soviet philosophical literature was logically divided along the following perspectives toward nature: theoretical, practical, aesthetic, and ethical (Tugarinov 1978). The practical attitude toward nature focuses on processes that produce goods and services and enable humans to live. Through these processes, the human forces at work in society are revealed. These practical and material economic arrangements determine, in accord with dialectical materialism, the theoretical and ideological structures of society, including society's attitude toward nature

The aesthetic attitude toward nature can be seen in the human need to observe the beauty of natural phenomena and various landscapes, and to engage in artistic reflection on observable forms, such as space and light. Natural elements enrich figurative artistic exploration of reality. In turn, the aesthetic attitude toward nature contributes to the formation in mass consciousness of the idea of an empathetic approach toward nature.

In the second half of the twentieth century, the importance of an ethical attitude toward nature grew significantly. Inculcating a love of nature and the need to preserve plant and animal species and their habitats became one of the most important tasks of moral education. The point was not to resurrect animistic feelings, but to understand that the surrounding flora and fauna are necessary for humans (for both their physical existence and spiritual enrichment) and that some limits on the transformation of nature are needed even under conditions of socialism. Ethical indifference toward nature was to be replaced with a clear understanding that destruction of nature is immoral and intolerable. More and more people understand that nature is a support for human life, a source of beauty, and the ground of an ultimate good. It thus became commonplace to include in the sphere of moral relationships some aspects of the relationship between humans and nature, along with traditional considerations of person-person and person-society relationships.

In the past, the specific content of these different human attitudes toward nature were considered to depend significantly on the socioeconomic structure of society and on the various social relations and value configurations that emerged in a given historical situation. Yet since the 1990s, through social connections, the surrounding environment has obtained universal significance for people, satisfying their different needs—from simple physical ones to spiritual ones. In Russia, East European countries, and the Commonwealth of Independent States, environmental ethics is now understood to depend on the formation of an environmental consciousness and an environmental culture, based on continuous environmental education and ecological enlightenment.

SUSTAINABLE DEVELOPMENT

Undoubtedly, the concept of sustainable development adopted by the 1992 U.N. Conference on Environment and Development (Rio de Janeiro) provided a new impulse for environmental ethics to go in that direction. In Russia, the concept of sustainable development reflects the increase of social, economic, and ecological problems on our planet, on the one hand, and also our overall level of scientific understanding of nature and society, on the other. The concept of sustainable development seems to be the point of convergence of several philosophical and sociological teachings and several theories in the natural sciences and mathematics.

Simply put, humankind has to transition to sustainable development for the survival of humanity and preservation of the biosphere—the thin sphere of life enveloping the geosphere. Needed is a transformation of all spheres of human activity in the direction of diminishing pressure on the biosphere. Without doubt, sustainable development should be characterized by economic effectiveness and social justice, as well as a general reduction of anthropogenic pressure on the biosphere.

The Russian approach to understanding ecological problems and sustainable development is connected with Vladimir I. Vernadskii's teachings about the biosphere and noosphere. The concept of a noosphere, a sphere of mind, reflects a belief in the power of reason and its unlimited creative capabilities. The notion was developed by the French scientists Edward Leroi and Pierre Teilhard de Chardin, as well as by Vernadskii, though their several ideas about the noosphere are significantly different from one another. According to Vernadskii, the epoch of the noosphere will be characterized by the following basic traits:

- Humankind settles the whole planet; connections between various peoples and countries intensify.
- The geological role of human activity prevails over natural geological processes taking place in the biosphere.
- The borders of the biosphere expand; humankind enters outer space.
- New sources of energy are found.
- People of all races and religions are discovered to be equal.
- The role of the masses in solving the problems of domestic and foreign policy grows.
- Wars are excluded from social life.
- The welfare of working people grows; a real opportunity emerges for the elimination of hunger, poverty, and disease.
- The original planetary biosphere is reasonably transformed to satisfy all the material and spiritual needs of a growing global population.
- Scientific thought and inquiry are liberated from the pressures of religion and politics, and state and society create conditions favorable to the realization of this intellectual freedom.

The philosophical connections between Vernadskii's vision and the later concept of sustainable development are striking. For instance, the concept of sustainable development means, among other things, a movement of humankind into the sphere of reason, where society and nature harmoniously interact with each other.

Since around 2005, there has been a generalization of the concept of sustainable development. Under this generalization, sustainable development is understood not simply as ecologically secure development, but also as stability in the social and political spheres of social life, based on respect for human rights, democratic principles, the rule of law, and norms of international law. Such sustainable development unites the environmental, economic, and sociopolitical domains into a single unified system. Moreover, it connects reason and morality. Historically, the achievements of reason, torn asunder from moral elements, have sometimes acquired a malevolent and antihuman character. To limit rational but inhumane social development, spiritual values and norms of morality are needed.

Transition to sustainable development means changing both the substance and form of the material and spiritual culture of civilization. This transition cannot be achieved by means of conventional thinking or traditional ideas and values. It requires working out new scientific and philosophical approaches to complex problems, corresponding not only to postmodern reality but also to the prospects for civilization in the third millennium.

Realizing sustainable development that affirms the intrinsic value of nature and ethical attitudes toward it is of utmost importance. Former strategies, oriented only to economic development, need to be replaced with a strategy of integrating the human economy into the natural economy and patterning the design of artificial systems of energy production and agriculture on the model of ecosystems, in accord with the concept of sustainable development. The modern way of life, with values formed in the previous half millennium, needs to undergo radical changes in order to meet the challenges of the new millennium.

In Soviet and post-Soviet cultural literature, the ecology of the biosphere is basically treated as the new means of reconciling humans and nature on the basis of knowledge about and understanding of nature. By knowing the modalities, depth, and scale of interaction between humans and nature, biospheric ecology can play a significant role in the formation of environmental thinking, reorienting all spheres of human life toward solving environmental problems. The main paradigm of the Russian environmental outlook and the main thrust of biospheric ecology is understanding that people do not exist on the earth simply for themselves, that they also perform specific ecological functions. The most important characteristic of Eastern European environmental ethics is a rejection of naive anthropocentrism and a transition toward a system of beliefs built on biospheric centrism. Such an environmental ethics is the conceptual basis of a new biosphere-centered culture. By concentrating attention on the problems of the biosphere, biospheric environmental ethics creates the preconditions for actions oriented toward preserving and developing the well-being of humans and nature.

Like most environmental ethics, Eastern European environmental ethics cares about the natural conditions of existence of future generations. Orientation toward the future, which presupposes also caring about the present, distinguishes environmental ethics from traditional ethics.

Sustainable development can be understood as development that satisfies the needs of the present without endangering the ability of future generations to satisfy their own needs. This orientation can be considered the main ethical regulator of interaction between people and the natural environment. In place of anthropocentrism, there emerges a new approach to reality based on biospheric centrism. The biospheric approach introduces a fundamentally new perspective into the moral experience of humankind, hitherto based on a traditional dichotomy of good and evil. Vernadskii (1988) claims that ethics can be approached scientifically if it proceeds from knowledge of the biosphere. Environmental ethics demands that we act responsibly toward future generations when making decisions affecting the environment.

Post-Soviet environmental literature suggests that sustainable development is possible, not under conditions of socialism, but only under a biosphere-centered culture that reproduces through sustainable development. Hence, the notions of biosphere-centered culture and sustainable development reflect different aspects of the same process: harmonizing the relations between society and nature.

GLOBALIZATION

The post-Soviet environmental literature since around 2005 increasingly focuses on issues of globalization. As Vernadskii predicted, humankind lives in an era of the formation of a new world outlook, the basic content of which is, to a large extent, global. The complex, multifaceted process of globalization is defined by a number of factors, including integrative trends in the economic and political spheres, synthesis of the natural sciences and humanities to present an integrated picture of reality, and development of information and communication technologies so as to change not only the human way of life and perspective on the world but also human consciousness. The emerging danger posed by global climate change requires reconsidering the entire system of our basic social values (such as ethical norms, attitudes toward nature, and the character of production to satisfy human needs and wants).

In relation to controlling the processes of globalization for sustainable development, the premises and conditions of the formation of global consciousness become extremely topical. The contents of global consciousness should be determined less through broadening traditional perceptions and thought than through forming fundamentally new perspectives based on emerging ideas for global integration.

In each era, humankind, relying on the current level of knowledge in various fields, regulated to varying degrees its relations with the natural environment. At the beginning of the twenty-first century, however, we face the greatest challenge of our time, perhaps of all time: harmonizing the development of the socioeconomic sphere and the biosphere. The task is to realize an environmentally sound revolution in human lifestyles, to environmentalize the human way of living.

EDUCATION

Meeting this challenge is unthinkable without a corresponding transformation of social consciousness, without each individual understanding the meaning of the coming changes. For this reason, the problems of education come to the foreground. Needed is a global change of orientation in educational systems toward the theory and practice of sustainable development. This transformation requires a search for forms and means of embedding environmental knowledge and imperatives into different levels of education.

In Russia, East European countries, and the Commonwealth of Independent States, a commonly accepted ideal is the need to deploy a multilevel system of environmental education that embraces all elements of preschool, primary, secondary, and higher education, as well as executive courses, retraining, and popularization of scientific knowledge. Such an education should be the main thrust for developing environmental consciousness and forming environmental responsibility. An environmental education is a precondition for the emergence of an environmental culture and for society's transition to sustainable development.

In Russia, environmental education should be a top priority in that country because one can especially feel a strong dependence on the material and spiritual forces of nature. Russia is not only the biggest country in the world and the richest country in terms of energy resources; it is also the coldest country, where permafrost covers more than half of its territory and an extreme natural environment dominates. Fortunately, in Russia the birthrate has been dropping catastrophically in recent years, thus relieving human pressure on the biosphere.

Education for sustainable development is integrative and multifaceted. It is directed toward solving a number of interrelated problems and thus cannot be reduced to ecological education alone. It must define strategic goals and develop the spiritual, intellectual, scientific, and technological means for achieving them. As is evident, education for sustainable development should ascend to the level of an integrated environmental education.

SEE ALSO *Anthropocentrism; Chernobyl; Conservation; Future Generations; Intrinsic and Instrumental Value; Schweitzer, Albert; Space/Place; Sustainable Development.*

BIBLIOGRAPHY

Attfield, Robin. 2003. *Environmental Ethics: An Overview for the Twenty-first Century.* Cambridge: Polity Press.

Carter, Francis William, and David Turnock. 1993. *Environmental Problems in Eastern Europe.* New York: Routledge.

Chumakov, Alexander N. 2004. *Globalizatzia: Kontury tzelostnogo mira* (Globalization: The outline of the holistic world). Moscow: Prospekt.

Chumakov, Alexander N. 2006. *Metafizika globalizatzii: Kulturno-tzivilizatzionnyi kontekst* (The metaphysics of globalization: Cultural and civilizational contexts). Moscow: Canon+.

Des Jardins, Joseph R. 2000. *Environmental Ethics: An Introduction to Environmental Philosophy.* Belmont, CA: Wadsworth Publishing/Thomson Learning.

Light, Andrew, and Holmes Rolston III, eds. 2003. *Environmental Ethics: An Anthology.* Oxford: Blackwell.

Mamedov, Nizami M. 2003. *Osnovy sotzialnoi ekologii* (The foundations of social ecology). Moscow: Stupeni.

Mamedov, Nizami M., ed. 2002. *Vvedenie v teoriu ustoichivogo razvitia* (Introduction into the theory of sustainable development). Moscow: Stupeni.

Mazour, I. I.; A. N. Chumakov; and W. C. Gay, eds. 2003. *Global Studies Encyclopedia.* Moscow: Raduga.

Pojman, Louis P., and Paul Pojman, eds. 2008. *Environmental Ethics: Readings in Theory and Application.* Belmont, CA: Wadsworth.

Tugarinov, Vasilii P. 1978. *Priroda, tzivilizatzia, chelovek* (Nature, civilization, person). Leningrad: Izd-vo LGU.

Vernadskii, Vladimir I. 1988. *Filosofskie mysli naturalista* (Philosophical thoughts of a natural scientist). Moscow: Nauka.

Alexander N. Chumakov
Nizami M. Mamedov
William C. Gay

S

SALMON RESTORATION

As anadromous fish dependent on the oceans for feeding and fresh water for spawning, salmon are vulnerable to a wide range of human activities. Pacific salmon were indigenous to the north Pacific in coastal regions and rivers from Japan, through Russia, and across the Pacific from Alaska south to California. Atlantic Salmon on the eastern coast of North America were indigenous from Long Island through the Atlantic Provinces of Canada to Greenland, Iceland, Scandinavia, the British Isles, western Russia and south in Europe as far as Portugal. In both the Atlantic and Pacific these geographic ranges are as of 2008 significantly reduced. Historically and ecologically, the long and precipitous decline of wild salmon stocks in the United States is traceable to a constellation of factors, the most important of which are habitat destruction and over-harvesting.

Both of these factors prevail in direct proportion to increased human population density along the coastal waterways and inland river basins where salmon migrated and spawned. On the East Coast of the United States, primarily Maine, the situation of Atlantic salmon is so dire that restoration of the species is highly dubious. According to the National Oceanic and Atmospheric Administration (NOAA) Fisheries and the U.S. Fish and Wildlife Service, "The number of wild Atlantic salmon in Maine rivers is at an all-time low, placing them in danger of extinction" (NOAA Fisheries 2007, p. 1).

On the West Coast of the United States, the status of salmon—which includes Chinook (King), Coho (Sliver), Sockeye, Chum, and Pink varieties—is much more variegated, depending on the region. Salmon stocks in Alaska, for example, are usually more stable and healthy, whereas salmon stocks in California are seriously threatened or endangered. A classic example of the complex and often contentious nature of salmon recovery is the early 2000s effort to restore salmon in the Columbia River Basin, which, according to some estimates, once had the largest number of returning salmon of any river in North America—between 10 and 15 million.

Salmon restoration as a formal strategic attempt to recover salmon on the West Coast began in earnest during the 1990s, when a number of wild seasonal spawning runs of salmon, identified as evolutionarily significant units (ESUs), were rapidly listed as either threatened or endangered under the federal Endangered Species Act (1973). An important tipping point was a report issued in 1991 by Nehlsen, Williams, and Lichatowich called "Pacific Salmon at the Crossroads: Stocks at Risk from California, Oregon, Idaho, and Washington." Their research showed that of 214 wild runs of salmon, steelhead, and sea-run cutthroat trout, 101 were in extreme risk of extinction, 58 at moderate risk, and 54 of concern. They identified 106 runs that had already become extinct. It was with this background that the National Marine Fisheries Service (later known as NOAA Fisheries) was mandated by the Endangered Species Act to develop salmon and steelhead recovery (or restoration) plans.

Salmon restoration encompasses a dizzying array of stakeholder groups that can add up to a bureaucratic and policy conundrum (Lackey, Lach, and Duncan 2006; Kolmes and Butkus 2006). Although it would be unwieldy to list all those involved, a short list includes the following: federal agencies, state governments, regional and municipal governments, Native American tribes, commercial and sport fishing associations, the hydroelectric power industry, the Northwest Power and Conservation Council, several environmentally oriented NGOs in favor of salmon recovery, and a number of industry and agriculture (e.g., irrigation) NGOs who favor the status quo. The primary entity responsible for creating salmon restoration plans and mediating the conflicts among various parties is NOAA Fisheries. The recovery process, however, is not beyond political maneuvering at the very highest levels of the federal government.

The contentious nature of salmon recovery was highlighted by the 2002 struggle in the Klamath River Basin, where competition over water between farmers and fishers (commercial and tribal) took a toll not only on community good will but also a significant number of adult salmon. In September 2002, nearly 34,000 adult salmon returning to spawn were killed when low water in the Klamath River created lethally warm conditions, causing the fish to succumb to fungal infection. The scenario in the Klamath Basin is an example of the clash of social values over salmon restoration and a notable case study in environmental ethics. It is also an example of how the abandonment of good science and the mismanagement of water resources in the Klamath River led to a conflict that nearly erupted into violence.

NOAA restoration plans have incorporated a vision of salmon recovery (reducing the risk of extinction for any population to 5% over the next hundred years) that is considered by some to be a narrow interpretation of the term *recovery*. Focused on preventing extinction and achieving the delisting of salmon as an endangered species, NOAA's approach to salmon restoration appears incongruous with the more generous vision of what "broad-sense recovery" means to most people. The shifting biological opinions and restoration plans produced by NOAA, struck down one after another by federal courts as inadequate, have raised questions about the capacity of NOAA Fisheries to carry out reliable salmon restoration planning. The attempt of NOAA Fisheries to count industrially produced hatchery fish, known to be genetically deficient and lacking the local evolutionary adaptations vital to population viability, has intensified the conflict between NOAA Fisheries on the one hand and environmental groups and locally based salmon-restoration efforts on the other (Meyers et al. 2004). In the view of some environmental activists, NOAA Fisheries has taken a view of salmon-recovery planning metrics that focuses excessively on fish population numbers to the exclusion of habitat issues or fish characteristics. Seeking to count hatchery fish to achieve stability of fish numbers, it has eschewed critical habitat requirements—and the attendant political implications for human land-use practices—that are essential for long-term salmon survival.

The issue of salmon restoration can also be addressed from the environmental-ethical perspective expressed in this question: Do human beings have a moral obligation to prevent the extinction of salmon and to commit time, money, and energy to the effort to restore the species? The answer to this question depends in large part on how one values salmon and salmon restoration. At one extreme there are those who do not value salmon at all and see salmon restoration as a major hindrance to the economic values of the industrial economy. In the middle—where the majority of people in the Northwest reside, based on opinion polls—are those who are in favor of salmon restoration and hold a variety of human values for salmon, including aesthetic, recreational, nutritional, culinary, cultural, scientific and spiritual At the other end of the spectrum are those, such as tribal peoples, who, for cultural and religious purposes, see salmon as sacred and others who believe salmon have intrinsic value apart from their utility to humans.

An interesting layer of analysis was added to the ethical assessment of salmon when, in 2001, the Roman Catholic bishops within the Columbia River Basin issued a pastoral letter entitled "The Columbia River Watershed: Caring for Creation and the Common Good." This international document surveyed the problems of the basin and provided a theological-ethical foundation for ecological and social responsibility. Grounding this idea in the concept of biblical stewardship and concern for the common good, the bishops of the region sought to create an ethical framework for action by offering ten ethical norms called "Considerations for Community Caretaking." On one of these principles, "Conserve and Protect Species of Wildlife," a critical norm for salmon restoration, the bishops state:

> The presence and health of wildlife is in many ways a sign of the health of our ecosystems, of the well-being of the people and communities dependent on the ecosystems for their livelihood, and our respect for God's creatures and creation. The presence and health of salmon and other species in the Columbia-Snake system, in particular, is a sign of the health of the entire region.

The ethical imperative to restore wild salmon is directly linked—scientifically, philosophically, and theologically—to the value of a species' life other than our own and to the generative, creative evolutionary process that produced it. The primary issue is that anthropogenic extinction inevitably curtails the evolutionary process that produces life. In the view of Holmes Rolston III, "Every extinction is an incremental decay in this stopping of the flow of life, no small thing. Every extinction is a kind of superkilling. It kills forms (species), beyond individuals. It kills 'essences' beyond 'existences,' the 'soul' as well as the 'body'" (1988, p. 144). Accordingly, the ethical duty to preserve and restore a species is, in Rolston's words, "a categorical imperative to living categories." In other words, the obligation to recover salmon from extinction is a moral duty without exception.

Prospects for the restoration of Atlantic salmon on the East Coast of the United States are bleak. Pacific salmon, however, have better prospects because of population density and the extent of commercial development in their historic breeding range. There are runs in California (e.g., the longstanding decline of the Sacramento River winter run of Chinook) where restoration is probably not realistic. The Sacramento River spring Chinook run suffered a dramatic population crash in 2008. Other runs of Pacific salmon, from Oregon to Alaska, can be restored or preserved in healthy condition if human activities detrimental to salmon survival are sharply curtailed to enhance water quality and reduce habitat alterations, riparian deforestation, genetic dilution of previously adaptive populations caused by hatcheries, and overfishing. If human beings in salmon-bearing areas are willing to embrace and implement an ethic of sustainability—a balance between the needs of humans and the legitimate ecosystem needs of wild salmon—then there are good prospects for restoration in areas where the habitat remains viable.

SEE ALSO *Ecological Restoration; Endangered Species Act; Hunting and Fishing: I. Overview; Hunting and Fishing: V. Commercial Fishing; Stewardship.*

BIBLIOGRAPHY

Augerot, Xanthippe et al. 2005. *Atlas of Pacific Salmon: The First Map-Based Assessment of Salmon in the North Pacific.* Berkeley: University of California Press.

Columbia River Pastoral Letter Project. January 8, 2001. *The Columbia River Watershed: Caring for Creation and the Common Good.* An International Pastoral Letter by the Catholic Bishops of the Region. Seattle, WA. Available from http://www.columbiariver.org

Kolmes, Steven A., and Russell A. Butkus. , 2006. "Got Wild Salmon? A Scientific and Ethical Analysis of Salmon Recovery in the Pacific Northwest and California." In *Salmon 2100: The Future of Wild Pacific Salmon,* eds. Robert T. Lackey, Denise H. Lach, and Sally L. Duncan. Bethesda, MD: American Fisheries Society.

Lackey, Robert T.; Denise H. Lach; and Sally L. Duncan. 2006. "Wild Salmon in Western North America: The Historical and Policy Context." In *Salmon 2100: The Future of Wild Pacific Salmon,* eds. Robert T. Lackey; Denise H. Lach; and Sally L. Duncan. Bethesda, MD: American Fisheries Society.

Lichatowich, James A. 1999. *Salmon without River: A History of the Pacific Salmon Crisis.* Washington DC: Island Press.

Myers, Ransom A., et al. 2004 "Hatcheries and Endangered Salmon," *Science* 303: 1980.

Nehlsen, W.; J. E. Williams; and J. A. Lichatowich. 1991. "Pacific Salmon at the Crossroads: Stocks at Risk from California, Oregon, Idaho, and Washington." *Fisheries* 16: 4–21. (Out of the 214 native stocks referred to in this paper, one population of Chinook Salmon in California had already been listed under the ESA.)

NOAA Fisheries Service. December 1, 2007. *Atlantic Salmon Recovery Plan, 1.* Available from http://www.nero.noaa.gov/prot_res/altsalmon/

Rolston, Holmes, III. 1988. *Environmental Ethics: Duties to and Values in the Natural World.* Philadelphia, PA: Temple University Press.

Williams, Richard N., ed. 2006. *Return to the River: Restoring Salmon to the Columbia River.* Boston: Elsevier Academic Press.

Russel A. Butkus
Steve A. Kolmes

SARO-WIWA, KEN

1941–1995

Kenule Beeson Saro-Wiwa, best known as Ken Saro-Wiwa, was born in Bori, an Ogoni village, on October 10, 1941. The Ogoni are an indigenous community of about 500,000 in the oil-rich Niger Delta of southeast Nigeria. He attended Government College, Umuahia, and later the University of Ibadan in western Nigeria. Saro-Wiwa became the community's spokesman in its battle against the environmental devastation caused by the long-time oil exploration and extraction activities of foreign oil companies, especially Shell.

Saro-Wiwa was a graduate assistant in English at Ibadan when a national political crisis broke out in 1966 in the form of a spate of coups d'état that pitted eastern and northern Nigerians against each other. As an easterner, Saro-Wiwa was obliged to leave Ibadan. He took up another assistantship at the University of Nigeria, Nsukka. When civil war erupted in 1967 and the Igbo declared the Republic of Biafra, he returned to his Ogoni homeland. There he taught secondary school, became a businessman, and campaigned for a federal Nigeria. In September 1967 he joined the University of Lagos as a teaching assistant.

In November 1967, at the age of twenty-six, Saro-Wiwa was appointed administrator for Bonny Island in Rivers State. After six months at that job, he returned to Lagos but soon went back to Rivers as commissioner, holding various portfolios for six years.

From 1973 until his death, Saro-Wiwa gradually moved from private life into the eye of a public storm. An Ogoni nationalist, Saro-Wiwa explained his outlook in his prison diary: "My worry about the Ogoni has been an article of faith, conceived of in primary school, nurtured through secondary school, actualized in the Nigerian civil war in 1967–70 and during my tenure as member of the Rivers State Executive Council, 1968–73" (1996, p. 75). From his youth onward, Saro-Wiwa had witnessed his people's plight as an oil-rich but exploited minority. He became committed as a leader of their struggle for survival against the exploitation of their natural resources and the destruction of their environment, with all the economic benefits accruing to the foreign oil companies. In 1977 he failed in his attempt to win election as a member of the Constituent Assembly that had been mandated by the Babangida military regime to fashion a new constitution that, in Saro-Wiwa's opinion, established an even stronger central government that further marginalized the minorities.

His writing output, both creative and critical, expanded as he became more and more engrossed in the Ogoni struggle. Among his over twenty books are: *Sozaboy* (1985); *Songs in a Time of War* (1985); *Prisoners of Jebs* (1988), and *Genocide in Nigeria* (1992). Penguin published his detention diary, *A Month and a Day*, in 1995. The Ogoni's oil, discovered in the 1950s, fast became a curse: Incidents of oil spillage, consistent gas flaring, and ecological devastation became the daily travails of the Ogoni fishing community. Saro-Wiwa believed in the political and economic autonomy of every nationality in the Nigerian federation, particularly the minorities such as the Ogoni, which were powerless in an overcentralized system controlled by the three major ethnic groups (Fulani-Hausa, Igbo, and Yoruba). It is therefore significant that Saro-Wiwa collaborated with other Ogoni leaders in 1990 to found the Movement for the Survival of the Ogoni People (MOSOP) that would soon become a major platform for his nationalist and environmental activities. As he traveled internationally to promote the movement and became popular and controversial, he drew the ire of both Ogoni elders and the military government.

Accusations of violence were rejected by Saro-Wiwa and his followers. He always preached nonviolence and claimed the late Martin Luther King Jr. as his model of activism even though King fought for inclusion and integration whereas Saro-Wiwa was an avowed separatist who sought "to establish a government of Ogoni people by Ogoni for Ogoni people in Ogoni within a confederal Nigeria" (Saro-Wiwa 1996, p. 111).

The Royal Dutch/Shell oil company was the main target of Ogoni protest; these protests led to the company's decision to abandon exploration and exploitation in the area. Significantly, the Nigerian government maintained that Shell was "a partner in progress" despite the environmental degradation caused by oil spillage and gas flaring, conditions that adversely affected the health of nearby residents. The Ogoni composed a song that featured the lyric, "The flames of Shell are hell, we bask beneath their light." The international community's failure to respond to issues alien to their economic interests did not help the Ogoni cause. For instance, the United States, Nigeria's main oil client, did nothing to interfere with Shell's oil extraction. Although Shell got away with a slap on the wrist, with official warnings and token fines, for its destructive actions in Nigeria, similar actions in Europe and North America were being heavily punished.

Protester in South Africa Recalls Ken Saro-Wiwa's Struggle for Justice. *Protesters in Johannesburg call for democracy in Nigeria, the native country of Ogoni activist Saro-Wiwa. Saro-Wiwa spoke out against human rights and environmental abuses inflicted on his people, most notably by Shell Oil Company. Saro-Wiwa was executed by the Nigerian military in 1995, along with eight other activists.* AP IMAGES.

On January 4, 1993, Saro-Wiwa organized Ogoni Day, announcing a bill of rights calling for the autonomy of the Ogoni people within a Nigerian confederation. Authored by Saro-Wiwa, the bill had been signed in 1990 by representatives of five of the six Ogoni kingdoms. In partial response the military began a "pacification" of the Ogoni that consisted of the repression or elimination of any dissidents. When four reactionary Ogoni leaders were murdered in May 21, 1994, the government fingered Saro-Wiwa as the hidden hand behind the act. He was arrested, with eight other associates, on May 22, 1994.

Despite the indifference of the United States government, Saro-Wiwa had many supporters among European and North American intellectuals and activists. The British author William Boyd advised him to contact Amnesty International and Greenpeace. Interaction with these organizations influenced him into drafting an addendum to the bill of rights titled "An Appeal to the International Community" and the establishment of the Ethnic Minority Rights Organization of Africa in 1993. Saro-Wiwa met with many nongovernmental organizations and filmmakers such as Kay Bishop who made the Ogoni documentary *The Heat of the Moment* (1992). Despite these gestures of support, on October 21, 1995, a special tribunal condemned Saro-Wiwa and his comrades to death by hanging. Notwithstanding calls from home and abroad for leniency, all were hanged on November 10, 1995.

Saro-Wiwa was nominated for the 1995 Nobel Peace Prize. In that year he won the Right Livelihood Award, known as the Alternative Nobel Prize. His exclusivist stance perhaps diminished the success of the Ogoni struggle. Perhaps Saro-Wiwa would have achieved more if he had championed the environmentalist and anti-exploitation cause on behalf of all or several Niger Delta groups. Nonetheless, his commitment to Ogoni rights and sovereignty as a minority, and to the well-being of the environment, still serves as a benchmark for activism in Nigeria. Since his execution, other groups, such as Movement for the Emancipation of the Niger Delta, have sprung up. Regarding the protection of the environment, one is tempted to believe that Saro-Wiwa's concern was more in terms of the particularity of the Ogoni, and less of the generality of nature's ecological well-being. Outside the Ogoni struggle, he would probably not have taken much interest in the environment.

SEE ALSO *Africa, Sub-Saharan; Environmental Activism.*

BIBLIOGRAPHY

WORKS BY KEN SARO-WIWA

Saro-Wiwa, Ken. 1985. *Songs in a Time of War.* Port Harcourt, Nigeria: Saros International Publishers.

Saro-Wiwa, Ken. 1985. *Sozaboy.* Port Harcourt, Nigeria: Saros International Publishers.

Saro-Wiwa, Ken. 1988. *Genocide in Nigeria: The Ogoni Tragedy.* London: Saros International Publishers.

Saro-Wiwa, Ken. 1988. *Prisoners of Jebs.* Port Harcourt, Nigeria: Saros International Publishers.

Saro-Wiwa, Ken. 1996. *A Month and a Day: A Detention Diary.* New York: Penguin.

WORKS ABOUT KEN SARO-WIWA

Grossman, Joe. 1996. *Hanged: Ken Saro-Wiwa in the American Print Media.* New York: Inner Image Ink.

Ikari, Ben Wuloo. 2007. *Ken Saro-Wiwa and Mosop.* Philadelphia, PA: Xlibris.

N'Allah, Abdul-Rasheed, ed. 1998. *Ogoni's Agonies: Ken Saro-Wiwa and the Crisis in Nigeria.* Trenton, NJ: Africa World Press.

Ojo-Ade, Femi. 1999. *Ken Saro-Wiwa: A Bio-Critical Study.* Brooklyn, NY: Africana Legacy Press.

Femi Ojo-Ade

SCANDINAVIA

This entry contains the following:

I. DENMARK AND SWEDEN

Scandinavia is a region of Northern Europe defined by geography, history, and language as including the countries of the Scandinavian Peninsula—Norway and Sweden—as well as Denmark. Some authorities advocate the inclusion of Finland because of economic and cultural connections, and Iceland and the Faroe Islands because their populations speak Northern European tongues closely related to Swedish and Norwegian. The region is often associated with nature-friendly attitudes and strong traditions of nature management, outdoor life, and environmental education. Danish and Swedish environmentalism share an ecological worldview, a critical view of technology, and a call for participatory democracy in environmental policy making.

Danish environmentalism, which encompasses a cosmological dimension, has emerged from a political culture of participatory democracy and a decentralized, small-scale economy. It speaks to civil society and encourages personal commitment to environmental goals. Swedish environmentalism is influenced by a centralized state administration emphasizing systemic solutions to environmental and developmental challenges.

Environmental controversies and debates in Denmark and Sweden have focused on genetically modified organisms (GMOs), animal welfare, the use and storage of pesticides and herbicides in the chemical and transport industry, the building of a fixed link across the Öresund (the strait that separates the Danish island Zealand from the southern Swedish province of Scania), the radioactive fallout from the nuclear accident in Chernobyl in 1985, and the question of environmentally friendly energy and nuclear power.

THE NUCLEAR ENERGY CONTROVERSY

Strong movements of opposition to nuclear energy have arisen in both Denmark and Sweden, but they took different forms in each country because of the countries' differing political cultures. Swedish antinuclear environmentalism has been mostly assimilated into previously existing governmental and nongovernmental institutions, although it has generated some new organizations. In 1980 the national nuclear-energy referendum resulted in the two more positive lines defeating the anti-nuclear line. Three months after the referendum the Swedish Parliament decided to phase out all nuclear reactors before the end of 2010. Many radical antinuclear activists considered this measure a setback because of its extended time frame. The antinuclear movement was scattered, and the issue was rapidly further institutionalized with the formation of the Swedish Green Party in 1981.

In Denmark the antinuclear movement engaged in critical debates with the political establishment, critical scientists, and journalists. As a result of these joint discussions, in the 1980s the Danish government repudiated nuclear energy, a decision made in a context of a small-scale, decentralized economy that aimed to develop alternative energy sources through increased research, local developments, and commercial incentives. The differences between Danish and Swedish environmentalism can be traced to the two countries' contrasting traditions of environmental law, policy making, environmental research, and environmental ethics and philosophy.

THE LEGAL AND POLICY DIMENSIONS OF SWEDISH AND DANISH ENVIRONMENTALISM

Allemansrätten (the right of public access) is a unique Scandinavan and Swedish regulation that protects the public's right to move freely in nature. This right also includes a responsibility not to destroy and disturb the environment. This is an example of how Swedish representative democracy, with its top-down institutionalized culture of environmental politics, governs public access to land.

In Denmark the culture of participatory democracy has engendered dialogues between the political establishment and the environmental movement. One important result of these dialogues has been the establishment of consensus conferencing that seeks to include the public in decisions concerning the environment and development issues. In 1986 one of the world's most rigorous precautionary laws on genetically modified organisms was passed as a result of this public-consultation procedure. In the late 1990s NGOs and the public argued that the issue of commercialization of GMO crops implied that precautionary principles should include ethics and value judgments. On the other hand, the experts and the industry continued to treat risk in a scientific and physical sense. Thus the issue of commercialization of GMO crops challenged this consensus.

In Sweden, because of the rapid incorporation of environmentalism into parliament (such as the Centre Party), administration (such as the Environmental Protection Board [1967]), and nongovernmental organizations (such as the Swedish Society for Nature Conservation) in the 1960s and 1970s, Swedish environmental policy making is largely the result of parliamentary work in which staff members from NGOs often serve as expert consultants. This process has resulted in predominantly large-scale solutions to environmental problems, including the incorporation of environmental concerns into the existing economic and corporate culture.

In Denmark, as previously noted, environmentalism has not been as easily channeled into official governmental bodies. Instead, the country's heritage of citizen participation and cooperation has played a vital role in creating alternative ways of enhancing public environmental awareness. The grass-roots movement NOAH, founded in 1969, refers to "the first environmental activist who fought for the species" (Jamison et al. 1990, p. 120). NOAH has played a significant role in discussions about alternative technologies. NOAH was formed by environmental activist students who dramatically revolted against NOA (a natural history society at the University of Copenhagen) at a NOA annual seminar in 1969.

Educational institutions have played an important role in the development of environmental awareness in both Sweden and Denmark. As far back as 1940, Elisabeth Tamm (1880–1958) and Elin Wägner (1882–1949) argued for the importance of integrating environmental and development concerns with education and women's emancipation—an argument that predates concepts such as sustainable development and education for sustainable development by thirty to fifty years.

The Swedish school system has long been responsible for instilling social values and instigating social change. The guidelines and regulations put forward by the United Nations Educational, Scientific and Cultural Organization's (UNESCO) Decade of Education for Sustainable Development (DESD) are, for example, implemented at all levels of the Swedish educational system, emphasizing critical reflection, pluralism, and democracy.

The People's High Schools have promoted the development of environmental activism and ecological awareness in the rural population of Denmark. The minister, poet, and educator N. F. S Gruntwig (1783–1872) is considered to be the father of these schools, which became arenas for alternative-technology workshops and the promotion of alternative ecological lifestyles in the 1970s and 1980s. Like the Danish People's High Schools of the 1980s, the Swedish implementation of the United Nations' Education for Sustainable Development (ESD) since 2000 has gone beyond traditional pedagogy to teach values and critical reflection

ENVIRONMENTAL RESEARCH AND PHILOSOPHY

The largest and most ambitious environmental research program in the humanities in Scandinavia, Man and Nature (Menneske og nature) (1992–1997), was based in Denmark at Odense University. In Sweden the best-known program is Roads to Sustainable Development (1996–2002) (Vägar till uthållig utveckling-beteenden, organisationer, strukturer).

The Swedish and Danish natural environments have influenced both countries' environmental research and philosophy. The lack of wildlife in mainland Denmark, the fact that Denmark is densely populated, and Denmark's history of decentralized and small-scale economy has inspired experimental and pragmatic environmental research and philosophy. This pragmatic take on nonhuman nature includes the view of nonhuman nature as a workshop. This view, in contrast to the Swedish theoretical view, was typical of Tycho Brahe (1546–1601) and Hans Christian Ørsted (1777–1851), who combined romantic and utilitarian views of nonhuman nature in his research.

In Sweden the heritage of Carl Linnaeus's (1707–1778) theoretical classifications approach to nonhuman nature, along with the need to understand and manage large areas of wilderness, has given birth to an almost imperialistic relationship to nonhuman nature. Nevertheless, empirical studies of Swedes' views of nonhuman nature reveal strongly nonanthropocentric, biocentric, and ecocentric attitudes.

SEE ALSO *Alternative Technology; Environmental Education; Environmental Policy; Genetically Modified Organisms and Biotechnology; Naess, Arne; Nongovernmental Organizations; Nuclear Power; Sustainable Development.*

BIBLIOGRAPHY

Jamison, Andrew; Ron Eyerman; and Jacqueline Cramer. 1990. *The Making of the New Environmental Consciousness: A Comparative Study of the Environmental Movements in Sweden, Denmark and the Netherlands.* Edinburgh: Edinburgh University Press.

Jamison, Andrew, and Erik Baark. 1999. "National Shades of Green: Comparing the Swedish and Danish Styles in Ecological Modernization." *Environmental Values* 8: 199–218.

Kronlid, David. 2005. *Miljöetik i Praktiken. Åtta Fall ur Svensk Miljö-och Utvecklingshistoria.* Lund, Sweden: Studentlitteratur.

Lundgren, Lars J. 2002. *Miljöns Mänskliga Dimension. En Studie av Humanistisk ochSamhällsvetenskaplig Miljöforskning.* Rapport 2002. Stockholm: Swedish Research Council.

Lundgren, Lars J., ed. 2003. *Vägar till Kunskap. Några Aspekter på Humanvetenskaplig och Annan Miljöforskning.* Stockholm: Symposion.

Melin, Anders. 2001. *Judgements in Equilibrium? An Ethical Analysis of Environmental Impact Assessment.* Linköping Studies in Arts and Science 235. Linköping, Sweden: Department of Water and Environmental Studies, Linköping University.

Merchant, Carolyn. 1995. *Earthcare: Women and the Environment.* New York: Routledge.

Sandell, Klas, et al. 2005. *Education for Sustainable Development: Nature, School and Democracy.* Lund, Sweden: Studentlitteratur.

Tamm, Elisabeth, and Elin Wägner. 1940. *Fred med Jorden.* Stockholm: Albert Bonniers Förlag.

Toft, Jesper. 2000. "Denmark: Potential Polarization or Consensus?" *Journal of Risk* 3(3): 227–235.

Uddenberg, Nils. 1995. *Det stora sammanhanget: moderna svenskars syn på människans plats i naturen.* Nora, Sweden: Nya Doxa.

Wägner, Elin. 2007 (1941). *Väckarklocka.* Stockholm: Albert Bonniers Förlag.

David Kronlid

II. NORWAY

In Scandinavia the ideas of ecophilosophy, ecosophy, and ecopolitics have been a staple of debates on issues related to environmental ethics. Norway has been a hotbed of these ecologically informed philosophies, which offer unique perspectives on living in harmony with the environment. Key Norwegian thinkers include the coauthor of *The Limits to Growth* (1972), Jørgen Randers; the founder of Deep Ecology, Arne Naess; the chair of the World Commission on Environment and Development, Gro Harlem Brundtland; the politician and diplomat Rolf Edberg (1912–1997); and the peace researcher Johan Galtung.

The development of environmental ethics and philosophy gained real momentum in Scandinavia only after the 1960s. It was environmental concerns that arose abroad that triggered Norwegians into action and reflection. Rachel Carson's famous warning against pesticides in *Silent Spring* (1962) was immediately translated into key Scandinavian languages, and it inspired many people to adopt an ecological perspective. Equally important were the environmental writings of Lynn Townsend White Jr. (1907–1987) and Jacques Yves Cousteau (1910–1997), and the reflections on the technological

Whalers off the Coast of Norway, 1999. *Whale blubber was once a hot commodity in Norway, used in a wide range of products. Today the price is so low that even hunters believe its not worth the cost of hauling to land. Environmentalists around the world criticize the whaling practice for violating animal rights; Norway defends it as being sustainable.* AP IMAGES.

standardization of human life and nature by the Finnish philosopher George Henrik von Wright (1916–2003).

Scholars active in the International Biological Program (IBP; an effort, from 1964 to 1974, to coordinate large-scale ecological and environmental studies) mobilized a series of students and philosophers throughout Scandinavia to rethink the human condition in the natural world. This program, initiated by members of the International Union of Biological Sciences, focused mainly on problems related to food production and management of natural resources in view of a rapidly increasing human population and widespread malnutrition in the world. In Scandinavia the program was active between the mid-1960s and the mid-1970s, and fully in effect between 1967 and 1972. The managerial benefits of ecological research were, at least initially, stressed in all the Scandinavian countries. The managerial perspective gave way to a biocentrism that accorded intrinsic value to the environment as a whole. The ecologists who dominated the program pledged to deliver scientific and philosophical methods that could generate useful knowledge about the consequences of various modes of land use. Most of their research sought to achieve an energy balance between species, as articulated by the American ecologist Eugene P. Odum (1913–2002). This methodology assumed the desirability of a steady-state economy of human beings that would be in balance with steady-state economy of nature.

ECOPOLITICS VS. ANIMAL RIGHTS

The hegemony of the ecological approach may explain why the protection of individual animals failed to capture public attention in Norway. The few vocal defenders of both animal liberation and animal rights have not been successful in translating philosophical points into political or legal achievements. A case in point is the issue of whaling, which is defended by Norwegian politicians, scientists, and environmental philosophers alike from an ecological point of view. The Norwegians have only been hunting mink whale, which, according to Norwegian marine ecologists, are not endangered. Despite criticisms from environmentalists around the world who more often than not focus on animal liberation or rights, Norwegian environmentalists thus tend to defend whaling as sustainable.

Many of the scientists involved in IBP were deeply concerned about environmental problems. For example, the Norwegian ecologist Ivar Mysterud argued that politics should be put on a secure ecological footing, and in 1970 he suggested the term *ecopolitics* to demarcate managerial environmentalism from an approach to politics based on the science of ecology. The term was quickly adopted not only by fellow ecologists but also by many scholars, activists, and students who questioned the practices of technocracy and industrialism. Much of this criticism had, since the mid-1960s, been informed by populist agrarian socialism, which persisted under the new label ecopolitics from 1970 onward.

Ecopolitics aims at developing a steady-state social economy that would mirror the steady-state balance of the economy of nature. One of many students inspired by steady-state reasoning was Nils Christian Stenseth, who later became a key figure in international ecological research. His first article, published at the age of twenty-three, was about ecopolitics. In it he argued that "all biologists should work for a *steady-state society* in place of the *growth society*," and one should limit the human population growth to zero (1972, p. 118). Ecological modeling represented the way forward, for simulation models could determine exactly when and how to achieve a steady state.

Ecologists began to arrange seminars and various outreach activities to engage scholars in the political and social sciences and philosophy. As a result, the environmental debates in Scandinavia were often framed in terms of ecological methodologies and perspectives. In Sweden, for example, the politician and diplomat Rolf Edberg wrote several popular books about the need to take care of the environment. He described the need to achieve a more harmonious relationship to nature through the science of ecology.

Beginning in the late 1960s, there was a growing concern in Sweden about the effect of airborne industrial pollution originating elsewhere in Europe on the nation's forest industry, a topic brought to international attention at the United Nations Conference on the Human Environment that took place in Stockholm in June 1972. At the heart of the Swedish agenda was not only raising international awareness but also presenting analytical tools that could address environmental issues. One such tool was socioeconomics, an academic field with strong intellectual ties to the socialist-inspired economies of all the Scandinavian countries.

ECOPHILOSOPHY

In Norway many philosophers were impressed with the ecologists, and they would attend their lectures and participate in their seminars. The Norwegian philosopher Sigmund Kvaløy organized a Coworking Group for the Protection of Nature and the Environment inspired by the ecologists. Those with a philosophical bent met in the Ecophilosophy Group, a subsection of this loose association. Kvaløy was granted a four-year Ph.D. scholarship in philosophy starting in 1967, which he used to explore ecological thinking. He had been Arne Naess's student and assistant since 1961. In 1969 he took over Naess's introductory seminar to epistemology, "Nature and Humans," and quickly turned it into a workshop for his group's work in "ecophilosophy" (a word Kvaløy coined).

Kvaløy's students and followers were fond of quoting Karl Marx's famous saying, "The philosophers have only interpreted the world in various ways; the point, however, is to change it." Turning words into action, the Coworking Group became an effective, hard-hitting student association that targeted hydropower developments. Most dramatic was their attempt to save the Mardøla River, which included Norway's highest waterfall, during the summer of 1970. Naess joined the Mardøla demonstration, and he decided afterward to resign his professorship so that he could devote himself fully to the environmental cause.

At the same time Naess started to attend his own "Nature and Humans" seminar, where Kvaløy was now in charge. Apparently, Naess was one of the few who took notes, and in the spring of 1971 he transformed them into a lecture series titled "Ecology and Philosophy." In these talks he introduced, for the first time, his

> ecosophy ... as a type of philosophy that takes its point of departure from an identification with all life in this life-giving environment. It establishes in a way a classless society within the entire biosphere, a democracy in which we can talk about justice not only for humans but also for animals, plants, and minerals. And life will not be conceived as an antagonism unto death but an interaction with surroundings, a life-giving environment. This represents a very strong emphasis on everything hanging together and emphasizes that we are only *fragments*—not even parts. (Naess 1971, p. 54)

ECOSOPHY

The Mardøla experience and the discussions at the "Nature and Humans" seminar radicalized the thinking of Kvaløy, the charismatic leader. After the Mardøla experience he adopted from ecology the idea that a complex ecosystem is more robust than a simple one in the face of environmental changes. Inspired by Herbert Marcuse, he argued that a complex human society would have a better chance of surviving the environmental crisis than one based on the "one-dimensional man" of industrial society.

Many of the ecophilosophers, including Naess and Kvaløy, were active members of the Norwegian Alpine Club, an organization devoted to technical climbing. Among their many activities, they made several visits to the high mountains of Pakistan. It was during a trip in 1964 that Naess first formulated what later became known as the ecosophy formula. He explained his "thriving" in Pakistan's mountains as a mixture of pain and excitement in the following mathematical terms: $T = G^2/(L_S + Å_s)$. In this equation T, *trivsel,* (thriving) equals G^2, *glød,* (excitement squared) divided by L_S, *legemlige smerter,* (bodily pains) plus $Å_S$, *åndelige smerter* (spiritual pains). This formula would later serve as a key explanation of the meaning of self-realization in Naess's "Ecosophy T," with the T standing for "thriving." The "T" could also be short for Tvergastein, the name of Naess's cottage, or also "Tolkning" (interpretation) as this was important to his early philosophy. There is, however, only circumstantial evidence for these readings.

Equally important to his ecosophy was Mohandas Gandhi's (1869–1948) teaching of nonviolence, which came to the forefront of Naess's thinking after his first visit to Pakistan in 1950. Back in Oslo he gave a lecture series about Gandhi's political ethics that resulted in a book he coauthored with the young sociologist Johan Galtung that was published in 1955. In 1960 Naess followed up with a popular version of this work, which was translated as *Gandhi and the Nuclear Age* in 1965. Here he argued that people from the Europe and North America had much to learn from Gandhi given the threat of nuclear Armageddon.

DEEP ECOLOGY

Naess introduced the concept of Deep Ecology in a paper at the World Future Research Conference in Bucharest in early September 1972. The conference was organized by the World Futures Studies Federation. What dominated future studies in 1972 was *The Limits to Growth* report for the Club of Rome written, among others, by the twenty-seven-year-old Norwegian solid-state physicist Jørgen Randers. At the time Randers was entirely unknown. It was therefore a shock to Norwegian environmentalists to see him rise to world fame through a report that came to dominate environmental debate at the United Nations' conference in Stockholm. Though *The Limits to Growth* predicted that there would be limits to natural resources, it did not predict limits to existing political systems. The MIT group behind the report was, in this respect, part of a larger trend of environmentalists looking for solutions to ecological problems within established social structures.

For Galtung and Naess the time was ripe in Bucharest to hit back at what they saw as a "shallow" technocratic analysis of the environmental situation. Galtung spoke first with his paper "*The Limits to Growth* and Class Politics" (1973), a head-on attack on the lack of social analysis in the report. When it was Naess's turn to mount the rostrum in Bucharest, he, too, took an anticlass posture in presenting "The Shallow and the Deep Ecology Movement" in Norway. Upon returning to Oslo Naess used his notes to compile a summary which he published as "The Shallow and the Deep, Long-Range Ecology Movements" in 1973.

Ironically, the long-range ecology movement Naess spoke of faded upon his return to Oslo, as the Coworking Group for the Protection of Nature and the Environment was infiltrated and taken over by Marxist-Leninists. The group dissolved in 1973 after a period of internal cleansings and futile debates about the value of democracy. Its last unified stand came with the national referendum on membership in the European Community at the end of September 1972. The group was decisively opposed to joining, arguing that "this industrial-serving mega-society seeks to break apart the established *diversity* of sturdy self-governed and heterogeneously, traditional-colored local communities and replace them with a uniform system of government that presupposes uniform social units and a uniform culture: a simplification that increases vulnerability, according to the science of ecology" (Samarbeidsgruppa 1972, p. 91). They made their case: Norway voted against EU membership but could not decide on what to do next. As a result, the ecophilosophers split into socialist and ecological wings. Kvaløy and Naess regretted this leftward turn in the politics of ecology because it undermined the broad science-based environmentalism they sought to mobilize. They subsequently continued with their activities outside the academic realm in various environmental organizations where they, among other things, mobilized people to save the Alta River in the north of Norway.

GRO HARLEM BRUNDTLAND

When the young and politically inexperienced feminist Gro Harlem Brundtland became minister of the environment in 1974, she faced the ecophilosophers in various heated debates and rough conflicts. They used every opportunity to show that the ecological steady-state society was not an herbal-tea party but a revolutionary break with industrial growth. As a medical doctor, she took a strictly anthropocentric stand against them and the ecologists claiming to speak on behalf of nature, arguing in favor of bureaucratic rules and democratic procedures instituted by humans to the benefit of humans and especially the working class she represented as the head of the Labour Party. In offering resistance to her views, both the ecophilosophers and the ecologists forced Brundtland to reflect more deeply on social aspects of environmental affairs, as she later did in *Our Common Future* (1987).

CONCLUSION

Norwegian ecologists' and environmental activists' concerns for the environmental future mobilized a series of students and philosophers to rethink the human condition in the natural world. Their innovative thinking about ecophilosophy, ecosophy, and eco-politics became part of the international debate.

SEE ALSO *Animal Ethics; Deep Ecology; Environmental Policy; Environmental Politics; Limits to Growth; Naess, Arne.*

BIBLIOGRAPHY

Anker, Peder. 2007. "Science as a Vacation: A History of Ecology in Norway." *History of Science* 45: 455–479.

Carson, Rachel. 1962. *Silent Spring*. Greenwich: Fawcett Crest.

Galtung, Johan. 1973. "*The Limits to Growth* and Class Politics." *Journal of Peace Research* 10(1, 2): 101–114.

Meadows, Donella H.; Dennis L. Meadows; Jørgen Randers; and William W. Behrens III. 1972. *The Limits to Growth: A Report for the Club of Rome's Project on the Predicament of Mankind*. New York: Signet.

Naess, Arne. 1964. *Opp stupet til østtoppen av Tirich Mir* (Up the cliff to the East Summit of Tirich Mir). Oslo: Gyldendal.

Naess, Arne. 1965. *Gandhi and the Nuclear Age*. New Jersey: Bedminster Press.

Naess, Arne. 1971. *Økologi og filosofi I* (Ecology and philosophy I). Oslo: Filosofisk Institutt.

Naess, Arne. 1973. "The Shallow and the Deep, Long-Range Ecology Movements: A Summary." *Inquiry* 16: 95–100,

Samarbeidsgruppa for natur og miljøvern (Coworking Group for the Protection of Nature and the Environment). 1972. *Dette bør du vite om EF* (This you should know about EC [the European Community]). Oslo: Pax.

Stenseth, Nils Chr. Stenseth. 1972. "En oppfordring til biologene om å utforme en økopolitikk" (A call to the biologists to shape an eco-politic). *Naturen* 96: 118–126.

World Commission on Environment and Development. 1987. *Our Common Future*. Oxford, UK: Oxford University Press.

Peder Anker

SCHUMACHER, ERNST FRIEDRICH
1911–1977

Ernst Friedrich Schumacher was an economist, policy adviser, and essayist who was born in Bonn, Germany, on August 16, 1911. His most famous work, *Small Is Beautiful* (1973), sold millions of copies and helped spark the "intermediate" or "appropriate technology" movement. Schumacher died in Switzerland on September 4, 1977.

EDUCATION AND EARLY WORK

E. F. Schumacher studied economics from an early age, eventually becoming a Rhodes Scholar, and traveled to England to study at New College, Oxford. Throughout the late 1920s and early 1930s Schumacher's life and work were overshadowed by the struggles in his Germany to cope with a series of economic and political crises that led to the rise of Adolf Hitler and the Nazis. At the outbreak of World War II Schumacher was interned in England as an enemy alien and later was forced to work as an agricultural laborer. That experience fostered his interests in farming, soil management, and Marxist socialism that were to influence his future activities.

At the end of the war Schumacher—now a British citizen—was made a member of the Strategic Bombing Survey, which examined the impacts of bombing on Germany; he later became a member of the British Control Commission in Germany. After his return to England he became the economic adviser to the National Coal Board, where he stayed until his official retirement in 1971. His early interests in the postwar period revolved around reconstructing Germany and the attempt to establish an international financial and trading system (the World Bank and the International Monetary Fund).

In 1955 Schumacher was invited to go to Burma for a short term as an economic adviser, an event that resulted in an essay published in 1955 titled "Economics in a Buddhist Country," later revised as "Buddhist Economics." The essay drew on his experience of a country whose core way of life was completely different from the world he had inhabited. Schumacher was astonished at the happiness of a people living in poverty. In that essay and in later works he ascribed their contentment to a series of approaches to life: (1) minimizing wants; (2) work as a means of enhancing life; and (3) a spiritual understanding of human beings. The enrichment of life was seen as the purpose of economics. In this way Schumacher rediscovered in Burma a buried British tradition (sometimes referred to as the Third Way) of economic socialism that derived from John Ruskin, William Morris, and R. H. Tawney and was instrumental in creating Gandhi's spiritual economics.

PHILOSOPHY AND LATER WORKS

Schumacher's new interest in the problems of developing countries led him to India, where the vast distance between the high-technology future planned by the government with the assistance of international economic advisers and the poverty of the population led him to push for intermediate technologies that would enhance the work people were doing through the application of appropriately considered improvements.

Small Is Beautiful: Economics as If People Mattered (1973) is a series of overlapping essays. "Nuclear Power—Salvation or Damnation" not only attacks the economics of nuclear power but also points out in detail the intransigence of the waste problem. "Development" articulates the emerging problem of rich elites in developing countries as enclaves of westernization in a sea of poverty. "Social and Economic Problems Calling for the Development of Intermediate Technology" puts forward a diagnosis and a work agenda that would be followed by the Intermediate Technology Development Group (ITDG). The most famous chapter, "Buddhist Economics," set out the essentials of ecological economics long before anyone gave it a name: Schumacher makes clear the distinctions between renewable and nonrenewable resources and the folly of living off capital rather than interest. "How to obtain given ends"—the dignity of the human—"with minimum means" is Schumacher's version of what a Buddhist economics should be; as opposed to a system driven by a goal of maximum consumption. *Good Work* (1979) is a complementary series of speeches and short essays published after Schumacher's death that includes an indictment of dependence on oil.

Guide for the Perplexed (1977), which was written after Schumacher's conversion to Catholicism, is an attempt to put forward a more extensive philosophy and theology. It is essentially an updated version of medieval Christianity based primarily on the thought of Saint Thomas Aquinas and argues that contemporary philosophies are "horizontal" (concerned only with the material world), whereas a "vertical" approach is more appropriate to beings who ascend from mere physical life, the mineral, through plants, to consciousness, and then to self-awareness. The book claims that people need an adequate level of response to the problems at each level.

SEE ALSO *Buddhism; Nuclear Power; Ruskin, John; Technology.*

BIBLIOGRAPHY

WORKS BY E. F. SCHUMACHER

Schumacher, E. F. 1973. *Small Is Beautiful: Economics as If People Mattered.* London: Blond and Briggs.

Schumacher, E. F. 1977. *Guide for the Perplexed.* London: Cape.

Schumacher, E. F. 1979. *Good Work.* New York: Harper & Row.

WORKS ABOUT E. F, SCHUMACHER

Wood, Barbara. 1984. *E. F. Schumacher: His Life and Thought.* New York: Harper & Row. Also published as *Alias Papa: A Life of Fritz Schumacher.* London: J. Cape, 1984.

Peter Timmerman

SCHWEITZER, ALBERT
1875–1965

Albert Schweitzer, theologian, philosopher, concert organist, and physician, was born on January 14, 1875, in Kaisersberg, Alsace-Lorraine. In 1912 Schweitzer left his position as professor of theology at the University of Strasbourg to become a mission doctor at Lambaréné in what later became Gabon.

Schweitzer is best known for his philosophy of reverence for life, which he developed in response to the sense of cultural crisis that was widespread in Europe in the late nineteenth and early twentieth centuries. Schweitzer argued that the failure of ethics to provide an orientation to life rests on trying to base ethics on a worldview that claims that the world itself is structured by and oriented toward ethical values. Rejecting any such metaphysical project, Schweitzer called for an ethics based on a view of life as practical action within the world.

Schweitzer's development of his ethical view of life is a form of life-philosophy (*Lebensphilosophie*), a nineteenth and twentieth century development that put life and will at the center of philosophical thought. His orientation is explicitly existential: How can an individual give positive meaning to life, starting with the will to live as each individual immediately experiences it. Ethics has the task of giving the human will to live a focus that frees it from alienation and pessimism. "Deepened affirmation of world and life consists in this: that we have the will to maintain our own life and every kind of existence that we can in any way influence, and to bring them to their highest value" (Schweitzer 1987 [1923], p. 278).

Schweitzer says that when human beings affirm both their own inner will to live and the will to live in all forms of life, they experience contact with something much deeper: "Reverence for life means to be in the grasp of the infinite, inexplicable, forward-urging will in which all Being is grounded" (Schweitzer 1987 [1923], p. 283). Schweitzer argues that this mystical experience is necessary for a genuinely positive orientation toward life.

The principle of reverence for life has important implications for human relations with animal and plant life (see Schweitzer 1988 [1919]). Our world is "a ghastly drama of the self-estrangement of the will to live. One existence makes its way at the cost of another; one destroys the other. … I can do nothing but hold to the fact that the will to live in me manifests itself as will to live that desires to come to harmony with other will to live" (Schweitzer 1987 [1923], p. 312).

Recognizing that injury to life is necessary for sustaining human life, Schweitzer asked how it is possible to lead a concrete life of reverence for life. While causing unnecessary suffering and harm is strictly forbidden, Schweitzer acknowledged that reverence for life has an unavoidably subjective or personal dimension. One person may conclude that reverence for life requires one to be vegetarian; another that eating animal flesh is permissible. Each, in making this subjective decision, may be showing reverence for life, but if one decides to eat meat, one has the objective obligation to ensure that animals live and die without unnecessary suffering. However one decides this question, one always lives at the expense of other life, and to this extent one is always guilty.

While Schweitzer's thought has been influential in animal-welfare movements, it has had less impact on environmental thought. But his insistence on reverence for all life, in whatever form, makes his philosophy a potentially important resource for biocentric thought, and reverence for life is a clear relative of Edward O. Wilson's notion of biophilia.

Schweitzer's philosophy of reverence for life presents two major problems for contemporary thought. First, his mysticism concerning the infinite will to live will be foreign to many people, though the experiential nature of this mysticism will make it accessible to some. He insists that his mysticism is not an abstract worldview, but rather a concrete attitude toward life. One may want to ask, however, whether this mysticism successfully avoids the problems Schweitzer sees in abstract mysticism. Second, Schweitzer's emphasis on the guilt that inevitably results when a moral agent lives in a world in which life lives at the expense of other life may strike some readers as a version of original sin. One may want to ask whether a moral life that lives at the expense of other life is possible without guilt.

SEE ALSO *Animal Ethics; Biocentrism; Biophilia; Intrinsic and Instrumental Value; Life: Respect/Reverence; Wilson, Edward O.*

BIBLIOGRAPHY

Evans, J. Claude. 2005. "Reverence for Life: The Philosophy of Albert Schweitzer." In his *With Respect for Nature: Living as Part of the Natural World.* Albany: State University of New York Press.

Schweitzer, Albert. 1988 (1919). *A Place for Revelation: Sermons on Reverence for Life,* trans. David Larrimore Holland. New York: Macmillan.

Schweitzer, Albert. 1987 (1923). *The Philosophy of Civilization,* trans. C. T. Campion. Buffalo, NY: Prometheus Books.

J. Claude Evans

SEED BANKS

Seed banks are facilities for conserving seeds, either by continually replanting and harvesting them in the places in which they were bred (in situ) or by placing them in some form of remote storage (ex situ). Seed banks play a key role in crop improvement and in the conservation of biodiversity. There are more than 1,500 government, private-sector, and nongovernment-organization seed banks around the world with more than 6 million accessions—different landraces or varieties—in storage.

Modern agricultural practices have dramatically reduced the number of landraces, or traditional regional varieties that are grown by local farmers. Farmers who continue to plant the varieties of their forebears conserve tremendous the genetic variation that has been created over thousands of years of artificially—as opposed to naturally—selected mutations, natural hybridizations, and chromosomal aberrations. Genetic diversity is also being destroyed by other human activity and global climate change. The Food and Agricultural Organization (FAO) estimates that in 1949 there were 10,000 wheat varieties grown in China. In the 1970s that number had shrunk to 1,000 varieties (1997).

Seed banks also provide a reservoir of genes and cultivars (a variety of plant that originates and persists under cultivation) that can be used by research scientists and farmers. There are instances in which catastrophic events such as wars or hurricanes have destroyed the seed of an important traditional variety and the seed bank material was used to reestablish the cultivar. Following the Rwandan civil war in 1994, several international organizations helped to reintroduce more than 275 bean varieties back to Rwandan farmers, thus helping to restore food security.

Plant breeders are the major users of the genetic material stored in seed banks. Cultivars, landraces, and wild populations stored in the seed banks provide the genes and genetic resources needed by plant breeders to develop new cultivars with increased yield, pest resistance, and other valuable traits. Plant breeders will screen large numbers of accessions from seed banks to identify new genes for biotic and abiotic stresses, including insects, diseases, temperature, and soil. They will also look for other genetic traits, including growth habit, nutrition, color and quality. Wild germ plasm can also be used by plant breeders to improve yield and other important traits of agricultural crops.

Seed banks only store the genetic material, whereas farmers who plant landraces or allow plants to grow in their natural habitat will continue to create new genes and gene combinations. Both biotic and abiotic stresses help to select new genes that will then be available to the plant breeders. Creating new genes through mutation is important as climate change continues to influence plant production. In situ cultivation over thousands of years is what has produced the genetic variation that we have today. The ideal way to store genetic diversity is in situ. This method allows the population to continue to create additional genetic diversity. The Convention on Biological Diversity (CBD) states that ex situ conservation should complement in situ conservation. The CBD (1993) enhanced global awareness of the importance of conserving, using, and sharing the benefits of genetic diversity.

Doomsday Vault, Norway. *The Svalbard Global Seed Vault, known as the "doomsday" vault, was officially opened in February 2008, to protect millions of food crops from climate change, wars, and natural disasters in the remote Arctic location. Several seed banks exist around the world, in order to protect against an accidental loss of genetic diversity.* AP IMAGES.

A seed bank (the Svalbard Global Seed Vault) has been built in northern Norway, 696 miles from the North Pole. It will hold more than 3 million samples at minus 18 degrees Celsius. Opened in February 2008, this seed bank will preserve seed from the world's major and minor food crops. It was built by the Norwegian government and will be operated by the Global Crop Diversity Trust.

SEE ALSO *Agriculture; Biodiversity; Convention on Biodiversity; Farms; Food; Global Climate Change.*

BIBLIOGRAPHY

Bioversity International. 2008. Available from http://www.bioversityinternational.org/Themes/Genebanks/index.asp#Seed_genebanks

Charles, Daniel. 2006. "Species Conservation: A 'Forever' Seed Bank Takes Root in the Arctic." *Science* 312 (5781): 1730–1731.

Consultative Group on International Agricultural Research. *25 Years of Food and Agriculture Improvement in Developing Countries, 1971–1996.* Available from http://www.worldbank.org/html/cgiar/25years/25cover.html

Gepts, Paul. 2006. "Plant Genetic Resources Conservation and Utilization: The Accomplishment and Future of a Societal Insurance Policy." *Crop Science* 46: 2278–2292

The State of the World's Plant Genetic Resources for Food and Agriculture. 1997. Rome: FAO. Available from http://www.fao.org/ag/agp/AGPS/Pgrfa/pdf/swrfull.pdf

Tanksley, Steven D., and Susan R. McCouch. 1997. "Seed Banks and Molecular Maps: Unlocking Genetic Potential from the Wild." *Science* 277 (5329): 1063–1066.

Russell Freed

SELF-AWARENESS

SEE *Consciousness.*

SELF-CONSCIOUSNESS

SEE *Consciousness.*

SHEPARD, PAUL
1925–1996

Paul Howe Shepard was born on July 12, 1925, in Kansas City, Missouri. His life-long fascination with natural history and evolution led to research in many fields of inquiry in the sciences and social sciences. He is credited with synthesizing this far-ranging view into a human ecology that explicated the relationship of humans to the whole of nature.

Shepard acknowledged that his theory was informed by his boyhood experiences, first in collecting and keeping animals as a child, and later in youth, through hunting, fishing, trapping, and falconry. His original view of human ecology was culminated, at the age of fifty-seven, in *Nature and Madness* (1982). In this psychohistory, Shepard explains attenuated development and maturation in humans, examines our puzzling estrangement from nature, and proposes the proper rearing of children close to nature.

Nature and Madness followed years of seeking the origins of nature perception in humans. In 1950, after graduation from the University of Missouri and a brief stint as field representative of the Missouri Wildlife Federation headed by Charles Callison, Shepard entered a Master's Conservation Program at Yale University, co-directed by renowned plant ecologist Paul Sears and evolutionist G. Evelyn Hutchinson. Focusing on the relationship of art to ecological perception in pioneer America, Shepard plunged into the study of the Hudson River School painters of the 1830s and 1840s. He broadened this study in an interdisciplinary doctoral program that culminated in his dissertation, "American Attitudes Towards the Landscape in New England and the West, 1830–1870."

During his graduate studies and his first academic appointment at Knox College, Shepard was a grassroots activist in his summer employment as a seasonal state and national park ranger. At Big Spring State Park in Missouri and at Olympic National Park in Washington, he was a whistleblower, publicizing the illegal logging of old growth forests, which led in both cases to administrative and policy changes. As conservation chairman of the National Garden Clubs of America, he initiated letter-writing campaigns supporting the creation of the Arctic National Wildlife Refuge and opposing the use of pesticides and the building of a dam in Dinosaur National Monument. In his first teaching position at Knox College, he transformed old mine tailings and pits into a biological field station, Green Oaks, restoring one meadow to tall-grass prairie. During this period of his life he was married to Melba Wheatcroft Shepard and they had three children, Kenton, Margaret, and Jane.

Shepard's studies on the influence of landscape aesthetics on nature perception were presented in 1967 in his book, *Man in the Landscape: A Historic View of the Esthetics of Nature.* This book, along with the anthology *The Subversive Science*, co-authored with Daniel McKinley, were important to developing ecologists and used as popular readers for burgeoning environmental studies programs. Published posthumously in 2003, *Where We Belong* is a collection of essays on landscape and place also written by Shepard during this period.

After living in Massachusetts and holding several teaching positions in the East, in 1973 Shepard and his second wife, Ruth Attwood Shepard, moved to California where he was named Professor of Human Ecology and Natural Philosophy at Pitzer College, one of the Claremont Schools.

Years of writing and research on nature perception led Shepard at this time to an unexpected conclusion: Landscape art and literature, rather than drawing us to nature, distance us from it and creates a world of observers rather than participants. It was then that he turned to anthropology. Soon after arriving at Pitzer College, Shepard published his much acclaimed *The Tender Carnivore and the Sacred Game* (1973), a celebration of our primal hunter and gatherer forebears. However, his assertion that domestication turned humans away from a harmonious foraging way of life drew strong criticism.

Enthralled by animals from an early age, Shepard believed that humanity emerged in close association with animals and cannot be complete without them. He explored this connection in three books. *Thinking Animals: Animals and Human Development* (1978) shows how animals have been and still are integral to human thought and language. *The Sacred Paw* (with Barry Sanders, 1985) establishes the significance of the bear to northern hemisphere peoples in art, mythology, and literature. And in *The Others: How Animals Made Us Human*, published posthumously in 1996, Shepard explores the relationship of humans to animals in diverse cultural contexts throughout history.

Shepard retired from teaching at Pitzer College in 1994. He died at home in Salt Lake City on July 16, 1996, as his last book, *The Only World We've Got*, a reader, was published. In *Coming Home to the Pleistocene* (1998)—edited by his third wife, Florence Krall Shepard, and published posthumously—he summarized and clarified his most important insights into our human ecology and development. He believed that carried deep in our DNA is an integrated person who knows how to live ethically and ecologically on Planet Earth.

SEE ALSO *Environmental Activism; Environmental Aesthetics; Environmental Art; Hudson River School; Hunting and Fishing: II. Recreational Hunting; Space/Place.*

BIBLIOGRAPHY

WORKS BY PAUL SHEPARD

Shepard, Paul. 1954. "American Attitudes Towards the Landscape in New England and the West, 1830–1870." PhD diss. New Haven, CT: Yale University.

Shepard, Paul. 1967. *Man in the Landscape: A Historic View of the Esthetics of Nature*. New York: Knopf.

Shepard, Paul, and Daniel McKinley, eds. 1969. *The Subversive Science: Essays toward an Ecology of Man*. Boston: Houghton Mifflin.

Shepard, Paul, and Daniel McKinley, eds. 1971. *Environmental Essays on the Planet as a Home*. Boston: Houghton-Mifflin.

Shepard, Paul. 1973. *The Tender Carnivore and the Sacred Game: The Human Past as the Key to Modern Man's Identity—and to his Future*. New York: Scribner.

Shepard, Paul. 1978. *Thinking Animals: Animals and the Development of Human Intelligence*. New York: Viking Press.

Shepard, Paul. 1982. *Nature and Madness*. San Francisco: Sierra Club Books.

Shepard, Paul, and Barry Sanders. 1985. *The Sacred Paw: The Bear in Nature, Myth, and Literature*. New York: Viking.

Shepard, Paul. 1996. *The Only World We've Got: A Paul Shepard Reader*. San Francisco: Sierra Club Books.

Shepard, Paul. 1996. *The Others: How Animals Made Us Human*. Washington, DC: Island Press.

Shepard, Paul. 1998. *Coming Home to the Pleistocene*, ed. Florence R. Shepard. Washington, DC: Island Press.

Shepard, Paul. 1999. *Encounters with Nature: Essays*, ed. Florence R. Shepard. Washington, DC: Island Press.

Shepard, Paul. 2003. *Where We Belong: Beyond Abstraction in Perceiving Nature*, ed. Florence Rose Shepard. Athens: University of Georgia Press.

Florence Shepard

SHIVA, VANDANA
1952–

Vandana Shiva, who was born in Dehra Dun, India, in November 1952, is one of the leading environmental scholars and activists in the world. Originally trained in physics, she completed a doctorate in the philosophy of science in 1978. She is the author and editor of fifteen books and has published more than three hundred articles in addition to lecturing worldwide. In 1982 she established the Research Foundation for Science, Technology and Natural Resource Policy in India, which supports local environmental struggles, promotes biodiversity, and protects indigenous knowledge. Shiva has developed an activist platform for "earth democracy" that is opposed to life-destroying corporate globalization and demands local sovereignty over food systems, water systems, and biodiversity systems.

Shiva's critique of corporate globalization identifies the multiple ways in which Western economic development strategies and technological incursions destroy not just nature but native knowledge and social relations as well. In *The Violence of the Green Revolution* (1989) she identified the destruction caused by first world corporations in their efforts to increase crop productivity and profits. For many, the Green Revolution brings to mind advances in agricultural development that allowed greater food production to feed an ever-increasing human population, particularly after years of catastrophic famines. Shiva's discussions reveal a darker side in which high-yielding seeds brought environmental, economic, and social disasters. Corporations that developed and owned the new technologies turned biologically diverse and sustainable local farms into monoculture plots that were dependent on corporate products and World Bank loans for funds to purchase the products. When a crop failed, there was nowhere to turn, and in communities that had changed their farming practices a bad year meant economic and social collapse. In addition, the toxic pesticides and chemical fertilizers that were necessary for "advanced" methods of farming destroyed native plants and depleted soils.

Building on her earlier criticisms of the development of industrialized agriculture, Shiva is an outspoken critic of genetically modifying foods and patenting life, which she addressed in *Biopiracy: The Plunder of Nature and Knowledge* (1997). When corporations claim ownership of the earth and the living things on it, the value of nature and those who work with nature is diminished. Bidiversity and sustainability are sacrificed for profit.

As both a writer and an activist, Shiva uses her scholarly analyses to develop ecologically sustainable alternatives. For example, to combat the problems associated with the intervention of corporate agribusiness in Indian farming and the increasing destruction of globalized food production, Shiva formed an organization, Navdanya, whose name means "nine crops" that represent India's collective source of food security. Navdanya supports local organic farming, rescues and protects seeds and plants that are threatened with extinction, and defends native knowledge and food security.

Most of the farmers who produce most of the food in the third world are women, and Shiva is aware of the way in which ecological destruction disproportionately affects women. From her early days as part of the Chipko movement, which was made up of Himalayan women dedicated to the protection of the environment, through her critique of "maldevelopment" in *Staying Alive: Women, Ecology and Development* (1988), to the formation of an international coalition of women to respond to globalization called Diverse Women for Diversity, Shiva's environmentalism has been linked to the struggles of women.

Shiva has argued that the masculine way of thinking, doing science, and defining progress as controlling wealth and property dehumanizes women as well as men and ultimately will destroy the earth. This destructive force of patriarchy requires harnessing the creative power of women and nature. To accomplish this, those in positions of economic and epistemic power relegate women and nature to the realm of the passive. Once something is made passive or inert, Shiva argues, it can be used and commodified more readily. The health of the planet demands that women's labor and knowledge and the earth itself not be used in these ways, and thus she sees feminism and environmentalism as powerful political partners. This partnership can provide liberation from the life-destroying systems that currently threaten women, particularly third world women, and the planet.

Some have criticized Shiva's stance against global corporate techno-culture as over-simplified. Critics claim that she has underestimated the value of technological advances in saving lives. Shiva's stance is not anti-technology, however. She advocates democratic, community participation in the decisions to use and control technology and raises cautionary concerns about the long-term effects of technological incursions into social relations and our delicate relationship with the Earth. Recently she has issued warnings about the dangers of rushing to replace oil with biofuels, or fuels derived from corn, soya, and other common food crops. She argues that this quick fix solution will negatively impact the lives of poor people who will not be able to afford food.

SEE ALSO *Agriculture; Berry, Wendell; Biodiversity; Borlaug, Norman; Chipko Movement; Ecological Feminism; Environmental Activism; Factory Farms; Farms; Food Safety; Genetically Modified Organisms and Biotechnology; India and South Asia; Jackson, Wes; Patenting Life; Sustainability.*

BIBLIOGRAPHY

WORKS BY VANDANA SHIVA

Shiva, Vandana. 1988. *Staying Alive: Women, Ecology and Survival in India.* New Delhi: Kali for Women.

Shiva, Vandana. 1992. *The Violence of the Green Revolution: Ecological Degradation and Political Conflict in Punjab.* London: Zed Press.

Shiva, Vandana. 1993. *Monocultures of the Mind: Perspectives on Biodiversity and Biotechnologies.* London and Atlantic Highlands, NJ: Zed Press.

Shiva, Vandana. 1997. *Biopiracy: the Plunder of Nature and Knowledge.* Boston: South End Press.

Shiva, Vandana.2000. *Stolen Harvest: The Hijacking of the Global Food Supply.* Cambridge, MA: South End Press.

Shiva, Vandana. 2002. *Water Wars: Privatization, Pollution, and Profit.* Cambridge, MA: South End Press.

Shiva, Vandana. 2005. *Earth Democracy: Justice, Sustainability, and Peace.* Cambridge, MA: South End Press.

Shiva, Vandana, and Maria Mies, eds. 1993. *Ecofeminism.* Atlantic Highlands, NJ: Zed Books.

Lori Gruen

SHRADER-FRECHETTE, KRISTIN
1944–

Kristin Shrader-Frechette was born on September 14, 1944, in Louisville, Kentucky. She got her B.A. in mathematics from Xavier University, earned a Ph.D. in philosophy from the University of Notre Dame, and held postdoctoral fellowships in biology (community ecology), economics, and hydrogeology. As of 2008 she is O'Neill Family Professor of Philosophy and Concurrent Professor of Biological Sciences at the University of Notre Dame. She is the author of more than 350 articles and 15 books, much of this work focused on ethical and methodological issues that arise in connection with technological risks to humans and the environment, the actual consequences of various technologies, and related governmental regulatory efforts. She also writes on ethical theory and the scientific method.

Shrader-Frechette's work has regularly addressed problems related to nuclear technology. She has written extensively on the proposed permanent nuclear-waste repository in Yucca Mountain, Nevada. Because of the nature and extent of the scientific uncertainty regarding whether such waste could be safely housed there (or anywhere) for 10,000 years—a number that is, in any case, problematically arbitrary—she claims that the Department of Energy should delay determination of the site's suitability and store waste for a century in numerous regional, monitored, retrievable facilities while the option of permanent interment is further studied. She has criticized federal regulations governing releases of radiation from the site, arguing that the U.S. Environmental Protection Agency's suggested radiation-exposure limits, which are twenty-three times higher for the distant future than for the near future, fail the demand for equal protection, indefensibly entailing that we merit more protection than our descendents in spite of the fact that we, not they, profit from the power for which the waste was generated. Furthermore, she holds that the agency should not assess compliance with these regulations according to mean and median doses of exposure across the affected population, since both of these approaches are capable of sanctioning lethal doses for many people. Instead, compliance should be assessed according to whether any individual faces an exposure dose over a certain amount. She has also written on the siting of uranium-enrichment facilities, the effects and cleanup of low-dose ionizing radiation from above-ground nuclear-weapons testing, and safety regulations in the nuclear-workplace environment.

Shrader-Frechette has consistently defended the use of cost-benefit analysis in environmental-policy decision making, albeit only where such analysis is conceived of and conducted in ways she sees as appropriate. She believes that while its use may be necessary for rational societal decision making, it is not sufficient, since considerations such as who is responsible for creating the risk, who benefits from the risk, whether the risk is involuntarily imposed or voluntarily chosen, what moral and legal rights affected parties have, and so on, can override narrow cost-benefit judgments. She maintains that assessments of policy-related gains and losses must be scientifically well informed—a demand to which economists have traditionally been somewhat insensitive. Such analysis must treat risk assessment as not purely subjective, that is, not as matters purely of value rather than fact. However, she argues, this does not mean that risk assessment or the resultant determination of costs and benefits should be conceived of as a purely scientific process in which only expert analysis counts. Rather, public deliberation is critically important. The appropriateness of lay persons' involvement is based in the stake that we all have in the outcomes of these policy decisions, in the very idea of democracy, and in the fact that scientific judgments are often unavoidably plagued by uncertainty. Thus, Shrader-Frechette makes a much smaller role for technocrats in regulatory decisions than do some others (perhaps most notably Cass Sunstein).

Shrader-Frechette has also produced writings central to the development of the environmental-justice movement and been involved on the ground in environmental-justice efforts around the world. In the Congo, for example, she worked with the World Council of Churches to advise locals on how to avoid having toxic waste shipped to their land from developed countries. And with her students, she helped the predominantly African-American community of Scarboro, Tennessee—one with high numbers of children suffering from respiratory and pulmonary ailments—assess local levels of exposure to pollutants, including beryllium, lead, ionizing radiation, mercury, and polychlorinated biphenyls. She has advised the United Nations, the World Health Organization, numerous foreign governments, as well as the president of the United States, the U.S. Congress, and various federal and state agencies. She was the first female president of several prestigious scholarly associations and societies.

SEE ALSO *China; Cost-Benefit Analysis; Environmental Justice; Environmental Policy; Global Climate Change; Nuclear Power; Pollution; Resource Management; Risk Assessment; Urban Environments; U.S. Environmental Protection Agency.*

BIBLIOGRAPHY

WORKS BY KRISTIN SHRADER-FRECHETTE

Shrader-Frechette, Kristin. 1991. *Risk and Rationality*. Berkeley: University of California Press.

Shrader-Frechette, Kristin. 1993. *Burying Uncertainty: Risk and the Case against Geological Disposal of Nuclear Waste*. Berkeley: University of California Press.

Shrader-Frechette, Kristin. 2002. *Environmental Justice: Creating Equality, Reclaiming Democracy*. New York: Oxford University Press.

Shrader-Frechette, Kristin. 2007. *Taking Action, Saving Lives: Our Duties to Protect Environmental and Public Health*. New York: Oxford University Press.

Ty Raterman

SIERRA CLUB

The Sierra Club has influenced environmental thought and policy in the United States for more than one hundred years. The club's original regional focus has gradually enlarged to the point where it now has a global presence, with more than 1 million members. This shift has been accompanied by an evolution of the organization's goals, methods, and overall philosophy. With a history intertwined with that of the National Park

Service, the Sierra Club has a strong tradition of both conservation and recreation.

As its name suggests, the Sierra Club was inspired by the Sierra Nevada mountain range of California. Revered today as a patron saint of not only the Sierra Club, but also of the conservation movement as a whole, John Muir (1838–1914) articulated an early statement of purpose. Even before the organization was founded, Muir, along with the editor Robert Underwood Johnson (1853–1937), lobbied Congress successfully for the creation of Yosemite National Park in 1890. On May 28, 1892, a coalition of prominent San Francisco Bay area residents, including professors from Stanford and the University of California, joined Muir and Johnson in founding the Sierra Club. They wanted to defend and enlarge Yosemite but also sought to formulate larger scientific and recreational goals for California.

The Sierra Club found its voice in Muir, whose writing was published by the organization in its journal, the *Sierra Club Bulletin*. Muir's enthusiasm was infectious, fostering opportunities for others to visit the mountains. By 1901 the Board of Directors approved summer excursions. That first year nearly a hundred people went to Tuolumne Meadows in Yosemite. The numbers of participants in outings grew rapidly, along with membership totals.

In 1914 the Sierra Club organized a final trip to Hetch Hetchy Valley. Considered by many to be Yosemite Valley's equal in beauty, Hetch Hetchy was the subject of a long fight between conservationists and the city of San Francisco, which sought to dam the valley for its municipal water supply. This controversy tested the young Sierra Club and drove a permanent wedge between two groups of conservationists that previously held a shaky truce: on one side were the "preservationists," whom the Sierra Club fully embraced, and on the other were the utilitarians, or "conservationists," headed by the first chief of the United State Forest Service, Gifford Pinchot (1865–1946). Preservationists argued for retaining natural lands with minimal management, while utilitarians believed that wild lands can be productively used for multiple purposes. The Sierra Club failed to preserve the valley: In 1913 Congress approved the damming of Hetch Hetchy.

Following that defeat the Sierra Club won a victory with the passage of the National Park Service Organic Act of 1916, which created this new government agency and appropriated federal funding in its support. Utilitarians such as Forest Service Chief Pinchot thus saw their influence eroded in the national parks. Efforts to develop and make accessible places such as Yosemite were a major focus of the next several decades. Hiking trails were built, and Sierra Club members began pioneering new ways to explore the mountains. Mountaineering and rock climbing techniques opened up previously inaccessible areas. Lightweight hiking and camping gear made these expeditions more enjoyable and popular. The 1920s saw major road-building efforts, and the automobile brought ever-increasing numbers of tourists to the expanding park system.

The recreational successes that the Sierra Club was enjoying began to clash with the group's ecological goals. Development of the most scenic areas in the country was so rapid and overwhelming that the parks were in danger of being destroyed by overuse. As Michael Cohen notes, the organization's statement of purpose was amended in 1951 to read "explore, enjoy, and preserve" instead of "explore, enjoy, and render accessible" (1988, p. 100). Working with other groups such as the Wilderness Society, the Sierra Club was no longer satisfied with national park designations unless they were backed by an ecological vision.

The early 1950s brought many changes and fights to the Sierra Club, both internal and external. David Brower (1912–2000) became the club's first executive director in 1952. His leadership, along with a proposal to dam the Colorado River in 1951 at Echo Park and flood Dinosaur National Monument, spurred the Sierra Club to employ more aggressive tactics. One strategy was full-page newspaper ads to lobby against the proposed dam; these ads brought the Sierra Club to the Supreme Court when their tax-exempt status was threatened. Meanwhile, the organization brought tens of thousands of new visitors to Dinosaur, many of whom experienced the area by rafting the Colorado. The club's tradition of making previously unknown places accessible proved successful in this case: The proposal to dam the river at Echo Park was thwarted in 1956, although construction began for the Glen Canyon Dam. The Sierra Club continued to publish books in support of its mission and branched out into filmmaking. The organization decided to avoid losing its tax-exempt status, however, by not directly lobbying the government.

The Sierra Club now had a national presence yet retained a regional focus through its local chapters. This local presence was enhanced when the club's tax-exempt status was revoked. Fighting against a proposal to dam parts of the Grand Canyon, the Sierra Club took out several full-page advertisements in national publications. Their most famous ad, published in 1966, proclaimed, "Should we also flood the Sistine Chapel so tourists can get nearer the ceiling?" Membership skyrocketed. April 22, 1970, the first Earth Day, sparked a rise in awareness of environmental issues that brought the Sierra Club still more members. However, this wave of members came without Brower, who resigned as executive director in 1969 after several years of clashes with board members.

Sierra Club Ceremony, Johnson City, Tennessee. *Members of a local Sierra Club chapter stage a "funeral" for four ancient beech trees cut down to make way for development in 2005. The organization, with John Muir as its patron saint, has been around since the early twentieth century.* AP IMAGES.

Through the 1970s the Sierra Club's platform became more wide-ranging. The group addressed issues such as overpopulation, arguing that birth rates in the United States are too high; the club, however, does not take a position on immigration or issues it sees as disconnected from conservation. Brower, after reconciling with the Sierra Club, led a faction that argued for a public position against immigration in 2000, but the club's moderate elements steered official club policy towards a more compromising stance. Notably, younger conservation and environmental organizations have criticized the Sierra Club as being too willing to compromise and make deals with Washington. These groups argue for grassroots activism and suggest that the Sierra Club has strayed too far from John Muir's originary vision.

Despite criticism, traditional conservation causes remain important to the Sierra Club. For example, it has repeatedly lobbied to keep the Arctic National Wildlife Refuge closed to oil drilling and has worked to create new wilderness areas. Global and urban issues also attracted the Sierra Club as the group passed its centennial in 1992. Climate change and renewable energy became major policy issues, as have clean urban spaces and environmental justice. The Sierra Club has evolved into a comprehensive conservation group that has managed to keep local chapters small and dedicated to excursions. Nevertheless, it maintains a national presence and wields considerable influence in Washington, D.C. These efforts to be local yet global, to protect yet promote, have created a flexible and lasting organization.

SEE ALSO *Conservation; Environmental Activism; Environmental Law; Environmental Policy; Hetch Hetchy; Mountains; Muir, John; Nongovernmental Organizations; Pinchot, Gifford; Preservation; U.S. Forest Service; Utilitarianism; Water; Wilderness.*

BIBLIOGRAPHY

Cohen, Michael P. 1988. *The History of the Sierra Club, 1892–1970.* San Francisco: Sierra Club Books.

Sierra Club. Available from http://www.sierraclub.org

Turner, Tom. 1991. *Sierra Club: 100 Years of Protecting Nature.* New York: H. N. Abrams.

Kyhl Lyndgaard

SINGER, PETER
1946–

Peter Singer was born in Melbourne, Australia, on July 6, 1946. He completed undergraduate studies in law, history, and philosophy at Melbourne University and did graduate work in philosophy at Oxford, where he received a bachelor of philosophy degree in 1971.

His primary appointments have been as a professor of philosophy at Monash University in Australia and a professor of bioethics at the Center for Human Values at Princeton University; he also has held visiting positions at universities around the world. Singer addresses timely and provocative moral issues, has a clear and accessible writing style, and is a socially engaged public intellectual. He is said to be among the most influential philosophers alive (Specter 1999).

PRACTICAL ETHICS

Singer's most important contributions have been in practical ethics. He helped establish the applied ethics movement in which philosophers turned their analytical and argumentative skills toward matters of moral significance and public interest. He has focused primarily on "relevant" issues that "any thinking person must face" (Singer 1993, p. v).

Singer has argued vigorously for challenging, usually unorthodox positions on the treatment of nonhuman animals; the obligations affluent people have to people living in absolute poverty; the ethics of globalization; environmental protection, including measures to reduce climate change; and major issues in bioethics, including euthanasia and abortion. His works offer a critique of the doctrine of the sanctity of life.

ANIMAL LIBERATION AND ANIMAL RIGHTS

In *Animal Liberation,* first published in 1975, Singer challenged the morality of using nonhuman animals for food and scientific experimentation. He maintained that it is inconsistent to use animals for purposes for which people would not consider using a human being, even an infant or a person whose mental level is similar to that of an animal. He argued that such uses are wrong because they result from an unjustified prejudice against other species, which, adopting a term coined by Richard Ryder, he called *speciesism.*

Singer has defended the "fundamental principle of equality," the demand that equal moral consideration be given to comparable interests regardless of whose interests they are, by arguing that that principle best explains why racism and sexism are wrong. In the human case it generally is agreed that women's and minorities' educational, political, and economic interests are comparable to those of white men, and so their interests should be given equal consideration.

In the case of animals Singer argues that many animals also have interests both in avoiding pain and suffering and in experiencing happiness. *Animal Liberation* uses photographs and vivid descriptions to show how animals' interests are given little or no consideration by agribusiness and research industries. That treatment violates the principle of equality and therefore is wrong.

Animal Liberation sold hundreds of thousands of copies and helped start a worldwide animal protection movement. It has been referred to widely as the Bible of the animal liberation movement. In *Practical Ethics,* first published in 1979, Singer supplemented his equality-based defense of animals with a utilitarian defense, arguing that the overall harms produced by these uses of animals typically outweigh the benefits.

In *Practical Ethics* Singer also discussed the major issues in biomedical ethics. His sanctity of life doctrine holds that all human life, regardless of its subjective quality, is equally valuable and thus deserves equal treatment. Singer has argued for an ethic sensitive to relevant details, for example, whether the individual is conscious, is in pain and suffering, has expressed a preference for how her or his life should end, the medical prognosis, and the family's wishes. He has argued that those factors can make it permissible or even obligatory to let a human being die and even be killed actively. Singer thus argues that abortion, voluntary euthanasia, assisted suicide, and even infanticide can be morally permissible and often are.

POVERTY AND MORAL OBLIGATIONS

Singer has applied the principle of equality and utilitarianism to questions about people's moral obligations toward the billion or more people living in and dying from absolute poverty. He has argued that affluent individuals and nations are morally obligated to provide generous assistance to people in poverty, with an immediate response to famines and natural disasters and, more important, long-term development aid. In his essay "Famine, Affluence, and Morality" (1972) Singer defended his conclusions by referring to a simple case: Someone sees a small child fall into a pond and realizes the child is in danger of drowning. She could save the child, but her clothes would get muddy. Is she morally obligated to save the child? Most people would think that she is. Singer argues that this principle best explains why if one can prevent something very bad from happening at an insignificant cost to oneself, one is obligated to do so. By analogy and by appeal to that principle Singer concludes that people are obligated to help prevent the daily needless deaths of tens of thousands of children: Those deaths are very bad, and people can prevent them at only minor costs to themselves.

In *One World,* first published in 2002, Singer applied the principle of equality to climate change and globalization. Affluent nations enjoy their prosperity at significant costs to the environment: Their development and energy-consumptive lifestyles contribute to climate change that threatens the future of life on the planet. There is a scientific consensus on predictions for increased hurricanes and storms, droughts and floods, tropical diseases, rising sea levels, and disruption in food production resulting from climate change. That change probably will affect poor nations the most because they lack the resources to move people from affected areas, store food, and fight disease.

Singer argues that industrialized nations must reduce their greenhouse gas and carbon emissions in an attempt to curb climate change. The principle of equal consideration of interests requires a new global ethic in which leaders of developed nations consider the effects of their policies on all the people affected by them, not just the people of their own nations. He argues that on a variety of plausible theories of fairness and justice, the refusal of the United States to commit to a plan to address atmospheric change, for example, the Kyoto Protocol, is self-serving and ethically indefensible because the United States has 5 percent of the world's population but produces 30 percent of the climate-changing emissions. Because the developing nations' contributions of greenhouse gases will not equal the built-up contributions of the developed nations until about 2038 and, per capita, for at least a hundred years,

Singer argues that industrialized nations should bear the initial burdens; he proposes a cap and trade program to give developing countries incentives to keep their emissions low.

ENVIRONMENTAL ETHICS

In regard to the environment Singer's ethic is an extension of the dominant Western ethics to include all sentient beings. It attributes no intrinsic value to nonsentient entities such as plants, ecosystems, wilderness, species (apart from individual members of a species), endangered species, and nature generally. Singer argues that if an entity lacks or is incapable of subjective experience, it lacks interests that need to be considered. He rejects developments in environmental ethics, such as holism, Deep Ecology, and the land ethic, that posit intrinsic value beyond conscious beings and their experiences.

Singer has argued that his ethic provides strong environmental protection. Many human beings value natural spaces for recreational, scientific, aesthetic, and spiritual purposes: Destroying wilderness, which is irreplaceable, thwarts those interests. Because future generations of humans probably will have such interests, short-term economic benefits from destroying nature should not outweigh this loss to an indefinite number of future humans.

SEE ALSO *Animal Ethics; Civil Disobedience; Environmental Activism; Environmental Philosophy: V. Contemporary Philosophy; Global Climate Change; Speciesism; Utilitarianism.*

BIBLIOGRAPHY

Jamieson, Dale, ed. 1999. *Singer and His Critics*. Malden, MA: Blackwell.

Singer, Peter. 1972. "Famine, Affluence, and Morality," *Philosophy and Public Affairs,* 1(1): 229–243.

Singer, Peter. 1993. *Practical Ethics,* 2nd edition. Cambridge, UK, and New York: Cambridge University Press.

Singer, Peter. 2002. *Animal Liberation,* 3rd edition. New York: Harper Collins.

Singer, Peter. 2004. *One World: The Ethics of Globalization,* 2nd edition. New Haven, CT: Yale University Press.

Specter, Michael. 1999. "Profile: The Dangerous Philosopher," *The New Yorker*, September 6. Available from http://www.michaelspecter.com/ny/1999/1999_09_06_philosopher.html

Nathan Nobis

SNYDER, GARY
1930–

Poet, essayist, teacher, Buddhist practitioner, and community activist, Gary Snyder has lived a colorful and storied life. Many readers became familiar with Snyder through his fictionalized representation in Jack Kerouac's 1958 novel *Dharma Bums,* in which his prowess as an outdoorsman and mountain climber is highlighted. Others are particularly fascinated with Snyder's ten-year immersion in Buddhist philosophy and practice from the late 1950s through the late 1960s, during which he lived in various Zen monasteries in Kyoto, Japan, developing habits and affinities to which he continued to adhere into the twenty-first century. And, of course, most people know Snyder through his many important works of poetry and nonfiction, ranging from *Riprap* (his first book of poetry, published in 1959), to his Pulitzer-prize-winning collection *Turtle Island* (1974), to more recent volumes such as *No Nature: New and Selected Poems* (1992), *Mountains and Rivers Without End* (1995), and *Danger on Peaks* (2005). The most compendious edition of his work is *The Gary Snyder Reader: Prose, Poetry, and Translations, 1952–1998,* which appeared in 1999.

Of particular relevance to environmental ethics and philosophy are Snyder's profound engagement with and articulation of the experience of physical labor, his application of Buddhist ideas to the conceptualization of the human relationship with nature, and his personal involvement with local, regional, and global political issues associated with environmental responsibility. The poems in *Riprap* and such later collections as *Myths & Texts* (1960), *Riprap and Cold Mountain Poems* (1965), and *The Back Country* (1967), among others, vividly evoke the presence of a human speaker immersed in the sensory qualities of outdoor experience. One poem from *Riprap,* titled "Mid-August at Sourdough Mountain Lookout," for instance, recalls the poet's seasonal work as a fire lookout in the North Cascades of Washington State, immersing the speaker in the physical realities of the job ("Down valley a smoke haze /... Swarms of new flies. // Drinking cold snow-water from a tin cup / Looking down for miles / Through high still air"). Philosopher Jack Turner writes bitterly in *The Abstract Wild* (1996) that "Science, including economics, tends to reduce nonhuman life to trash....We need to find another way of describing the world and our experience in it" (p. 65). He points specifically to Snyder's work—using an excerpt from *Mountains and Rivers Without End* as an example—to demonstrate the language of engaged non-abstraction.

Many of Snyder's poems resonate with echoes of his Buddhist training, but perhaps one of the most powerful applications of this philosophical tradition to the poet's artistic vision occurs in "Ripples on the Surface," the final poem in *No Nature.* This poem explores how the human mind knows nature, seeking to read it as a text, to apprehend it as if it were a "performance." He concludes by erasing the distinction between nature and human culture, stating "No nature // Both together, one big empty house." Despite the frequent immersion in the physical world that occurs in Snyder's poetry, this Buddhist angle presents an opposing viewpoint: The phenomenal world is not the essence of reality and essential Truth cannot be perceived or articulated. This idea directly echoes the *Diamond Sutra*. This sacred text, dating back to ninth-century China, explores the fundamental notion of reality and the importance of nonattachment and is a central text in the Mahayana Buddhist tradition.

Despite his Buddhist skepticism toward the ability of humans to apprehend nature, Snyder took a practical stance in late-twentieth-century debates about the meaning of wilderness in American culture. In the essay "Is Nature Real?" (collected in *The Gary Snyder Reader* [1999]), he both acknowledged the validity of contemporary deconstructions of "wilderness" by scholars such as historian William Cronon ("So we understand the point about wilderness being in one sense a cultural construct, for what isn't?") and sharply criticized the political naiveté of such critiques. He states: "The attacks on nature and wilderness from the ivory towers come at just the right time to bolster global developers, the resurgent timber companies ... and those who would trash the Endangered Species Act" (p. 388).

Patrick D. Murphy begins his 2000 study, *A Place for Wayfaring: The Poetry and Prose of Gary Snyder*, by recalling that when Snyder was a small child, his family moved to Lake City, Washington, an area that had been devastated by clear-cutting. "The significance of this land having been clear-cut for Snyder's psychological and environmental-ethical development should not be underestimated," Murphy argues (p. 1). The poet himself remarked in *The Practice of the Wild* (1990): "I suspect that I was to some extent instructed by the ghosts of those ancient trees as they hovered near their stumps" (p. 118). Later, Snyder directed his social concerns toward the militaristic ambitions of the U.S. government and its allies, noting in his 2003 essay "Ecology, Literature, and the New World Disorder" in *Back on the Fire* (2007) that "we have entered a period in which global relations are defined by new nationalisms, religious fundamentalism, developed world hubris, stepped-up environmental damage, and everywhere expanding problems of health and poverty" (p. 22). Snyder's poetry, too, reflects his sense of social commitment—commitment, that is, to the larger community of life. In "For All," collected in *Axe Handles* (1983) and later in *No Nature*, he laughingly declares his true loyalty: "I pledge allegiance // I pledge allegiance to the soil / of Turtle Island, / and to the beings who thereon dwell / one ecosystem / in diversity / under the sun / With joyful interpenetration for all." This single brief poem captures the essence of the essays collected in the opening section, titled "Ethics," of his 1995 collection of essays, *A Place in Space: Ethics, Aesthetics, and Watersheds*. The "Watersheds" section of that volume includes several essays that explain in detail what it means to live responsibly in a particular community in a particular place, with a long-term (sustainable) vision of that place; essays such as "Reinhabitation" and "Coming into the Watershed" probe the author's essential concepts of community and bioregional governance.

In his 2006 essay "Writers and the War Against Nature" (collected in *Back on the Fire*), Snyder points to "Song of the Taste" (initially published in *Regarding Wave* in 1969) as his "first truly ecological poem" and an exploration of "the essential qualities of human foods" (p. 68). As he explains in this essay, the poem goes "straight to the question of conflict between the ethics of *ahimsa*, nonviolence, 'respect for all beings,' and the lives of necessity and subsistence of indigenous peoples and Native Americans I had known." People who derive their existence from the bounty of nature, he continues, "can enter into the process with gratitude and care, and no arrogant assumptions of human privilege. This cannot come from 'thinking about' nature; it must come from being *within* nature" (p. 69). From his trim, haiku-like meditations on the physicality of human engagement with nature to his Buddhism-inspired reflections and his political critiques, Gary Snyder's work resonates with the implications of a life lived "within nature" and, at the same time, within community.

SEE ALSO *Buddhism; Environmental Activism; Wilderness.*

BIBLIOGRAPHY

WORKS BY GARY SNYDER

Riprap. 1959. Ashland, MA: Origin Press.

Myths & Texts. 1960. New York: Totem.

Riprap; and Cold Mountain Poems. 1965. San Francisco: City Lights.

The Back Country. 1967. London: Fulcrum.

Regarding Wave. 1969. Iowa City, IA: Windhover Press.

Turtle Island. 1974. New York: New Directions.

Axe Handles. 1983. San Francisco: North Point Press.

The Practice of the Wild. 1990. San Francisco: North Point Press.

No Nature: New and Selected Poems. 1992. New York: Pantheon.

A Place in Space: Ethics, Aesthetics, and Watersheds. 1995. Washington, DC: Counterpoint.

Mountains and Rivers Without End. 1996. Washington, DC: Counterpoint.

The Gary Snyder Reader: Prose, Poetry, and Translations, 1952–1998. 1999. Washington, DC: Counterpoint.

Danger on Peaks. 2005. Emeryville, CA: Shoemaker & Hoard.

Back on the Fire. 2007. Emeryville, CA: Shoemaker & Hoard.

WORKS ABOUT GARY SNYDER

Kerouac, Jack. 1958. *Dharma Bums*. New York: Viking.

Murphy, Patrick D. 2000. *A Place for Wayfaring: The Poetry and Prose of Gary Snyder*. Corvallis: Oregon State University Press.

Turner, Jack. 1996. *The Abstract Wild*. Tucson: University of Arizona Press.

Scott Slovic

SOCIAL CONSTRUCTIVISM

Social constructivism is a group of connected claims about how perspectives, knowledge, and perhaps more concrete objects arise through social processes. Most important for environmental ethics, though, are four similar ideas:

1. First, and most important, social constructivists draw attention to the ways in which scientific knowledge does not simply reflect nature, but instead stems from work done by groups of people to describe nature. That work shapes the knowledge produced; therefore, our ideas have social histories. There is no direct path from nature to our ideas about it.
2. Second, not only is scientific knowledge socially constructed, but so are values and meanings (Macnaghten and Urry 1998). People's understandings of nature and wilderness are deeply shaped by their cultures, by their economic positions, by political struggles over environmental issues, and by more idiosyncratic factors such as personal histories. To one person a forest might be a beautiful and fragile piece of wilderness needing protection. To another it might be a resource waiting to be sustainably exploited. To yet another it might be a forbidding and terrifying source of dangers to be contained. All of these meanings stem from particular cultural, economic, and political contexts. If meanings of particular natural objects are socially constructed, so are general environmental values, especially those for which philosophers argue.
3. Third, more materially, much of physical nature itself is socially constructed because many of the environments and landscapes that we take to be natural have themselves been affected by many generations of human actions (Cronon 1995). Forests in North America once looked like pristine wilderness, but even before Europeans logged and cleared them, Native Americans were managing some of those forests through controlled fires. Many of the places that modern urban dwellers most appreciate as natural refuges from the built world have been profoundly affected by human activity and even deliberately shaped for recreation.
4. Finally, what we might call "strong" social constructivism is the claim that how people understand the world *is* the world. This is the claim that with their knowledge of it people shape the world directly. Although there is a philosophical case to be made for strong social constructivism, it is controversial, and we can set it aside here.

THE CASE FOR THE SOCIAL CONSTRUCTION OF SCIENTIFIC KNOWLEDGE

The case for the social construction of scientific knowledge has been made repeatedly in science and technology studies (STS) (Sismondo 2004), and in postmodern philosophy. There is a common saying that "data do not speak for themselves." This point, which denies that data are a sufficient foundation of knowledge, might illustrate justifications for social constructivism about scientific knowledge. Data need to be interpreted before they can have any meaning. There are no absolute rules for interpretation. Instead there are practices, tools, and innovations, all of which have to be acceptable or justified to expert communities. More deeply, the choice of data to be gathered is determined by the questions posed or the hypotheses to be tested; by the means or lack thereof for acquiring data, such as satellite imaging technology or transportation to a remote site; and by the resources for purchasing equipment, conducting surveys, or paying graduate students.

The widely held view that knowledge cannot be absolutely or even solidly grounded is known as *antifoundationalism*. Antifoundationalism is important not only to STS, but also to postmodern philosophy—and even to much modern philosophy. Postmodernists argue and emphasize that people's access to nature is always mediated because there is no direct way to know nature as it is in itself (Smith 1999). They further argue that there are no stable foundations for any values or institutions; therefore, knowledge and values are constructed.

Science and technology are thoroughly social activities. They are social because scientists are always members of communities, enculturated into those communities and necessarily working within them. Communities, among other things, set standards for inquiry and evaluate knowledge claims; there is no abstract, ideal, and logical scientific method apart from the best practices of scientists. In addition, science is an arena in which rhetorical work is crucial because scientists always have to convince their peers and others (including funding sources) of the value of their ideas and plans. Scientific actors have investments in skills, prestige, knowledge, and specific theories and practices. Thus many different types of ideologies and values are important components of research; even conflicts in a wider society may be connected to conflicts within science and vice versa.

Thousands of empirical case studies have shown that scientists have choices about how and where to look for data, how to interpret that data, what can count as good models and theories, and how to interpret and use those models and theories. To take only one important and well-studied example: The general circulation models (GCMs) that provide our best evidence of human-induced climate change are extremely complicated computer programs that combine models and data in innovative ways (Demeritt 2001; Edwards 2001). The models, and the ways in which the models are applied and programmed, differ from one GCM to another, although they are often related by descent and by cross-fertilization (models typically share ancestors; when researchers can, they use good ideas that they see in other GCMs). The models are corrected by factors that bring them into line with existing data and expectations; this kind of correction is an art. Even the data are not raw, but corrected and smoothed, to take account of the fact that they are collected differently at irregularly spaced weather stations and satellites. Every aspect of a GCM is shaped by choices; communities of experts decide what the acceptable choices are.

Scientific knowledge is produced when the socially sanctioned expert communities, or important parts of them, are satisfied with claims. For uncontroversial claims, the experts might agree very quickly, accepting the authority of the researchers making those claims or the reasonableness of the claims given other things known. Controversies can arise, however, prompting the examining and challenging of many of the choices and arguments, the competence of researchers, and the reliability of equipment. Such controversies can expose the social construction of scientific knowledge. Eventually, minority views are marginalized and excluded from the debates, and the result is a rough consensus.

OPPOSITION TO SOCIAL CONSTRUCTIVISM

The idea that our knowledge is socially constructed may appear obvious. There is, however, another set of connected claims and perspectives, typically going under the label of *realism,* that opposes the views of social constructivism. For realists, our best scientific knowledge is true to nature. Realists claim that social processes do not significantly shape scientific knowledge but instead faithfully record nature's nature and that the emphasis of social constructivism is therefore misplaced (Crist 2004). On this view, scientific knowledge, the meaning of nature, environmental values, and even "natural" spaces may be shaped socially, but they are also shaped by nature. Not only is science shaped by nature, but it must also be approximately true to nature because it is so successful—at contributing to technology, for example. In other words, scientific knowledge passes the pragmatic test of truth: It works.

A further case against social constructivism is that its emphasis makes environmental politics difficult (Soulé and Lease 1995). A focus on the social processes by which scientific knowledge is made tends to undercut that knowledge because it implies that other social circumstances would have produced different knowledge and because confidence in science is based on its reliance on formulaic methods for uncovering truths of nature. Social constructivism, in this view, cripples the ability of science to serve as a solid foundation for environmental politics. Likewise, somebody who claims that wilderness is valuable for its own sake wants to assert that claim as objectively true, not merely as one culturally bound "story" among many possible "stories." Social constructivism, however—in this case, the social construction of values and meanings—does not allow environmental values to trump other factors in political disputes. Even the third, material, form of social constructivism—the observation that some landscapes normally thought to be natural are shaped by people—poses problems for environmental politics because it raises questions about how much and why we should value those landscapes.

SOCIAL CONSTRUCTIVISM AND ENVIRONMENTAL POLITICS

We might think that at least part of the debate about social constructivism and environmental politics hinges on a theoretical clash between social constructivism and realism. If realism is right, and science and philosophy provide natural and ethical truths that mirror the way things are, then environmental politics can rest securely on those truths. If social constructivism is right, and science and philosophy provide only contingent natural and ethical truths, then environmental politics can rest only insecurely on those truths.

In practice, however, these theoretical debates are not important to environmental politics or even to most environmental arguments—not just because the theoretical debates are far removed from the action but also because practical and down-to-earth versions of social constructivism and realism often play roles in environmental disputes. The practical versions make the theoretical ones less relevant.

Environmental politics often pits experts against one another. On the one hand, experts typically try to present their own views as entirely constrained by nature and rationality so that there is no room for disagreement. On the other hand, those same experts find ways in which opposing arguments are open to challenge. Scientific knowledge laden with choices is not hidden, a truth seen not only by people working in STS or postmodern philosophy; it is regularly rediscovered in disputes. For example, the authority of studies on climate change through GCMs has been continually challenged, both substantively and as a basis for policy (Demeritt 2001). The result is that political expertise in other domains has often played a larger role in energy policy than has climate science.

Values are even more open to challenge. Therefore, environmental politics (and any other kind of politics) is already a constructivist arena. If the stakes are high enough, the pronouncements of experts do not go unchallenged. It is well established in studies of public controversies that as long as the social and political issues are being actively disputed, so will be the scientific issues. In rough terms, all the issues have to be resolved at once.

If the above argument is right, then for science (and philosophy) to play a larger role in politics, they should become more engaged, not less. Since expertise will be "deconstructed" during controversies; knowledge should be constructed with controversy already in mind. The authority of science cannot depend on an incorrect formal picture of itself that is open to periodic debunking. The social-constructivist view brings to the fore the complexity of real-world science and therefore can contribute to its success. Successful science in the public sphere can be the result of the "coproduction" of science and politics (Jasanoff 2004); science can more easily solve problems in the public domain if scientific knowledge is carefully adjusted to its public contexts and attuned to knowledge that non-scientists have. Perhaps the third, material, form of social constructivism can give us a hint about how environmental politics can thrive in the absence of foundations. A landscape does not have to be a wilderness to have value—although enough people have to be convinced of its value, even in the absence of pure naturalness, for this claim to carry weight. Similarly, scientific and philosophical claims do not have to be indisputable to gain respect—although their authority hinges at least in part on a sufficient number of people being convinced of their value. Social constructivism, then, tells us we have to live without foundations.

SEE ALSO *Environmental Philosophy: VI. Postmodern Philosophy; Environmental Politics; Forests; Global Climate Change; Wilderness.*

BIBLIOGRAPHY

Crist, Eileen. 2004. "Against the Social Construction of Nature and Wilderness." *Environmental Ethics* 26: 5–23.

Cronon, William, ed. 1995. *Uncommon Ground: Toward Reinventing Nature.* New York: W. W. Norton.

Demeritt, David. 2001. "The Construction of Global Warming and the Politics of Science." *Annals of the Association of American Geographers* 91: 307–337.

Edwards, Paul N. 2001. "Representing the Global Atmosphere: Computer Models, Data, and Knowledge about Climate Change." In *Changing the Atmosphere: Expert Knowledge and Environmental Governance*, ed. Paul N. Edwards and Clark A. Miller. Cambridge, MA: MIT Press.

Jasanoff, Sheila. 2004. *States of Knowledge: The Co-Production of Science and the Social Order.* New York: Routledge.

Macnaghten, Phil, and John Urry. 1998. *Contested Natures.* London: Sage.

Sismondo, Sergio. 2004. *An Introduction to Science and Technology Studies.* Oxford, UK: Blackwell.

Smith, Mick. 1999. "To Speak of Trees: Social Constructivism, Environmental Values, and the Future of Deep Ecology." *Environmental Ethics* 21: 359–376.

Soulé, Michael, and Gary Lease, eds. 1995. *Reinventing Nature? Responses to Postmodern Deconstruction.* Washington, DC: Island Press.

Sergio Sismondo

SOCIAL CONTRACT THEORY

Social contract theory is generally considered to be one of the most significant contemporary political theories. While early forms of the theory are found in views attributed to Protagoras in the fifth century BCE and expressed in Plato's dialogue *Crito* (c. 350 BCE) and summarized in his *Republic*, as well as by Manegold of Lautenbach (1080), most commentators consider Thomas Hobbes's *Leviathan* (1651) to be the first fully developed statement of social contract theory. For Hobbes, a fictional state of nature illustrated the necessity of political authority. People are equal, in the sense that everyone can be killed by others, and therefore some kind of agreement is necessary in order for people to live together in peace: The social contract is essentially a life preserver. After Hobbes, John Locke, and Jean-Jacques Rousseau were the most prominent proponents of this theory. These thinkers differed from Hobbes in asserting that there are certain irreducible moral notions, based on a natural equality of moral status. Rousseau's *Discourse on Inequality* (1754) affirmed that self-preservation and pity are prior to reason and that the duty not to harm others is based not on rationality but on sentience, the capacity to feel. Thus animals have the natural right not to be mistreated by human beings. Locke, in his *Second Treatise of Government* (1689), saw humankind in the state of nature as inherently social and relatively peaceful, but as prone to conflict over property. Locke's view, in which private ownership of property preceded the social contract, foreshadowed many conflicts between private property and the public good.

In the twentieth century, John Rawls's *A Theory of Justice* (1971) emerged as the most influential statement of social contract theory. Rawls adopts Immanuel Kant's position that persons are able to reason from a universal point of view. Rawls's premise is that principles of justice are those which rational self-interested persons would agree to in an ideally fair situation, which he termed the "original position." Rawls's original position, unlike the state of nature in some earlier contractarian theories, is not supposed to be an actual historical situation, but rather a fiction that can help us think well about justice. Rawls proposed that those in the original position were behind the "veil of ignorance," meaning that they do not know their gender, race, and socioeconomic position in society, their natural talents or endowments, or their own conceptions of the good.

Rawls argued that those in the original position would agree on two principles of distributive justice. The first principle is that each person is to have a right to the most extensive liberties compatible with the same liberties for all others. The second principle is that social and economic inequalities are to be arranged so that they are to the greatest benefit of the least advantaged and attached to offices and positions open to all under conditions of fair equality of opportunity.

Rawls's theory has evoked opposition, due in part to the exclusion of nonhuman animals and the environment from the sphere of those entitled to justice. Tom Regan argues that Rawls's theory is not impartial, since while at key points Rawls indicates that mentally disabled humans, who are relevantly like animals, are owed duties of justice, he maintains that we have only indirect duties to animals. For Regan, the veil of ignorance should veil knowledge of the eventual species of the contractor. Regan is joined in his view by Paola Cavalieri and Will Kymlicka in "Expanding the Social Contract" (1996). Mark Rowlands in "Contractarianism and Animal Rights" (1997) makes a similar point in claiming that rationality is an undeserved quality, just as are race and economic class. Peter Singer, in the revised *Practical Ethics* (1993), argues that social contract theory should be rejected for the above reasons and also for its exclusion of future generations and of poor nations from the sphere of those to whom justice is owed. J. Baird Callicott maintains that positing reason as the origin of the moral community, even in a hypothetical manner, is contradicted by what is known of the evolutionary development of ethics.

Some commentators have offered revisions that could be incorporated into Rawls's theory. For example, the idea of what constitutes the basic goods to which everyone is entitled could be expanded to include clean air and water. Another proposal is to include behind the veil of ignorance to what generation the contractor will in fact belong.

Other revisions involve the basic premises of Rawls's theory. For example, many feminist thinkers reject, as male-biased, the idea of the original position as made up of rational and self-interested humans. They argue on the contrary that humans are fundamentally social, concerned for others as well as themselves, and mutually dependent. Avner de-Shalit points out in *The Environment: Between Theory and Practice* (2000) that Rawls's theory requires those in the original position to be without identity, including without a place in the world and without moral values that affirm the environment.

Many thinkers reject reciprocity between equals as a necessary component of justice. For these thinkers, rather than being limited to agreements between equals, justice is owed to all individuals who have morally weighty interests that are in need of protection. Mary Midgley, in *Animals and Why They Matter* (1983), argues that in many cases, such as those of the mentally handicapped, animals, plants, and rivers, our duties are noncontractual.

Rawls addresses this issue by stating that we have direct moral duties to animals, but these are duties of compassion and humanity rather than duties of justice. Rawls also states that social contract theory does not comprise all of moral theory. Martha Nussbaum and others object that this strategy of separating justice from compassion tends to trivialize animals, mentally disabled humans, and the environment, due to the heavy emphasis on justice in Western moral theory.

In contrast, some have held that nonhuman animals and the natural world are active communicators who are in fact participants in the contract with humanity. Bernard Rollin (1992) argues that domestication is an implicit social contract humans have made with animals, which is violated by factory farming. Michel Serres, in *The Natural Contract* (1995), acknowledges the earth itself as a partner. He interprets a contract in terms of the relationships, the "cords" between humans and the world. For Serres, we live in a play of energies, as participants in our entire community.

SEE ALSO *Callicott, J. Baird; Environmental Citizenship; Environmental Justice; Environmental Politics; Intergenerational Justice; Regan, Tom; Singer, Peter.*

BIBLIOGRAPHY

Cavalieri, Paola, and Will Kymlicka. 1996. "Expanding the Social Contract." *Etica & Animali* 8: 5–33.

Midgley, Mary. 1983. *Animals and Why They Matter*. Athens: University of Georgia Press.

Nussbaum, Martha C. 2004. "Beyond 'Compassion and Humanity.'" *Animal Rights: Current Debates and New Directions*, ed. Cass R. Sunstein and Martha C. Nussbaum. New York: Oxford University Press.

Rawls, John. 1971. *A Theory of Justice*. Cambridge, MA: Belknap Press of Harvard University.

Regan, Tom. 1983. *The Case for Animal Rights*. Berkeley: University of California Press.

Rollins, Bernard. 1992. *Animal Rights and Human Morality*, rev. edition. Buffalo, NY: Prometheus Books.

Rowlands, Mark. 1997. "Contractarianism and Animal Rights." *Journal of Applied Philosophy* 14(3): 235–247.

Seres, Michel. 1995. *The Natural Contract*, trans. Elizabeth MacArthur and William Paulson. Ann Arbor: University of Michigan Press.

Singer, Peter. 1993. *Practical Ethics*, rev. edition. Cambridge, UK: Cambridge University Press.

Thero, Daniel P. 1995. "Rawls and Environmental Ethics." *Environmental Ethics* 17(1): 93–106.

Susan J. Armstrong

SOCIAL ECOLOGY

It is difficult to separate Social Ecology from the work of the social theorist Murray Bookchin. Having elaborated its philosophical foundations in a dozen books and many essays, he is considered by many to be the founder of the field. Bookchin drew on history, anthropology, philosophy, political theory, and ecology to formulate a comprehensive analysis of the relationship between humanity and nature, the causes of the ecological crisis, and the pathways humanity could reinstate to create an ecologically sustainable and just world (Bookchin 1982, 1986, 1995, 2003).

DEFINING THE TRADITION

The Social Ecologist John Clark has argued that although Social Ecology is associated closely with Bookchin's work, it is part of a long philosophical tradition (Clark 1998). Clark creatively traced—or, as Andrew Light (1998) described it, creatively invented—the intellectual tradition that preceded Social Ecology. Among its forerunners was the anarchist Petr Kropotkin with his emphasis on the significance of "mutual aid" among animals and humans and model of a human society organized in communities founded on cooperation and free from central government (Kropotkin 1987). The geographer and anarchist Elisée Reclus, a colleague of Kropotkin's, prefigured ideas of Social Ecology in envisioning the reorganization of societies into communities embedded within their ecological and geographic regions (Clark and Martin 2004). The historian and social theorist Lewis Mumford may be regarded as an early Social Ecologist for his analysis of the way mechanization and domination lead to the dissolution of human ties to the natural world (Mumford 1967). Bookchin contributed the most elaborate articulation of Social Ecology: His argument, which continues to stand as the central pillar of Social Ecology, is that the destruction of nature originated in hierarchical and class-structured social domination among humans (Bookchin 1982).

WHAT IS SOCIAL ECOLOGY?

Clark's description of Social Ecology as "the awakening earth community reflecting on itself, uncovering its history, exploring its present predicament, and contemplating its future" highlights its general threads and calls attention to its predilection for theorizing nature and society as a unity (Clark 1998, p. 416). More specifically, key themes and arguments of Social Ecology include the following:

- Viewing nature and society as emerging through an evolutionary unfolding toward increasing diversity, complexity, freedom, and consciousness by means of processes that foundationally involve interconnection, complementarity, and cooperation;
- Understanding the relationship between nature and society as a holistic unity in diversity and seeking to discover why this relationship has gone awry; regarding social conditions and structures as the causes of the detrimental impact of humanity on nature;
- Critiquing institutionalized forms of dominance, both hierarchical and class-based, not only from a social-justice perspective but also for being causally implicated in ecological destruction;
- Privileging social-structural explanations of ecological disruptions over biological and/or psychospiritual explanatory frameworks such as human population growth and human chauvinism;
- Assessing the capitalist market economy as the major force behind intensifying ecological problems;
- Identifying capitalism as an economy, way of life, and thought style that has colonized every aspect of human life and the natural world;
- Agitating for the revolutionary abolition of all forms of domination rather than seeking reformist solutions to social and ecological problems or encouraging individual spiritual transformation;
- Urging the realization of freedom for both people and nature;
- Providing a vision of the ecological society to counter the dominance of the economism (the hegemony of the market economy) that is destroying the biosphere.

Slum District in Manila, Philippines. *A group of children in the district of Baseco in Manila play near their shanty house, which contrasts sharply with the homes being built by Habitat for Humanity aid workers behind them. According to principles of Social Ecology, ecological problems cannot be separated from social inequality and economic exploitation. In a similar vein, Social Ecologists would say that ecological destruction cannot be evaded until hierarchy and class structure are abolished.* JOEL NITO/AFP/GETTY IMAGES.

According to Social Ecologists, the human-nature relationship is formed through the structural and conceptual relations that predominate in any society. The drive to dominate nature originated in and is perpetuated by the human domination of humans. Social domination is organized along lines of hierarchy and class stratification: Hierarchy involves the valorization and institutionalization of human differences (gender, race, ethnicity, etc.), and class divisions are based on unequal ownership or control of material wealth and means of production. Both forms of domination underlie the destruction of nature, for it is only when differential status, master-servant relationships, and economic exploitation emerge in the social world that human beings direct related ideas and actions to the world of landscapes, animals, and plants. Ecological problems never have been separate from social inequity and economic exploitation, and the ecological crisis cannot be resolved without a revolutionary restructuring of society on the economic, political, cultural, and value levels.

In making the case for the causal primacy of social structure in the way nature is treated, Social Ecologists echo a long-standing sociological predilection for viewing social patterns as being projected onto nonsocial domains, especially the realms of gods and nature. Within any society all its dimensions are aligned through the structural and ideological mappings that the sociologist Max Weber characterized as manifesting "elective affinity." Thus, societies stratified through systems of domination project a blueprint of stratification onto the natural world, representing it as a domain inferior to humanity. That projection makes nature available for many forms of physical domination: destruction of habitats, conversion of ecosystems, ownership of land, exploitation of life forms, and experimentation on animals as well as the overarching constitution of the nonhuman world as a realm for instrumental use.

The upshot of this analysis is that ending ecological destruction hinges on abolishing domination in society. This analysis may explain why the contemporary environmental movement, along with the earlier work of Henry David Thoreau and John Muir, has been unable to turn things around. From a social-ecological standpoint, the creation of an ecological society requires nothing less than the emergence of an emancipated humanity that abandons hierarchical valuations and economic inequalities, charting instead a new historical course for both humans and nature into realms of creativity and freedom.

Although the destruction of nature did not originate with industrialism but has roots in the earliest forms of hierarchy (especially patriarchy), Social Ecologists indict the market economy as the major force behind the ecological crisis. Steven Best stated that for Social Ecology "environmental problems emerge from a long history of hierarchical social relations that culminate in a class-ridden, profit-driven, accumulation-oriented capitalist society" (Best 1998, p. 337). In its addiction to limitless growth the market economy, especially "the horror of economistic-technocratic globalism" (Clark 1998, p. 429) that it has turned into, is jeopardizing the integrity of the biosphere as a whole.

From the point of view of Social Ecology, economic expansionism is leading to the colonization of all worlds: natural, social, cultural, and personal. Economism homogenizes and impoverishes the natural world while degrading human relations and experiences into commodities. Economism also has co-opted the Enlightenment concept of progress as social development that unfolds through competition and expansionism rather than through cooperation and balance. Social Ecologists do not regard the negative impact of industrialism as stemming from either technological development or cultural-ideological contrivances such as commodity fetishism that sustain overproduction but instead from an economic system founded on the "the universal reign of limitless buying and selling, indeed, of limitless growth and expansion" (Bookchin 1986, pp. 28–29). This imperative renders capitalism nearly impervious to ethical considerations and unmasks the idea of "greening capitalism" as an oxymoron if not an Orwellian smoke screen (Bookchin 1993).

Diagnosing socioeconomic forms of domination as the source of ecological destruction presents Social Ecology with the task of envisioning an alternative way of life: the ecological society. The future ecological society is portrayed as organized in ecocommunities that will be egalitarian, democratic and participatory, and semiautonomous but interconnected. Such networked communities will live in balance—both knowledge-based and heartfelt—with their ecological regions. In the ecological society people will integrate ethical considerations into their energy choices, forms of land use, and treatment of animals. Economies will be human-scaled. In the creation and exchange of objects craft will be valued over mass production, durability over constant turnover, and simple lifestyles over consumption (Biehl 1998).

The ethos of the ecological society is envisioned as cooperative with respect to people, animals, and the land. Because cooperative relationships are ontologically primary in evolutionary, ecological, and social processes, the ecological society is conceived as a realizable and actionable vision rather than a utopian will-o'-the-wisp. The creation of a social world rooted in the praxis and ethic of mutualism is theorized as restoring the primal and ever-present, even when repressed and marginalized, ground of being (Clark 1998).

THE COSMOLOGICAL FOUNDATIONS OF SOCIAL ECOLOGY

Even as Social Ecology has problematized material relations within economic, political, and sociocultural systems of domination, it has articulated a cosmological context for the emergence of humanity within an evolving universe and biosphere; the rupture between humanity and nature that has culminated in the ecological crisis is scrutinized in the same context. The materialism of Social Ecology is thus a far cry from the materialism of European and American Marxist and anarchist traditions, which shy away from theorizing the universe at large and the Earth in particular as manifestations of a cosmos of intrinsic integrity, beauty, and order. Bookchin in particular elaborated those cosmological foundations for Social Ecology, relying on dialectical naturalism to represent society as an emergent of nature and redeem the human-nature relationship (in its original unity and future potential) as complementary, harmonious, and mutually supportive. The framework of dialectical naturalism allowed him to tap into an established philosophical tradition while signaling his divergence from Hegel's Christian idealism and Marx's anthropocentric materialism.

Bookchin portrayed natural history as grading into social history without sacrificing the distinctive qualities of either domain. Everything that is characteristically or quintessentially human, from the development of science and technology to the creation of cities, the invention of writing, and the composition of music, has been eons in the making (Bookchin 1993). The peculiar qualities of human beings, such as the capacity for reason and self-consciousness, sophisticated language, the aspiration to freedom, and the power to innovate and intervene, have emerged through and as a form of biological evolution. Social Ecology thus affirms nature and society as a continuum, or a "differentiated unity" (Bookchin 1996): This perspective opposes lifting humanity into a realm above the natural world but also objects to conflating society and nature by discounting the unique features of humanity.

Bookchin coined the terms *first nature* (the natural world) and *second nature* (human society) to underscore their similarity and divergence. The rise of domination within second nature has had dire repercussions for first nature and for society's relationship with it: The schisms and alienation created within society echo a human schism and alienation from the natural world, and the abuses inflicted on people are all easily directed at nonhumans. By overcoming the distortions arising from social domination, a dialectical unfolding can lead humanity and nature into a higher synthesis, a liberated coexistence that Bookchin called free nature (Bender 2003).

In remaining vague about the meaning of free nature, Bookchin skirted an important issue: the substantive contours of the relationship between second nature and first nature in the ecological society. Clark endeavored to correct this by unpacking Bookchin's free nature in connection with the issues of biodiversity and wilderness protection:

> The social-ecological conception of freedom as spontaneous creative order points to the need for a larger sphere of wild nature so that biodiversity can be maintained and evolutionary processes can continue their self-expression, not only in human culture and humanized nature, but also in the natural world substantially free of human influence and control. A Social Ecology therefore implies the necessity not only for wilderness preservation but also for an extensive expansion of wilderness (and relative wilderness) areas where they have been largely destroyed. (Clark 1998, p. 430)

This passage points to the attempt to harmonize free nature with key themes of environmental thought; it also represents an attempt to begin reconciling social-ecological and deep-ecological perspectives.

Human society is born out of the natural world in a biological sense but also in the ancient and originally egalitarian structuring of human relations along kinship lines, a sexual division of labor, and rights and obligations of different age groups (Bookchin 1982). Although all human qualities exist in inchoate form in first nature, second nature has evolved an unparalleled potential aptitude for rationality, self-consciousness, and intentionality. For Bookchin the deepest realization of these abilities requires freedom, but freedom has been stifled by social domination in all its forms: status distinctions, patriarchy, racism, tribalism, economic exploitation, discrimination against minorities, and state oppression. The future of both humanity and the biosphere depends on establishing the freedom within which the highest human qualities can become actualized. Bookchin tends to echo a Nietzschean assessment of human beings as an unrealized but promise-filled species hovering between ape and superman: grown beyond unselfconscious animal but not yet nature rendered self-conscious.

Bookchin attempted to sustain first nature and second nature as a continuum while honoring the qualities that make the human phenomenon unique. By underscoring positive features of humanity such as the celebration of freedom, the cultivation of reason, and the aspiration to self-consciousness, he seems to have had a twofold goal: to preempt a conscious or subterranean response of misanthropy to the ecological damage human beings have wrought and to highlight the human qualities that can be actualized after the relational and mental shackles of domination are undone and the way to the ecological society is opened.

CRITICISMS

The social-ecological preoccupation with human uniqueness has been criticized, especially by Deep Ecologists, as an expression of human chauvinism or anthropocentrism (Bender 2003). Passages in Bookchin's work in which he draws sharp lines between human nature and all other animals invite that critique. To get a theoretically tidy distinction between first nature and second nature, Bookchin tended to oversimplify animal life as one of fixed instincts and genetic programs while exulting humanity as epitomizing the achievements of reasoning, self-consciousness, intentional planning, and behavioral plasticity. However, a dualistic frame of this type is empirically problematic because it ignores advances in behavioral ecology and cognitive ethology that reveal the complexities of animal life and ethically problematic because it underwrites a human-supremacist argument.

The motive behind such dualistic maneuvers is to avoid naturalizing the ecological crisis by anchoring it in biological programs or regarding it as a consequence of Darwinian processes. After hypostatizing the distinction between human beings and animals, Bookchin and other Social Ecologists exorcise terms such as *hierarchy, domination, competition,* and *slavery* from animal relations. When that terminology is applied to the natural world, domination of people and nature (and ultimately human colonization of the biosphere) can appear legitimated as an extension of biological basics. Thus, Bookchin's attempt to distinguish humanity from the rest of the animal kingdom allowed him to pathologize domination as a pure sociocultural phenomenon and at the same time exonerate first nature from the vices of inequity, exploitation, oppression, and subservience.

Although Bookchin's critics often deplore the sharp line he drew between humans and animals, they rarely give him and other Social Ecologists credit for defending the natural world against its historical and recent constructions as blind, mute, cruel, selfish, competitive, and stingy. Countering the ideological fiction of nature as "demonic and hostile" (Bookchin 1986), Bookchin insisted on an understanding of the natural world as creative, pregnant, fecund, participatory, relational, and wondrous (Bookchin 1986, Clark 1998).

Social Ecology propounds a philosophy of all phenomena as interrelated, jointly enhancing, and mutually forming through symbiotic and cooperative processes. Within that framework the destructiveness of domination is conceived of as fortuitous, a possible historical trajectory but not an essential or necessary characteristic of the world. The cosmology of Social Ecology is thus openly spiritual in offering a vision of the good and goodness as primary forces and in affirming evolutionary emergence in the universe and the biosphere as a creative, awe-inspiring process, though not one conceived as either supernaturally designed or with a predetermined telos. As Clark noted about the implications of this cosmology for a possible future, "the ecological society that is the goal of Social Ecology is found to be rooted in the most basic levels of being" (Clark 1998, p. 421). For Social Ecologists, in a just and ecologically harmonious world, humanity will return to a primordial condition, but this will involve the restoration of the original essence and potential of humankind, not the reinstatement of the Stone Age or the Pleistocene.

THE CLASH BETWEEN SOCIAL ECOLOGY AND DEEP ECOLOGY

Despite Bookchin's theoretical brilliance, decades of activism and writing, and comprehensive formulation of Social Ecology, his divisiveness marred his contribution and damaged the ecological cause. His sectarianism manifested itself in his attack on Deep Ecology during the 1980s and 1990s. The ensuing conflict between Deep Ecology and Social Ecology contributed to polarizing the environmental movement into nature and social justice camps, the very schism that Bookchin was striving to supersede (Best 1998, Light 1998, Clark 2000).

Deep Ecology emerged with the work of the Norwegian ecophilosopher Arne Naess and evolved into the Deep Ecology movement through the contributions of American and Australian environmental thinkers and activists. The underlying cause of ecological destruction and of the alienation of the human species from the biosphere is identified by Deep Ecologists as anthropocentrism: the self-exultation of human beings, whose ostensible superiority and perceived entitlements sanction dominion over nature. In the culture at large anthropocentrism manifests itself as the pervasive and often unquestioned belief that nature is a domain to be used, a domain primarily of instrumental value for people.

In an effort to recover traditions of thought and practice that transcend anthropocentrism, foster biospheric egalitarianism, and celebrate the intrinsic value of nature, Deep Ecologists have formulated a syncretic platform that has integrated elements of Western philosophy, scientific ecology, conservation biology, humanistic psychology, primitive ritual, and Eastern spirituality. Those syntheses are used in an attempt to recover a biocentric sensibility that counters the supremacist tendencies of *Homo sapiens* with a vision of harmonious coexistence with all beings. In the idiom of Aldo Leopold, Deep Ecologists agitate for the day when human beings will be "plain members and citizens" rather than conquerors of the land community.

Bookchin disparaged deep-ecological thought as a random collage of ideas that was deplorable in its theoretical incoherence, mystical inclinations, inattention to social-justice issues, and denial of social domination as the root of ecological exploitation and destruction. He considered Deep Ecological philosophy a hodgepodge of "Taoist moods, Buddhist homilies, and New Age platitudes," with Spinoza, Whitehead, and Heidegger thrown into the mix, that displaces the "vigorous reasoning" and "muscularity of thought" of Western philosophical and political traditions (Bookchin 1996, p. 98ff.). Bookchin's criticisms might have inspired a dialogue between Deep Ecology and Social Ecology, but his invective proved more polarizing than constructive. The tendentious tone that Bookchin set can be seen in a deep-ecological backlash that finds little if anything to learn from his ideas.

In focusing on what repelled him in deep-ecological literature—especially appeals to spiritual transformation, invocations of mystical unity, and antirationalism—Bookchin failed to acknowledge important convergences between Deep Ecology and Social Ecology (Clark 2000). Deep Ecology has always stressed the idea that creating a balanced world will necessitate profound changes at the economic, political, cultural, and ideological levels. Social Ecologists similarly insist that real transformation will not occur without reimagining and revolutionizing all dimensions of society. Concurring with social-ecological analyses, Deep Ecologists have not shied away from criticizing capitalist wasteful production as well as the consumer culture that both is spawned by overproduction and feeds it (Roszak, Gomes, and Kanner 1995). Social Ecology and Deep Ecology share a broad vision of the ecological society, highlighting the desirability of local governance and democratic decision making, small-scale economies of production and consumption, community spirit, lifestyles that respect cultural and natural diversity, ecological embeddedness, and care of nonhuman species both for the ways they enhance human life and for their intrinsic value.

The legacy of Social Ecology for the environmental movement and the future directions of social-ecological thought are not known. It is unclear whether Social Ecology will thrive as an ecophilosophy in its own right, whether its insights will be absorbed into new syntheses in environmental thought and activism, or whether the term *Social Ecology* will end up being applied only to analyses narrowly focused on social-justice concerns in environmental and ecological contexts. Murray Bookchin died in 2006. In the years before his death his polemics alienated him from many of his contemporaries. However, the key insights of Social Ecology continue to command attention: Social and ecological problems are inseparable, and social domination has long been implicated in the destruction of the biosphere.

SEE ALSO *Animal Ethics; Biocentrism; Bookchin, Murray; Darwin, Charles; Deep Ecology; Ecological Feminism; Ecology: II. Community Ecology; Environmental Justice; Environmental Philosophy: V. Contemporary Philosophy; Land Ethic; Leopold, Aldo; Naess, Arne.*

BIBLIOGRAPHY

Bender, Frederic L. 2003. *The Culture of Extinction: Toward a Philosophy of Deep Ecology.* Amherst, NY: Humanity Books.

Best, Steven. 1998. "Murray Bookchin's Theory of Social Ecology." *Organization & Environment* 11(3): 334–353.

Biehl, Janet. 1998. *The Politics of Social Ecology: Libertarian Municipalism.* Montreal, Canada, and Buffalo, NY: Black Rose Books.

Bookchin, Murray. 1982. *The Ecology of Freedom: The Emergence and Dissolution of Hierarchy.* Palo Alto, CA: Cheshire Books.

Bookchin, Murray. 1986. *The Modern Crisis.* Philadelphia: New Society Publishers.

Bookchin, Murray. 1996. *The Philosophy of Social Ecology: Essays on Dialectical Naturalism.* 2nd rev. edition. Montreal, Canada: Black Rose Books.

Bookchin, Murray. 1993. "What Is Social Ecology?" In *Environmental Philosophy: From Animal Rights to Radical Ecology*, ed. Michael Zimmerman. Englewood Cliffs, NJ: Prentice Hall. Available from http://www.communalism.org/Archive/4/wise.html

Clark, John. 1995. "Not Deep Apart." *The Trumpeter* 12(2): 98–104.

Clark, John. 1998. "A Social Ecology." In *Environmental Philosophy: From Animal Rights to Radical Ecology,* 2nd edition, ed. M. E. Zimmerman, J. Baird Callicott, John Clark, et al. Upper Saddle River, NJ: Prentice-Hall.

Clark, John, 2000. "How Wide Is Deep Ecology?" In *Beneath the Surface: Critical Essays in the Philosophy of Deep Ecology,* ed. Eric Katz, Andrew Light, and David Rothenberg. Cambridge, MA: MIT Press.

Clark, John P., and Camille Martin, eds. 2004. *Anarchy, Geography, Modernity: The Radical Social Thought of Elisée Reclus.* Lanham, MD: Lexington Books.

Kropotkin, Petr. 1987. *Mutual Aid: A Factor of Evolution.* London: Freedom Press. (Orig. publ. in 1902.)

Light, Andrew. 1998. "Bookchin as/and Social Ecology." In *Social Ecology after Bookchin,* ed. Andrew Light. New York: Guilford Press.

Mumford, Lewis. 1967. *The Myth of the Machine: Technics and Human Development.* London: Secker & Warburg.

Roszak, Theodore; Mary E. Gomes; and Allen D. Kanner, eds. 1995. *Ecopsychology: Restoring the Earth, Healing the Mind.* San Francisco: Sierra Books.

Eileen Crist

SOCIETY FOR CONSERVATION BIOLOGY

The Society for Conservation Biology (SCB), established in 1986, seeks to promote the scientific study of issues pertaining to the loss, maintenance, and restoration of biodiversity. The SCB and its flagship journal, *Conservation Biology,* bring together scientists, scholars, policy makers, and members of nongovernmental organizations who share the goal of protecting and perpetuating the earth's biological diversity. Since its inception the SCB has recognized an essential role for environmental ethics in informing, shaping, and applying the science of conservation biology. Conversely, ideas and insights from conservation biology have contributed to the development of environmental philosophy and ethics.

ORIGINS AND DEVELOPMENT OF CONSERVATION BIOLOGY

The SCB was founded in 1986 in response to the increasingly urgent concern over global threats to biological diversity in the late twentieth century. More broadly, however, the emergence of conservation biology as a new interdisciplinary field reflected long-term trends in conservation science and practice. A concern with biological diversity has deep roots in the worldviews of native cultures around the world; in the scientific tradition of Europe and North America (in the fundamental contributions, for example, of Linnaeus, Charles Darwin, and Alfred Russel Wallace); and in the work of naturalists and protoconservationists of the 1800s (the writings, for example, of Alexander von Humboldt, Henry David Thoreau, and George Perkins Marsh). With the rise of the Progressive-Era conservation movement in the United States in the early 1900s, science became more intimately tied to conservation policy and practice. It was, however, fragmented into varied disciplines, preecological in content and narrowly utilitarian in application. This scientific content both reflected and reinforced the philosophical split in the early conservation movement between the utilitarian resource-conservation ethic (often associated with forester Gifford Pinchot) and the nature-preservation ethic (often associated with naturalist John Muir).

By the 1930s ecological and evolutionary science had begun to influence various resource-management fields (including agriculture, forestry, wildlife management, range management, and fisheries management), easing this long-standing tension within the conservation movement. Early biogeographers and ecologists such as Henry C. Cowles, Frederic Clements, Henry Gleason, Victor Shelford, Charles Elton, and Ernst Mayr developed basic concepts of community ecology, ecological change, population dynamics, and plant and animal distribution, giving greater emphasis to the role of biological diversity in the structure, composition, and function of biotic communities. Aldo Leopold, applying these concepts to land management and stewardship in the 1930s and 1940s, redefined conservation as "a state of health in the land," which he further described as "the capacity for self-renewal in the soils, waters, plants, and animals that collectively comprise the land" (1991, p. 318). In thus recasting conservation's goals—most explicitly in his influential essay "The Land Ethic," published posthumously in the environmental classic *A Sand County Almanac* (1949)—Leopold wedded conservation science and conservation ethics. Leopold's land ethic implied that conservation was no longer just the purview of professional resource managers charged with the efficient production of goods from the earth, but also of individuals and institutions assuming responsibility for the health of the land. A generation later this coupling of science and ethics in the service of an ecologically robust approach to conservation provided an important cornerstone for the field of conservation biology.

In the decades following World War II, the resource-management professions faced mounting environmental and philosophical challenges in the United States and around the world. An expanding and increasingly globalized economy increased the burdens on natural communities through overexploitation; pollution; the spread of invasive species; the early indications of climate change; and widespread habitat loss, alteration, and fragmentation. These environmental changes engendered ever-lengthening lists of threatened and endangered species (later legally defined and protected) and alarm over the loss of biological diversity at various geographical scales. The world's species-rich tropical forests, for example, became a focal point of global environmental concern by the late 1970s. The traditional resource-management fields, with their inherited disciplinary boundary lines, reductionist tendencies, and commodity-dominated priorities, were ill-equipped to address these systemic challenges.

Conservation biology emerged in response to these trends. It was a part of the same process of intellectual cross-pollination that spawned such fields as environmental ethics, environmental history, ecological economics, landscape ecology, agroecology, and restoration ecology in the late 1970s and 1980s. Conservation biology was the product of a fusion of several overlapping spheres of scientific inquiry: coevolution and population biology (as developed in the 1960s by Peter Raven and Paul Ehrlich, among others); island biogeography (grounded in the landmark research of E. O. Wilson and Robert MacArthur); conservation genetics (especially as synthesized by Otto Frankel and Michael Soulé); and inquiry into the social dimensions of conservation policy and international development (through key contributors such as Thomas Lovejoy, Bruce Wilcox, and Norman Myers).

THE FOUNDING OF THE SCB

Conditions were ripe for the creation of the new field and its namesake professional society. A series of scientific workshops and conferences held between 1978 and 1985 gave the field increasing definition. In 1986 a major forum on the conservation of biodiversity—a neologism adopted in the course of planning the forum—was convened in Washington, D.C., under the auspices of the Smithsonian Institution and the U.S. National Academy of Sciences. Many of the forum's scientific leaders helped to establish the new group. In 1987 the SCB, under its first president Michael Soulé, organized its first annual meeting and published the first issue of *Conservation Biology*.

A close and continuing relationship between environmental ethics and conservation biology was forged in these early years of the SCB. The 1986 forum on biodiversity included not only a wide range of scientists but also environmental ethicists and scholars from other fields. Soulé, the central figure in defining conservation biology and organizing the SCB, credited the influential Norwegian environmental ethicist Arne Naess with shaping his priorities for the field; it was to be a "mission-driven" and "value-laden" field that accepted the moral responsibility of humans to safeguard and sustain the community of life. The bylaws required that one seat on the society's board of directors be reserved for a scholar in the humanities and environmental ethics. The first editor of *Conservation Biology*, David Ehrenfeld, had published his own contribution to environmental philosophy, *The Arrogance of Humanism*, in 1981. Beginning with Ehrenfeld's editorship, the journal regularly featured articles on environmental ethics alongside its more customary scientific studies.

Even as conservation biology has sought to unify the disparate domains of natural-resource management, it has also significantly influenced the discourse of environmental ethics. It has raised vital issues such as the human role in shaping "natural" ecosystems; the role of biological diversity in conservation strategy; the movement toward more community-based, participatory approaches to conservation decision making; the philosophical rationales and ecological foundations of sustainability; and the role of the conservation biologist as scientist in an explicitly values-driven undertaking.

SEE ALSO *Biodiversity; Conservation; Conservation Biology; Globalization; Land Ethic; Leopold, Aldo; Naess, Arne; Resource Management.*

BIBLIOGRAPHY

Callicott, J. Baird. 1990. "Whither Conservation Ethics?" *Conservation Biology* 4(1):15–20.

Ehrenfeld, David. 1981. *The Arrogance of Humanism.* New York: Oxford University Press.

Leopold, Aldo. 1991. *The River of the Mother of God and Other Essays by Aldo Leopold,* ed. Susan L. Flader and J. Baird Callicott. Madison: University of Wisconsin Press.

Mayr, Ernst. 1982. *The Growth of Biological Thought: Diversity, Evolution, and Inheritance.* Cambridge, MA: Harvard University Press.

Meine, Curt; Michael Soulé; and Reed F. Noss. 2006. "A Mission-Driven Discipline: The Growth of Conservation Biology." *Conservation Biology* 20(3): 631–651.

Quammen, David. 1996. *The Song of the Dodo: Island Biogeography in an Age of Extinctions.* New York: Simon and Schuster.

Soulé, Michael. 1987. "History of the Society for Conservation Biology: How and Why We Got Here." *Conservation Biology* 1(1): 4–5.

Curt Meine

SOCIOBIOLOGY

SEE *Evolutionary Psychology.*

SOILS

Soil is the portion of the earth's surface that consists of a mixture of disintegrated rock and humus, or dead organic matter. Soil science, a branch of agronomy, has categorized thousands of soil types according to their physical and biochemical characteristics. In addition to its mineral substrate and humus tilth, a soil may contain 30,000 species of organisms, with a thimbleful containing billions of bacteria, fungi, algae, protozoa, and nematodes, plus virus particles. The soil microbiologist Selman Waksman received the Nobel Prize for discovering soil actinobacteria that produce lifesaving antibiotics. Soil may be critical in preventing the spread of antibiotic resistance to preserve the medical value of these soil-borne antibiotics.

VALUE AND DEGRADATION

Soil has both intrinsic and instrumental value, but modern agriculture has allowed soil degradation through erosion and contamination of soils and waters with chemicals. Though considered by some medical authorities to be a pathology, eating soil (geophagy or pica) can improve human or animal health by dissolving micronutrients. The clay mineral surfaces adsorb harmful bacteria, viruses, and toxic organic compounds and are eliminated from the body. Soil also performs a variety of ecosystem services. Soil microbes enzymatically digest the complex organic compounds of dead plants, animals, and other microbes, producing simple inorganic ions by mineralization (rotting or composting). Roots absorb those ions, and the carbon dioxide gas released enables photosynthesis. This action constitutes nutrient cycling in all terrestrial ecosystems, the study of which is a major branch of ecosystem ecology. A wide variety of toxic and hazardous organic compounds are bioremediated into harmless or beneficial inorganic substances by soil microbes. The soil biota provides biological resilience to chemical insults if it is not overwhelmed.

The "spirit of the soil" includes several extrascientific concepts. The idea of native soil has inspired patriotism for centuries, which results from the concept of soil as a symbol for the place of a person's birth and, before world trade and transport, the source of a person's nutrition. Centuries ago the apparently spontaneous appearance of mushrooms caused humans to develop the idea of vitalism, by which the soil was said to transmit a vital force from animals, through their manure, to the soil and then into the newly developing plants, once more providing sustenance for animals. The development of soilless hydroponic horticulture—plants are grown in water and supplied by dissolved inorganic fertilizers (nutrients)—developed by Justus von Leibig disproved the idea of vitalism. However, Wendell Berry and others have suggested that soil nevertheless retains a memory of its past management.

Aldo Leopold proposed a land ethic that would promote biological diversity, ranging from the humble earthworm to all other life-forms, with each one having a right to maintain a habitat within the soil. Leopold's early papers, written while he was a U.S. Forest Service employee in the Southwest, indicate that his land ethic was rooted in concern about the soil erosion caused by the advent of fire suppression and cattle ranching in that region. Farmers holding soil in their hands often have a reverence for the plant and animal productivity developing from soil. The ancient Jewish stewardship environmental ethic emphasizes human responsibility to conserve the soil by periodically resting it because the land or soil belongs to God, not to humans. Thus, humans are caretakers of the earth.

PROBLEMS AND REMEDIES

Contemporary nonsustainable industrialized crop-production practices create serious threats to water quality, in part because they pose serious threats to soil, which modulates the flow of water and purifies it. Those nonsustainable practices also cause soil erosion, which, in addition to the loss of productive, life-sustaining soil, loads streams and reservoirs with sediments and excess nutrients, which also diminish water quality. Phosphates and pesticides move away from fields with the soil into rivers, lakes, and streams, resulting in algal blooms and fish kills or the death of a wider variety of organisms and pose threats to human health. Nitrates and herbicides are moving downward into groundwater through depleted soils, contaminating sources of drinking water.

Because of the loss of soil as a result of the use of modern tillage equipment, soil conservation practices have been a major focus of U.S. public policy. However, growers have ignored these practices in the interest of expediting farming operations and increasing short-term profits. Soil loss of one to five tons per acre per year is considered acceptable, but accelerated soil loss may occur at ten or more times that rate. A sustainable society would ensure that soil is preserved and further soil-related environmental degradation does not continue.

A major controversy surrounds the goal of nutrient replenishment. Although synthetic industrial chemical fertilizers provide the base for modern plant nutrition, more environmentally friendly sources might be reintroduced. Commercial fertilizers require extensive petroleum energy to produce, transport, and apply compared with more sustainable fertilizers, such as animal manures, that are an agricultural by-product that can be produced locally. Also, legumes used in crop rotation provide free low-energy nitrogen not only to themselves but to other plants grown on the same soil. Biodynamic microbial preparations, compost teas (composted material dissolved or suspended in water so that it can be sprayed), and greater use of animal wastes, including human waste, are key ingredients in the recipe for a postindustrial sustainable agriculture. Using more organic materials greatly enhances phosphate (the second most limiting plant nutrient after nitrogen) availability to plants. This is the key to organic agricultural production systems, including long-term sustainable practices to prevent soil and water erosion.

A holistic ecological approach to soil management requires viewing the health of the soil and the conservation of soil as a vital component of human health. Soil fertility promotes regenerative sustainable crop production, the most salient elements of which are biological nitrogen-fixing bacteria functioning inside root nodules of legumes, mycorrhizal fungi living on plant roots that serve as extra root hairs that enhance plant nutrition, and bacteria, fungi, and nematodes serving as natural biological control agents by killing insects and other plant pathogens.

The loss of soil microbial diversity may threaten future generations through loss of soil-based ecosystem services and basic biological processes and through loss of potential lifesaving antibiotics and medicines.

SEE ALSO *Agriculture; Berry, Wendell; Biodiversity; Conservation; Ecosystem Health; Future Generations; Judaism; Land Ethic; Sustainable Agriculture; U.S. Forest Service; Water.*

BIBLIOGRAPHY

Bennett, Hugh Hammond. 1939. *Soil Conservation.* New York, London, McGraw-Hill.

Mahaffee, W. F., and S. Scheuerell. 2006. "Compost Teas: Alternatives to Traditional Biological Control Agents." In *Microbial Ecology of Aerial Plant Surfaces*, eds. M. J. Baily, A. K. Lilley, T. M. Timms-Wilson, and P. T. N. Spencer-Phillips. Oxford, U.K.: CABI.

McNeill, J. R., and V. Winiwarter, V. 2004. "Breaking the Sod: Humankind, History and Soil." *Science* 304(5677): 1627–1629.

McNeill, J. R., and V. Winiwarter, eds. 2006. *Soils and Societies: Perspectives from Environmental History.* Isle of Harris: The White Horse Press.

Thomas A. Ruehr

SONTOKU, NINOMIYA
1787–1856

Ninomiya Sontoku, also known as Ninomiya Kinjiro, a Japanese farmer-sage, was born in a small village in Odawara Han (the contemporary Kanagawa prefecture, near Tokyo) in the Edo era (1603–1867). As an orphan living in poverty, Sontoku restored his father's house and became a landowner, enlarging his estate through hard work and rational management. He served a samurai family and rescued it from debt, after which he was sought out to restore ruined villages. He later was appointed a samurai officer and helped the government restore the Nikko area, where he died and was enshrined.

Sontoku's environmental thoughts and practices were based on the premodern ecological worldview that was characteristic of preindustrial Japanese society; it consisted mainly of Shintoism mixed with Japanese Confucianism and Buddhism. Sontoku believed that all humans and nature exist in a circle within which everything exists in fusion and unity (*ichi-en yu-go*). This is reminiscent of the Great Ultimate of Chu Hsi, the synthesizer of Neo-Confucianism in the Sung Dynasty in China, which also took the form of a circle as the monistic origin or ground from which everything was brought forth. The circle that represents the Confucian unification of humans and nature might be considered a symbol of the ecoholistic view.

The circle expresses the unity of all things. Nothing can be outside the circle, and everything is contained within it. The circle also expresses the limits of humans and nature as a whole, in which reside all dualistic principles such as yin and yang, masculine and feminine, light and dark, life and death, good and evil, and rich and poor. A circle also expresses the idea that everything within the circle will change not in the direction of progress but in a cyclical way. This holistic way of thinking leads to the conclusion that a one-sided increase in the welfare of humans or nature will destroy the balance; that is, an increase in the welfare of humans will lead to a decrease in the welfare of nature.

Sontoku's practical solution to the problem of the relationship of human beings to nature lay in the symbiotic coelaboration of both. Human beings owe gratitude (*on*) to heaven and earth (the great father and mother), to their ancestors, parents, and lords, and are obliged to repay the debt. The main virtues Sontuku practiced and recommended were diligent labor (*kin*), frugality (*ken*), and concession (*jo*) in agriculture and economics to increase natural produce by "assisting the transforming and nourishing process of Heaven and Earth" (Tu 1989, p. 86). This is part of Sontoku's teaching of "reward for virtue" (*hotoku*).

Sontoku eventually rescued more than six hundred villages and tens of thousands of people. He not only restored devastated farms but also saved people from mental collapse by helping them be financially and morally independent. To help distressed people, he used large sums of money from his successful projects, leaving none for his own family. The voluntary credit union organized by his followers, the Society for Rewarding Virtue (*Hotoku-sha*), was highly successful, with many branches throughout Japan; it continues to exist. Sontoku's achievements testified to his belief that Confucian morals (*'jen*), politics, and economics rather than Western power politics and self-interested economics can both make people happy and restore nature.

In prewar Japan, Sontoku was a national hero who was cited in elementary school textbooks on moral education called *shushin* ("cultivating oneself") as a model of Confucian and other virtues. All elementary schools had a bronze statue of him on a pedestal near the main entrance in which he carried on his back a bundle of firewood gathered in the mountains, reading a book while walking. Boys and girls were encouraged to work hard and study diligently so that they would become decent citizens and successful people. Thus, his influence was felt not only in agriculture and economics but also in moral education in Confucian virtues. In postwar Japan, he was neglected and his school of thought was almost forgotten. However, his thoughts and practices have been revived by the International Ninomiya Sontoku Association, which was founded in 2003.

SEE ALSO *Confucianism; Japan.*

BIBLIOGRAPHY

Sontoku, Ninomiya. 1970. *Sage Ninomiya's Evening Talks*, trans. Isoh Yamagata. Westport, CT: Greenwood Press.

Tu, Weiming. 1989. *Centrality and Commonality: An Essay on Confucian Religiousness*. Albany: State University of New York Press.

Uchimura, Kanzo. 1908. *Representative Men of Japan: Essays.* Tokyo, Japan: Keiseisha.

Yamauchi, T. 2004. "Sontoku's Environmental Ethics." In *Hotokugaku*. International Ninomiya Sontoku Association.

T. Yamauchi

SOUTH AMERICA

The cultural and biogeographic identity of South America, as well as the history of its environmental philosophy, is embodied in the Andes Cordillera, the great mountain system that crosses the continent from south to north and is home to the emblematic Andean Condor. According to Tiahuanaco pre-Inca culture, in ancestral times *Viracocha* (one of the most important deities in the Inca pantheon) emerged from Lake Titicaca in the heights of the Andes and created the sun with his light; the rain and water with his tears; the heavens, the stars, the humans and the other living beings that inhabited the region (Kusch 1962). Today, along this Andean backbone, there is a mosaic of altiplanic, rain-forest, wetland, desert, coastal, glacial, steppe, and prairie ecosystems hosting most of the world's plant and animal biodiversity (Mittermeier et al. 2003). The highest mountain peaks in the Americas, as well as the vast Amazonian basin, Pantanal wetlands, Chaco savannas, and Patagonian high latitudes

Gateway of the Sun. *The figure of* Viracocha *in the center of the Sun Gate in Tiahuanaco in the highlands of Bolivia, surrounded by human-condor guardians illustrates how deities, humans, and nature have been and are still united in Amerindians' worldviews and lives.* © TINA MANLEY/ARCHEOLOGY/ALAMY.

are still inhabited by endemic biological species, cultures, and languages. Amerindian people have coevolved with each of these unique and diverse landscapes, ecosystems, and biota, developing a plethora of environmental worldviews and practices that have come under consideration by South American environmental philosophers since the 1980s (Rozzi 2001).

ENVIRONMENTAL ISSUES

The principal environmental concern in South America is the threat to this, the world's treasure trove of biodiversity. The drivers of biodiversity destruction and losses of cultural diversity in South America are many (Primack et al. 2006). Global climate change is provoking a rapid retreat of high Andean glaciers. Sine 1970, glaciers in the Andes have lost 20 percent of their volume, with drastic water supply consequences that are affecting mountain communities, agriculture, and ecosystem integrity. Ozone depletion in the Earth's stratosphere has its strongest expression in the Antarctic ozone hole. At the beginning of the 2000s, in this area of the Antarctic stratosphere ozone levels have dropped 33 percent of their pre-1975 values, with a variety of human health and ecological consequences, such as increases in skin cancer, damage to plants, and reduction of plankton populations and photosynthetic capacity. Exotic invasive species represent another major threat to South American biodiversity. Exotic mammals (such as feral goats, European rabbits, or North American beavers and minks) are especially harmful in archipelagoes such as Galapagos, Juan Fernandez, and Cape Horn. Exotic predatory fish, such as trout and salmon, have a high impact on Andean and temperate freshwater ecosystems.

Mining is also a main driver of rapid and intensive changes in freshwater, marine, and terrestrial biological diversity, as well as cultural diversity involving displacement of communities from their ancestral territories, and destruction of their habitats. In addition to direct habitat destruction, mining frequently involves pollution. For example, mercury pollution caused by the amalgamation of gold in tropical regions such as the Amazon affects the health of aquatic invertebrates, fish, and humans downstream from gold-mining activities. Dams and construction of waterways represent a frequent source of social conflicts and environmental impacts. One of the most debated mega-projects in South America is the Hidrovia, in which the Paraguay-Parana River would be dredged to let large ships carry cargo from Buenos Aires on the Argentinean coast 3000 km north to Bolivia, Paraguay, and Brazil. This project could cause significant drainage of the Pantanal, the world's largest wetland, which is the habitat of endangered jaguars, giant otters, thousands of invertebrates, and tens of Indian tribes. The latter have joined scientists, artists, humanists, and numerous non-governmental organizations (NGOs) in their protest. This type of mega-project often involves serious and rapid environmental degradation, but they have a widespread presence throughout South American geography and post-Columbian history. Currently, extensive monospecific plantations of eucalyptus are replacing native forests in Colombia, southern Brazil, Uruguay, and Chile, and large-scale plantations of genetically modified sugarcane and soybean are replacing vast areas of tropical forests. At the same time, tropical and temperate coastal marine biodiversity of South America is threatened by shrimp and salmon aquaculture, respectively.

As the process of globalization accelerated over the last quarter of the twentieth century, South America's agriculture industrialized—one element of which is an economy of scale—forcing subsistence farmers off their smallholds, for which they had no legal title. As the twenty-first century unfolds, South America not only continues to supply a hungry world with mineral, wood, and food resources, but also an energy-starved world with biofuels—more especially ethanol derived from sugar cane, further exacerbating social dislocation and environmental degradation. In spite of this pressing scenario, South America is still the home of vigorous peasant, indigenous, and diverse rural and urban communities, who value and defend their biological and cultural heritage. Thus South American environmental philosophy has from its start integrated social and often political analysis with environmental concerns. Today, they are frequently involved in conservation, ecological and social justice, and sustainable development initiatives.

SOUTH AMERICAN ENVIRONMENTAL PHILOSOPHY

There are two main sources of South American environmental philosophy (Rozzi 2001). The first is rooted in the ancient ethos and biocultural landscapes of Amerindian people, as well as African American, old peasant, and contemporary rural and urban communities. Since the 1960s these rich biocultural landscapes have been increasingly studied and valued by academics through historical and critical thinking (Mignolo 1995, Escobar 1996), liberation philosophy and theology (Boff 1995), ecofeminism (Parentelli 1996, Gebara 1999), and biocultural conservation (Rozzi 1997, 2001). The second source is the incorporation of international environmental thinking and philosophy in South American universities. This trend was first sparked in the 1970s by the United Nations Environment Programme that focused on sustainable development and interdisciplinary education (PNUMA/UNESCO 1985, Carrizosa 2001, Leff 2002, Porto-Gonçalves 2006). Second, in the 1990s, the spontaneous interest of South American scholars initiated the discussion and translation into Spanish the work of environmental philosophers in Europe, Australia and the United States, thereby introducing Deep Ecology, Social Ecology, ecocentric ethics, and animal liberation into Latin America (Sosa 1990; Bugallo 1995; Kwiatkowska and Issa 1998; Valdés 2004; Rozzi 1997, 2007).

Roots of South American Environmental Philosophy The Argentine philosopher Rodolfo Kusch was a pioneer in understanding the links between regional landscapes and Amerindian cultures in the Andean region. He realized that no genuine philosophy in the Americas can be conceived without incorporating the Amerindian cultures (1962). Kusch's efforts pitted him against the almost determined ignorance of indigenous thought and lives in South American academic philosophy.

In the1960s Kusch began to develop studies in comparative ethno-philosophy while working in northern Argentina at the University of Salta. Kusch was interested in learning how much of the Inca legacy persisted in twentieth-century peasant communities in Bolivia and northwest Argentina. In "Geoculture of the American Man" (1976) Kusch coined the term *geoculture*; thanks to the insights yielded by this concept, South American geography was no longer seen merely through "colonial lenses" as a virgin territory to be conquered and used. Instead, it was understood as a source of cultural meanings. Kusch disclosed the embeddedness of various South American ethos in the environment, "always situated, always grounded" (Gutierrez 2008, p. 2). Kusch's philosophical inquiry was motivated by a need, felt by Latin-American intellectuals since the second half of the nineteenth century, to discover or renew the cultural identity of the Americas (Mignolo 1995). The Andean biocultural geography inspired this search for identity in the southern cone of South America.

Decolonization Research Program and Environmental Philosophy In the 1980s the Argentine-Mexican philosopher Enrique Dussel developed the notion of transmodernity. This notion, combined with Kusch's geocultural approach, inspired another Argentine philosopher, Walter Mignolo, to develop the notions of *border thinking,*

An Indigenous Community in Bolivia. *Indigenous Quechua and Aymara communities in the high Andes have many celebrations and rituals to pay (*pagos*) and thank (*despachos*) the Earth or* Pachamama. *These rituals seek renovation and purification, fostering reciprocity among people, the land, and deities. The picture illustrates a celebration at Sajama National Park in Bolivia.* PHOTO BY JUERGEN CZERWENKA. COURTESY OF RICARDO ROZZI.

border epistemology, and *pluritopic hermeneutics* (Mignolo 1995). Dussel's liberation philosophy aims to overcome eurocentric modernity not simply by negating it "but by thinking it from the underside, from the perspective of the excluded other": the colonized indigenous people, poor peasant communities, and urban marginal citizens and workers (1996, p. 14).

At Duke University Mignolo has further developed this project through the Latin American Modernity/Coloniality Research Program (LAMCRP). He affirms "the 'West and the rest,' in Huntington's phrase, provides the model to overcome, as the 'rest' ... emerges in its diversity" (Mignolo 2000, p. 310). In contradistinction to eurocentric abstract universals, the alternative proposed by Mignolo is a kind of border thinking that "engages the colonialism of Western epistemology from the perspective of epistemic forces that have been turned into subaltern (traditional, folkloric, religious, emotional, etc.) forms of knowledge" (Mignolo 2001, p. 11). He emphasizes the need to permit the expression of "pluriversal" epistemologies and local histories and communities that exist at the borders of globalization (Mignolo 1995).

At the University of North Carolina at Chapel Hill, another LAMCRP researcher, the Colombian philosopher Arturo Escobar, has developed a geopolitical perspective by working closely with African American communities in the Colombian Pacific on projects about globalization, culture, women, environment, and place. In these communities Escobar finds powerful elements of ecological sustainability in the reinterpretation of anthropological practices related to mythical and symbolic traditions and ecosystemic contexts (Noguera 2007). However, growing violence, poverty, and environmental and social degradation in Latin America are replacing these realities. In "Invention of the Third World," Escobar affirms that "it suffices to take a quick look to the biophysical, economic, and cultural landscapes of the Third World to realize that the Development Project is in crisis" (1996, p. 9). Against this background Escobar calls for a postdevelopment era, which emphasizes local economies rather than global markets and consumerist lifestyles.

Liberation Theology and Environmental Philosophy In contrast to Escobar, Leonardo Boff, a Brazilian liberation theologian and leading environmental philosopher, asserts in his landmark book *Ecology and Liberation: A New Paradigm* (1995), that "what is today in crisis is not so much the development model, but the model of society that dominates the world" (p. 31). He proposes a holistic, ecosocial approach to environmental ethics, affirming that "the new model of society has to aim at a reconstruction of the social fabric, starting from the multiform potentialities of human beings and society" (1995, p. 36). At the same time Boff calls for broadening the spectrum of environmental ethics to encompass a notion of environmental justice that would incorporate the interests of poor people. In *Cry of the Earth, Cry of the Poor* (1997), he situates the social and political concerns of liberation theology in broader ecological frameworks: "We need to refine the concept of ecological justice, but without a minimum of social justice it is impossible to make ecological justice fully effective. The one involves the other" (p. 45). Boff's concept of ecological justice demands an overcoming of both anthropocentrism and ethnocentrism; social well-being requires consideration of biological and ecological diversity as much as of social classes, native languages, worldviews, and cultural practices.

Boff demands integration at multiple levels of relations, repeatedly calling for an interior ecology (psychological, spiritual) and a reconnection with the earth as a whole, a *dignitas terrae.* He highlights the importance of revering the masculine and feminine, a position that echoes the philosophy of leading South American ecofeminists and liberation theologians Gladys Parentelli (Uruguay and Venezuela) and Ivone Gebara (Brazil). Based on their work with poor women, Parentelli (1996) and Gebara (1999) have inaugurated a Latin American theology from the "optic" of women, pointing out that poverty is not a gender-neutral category. Frequently, poor women are victims of physical and sexual violence; they also lack control over maternity and often are left to provide the primary economic support for their children. Since the 1980s ecofeminists have created new networks and journals that explore the relationship between the oppression of women, indigenous people, and nature in Latin America. South American ecofeminists have called attention to the daily lives of women in slums, showing the ways that the exclusion of the poor is linked to the destruction of their lands. This approach concurs with the perspective that in Latin America the most negative impacts of environmental degradation affect poor people; they are the main victims, not the agents of most degradation (Rozzi 2001).

Biocultural Conservation and Environmental Philosophy Working with indigenous communities in southern South America, the Chilean ecologist and philosopher Ricardo Rozzi has integrated ecological sciences and environmental philosophy. He has developed epistemological and ethical frameworks that are designed to gain a better understanding of the differences and similarities between contemporary scientific knowledge and indigenous ecological knowledge. His work has two main goals: (1) enhancing communication and respect among different sociocultural actors and (2) promoting biocultural conservation. In South America numerous case studies show that indigenous and other local communities agree with scientists and other scholars that, where biodiversity has been protected, local communities enjoy higher levels of autonomy and social well-being (Primack et al. 2006). This convergence between indigenous and scientific views is also supported by the concept of "environmentalism of the poor" developed in South America by the Spanish ecological economist Juan Martinez Alier (2002). Likewise, Rozzi (2001) has called for a "field environmental philosophy" approach whereby philosophers participate in long-term interdisciplinary biocultural conservation projects that involve "direct encounters" with human and nonhuman beings living in their habitats. This field environmental philosophy approach aims to prevent the imposition of global epistemological and development models and to promote instead the expression of diverse ethos and traditional ecological knowledge by local communities.

ENVIRONMENTAL PHILOSOPHY AND SOUTH AMERICAN ACADEMIA

There are two main sources of the influence that environmental philosophy has come to have in South American academia: (1) UNEP's support for environmental academic programs and (2) the efforts of individual scholars.

United Nations Environment Programme (UNEP) At the beginning of the 1970s, the United Nations launched two important programs that promoted environmental thought in South American universities. In 1970 the United Nations Educational, Scientific, and Cultural Organization (UNESCO) created the Man and Biosphere Programme (MAB), which "proposes an interdisciplinary research agenda and capacity building aiming to improve the relationship of people with their environment globally" (UNESCO 2008). In 1972 UNEP was founded, and it immediately proposed to establish "an international program on interdisciplinary environmental formal and informal education" (PNUMA/UNESCO 1985).

In 1977 UNESCO and UNEP organized the International Conference on Environmental Education in Tbilisi, Russia; the conference issued a call for each continent to establish a regional network devoted to environmental thought and education. The most successful such undertaking was the Latin American and Caribbean network, which was consolidated in 1985 at the University and Environment Conference held at the Universidad Nacional de Colombia (UNC). Three central concepts emerged from this meeting: (1) The environment includes not only biotic-physical elements but also sociocultural ones; (2) environmental problems are associated with human development; and (3) university education requires interdisciplinary approaches to address the interactions among nature, technology, and society (PNUMA/UNESCO 1985).

The University and Environment Conference triggered the creation of the Institute on Environmental Studies (IDEA) at UNC, which in 1987 gave rise to the *Pensamiento Ambiental* (Environmental Thought) working group. Since the 1990s this group has researched the relationships between environmental ethics, epistemology, and politics, questioning the technocratic character that prevails in public administration and environmental sciences (Noguera 2005). Since 2000 IDEA has collaborated with the Mexican environmental economist Enrique Leff in publishing the *UNEP Series on Environmental Thought*, an essential bibliographic source on South American environmental philosophy.

Among the philosophers from IDEA, Augusto Angel-Maya stands out for having pioneered an influential school of environmental thought in Colombia. He criticizes the rationalist tradition of philosophy that separates humans from nature. Angel-Maya affirms that "Platonism has drowned Western philosophy" (2002, p. 85). Angel-Maya urges philosophers to turn away from Platonic metaphysics by rediscovering the work of the Ionian pre-Socratic philosophers, with their immersion in the multidimensional complexities of immanent, here-and-now reality, a task he believes is best accomplished through interdisciplinary approaches.

Influences of Anglo-Saxon Environmental Philosophy As an academic subdiscipline, environmental ethics emerged during the early 1970s, mainly in U.S., British, and Australian universities (Callicott and da Rocha 1996). Since the 1990s a few South American scholars have begun to translate and discuss Anglo-Saxon environmental philosophers. In Argentina Alicia Bugallo did extensive research in Deep Ecology and published *De Dioses, Pensadores, y Ecologistas* (Of Gods, Thinkers, and Ecologists) (1995). In Uruguay Eduardo Gudynas introduced the work of the American anarchist ecologist Murray Bookchin and created the Latin American Center of Social Ecology in 1989. In Brazil Sonia Felipe has adapted the work of the animal-liberation philosophers Peter Singer and Tom Regan, participating in the journal *Revista Brasileira de Direito Animal* (Brazilian Animal Rights Review). In Chile Ricardo Rozzi has worked on ecological ethics and introduced the work of Baird Callicott and Eugene Hargrove through a collection of articles in the journal *Ambiente y Desarrollo* since 1996. To enhance the dialogue between South American and Anglo-Saxon environmental philosophers, the University of North Texas and Chilean universities have collaborated on a number of programs in environmental philosophy and biocultural conservation, including a series of occasional papers published online by the International Society of Environmental Ethics (Rozzi 2007).

Social Movements and Environmental Philosophy Since 2000 social movements have played an increasingly influential role in South American environmental philosophy. "Another world is possible" is the slogan of the World Social Forum (WSF) first held in Porto Alegre, Brazil, in 2001. The WSF has brought together countless social movements and leading environmental philosophers and scholars, helping to forge an approach to environmental philosophy that builds on the knowledge of indigenous, peasant, and other subaltern or minority groups to envision other ways of coexisting with a multiplicity of human and nonhuman beings. In the words of the "Manifesto for Life: Ethics for Sustainability," "The ethic for building a sustainable society leads to an emancipation process which, as Paulo Freire taught, recognizes that no one frees anyone else and no one frees himself alone; human beings are only freed in communion" (in Reichman 2004, p. 18). This manifesto, prepared by 35 distinguished Latin American scholars, was signed during the Thirteenth Forum of Ministers of the Environment of Latin America held in Rio de Janeiro, Brazil, in October 2001 and ratified by leading South American environmental philosophers at the Symposium on Environmental Ethics and Sustainable Development in Bogota, Colombia, in 2002; it suggests that a future of South American environmental philosophy is oriented toward a dialogue among the multiplicity of human and nonhuman forms of life.

SEE ALSO *Antarctica; Biocultural and Linguistic Diversity; Biodiversity; Caribbean; Dams; Exotic Species; Fish Farming; Forests; Global Climate Change; Invasive Species; Mining: I. Overview; Ozone Depletion; Traditional Ecological Knowledge; United Nations Environment Programme.*

BIBLIOGRAPHY

Angel-Maya, Augusto. 2002. *El Retorno de Icaro*. Bogota, Colombia: PNUD.

Boff, Leonardo. 1995. *Ecology and Liberation: A New Paradigm*. New York: Orbis Books.

Boff, Leonardo. 1997. *Cry of the Earth, Cry of the Poor*. New York: Orbis Books.

Bugallo, Alicia. 1995. *De Dioses, Pensadores, y Ecologistas.* Buenos Aires, Argentina: Grupo Editor Latinoamericano.

Callicott, J. Baird, and Fernando J. R. da Rocha. 1996. *Earth Summit Ethics: Toward a Reconstructive Postmodern Environmental Philosophy on the Atlantic Rim.* Albany, NY: SUNY Press.

Carrizosa, Julio. 2001. *Que es Ambientalismo?* Bogota, Colombia: IDEA-PNUMA.

Dussel, Enrique. 1996. *The Underside of Modernity: Apel, Ricoeur, Rorty, Taylor, and the Philosophy of Liberation*, ed. and trans. Eduardo Mendieta. New York: Humanity Books.

Escobar, Arturo. 1996. *Encountering Development: The Making and Unmaking of the Third World.* Princeton, NJ: Princeton University Press.

Gebara, Ivonne. 1999. *Longing for Running Water: Ecofeminism and Liberation.* Minneapolis, MN: Augsburg Fortress Press.

Gutierrez, Daniel E. 2008. "Environmental Thought in Argentina: A Panoramic View." South American Philosophy Section, International Society of Environmental Ethics Occasional Papers 3. http://www.cep.unt.edu/iseepapers/gutierrez-engl.pdf

Kusch, Rodolfo. 1962. *América Profunda.* Buenos Aires: Hachette.

Kusch, Rodolfo. 1976. *Geocultura del Hombre Americano.* San Antonio de Padua, Argentina: Editorial Castañeda.

Kwiatkowska, Teresa, and Jorge Issa, eds. 1998. *Los Caminos de la Ética Ambiental: Una Antología de Textos Contemporáneos.* Mexico City: Plaza y Valdés.

Leff, Enrique. 2002. *Etica, Vida, Sustentabilidad.* Bogota, Colombia: IDEA-PNUMA.

Martinez-Alier, Joan. 2002.*The Environmentalism of the Poor.* Cheltenham, UK: Edward Elgar.

Mignolo, Walter. 1995. *The Darker Side of the Renaissance: Literacy, Territoriality, and Colonization.* Ann Arbor: University of Michigan Press.

Mignolo, Walter. 2000. *Local Histories/Global Designs.* Princeton: Princeton University Press.

Mignolo, Walter. 2001. "Local Histories and Global Designs: An Interview with Walter Mignolo," *Discourse* 22(3): 7–33.

Mittermeier, Russell A., Cristina Goettsch Mittermeier, Patricio Robles Gil, and John Pilgrim 2003. *Wilderness: Earth's Last Wild Places.* Arlington, VA: Conservation International.

Noguera, Ana Patricia. 2005. "Editorial." *Ideas Ambientales* 1: 2–3.

Noguera, Ana Patricia. 2007. "Horizons of Environmental Ethics in Colombia: From Anthropocentric Environmental Ethics to Complex Environmental Ethics." International Society for Environmental Ethics Occasional Papers 2. http://www.cep.unt.edu/iseepapers/noguera-eng.pdf

Parentelli, Gladys. 1996. "Latin America's Poor Women." In *Women Healing Earth: Third World Women on Ecology, Feminism, and Religion,* ed. R. R Ruether. Maryknoll, NY: Orbis Books.

PNUMA/UNESCO. 1985. *Universidad y Medio Ambiente En America Latina y EL Caribe.* Bogota: ICFES.

Porto-Gonçalves, Carlos Walter. 2006. El Desafío Ambiental. PNUMA, Mexico City, Mexico.

Primack; Richard; Ricardo Rozzi; Peter Feinsinger; Rodolfo Dirzo; and Francisca Massardo. 2006. *Fundamentos de Conservación Biológica: Perspectivas Latinoamericanas.* Mexico City: Fondo de Cultura Económica.

Reichman, Jorge. 2004. *Etica Ecologica.* Madrid: Nordan.

Rozzi, Ricardo. 1997. "Hacia una Superación de la Dicotomía Antropecentrismo/Biocentrismo." *Ambiente y Desarollo* XIII (3): 48–58.

Rozzi, Ricardo. 2001. "Éticas Ambientales Latinoamericanas: Raíces y Ramas." In *Fundamentos de Conservación Biológica: Perspectivas Latinoamericanas,* eds. R. Primack; R. Rozzi; P. Feinsinger; R. Dirzo; and F. Massardo. Mexico City: Fondo de Cultura Económica.

Rozzi, R. 2007. *Environmental Ethics: South American Roots and Branches.* International Society for Environmental Ethics Occasional Papers 1. http://www.cep.unt.edu/iseepapers/introduction.pdf

Sosa, Nicholas. 1990. *Etica Ecológica.* Madrid, España: Libertaria.

UNESCO. 2008. Man and Biosphere Programme. http://www.unesco.org/mab/mabProg.shtml

Valdés, Margarita. 2004. *Naturaleza y Valor: Una Aproximación a la Ética Ambiental.* Mexico City: Fondo de Cultura Económica.

Ricardo Rozzi

SOUTHEAST ASIA

Southeast Asia forms a major part of the Indomalayan bioregion. Myanmar (Burma), Thailand, peninsular Malaysia, Singapore, Laos, Cambodia, and Vietnam belong to the Asian mainland, and have a biota that is similar to that of the Indian subcontinent. The insular Sabah and Sarawak (Malaysia), Brunei Darussalam, the Philippines, Indonesia, Timor-Leste, and Papua New Guinea make up most of the great Malay Archipelago, linking Indomalaya to the Australian bioregion. Southeast Asia is also usually considered to include the Andaman and Nicobar islands (India), and the Australian islands of Christmas and Cocos (Keeling).

POPULATION AND POLITICS

Geologically complex and biodiversity rich, this region is home to more than 8 percent of the world's human population and has a land area of about 5 million square kilometers (comparable to that of the European Union). With an average of about 125 people per square kilometer, its population density is similar to that of China and Europe, less than half that of India but four times greater than that of the United States. Although the two major divisions (mainland and archipelago) have many biotic and ethnic connections, the differences give rise to a variety of environmental challenges compounded by cultural diversity and a wide range of political systems: a military dictatorship (Myanmar), a communist state (Laos), a socialist republic (Vietnam), democracies (constitutional monarchies in Thailand, Malaysia, Cambodia, and Papua New Guinea; and presidencies in Singapore, Philippines, Indonesia, and Timor-Leste), and a hereditary kingdom (Brunei). There are also pockets of unassimilated people, such as the Mangyan tribes of Mindoro (Philippines) and the totally separate Sentinelese (Andaman Islands). The recent histories of these states offer remarkable contrasts, ranging from stability to violent struggle (internal and external), including ecocide and even genocide—despite which there is evidence that both people and nature can achieve significant recovery. Most of these countries now belong to the Association of Southeast Asian Nations (ASEAN).

Although all major nations in the region are designated as "developing countries" (in terms of gross national product and population, some have had dramatic growth rates over recent decades), there are wide variations in human wealth and well-being and associated environmental problems. Singapore has attained a European standard of living, whereas in Myanmar, beyond the ruling elite poverty is widespread. Many Javanese are prosperous compared to the rural poor on other Indonesian islands. Although most areas are safe, the islands of the north Celebes Sea (southwestern Philippines) harbor many pirates and dissidents. Social unrest occurs locally, including southern Thailand, Ambon (Indonesia), Kalimantan (transmigrant and Dayak conflicts), Myanmar, and Timor-Leste (when it achieved nationhood in 2002, by some accounts it was the poorest country in the world), and many of these struggles seem likely to continue.

ENVIRONMENTAL CHALLENGES

The region faces many environmental problems. The biodiversity of Neomalaya (Malay Peninsula, Sumatra, and Borneo) has been greatly influenced by sea-level change over the last 2.5 million years, and the projected ocean rise associated with global warming would have a major impact on the densely populated coastal areas of Southeast Asia. Although much of the region has abundant rainfall, access to freshwater is threatened by climate change, agribusiness, and pollution. Loss of Tibetan glacial feed water for the great rivers of Indochina, including the Irrawaddy, the Mekong, and the Red River, will have massive impacts on humans and wildlife alike. Local rainfall patterns are often seasonal, as in central Philippines and Nusa Tenggara (Indonesia), with the effects of climate change difficult to predict. Inshore marine pollution, such as eutrophication caused by conversion of mangroves for aquaculture, is extensive. Since 1975 about half the mangrove of the Irrawaddy Delta has been cut, and the exceptional flooding caused by Cyclone Nargis in 2008 was at least partly due to this rapid and uncontrolled change.

Many areas have been affected by smoke drifting hundreds of kilometers when peat and forest fires burn out of control. In Kalimantan the so-called "El Niño fires" of 1982–83 destroyed more than 3.5 million hectares of degraded forest, and at least 200,000 additional hectares burned in 1998. These huge fires cause serious health problems, especially in megacities (where the smoke combines with vehicle and factory fumes), and economic loss (including reduced crop growth). Controversy surrounds their cause: Developers blame climate change, whereas environmentalists argue that a mixture of local uncontrolled and internationally financed forest clearance is to blame.

If all major regional forests are destroyed, these fires will eventually come to an end. But long before that point is reached, the impacts of forest destruction on local and global climate regimes, nutrient cycles, erosion, and sediment discharge into the sea are likely to be profound. Following deforestation many local soils rapidly become infertile and are abandoned, after which they are often colonized by dense siliceous grasses (notably "lalang," *Imperata cylindrica*) that inhibit natural tree regeneration. Former forest becomes anthropogenic grassland of very limited utility and greatly reduced biodiversity—as evident in the central Philippines.

UNIQUE BIODIVERSITY UNDER THREAT

These issues raise an acute challenge to the region's biodiversity. The impact of wholesale forest destruction will be massive because the region is one of the greatest biodiversity hot spots in the world. Indonesia covers less than 1.5 percent of the Earth's land surface but supports about 17 percent of all species. If there are roughly 15 million species worldwide, then probably some 2.5 million exist in Indonesia alone. Assuming a conservative 10-percent level of endemism, this means that the area is home to 250,000 species found nowhere else. For Southeast Asia as a whole, the number of endemics probably far exceeds 1 million, most of which depend on closed forest. New Guinea has the third-largest rain forest on Earth, but pressures for timber extraction and land conversion mount daily.

Myanmar, Thailand, the Malay Peninsula, and New Guinea are probably the most species-rich areas of Southeast Asia, closely followed by Vietnam and the islands of Borneo and Sumatra. If one considers marine biodiversity, the Malay Archipelago is very likely the richest on Earth, with one-third of the world's coastline and coral reefs. Nowhere is this biodiversity homogeneous—every mountain range, valley system, and shallow sea harbors a unique mix of species and, in many cases, unique endemics as well. This lack of homogeneity is shown most dramatically in the suture zone between the Asian and Australian tectonic plates occupied by the Philippines, Sulawesi, Maluku, and Nusa Tenggara (Indonesia). Biodiversity measured as species richness per unit area is lower here, but many of the species are unique to local areas. The Philippine islands, for example, are divisible (on the co-occurrence of endemic species) into six subregions, four of which make up the Philippines "proper," with many thousands of species of organisms found only there. It seems certain that many of these have gone extinct since Magellan arrived in 1521, with the rate of disappearance accelerating sharply since 1900. Although it is a single landmass, Sulawesi offers a comparable picture of relatively low richness coupled with high endemism, including endemics restricted to particular parts of the island. To a lesser extent the same is true of northern and central Maluku, where islands such as Buru, Obi, and Seram all have a good number of endemics. The island of New Guinea (politically divided between Indonesia and Papua New Guinea) is another area of exceptional endemism.

This very high rate of change in species composition across habitats and whole regions (beta and gamma diversity, respectively) is a major challenge for conservation, especially given the wholesale conversion of lowlands for cash-crop monocultures and lumber, displacement of subsistence agriculture into montane regions, mechanized and illegal fishing, and capture of mountain water for irrigation. To conserve the region's biodiversity requires, *in extremis*, what has been called a "structural solution"; actions are required that are unique to every local ecosystem coupled with general restraint to prevent wholesale conversion to anthropogenic landscapes (already far progressed in many areas such as the Philippines, which has only 7 percent of its original forest cover left). In the archipelago in particular, because of the vast number of medium and small islands, where there is a strong tendency toward the evolution of local forms and subspecies, the threat to such "evolutionary significant units" is massive. There is every reason to expect that extensive local genetic diversity affects the majority of organisms throughout this, the greatest archipelago on Earth.

Protest Against New Forest Access Regulations. *Indonesian women protest about use of the nation's forests by mining corporations, in Jakarta on March 10, 2008. Driven by global demand for commodities, Indonesia has one of the highest rates of deforestation in the world—but public concerns about the likely consequences are growing.* BAY ISMOYO/AFP/GETTY IMAGES.

WHAT CAN BE DONE?

If climate change is driven by fossil fuel use, then, regardless of its own emissions, the region is largely at the mercy of China, India, Europe, and the United States. An equally potent threat to biodiversity is poverty, coupled with rising population, over which nation states have more control; Thailand, for example, has achieved some measure of population restraint through the work of the family-planning activist Mechai Viravaidya. If, however, escape from poverty depends on increased use of fossil fuel, then this policy may be self-defeating.

For many people now living in the central areas of the archipelago, such as Indonesians resettled by World Bank-funded transmigration schemes during the 1980s, primary forests appear hostile and frightening. Education, especially regarding ecological literacy, will be vital if this is to change, but opportunities are limited by persistent poverty. Many rural people now engaged in agriculture and fishing have little local or traditional ecosystem knowledge. Even though there are encouraging developments, such as dissemination of local-language guides about sustainable fisheries, these have yet to overcome the use of blast (explosive) fishing for local food needs and cyanide to collect reef fish for the international aquarium trade—both hideously destructive and all-too-common practices.

For the conservation of biodiversity, international nongovernmental organizations (NGOs) play a growing role in raising awareness and providing funds for action, although organizations such as the Nature Conservancy have been criticized for dependence on multinationals as major donors. There is increasing collaboration between Southeast Asian and international scientists concerning environmental projects, including the Millennium Ecosystem Assessment of 2001–2005.

However important NGOs, academic research, and the business world may be, any shift to a sustainable Human-Earth relationship will depend on the understanding and willingness of the mass of people. Worldviews and social aspirations, coupled with better education and relief from poverty, will largely determine the future environmental impact of *Homo sapiens* on all parts of the Earth. In Southeast

Asia there are many cultural divisions as well as demographic and economic differences, although in one respect the overall picture appears simple: On the mainland the dominant cultural outlook has been shaped by Buddhism, and in the archipelago (including the Malay Peninsula) by Islam. Christianity is a lesser but important influence throughout, with notable Roman Catholic dominance in the northern and central Philippines and in Timor-Leste. The archipelago, now dominated by monotheistic religions, has mostly had a poorer environmental record than the Buddhist-influenced mainland. However, wherever a consumer-orientated middle class has emerged, as in parts of Thailand, increasing materialism is also evident, with concomitant impacts even on areas formerly held sacred.

ENVIRONMENTAL ACTION AND PHILOSOPHY

Cultural differences, coupled with wide variations in political and economic reality, mean that structural solutions to biodiversity loss are not just desirable—they are essential. Environmental action has to deal with differences not only between but also within countries, especially in large and culturally diverse nations such as Indonesia and Vietnam. International NGOs and national and local government agencies that fail to recognize this reality (including wide variations in land tenure) are unlikely to meet with success—and could well do more harm than good.

Against this background the development of environmental philosophy within the region, including courses taught and leading figures, is vitally important as a source of insight and future leadership. Influential figures in the conservation movement include Angel Alcala (Silliman University, Philippines, a key person in marine protection), Maryati Mohamed (director of the Institute for Tropical Biology and Conservation, University of Malaysia Sabah, Kota Kinabalu), and Chris Margules (Australian team leader for Conservation International's Indo-Pacific Region program). Establishment of the ASEAN Center for Biodiversity (University of Philippines, Los Baños), and the Conservation Training and Resource Center (Bogor, Indonesia) are important steps. BirdLife International in Indochina is active in Vietnam, Cambodia, Laos, and Myanmar. There are also valuable national organizations, such as Education for Nature Vietnam; Wildlife Conservation Society of the Philippines (founded in 1992 and later a key meeting ground for government agencies, NGOs, and academics); the Malaysian Nature Society; and the Center for Environment, Technology, and Development Malaysia (founded by the renewable-energy activist Gurmit Singh).

Most of these initiatives are oriented toward technology and action, not philosophy. Few universities in Southeast Asia teach general philosophy, much less environmental values, although ethical issues are addressed within ecological and environmental courses. The International Society for Environmental Ethics does not list a single representative in Southeast Asia. The philosophy course at the National University of Singapore does include one module, taught by Cecilia Lim. International academic collaboration is likely to be critical to foster growth in environmental philosophy in the region, such as the recently established International Network of Environmental Education, led by Fumiaki Taniguchi from Konan University (Kobe, Japan) in partnership with Phranakhon Rajabhat University (Bangkok, Thailand), Peking University (Beijing, China), and University of Malaya (Kuala Lumpur, Malaysia). The Earth Charter Initiative should also play a valuable role. The effectiveness of such initiatives will hinge on sensitivity to cultural differences, a point that has been explored by Hana Panggabean, from the psychology faculty, Atma Jaya Catholic University, Jakarta (Fremerey and Panggabean 2004).

SEE ALSO *Biodiversity; Buddhism; Christianity; Conservation; Earth Charter; Forests; Global Climate Change; Islam; Nongovernmental Organizations; Population; Species; Sustainability.*

BIBLIOGRAPHY

Dearden, Philip. 2002. "Development and Biocultural Diversity in Northern Thailand." In *Environmental Protection and Rural Development in Thailand*, ed. Philip Dearden. Bangkok, Thailand: White Lotus Press.

Fremerey, Michael, and Hana Panggabean. 2004. "Between Difference and Synergy: Cultural Issues in an International Research Scheme." In *Land Use, Nature Conservation, and the Stability of Rainforest Margins in Southeast Asia,* eds. Gerhard Gerold; Michael Fremerey; and Edi Guhardja, pp. 523–533. Berlin, Germany: Springer.

Gerold, Gerhard; Michael Fremerey; and Edi Guhardja, eds. 2004. *Land Use, Nature Conservation, and the Stability of Rainforest Margins in Southeast Asia.* Berlin, Germany: Springer.

Kemf, Elizabeth. 1990. *Month of Pure Light: The Regreening of Vietnam.* London: The Woman's Press.

Myers, Norman; Russell A. Mittermeier; Cristina G. Mittermeier; et al. 2000. "Biodiversity Hotspots for Conservation Priorities." *Nature* 403: 853–858.

Nature Conservancy. 2008. Indonesia. http://www.nature.org/wherewework/asiapacific/indonesia/

Primack, Richard B., and Thomas E. Lovejoy, eds. 1995. *Ecology, Conservation, and Management of Southeast Asian Rainforests.* New Haven, CT: Yale University Press.

Sterling, Eleanor.; Martha M. Hurley; and Le Duc Minh. 2006. *Vietnam: A Natural History.* New Haven, CT: Yale University Press.

Swearer, Donald K.; Sommai Premchit; and Phaithoon Dokbuakaew. 2004. *Sacred Mountains of Northern Thailand and their Legends.* Chiang Mai, Thailand: Silkworm Books.

R. I. Vane-Wright

SPACE/PLACE

Humans see things and events not only as they are, but also as signs or symbols that point beyond themselves to other associated things, events, or ideas. These symbols point to meanings that might otherwise be overlooked because they are obscure, intangible, or far removed in time or space. A sign in the park warning of poison ivy points out a hazard to visitors who can read English, but not the shapes of leaves; a war memorial points back to a past event; a steeple points beyond this world to a transcendent hope.

UNDERSTANDING A SENSE OF PLACE

Such symbols are part of a public and a local vocabulary. Every day each person uses symbols in a locality to take and keep his or her physical and spiritual bearings. Understanding this local symbolic vocabulary—and the intangibles to which it points—is the first step toward sensing the locality as a place, toward feeling "in place." We can appreciate this feeling when we travel to foreign localities that are, to us, largely unintelligible, scarcely places at all. The commonalities of human life ensure that no human locality is to any other human wholly unintelligible, but to be deeply uncertain as to the meaning of many of the things and activities by which one is surrounded is to feel "out of place."

EMOTIONS, MEMORY, AND A SENSE OF PLACE

It is, of course, possible to comprehend meanings while feeling toward them intense repugnance, disapproval, or dislike. We have all understood and yet disliked a book. To understand and detest a place is not, perhaps, so common as to understand and detest a book; but there is, commonly enough, what the philosopher Roger Scruton (2007) calls *oikophobia:* antipathy toward one's home place. The oikophobe is of the place and so understands the place (detachment may enhance insight), but he is not altogether in the place. He is also, in his own way, "out of place." This is because sense of place entails empathy, affection, and understanding. The oikophobe is cut off from full understanding of the place just as the teetotaling prohibitionist, notwithstanding degrees in organic chemistry and physiology, is cut off from full understanding of wine. To understand Austria, for instance, one must not only understand the meaning and recognize the manifestations of gemütlichkeit (cordiality, friendliness)—one must also enjoy gemütlichkeit.

This identification with a place is often tacitly present in natives who cannot imagine themselves living anywhere else. But visitors to places with a distinctive ethos and way of life—places like New York City, New Orleans, Santa Fe, or San Francisco—sometimes feel a surge of similar affinity. Some who do will stay. Chronic homesickness afflicts anyone forced to leave a place with which they identify, because a sense of place implies desire to be in that place.

Empathy and affection may rise to the level of love. Indeed, a sense of place is often described in the language of love. Love of a place is the root meaning of *patriotism* (a word that should not be surrendered to jingoists and chauvinists or to the oikophobes who oppose them). People say that they have "fallen in love" with a place. In this, as in any loving relation, there is an interpenetration and mutual involvement that progressively knits the partners together. A husband, for instance, comes to see something of himself in his wife, as she presumably comes to see something of herself in him. The expression "grown together" is, in such cases, more than mere metaphor. Persons with a sense of place are knitted to the place in an analogous fashion. Looking at the place, they are reminded of the life they have lived there; thinking about the life they have lived, they are reminded of that place.

Memory is, therefore, the third component in a sense of place. This is not reminiscence, for sense of place does not mean that the mind is forever asking itself, "Do you remember when?" Memory is, rather, recognition that, for better or for worse (recurring to the language of love and marriage), here and nowhere else is where my life has been and is being lived. It is, in other words, love completed by conscious and demonstrated commitment to the place.

PLACELESSNESS

Understanding, affection, and memory are the three components in a sense of place. Possessing any one of these in even a small degree, one may lay claim to a sense of place, but it is doubtful that a person utterly devoid of any one of them really feels a sense of place. Localities are, for such persons, only more or less satisfactory. They are enigmas that arouse no curiosity—disposable instruments, like a parking place, a motel room, or a seat in an airport waiting area.

The geographer Edward Relph called this attitude *placelessness* (1976). Placelessness is the absence of understanding, affection, or memory. It may arise in placeless individuals who are incapable of developing a sense of place, or in placeless localities that are impossible to understand, love, or remember. Relph argued that placelessness is increasing for both reasons. There are more drifting nomads who cannot put down roots, and more impenetrable localities that do not provide fertile soil for the sinking of roots.

This argument is easily criticized as a mask for xenophobia; it is not at all clear, however, that people with a secure sense of place are prone to xenophobia. Xenophobia is, if anything, a pathology of persons who resent unsought placelessness and blame foreigners and immigrants for this unsettled condition. This argument has also engendered a postmodern style of architecture that seeks to recover the unique and local, but the rearguard of modernism routinely denounces it as ersatz, eclectic, and kitschy. Perhaps there is something fundamentally phony about much of this superficially idiosyncratic architecture, but one can admit this and still agree with its protest against the placeless "geography of nowhere" imposed by the "international style."

THE ENVIRONMENT AND A SENSE OF PLACE

The connection between sense of place and environmental values and attitudes is complex. Clearly, persons with a sense of place will not wish to see that place destroyed; they can be expected to show more concern for stewardship of the local environment than, say, nomadic managers of global corporations. But the natural environment will be, for such proud locals, only one facet of the place, joined in complicated ways to its economy, society, and culture. The conflicts between logging communities and environmentalists in the Pacific Northwest show that residents with a sense of place may not prize environmental values. Whether understood as scenery or as natural resources, the natural environment is just one of the things these people understand, love, and remember when they experience a sense of place. They will, therefore, often accept environmental costs as tradeoffs for economic, social, or cultural benefits. Environmentalists who would harness the powerful sentiment that we call a sense of place must, therefore, learn to see and describe the ecosystem as part of the larger social and symbolic ecology of the locale. When environmental degradation can be shown to threaten a way of life and the identities of individuals who are part of it, environmentalism and a sense of place are in harmony. When environmentalism appears to threaten an established local way of life, however, environmentalism and sense of place will clash.

Regard for the viability of a place should include concern for the natural environment, as well as the larger social and symbolic ecology. Such concern is, however, sometimes coupled to indifference to the viability of the natural environment in other places. If my drinking water is safe, I may too easily grow indifferent to the quality of drinking water in other places, even when this has been rendered unsafe by my behavior as a consumer. Thus, to some ethicists attaching special value to one's local environment is a vice akin to the vice that other ethicists find in preferential regard for one's neighbors and kin. The answer to both of these challenges is the same: Abstract regard (for humanity or the environment) normally issues in a censorious attitude, whereas concrete regard (for persons and places) more often issues in constructive action.

SHRINKING SPACE AND TIME

For humans and presumably all creatures endowed with sight, the external world appears to consist of objects that occupy space. Philosophers disagree on this point: Some argue that space is itself a feature of the external world, whereas others regard space as a concept whereby the mind imagines something that is, in fact, quite different from space. Space for the second group of philosophers is analogous to color. The external world has no color, just light with different wavelengths bouncing off or being absorbed by objects: but color is the way our mind imagines these wavelengths. Whatever the answer may be to this abstruse question, there can be no doubt that it is highly practical to understand the world as objects arrayed in space. Spatial maneuvers are, after all, our most effective means of manipulating and controlling the external world.

This is why space is one of the first things the human infant learns to understand. At a young age she discovers that she can have much that she desires by reaching out for it, rolling toward it, crawling after it. This primary spatial education continues in the spontaneous play of childhood and the organized sport of adolescence. Children learn the basic nature of space (along with other things) when they run, jump, and fall; when they catch and throw balls; when they frighten their parents with tricks on their bicycles. In adulthood such exercises may continue in the spatial virtuosity of the athlete, acrobat, or dancer but are more commonly transformed into practical geometries such as navigation, engineering, and ballistics.

The mention of those three sciences should make it clear why humans seek spatial understanding. Practical geometries are instruments of power. It is no coincidence that revolutions in navigation and ballistics (as well as cartography and perspectival drawing) accompanied the birth of the modern age. Successful navigation, engineering, drawing, and ballistics demand advanced spatial understanding, but they are also means of shaping space. Navigation and ballistics, for example, shrink space. Engineering has contributed mightily to what nineteenth-century writers, thinking of trains and the telegraph, called the "annihilation of distance." Since the fifteenth century distance annihilating innovations such as highways, airplanes, and microwave transmitters have radically reduced the effective size of the earth.

The geographer Donald G. Janelle (1969) has described this shrinkage as time-space convergence, because the convergence of locations in space was a reduction in transportation time. Time-space convergence is a pervasive fact of society. People, goods, capital, images, and ideas move through space with astonishing speed, as do pathogens, pollutants, terrorists, and intercontinental ballistic missiles. (Efforts to channel or stem these flows are another class of space-shaping technologies).

Time-space compression has profoundly transformed the global environment: first by bringing together organisms that had evolved separately—with disastrous consequences for some of those organisms (Native American populations were devastated by Old World diseases such as smallpox)—and second by bringing the farthest corners of the earth into the reach of metropolitan markets. Many critics note that the spatially extensive economy also hides from consumers the consequences of their choices, putting resource extraction, polluting factories, and waste out of sight and out of mind. Others have countered that this annihilation of distance also permits the dissemination of images of environmental degradation. Indeed, because such images often depict vivid extremes, consumers may in some instances overestimate environmental degradation.

The greatest effect of time-space compression on consumer perception of the environment is to separate degradation from the intentions of any single human actor, thereby diffusing the sense of responsibility. If someone cuts down a tree and builds a table, she has no doubt that it was she who cut the tree. If a shopper buys a table, it is much harder for her to connect her intention with the felling of any particular tree. Indeed, she never forms the intention to fell a tree—only the intention to buy a table so that she may give a dinner party for twelve. At the other end of the commodity chain, the lumberjack likewise works in a fog of abstractions. He cuts down the tree not necessarily because he wishes to but because he is impelled to do so by the market demand that reaches him through the orders and inducements of his employer. Thus, a tree is cut and no one feels that it is he or she who bears responsibility for the act. Hence environmental consciousness is not everywhere identical to a sense of place, and time-space convergence is sometimes—but not always—harmful to the environment.

SEE ALSO *Globalization; Land Ethic; Native Americans; Regionalism; Stewardship.*

BIBLIOGRAPHY

Cresswell, Tim. 2005. *Place: A Short Introduction.* Malden, MA: Blackwell.

Cronon, William. 1991. "The Busy Hive." In *Nature's Metropolis: Chicago and the Great West.* New York: Norton.

Curry, Michael R. 1996. "On Space and Spatial Practice in Contemporary Geography." In *Concepts in Human Geography*, ed. Carville Earle; Kent Mathewson; and Martin Kenzer. Lanham, MD: Rowman and Littlefield.

Janelle, Donald G. 1969. "Spatial Reorganization: A Model and Concept." *Annals of the Association of American Geographers* 59: 348–364.

Relph, E. C. 1976. *Place and Placelessness.* London: Pion.

Sack, Robert, *A Geographical Guide to the Real and the Good.* London: Routledge, 2003.

Scruton, Roger. 2007. "Conserving Nations." In *A Political Philosophy.* Rev. edition. London and New York: Continuum.

Tuan, Yi-fu. 1977. *Space and Place: The Perspective of Experience.* Minneapolis: University of Minnesota Press.

Jonathan M. Smith

SPECIES

Concern for the protection of endangered and threatened species is central to environmental ethics, and is embodied in public policy. Our moral and legal deliberations about species ought to be grounded in the best current biological and philosophical thinking about species. Unfortunately, there is no single unambiguous definition of *species* in biology, but rather a host of competing species concepts and definitions. Biologists and philosophers disagree about which, if any, is the "correct" species concept, whether we should accept a plurality of species concepts, and even whether *species* are real natural categories. These difficulties are known as "the species problem." This entry discusses various aspects of the species problem and how they impact our moral and legal thinking about species. It argues that there is a sufficiently clear account of the entities our species protection laws aim to protect, and the rationale for protecting them.

PROBLEMS WITH SPECIES

A species concept is an account of the species category. It spells out the special features that distinguish species from other groups of organisms. A species concept indicates where the boundaries are between distinct species and what makes two organisms conspecific. There are many species concepts in the biological literature; four of the most important in current use are discussed below.

The Biological Species Concept: A species is a group of interbreeding natural populations that is reproductively isolated from other such groups (Mayr and Ashlock 1991, p. 26).

The Evolutionary Species Concept: A species is a single lineage of ancestral descendant populations of organisms which maintains its identity from other such lineages and which has its own evolutionary tendencies and historical fate (Wiley 1978, p. 18).

The Ecological Species Concept: A species is a lineage (or a closely related set of lineages) which occupies an adaptive zone [ecological niche] minimally different from that of any other lineage in its range and which evolves separately from all lineages outside its range (Van Valen 1976, p. 233).

The Phylogenetic Species Concept: A species is a group of organisms, including a common ancestor and all of its descendants (a monophyletic group), that is the smallest diagnosably distinct such group. (See Cracraft 1983; Mishler and Brandon 1987.)

These species concepts give different accounts of what makes a group of organisms a species, and generate different classifications of organisms. For example, the Biological Species Concept (BSC) does not recognize as species groups of organisms that do not interbreed, and consequently does not recognize asexual organisms as forming species. The other three species concepts recognize asexual organisms as species so long as they form ancestor-descendant lineages that meet their other criteria. The Ecological Species Concept and the Evolutionary Species Concept will lump together as a single species populations that do not exchange genetic material due to geographical isolation (unlike the BSC), so long as those populations occupy the same ecological niche, or maintain the same evolutionary tendencies, respectively. The Phylogenetic Species Concept (PSC) is the most fine-grained species concept in that it splits such populations into distinct species, so long as they are diagnosably discernible. The PSC recognizes distinct species wherever a monophyletic group of organisms is recognized as a distinct group for a variety of reasons. Subspecies, distinct forms of species often recognized by geographical or minor character differences, may be regarded as distinct species by the PSC.

Species monism is the view that there is one correct species concept, and a task of biological systematics is to determine which species concept gives the correct account of species and generates a classification that reflects true divisions in nature. Given the ongoing and unresolved dispute concerning the species concept, and the fact that biologists adopt different species concepts in different contexts, an increasingly popular alternative is *species pluralism*. Species pluralists (John Dupré 1993; Marc Ereshefsky 2001; Philip Kitcher 1984) maintain that there is more than one kind of species, and we can accept a number of different species concepts, which need not conflict. Biologists with different concerns may simply be talking about different things. For example, taxonomists employing the PSC generally follow the cladistic approach to taxonomy, which aims to reconstruct evolutionary history, while de-emphasizing the importance of similarities in classification. Those adopting the other species concepts have additional concerns and wish to retain a more intuitive classification. Pluralists have suggested we adopt explicit terminology ("biospecies," "ecospecies," "phylospecies") to make clear which things we are talking about by *species*.

Another area of contention regarding species concerns their ontological status. The traditional Platonic view considered species to be universals—abstract entities distinct from the concrete individual organisms that are their instances. The view that species are universals—eternal, unchanging, abstract entities—is thought to have no place in contemporary evolutionary biology, where species are changing, evolving units that have beginnings and go extinct. According to the class view, a species is a class or set of organisms with certain properties in common. A species concept tells us which qualitative properties (morphological, genetic, etc.) distinguish one species class from another. On one version of the class view, there are properties shared by all and only the members of a species, in virtue of which organisms belong to that species. The other version rejects precise species essences in favor of clusters of features, such that organisms belong to a species if they have a sufficient number of them. Much contemporary thinking about species has de-emphasized qualitative similarities between organisms, and focused instead on populations and the biological relations that unite them into species. This shift has caused many to adopt the species-as-individuals view—that species are concrete, persisting, particular individuals, with organisms as parts rather than members. (See works by Judith Crane 2004, Michael Ghiselin 1974, and David Hull 1978.)

A final aspect of the species problem concerns the reality of species. The difficulties in settling on a species concept and understanding the ontology of species invite this worry. Perhaps species are merely artificial categories drawn up for the sake of convenience. The idea that species gradually evolve into one another suggests that the boundaries between species are quite arbitrary, which is what led Charles Darwin to doubt the reality of species. For species to be real is for there to be an objective fact as to what the different species are, and where the boundaries are between species. Philosophers sometimes talk of "carving nature at the joints," echoing Plato's expression from *The Sophist*. If species are real, there are natural joints to be discovered, and the quest for the correct species concept is the attempt to carve nature at its joints.

CONSERVATION SPECIES

What kind of "species" matter in our efforts to protect endangered species? Consider the U.S. Endangered Species Act of 1973 (ESA). The "Definitions" section contains the following: "The term 'species' includes any subspecies of fish or wildlife or plants, and any distinct population segment of any species of vertebrate fish or wildlife which interbreeds when mature" (Section 3[16]).

The inclusion of "subspecies" and "distinct population segments" shows that the ESA is using a finer-grained notion of species than the Biological, Ecological, or Evolutionary Species Concepts. Many of the endangered and threatened species listed under the ESA are in fact classified as subspecies in standard classifications. Perhaps the ESA is using the fine-grained Phylogenetic Species Concept (PSC), which treats many "subspecies" as species. However, the PSC's strict adherence to only monophyletic groups as species is clearly not a concern of the ESA. According to the PSC, an ancestral population that buds off a new species but persists without any noticeable change is not considered a species, since it does not include all of its descendants. The ESA's appeal to interbreeding, at least for vertebrate animals, suggests that interbreeding relations would be sufficient to unify such a group of organisms into a species, even if it were not monophyletic. That interbreeding is mentioned only in the case of vertebrate animals suggests that other biological processes, such as ecological forces, could unify a population into a species. The "species" of the ESA are finely divided populations, but the ESA does not specify which biological processes unify populations into species.

Robin Waples (1991) suggests that the protectable populations of the ESA be understood as "evolutionarily significant units" (ESUs). ESUs were introduced to capture the idea of a minimal unit of conservation, which avoids the controversial taxonomic categories of species and subspecies. This would allow ESA species to be more fine-grained than many standard species concepts. However, ESUs are identified by morphological or genetic distinctness from related populations. David Pennock and Walter Dimmick (1997) note that many populations listed under the ESA do not qualify as ESUs. Protected populations of grizzlies and grey wolves are identified by region, and do not differ genetically or morphologically from unprotected populations in other regions.

The fine-grainedness of ESA species appears to flow from the goals of slowing the current rate of extinctions, and preserving ecosystems on which species depend. Section 2(b) reads: "The purposes of this Act are to provide a means whereby the ecosystems upon which endangered species and threatened species depend may be conserved ..." The emphasis on ecosystems accompanies a concern with regional populations. In some cases regional populations are classified as subspecies. For example, the ESA lists six subspecies of beach mice. The mice occupy similar habitats in the Southeastern United States, are closely related biologically, and are distinguished only by region and minor morphological differences. Where biological classifications do not differentiate populations as subspecies, the ESA may differentiate them nonetheless. Grizzly bears are listed as a single species (*Ursus arctos horribilis*, a subspecies of brown bear), but the ESA distinguishes and treats separately five different regional populations of grizzlies. In March 2007, the Yellowstone grizzly was removed from the threatened list, while the remaining four populations of grizzlies remained protected.

Grizzly Bear Peering Over the Grass, Yellowstone National Park. *The grizzly bear (*ursus arctos horribilis*) is a symbol of the American wilderness; it is the largest species of bear found in North America. Between 1800 and 1975, the number of grizzlies in the United States decreased from around 50,000 to fewer than 1,000. The effort to recover the number of grizzly bears, initiated in 1975, has met with some success.* **NPS.**

The ESA uses a species concept based on the idea of a regional population. How fine-grained is it supposed to be? Where do we stop dividing populations? Consider the ESA definition of "endangered species": "The term 'endangered species' means any species which is in danger of extinction throughout all or a significant portion of its range ..." (Section 3[6]). The range of a species may be broader than an area occupied by a protected population, as in the case of grizzlies. The range of grizzlies includes Washington, Idaho, Montana, and Wyoming, but grizzlies are endangered in a portion of their range, as indicated by the four out of five regional populations protected under the ESA. But portions of ranges used to divide populations for protection must be "significant." This language is vague, as is often the case with legal language, and allows room for judgment. Clearly the status of the American red squirrel should not be based on the fate of one backyard population. Given the concern to protect ecosystems, it is plausible that the limit of a "significant" portion of a range is an ecosystem that can be identified and discerned for the purposes of conservation. Thus the following species concept appears to be implicit in the ESA and in our practical thinking about species conservation: *The Conservation Species Concept:* A (conservation) species is a population of organisms (unified by interbreeding relations, occupation of a common ecological niche, or other biological process) that depends on a discernable ecosystem for its survival.

The Conservation Species Concept recognizes as species biologically unified populations that can be discriminated on the basis of protectable ecosystems. It is such populations that species protection laws aim to protect.

Because the Conservation Species Concept is population-based, it suggests a species-as-individuals ontology. Perhaps understanding species as individuals can help to explain the value of species and why they ought to be protected. It has been suggested that species as individuals have interests in addition to any interests of their members, and so deserve moral consideration independently of their value to humans. (See Lawrence Johnson [2003] for an attempt to show this.) But even if species are the same type of entity ontologically as organisms, given the comparative lack of cohesion and other significant differences between species and organisms, trying to establish a meaningful account of species interests still poses a substantial challenge (see Sandler and Crane 2006).

The ESA provides another account of the value of species: "These species of fish, wildlife, and plants are of esthetic, ecological, educational, historical, recreational, and scientific value to the Nation and its people" (Section 2[3]). This account of the value of species is not only human-centered, it is nation-centered, which many environmentalists may find unsatisfactory. It provides a rationale for protecting endangered populations that avoids the difficulties in giving an account of species value independently of human interests. In fact, the ESA explicitly does not recognize as endangered any species of insect that is considered a "pest" (Section 3[6]). If we rely on such an account of species value, there are no true conflicts between human interests and those of endangered species. Such apparent conflicts are between competing human interests.

The question of species realism would appear to have profound implications for efforts to preserve species. If species are not real, but merely convenient categories, are efforts to protect species misguided? No. Even biologists and philosophers who reject the reality of species can accept that populations are real. A population is a group of organisms that is integrated in some biologically interesting way—for example, by interbreeding or ecological forces. The Conservation Species Concept is based on this idea of a population. If we take a pluralist approach to species, we can accept such species concepts as legitimate. Species monists who believe there is one correct scientific concept can allow that conservationists are not talking about biological species, but populations that can be identified by the ecosystems on which those populations depend, and which we have an interest in protecting.

SEE ALSO *Conservation; Defenders of Wildlife; Ecosystem Health; Endangered Species Act; Environmental Law; Speciesism.*

BIBLIOGRAPHY

Cracraft, Joel. 1983. "Species Concepts and Speciation Analysis." *Current Ornithology* 1: 159–187.

Crane, Judith K. 2004. "On the Metaphysics of Species." *Philosophy of Science* 71(2): 156–173.

Dupré, John. 1993. *The Disorder of Things: Metaphysical Foundations of the Disunity of Science.* Cambridge, MA: Harvard University Press.

Ereshefsky, Marc. 2001. *The Poverty of the Linnaean Hierarchy.* Cambridge: Cambridge University Press.

Ghiselin, Michael T. 1974. "A Radical Solution to the Species Problem." *Systematic Zoology* 23(4): 536–544.

Hull, David L. 1978. "A Matter of Individuality." *Philosophy of Science* 45(3): 335–360.

Johnson, Lawrence E. 2003. "Future Generations and Contemporary Ethics." *Environmental Values* 12(4): 471–487.

Kitcher, Philip. 1984. "Species." *Philosophy of Science* 51(2): 308–333.

Mayr, Ernst, and Peter D. Ashlock. 1991. *Principles of Systematic Zoology.* 2d edition. New York: McGraw-Hill.

Mishler, Brent D., and Robert N. Brandon. 1987. "Individuality, Pluralism, and the Phylogenetic Species Concept." *Biology and Philosophy* 2: 397–414.

Pennock, David S., and Walter W. Dimmick. 1997. "Critique of the Evolutionary Significant Unit as a Definition for 'Distinct Population Segments' under the U.S. Endangered Species Act." *Conservation Biology* 11: 611–619.

Ryder, Oliver A. 1986. "Species Conservation and Systematics: The Dilemma of Subspecies." *Trends in Ecology and Evolution* 1: 9–10.

Sandler, Ronald, and Judith K. Crane. 2006. "On the Moral Considerability of *Homo sapiens* and Other Species." *Environmental Values* 15(1): 69–84.

Van Valen, Leigh. 1976. "Ecological Species, Multispecies, and Oaks." *Taxon* 25(2/3): 233–239.

Waples, Robin S. 1991. "Pacific Salmon, Oncorhynchus ssp., and the Definition of 'Species' under the Endangered Species Act." *Marine Fisheries Review* 53: 11–22.

Wiley, Edward O. 1978. "The Evolutionary Species Concept Reconsidered." *Systematic Zoology* 27(1): 17–26.

Judith K. Crane

SPECIESISM

Richard Ryder is credited with coining the term *speciesism* in 1970 (see Ryder 1975, 1989). As Ryder himself observed, "In 1985 the *Oxford English Dictionary* defined speciesism as 'discrimination against or exploitation of certain animal species by human beings, based on an assumption of mankind's superiority.' This definition marked the official acceptance of 'speciesism' into the language" (1998, p. 320). Ryder goes on to refer to the pioneering work of Peter Singer, who has done much to both popularize the concept of speciesism and to present arguments justifying claims that speciesism is a morally odious practice.

The charge of speciesism, as it occurs in debates about the moral status of nonhuman animals, usually applies to people who attempt to justify different treatment of nonhuman animals (or who attribute to such animals different value) using the criterion of species membership. Speciesism has been compared to both racism and sexism. Racists attempt to justify different treatment and different attribution of value using race membership as a criterion, and sexists do the same using gender as a criterion. Opponents of speciesism argue that just as racism and sexism are morally odious as applied to our fellow humans, so, too, is speciesism as applied to our fellow creatures. Opponents of speciesism believe that the moral community—the community of beings that count and have moral worth—needs to be expanded to include both humans and nonhuman animals.

The debate about speciesism has been particularly intense in connection with the use of millions of nonhuman animals in biomedical experiments. As Singer has pointed out:

> The experimenter, then, shows bias in favor of his own species whenever he carries out an experiment on a non-human for a purpose that he would not think justified him in using a human being at an equal or lower level of sentience, awareness. ... No one familiar with the kind of results yielded by most experiments on animals can have the slightest doubt that if this bias were eliminated the number of experiments performed would be a minute fraction of the number performed today. (1989, p. 80)

Bias against animals in the context of animal experimentation in turn has led to intense debates about (a) the cognitive status of nonhuman animals; and (b) the problem of cognitively marginal humans. In connection with the latter issue, some have argued that many nonhuman animals are cognitively more sophisticated than young infants, those with severe mental retardation, or advanced senility. Because these latter are included in the moral community of beings that count, why not include nonhuman animals—unless you are guilty of speciesism? Tom Regan (1979) labeled this "the argument from marginal cases," and it appears in nearly all appeals for extending the same ethical regard we accord so-called "marginal" members of our own species to cognitively equal or superior members of other species.

LaFollette and Shanks (1996) have distinguished between *bare speciesism* and *indirect speciesism*. Bare speciesism involves differential treatment of organisms simply on the basis of a biological species difference. It is hard to see how a mere species difference can possibly be morally relevant. By contrast, indirect speciesism is the view that the evolutionary changes giving rise to species differences were accompanied by changes in morally relevant cognitive properties that restrict the moral community to the human species. Issues concerning the mental lives of animals have been reviewed in the work of Niall Shanks (2002).

In an important early discussion of these matters, laid out in a footnote in *The Principles of Morals and Legislation* (1789), the utilitarian philosopher Jeremy Bentham observed:

> The day *may* come, when the rest of the animal creation may acquire those rights which could not have been withholden from them but by the hand of tyranny. The French have already discovered that the blackness of the skin is no reason why a human being should be abandoned without redress to the caprice of a tormentor. It may come one day to be recognised, that the number of legs, the villosity of the skin, or the termination of the *os sacrum* are reasons equally insufficient for abandoning a sensitive being to the same fate. What else is it that should trace the insuperable line? Is it the faculty of reason, or, perhaps, the faculty of discourse? But a full-grown horse or dog is beyond comparison a more rational, as well as a more conversable, animal than an infant of a day, or a week, or even a month old. But suppose the case were otherwise, what would it avail? The question is not, Can they *reason*? nor, Can they *talk*? but, Can they *suffer*? (1879, p. 311)

In this passage we find the earliest statement of the argument from marginal cases, in this instance "an infant of a day, or a week, or even a month old." Peter Singer, in his book *Animal Liberation* (1975), elaborated a view of the moral status of nonhuman animals along lines sketched here by Bentham, using the same utilitarian arguments. Singer focuses, as Bentham does, on the moral significance of animals' pain and suffering. As Singer observed in *Animal Liberation*, "If a being suffers, there can be no moral justification for refusing to take their suffering into consideration" (1990, p. 8). For Singer any being capable of suffering has a place in the community of beings that have moral standing. He argues that just as racists and sexists wrongly treat certain individuals differently on the basis of morally irrelevant traits, so speciesists are guilty of a similar error. Though Singer sometimes uses the phrase "animal rights," he has made it very clear that, as a utilitarian moral theorist, he prefers not to talk about rights at all, whether animal or human.

By contrast, Tom Regan, in *The Case for Animal Rights* (1983), has developed an explicitly rights-based approach to the moral status of animals. If, as Regan argues, nonhuman animals have rights, then their interests cannot be sacrificed even if such sacrifice would greatly benefit human beings. Regan's basic idea is that creatures should be treated the same unless there is a morally relevant reason to justify a difference in treatment. A mere difference in species membership is not, in his view, a morally relevant difference. For Regan nonhuman animals have an inherent worth that trumps their instrumental value to us as subjects of painful experiments and as farm animals.

Defenders of animal experimentation such as Carl Cohen (1986) object to the comparison of speciesism with racism and sexism. For Cohen the capacity for moral judgment is what distinguishes humans from other animals. By contrast, Peter Carruthers (1992) has argued that the experiences of animals (if any) are not sufficiently like ours to confer moral standing on them.

SEE ALSO *Animal Ethics; Regan, Tom; Singer, Peter.*

BIBLIOGRAPHY

Bentham, Jeremy. 1879. *An Introduction to the Principles of Morals and Legislation*. Oxford, UK: Clarendon Press.

Carruthers, P. 1992. *The Animals Issue*. Cambridge, UK: Cambridge University Press.

Cohen, Carl. 1986. "The Case for the Use of Animals in Biomedical Research." *New England Journal of Medicine* 315: 865–870.

Lafleur I. J., ed. 1948. *Bentham: An Introduction to the Principles of Morals and Legislation*. New York: Hafner.

LaFollette, H. H., and Niall Shanks. 1996. *Brute Science: The Dilemmas of Animal Experimentation*. London: Routledge.

Regan, Tom. 1979. "An Examination and Defense of One Argument Concerning Animal Rights." Inquiry 22: 189–219.

Regan, Tom. 1983. *The Case for Animal Rights*. Berkeley: University of California Press.

Ryder, Richard D. 1975. *Victims of Science: The Use of Animals in Research*. London: Davis Poynter.

Ryder, Richard D. 1989. *Animal Revolution: Changing Attitudes towards Speciesism*. Oxford, UK: Basil Blackwell.

Ryder, Richard D. 1998. "Speciesism." In *Encyclopedia of Animal Rights and Animal Welfare*, ed. M. Bekoff and C. A. Meaney. Westport, CT: Greenwood.

Shanks, Niall. 2002. *Animals and Science: A Guide to the Debates*. Santa Barbara, CA: ABC-CLIO.

Singer, Peter. 1989. "All Animals are Equal." In *Animal Rights and Human Obligations*, ed. Tom Regan and Peter Singer. Englewood Cliffs, NJ: Prentice Hall.

Singer, Peter. 1990. *Animal Liberation*. 2nd edition. New York: Avon Books.

Niall Shanks

SPINOZA, BARUCH
1632–1677

Baruch (Benedict de) Spinoza scandalized the western European world and his own Jewish community in Holland with his declaration that God is nothing other than nature. His philosophy maintains that one substance comprises all of existence. God, the force of creation, cannot be outside, above, or beyond the created world. Because God is perfect, infinite, eternal, and coextensive with all of existence, the world itself must be identical with divine perfection. God is in everything, and everything is in God (both "pantheism" and "panentheism").

According to Spinoza, God expresses himself in infinitely many ways; he has infinitely many "attributes," which are further divided into infinitely many "modes." Humans conceive two attributes, "thought" and "extension," each of which is perfect, complete, infinite, and eternal. Thus, God is not merely spiritual or mental, but equally corporeal. Infinitely many ideas comprise thought, whereas "extension" names the interactive community of infinitely many bodies. Bodies and ideas express the same order and connection of causes. This means that any individual thing in nature exists, at the same time and to the same degree, as a thinking and a bodily power. Human minds are nothing other than the ideas that correspond to human bodies. All things, including rocks, mice, and computers, have "minds," ideal powers that correspond precisely to the powers of their bodies ("panpsychism").

Such metaphysical principles inspired the Deep Ecologist philosopher Arne Naess to declare, "No great philosopher has so much to offer in the way of clarification and articulation of basic ecological attitudes as Baruch Spinoza" (1977, p. 54). Deep Ecologists are also attracted to Spinoza's doctrine of *conatus* (striving), which contends that each and every thing strives to persevere in being and enhance its existence. The recognition that all beings aim at "self-realization," in an ecological interpretation, fosters an appreciation of other beings' integrity in a way that might restrain human exploitation. Spinoza's principles and his own attitudes condemn anthropocentrism, the view that the nonhuman world exists for the use and pleasure of people (see also Sessions 1977, Mathews 1991).

Deep Ecological interpretations of Spinoza have met with criticism from several theorists, notably the philosopher Genevieve Lloyd. Lloyd notes that, although Spinoza propounds a nonanthropocentric metaphysics, his "morality" is entirely human-centered and excludes any moral consideration of nonhuman beings, affirming that "other species can be ruthlessly exploited for human ends" (1980, p. 294). Her criticism centers on Spinoza's assertion that the "rational principle of seeking our advantage teaches us to establish a bond with men, but not with the lower animals, or with things whose nature is different from human nature" (Spinoza 1985, p. 566).

Naess rebuts Lloyd's criticism by pointing out that the term "moral" is never used by Spinoza and is entirely inappropriate to his philosophy. He contends that Lloyd misleads her readers by suggesting that there is a uniquely human moral community that enjoys special "rights" separate from "natural right," the capacity to preserve and enhance one's life. The "rights" a community institutes and upholds are simply techniques humans have within their power to live well. Such forms of association are different in character, but not in kind, from, for example, packs of wolves. Although Spinoza cannot support a program of animal rights, his views allow nonhuman animals to be "members of *life communities* on par with babies, lunatics, and others who do not cooperate as citizens, but are cared for in part for their own sake" (Naess 1980, p. 319).

Nevertheless, Lloyd indicates an ambiguity in ecological philosophy; although ecological thought aims above all to undermine anthropocentrism, it often does so by "humanizing the nonhuman" (1980, p. 307). Anthropocentrism is not obviously weakened by affirming that all beings resemble us in some ways, or by viewing the universe as a great Self. Spinoza himself consistently criticized the theological tradition for imagining God in human terms. Lloyd's critique and Naess's attraction to Spinoza's antimoralism points to new directions in which Spinozism might go in support of an ecological perspective. Rather than expanding human categories to foster respect for nonhuman reality, Spinoza's philosophy suggests that we naturalize humans as thoroughly as possible. Spinoza encourages us to see that we are constituted deeply by the myriad powers we depend upon to survive and flourish, whether they are our families, the atmosphere, the military-industrial complex, or our nonhuman animal companions. Our constitutive connections complicate any boundary between the human and nonhuman, whereas a moral law upheld in the name of humanity as a special genre of being might falsely elevate people out of the rest of nature. Spinoza affirms our necessary dependence upon the human and nonhuman world, and thereby points to the importance of caring for and transforming the quality of such relationships to enjoy more vital ways of being together.

SEE ALSO *Deep Ecology; Naess, Arne; Pantheism.*

BIBLIOGRAPHY

Lloyd, Genevieve. 1980. "Spinoza's Environmental Ethics." *Inquiry* 23: 293–311.

Mathews, Freya. 1991. *The Ecological Self.* London: Routledge.

Naess, Arne. 1977. "Spinoza and Ecology." *Philosophia* 7(1): 45–54.

Naess, Arne. 1980. "Environmental Ethics and Spinoza's Ethics: Comments on Genevieve Lloyd's Article." *Inquiry* 23: 313–325.

Sessions, George. 1977. "Spinoza and Jeffers on Man in Nature." *Inquiry* 20: 481–528.

Spinoza, Benedictus de. 1985. *The Complete Works of Spinoza,* Vol. 1, ed. and trans. Edwin Curley. Princeton, NJ: Princeton University Press.

Hasana Sharp

ST. FRANCIS OF ASSISI
1181 or 1182–1226

St. Francis of Assisi, patron saint of the environment and animals, was born in either 1181 or 1182 in the town of Assisi in Umbria, Italy, and was baptized as Giovanni de Bernardone. His wealthy merchant father, who was in France on business at the time, renamed him Francesco, in honor of his maternal ancestors from France. As a young man, Francis pursued an education while engaging in carousing with his friends from the nobility. Never interested in entering into his father's commercial ventures, Francis experienced a spiritual crisis in his early twenties. He began spending time as a hermit, while seeking guidance from God. Much to his father's chagrin, Francis rebuilt dilapidated churches, aided those without adequate clothes or food, nursed lepers, and abandoned his own fine garments and possessions, to cleanse himself of worldly attachments. In 1209, Francis founded the Order of Friars Minor, bound by vows of poverty, chastity, and obedience. Three years later, St. Clare of Assisi, supported by Francis, organized an order for women, the Poor Clares, based on the same principles. In 1221, St. Francis founded the Third Order open to the laity. Francis died in 1226, and was canonized a mere two years later in 1228 by Pope Gregory IX.

St. Francis of Assisi is often misunderstood as the first saint to encourage Christians to care for nature, when he is actually a participant in a long tradition of Christian protection of nature beginning with Christian monks in the third century. His medieval biographers, such as Thomas of Celano, report Francis fed bees in winter, released animals from traps, and allowed native wildflowers to remain around the edges of a cultivated garden. These practices, however,

St. Francis of Assisi. *St. Francis Preaches to the Birds, a late thirteenth-century painting by the Italian artist Giotto. In Christianity, St. Francis is considered to be the patron saint of the environment and animals.* THE ART ARCHIVE/MUSEE DU LOURVE PARIS/ALFREDO DAGLI ORTI.

continued the values of late ancient Celtic and desert monastics, who protected wildlife from hunting and appreciated regional flora. Francis's original contributions lie instead in his resistance to Christian treatment of nature as a mere material possession; his concern for the marginalized, whether human or in nature; and his practical approach to the social issues presented by the growth of towns and cities as Europe slowly emerged from feudalism. In his "Canticle to Brother Sun," for example, Francis exhorted Christians to a non-hierarchal valuation of nature, by identifying the planets as brother and sister. When Francis preached his well-known sermon to the birds, he not only acknowledged the worth before God of all non-human creatures, he also subtly argued for the spiritual equity of humans who were economically marginalized or socially ostracized, such as the lepers Francis physically embraced. The Franciscan friars were mendicants, which meant they were not necessarily tied to the church's estates and property holdings, and could reach out to the urban poor—who, unlike the feudal peasants, had no access to the land. Franciscans ministered to the victims of natural disasters, like earthquakes, rather than adopting the fatalistic attitude that such tragedies were God's judgment on the unrighteous.

Many stories about St. Francis's environmental ethos can be interpreted in two ways. According to legend, St. Francis tamed the wolf who had been terrifying people who ventured outside the town walls of Gubbio by striking an agreement between the wolf and the citizenry: If the wolf stopped his attacks, the town's residents would feed him. On one level this tale calls for the protection of all God's creatures, even the large predators. On another, it argues that those unable to easily feed themselves will eventually attack or harm others. Sharing resources with those who have no honest way to make a living thus will prevent the hungry from becoming outlaws and threats to the greater society.

Francis withdrew to Mount Alverna for prayer, and in doing so received the stigmata or the wounds of Christ from a seraph, an angelic being. For Christians, Christ is the Creator of the universe and his body is an analog for the cosmos. At the time, avaricious churchmen utilized the centrality of the body and blood of Christ, in the bread and wine of the Eucharist, as a rationalization for church ownership of vineyards or collection of offerings of bread or grain; Francis received the environmentally critical symbolism of Christ's suffering and death as a result of his unselfish service to others. Francis thus reminded Christians that care for the poor and needy, and for God's Creation, should be the expression of their religious faith, rather than with feudal hegemony over natural resources. In experiencing a vision at an isolated and uncultivated locale, Francis continued a Biblical tradition of encounters with the divine in wilderness settings.

After the death of Francis, Franciscan appreciation of the environment contributed to the development of modern science through the investigations of friars such as Roger Bacon, who during the thirteenth century utilized the scientific method and advocated detailed observation of nature. Today, both the Roman Catholic Church and the Anglican Communion (Episcopalians) continue to sanction Franciscan orders that champion environmental causes, work to relieve poverty, sponsor retreats in natural settings, and encourage Christians to simple living and care of the earth.

SEE ALSO *Animal Ethics; Bible; Christianity; Ecotheology.*

BIBLIOGRAPHY

Armstrong, Edward. 1973. *Saint Francis: Nature Mystic; The Derivation and Significance of the Nature Stories in the Franciscan Legend.* Berkeley: University of California Press.

Bratton, Susan. 2007. *Environmental Values in Christian Art.* Albany: State University of New York Press.

Sorrell, Roger D. 1988. *St. Francis of Assisi and Nature: Tradition and Innovation in Western Christian Attitudes toward the Environment.* New York: Oxford University Press.

Susan Power Bratton

STEWARDSHIP

The term *stewardship* refers to a way of thinking about environmental responsibility that is based on the metaphor of human beings as stewards: persons who are responsible to an owner for the care or management of that person's household and goods. Environmental stewardship extends the concept of a household to include the whole earth or some part of it. It also extends the role of a steward to the human race, the present generation, an organization or agency, or an individual landowner. Synonyms or near synonyms for *steward* include *caretaker*, *curator*, *custodian*, *guardian*, and *trustee*.

THE STEWARDSHIP CONCEPT

Steward comes from the Old English word *stīweard*, from *stī*, meaning "hall," and *weard*, meaning "ward," or "guard." Its primary meaning is an official or servant who is responsible for the domestic affairs of a household, including supervising the other servants, managing money and keeping the books, and directing the serving of meals. The term also has been used to denote a ruler or highly placed noble serving as a minister to a ruler and to offices and occupations such as magistrate, labor union representative, flight attendant, shipboard caterer, and financial manager.

The two key elements of the stewardship concept are the ability to care for, manage, or control persons or things and accountability for the proper exercise of that ability. A steward exercises power and authority but does not have license to do so in a self-serving or careless manner.

Although the term *stewardship* is used widely in environmental writing, its implications and appropriateness have been debated. Critics charge that stewardship is based on the same problematic assumptions and values that underlie the environmental crisis: the idea that humans are separate from and superior to the rest of nature, which is seen as a pool of resources to be managed and controlled by the rational application of science and technology.

Defenders counter that the stewardship model recognizes that humans are unique in their power to transform, degrade, or destroy the earth and their ability to make individual and collective decisions about ways to use that power. For them, the concept of stewardship expresses a sense of responsibility for one's actions as they affect, directly or indirectly, other people and the natural environment over great distances and far into the future, a sense that has become rare in a competitive, individualistic, shortsighted, profit-oriented, and anthropocentric consumer culture.

RELIGIOUS TRADITIONS OF STEWARDSHIP

In many Western religious traditions God is the true owner of the earth and the one to whom human beings are responsible in every aspect of their lives. The stewardship concept is less prominent in or largely absent from Eastern and indigenous religious traditions. Some regard stewardship as a weak or marginal theme in Western religion, but others see it as firmly rooted in Jewish, Islamic and Christian scripture and tradition.

Psalm 24:1 declares, "The earth is the Lord's, and the fullness thereof; the world, and they that dwell therein." God tells the people of Israel, "The land is mine; with me you are but aliens and tenants" (Leviticus 25:23). Islam teaches that Allah has appointed humans as caliphs or viceroys in the earth (Qu'ran, 6:165). In Christianity the overt use of the term *stewardship* for a person's responsibility to God for the use of created things began with the sixteenth-century theologian John Calvin: "Let everyone regard himself as the steward of God in all things which he possesses. Then he will neither conduct himself dissolutely, nor corrupt by abuse those things which God requires to be preserved" (*Commentary on Genesis,* 1554). In the seventeenth-century English chief justice Matthew Hale extended the concept to human responsibility for the natural environment, writing, "In relation therefore to this inferior world of Brutes and Vegetables, the End of Man's Creation was, that he should be the vice-roy of the great God of Heaven and Earth in this inferior world; his Steward, *Villicus,* Bayliff, or Farmer of this goodly Farm of the lower World" whose charge is "to preserve the face of the Earth in beauty, usefulness, and fruitfulness" (*The Primitive Origination of Mankind* 1677).

In twentieth-century Protestantism the most common meaning of the term entailed giving a portion of one's income, volunteering one's time, or lending one's talents to the work of the church. With the rise of environmental concerns in the second half of the century, *environmental stewardship*, *stewardship of the environment*, and similar usages became increasingly common in religious and ethical literature. The concept and its scriptural roots have been used to refute the charge that rapacious domination is the orthodox Jewish and Christian attitude toward nature. It also has been used to define an ethical position between exploitive domination and subordination to nature.

Some theologically conservative Christians, especially those who expect the imminent end of the world, regard environmental stewardship as at best irrelevant to the church's mission to spread the message of salvation or at worst heretical earth worship. Political and economic conservatives criticize religious communities' advocacy for religious environmental concern as overly influenced by "radical environmentalism." Although they may characterize their own position as stewardship (e.g., the Interfaith Stewardship Alliance), their skepticism about ecological threats and opposition to environmental regulation puts them at odds with most people who embrace that term.

In contrast, some religious writers, especially more liberal and radical theologians, have criticized stewardship language for being too hierarchical and managerial to express human solidarity with the rest of creation or God's active presence within nature.

SECULAR VERSIONS OF STEWARDSHIP

There are also significant uses of the term in secular contexts, suggesting that it need not assume a theistic or God-centered worldview. For example, there are organizations such as the Forest Stewardship Council, the Land Stewardship Project, and the Alberta Stewardship Network. Conservation programs that encourage voluntary efforts by landowner communities to promote ecological preservation and restoration on their property often are called stewardship programs, reflecting their reliance on a sense of obligation on the part of the property owner that transcends short-term economic self-interest.

What could take the place of God in a secular stewardship ethic? There are a number of human communities that environmental stewards might serve and to which they might be responsible:

1. Future generations: The present generation of human beings holds the earth in trust for those who will come after it, who are entitled to their fair share of the planet's bounty. Persons alive today thus have a responsibility to live sustainably, to use the earth's resources in ways that will not deplete them or impair the functioning of the earth's life-support systems.
2. The international community: Nations may be seen as responsible to the wider international community, including its most impoverished members, for the fair distribution of the benefits of economic development and for maintaining the quality and integrity of the global environment.
3. Citizens: Government agencies have stewardship responsibility for public lands and resources (the U.S. Bureau of Land Management, the National Park Service, etc.) on behalf of the nation's citizens.
4. Local or regional communities: Individual landowners may be held responsible to their neighbors for maintaining the ecological values (e.g., wildlife habitat, water quality, soil fertility) of their property.

Some secular stewards may see themselves as responsible not only for some portion of nature but also to nature as the whole community of life or as a cosmic creative process.

MANAGEMENT OR CARE?

Stewardship sometimes is presented as wise use of natural resources for the benefit of human beings. In contrast to such an anthropocentric, or human-centered, view, nonanthropocentric understandings of stewardship regard humans as accountable for the welfare of nonhuman individuals or communities as well. The approach that is adopted affects the characterization of stewardship: as the management of a tool or resource or as care for a loved and respected fellow being.

The term *management* suggests an effort to control someone or something to serve a purpose that is outside or separate from that entity, imposing order on activities that otherwise would be chaotic or directionless. Although one person may manage another in certain circumstances, to treat people only as tools or resources to be directed or controlled violates their inherent dignity as beings who have intrinsic value. For some, to treat animals, plants, ecosystems, or other natural entities solely as means to an end is also morally objectionable.

Whether the object of care is valued for its usefulness (a car or a tool) or loved or respected for its own sake (a child, elderly parent, or medical patient), the concept of caring implies an attempt to maintain or restore the proper functioning of a machine or the health and well-being of an organism in accordance with its inherent norms or requirements. In a nonanthropocentric definition of stewardship, *care* is a more appropriate term than *management* in defining human beings' relationship to nonhuman beings. Nonetheless, some environmental philosophers deny that humans have the right or competence to exercise power over other species. For them, care may imply a condescending, presumptuous, or paternalistic attitude toward "inferior" creatures that undercuts a proper regard for fellow beings.

INTERVENTION OR RESTRAINT?

The definition of stewardship as acting upon its object as caring or as management gives rise to the objection that stewardship is inappropriately interventionist when applied to nature. For some critics of stewardship the idea that humans have a moral responsibility to intervene in natural systems assumes that humans know what is best and that nature cannot flourish without them. Those who assume the role of steward overestimate their ability to understand and control vast, complex webs of environmental relationships and overlook the ability of ecosystems to be self-regulating. A totally managed and humanized world also would deprive nature of its mysterious otherness and independence.

In response, advocates of stewardship argue that a hands-off policy toward nature is unrealistic and irresponsible. Very little of nature is untouched by humans, and not everything that once was thought of as "unspoiled wilderness" really was.

Humans must intervene in nature to meet their needs for survival and self-realization; stewardship means doing so in ways that will not waste resources or damage the environment to the detriment of human and other forms of life. Because so much of human activity has had destructive consequences, people have a responsibility to protect nature actively from those consequences and to reverse them where they can.

It also can be argued that humans are by nature creative transformers of their surroundings; denying them the right to do that violates their integrity and implies that they are unnatural. There is also a long-standing tradition, not necessarily embraced by all advocates of stewardship, that humans have a role in perfecting or completing an unfinished creation, carrying on the continuing work of the Creator or nature's creative powers.

The most plausible interpretations of stewardship fall between the extremes of total disengagement and total management. A restrained approach to stewardship would allow the minimum intervention necessary to maintain a sustainable, just, and dignified human civilization and protect and restore the species and ecosystems that human activity threatens or has harmed. Although human responsibility would be global in the sense that no part of the earth can be immune from being threatened or influenced by human activities, different degrees of intervention would be appropriate for different types of landscape, for example, more intervention in domesticated landscapes and less in wilderness areas.

CONTRIBUTIONS AND SHORTCOMINGS

Stewardship has contributed to the evolution of environmental ethics. It has helped draw questions about human impacts on the environment into the realm of moral and religious responsibility rather than leaving them to economists, industrialists, and engineers. It also has focused attention on the importance for environmental ethics of the choice of metaphors for defining the human place and role in nature.

As is the case with any moral model or metaphor, stewardship also has shortcomings. By itself, it does not provide criteria for good stewardship; stated in another way, its susceptibility to widely divergent interpretations offers too many possible types of criteria. It does not present a direct challenge to many of the cultural assumptions that for some philosophers and theologians lie at the roots of the environmental crisis. Even if stewardship does not require or endorse those assumptions, this ambiguity leaves it open to being reduced to more or less enlightened anthropocentric utilitarianism.

However, it is clear that for some who advocate a stewardship ethic such a reductive understanding is not adequate. To grasp how a particular stewardship ethic interprets environmental responsibility, one must look beyond its use of the word to its underlying convictions about humanity, nature, and the ultimate sources of moral obligation.

SEE ALSO *Animal Ethics; Christianity; Future Generations; Islam; Judaism; Land Ethic; Utilitarianism.*

BIBLIOGRAPHY

Attfield, Robin. 1999. *The Ethics of the Global Environment.* Edinburgh, Scotland: Edinburgh University Press.

Black, John. 1970. *The Dominion of Man: The Search for Ecological Responsibility.* Edinburgh, Scotland: Edinburgh University Press.

DeWitt, Calvin B. 1998. *Caring for Creation: Responsible Stewardship of God's Handiwork.* Grand Rapids, MI: Baker Books.

Hall, Douglas John. 1986. *Imaging God: Dominion as Stewardship.* Grand Rapids, MI: W. B. Eerdmans; New York: Friendship Press.

Palmer, Clare. 1992. "Stewardship: A Case Study in Environmental Ethics." In *The Earth Beneath: A Critical Guide to Green Theology*, ed. Ian Ball, Margaret Goodall, Clare Palmer, et al. London: SPCK.

Passmore, John. 1974. *Man's Responsibility for Nature: Ecological Problems and Western Traditions.* New York: Scribner.

Paterson, John L. 2003. "Conceptualizing Stewardship in Agriculture within the Christian Tradition." *Environmental Ethics* 25(1): 43–58.

Roach, Catherine M.; Tim I. Hollins; Brian E. McLaren, et al. 2006. "Ducks, Bogs, and Guns: A Case Study of Stewardship Ethics in Newfoundland." *Ethics and the Environment* 11(1): 43–70.

Peter W. Bakken

STRIP MINING

SEE *Mining: III. Mountaintop Removal.*

SUBSTANTIAL EQUIVALENCE

Substantial equivalence is a concept that was developed to aid in the evaluation of the safety of new food products, particularly genetically modified foods. Many traditional foods, such as tomatoes and potatoes, contain some toxins but nevertheless have a long history of safe consumption. In evaluating the safety of novel foods, the question is not whether certain foods are completely safe but whether they are at least as safe as traditional foods.

Substantial equivalence is the principle that novel foods should be compared with traditional foods to determine whether a novel food is likely to present a new risk or a greater risk than the conventional food. The comparison involves both the biological composition of the foods and their nutrient, antinutrient, and toxin levels. Substantial equivalence is a doctrine of relative safety, not absolute safety.

If substantial equivalence is established for a novel food product, that product generally is considered to be as safe as its conventional counterpart. For many genetically modified foods, substantial equivalence can be established with the exception of the one or two specific new traits that the genetically modified food was engineered to contain. In these cases, risk assessment is directed toward those new traits. In rare instances, a novel food may be wholly unlike any traditional food and no substantial equivalence will exist, requiring extensive safety testing of the new product.

As genetically modified food products were being developed and tested in the late 1980s and early 1990s, concern arose about finding ways to evaluate their safety. The Organization for Economic Cooperation and Development (OECD) brought together a team of experts from many countries to consider how to assess the safety of genetically modified foods. One of the recommendations was the substantial equivalence concept. The principle of substantial equivalence later was endorsed by a United Nations Food and Agriculture Organization (FAO) and World Health Organization (WHO) joint expert consultation. The adoption of the substantial equivalence concept by the OECD, FAO, and WHO was not binding on any countries but did lead to the adoption of the doctrine by many member countries. The manner in which the doctrine was implemented in different countries, however, varied significantly and led to ongoing disputes about genetically modified food products.

In the United States, the Food and Drug Administration (FDA) is responsible for ensuring that all marketed food products other than meat and poultry are safe. The FDA treats genetically engineered food products the same way as traditional food products; no special requirements apply.

In general, FDA approval is necessary before a food product that contains a new additive is commercialized. This could include genetic material added to a conventional food to produce a desired trait. There is an exception to the FDA approval requirement for foods that are "generally recognized as safe" (GRAS) by experts. With respect to genetically modified foods, the FDA relied on the concept of substantial equivalence to determine that "[i]n most cases, the substances expected to become components of food as a result of genetic modification of a plant will be the same as or substantially similar to substances commonly found in food, such as proteins, fats and oils, and carbohydrates" and therefore will be GRAS (Statement of Policy 1992, 22984, 22985).

Although the FDA sets standards, the manufacturer of a food additive, not the FDA, is responsible for determining whether that additive is GRAS. A manufacturer does not need to report to the FDA that it has made a GRAS determination. As a result, use of the substantial equivalence doctrine by the FDA has produced a system in which genetically modified foods generally do not need regulatory approval prior to their commercialization.

Many European countries have adopted the substantial equivalence concept but have applied it much more stringently because of a high level of concern about the safety of genetically modified products. Many European countries generally do not permit the commercialization of most genetically engineered food products. This difference has led to ongoing international disputes between the United States and the European Union about the international shipment of and trade in genetically modified products.

The substantial equivalence concept has been criticized for lacking a definition of what precisely to compare or how similar items need to be in order to be substantially equivalent. This vagueness allows for a wide degree of flexibility in applying the doctrine, which can result in less rigorous regulation or safety evaluation than is appropriate. There are also concerns about the ability of substantial equivalence to provide an adequate basis for safety assessment for more complex next-generation genetically modified products. In 2007 the FDA issued a draft risk assessment that applied a substantial equivalence analysis to conclude that food products from cloned animals, such as meat and milk products, were as safe to eat as products from conventionally bred animals.

SEE ALSO *Animal Cloning; Food; Genetically Modified Organisms and Biotechnology; Risk Assessment; U.S. Food and Drug Administration.*

BIBLIOGRAPHY

Levidow, Les; Joseph Murphy; and Susan Carr. 2007. "Recasting 'Substantial Equivalence': Transatlantic Governance of GM Food." *Science Technology Human Values* 32(1): 26–64.

Mandel, Gregory N. 2004. "Gaps, Inexperience, Inconsistencies, and Overlaps: Crisis in the Regulation of Genetically Modified Plants and Animals." *William and Mary Law Review* 45: 2167–2259.

McGarity, Thomas O. 2002. "Seeds of Distrust: Federal Regulation of Genetically Modified Foods." *University of Michigan Journal of Law Reform* 35: 403–510.

"Statement of Policy: Foods Derived from New Plant Varieties." 1992 *Federal Register* 57: 22,984, 22,985.

Gregory N. Mandel

SUSTAINABILITY

The word *sustainability* has become ubiquitous in environmental affairs since the 1987 World Commission on Environment and Development (WCED) report *Our Common Future* popularized the concept of sustainable development. The idea, however, has a long history. The term *sustainability* has a range of definitions running into hundreds, making any preliminary definition necessarily highly abstract, but all cluster around the core idea that some system, process, range of welfare, or set of items can be maintained at a certain rate or level for the long term; the ingredients of this formulation and its applications, however, vary widely, as do their disciplinary roots and practical implications.

POLITICAL ECONOMY

The earliest clear example of the concept of sustainability in economic thought is in John Stuart Mill's (1806–1873) treatment of the "stationary state" in Book IV of his *Principles of Political Economy* (1848). In this work Mill argues that an end to economic growth is ultimately unavoidable but that this limitation need not imply a rejection progress; rather, he anticipated significant moral and emotional human improvement through a more egalitarian distribution of wealth and reduced economic competition. Although this prescription was original to Mill, in making it he acknowledged debts to Thomas Malthus's (1766–1834) earlier writings on natural limits, especially "An Essay on the Principle of Population," which had gone through six editions between 1798 and 1826 and significantly influenced opinion among Mill's utilitarian philosophical bedfellows in England. Malthus's argument, however, did not share Mill's optimism about prospects for social improvement, and was originally motivated precisely by Malthus's opposition to doctrines of human progress advanced in the wake of the French Revolution.

Malthus argued that unchecked population increases geometrically (e.g., 1,2,4,8) whereas food supply increases only arithmetically (e.g., 1,2,3,4); hence there is a constant tendency for demand on necessities to outstrip supply when population rises, along with a permanent likelihood of poverty and starvation for some section of the population, a circumstance that undercuts arguments for social improvement. This focus on population rather than differences in wealth and consumption was underscored by Malthus's opposition to contraception and was further emphasized by his supporters' tendency to concentrate on (possibly compulsory) birth control, but only for the poorer classes, priorities that were sharply condemned by the nineteenth-century radical left (e.g., Friedrich Engels's 1844 work *Outlines of a Critique of Political Economy*) and that still fuel suspicions of Malthusian influence on thinking about sustainability today.

Mill's idea of the stationary state presented a contrast to Malthus's views not only in its optimism and advocacy of contraception—Mill served two nights in prison for distributing advocacy literature on birth control methods in 1823—but also in the conditions he envisioned. Whereas Malthus saw the changes of population and resource base as a potential source of chronic instability, Mill's stationary state is stable and loosely egalitarian, and thus a progenitor of notions of a "steady-state economy" that have been popular among contemporary advocates of sustainability.

Both Malthus's outlook and Mill's utilitarian schemes grew increasingly irrelevant to mainstream political economy in the latter nineteenth and early twentieth centuries as technological improvements, along with increased agricultural and industrial productivity, appeared to dispel Malthusian gloom about an unavoidable disparity between a sprinting population and a crawling resource base. Cornucopian technological and productionist optimism were the prevailing ideas in economics at the turn of the twentieth century. Developments within economic theory also contributed, for the marginal-utility theory that arose in neoclassical economics at the nineteenth century involved a new scarcity postulate worked out simultaneously by Carl Menger, W. S. Jevons, and Leon Walras. This postulate saw need in radically subjective terms, as an internal human state rather than as the naturalistic, interactive phenomenon postulated by Mill and Malthus, and maintained that internal human desires defined need, and that satisfying these desires/needs, which themselves are stimulated by seeing desirable objects, drives human activity. This in turn means that individuals choose between satisfying various needs, with each person having an internal hierarchy of needs and endeavoring to calculatively obtain the best possible result in relation to goods that are in short supply. The existence of an infinite number of these needs places limits on any given single need at any particular moment; accordingly aggregate needs are infinitely expandable, but are rendered calculable by individuals making hierarchical choices that limit particular needs, This desocialized model of need also incorporated optimism about the functional substitutability of goods, assuming on the basis of this model that scarcity of a given good (e.g., oil) would generate incentives to develop resource substitutes for that good (e.g., ethanol) and so absolute external scarcity could be kept at bay. This new theory came to dominate twentieth-century academic economics, pushing the consideration of external limits into the background (Xenos 1989).

The concept of hard external limits to economic expansion of a sort not amenable to technological fixes or resource substitution resurfaced with the Club of Rome's 1972 *Limits to Growth* report (Meadows et al. 1972). This report examined five variables—world population, industrialization, pollution, food production, and resource depletion—and ran these through successive computer simulations to explore possible outcomes of exponential growth combined with finite resources. The simplified models, though not aimed at explicit predictions, consistently manifested feedback loops producing dire consequences before 2100, suggesting a rate of nonrenewable resource depletion rapid enough to portend exhaustion within a little more than a century, with no likelihood of any technological rescue. The book popularized the idea of physical limits to growth and paved the way for concepts of sustainability based on that prospect. This work was still haunted by the specter of population growth, but since that time most sustainability-oriented environmentalists have increasingly emphasized the dangers of overconsumption and downplayed those of overpopulation. Nevertheless, the argument that continuous population increase will eventually place strains a depleting natural resource base, even in the most egalitarian social arrangements, remains part of environmentalist discourse. Accordingly, cornucopian and market-based critics of environmentalism such as Julian Simon (1996) and Bjorn Lomborg (2001) have characterized sustainability arguments as neo-Malthusian.

FORESTRY AND SUSTAINABLE YIELDS

Forestry has also informed modern ideas about sustainability. The work of the American forester-conservationist Gifford Pinchot (1865–1946) has been a major influence. Echoing Mill, Pinchot combined theories of resource scarcity with an anthropocentric utilitarian moral concern for human welfare. For Pinchot the forester's mission was "based on the elimination of waste, and directed toward the best use of all we have for the greatest good of the greatest number for the longest time" (Pinchot 1914, p. 25). In his autobiography *Breaking New Ground* (1947) Pinchot recounts that, upon his return to the United States in 1890 after a period of forestry training in Europe, he was horrified at American lumbermen's wastefulness. He and his allies, pointing to the dangers of timber famine, established a national U.S. Forest Service based on principles of efficient harvesting of resources through scientific forest management and replanting, and the prevention of fire, theft, improper use, and destruction. These practices aimed at preserving the resources in perpetuity. This mandate came to include economic and longterm social-justice concerns, as manifested in Pinchot's concerns about the theft of timberland land from Native Americans and his campaign in 1908 and 1909 to introduce systematic forestry on American Indian reservations. He claimed that this last measure, within eighteen months, "saved large sums of money to the Indians, gave many of them profitable employment, and by the introduction of Forestry promised to make that employment permanent" (Pinchot 1947, p. 412). Although these arrangements were truncated in 1909 by political dispute, they were resurrected in the mid-1930s under Forest Service head Ferdinand A. Silcox as the Indian New Deal, reviving ideas of social service in forestry that are still influential.

Thanks in large measure to the precedents set by Pinchot's work, the range of functions included as legally mandatory in forest planning have expanded. The Multiple Use Sustained Yield Act of 1960 formalized the U.S. Forest Service range of duties by requiring forest planning to consider issues such as outdoor recreation, location in relation to human settlements, watersheds, and fish and wildlife preservation in addition to the more familiar concerns about timber and grazing. In each case the operative principle is "sustainable yield"—the amount of a resource that can be extracted without undermining the natural system's core capacities to maintain or improve upon its full range of services.

Pinchot's original conception of forestry was anthropocentric and geared to economic development; a raft of other issues, however, has arisen in the past forty years. One major source of controversy is clear cutting, the clearing and replanting of an entire area of forest as opposed to selective felling in a given area. This practice, which had become dominant in the U.S. Forest Service by the late 1950s, is supported by timber interests (for whom it can be more profitable) and by many foresters, but most environmentalists regard it as abusive to forestlands, especially because of habitat loss, even if the species affected may be ecologically unimportant to the system's productivity. This controversy is an example of how anthropocentric and nonanthropocentric conceptions of nature's value can result in practical differences even when there is agreement among the parties about the goal of sustaining the long-term use of natural resources. The problem is most pronounced in areas where clear cutting might affect vulnerable species; not surprisingly, then, the first wave of organized opposition to Forest Service clear cutting occurred around the same time as the passage of the Endangered Species Act (1973) and the 1975 Convention on International Trade in Endangered

Species of Wild Fauna and Flora (CITES), which sought to integrate balanced species use with conservation. The latter was the first major international agreement in which the idea of sustainable use was implicit, though the phrase was not used; the convention does not expressly promote sustainable use by defining the term or demanding particular practices, but it does seek to prevent destruction and unsustainable use.

SUSTAINABILITY AND SUSTAINABLE DEVELOPMENT

Both political economy and forestry were prominent influences on the ideas sustainable development discussed in the 1987 WCED Report *Our Common Future* (often known as the Brundtland Report after its chair, the former Norwegian prime minister Gro Harlem Brundtland). Though it did not coin the phrase "sustainable development," the report furnished its basic definition as development that "meets the needs of the present without compromising the ability of future generations to meet their own needs" (WCED 1987, p. 8). It drew upon earlier precedents by linking sustainable use of resources to intergenerational, intragenerational, and international distributive justice and poverty relief, noting the extent to which poverty causes ecological depletion and linking these points to conservation concerns. The WCED sustainable-development model is, however, clearly anthropocentric, embracing technological optimism and suggesting a new kind of economic growth rather than questioning or rejecting the very idea of growth. The 1992 United Nations Conference on Environment and Development in Rio de Janeiro built upon this report in forging the Convention on Biological Diversity, the first treaty to expressly promote the idea of sustainable use as an international ideal.

A concern with yield is an important but not sufficient element of a practice of ecologically sustainable development. Ecological sustainability implies the satisfaction of three conditions in human interactions with nature: (1) Rates of use of renewable resources must not exceed their rates of regeneration; (2) rates of use of nonrenewable resources must not exceed the rate at which renewable substitutes can be developed; and (3) rates of pollution emission must not exceed the assimilative capacity of the environment (Jones 2003). Human impacts in these areas may be measured by using ecological footprint analysis, as developed by Wackernagel and Rees (1996).

VARIETIES OF SUSTAINABILITY

The concept of sustainability poses two major questions: What is to be sustained? Who or what should be the beneficiary of sustainability? In anthropocentric theories the sustaining of ecological systems aims at the flourishing of humans; indeed, some argue that a sufficiently broad conception of human fulfillment coupled with a recognition of human ignorance may lead to a policy convergence between anthropocentric and nonanthropocentric views (Norton 1991). For example, the loss of a species in a given ecosystem might superficially appear unimportant within an anthropocentric view of policy, but if the complexity of ecosystems and the possibility of human error are used to mandate caution, the sensible policy course may still be to avoid risking any possible unforeseen impacts of the loss, thus mandating the same policy as that which would come from a nonanthropocentric perspective. Alternatively it can be argued that, if some species are unnecessary to human continuity and one allows some resource substitutability (for example, moving away from consuming scarce Atlantic cod supplies and towards using more plentiful and functionally equivalent European haddock), then only natural capital critical to human survival need be sustained for future generations (Dobson 2000); such a view might call for the complement of a nonanthropocentric perspective to justify the protection of areas of nature not critical to human well-being.

A quandary of intergenerational justice is that granting equal resource access to every generation without calculating an endpoint yields absurd conclusions: Finite resources must be divided among an infinite number of claimants, and so "no one gets anything at any time" (Laslett and Fishkin 1992, p. 6). Some argue, therefore, for a compromise between discounting the future (i.e., measuring the entitlements of future people as becoming progressively smaller and less important the further away they are from us in time) and the need to impose legitimate limits on the present generation.

One option is a "just savings" solution in the manner of John Rawls's justice theory (Wissenburg 1998), whereby people of all generations are regarded as morally equal and equally entitled to a particular basic set of opportunities, thus creating an obligation for each generation to pass on that set of opportunities to the next generation. Alternatively, a moral appeal may be made to future generations' vulnerability, arguing that this vulnerability creates obligations for the current generation (Goodin 1985; Cowen and Parfit 1992; Dobson 2000). Even in purely anthropocentric terms the details of such options still need calculation, and here the distinction between weak and strong sustainability becomes significant. Weak sustainability espouses the substitutability of natural capital (i.e., naturally occurring goods that have beneficial economic effects, such as the ability of forests to produce oxygen and absorb carbon dioxide) for human-made capital (i.e., human products that may have functionally similar economic effects to such naturally occurring goods). Weak sustainability maintains that so long as an even stock of total capital is maintained, economic growth can be beneficial and consumption rates maintained. In some formulations an even stock of welfare functions is to be maintained, and so the issue becomes still clearer: a choice between sustaining either a particular list of goods or a particular level of human welfare.

Strong sustainability insists on treating natural capital independently of human-made capital, rejecting the idea that natural capital (i.e. naturally occurring economically beneficial goods) can always in principle be substituted by man-made equivalent goods, and so strong sustainability theory advocates sustaining particular natural goods and processes (i.e., physical "stuff") rather than undifferentiated total capital or welfare (i.e., abstract measurements of welfare held at a particular level). Although weak-sustainability has been more popular among thinkers stressing the range of future individual choices, Bryan Norton has supported the strong-sustainability perspective by a series of highly ingenious arguments concerning future human options and collective goods, maintaining that future people's opportunities for living fulfilling lives mandates strong rather than weak sustainability (Norton 2005).

SEE ALSO *Brundtland Report; Convention on Biodiversity; Environmental Law; Future Generations; Intergenerational Justice; Limits to Growth; Norton, Bryan; Pinchot, Gifford; Population; Resource Management; Sustainable Development; U.S. Forest Service.*

BIBLIOGRAPHY

Beckerman, Wilfred. 1995. "How Would You Like Your 'Sustainability' Sir? Weak or Strong? A Reply to My Critics." *Environmental Values* 4: 169–179.

Cowen, Tyler, and Derek Parfit. 1992. "Against the Social Discount Rate." In *Justice between Age Groups and Generations*, eds. Peter Laslett and James Fishkin. New Haven, CT: Yale University Press.

Daly, Herman E. 1992. *Steady State Economics*. 2nd edition. London: Earthscan.

Daly, Herman E. 1995. "On Wilfred Beckerman's Critique of Sustainable Development." *Environmental Values* 4: 49–55.

Dobson, Andrew. 2000. *Justice and the Environment*. Oxford: Oxford University Press.

Engels, Friedrich. 1973 (1844). "Outlines of a Critique of Political Economy." In *Collected Works of Karl Marx and Friedrich Engels, 1843-44*. Vol. 3. London: International Books.

Goodin, Robert E. 1985. *Protecting the Vulnerable*. Chicago: University of Chicago Press.

Jones, Alan. "Ecological Sustainability" at Hunter and Central Coast Region: Education for Environmental Responsibility. Available from http://www.rumbalara.eec.education.nsw.gov.au

Laslett, Peter, and James Fishkin, eds. 1992. *Justice between Age Groups and Generations*. New Haven: Yale University Press.

Lomborg, Bjorn. 2001. *The Skeptical Environmentalist: Measuring the Real State of the World*. London: Cambridge University Press.

Malthus, Thomas R. 1999 (1798). *An Essay on the Principle of Population*. London: Oxford University Press.

Meadows, Donella H.; Jorgen Randers; Dennis L. Meadows; and William W. Behrens. 1972. *The Limits to Growth: A Report for the Club of Rome's Project on the Predicament of Mankind*. New York: Universe Books.

Mill, John Stuart. 2004 (1848). *Principles of Political Economy*. New York: Prometheus Books.

Norton, Bryan G. 1991. *Toward Unity among Environmentalists*. London: Oxford University Press.

Norton, Bryan G. 2005. *Sustainability*. Chicago: University of Chicago Press.

Pinchot, Gifford. 1914. *The Training of a Forester*. Philadelphia: Lippincott.

Pinchot, Gifford. 1947. *Breaking New Ground*. New York: Harcourt, Brace.

Simon, Julian L. 1996. *The Ultimate Resource 2*. Princeton, NJ: Princeton University Press.

Wackernagel, Mathis, and William Rees. 1996. *Our Ecological Footprint*. London: New Society.

Wissenburg, Marcel 1998. *Green Liberalism*. London: UCL Press.

World Commission on Environment and Development (WCED). 1987. *Our Common Future*. New York: Oxford University Press.

Xenos, Nicholas. 1989. *Scarcity and Modernity*. London: Routledge.

Piers H. G. Stephens

SUSTAINABLE AGRICULTURE

At the beginning of the twenty-first century there are renewed threats of the starvation of millions, not due to warfare but to straightforward imbalance between food production and consumption. As seen frequently in Bangladesh, starvation will occur not first in overcrowded inner cities but in the very fields where food is grown. *Sustainable agriculture* is the technical name given policies and agricultural systems whose bottom-line goal is the prevention of such systemwide failures of agriculture.

Agricultural sustainability is defined as the ability to provide sufficient, healthful, and accessible food supplies into the indefinite future for the populations that depend on the systems. Agricultural sustainability has two more particular meanings: 1) sustainability in the goal of agriculture, where it implies a permanent ability to feed its constituent populations; and 2) sustainability in the means (or tools) that actual agricultural systems use to attain the goal of sustainable agriculture. The first goal for sustainability has a morally obligatory property: "We cannot responsibly establish a system of agriculture that is doomed eventually to fail to feed those whom the system is designed to feed." The second is a complex of technically measurable factors in the means (tools), to the goal, that enables us to confirm their success. The tools are various agricultural methods and integrated systems of agriculture. Among the factors determining the success of the means must be their ability to withstand predictable shocks such as droughts, pest invasions, and plant diseases. Some factors are directly measurable such as crop yields and rates of soil erosion (or the lack thereof). Others, noted by Richard Harwood (1990), may be more elusive such as the ability of a system to allow the nondisruptive evolution to new systems better fit to future environments and needs.

A key requirement for such future practices is that the natural and human resources needed for food production be prevented from becoming exhausted. Massive and constantly expanding use of current agricultural resources for the manufacture of biofuels is a prima facie violation of this requirement and its impact on the poor is already being felt.

OBLIGATIONS OF POLICY MAKERS

No plausible disagreement exists among agricultural professionals about the moral obligation to pursue "goal sustainability." The acrimonious debate is about whether different and often rival agricultural systems are capable of providing the means to the goal of sustainability. A constant review of agricultural systems to ascertain their current and future contribution to sustainability is a principal ethical duty of agricultural research and policy professionals. Because use of unsustainable means (tools/systems), especially on a national or large regional scale, will have tragic impacts on humans and the environment, such as those predicted for Bangladesh, the review of agricultural systems is ethically demanding. An ideal review is free of all bias, such as an unreflective preference for one's "usual way of doing things." Such a review of tools/means to a goal might be called simply "practical wisdom," a multifaceted virtue long ago identified by Aristotle. In the *Nichomachean Ethics*, Aristotle considered the nature of the intellectual virtues as well as of the moral virtues. And among the intellectual virtues, he sharply distinguished practical wisdom from scientific knowledge. The latter concerns the universal and changeless laws of nature, such as the law of gravity and the several laws of motion in physics. Practical wisdom is concerned more with doing than knowing in reference to things that are particular, not universal, and might be one way rather than another, this way, rather than that. The science of agriculture is called "agronomy." Adding the suffix "nomy"—law in Greek—to "agro" suggests that the science of agriculture is similar to astronomy, the science that discovers the universal laws of the motions of the stars and planets. But agricultural science should be conceived less like a type of Aristotelian scientific knowledge and more like a type of Aritstotelian practical wisdom. In addition to avoiding bias, the intellectual virtue of practical wisdom requires "memory," which in this context means memory of past failures that have burdened past agricultural systems (King 2004, p. 18). The function of "practical wisdom" in attaining a sustainable agriculture may be best illustrated by some history.

FAILURES IN CHOICE OF AGRICULTURAL SYSTEMS

Expensive agricultural projects by the U.S. Agency for International Development have repeatedly failed to be sustainable for the targeted populations in Asia, Africa, and Latin America because the farmers lacked access to and control of the resources needed for food production by means of those systems—such as artificial fertilizers, chemical pesticides and herbicides, and irrigation equipment. A careful review of the history of successful projects versus those that failed to provide sustainable benefits would have revealed such problems and the ways to avoid them. Further, involvement of the recipient communities in the design of their agricultural systems is vital to the long-term success of those systems (Korten 1980, Ingle 1982). The wise reviewer of tools needs to cultivate institutional memory as an essential component of the review of means and methods for sustainability.

Students of the applied sciences have noted that standard paradigms, of considerable value for efficiency in the exact sciences, like astronomy, exist also in applied sciences, like agronomy, where their function is to establish a priori categories of the "right way of doing things." What such paradigms provide in efficiency, however, they lose in efficacy, because other, more effective means lie beyond their purview. Further, prevailing agricultural paradigms can prevent a genuinely open search for optimum agricultural methods and means because, as elaborated by Thomas Kuhn (1970), the ruling paradigm shapes what subfields and competencies researchers will pursue, what equipment will be in their laboratories, and what courses they will have taken in learning to use that equipment. They will find it extremely difficult to give a fair hearing to alternative agricultural systems that use different tools and require different competencies.

The prevailing agricultural paradigm in the early 2000s is based on an industrial model and requires, among other things, the use of machinery (tractors, harvesters), fossil fuel energy (gasoline, diesel fuel), artificially manufactured mineral fertilizers (nitrogen, potassium, phosphorous), chemical pesticides and herbicides (atrazine, Round-up), and energy- and capital-intensive irrigation systems (pumps, pipes). A paradigm suited to such high resource use shortens the amount of time it takes to produce a publishable paper because less energy-intensive, low-input systems are more time consuming to develop. Thus, the low-input systems, while plausibly more sustainable and accessible to resource- and capital-poor farmers, receive little or no attention from agricultural "experts" (agronomists) because developing them threatens the careers of young researchers. In short, it can be professionally very risky to think outside the business-as-usual agriculture box.

Genuine "practical wisdom" must consider that an eclectic tool kit assembled from elements taken from various paradigms of agricultural systems may be best to assure sustainability. For example, an "agronomist," who becomes an agro-ecologist in designing an agricultural system for an impoverished community of farmers, might substitute oxen for tractors and gravity-driven for pump-driven irrigation to reduce a local farmers' need for capital (money) while still relying on genetically modified crops engineered to be resistant to pests. Such an open review of potentially sustainable agricultural tools assembled from apparently competing paradigms implies that in the agricultural sciences there is less need for universal and unanimously accepted first principles. This is clearly abhorrent in theoretical physics but is perfectly acceptable in an applied science. The "useful good" is the object of applied science whereas physics, like astronomy, pursues something close to universal, mathematical "truth," one feature of which is self-consistency and elegant coherence. In agricultural sciences the paradigm cultivated in American agricultural schools was variously named conventional, industrial, energy-intensive, capital-intensive, or production-oriented agriculture. With the goal of increasing plant nutrition and yield improvement, influenced by the nineteenth-century chemist Justus von Liebig, the paradigm took on pretensions of an exact science. In the late twentieth century an "opposing" camp, not yet having coalesced enough to have created a recognizable paradigm of its own, began to emerge. The competing paradigm-in-the-making was variously labeled alternative agriculture, low-input agriculture, agroecology, and organic agriculture, each variant claiming to be more conservative in the use of resources, especially artificial fertilizers, chemical pesticides and herbicides, and fossil-fuel energy. Advocates of the newly emerging paradigm thus also claimed that alternative, low-input, agroecological, organic agriculture was also "sustainable agriculture." The name was offensive to conventional agronomists because it implies that conventional agriculture is unsustainable and thus in fundamental violation of a basic obligation of agriculture.

OBJECTIVE SIGNS OF SUSTAINABILITY

In any case, both conventional and alternative agriculture need something closer to the ideal of "practical wisdom." A policy maker reviewing candidate agricultural systems must require that they meet concrete criteria of sustainability. J. W. Hansen (1996), applying the rationality of conventional production agriculture, indicates that any prudent review of means to sustainability assumes that we know what we are looking for. A positive determination of "means sustainability," in other words, requires explicitly specifying objective and measurable characteristics of systems that promise goal sustainability. An example of such criteria would be the capacity to produce the same yield, year after year, without loss of soil fertility or increase in soil salinity. In the absence of such explicitly stated, objective, measurable criteria, no agricultural system can be declared to be sustainable with any certainty. And because the history of natural shocks to agricultural production, such as drought and pest invasions, warns against relying on the stability of environmental conditions, quantitative probabilities of the system's ability to withstand shocks must be determined. To this should be added Harwood's criterion: the capacity to support the evolution of new, even better, more sustainable agricultural systems, under changing environmental and cultural circumstances.

CAUTIONARY COMPLETION OF THE DEFINITION

These signs of "means sustainability" are not doctrinaire and seem intuitively obvious. But they do not deal directly with the kind of properties found in a definition coming from the alternative agriculture paradigm in which sustainable agriculture systems are defined qualitatively as ecologically sound, economically viable, socially just, culturally appropriate, and based on holistic scientific approaches, including indigenous and community-based knowledge systems. This conglomerate, qualitative definition is criticized by defenders of conventional agriculture for including extraneous, nonagricultural functions like continuity or compatibility with local cultural traditions, but above all, "community food security"—reliable access to the food produced in a region by most everyone who lives there, whether they themselves are farmers or not. Defenders of the definition respond that conventional, industrial systems have led to the worsening of poverty and hunger in less developed countries precisely because, even though such systems increased regional food production, regional increases in food production did not lead to increased local access to food. Because industrial production systems involve high costs to producers, their commodities must fetch high prices in the global marketplace, often resulting

Bangladeshi Farmers Cleaning Paddy, 2008. *For Bangladesh and other Asian countries, paddy is one of the main sources of agriculture. Researchers have been trying to promote new forms of sustainable agriculture to these and other regions around the world. Although the definition is not yet concrete, most agree that sustainable agriculture should be ecologically sound, economically viable, socially just, culturally appropriate, and based on holistic and scientific approaches.* FARJANA KHAN GODHULY/AFP/GETTY IMAGES.

in the export of the foods produced, which are thus unavailable, at affordable prices, to local people who are not themselves food producers. This circumstance is exacerbated by the economy-of-scale principle of conventional agriculture, which requires farmers to "get big or get out." As some farmers "get big" by acquiring the lands of their neighbors, those who are driven out fall victim to poverty and ultimately to hunger.

The qualitative definition of sustainability proffered by advocates of alternative agricultural systems, however, does not provide the quantitative measures that are vital to genuine sustainability and future food security. Organic food systems have an appeal in their conservative use of nonrenewable resources, but their capacity to provide adequate and reliable future yields needs substantiation. For there may be an unavoidable trade-off: relatively low in-put agriculture is often also relatively low-output agriculture. The inability as of the early 2000s of organic agriculture to provide for entire populations in European and North American societies does not, however, make it irrelevant to sustainability. If the failures already mentioned teach us anything, it is that in agriculture one size—that is, one system, one paradigm—does not fit all. What may work well, at least for now, in North America and Europe may be a disaster in Africa and Latin America. Today's practitioners of organic agriculture may, furthermore, be learning the arts—for practical wisdom in the area of agriculture is more an art (in the Aristotelian sense, a *techne*) than a science—needed for resilient future agricultural systems.

Agroecological systems involve harnessing ecological knowledge to food production and enhancing sustainability by reducing the need to overwhelm local ecosystems with energy and synthetic chemical inputs—for example, in weed and pest control. Such systems depend on an intimate knowledge of the local ecology and hence require a community-based approach, because local people may have a wealth of such knowledge, known as "vernacular ecological knowledge" or, more commonly, as "traditional ecological knowledge" (TEK). Traditional agricultural societies are well suited to utilize agroecological approaches, because many have been doing so for many generations past. These approaches are not burdens on either the surrounding ecosystems or the host communities but are simply more sustainable augmentations of traditional practices or replacements for the conventional systems that involve "universal" prescriptions after they have catastrophically failed. Supporters of agroecology defend the dependence on community wisdom and labor as evidence that agroecological systems will be more responsive to community food security, the ultimate test, in their view, of true sustainability as a goal.

DEVELOPING-WORLD APPLICATIONS

Nadia Scialabba's United Nations 2007 FAO report, *Organic Agriculture and Food Security,* supplies some of the quantitative work needed on the theoretical capacity of organic agriculture to feed the world's population. She notes global models suggesting a capacity to feed the world at levels comparable to current conventional agriculture without an increase in land under cultivation and with reduced environmental impacts.

The alternative definition of "sustainable agriculture" differs from the definition given at the outset in part because it focuses explicitly on the human and environmental aspects of food production. This focus is in part due to the memory that U.S. agriculture became "unsustainable" from an ethical viewpoint at the turn of the twentieth century and again in the 1920s, when poverty, subhuman living conditions, and hunger were common in the farming communities. In efforts to find sustainable food systems, nothing could be more dangerous than leaving out the condition of the food producers and treating sustainability as a resource-input and food-output equation. Developing countries in the early 2000s provide evidence of the pervasive harm done to every kind of agricultural and environmental value when the human agents and their communities are not included in the selection of means to sustainability (Korten 1980).

It is scarcely conceivable that a global collapse of food adequacy would occur suddenly. The flip side of locally determined agricultural systems is that agricultural failure has its first visible impacts in rural locales among poor farmers who cannot feed their families. This is illustrated tragically again in Bangladesh where near famine strikes first among farm families in the months just before harvest. Neglect of political and socioeconomic factors in pursuit of sustainability will again create the environmental chaos observed by the soil scientist Charles Kellogg during one of America's brushes with unsustainability. He wrote, "The final exhaustion of the land follows, not precedes, the exhaustion of the people. In a final effort, exploited people pass their suffering to the land" (1938, p. 878).

SEE ALSO *Agriculture; Organic Farming; Sustainability.*

BIBLIOGRAPHY

Hansen, J. W. 1996. "Is Agricultural Sustainability a Useful Concept?" *Agricultural Systems* 50: 117–143.

Harwood, R. R. 1990. "A History of Sustainable Agriculture." In *Sustainable Agriculture Systems,* eds. C. A. Edwards et. al. Akeny, Iowa: Soil and Water Conservation Society.

Ingle, M. D. 1982. *Reaching the Poor through Development Assistance: An Overview of Strategies and Techniques.* Washington, DC: Development Project Management Center, USDA with USAID.

Kellogg, Charles E. 1938. "Soil and Society." Washington, DC: U.S. Department of Agriculture Yearbook.

King, J. L. 2004. "Thomas Aquinas on Prudence." M.A. thesis. San Francisco: San Francisco State University.

Korten, D. C. 1980. "Community Organization and Rural Development: A Learning Process Approach." *Public Administration Review* 40: 480–511.

Kuhn, Thomas S. 1970. *The Structure of Scientific Revolutions.* 2nd ed. Chicago: University of Chicago Press.

Scialabba, Nadia, 2007. *Organic Agriculture and Food Security.* Rome: United Nations FAO.

Stanislaus J. Dundon

SUSTAINABLE ARCHITECTURE AND ENGINEERING

In the late twentieth century many citizens in Western societies began to recognize that if other societies consumed resources at the same rate that they did, the ecosystem soon would be exhausted and unable to reproduce itself. That recognition motivated the United Nations to charter the World Council on Environment and Development (WCED) to investigate ways in which the apparent conflict between economic development and environmental degradation might be reconciled. Under the auspices of the Brundtland Commission, the WCED published its findings as *Our Common Future* in 1987. That document, long considered the seminal text on sustainable development, defined sustainable development as "development that meets the needs of the present without compromising the ability of future generations to meet their own needs" (World Council on Environment and Development 1987, p. 8). In the conflict between economic development and environmental protection unsustainable development usually is associated with the industrial and transportation sectors. Although industrial wastes and automobile emissions contribute substantially to degraded environmental

Earthship in Taos, New Mexico. *One modern example of sustainable architecture is the "earthship," designed by architect Michael Reynolds since the 1970s. The low-tech designs range from multi-million dollar luxury homes to small, communal living environments.* AP IMAGES

and social conditions, they are not the largest source of the problem.

RESPONSES TO THE CHALLENGE

According to the U.S. Department of Energy (DOE), the production and operation of the built environment accounts for almost half of all greenhouse gas (GHG) emissions and more than half of annual energy consumption in North America. These general statistics were put in a critical context by a Brookings Institution study that projected that in 2030 about half of the buildings in which Americans live, work, study, and shop will have been built after 2000 (Nelson 2004). If these projections are accurate, the construction and operation of buildings could become the single largest threat to public health, safety, and welfare as well as the major cause of environmental degradation and threats to other species. These statistics present a serious ethical challenge to architects, engineers, and those who commission their services.

As the public conversation about sustainable development has matured, three responses to this challenge have developed: resignation, denial, and hope. If one takes these statistics as inevitable, as do neo-Malthusians such as Paul Ehrlich (1971), one tends toward resignation. If one considers them overblown or unreliable, as do those with an interest in maintaining the status quo, one tends to deny the mounting evidence that shows that people's habits are unsustainable. However, if one takes these statistics seriously but rejects historical determinism, one tends to be hopeful that environmental collapse can be avoided. This is the story line of sustainable development, a modern narrative in which people create hope for the future by taking collective actions that may alter the path of history.

MODERN ARCHITECTURE

The popular thesis among many environmentalists, that ancient architecture is the best model for sustainable development in the future, generally is accompanied by the corollary that modern architecture has been relentlessly antinature. Although many modern buildings consume vast amounts of energy because they ignore the natural energy flows of their locations, there also are buildings that look stylistically modern but act in harmony with the ecologies and cultures for which they were designed. It is inaccurate to imagine that modern and sustainable architectures embody singular and opposed sets of values; reality is far more complex.

A good example of midcentury modern architecture that works with natural forces is the Tremaine House in Santa Barbara, California (1947–1948), designed by the Austrian-American architect Richard Neutra (1892–1970). The deep overhangs of the roof shield the large expanses of glass from unwanted solar heat gain, and the operable transom panels above the sliding glass doors allow for natural cross-ventilation even when the doors are closed. Neutra (1948) referred to this system as CSSA/LS, or continuous sub-soffit airchange over a lowered spandrel. What makes Neutra's work a particularly good example of protosustainability (the initial ideas required for sustainability to emerge in the form articulated by the Bruntland Report) is the fact that he employed energy-saving technologies not only in houses for people of substantial means such as the Tremaines but also for people of modest means. This is demonstrated by his 1948 proposal that Brazilian schools be equipped with his CSSA/LS system along with technologies designed to harvest rainwater and produce electricity on site. Neutra's work demonstrates a balanced sensitivity to the competing interests of economic development, environmental preservation, and social equity—commonly referred to as the three Es—long before the Brundtland Commission declared them to be the core values of sustainable development.

Neutra was not alone in the search for environmentally and socially responsible architecture. Among his peers were the theorists Patrick Geddes (1854–1932), Lewis Mumford (1895–1990), and Frederick Keisler (1890–1965), as well as a diverse group of practitioners that included Frank Lloyd Wright (1867–1959), Alvar Aalto (1898–1976), Harwell Hamilton Harris (1903–1990), and the many practitioners of regionalism in the Bay Area of California, Texas, and Mexico, among other places. In his collection of regionalist texts Vincent Canizaro (2007) demonstrated that some modern architects have championed the values that are associated with sustainability. In the world of engineering the appropriate technology movement that emerged in the 1960s demonstrates similar values.

After the appearance of *Our Common Future* in 1987 pressure from outside the building professions to build sustainably increased slowly; after the turn of the millennium the subject of sustainable design became a dominant discourse in both architecture and engineering. This does not imply that the legacy of Neutra and his peers dominated the theorizing and designing of sustainable environments. The sociologist Simon Guy and the architect Graham Farmer (2001) found six distinct traditions within architectural discourse, each based on differing and sometimes conflicting assumptions yet all claiming to define what sustainable design must be. Guy and Farmer's categories would include buildings as diverse as the low-tech back-to-the-land earth ships built in New Mexico by Mike Reynolds (1970), the technocratic and energy-efficient Commerzbank Tower designed for Frankfurt by Sir Norman Foster and Associates (1995–1998), and the communal solar kitchen developed for Jiutepec, Mexico, by the BaSiC Initiative (2004). Although some technocrats would prefer to standardize sustainable technologies as lists of best practices or universal technical codes, the practice of sustainable design has continued to diversify.

DISCIPLINARY APPROACHES

Another popular conception involves the division of work between architects and engineers in the planning of sustainable built environments. People are predisposed to associate architects with space planning and the superficial embellishment of building facades and engineers with technical problem solving. Although there is some truth in this categorization, reality is far more complex. Although these two modern professions emerged from a common European origin, over the last five centuries they have developed distinct cultures. Where architects tend to be expansive in their approach to problem solving—as in the arts—engineers tend to be reductive—as in the sciences. Where architects tend to emphasize the visual or communicative qualities of the environments they design, engineers tend to emphasize economic efficiency. However, the tendency to essentialize the values of the disciplines is as erroneous as it would be to characterize the visual characteristics of modern architecture as inherently antinature. The building professions are far more pluralistic than many critics suppose.

Among the diverse design practices developed in the early twenty-first century were those which derived from the critique of modern technology offered by science and technology studies (STS). Rather than focus on the expressive possibilities of building form, economic efficiency, or the artifacts themselves, STS scholars study the relationship of particular material technologies to the societies that develop, maintain, and operate them. The empirical study of the built environment as a socio-technical system is a promising area of analysis that may influence not only the teaching of design but also the consequences of design practice.

SEE ALSO *Alternative Technology; Brundtland Report; Built Environment; Energy; Industrial Ecology; Sustainability; Sustainable Development.*

BIBLIOGRAPHY

Canizaro, Vincent B., ed. 2007. *Architectural Regionalism: Collected Writings on Place, Identity, Modernity, and Tradition.* New York: Princeton Architectural Press.

Ehrlich, Paul R. 1971. *The Population Bomb.* New York: Ballantine Books.

Energy Information Administration. 2007. *Households, Building, Industry, & Vehicles: End-Use Consumption Data and Analysis.* Available from http://www.eia.doe.gov/emeu/consumption/index.html

Guy, Simon, and Graham Farmer. 2001. "Re-Interpreting Sustainable Architecture: The Place of Technology." *Journal of Architectural Education* 54(3): 140–148.

Guy, Simon, and Steven A. Moore, eds. 2005. *Sustainable Architectures: Cultures and Natures in Europe and North America.* New York: Spon Press; London: Taylor & Francis Group.

Mazria, Edward. 2007. *Architecture 2030: Global Warming, Climate Change, and the Built Environment.* Available from http://www.architecture2030.0rg/home.html

Moore, Steven A. 2007. *Alternative Routes to the Sustainable City: Austin, Curitiba, and Frankfurt.* Lanham, MD: Lexington Books.

Nelson, Arthur C. 2004. *Toward a New Metropolis: The Opportunity to Rebuild America.* Washington, DC: Brookings Institution. Available from http://www.brookings.edu/reports/2004/12metropolitanpolicy_nelson.aspx

Neutra, Richard. 1948. *Architecture of Social Concern in Regions of Mild Climate.* São Paulo, Brazil: Gerth Todtmann.

World Council on Environment and Development. 1987. *Our Common Future.* Oxford and New York: Oxford University Press.

Steven A. Moore

SUSTAINABLE DEVELOPMENT

Sustainable development is designed to meet present needs without compromising the needs and aspirations of future generations. Such goals inherently pose issues of ethics and values through discussions of both intragenerational and intergenerational equity; if current standards of living in some societies are unsustainable, is it possible to justify their existence in the present, much less their extension into the future?

HISTORY

Although the term *sustainable development* is often traced to the World Commission on Environment and Development's publication of *Our Common Future (1987)*, this report was the outcome of decades of concern about the impact of human beings on the natural environment. Notions of sustainable development emerged from the birth of the modern environmental movement, sparked by the work of authors such as Rachel Carson (1962), who focused on the impact of chemical pollution on the environment. These environmental concerns were linked to development as authors such Paul Ehrlich (1971) and organizations such as the Club of Rome (Meadows et al. 1972) highlighted the growth of the global human population and the stresses that growth was placing on the natural-resource base of the planet. This initial environment-development linkage was focused on the challenges human activities posed for the earth's natural-resource base. Consequently, initial efforts to address these challenges, including the United Nations Conference on the Human Environment in 1972, focused on the protection and preservation of that resource base.

Our Common Future (World Commission on Environment and Development 1987), also known as *The Brundtland Report*, reflects a shift in thinking about environment and development. By the mid-1980s, years of experience in development and international aid had demonstrated that the complexity of the connections between the environment and human well-being far exceeded the compass of analyses that were confined to the protection of natural resources. For example, it was in this period that the analysis of famine shifted from purely environmental causes (e.g., the absence of sufficient rainfall) to environmental causes as they intersect with local sociocultural factors and national and global political economies (e.g., Watts 1983). *Our Common Future* reflected this shift, turning from the previous environment-first approaches to the environment/development interface and thereby embracing a more holistic understanding of this interaction and its impact on human well-being.

The 1992 United Nations Conference on Environment and Development (UNCED) in Rio de Janeiro carried this new focus into an institutional context by producing Agenda 21, an action plan to address the interconnections between environment and development. The United Nations Commission on Sustainable Development (UNSCD) was founded to ensure appropriate follow-through after the conference. Nevertheless, the outcomes of UNCED mark something of a return to a resource-base-first approach to sustainability. For example, the Convention on Biological Diversity (CBD), a product of UNCED, focuses on the preservation of a particular key resource with the assumption that such preservation would preserve the resource in the context of development and, in so doing, improve human well-being.

The Extent of Urban Sprawl. *A tractor working in the fields near Portland, Oregon, seems dwarfed by a large tract of new apartments in the background. A study conducted by the Sierra Club reported that Oregon was ranked first in land-use planning.* AP IMAGES.

In later years major environmental assessments tried to better articulate the links between environmental preservation, development, and human well-being. The conceptual framework of the Millennium Ecosystem Assessment (MA) (2003) employed an ecosystem-services approach to draw this connection. Ecosystem services are the rewards human beings obtain from ecosystems, ranging from water filtration to raw materials to cultural/spiritual benefits. A change in a particular ecosystem will change the services that ecosystem provides to human beings—it will provide more or less of a particular service and more or fewer services overall. Through this approach the MA evaluated the earth's ecosystems from an anthropocentric perspective, evaluating the current state and trends of these ecosystem services in terms of the benefits humans derive from them and projecting the future benefits that would be available from these ecosystems under current and other hypothesized policy and technological regimes. Hence the MA marks something of a return to a WCED-style understanding of sustainable development.

CONTEMPORARY TENSIONS BETWEEN ENVIRONMENT AND DEVELOPMENT

The most commonly discussed tension in the contemporary sustainable-development literature is that of intergenerational equity. The definition of *sustainable* cannot be separated from a sense of ethical responsibility to future generations. For example, does *sustainability* mean the maintenance of the basic conditions necessary for human life on earth or the maintenance of the current standard of living for future generations? The answer to this question informs choices about how to live in the present. Maintaining human life is a fairly low standard that might be achieved even with a substantially irresponsible use of natural resources. However, as the MA illustrates, the maintenance of our current standard of living into the future without substantial technological change is likely impossible because we are degrading 60 percent of the (broadly categorized) services we receive from ecosystems. If sustainable development is a worthy goal, it demands consideration of how current behavior prevents the attainment of that goal.

Less commonly discussed are the ethical dimensions of *intra*generational equity raised by contemporary thought and practice in sustainable development. Various ecological-footprint studies (e.g., Rees and Wackernagel 1995) suggest that current levels of consumption in the global north are possible only because very few people in the global south enjoy this resource-intensive standard. Furthermore, these studies make clear that it would require more than three planet earths' worth of resources to allow everyone in the world to enjoy this standard of living. This analysis points to an inherent incompatibility in the goals of environmentalists seeking the sustainable use of natural resources and those working on development efforts intended to raise the standard of living of those in the global south. Such observations have led some (e.g., Esteva 1992, Banerjee 2003) to question whether sustainable development represents a form of neocolonialism whereby those in the global south are encouraged to develop in a sustainable manner (a goal impossible under current technological regimes and societal values) so as to slow their growth and preserve the lifestyles and prerogatives of those living in the global north.

Issues of intragenerational equity are further complicated by the mainstreaming of environmental concerns into development thought and practice. Although such mainstreaming maintains environmental quality and preserves key environmental resources for human well-being, the frameworks that are emerging out of this effort may create or perpetuate challenges that hinder development efforts and maintain the difficult circumstances facing many of the world's poor. The United Nations Environment Programme's Fourth Global Environment Outlook (GEO-4), which employs a conceptual framework similar to that of the MA, is an example of the dangers of this mainstreaming. Titled *Environment for Development (UNEP 2007)*, GEO-4 uses an environmental reporting framework to evaluate sustainable development initiatives. Carr and his coauthors (2007) argue that the use of such frameworks in the evaluation of development initiatives inadvertently produces situations that disempower local communities and local knowledge in the development process, thus squandering key resources for development and losing sight of the intended beneficiaries of such efforts.

The framework employed by GEO-4 divides linked environment/development challenges into a series of causes, ranging from large-scale ultimate causes like population and economic growth to proximate causes ranging from increased sewage output into the environment to ecosystem-specific changes that affect human well-being. The responses to these changes purport to address these challenges at each of these levels. However, local actors are usually constrained in their individual efforts to responses at the ecosystem level, whereas governments and development agencies are privileged in their ability to address ultimate and proximate causes of environmental changes. This general set of outcomes is consonant with development studies that argue that contemporary development practice does not address the ideas or needs of the poor unless those ideas and needs fit into and support the ideas, values, and careers of those working for development agencies (Ferguson 1994, Escobar 1995, and Easterly 2006). For example, local (indigenous) knowledge, today a popular buzzword in development, rose to prominence only when the development community deemed such knowledge useful and/or appropriate. Before this shift in the outlook of the development community, this same local knowledge was largely ignored.

AN ETHICAL FUTURE FOR DEVELOPMENT

The mainstreaming of environmental concerns in development thought and practice remains a work in progress, and among the most significant challenges raised by this effort are the ethical issues surrounding intergenerational and intragenerational equity. In debates about tradeoffs between the present and future, work in sustainable development has provided new openings for thinking about the ethical dimensions of environmental protection and development, and the value systems that allow unsustainable (and therefore unjust) practices to continue.

SEE ALSO *Brundtland Report; Convention on Biodiversity; Future Generations; Millennium Ecosystem Assessment; Sustainability; United Nations Environment Programme.*

BIBLIOGRAPHY

Banerjee, Subhabrata Bobby. 2003. "Who Sustains Whose Development? Sustainable Development and the Reinvention of Nature."*Organizational Studies* 24(1): 143–180.

Carr, Edward R.; Philip M. Wingard; Sara C. Yorty; et al. 2007. "Applying DPSIR to Sustainable Development." *International Journal of Sustainable Development and World Ecology* 14(6): 543–555.

Carson. Rachel. 1962. *Silent Spring*. Boston: Houghton Mifflin.

Easterly, William, 2006. *The White Man's Burden: Why the West's Efforts to Aid the Rest Have Done So Much Ill and So Little Good.* New York: Penguin.

Ehrlich, Paul R. 1971. *The Population Bomb.* New York: Ballantine Books.

Escobar, Arturo. 1995. *Encountering Development.* Princeton, NJ: Princeton University Press.

Esteva, Gustavo. 1992. "Development." In *The Development Dictionary: A Guide to Knowledge and Power*, ed. W. Sachs. London: Zed Books.

Ferguson, James. 1994. *The Anti-Politics Machine: Development, Depoliticization, and Bureaucratic Power in Lesotho.* Minneapolis: University of Minnesota Press.

Meadows, Donella H.; Dennis L. Meadows; Jorgen Randers; and Willam W. Behrens. 1972. *The Limits to Growth.* London: Earth Island.

Millennium Ecosystem Assessment Conceptual Framework Working Group. 2003. *Ecosystems and Human Well-Being: A Framework for Assessment.* Washington, DC: Island Press. Available from http://www.millenniumassessment.org/en/Framework.aspx

Rees, Williams E., and Mathis Wackernagel. 1995. *Our Ecological Footprint: Reducing Human Impact on the Earth.* Gabriela Island, BC: New Society Publishers.

United Nations Environment Programme. 2007. *Fourth Global Environment Outlook.* Valetta, Malta: United Nations Environment Programme. Available from http://www.unep.org/geo/geo4/media/

Watts, Michael. 1983. *Silent Violence: Food, Famine and Peasantry in Northern Nigeria.* Berkeley: University of California Press.

World Commission on Environment and Development. 1987. *Our Common Future.* Available from http://www.un-documents.net/wced-ocf.htm

Edward R. Carr

SWAMPS

SEE *Wetlands.*

SYLVAN, RICHARD

1935–1996

Richard Sylvan was a seminal thinker in environmental philosophy. He also was known for pioneering work in logic, metaphysics, the philosophy of language, semantics, epistemology, social philosophy, political philosophy, ethics, the philosophy of science, the philosophy of mind, and computation theory.

EDUCATION AND EARLY WORKS

Sylvan was born Francis Richard Routley in Levin, New Zealand, on December 13, 1935. He met and later married the philosopher Val McCrae in 1963, and they jointly authored several works in environmental philosophy under the names R. and V. Routley. Following their separation, Val changed her name to Plumwood. Richard changed his name to Sylvan when he remarried in 1983. He studied at Victoria University, Wellington, and then Princeton before taking positions at the University of Sydney, the University of New England, and Monash University. From 1971 until his death he was a fellow at the Research School of Social Sciences at the Australian National University. Sylvan died on June 16, 1996, in Bali, Indonesia. He is buried in New South Wales at the edge of one of the forests which he cherished.

Sylvan was responsible for much of the development of environmental philosophy in Australasia. An example of his provocative style and controversial choice of topics can be found in the 1982 paper "In Defence of Cannibalism," which was published in Sylvan's *Discussion Papers in Environmental Philosophy*, one of several preprint series which he edited.

The content of "In Defence of Cannibalism" is less sensational than the title. The essay addresses the ethics of killing, in particular killing humans, and the ethics of eating dead animals, including dead humans. Sylvan carefully separated those questions. The title of the paper generated alarm among some members of the philosophical community, and Sylvan may have derived satisfaction from the unsettling effects of the paper.

DEEP GREEN THEORY

Sylvan's research program was much broader than the ambit of environmental philosophy; it was systematically linked with his and others' work in metaphysics, semantics, logic, epistemology, and value theory. He also connected it with work outside mainstream Australasian and Anglo-American philosophical inquiry as well as with work in other disciplines, including Taoism, Buddhism, nihilism, cosmology, demography, politics, and economics. Sylvan also addressed environmental policy issues in his monograph *The Fight for the Forests* (Routley and Routley 1973), written with Val Plumwood.

Sylvan's work in environmental philosophy goes back to the early 1970s. His 1973 paper "Is There a Need for a New, an Environmental, Ethic?" is a landmark in the discussion of anthropocentrism. The last man argument presented in that paper remains an important locus of that discussion. Another important paper written in collaboration with Plumwood (Routley and Routley 1980) elaborates the argument. Sylvan's critique of anthropocentrism, or human chauvinism, aligned him with the central concerns of Deep Ecology, though he distanced himself from much of Deep Ecology (Sylvan 1985). Sylvan went on to develop his own environmental philosophy, which he called Deep Green Theory (DGT) in *The Greening of Ethics* (1994). At the time of his death he was working on a fuller explication of DGT that was never completed. A manuscript published posthumously in 1997, *Metaphysics: From Radical to Deep Plurallism* [sic], addressed metaphysical and associated logical issues with a strong pluralist twist, which Sylvan believed was essential to providing a satisfactory foundation for environmental philosophy. Sylvan developed his views in opposition to those of his university colleague John Passmore, whose very different views were published in *Man's Responsibility for Nature* (1974), which rejected the need for radical revision in ethics to accommodate concern for the environment.

DGT aims to clear the "garbage" (Sylvan's description) from environmental philosophy and replace it with theoretical rigor. It is a pluralistic position that shares with Deep Ecology a rejection of the prevailing technocentric approach to the environment of industrial society. Although it shares a number of features with Arne Naess's eight-point platform for Deep Ecology (Naess 1989), its focus and emphasis are different. A central theme is that many environmental items are valuable in themselves; that is, their value does not depend on human values and interests. In developing his alternative position, Sylvan introduced a number of distinctions marked by a plethora of neologisms: *non-jective, gre-een, extranalities, expiricism, intraneous, extitution* (Grey 2000). Sylvan's DGT is a complex articulation of reflections which help to clarify central problems in environmental philosophy.

SEE ALSO *Australia and New Zealand; Deep Ecology; Environmental Philosophy: V. Contemporary Philosophy; Plumwood, Val.*

BIBLIOGRAPHY

Grey, William. 2000. "A Critique of Deep Green Theory." In *Beneath the Surface: Critical Essays in the Philosophy of Deep Ecology,* ed. Eric Katz, Andrew Light, and David Rothenberg. Cambridge, MA: MIT Press.

Naess, Arne. 1989. *Ecology, Community, and Lifestyle: Outline of an Ecosophy,* trans. David Rothenberg. Cambridge, UK, and New York: Cambridge University Press.

Passmore, John. 1974. *Man's Responsibility for Nature: Ecological Problems and Western Traditions.* London: Duckworth.

Routley, Richard. 1973. "Is There a Need for a New, an Environmental, Ethic?" In *Proceedings of the XVth World Congress of Philosophy,* Sophia, Bulgaria, 1: 205–210.

Routley, Richard. 1982. "In Defence of Cannibalism." *Discussion Papers in Environmental Philosophy,* No. 2. Canberra: Australian National University.

Routley, Richard, and Val Routley. 1973. *The Fight for the Forests: The Takeover of Australian Forests for Pines, Wood Chips, and Intensive Forestry.* Canberra: Research School of Social Sciences, Australian National University.

Routley, Richard, and Val Routley. 1978. "Nuclear Energy and Obligations to the Future." *Inquiry* 21: 133–179.

Routley, Richard, and Val Routley. 1979. "Against the Inevitability of Human Chauvinism." In *Ethics and Problems of the 21st Century,* ed. K. E. Goodpaster and K. M. Sayre. Notre Dame, IN: University of Notre Dame Press.

Routley, Richard, and Val Routley. 1980. "Human Chauvinism and Environmental Ethics." In *Environmental Philosophy,* ed. D. S. Mannison, M. A. McRobbie, and R. Routley. Canberra: Research School of Social Sciences, Australian National University.

Sylvan, Richard. 1985. "A Critique of Deep Ecology." *Radical Philosophy* 40: 2–12 and 41: 10–22. Also published as *Discussion Papers in Environmental Philosophy,* No. 12. Canberra: Australian National University.

Sylvan, Richard. 1986. "Three Essays on Deeper Environmental Ethics." *Discussion Papers in Environmental Philosophy,* No. 13. Canberra: Australian National University.

Sylvan, Richard. 1990. "In Defence of Deep Environmental Ethics. "*Discussion Papers in Environmental Philosophy,* No. 18. Canberra: Australian National University.

Sylvan, Richard. 1997. *Metaphysics: From Radical to Deep Plurallism.* [sic] Cambridge, UK: White Horse Press.

Sylvan, Richard, and David Bennett. 1994. *The Greening of Ethics.* Cambridge, UK: White Horse Press; Tucson: University of Arizona Press.

William Grey

T

TAKINGS

The Fifth Amendment to the U.S. Constitution states that no person "shall be deprived of life, liberty or property without due process of law; nor shall private property be taken for public use, without just compensation." This last clause, the "takings clause," prohibits the U.S. government from seizing property without just compensation. In the United States it is this amendment that authorizes the federal government to seize properties for common use, a legal concept known as *eminent domain*. Many other nations have similar laws. The United Kingdom, New Zealand, and the Republic of Ireland have "compulsory purchase" laws; Australia has "resumption or compulsory acquisition" laws; and South Africa has "expropriation" laws. Takings restrictions date as far back as the Magna Carta, issued in 1215 as a curb on the absolute will of the king of England. All such laws protect property owners from unjust seizure but also allow governments to seize property in the common interest.

The requirement of compensation for the physical taking of property is uncontroversial. As regards regulations, however, the requirement is more complex. The environmental issue referred to as "takings" or "regulatory takings" relates to a property holder's claim to compensation for damages incurred or benefits forestalled, as a result of regulations—restrictions, prohibitions, or requirements—placed on the use or lease of a given parcel of property. For instance, if an owner of a piece of land is restricted from building on his or her property, say, because of an endangered-species regulation, the property owner may argue that this regulation has effectively "taken" the property, or taken a significant aspect of value on the property. The central theoretical question is the extent to which a given government action—or, in this case, a given regulation—constitutes a taking.

Until the late 1970s controversies over regulatory takings in the United States had been addressed chiefly by the courts. One of the earliest takings cases was *Pennsylvania Coal Co. v. Mahon* (1922). In this case the Pennsylvania legislature had prohibited the mining of coal underneath streets and houses. The controversy emerged because Pennsylvania Coal had, forty years earlier, granted strict surface rights to H. J. Mahon, under the express agreement that they would eventually mine coal under his dwelling. Pennsylvania Coal argued that the legislature's prohibition of mining under streets and houses constituted a taking because the coal company was no longer permitted to mine coal in these areas. The U.S. Supreme Court found in favor of Pennsylvania Coal, stating that "[W]hile property may be regulated to a certain extent, if regulation goes too far it will be recognized as a taking." This case established the precedent that regulations, not only physical seizures of property, can constitute takings.

Since then two cases have emerged as central to modern regulatory takings law (Squillace 2008). In the first case, *Penn Central Transportation Co. v. New York City* (1978), Penn Central petitioned the city for the right to develop a high-rise tower above Grand Central Terminal. Because Grand Central had been designated by the city as an historic landmark, it was subject to zoning restrictions that prohibited such construction. The Court rejected Penn Central's regulatory takings claim, focusing on two issues: the character of the government action and whether the regulation interfered with distinct and reasonable "investment-backed expectations." No compensation was paid.

In the second case, *Lucas v. South Carolina Coastal Council* (1992), David H. Lucas purchased two properties on the coast of South Carolina for $975,000. In 1988 the state passed the Beachfront Management Act, which restricted Lucas from developing his two properties. Lucas sued and won several critical cases, leading to a Supreme Court hearing. The Supreme Court held that the regulation had effectively deprived Lucas of all economically beneficial use of his property and therefore amounted to a "total taking." By appeal to the character of government action, the concept of reasonable "investment-backed expectations," and the notion of a "total taking," the Penn Central and Lucas cases provide a sound analytical framework for analyzing most regulatory takings cases.

Aside from this judicial history, the concept of takings has sparked heated political controversy. By the late-1970s, the United States had adopted broad-reaching environmental legislation such as the U.S. Endangered Species Act (ESA), the U.S. National Environmental Policy Act (NEPA), and the U.S. Federal Land Policy and Management Act (FLPMA). As these acts came into effect, some property owners, particularly in the western United States, began to question their legitimacy, citing as a precedent the conservationist platform of Gifford Pinchot, the first head of the U.S. Forest Service. Out of these concerns the so-called "sagebrush rebellion" or "wise use movement" was born. This political movement has influenced policy in all branches of government, but most notably in the executive and legislative branches.

For instance, when Ronald Reagan became president in 1981, he appointed James G. Watt, a central figure in the sagebrush rebellion, as secretary of the interior. Watt's appointment resulted in a series of controversial administrative decisions, the ostensible aim of which was to shore up property rights; the real aim, according to critics, was to dismantle environmental law. One of the central decisions came in 1988, when Reagan introduced Executive Order 12630, otherwise known as "Reagan's Order." This order, formally titled "Governmental Actions and Interference with Constitutionally Protected Property Rights," required all agencies in the executive branch to determine whether their proposed action may imply a taking and, if so, to conduct a takings impact assessment (TIA). If any takings implications were found, the order restricted an agency's ability to carry out that proposed action.

During the same decade, state legislatures battled over a variety of takings bills. By 1991 every state in the United States had considered some form of environmental takings legislation. The first federal takings bill had been introduced a year earlier, in 1990, by Senator Steve Symms of Idaho. Although particular bills vary widely in details, takings legislation usually seeks to establish guarantees of compensation for regulation or at least to assure that some assessment of costs and the possibility of compensation is put into place prior to the establishment of a regulation.

Regulatory takings poses complex philosophical and ethical issues, ranging from questions about the nature of private property to observations about harms to nature or to humans. At any point in the history of the "takings debate"—whether in the court decisions, in the legislation, in the policy of sitting administrations, or even in the more generalized civil sphere—philosophical issues regarding rights, harm, freedom, participation, representation, identity, self-actualization, moral status, public good, and so on, intertwine and overlap.

SEE ALSO *Environmental Law; Environmental Philosophy: V. Contemporary Philosophy; Land Ethic; Private Property.*

BIBLIOGRAPHY

Emerson, Kirk, and Charles R. Wise. 1997. "Statutory Approaches to Regulatory Takings: State Property Rights Legislation Issues and Implications for Public Administration." *Public Administration Review* 57(5): 411–422.

Epstein, Richard. 1985. *Takings: Private Property and the Power of Eminent Domain.* Cambridge, MA: Harvard University Press.

Marcellino, Carl P. 1998. "The Evolution of State Takings Legislation and the Proposals Considered During the 1997–1998 Legislative Session." *Journal of Legislation and Public Policy* 2(1): 143–170.

Squillace, Mark. 2008. Professor of Law and Director of Natural Resources Law Center, University of Colorado School of Law. Interview by Benjamin Hale, March 2008.

Sterk, Stewart E. 2006. "The Demise of Federal Takings Litigation." *William and Mary Law Review* 48(1): 251–302.

Benjamin Hale

TAOISM

SEE *Daoism.*

TAYLOR, PAUL
1923–

Paul Warren Taylor was born in Philadelphia on November 19, 1923. He is emeritus professor of philosophy at Brooklyn College, where he specialized in normative and applied ethics. The author of several works in ethics, including *Normative Discourse* (1961) and *Principles of Ethics* (1975), Taylor is probably best known for *Respect for Nature* (1986), in which he develops and defends a sophisticated biocentric (life-centered) environmental ethic.

Taylor's egalitarian biocentric ethic (1986) synthesizes elements of classical virtue ethics with Albert Schweitzer's ethic of reverence for life, Peter Singer's egalitarianism, and Kenneth Goodpaster's account of moral considerability. Taylor contends that one who adopts the ultimate moral attitude of respect for nature will become an environmentally virtuous person. He identifies environmentally ethical conduct with conduct motivated by respect for nature. Such environmentally virtuous conduct seeks to promote the flourishing of all living organisms. In Taylor's words, "Ethical action and goodness of character naturally flow from the attitude [of respect for nature], and the attitude is made manifest in how one acts and in what sort of person one is" (1986, p. 81).

Taylor admits that "we cannot see the point of taking the attitude of respect" until we understand and accept the biocentric outlook, but he insists that "once we do grasp it and shape our world outlook in accordance with it, we immediately understand how and why a person should adopt that attitude [of respect] as the only appropriate one to have toward nature" (1986, p. 90). The biocentric outlook, a scientifically grounded view of humanity's place in the natural order, consists of the following four theses:

1. *Homo sapiens*, like all other species, emerged as a result of random genetic drift and natural selection. As such, humans are members of the earth's biotic community on a par with all other living organisms.
2. The earth's biotic community forms a complex web of functionally interdependent organisms. In this web, the survival of each organism is determined in part by its relations to other organisms.
3. Each individual living organism is a "teleological center of life" pursuing its own good in its own way.
4. Humans are not superior to other living things. Their inherent worth is no greater than that of any other living organism. (1986, pp. 99–100).

Theses 1 to 3 are solidly supported by the sciences of biology and ecology. Taylor argues that those who accept these theses are rationally committed to thesis 4, which, together, support and make intelligible the attitude of respect for nature (1981, p. 206). This outlook sees living things "as *the appropriate objects of the attitude of respect* and are accordingly regarded as entities possessing inherent worth" (1981, p. 206).

Taylor derives his biocentric egalitarianism as follows. First, he argues that all living organisms are biologically goal-directed toward goods of their own. Next, following Goodpaster, he argues that any being with a good of its own deserves moral consideration. Coupling the latter conclusion with Singer's egalitarianism, Taylor concludes that every living organism possesses equal inherent worth and deserves equal moral consideration.

Numerous objections have been raised against Taylor's biocentric ethic. Principal among them are challenges to its account of moral considerability, its egalitarianism, its individualism, and its demands, along with a worry that Taylor commits the naturalistic fallacy.

To treat these challenges in order, some critics (e.g., Singer 1975, pp. 8–9) maintain that only sentient beings have interests and that only beings with interests deserve moral consideration. Taylor argues that it is arbitrary to restrict the class of morally considerable beings to sentient beings. Since all living organisms can be harmed or benefited and what benefits them promotes their good, Taylor insists that there is no nonarbitrary reason not to extend moral consideration to all living organisms. Mary Anne Warren (1997, p. 48) rejects Taylor's reasoning on the grounds that since lower organisms do not care whether their biological interests are satisfied, neither should humans.

Some biocentrists (Goodpaster 1978, Varner 2002) take issue with Taylor's egalitarianism. They agree that all living organisms deserve moral consideration, but deny that being morally considerable entails having *equal* moral significance. These critics reject Taylor's egalitarianism in favor of a hierarchical account of moral significance.

Other critics object to Taylor's stated view that "it is the good (well-being, welfare) of individual organisms ... that determines our moral relations with the Earth's wild communities of life" (1981, p. 198). These critics contend that Taylor's focus on individual welfare fails to address the actual concerns of environmentalists. Most environmentalists are concerned not with the welfare of individual mosquitoes, dandelions, and microbes, but rather with species preservation, ecological integrity, and pollution. These critics insist that a holistic ethic can better address these environmental concerns.

The idea of extending equal moral consideration to every living organism, including every insect and plant, strikes most people as not only too demanding, but outright absurd. How can people live their lives if they must give plants and insects the same moral consideration owed humans? Taylor tries to mitigate this objection by formulating a complex set of principles (self-defense, proportionality, minimum harm, distributive justice, and restitutive justice) for fairly resolving the conflicts that inevitably arise between humans and other equally considerable organisms. Even with these principles in place, however, Taylor's biocentric ethic remains extremely demanding, since the principle of proportionality dictates that the basic interests of plants trump the nonbasic interests of humans.

A final worry is that at some point Taylor must move from the purely descriptive aspects of his biocentric outlook to the moral attitude of respect for nature, and that in doing so he must either commit the naturalistic fallacy or beg the question by smuggling a normative judgment (thesis 4) into his biocentric outlook. Taylor recognizes, however, that the biocentric outlook "is not wholly analyzable into empirically confirmable assertions" and thus is best viewed as "a philosophical worldview" (1981, p. 205). Moreover, he never attempts to *derive* an *ought* from an *is*; rather, he seeks to provide us with a rational, coherent perspective on nature that will allow us to accurately perceive (not deduce) the inherent worth of all living beings.

Whether these objections to Taylor's biocentric egalitarianism prove insuperable remains to be seen. Regardless of whether his ethic prevails in the end or ultimately forces us to look elsewhere for an adequate environmental ethic, Taylor's biocentric outlook helps those who accept it to have a greater appreciation and respect for nature.

SEE ALSO *Biocentrism; Nuclear Power; Risk Assessment.*

BIBLIOGRAPHY

Goodpaster, Kenneth. 1978. "On Being Morally Considerable." *Journal of Philosophy* 78(6): 308–325.

Schweitzer, Albert. 1923. *Civilization and Ethics*, trans. John Naish. London: A. and C. Black.

Singer, Peter. 1975. *Animal Liberation.* New York: New York Review.

Taylor, Paul. 1961. *Normative Discourse.* Engelwood Cliffs, NJ: Prentice-Hall.

Taylor, Paul. 1975. *Principles of Ethics: An Introduction.* Belmont, CA: Dickenson.

Taylor, Paul. 1981. "The Ethics of Respect for Nature." *Environmental Ethics* 3(3): 197–218.

Taylor, Paul. 1986. *Respect for Nature: A Theory of Environmental Ethics.* Princeton, NJ: Princeton University Press.

Varner, Gary. 1998. *In Nature's Interests.* Oxford: Oxford University Press.

Varner, Gary. 2002. "Biocentric Individualism." In *Environmental Ethics: What Really Matters, What Really Works,* ed. David Schmidtz and Elizabeth Willott, pp. 108–120. Oxford: Oxford University Press.

Warren, Mary Anne. 1997. *Moral Status: Obligations to Persons and Other Living Things.* Oxford: Oxford University Press.

Mylan Engel Jr.

TECHNOLOGY

Environmental ethics often deals with the ethical and philosophical implications of human powers over the natural world. Because those powers nearly always are manifested technologically, environmental ethics is in large measure a philosophy of technology. The historical genesis of contemporary environmental ethics coincides with the unprecedented environmental impacts brought about by the advent of advanced industrial technology in the mid-twentieth century. It was also at that time that the philosophy of technology developed as a variegated research field in the English-speaking world, aspiring to comprehensive reflection on the making and using of artifacts. Although they occasionally have entered into dialogue (Ferrè 1992), these two areas of philosophy have remained largely separate despite their potential for support and synthesis.

HISTORICAL BACKGROUND

From its beginnings philosophy has included some attention to technology. However, it was only with the emergence of the industrial technosciences in the late nineteenth century that philosophers systematically turned their attention to the conceptualization and evaluation of technology (Kaplan 2004).

Premodern Socrates questioned those who pretended to wisdom: not just poets and politicians but also artisans. In Socrates's account it was the artisan practitioners of *techne* who came closest to having legitimate knowledge. The term *techne* is often translated as craft or art, but can also be seen as a practice that is grounded in theoretical knowledge, or "an account," thus linking it closely to notions of expertise or know-how. In another dialogue (*Phaedrus*) Socrates indicated that he had nothing to learn from nature. In contrast, Aristotle's philosophy included nature, or *physis,* as a fundamental theme. Aristotle did, however, maintain a strict division between natural and artificial entities. Natural entities are self-generating substantial unities of form and matter. An acorn sprouts into an oak tree, fulfilling its essence. Artifacts, by contrast, never achieve that substantial integration because the source of their being is external to them; if a bed were to sprout, it would give rise to a tree, not a bed (*Physics*). The ancient theme of *techne* and *physis* underpins contemporary work in both philosophy of technology and environmental ethics on the meaning of and proper relationships between technology and nature (McKibben 1989, Haraway 1991, Latour 1993, Rothenberg 1993, Strong 1995).

In Christian adaptations of philosophy, nature is thought of as a creation of God, who also is disclosed by means of supernatural revelation. Christian theology thus identifies two paths to insight into the ultimate nature of reality: the word of God in Scripture and the book of nature. On neither path, however, does technical thought make a significant appearance, although historians of technology such as Lynn White, Jr. (1967) have argued that Christian theology prepared the way for modern science and technology. White claimed that the Judeo-Christian teleology—God made nature solely to serve humanity—underpins the Western technological mastery of nature.

Modern Many of the founders of modern science and technology, such as Francis Bacon, René Descartes, and Isaac Newton, were motivated partly by the aim of natural theology: As Johannes Kepler phrased it, "to think God's thoughts after him." However, the modern era also signaled a radical break from ancient and medieval religious views that valorized contemplation and the control of one's desires and maintained that technology should be subject to ethical constraints. From the Renaissance through the Enlightenment, by contrast, traditional restraints were replaced with an ethical commitment to the unfettered pursuit of technology. Bacon argued that the production of knowledge would culminate in "the conquest of nature for the relief of man's estate" (*Novum Organum*). The modern approach to the world is essentially technological because it upholds an intimate alliance between knowing and changing the world. For example, Bacon wrote that "the nature of things betrays itself more readily under the vexations of art than in its natural freedom" ("Plan of the Work," paragraph 21). It is this commitment to technology, along with its subsequent questioning in response to problems, that frames the prominence of ethical issues both in the philosophy of technology and in environmental philosophy.

Steel Mills, Pittsburgh, Pennsylvania, ca. 1920–1950. *Beginning in the nineteenth century, industrial technology, such as steel production, brought about important changes in both environmental ethics and philosophy. Some have suggested that the powerful role of technology in modern times allows for a reduction in the sense of ethical responsibility for many individuals.* **THE LIBRARY OF CONGRESS.**

In the nineteenth century, technics and science began to merge and the machine-powered industrial technology employed in capitalist systems of trade and production brought massive changes. Systematic reflection on the social organization and ethical import of technology began at that time. Social theorists such as Karl Marx and Max Weber theorized technology in terms of human activity and the causal relations between technical change and social change (e.g., technology as an autonomous determinant of human affairs or as a social construction). Weber analyzed technicalization: the movement from traditional societies, in which techniques are situated within and delimited by nontechnical values, to modern societies, in which techniques are evaluated solely in technical terms. Marx argued that material culture largely determines the character of society because it is driven by the increasing capacities of machines rather than the needs of people. Yet Marx also maintained that industrial technology could support a just and fulfilling utopia if the social organization of labor, especially ownership of the means of production, was altered. Marxism has inspired reflections on technology within critical theory, especially in the Frankfurt School and among contemporary philosophers such as Andrew Feenberg (1991).

Contemporary The late nineteenth and twentieth centuries witnessed increasingly rigorous and diverse philosophical reflection on technology. Carl Mitcham (1994) organized those works into two groups: Engineering philosophy of technology argues that technology is central to human life, and humanities philosophy of technology is concerned with the moral and cultural boundaries of technology. Representatives of the first group include Ernst Kapp (1877), the first to write a book with *philosophy of technology* in the title, who pictured all technologies as "organ projections" (e.g., the telegraph extends the human nervous system). Friedrich Dessauer (1972) viewed technological activity as a mystical experience involving supreme participation in reality. Dessauer argued that technological invention offers direct contact with things-in-themselves. Insofar as they make ethical judgments of technology, members of this group tend to concur with Julian Simon (1995) and other contrarians in environmental ethics who emphasize the benefits of technology and are optimistic about the prospects of managing natural systems on a large scale.

Representatives of the second group tend to adopt a more historical approach to and mixed evaluation of technology, often motivated by a concern to preserve the harmonies of nature. This viewpoint allies humanities philosophers of technology with the romantic influences in environmental ethics, including Ralph Waldo Emerson and Henry David Thoreau (Marx 1964). Lewis Mumford (1934), arguing that humans are essentially linguistic rather than engineering animals, distinguished life-oriented technologies from the bureaucratic, systemic megamachine, which diminishes human meaning and constricts personal existence. Jacques Ellul (1964) argued that *la technique*—the whole ensemble of modern technologies—operates autonomously in the modern world to reduce life to the narrow demands of efficiency. José Ortega y Gasset saw technology as central to humans' ability to transcend nature and adapt it to their needs but also expressed concern that the unlimited power of technology will lull people into a flattened existence.

Bernard Charbonneau (Cèrèzuelle 2004), who influenced Ellul's thought, was a founder of the French environmental or political ecology movement who argued that the traditional warring ideologies (e.g., liberalism and communism) were insufficient responses to industrial society because they took its basic logic of production for granted. Charbonneau's work bridges environmental and technological ethics. He argued that World War I brought about a "great mutation" in which human freedom has been subordinated to the logic of an ever-accelerating technological industrialism. Humans seek freedom from nature's necessities in society, but those freedoms are paid for with the additional constraints of an impersonal, bureaucratic "societal frame." Charbonneau argued that embodied and personal contact with nature is as essential to human freedom as is technological escape from nature.

Many philosophers of technology follow Heidegger in treating technology as a unified system with a defining essence. This complements the thoughts of White and others in environmental ethics who attempt to situate the contemporary human condition within a broader sweep of history and a worldview that is in need of revision. Aldo Leopold, for example, argued that people need to evolve from a self-image that pictures humanity as conqueror of nature to one that pictures humanity as citizen of the biotic community. By contrast, those influenced by the more recent "empirical turn" in the philosophy of technology—mostly in the United States but increasingly in the Netherlands as well—tend to focus on individual technologies as they coevolve with society and express their potential uses in specific contexts (Achterhuis 2001). This approach relates to the context sensitivity of the pragmatist and policy-turn influences in environmental ethics, which are both approaches that begin inquiry with actual environmental controversies and seek to offer practical advise.

KEY CONCEPTS AND THEMES

The philosophy of technology can be discussed in terms of a number of basic issues associated with the standard branches of philosophy: logic, epistemology, metaphysics, and ethics. In both philosophy of technology and environmental philosophy, ethical and political concerns have had primary importance.

Responsibility and Precaution Hans Jonas (1984) argued that responsibility was not a central concept in premodern ethics because of the narrow compass of scientific knowledge and technological power. For Jonas, "responsibility ... is a function of power and knowledge," which "were formerly so limited" that consequences distant in time and space had to be left to fate as attention was focused on the present (Jonas 1984, p. 123). Modern technology, by contrast, introduces such novel actions, objects, and consequences that the old ethical frameworks are not appropriate. With the extended powers of technology, modern people face the unprecedented task of considering the global condition of human life, the well-being of future generations, and the existence of entire nonhuman species. This demands "a new conception of duties and rights" (Jonas 1984, p. 8).

Science and technology lengthen the human reach. As Jonas noted, this means that people need "lengthened foresight" to guide their actions. However, foresight is blurred by uncertainty about the consequences of people's actions. For Jonas, the correct reaction to uncertainty in the context of potentially dangerous technologies is precaution. People must apply a "heuristic of fear" that will replace predictions of hope and inform themselves about what is at stake. In this view, precaution is a noble fear grounded in an appreciation of the fragility of human existence in light of technological dangers. This position clearly situates him in the precautionary principle debates within environmental ethics.

Technosocial systems are so complex that a new ethical imperative of responsibility is not easy to implement. Moral responsibility is distributed across multiagent systems, and technologies serve as moral aggregators that turn individual morally negligible acts (driving a car) into major moral consequences (global climate change). Garrett Hardin (1968) put this in terms of the "tragedy of the commons," in which individuals immediately experience short-term benefits while creating longer-term, more diffuse costs.

Hardin's insight relates to Albert Borgmann's 1984 analysis of modern technology as a "device paradigm." Through the example of a central heating system, Borgmann demonstrates how the machinery of devices fades into the background as people increasingly confront commodities (heat, in this case) isolated from the conditions that make them possible. People flip a light switch, pull the lever on a gasoline pump, or press a button on the thermostat but do not experience the full effect of those actions. As users, people do not engender the sociotechnical systems that "lie behind" and make possible these simple actions at the user interface: People click the print icon on the computer screen but do not fell the tree, process the paper, manufacture the printer, and so on.

Thus, the conveniences afforded by technology create situations in which responsibility for the aggregate or emergent consequences seems both nowhere and everywhere. Ulrich Beck (1992) argued that many predominant techniques of risk management block out responsibility. Society is a "laboratory" in which no one has to answer for the negative effects of technological experimentation. The institutions of modern society recognize the existence of risk but permit an "organized irresponsibility."

Environmental ethics and the ethics of technology have generated the same responses to this problem, which include limiting individual freedom through expanded governmental regulations and bureaucracies and altering markets to eliminate externalities. Other proposals include enhancing individual ethical responsibility through improvements in education that incorporate expansions of affective sensibility and the cultivation of more virtuous or less consumerist character traits. Also, role responsibilities can be broadened. For example, corporations and manufacturers can adopt extended responsibility for their products and users can practice more responsible consumption. In recognition of engineering as social experimentation, engineers have shifted the interpretation of their primary responsibility from loyalty to a company or client to responsibility for public health, safety, and welfare. This is demonstrated in the growing emphasis on cradle-to-cradle design, industrial ecology, and sustainable architecture (McDonough and Braungart 2002).

Democracy and Justice Modern liberal democracy is about limits to government, yet technology has at least as much power as governments to shape people's lives—to constrain people, free them, and even constitute their identities. Richard Sclove (1995) and others argue that democracy also should be about limiting science and technology, especially by consciously guiding those increasingly powerful forces through democratic processes of foresight, reflection, participation, and discussion. This argument underpins many practical attempts to reintegrate ethics and other values into technical production processes that have become dissociated from society.

Many examples of these practices come from environmental contexts as governments routinely mandate environmental assessment processes—including public comment periods—for proposed engineering projects with environmental risks. Pointing out the inefficiencies of these processes as well as their often undesirable outcomes, others argue for increased control by those with the technical

Windfarm in East Frisia, Germany. *Sheep graze among wind turbines in East Frisia, Germany. With the extended powers of technology, modern people face the unprecedented task of considering the global condition of human life, the well-being of future generations, and the existence of entire nonhuman species.* OTMAR SMIT, 2008. USED UNDER LICENSE FROM SHUTTERSTOCK.COM

knowledge required to ensure sustainable development or other environmental goals; this is a technocratic vision long debated in the philosophy of technology.

Proposals for the democratization of technology often are motivated by the desire to bolster human autonomy in a world largely driven by technological and growth imperatives. The human-built world is displacing and despoiling nature, prompting many people to seek greater control over technology through political processes. Such proposals often are advanced to rectify injustices stemming from the ways in which technologies shift the distribution of power. In particular, environmental burdens and benefits often are distributed unequally across lines of race, class, and gender, for example, when polluting industries are placed in impoverished neighborhoods. The Green Revolution demonstrated the ethical ambiguities of technology transfer to developing countries. The populations of those countries are also the most vulnerable to the impacts of global climate change even though they are responsible for a relatively small proportion of greenhouse gas emissions. Finally, technologies may carry certain political characteristics by their nature. A nuclear power plant, for example, requires a hierarchical distribution of power and advanced security superstructures, whereas distributed solar power generation entails far different political consequences (Winner 1986).

The Good Life The way people relate to the natural world depends on what they believe about themselves and their relationship with the environment. Thus, visions of human nature and the good life are ultimately at work—albeit implicitly—behind modern technological society. The predominant vision is that of material comfort and abundance espoused by Bacon and encapsulated in consumerism. At least for those in the developed world, modern technology has brought vast improvements in material well-being. As more people aspire to this lifestyle, environmental ruin may follow. However, fertility declines with increasing development, and greater wealth brings with it the ability to afford cleaner technologies and the possibility of caring for nature for its own sake.

However, more fundamental to this vision is the belief—coeval with human existence but now magnified beyond previous proportions—that the world as it is does not provide a suitable home for human beings; humanity must construct a home for itself. Human beings often do not find themselves at home within the worlds they create. Levels of happiness do not rise after people cross a certain income threshold, and antidepressants are among the most frequently prescribed drugs in some developed countries. Disconnected from nature and overwhelmed by the pace of modern media culture, children as well as adults tend to lead more hyperactive and desultory lives. Both romanticism and socialism critique these kinds of technological alienation. Jean-Jacques Rousseau focused on how technology alienates the individual from feelings and sentiments. Karl Marx analyzed the power of industrial capitalism to alienate individuals from their full potential and separate people from control over the tools and products of their labor.

Thus, both environmental ethics and philosophy of technology debate the merits and feasibility of alternative visions of the good life, including various types of post-materialism, communalism, and bioregionalism. Such visions often are criticized as nostalgic idealizations of primitive existence when they are elaborated on a large scale rather than used to justify marginal or individual shifts to alternative technologies. Certainly, one would face strong humanitarian arguments against advocating anything as radical as giving up indoor plumbing or electricity as a social policy. Nonetheless, deep ecologists point out a fundamental human need when they speak of the ennobling power of wilderness. There is more to living well than material comfort, and not all burdens are odious. Indeed, in relinquishing technological aids and meeting nature on its own terms, people replenish a vital part of the human spirit. Debating the proper scope and character of technology, then, bears not just on matters of sustainability, survival, and justice but also on what it means to live well.

PRACTICAL CONTEXTS

Much of environmental ethics falls within the purview of the philosophy of technology because many of its central questions relate to the development, use, and impacts of artifacts. They also relate to the appropriate reach of technical knowledge and activities. For example, debates over sustainability in environmental ethics often stem from differences between those who picture it as a matter of technologically managing nature and those who picture it as an essential limitation on technology. This and other issues are manifested in practical contexts in which individuals and groups face decisions about technology. These contexts range from the construction of dams and power plants by societies, to business investments in new prototypes, to decisions to drive cars and run air conditioners by individuals. Thus, a focus on technology can help move environmental ethics out of academia and into the public, corporate, and private arenas in which such decisions are being made.

SEE ALSO *Bacon, Francis; Christianity; Descartes, Rene; Emerson, Ralph Waldo; Environmental Philosophy: I. Ancient Philosophy; Environmental Philosophy: III. Early Modern Philosophy; Environmental Philosophy: V. Contemporary Philosophy; Industrial Ecology; Ortega y Gasset, José; Precautionary Principle; Risk Assessment; Sustainability; Sustainable Agriculture; Thoreau, Henry David; Tragedy of the Commons; White, Lynn, Jr.*

BIBLIOGRAPHY

Achterhuis, Hans, ed. 2001. *American Philosophy of Technology: The Empirical Turn*, trans. Robert P. Crease. Bloomington: Indiana University Press.

Beck, Ulrich. 1992. *Risk Society: Towards a New Modernity.* Newbury Park, CA: Sage Publications.

Borgmann, Albert. 1984. *Technology and the Character of Contemporary Life: A Philosophical Inquiry.* Chicago: University of Chicago Press.

Cèrèzuelle, Daniel. 2004. "Nature and Freedom: An Introduction to the Environmental Thought of Bernard Charbonneau." In *Rethinking Nature: Essays in Environmental Philosophy*, ed. Robert Frodeman and Bruce V. Foltz. Bloomington: Indiana University Press.

Dessauer, Friedrich. 1972. "Technology in Its Proper Sphere." In *Philosophy and Technology: Readings in the Philosophical Problems of Technology*, ed. Carl Mitcham and Robert Mackey. New York: Free Press.

Ellul, Jacques. 1964. *The Technological Society*, trans. John Wilkinson. New York: Knopf. Translation of *La technique ou l'enjeu du siècle.* Paris: Armand Colin, 1954.

Feenberg, Andrew 1991. *Critical Theory of Technology.* Oxford: Oxford University Press.

Ferrè, Frederick, ed. 1992. *Research in Philosophy and Technology.* Greenwich, CT: JAI Press.

Haraway, Donna J. 1991. *Simians, Cyborgs, and Women: The Reinvention of Nature.* New York: Routledge.

Hardin, Garrett. 1968. "The Tragedy of the Commons." *Science* 162: 1243–1248.

Heidegger, Martin. 1977. *The Question Concerning Technology*, trans. William Lovitt. New York: Harper & Row. Translation of *Die Frage nach der Technik*, 1954.

Ihde, Don. 1990. *Technology and the Lifeworld: From Garden to Earth.* Bloomington: Indiana University Press.

Jonas, Hans. 1984. *The Imperative of Responsibility: In Search of an Ethics for the Technological Age.* Chicago: University of Chicago Press.

Kaplan, David M., ed. 2004. *Readings in the Philosophy of Technology.* Lanham, MD: Rowman and Littlefield.

Kapp, Ernst 1877. *Grundlinien einer Philosophie der Technik.* Braunschweig. Republished by Stern-Verlag Janssen, Dusseldorf, Germany, 1978.

Latour, Bruno. 1993. *We Have Never Been Modern*, trans. Catherine Porter. Cambridge, MA: Harvard University Press.

Marx, Leo. 1964. *The Machine in the Garden: Technology and the Pastoral Ideal in America*. New York: Oxford University Press.

McDonough, William, and Michael Braungart. 2002. *Cradle to Cradle: Remaking the Way We Make Things*. New York: North Point Press.

McKibben, Bill. 1989. *The End of Nature*. New York: Random House.

Mitcham, Carl. 1994. *Thinking through Technology: The Path between Engineering and Philosophy*. Chicago: University of Chicago Press.

Mitcham, Carl. 2004. "Technology: II. Philosophy of Medical Technology." In *Encyclopedia of Bioethics*, 3rd edition, ed. Stephen Post (New York: Macmillan Reference USA) vol. 5, pp. 2503–2511.

Mumford, Lewis. 1934. *Technics and Civilization*. New York: Harcourt, Brace and Company.

Rothenberg, David. 1993. *Hand's End: Technology and the Limits of Nature*. Berkeley: University of California Press.

Sclove, Richard E. 1995. *Democracy and Technology*. New York: Guilford Press.

Simon, Julian L., ed. 1995. *The State of Humanity*. Oxford, UK, and Cambridge, MA: Blackwell.

Strong, David. 1995. *Crazy Mountains: Learning from Wilderness to Weigh Technology*. Albany: State University of New York Press.

White, Lynn, Jr. 1967. "The Historical Roots of Our Ecologic Crisis." *Science* 155(3767): 1203–1207.

Winner, Langdon. 1986. *The Whale and the Reactor: A Search for Limits in an Age of High Technology*. Chicago: University of Chicago Press.

Adam Briggle

TELEOLOGY

The word *teleology* comes from the Greek words *telos*, meaning "end or purpose," and *logos*, meaning "word, thought, speech, principle." Teleology, as a specialty of modern philosophy, is the study of ends or purposes. With roots in Aristotle's philosophy, teleology refers both to a traditional model of scientific explanation—one that encompasses final causes—and an approach to ethics. Teleological explanations and teleological ethics have played a role, albeit a controversial one, in modern environmentalism.

ARISTOTELIAN AND MEDIEVAL ORIGINS

Based on his observations of nature, the ancient Greek philosopher Aristotle (384–322 B.C.E.) concluded that all natural objects have a characteristic and distinctive activity. Aristotle's teleological science differs from modern science in holding that one does not have a complete scientific understanding of an object until one understands this characteristic activity. The goal of this activity, what can be called its purpose or function, is identified as the object's *telos*. For example, in Aristotle's view one does not fully understand any artificial object, such as a house, until one understands the purpose it serves (shelter in the case of a house). Nor does one fully understand any natural object—including the stars, plants, and animals—until one understands the purpose it serves or the goal it strives to reach.

These brief examples demonstrate the close connection between teleological science and teleological ethics. Once one has understood the *telos* of a house or a plant or an animal (including the human animal), something one discovers through Aristotelian science, one also comes to understand what a *good* house is or what a good specimen of its species is—something with ethical implications. A good pine tree is one that grows straight and tall and produces much viable seed from which new pine trees might grow. A good pine tree is one that achieves its *telos*. Aristotle believed that this teleological framework could be applied to all natural objects, including humans. All natural objects achieve their good when they fulfill their function or achieve their *telos*.

This approach was further developed in the Middle Ages, when Christian philosophers synthesized Christian theology with Aristotle's science and ethics as the basis of natural-law philosophy. According to this perspective, as science comes to understand the natural *telos* of each living thing, the "laws of nature," it helps us understand God's "laws." Because the *telos* and purposes discovered in nature are God's purposes, the natural order can be equated with the moral order. Natural law thus has both a descriptive meaning, as the scientific laws of nature, and a prescriptive meaning, as the normative rules that we ethically ought to follow. In this ethical tradition fulfilling one's natural potential—a potential implicitly in harmony with the rest of nature—is the highest form of ethical activity.

APPLICATIONS TO ENVIRONMENTAL PHILOSOPHY

Several themes in contemporary environmental philosophy have parallels to this ethical tradition. Some environmentalists believe that ecosystems are naturally well ordered and harmonious. All parts of an ecosystem have a distinctive place and function in the overall scheme, each contributing to the natural order in its own way. Predators control the populations of their prey, bees pollinate flowering plants, and earthworms aerate the soil; hence each species has its telos in the ecological order. Undisturbed nature is thus good; ecological problems arise only when humans interfere with the natural order.

Other environmentalists have argued that all living things have moral standing because every living thing individually has a telos of its own—the end it strives to achieve—and thus a good of its own. For example, Paul Taylor argues that all living beings are "teleological centers of life," and possess a good of their own that is independent of human interests or, for that matter, independent of any ecological function they may (or may not) perform in the larger ecological order (1986).

Major objections to the teleological tradition challenge its relevance to contemporary debates. Most tellingly, modern evolutionary science provides a significant and perhaps insurmountable challenge to the teleological tradition in both science and ethics. The process of evolution by natural selection offers an account of the apparent design found in nature without appealing to any purpose or *telos*. On this view the order and characteristic activity that is found in nature is not inherent in nature itself, as Aristotle thought, nor does it come from a divine plan, as the Christian Aristotelians in the Middle Ages believed; it results from random genetic mutation and natural selection.

For example, it is tempting to offer a teleological explanation for the long neck of the giraffe by claiming that the long neck exists (or was designed) in order to allow the giraffe to reach food high off the ground. Yet evolutionary biology explains that the giraffe did not develop a long neck *in order to* reach the leaves but that the longer-necked ancestors of giraffes survived and reproduced similarly longer-necked progeny because, having longer necks, they were better able to reach food high off the ground than competing organisms with shorter necks. On this view nature is aiming at nothing, and species have no assigned function in the natural order; nature is headed nowhere in particular. It is neither good nor bad; it just is.

But the appeal of the teleological tradition persists. Contemporary biologists are comfortable using teleological categories when speaking about the natural world. Even within a Darwinian framework, such teleological concepts as *function, purpose, goal,* and *design* are used regularly by scientists and philosophers. Consider the following examples: "The purpose of the kidney is to remove waste from the blood"; "the goal of brightly colored plumage on male birds is to attract females"; "wetlands function as flood-control and water-filtration systems." In the teleological tradition, it is a reasonable inference from such purposive and functional ascriptions to a value or normative conclusion: "This is a healthy kidney"; "this is a successful male"; "wetlands should not be destroyed." The contemporary challenge is whether such inferences are legitimate and, if so, what normative meanings inhere in them.

Many observers continue to resist any inference from natural facts to value claims. One approach is to claim that the functional language that is common and appropriate in the biological sciences is a kind of convenient shorthand and that all such ascriptions can be fully explained, albeit more awkwardly, in terms of antecedent causes. For convenience navigators treat the Earth as an object at rest, the Sun as a moving object, and the pole star as a motionless point of reference even though they know full well that the Earth orbits the Sun and turns on its axis. The science of navigation could be expressed in the language of modern astronomy, but with great sacrifice of economy and simplicity of expression. Thus, in the view of modern science, although kidneys, plumage, wetlands, and even long necks might perform a function, they do not do so out of a prior built-in purpose, as Aristotle supposed, or a divine design as his Christian exponents supposed. The functions themselves are simply the result of previous evolutionary processes. Inferring a value conclusion from these facts would require an implicit value component already assumed. For example, only by assuming that reproductive success is good could one infer that bright plumage is a good thing for male birds; and only by assuming that survival is good could one infer that long necks are good for giraffes.

Some philosophers argue that it does seem reasonable to assume that adaptation for survival and reproductive success (what biologists call "inclusive fitness") *is* good and therefore that it is reasonable to reach normative conclusions from biological facts. It seems reasonable to conclude that the adaptive capacity that a species has developed to outrun or hide from its predators is good for that species. It is good for the giraffe to have a long neck, and a kidney that does not filter blood effectively is bad kidney.

Assessing these debates would require carefully distinguishing between such concepts as *function, purpose, design,* and *goal.* When used in science, all of these concepts involve explaining some phenomena (kidneys, long necks, bright plumage, predators, wetlands) in terms of some future state or activity (filtering blood, reaching high food, attracting a mate, controlling prey populations, absorbing flood water). This is the essence of any teleological explanation. Aristotle's teleological science seems right in this: One has not fully understood such natural objects until one has understood how they function. But the questions remain: Are such forward-looking explanations scientifically valid? If so, is it always a good thing to attain this future state or perform this activity? Are functional explanations truly teleological? Is value built in to any notion of aiming for and attaining some future state? If so, is this value necessarily an ethically good thing?

Much modern science, particularly as it developed under the influence of physics and mechanics, argues that the only legitimate scientific explanations are those that refer back to antecedent causes, not forward to future goals, or final causes. Thus, some critics challenge the legitimacy of any teleological explanations at all in the biological sciences. Others argue that, although some teleological explanations are legitimate, others are not and that no value conclusions can be drawn from them in any case. For example, although it may be legitimate to think that the function of the kidney is to filter blood, it is not legitimate to think that the function of earth worms is to aerate soil; rather, as their digestive tracts extract nourishment from detritus, a fortuitous side effect is soil aeration. Still others argue that, although value conclusions might be drawn from biological facts, the values are always qualified and conditional. *If* you assume that survival and reproductive success is good, or *if* you assume that adaptive fit is good, then one can conclude that certain traits are good for certain species. But inclusive fitness and adaptability are not always an *ethically* good thing. That which is good for a species or an individual is not always identical with an ethical good, a distinction foreign to Aristotle. Thus, although the teleological tradition provides a framework for thinking and reasoning about relations between nature and ethics, it is fraught with philosophical and scientific controversies.

SEE ALSO *Agricultural Ethics; Christianity; Darwin, Charles; Ecosystem Health; Evolution; Natural Law Theory; Species; Taylor, Paul.*

BIBLIOGRAPHY

Allen, Colin, Marc Bekoff, and George Lauder, eds. 1998. *Nature's Purposes: Analyses of Function and Design in Biology.* Cambridge, MA: MIT Press.

DesJardins, Joseph R. 2006. *Environmental Ethics. An Introduction to Environmental Philosophy.* 4th edition. Belmont, CA: Thomson.

Taylor, Paul W. 1986. *Respect for Nature: A Theory of Environmental Ethics.* Princeton, NJ: Princeton University Press.

Joseph DesJardins

THEORY

One of the principal tasks of environmental ethics and philosophy is to posit and defend an adequate normative ethical theory. This agenda was set by two seminal essays: In "Historical Roots of Our Ecologic Crisis" (1967), Lynn White Jr. blamed the environmental crisis on the Judeo-Christian worldview, claiming that Christianity was the most anthropocentric of world religions. An environmentally friendly worldview, he implied, would have to be nonanthropocentric. But, he argued, traditional European and North American ethical theory is anthropocentric, requiring an effort at building a new ethical theory. In the 1973 essay "Is There a Need for a New, an Environmental Ethic?" Richard Routley constructed the now-famous "Last Man" thought experiment, in which the last human being "lays about him" destroying everything within reach. Routley correctly expected that most of his readers would judge the last man's behavior to be morally reprehensible, but standard European and North American ethical theory could not support such an intuition. The Last Man thought experiment claims to demonstrate that the foundations of environmental ethics must be nonanthropocentric.

This interest in theory marks environmental ethics and philosophy as distinct from the more immediate practical work of ecological restoration, the development of sustainable technologies, or the institution of ecologically informed environmental policy. In addition to the implementation of environmentally sound practices, environmental ethicists and philosophers focus on fundamental questions concerning the types of values attributed to nature, what it would mean to actually restore a landscape, what it means to engage in a sustainable technology, or what constitutes an ecologically informed policy. Underlying and motivating all of these more practical environmental aims are implicit theoretical, environmental, ethical, and philosophical assumptions about the value of the environment itself. Theoretical environmental philosophy exposes and critically engages such assumptions.

Theorists in environmental ethics and philosophy have historically been interested in both normative and metaethical theoretical questions. Metaethics addresses questions *about* ethics, whereas normative ethics focuses on questions *within* ethics. Metaethicists are interested, for example, in whether or not environmental ethical claims can be true or false, whereas normative ethicists work to formulate and defend particular systems or theories of environmental ethics.

NORMATIVE ETHICAL THEORY IN ENVIRONMENTAL ETHICS AND PHILOSOPHY

An ethical theory is an attempt to determine which entities are worthy of direct moral standing, which are worthy of only indirect moral standing, and which do not matter morally. Consider, as an illustration, a circle. If something lies within this circle of moral concern (or within the moral community), it possesses direct moral standing. Things with direct moral standing count, period. If something lies outside of the moral community, it might count, but only indirectly at best. That is, something outside of the moral community might be tethered to (i.e., somehow important for or to) something within the moral community. It might also be the case that those things outside of the moral community possess no moral standing at all if it can be demonstrated that they serve no end for things within the moral community. So, for example, if our moral community included only human beings, then, although we would not necessarily be concerned with the loss of tropical plant species per se, we might still be concerned with them if their well-being somehow served a human end (e.g., provided chemical extracts that could treat a human illness). If, however, our moral community included all living things, then, in addition to being important as a source of medicine for humans, plants would also count directly.

Moral standing—whether direct or indirect—depends largely on what a given theory presupposes as the key to this inclusion. Typically the key to moral standing is a quality that entities possess or fail to possess. Because an entire moral-community structure depends on an established key to moral standing, normative environmental ethical theory has focused a great deal of energy on determining the nature of this key.

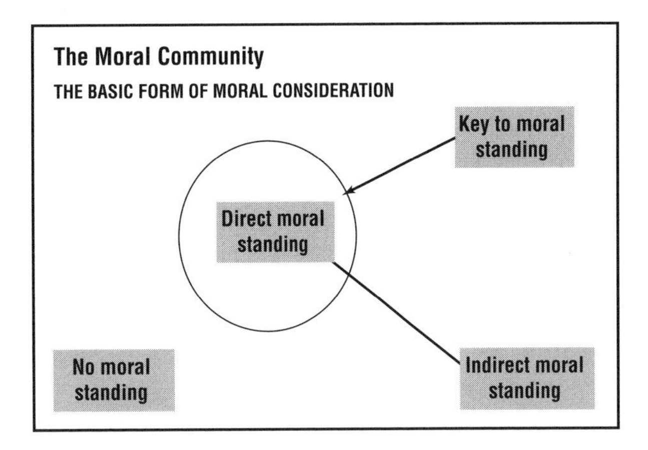

Figure 1. *Environmental ethical theories vary with regard to who or what possess direct, indirect, or no moral standing. For example, for some theories species and ecosystems might be members of the moral community directly, for others they might matter only because they support that which does count directly (i.e., they might merit only indirect moral standing), and for still others they might not matter at all. Different theories propose different relevant qualities (or keys) or moral consideration which then determines which things count (or do not count) and in what way.* CENGAGE LEARNING, GALE.

Although some environmental scholars downplay or dismiss the value of environmental-ethical theorizing, others point out the advantages to such theory building. First, they note that all actions and polices presuppose a theoretical foundation. Absent coercion, we are usually willing to perform only those actions that are consistent with our value assumptions. Hence, to engage in environmental ethical theory building is to engage at the same time in environmental policy making (albeit indirectly). Furthermore, given the inescapability and force of environmental ethical theory, it seems wiser to attend to such theory building than to ignore it. Second, the theoretical foundations provided by environmental ethical theory might be empowering. As opposed to environmental discourse that appears ungrounded, or environmental decision-making that happens only as the result of political maneuvering and power struggle, environmental ethical theorizing allows us to create a solid foundation from which to judge and defend a certain course of action and to understand the roots of other courses of action. Hence, even those without political power can participate in environmental discourse and possibly even in environmental decision-making. Third, environmental ethical theory provides us with at least a rough idea—but not the specific details—about how we ought to live. The application of theory, as opposed to concrete rules or policies, requires us to think for ourselves, allows us to adjust to novel situations and to consider how a given theory might manifest itself in different ways under different conditions. Finally, the establishment of an environmental ethical theory, and its corresponding notions of direct and indirect moral standing, establishes a powerful and important burden of proof. Given that those with direct moral standing would be "innocent until proven guilty," whereas those without direct moral standing would be "guilty until proven innocent," and given that the establishment of burden of proof is no small matter to those entities whose fate is thus decided, environmental ethical theory building takes on enormous importance.

Like many disciplines, environmental ethics has its own vocabulary. Two key terms in environmental ethics are *instrumental value* and *intrinsic value*. Something is said to have instrumental value if it is a means to some other end (e.g., a child can have instrumental value if she can mow the lawn). Something is said to have intrinsic value when it possesses value that transcends its instrumental value (e.g., a child, even if she will not or cannot mow the lawn). The views of environmental ethicists differ most markedly in their attribution of either instrumental or intrinsic value to various nonhuman things in the world. These differences produce profound differences in how and to what extent one sees and likewise how one feels we ought to work to solve environmental "problems."

KEY PERSPECTIVES IN ENVIRONMENTAL ETHICS

The key perspectives in environmental ethics are the following:

1. *Anthropocentrism* is the position that all humans—and only humans—possess intrinsic value and direct moral standing. In this view, nonhumans have only instrumental value to the extent that human well-being may in some way depend on them. For the anthropocentrist, environmental ethics and policies are motivated and justified solely on the basis of their effect on humans, without regard for the nonhuman world. An anthropocentrist, for example, would be concerned about rapid global climate change only insofar as it affects the welfare of human beings. Anthropocentrists argue variously that it is either unintelligible or unnecessary to extend direct moral standing to the nonhuman world. The philosopher John Passmore (1974) represents the anthropocentric camp.
2. *Nonanthropocentrism* attributes intrinsic value to humans and to at least some nonhuman entities. Nonanthropocentrists vary in how inclusively they view the moral community. The U.S. Endangered Species Act (1973), as an example, is nonanthropocentric to the extent that it dissociates the value of a species from its economic and narrowly human-centered value. Each of the perspectives described below are types of nonanthropocentrism (except that extensionism is more general than nonanthropocentrism):
   a. *Extensionism* is exemplified by zoocentrism and biocentrism (see below). These perspectives vary according to the extent to which they argue moral consideration ought to be attributed to various kinds of other individuals. Extensionism attempts to extend traditional moral theories (such as utilitarianism or rights theory) to entities that have not traditionally been considered worthy of direct moral standing.
   b. *Zoocentrism* attributes intrinsic value only to humans and certain nonhuman animals, although adherents to this view differ about which animals possess direct moral standing and intrinsic value. A zoocentrist could, for example, be concerned about the loss of biodiversity insofar as it harms humans and nonhuman animals that possess clear indications of self-consciousness (e.g., primates). Peter Singer (1975) and Tom Regan (1983) are major proponents of zoocentric philosophy.
   c. *Biocentrism* attributes intrinsic value and direct moral standing to all individual living creatures. It takes "being alive" as the key to moral inclusion. Nonliving things (e.g., lakes or rocks) and collectives (e.g., species and ecosystems) possess only instrumental value or no value at all. Biocentrists would care, for example, about biodiversity loss because of its effect on all individual living things. The philosopher Paul W. Taylor (1986) defends this position, as do Kenneth Goodpaster, Robin Attfield, and James Sterba.
   d. *Universal consideration* is a position that attributes intrinsic value and moral standing to everything (living or not). Hence, from this perspective, biodiversity loss would be decried not only for its potential harm to all living things, but also for its negative impact on even nonliving things such as mountains, rivers, or rocks. The philosopher Thomas Birch (1993) has championed this position.
   e. *Ecocentrism* is a reaction against the atomism or individualism represented by extensionism. Adopting Charles Darwin's analysis of ethics as generated by community membership and inspired by principles of ecological science, ecocentrism reflects the social-like connectedness among individuals in nature. Emergent properties of biological wholes—such as species, biotic communities, and ecosystems—transcend the properties of the individuals that compose such collectives. Ecological collectives, ecocentrists argue, merit moral standing because of their emergent properties and connectedness. Ecocentrism thus focuses moral concern on the maintenance of biotic communities, species, and ecosystems and less on the welfare of animals and other organisms. Aldo Leopold represents ecocentrism, especially when he writes, "A thing is right when it tends to preserve the integrity, stability, and beauty of the biotic community. It is wrong when it tends otherwise." (Leopold 1949, pp. 224–225). The philosopher J. Baird Callicott (1989, 1999) is the most noted defender of Leopold's land ethic. The philosopher Arne Naess (1989) is the originator and noted defender of the ecocentric environmental ethic known as Deep Ecology, which is based on a mystical sense of self intimately connected with all of nature. Some argue that ecocentrism, taken to its logical conclusion, is equivalent to James Lovelock's Gaia hypothesis (1979), in which the entire Earth merits moral consideration.
   f. *Environmental virtue theory* began to emerge in the early part of the twenty-first century. Harking back to Aristotle's approach to ethical reasoning, environmental virtue ethicists suggest that we should focus our energies on the creation of virtuous people, or people of appropriate character, instead of on working to determine the proper key to moral consideration and membership in

the moral community. Traits such as respect, humility, caring, and attentiveness are often advanced as the key virtues. The philosophers Phil Cafaro (2001) and Ron Sandler (2007) represent this trend in environmental ethics.

METAETHICAL THEORY IN ENVIRONMENTAL ETHICS AND PHILOSOPHY

Although many metaethical questions surround environmental ethics and philosophy, three of them have been at the center of particularly robust debates. Environmental ethics is often regarded as one among several types of "applied ethics," such as biomedical ethics, engineering ethics, and business ethics. Applied ethicists rely on the prevailing European and North American ethical theories—variations on utilitarianism and Kantian deontology to new ethical questions that the eighteenth- and nineteenth-century authors of these theories could never have imagined or anticipated. Accordingly, some philosophers view environmental philosophy as work that takes traditional ethical theory and examines environmental concerns through the lens of these theories. These philosophers might therefore be concerned with what the prevailing European and North American ethical theories of utilitarianism or deontology might have to say about the rightness or wrongness of factory farming or sport hunting.

Other environmental philosophers, however, view the work of environmental philosophy and ethics as something quite beyond mere applied ethics—as work that explores new ideas about ethics and even metaphysics and that, although practical in its application, is also freshly theoretical. These philosophers might point to the unique nonanthropocentric ethical theories created by environmental philosophers or the work done in policy and philosophy or between various sciences (e.g., ecology, geography, geology, and biology) and philosophy as examples of uniquely theoretical environmental ethics. The work of Robert Frodeman has taken this approach (2003).

Second, there has been a debate between those who operate as if environmental ethicists, like classical European and North American ethical theorists, should pursue a unified ethical theory and those who recommend embracing several theories at once. The former favors ethical monism, the belief that there is only one proper ethical theory. Ethical pluralists, on the other hand, believe that there may be more than one legitimate ethical theory and a plurality of ethical truths. Pluralists worry about the homogenizing and totalizing effect of the pursuit of monism. Monists worry that pluralism is little more than ethical relativism. Pluralists deny this reduction to relativism and instead suggest that the standard of an acceptable ethic ought to shift from a focus on truth to a focus on reasonableness, pointing out that any number of ethical prescriptions can be reasonable. Are pluralists really suggesting that a number of incommensurable ethical theories can be coherently held at the same time, or are they suggesting that different persons implicitly or explicitly hold different ethical theories and that there is, therefore, no decisive way to declare that one or the other is the one true theory? Are monists really suggesting that there is only one true ethical theory or merely demanding that each person hold an internally consistent ethical point of view while allowing that there may be many self-consistent ethical theories? Synthesizing, in Hegelian fashion, monism and pluralism in environmental ethics, could a monistic theory be constructed that is sufficiently general to allow for a plurality of approaches and applications in the real world? Within the literature of environmental ethics, Christopher Stone (1987) advocates an extreme version of pluralism, whereas Peter Wenz (1993) advocates a more moderate pluralism. J. Baird Callicott (1999) has defended a moderate version of monism; Peter Singer (1990) represents a more extreme monism in his steadfast commitment to utilitarianism.

Third, there is a debate between those environmental philosophers who have focused primarily on the creation and defense of ethical theories (theories that defend the intrinsic value of nature) and environmental pragmatists who are motivated primarily by effecting environmental change in the "real world." Whereas the more theoretically motivated environmental philosophers contend that environmental philosophers should continue to create and defend abstract theories of values that underpin environmental attitudes and decision making, pragmatists contend that they should instead focus on variable and context-dependent accounts of value and truth (i.e., on solving real-world environmental problems). Pragmatists often assert that environmental ethical theorizing has had no real impact on environmental problems or policy formation and that we ought to make philosophy more practical. Although some pragmatists assert that environmental ethical theorizing is simply irrelevant or useless, some assert that it is actually counterproductive. Commonly taking a pluralist approach, some environmental pragmatists strive to marshal the values and ethical commitments of ordinary people to support of environment-friendly policies; others recommend suppressing the often conflicting values and ethical commitments of ordinary people—because they can be divisive—and focus on more situation-dependent solutions that all sides can live with. Ethical theorists might, however, argue that any such solutions cryptically rest on implicit values and ethical commitments and that the very notion of an environmental problem presupposes the significance of environmental values and ethics. Finally, although ethical theorizing has not yet had the impact that environmental philosophers had originally hoped, it is also not clear why a pragmatist would necessarily care what ethical theorists do. It would seem that, as self-avowed pluralists, pragmatists would be content to let theorists theorize, while they, the pragmatists get on with their problem-oriented, situation-dependent solutions, What could be more unpragmatic—that is, impractical—than spending thousands of hours writing dozens of books and articles pointing out the unpragmatic product of the unpragmatic theoreticians?

SEE ALSO *Callicott, J. Baird; Gaia Hypothesis; Last Man Arguments; Leopold, Aldo; Naess, Arne; Passmore, John Arthur; Pragmatism; Singer, Peter; Sylvan, Richard; White, Lynn, Jr..*

BIBLIOGRAPHY

Attfield, Robin. 1991. *The Ethics of Environmental Concern.* 2nd edition. Athens: University of Georgia Press.

Birch, Thomas H. 1993. "Moral Considerability and Universal Consideration." *Environmental Ethics* 15:313–332.

Cafaro, Philip. 2001. "Thoreau, Leopold, and Carson: Toward an Environmental Virtue Ethics." *Environmental Ethics* 23:3–17.

Callicott, J. Baird. 1989. *In Defense of the Land Ethic: Essays in Environmental Philosophy.* Albany: State University of New York Press.

Callicott, J. Baird. 1994. "Moral Monism in Environmental Ethics Defended." *Journal of Philosophical Research* 19: 51–60.

Callicott, J. Baird. 1999. *Beyond the Land Ethic: More Essays in Environmental Philosophy.* Albany: State University of New York Press.

Frodeman, Robert. 2003. *Geo-Logic: Breaking Ground Between Philosophy and the Earth Sciences.* Albany: State University of New York Press.

Goodpaster, Kenneth E. 1978. "On Being Morally Considerable." *Journal of Philosophy* 75(6): 308–325.

Hargrove, Eugene C., ed. 1979–2008. *Environmental Ethics* (the journal). Denton: The Center for Environmental Philosophy and the University of North Texas.

Jamieson, Dale. ed. 2001. *A Companion to Environmental Philosophy.* Malden, MA Blackwell.

Leopold, Aldo. 1949. *A Sand County Almanac, and Sketches Here and There.* New York: Oxford University Press.

Lovelock, James E. 1979. *Gaia: A New Look at Life on Earth.* Oxford and New York: Oxford University Press.

Marietta, Don E., and Lester Embree. 1995. *Environmental Philosophy and Environmental Activism.* Lanham, MD: Rowman and Littlefield.

Naess, Arne. 1989. *Ecology, Community, and Lifestyle: An Outline of an Ecosophy,* trans. and rev. David Rothenberg. Cambridge, UK, and New York: Cambridge University Press.

Passmore, John. 1974. *Man's Responsibility for Nature.* New York: Scribners.

Regan, Tom. 1983. *The Case for Animal Rights.* Berkeley: University of California Press.

Routley, Richard [later Richard Sylvan]. 1973. "Is There a Need for a New, an Environmental, Ethic?" In *Proceedings of the Fifteenth World Congress of Philosophy,* Vol. 1, 205–210. Sophia, Bulgaria: Sophia Press.

Sandler, Ronald L. 2007. *Character and Environment: A Virtue-Oriented Approach to Environmental Ethics.* New York: Columbia University Press.

Singer, Peter. 1990. *Animal Liberation: A New Ethics for Our Treatment of Animals.* 2nd edition. New York: Avond.

Sterba, James P. 2005. "Kantians, Utilitarians and the Moral Status of Nonhuman Life." In *The Triumph of Practice Over Theory in Ethics.* New York: Oxford University Press.

Stone, Christopher D. 1987. *Earth and Other Ethics: The Case for Moral Pluralism.* New York: Harper and Row.

Taylor, Paul W. 1986. *Respect for Nature: A Theory of Environmental Ethics.* Princeton, NJ: Princeton University Press.

Wenz, Peter. 1993. "Minimal, Moderate, and Extreme Moral Pluralism." *Environmental Ethics* 15:61–74.

White, Lynn, Jr. 1967. "The Historical Roots of Our Ecologic Crisis." *Science* 155:1203–1207.

Michael P. Nelson

THOREAU, HENRY DAVID

1817–1862

Henry David Thoreau was born in Concord, Massachusetts, and lived there all his life. A writer, naturalist, and philosopher, he was an important forerunner of American environmentalism and remains a key source of insight and inspiration for millions of environmental and political activists around the world. *Walden* is his most famous work. His influence on the development of environmental ethics has been profound.

NONANTHROPOCENTRIC ETHICS

Thoreau was one of the earliest and strongest critics of anthropocentrism: the view that only human beings have rights or "intrinsic value" and that other creatures may be used in any way people see fit. For example, his first book, *A Week on the Concord and Merrimack Rivers* (1849), discussed the plight of anadromous fishes formerly found in great numbers in New England's rivers but by that time mostly blocked by dams.

One hundred thirty years before Earth First! Thoreau suggested that unjust treatment of the shad was grave enough to justify civil disobedience. This was a method of social protest which Thoreau pioneered, influencing both Mohandas Gandhi and Martin Luther King, Jr.

Thoreau made one of the earliest explicit calls for a nonanthropocentric ethic, writing: "Away with the superficial and

Henry David Thoreau. *A famous writer and philosopher and a dedicated critic of anthropocentrism, Thoreau pioneered the method of civil disobedience as a pro-environmental technique. Many of his works extoll the value of nature and the wilderness for the human experience.* © BETTMANN/CORBIS.

selfish phil-*anthropy* of men [emphasis and "fish" puns in the original]—who knows what admirable virtue of fishes may be below low-water mark, bearing up against a hard destiny, not admired by that fellow creature who alone can appreciate it!" (Thoreau 1849, p. 37). Reserving all love and concern for humans is both superficial, based on ignorance of what is below the surface, and selfish, an excuse for unjustified self-partiality.

WALDEN AND LATER WRITINGS

Walden represents a more searching, sustained attempt to specify and live a nonanthropocentric ethics. In that book Thoreau repeatedly asserted the intrinsic value of nonhuman nature—whether in trees, woodchucks, or Walden Pond itself—and tried to justify those assertions. However, *Walden* also discusses the benefits to people of recognizing the value of nature and living in harmony with it. Like Emerson's *Nature* (1836), but more practically and with a greater emphasis on wild nature, *Walden* teaches that nature is humankind's greatest resource. It provides all that humanity needs to flourish if people protect it and use it wisely.

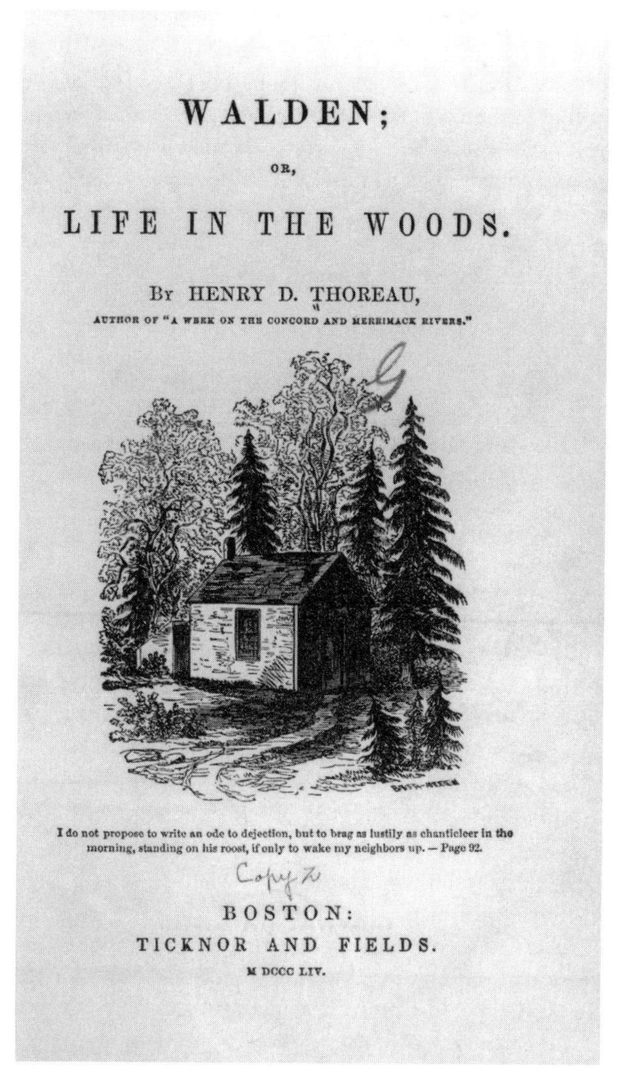

WALDEN;

OR,

LIFE IN THE WOODS.

BY HENRY D. THOREAU,

AUTHOR OF "A WEEK ON THE CONCORD AND MERRIMACK RIVERS."

I do not propose to write an ode to dejection, but to brag as lustily as chanticleer in the morning, standing on his roost, if only to wake my neighbors up. — Page 92.

BOSTON:

TICKNOR AND FIELDS.

M DCCC LIV.

"Walden" Title Page, 1854. *One of Henry David Thoreau's most famous works, "Walden" speaks to the benefits of valuing nature and living in harmony with it. This original title page features the cabin near Walden Pond in which Thoreau lived for two years, two months, and two days while writing the book.* LIBRARY OF CONGRESS.

However, wise use does not mean managing the whole of nature, and Thoreau's late essay "Walking" further develops the claims in *Walden* for the high value of wild nature. "In Wildness is the preservation of the World," Thoreau exclaimed there (Thoreau 1980, p. 112). "From the forest and wilderness come the tonics and barks which brace mankind" (p. 114). For Thoreau this vital connection was literal, physical, and sensual. "I think that I cannot preserve my health and spirits," he wrote, "unless I spend four hours a day at least . . . sauntering through the woods and over the hills and fields" (p. 95). This physical contact with nature revitalizes the mind, stimulating that "uncivilized free and wild thinking" at the heart of all human creativity (p. 96).

Walden and "Walking" give detailed advice for deepening one's experience of nature. However, it would be foolish to focus solely on improving the walker while failing to preserve the landscape he or she walks through, just as it would be a mistake to locate all value in the human experience and none in wild nature. The greatest value comes when one brings a lively mind to a vital place. Thoreau made much of the Concord countryside; still, as he reflected in his journal:

> I spend a considerable portion of my time observing the habits of the wild animals, my brute neighbors. ... But when I consider [that] the nobler animals have been exterminated here,—the cougar, panther, lynx, wolverene, wolf, bear, moose, deer, the beaver, the turkey, etc., etc.,—I cannot but feel as if I lived in a tamed, and, as it were, emasculated country. (Thoreau 1906, pp. 220–221)

Thoreau wanted "to know an entire heaven and an entire earth" rather than "a maimed and imperfect nature." In the end, humanity cannot separate its own flourishing from that of nature.

One of Thoreau's major insights was that human flourishing no longer requires the further taming of nature but rather the preservation of what wildness remains. "I would not have every man nor every part of a man cultivated, any more than I would have every acre of earth cultivated," he wrote (1980, p. 126). A "true culture" must respect spontaneity as well as order, creativity and daring as well as safety and good behavior (p. 124). It also must preserve wilderness landscapes so that it can preserve wildness in people's hearts and minds: "To preserve wild animals implies generally the creation of a forest for them to dwell in or resort to. So it is with man" (p. 117).

SEE ALSO *Civil Disobedience; Earth First!; Emerson, Ralph Waldo; Environmental Activism.*

BIBLIOGRAPHY

Buell, Lawrence. 1995. *The Environmental Imagination: Thoreau, Nature Writing, and the Formation of American Culture.* Cambridge, MA: Belknap Press of Harvard University Press.

Cafaro, Philip. 2004. *Thoreau's Living Ethics: Walden and the Pursuit of Virtue.* Athens: University of Georgia Press.

Richardson, Robert D., Jr. 1986. *Henry Thoreau: A Life of the Mind.* Berkeley: University of California Press.

Thoreau, Henry D. 1849. *A Week on the Concord and Merrimack Rivers.* Boston: James Munroe and Co.

Thoreau, Henry David. 1906. *The Journal of Henry D. Thoreau,* vol. 8 (March 23, 1856). Boston: Houghton Mifflin, 220–221.

Thoreau, Henry David. 1971. *Walden,* ed. J. Lyndon Shanley. Princeton, NJ: Princeton University Press.

Thoreau, Henry David. 1980. *The Natural History Essays.* Salt Lake City, UT: Peregrine Smith.

Philip Cafaro

TRADITIONAL ECOLOGICAL KNOWLEDGE

Humans have understood their environments in terms of traditional (or indigenous) ecological knowledge (TEK). TEK has been a major determinant of the ways in which people have interacted with their environment—for all peoples at some point in the past, and for many today as well, though much TEK has been lost via the diminution or disappearance of cultural groups, or greatly modified by interaction with scientific ecological knowledge (SEK), the knowledge generated by the more formal, organized investigation of the world that has become increasingly dominant.

TEK is central to environmental ethics and philosophy because understanding it opens up new perspectives on the ontology, axiology, epistemology, and praxis of ecological knowledge in general, and because understanding TEK and its similarities to and differences from modern SEK could be critical for developing and managing more sustainable ecosystems. (*Knowledge* herein is defined as consisting of values, descriptive data, and theory in a person's mind, which is shared to differing degrees within groups of different sizes to which individuals belong.)

HOW CAN WE UNDERSTAND TEK?

A major difficulty in defining and discussing TEK is that the TEKs of different indigenous groups are different because of variations in local cultural and environmental contexts; therefore, only those brought up in a specific culture and environment are likely to fully understand that culture's TEK. This insider's (*emic*) perspective contrasts with an outsider's (*etic*) perspective. Although some social scientists believe that it is fruitless for outsiders to attempt to understand TEK, others believe that it is possible to gain useful etic understanding and to generalize about TEKs, and that such work may be critical for their survival. A major challenge to generalization is the variation in TEK among individuals within local groups and in SEK within groups of scientists, e.g., plant breeders (Cleveland 2001), so that to compare TEK and SEK requires a comparison of variances and central tendencies—monolithic TEKs and SEKs do not exist.

Understanding TEK etically, however, requires a baseline for describing similarities and differences among

A San Bushman Teaching His Son to Hunt in Namibia. *Traditional Ecological Knowledge, or TEK, is an integral part of environmental philosophy. TEK is commonly understood to be based on direct interactions between humans and their environment, but what remains to be decided is whether the effects of indigenous peoples on their local ecosystems has been positive (conservationist) or negative (destructive). The UN Declaration on the Rights of Indigenous Peoples grants such native populations the right to pursue development and maintenance of their land as they see fit.* JOY TESSMAN/NATIONAL GEOGRAPHIC/GETTY IMAGES.

TEKs (and between TEKs and SEKs, as discussed below). Because most who have attempted this comparison have SEK, it is SEK that has provided the methodological base for comparison. (Therefore, attempts to understand TEK include the following assumptions: (a) There is an external objective reality that both TEK and SEK are based on; and (b) SEK can provide a description of this reality that can serve as an ontological comparator.)

Indeed, in an increasingly crowded and interconnected world, it is not possible to assume that each local group and its TEK and natural-resource management regimes can be understood only emically, because all activities affect other groups with different values and different management strategies. Therefore, negotiating conflicts based on some etic standards is required for the survival of TEKs and the external ecological reality they refer to.

HOW IS TEK CREATED, AND HOW IS IT RELATED TO ECOLOGY AND ETHICS?

A major controversy in the study of TEK has been whether indigenous classification of the environment is the result of the universal structure in nature that imposes itself on the human mind, perhaps facilitated by universals in human cognition (the intellectualist view), or whether it is the result of culture-dependent differences in goals, values, and theories (the utilitarian view) (Medin and Atran 1999). Boster's research with Aguaruna farmers in the Amazon is an example supporting the first view: Their cassava classification tends to classify the smallest distinct taxonomic unit in patterns similar to those of scientists. Support for the utilitarian view is a more common research finding, however: For example, the Mende of Sierra Leone use growth duration as a major criterion for classifying African rice varieties, with a mixture of varieties of different durations managed and planted to avoid labor bottlenecks and interharvest food shortages; Hopi and Quechua farmers classify and choose maize varieties based on ceremonial and religious values (Cleveland et al. 2000).

Differences in conclusions about the basis for classifying the environment may be due in part to differences in the nature of the plants or other organisms and environments involved. For example, the pattern of phenotypic expression (the appearance of an organism) of qualitative traits (e.g. seed or leaf color) in a clonally propagated crop (cassava) is much different than for quantitative traits (e.g. plant height or yield) in sexually propagated crops (rice), especially cross-pollinated ones (maize). Indigenous people may simply enjoy "playing" with diversity, yet their perceptions of genetic variation (to the extent revealed in plant phenotypes) depend on their ability to observe it and are determined in turn by the scale at which it occurs, the extent to which it is hidden by environmental variation, and on how important it is to them.

The preceding discussion assumes that TEK is based on direct interaction of individuals with the environment. However, TEK can also be learned indirectly—through teaching or imitating a respected individual, which introduces additional challenges for understanding TEK

TEK PRACTICE AND THE ENVIRONMENT

There is evidence that indigenous peoples have had massive effects on their environments, often in ways that increased useful production for humans, as summarized for the Americas by Mann (2005). But what criteria should be used to judge whether an effect is positive in the sense of conserving ecosystem functions, or negative in the sense of disrupting them? Change in biodiversity is often used as a criterion, and it is sometimes assumed that evidence that indigenous people and biodiversity coexist in space and time means the actions of the former are responsible for the latter. For example, the Global Plan of Action for crop genetic resources of the Food and Agriculture Organization of the United Nations (FAO) calls for more emphasis on in situ conservation based on evidence that "the rich diversity that exists today offers ample testimony of what has already been achieved" through farmer management and development of their crop genetic resources (FAO 1996, para. 26). Similarly, data showing a correlation between increased human presence and loss of biodiversity support the assumption that indigenous peoples tend to destroy their environments—for example, in the massive extinctions of large mammals with the exception of Africa during the last 10,000 to 50,000 years that followed indigenous peoples movements (Koch and Barnosky 2006).

Regardless of how the effects on the environment are judged, the question remains as to the connection between peoples' action and their TEK. Those who accept the conclusion that indigenous peoples conserve their environments often assume that their TEK is accurate and "ethical" because it leads to action that conserves the environment. On the other hand, those who accept the conclusion that indigenous peoples destroy their environments often assume that their TEK relatively is inaccurate and "unethical" because it leads to action that destroys the environment. Research on the relationships among TEK, behavior, and the effects of behavior on the environment is difficult to do and has produced conflicting results, making generalizations problematic and suggesting that these relationships are often contingent on local histories, cultures, and environments.

HOW DOES TEK COMPARE WITH SEK, AND CAN THEY WORK TOGETHER?

Social scientists often contrast SEK and TEK, seeing the former as rationalistic, reductionist, theoretical, generalizable, objectively verifiable, abstract, and imperialistic—in sharp contrast to the latter, which they see as organic, holistic, intuitive, local, socially constructed, practical, and egalitarian. On the other hand, there is evidence that SEK and TEK are more similar than different (Agrawal 1995). For example, since the 1920s, work by social scientists, historians, and philosophers on the nature of SEK has explicitly explored the ways in which it is shaped by personal psychology, historical contingencies, and social context, and some current research on TEK shows that it can be theoretical and objectively verifiable. One difficulty, discussed above, is how outsiders understand TEK if communication is structured so that indigenous people cannot explain the abstract or generalized basis for their specific TEK; outsiders may assume their practices are untheorized responses to changing, unpredictable circumstances.

Soleri and colleagues (2002) used scenarios based on basic biological principles to elicit TEK from traditional farmers in different parts of the world about genotype-by-environment interaction (i.e., the relationship between nature and nurture) and heritability based on a fundamental biological model. They found patterns in TEK across different crops and countries, and between TEK and SEK that supported the hypothesis that empirical and theoretical TEK and SEK consistently reflect similar environmental patterns and relationships. However, they also found differences among farmers, and among scientists, and between TEK and SEK, which could often be explained in terms of differences in crop varieties, environments, or cultural values.

Similarly, Ellen concluded that indigenous rain-forest peoples' observations of many individual species leads inductively to ecological models that are "privileged over accumulated inductive knowledge" (1999, p. 106). In some cases TEK may even be more complete than SEK—Malawian farmers' taxonomy of cassava varieties based on plant morphology visually distinguishes varieties between which scientists can see no differences, but whose distinctness was supported by molecular analyses for cyanogenic glucoside levels and genetic analysis (Mkumbira et al. 2003). TEK may also be less complete than SEK—Wola farmers of New Guinea are aware of geomorphological forces that destroy and renew their soil but not of processes on a geological time scale (Sillitoe 1996).

TEK IN RESEARCH, DEVELOPMENT AND POLITICAL CONTEXTS

All of the aspects of understanding TEK discussed above can be influenced by the research, development, and political contexts in which TEK is used. For example, an outside researcher's personal values and relationships with an indigenous community may affect her or his research and conclusions about TEK. Definitions of TEK in applied contexts often depend on the assumptions and goals of those in control. For example, the way in which "sustainable agriculture" is defined affects the way in which TEK is defined, which in turn leads to different roles for indigenous peoples in their development as controlled by outsiders (Cleveland and Soleri 2007).

The use of TEK in improving the well-being of local communities was pioneered by local and nongovernmental organizations and by indigenous communities themselves, but it has become institutionalized in the last two decades in mainstream economic development—for example by the World Bank and many national governments. This institutionalization has been criticized by many indigenous groups and their supporters for decontextualizing TEK and co-opting it for the goals of mainstream development, which may result in destroying TEK or even entire cultures.

Success in using TEK in "development" and applied science depends on its long-term results in social, cultural, economic, and environmental terms for local people and the world. This use of TEK challenges the cultural-relativist viewpoint that each local people's TEK is valid and should be respected by outsiders. In an increasingly crowded and interconnected world, however, TEK and natural-resource management practices cannot be judged only emically, because all activities affect other groups with different TEK and different management strategies. Therefore, we need to evaluate local solutions in global contexts of social, economic, and environmental sustainability.

TEK AND INDIGENOUS RIGHTS

Part of the problem with understanding the relationship between rights and TEK is that TEK often includes different concepts of rights than those of outsiders. It is possible, however, for outsiders to elicit indigenous peoples' knowledge of rights. For example, interviews using scenarios of potential conflicts over rights to crop genetic resources elicited consistent concepts of rights from Zuni community members (Soleri et al. 1994). There is a wide range of concepts of rights in TEK among various indigenous groups; they usually place more emphasis on community rights and on individual rights that benefit the community than SEK, which emphasizes individual rights to knowledge for personal gain and their protection through state-enforced legal systems such as patents.

Claims of indigenous farmers' rights to resources are often based on assumptions that indigenous farming is environmentally sustainable and that farmers' conservation of resources is based on accurate ecological knowledge and/or ethical principles of natural resource conservation. For example, Article 8(j) of the 1992 Convention on Biological Diversity (CBD) on in situ conservation calls for signatories to "respect, preserve and maintain knowledge, innovations and practices of indigenous and local communities embodying traditional lifestyles relevant for the conservation and sustainable use of biological diversity" and to "encourage the equitable sharing of benefits" arising from the use of those resources (CBD 1992). Yet, as discussed above, there are variations in sustainable resource use within communities and between indigenous communities.

Some arguments for indigenous peoples' human rights to land and biological resources rest on an assumption that these peoples are inherently conservationist; such arguments often mix value judgments about human rights with empirically testable hypotheses about the extent and efficacy of indigenous peoples' conservation of biodiversity. For example, indigenous rights and environmental conservation advocates may try to portray indigenous peoples in terms of European and North American environmentalist stereotypes—dubbed by some as "green Orientalism." If the empirical data do not support the claim that an indigenous people are conservators, then those who do not share the belief in their human rights—to which indigenous conservation has been linked—may deem this as sufficient justification for not recognizing indigenous rights to their traditional environments (Cleveland and Murray 1997).

On September 13, 2007, the United Nations General Assembly voted 143 to 4 (Australia, Canada, New

Zealand, United States; eleven abstentions), to adopt the nonbinding "United Nations Declaration on the Rights of Indigenous Peoples," which "sets out the individual and collective rights of the world's 370 million native peoples, calls for the maintenance and strengthening of their cultural identities, and emphasizes their right to pursue development in keeping with their own needs and aspirations," thereby ending nearly twenty-five years of "contentious negotiations" (United Nations General Assembly 2007). As this entry has shown, implementing these rights will not be easy. Successful implementation will depend on continuing research on the factors affecting variation in TEK within and among indigenous groups, and on the relationship of TEK to SEK. It will also depend on resolving conflicts over rights between individual indigenous groups and between indigenous groups and the more industrialized modern societies within which they exist. Such resolution in turn will require dealing with the reality that in an increasingly crowded and globalized world, consensus on dealing with common environmental resources will depend on some groups modifying their knowledge, including claims to rights.

SEE ALSO *Agriculture; Biodiversity; Convention on Biodiversity; Environmental Philosophy: V. Contemporary Philosophy; Sustainability.*

BIBLIOGRAPHY

Agrawal, Arun. 1995. "Dismantling the Divide between Indigenous and Scientific Knowledge." *Development and Change* 26: 413–439.

Cleveland, David A. 2001. "Is Plant Breeding Science Objective Truth or Social Construction? The Case of Yield Stability." *Agriculture and Human Values* 18(3): 251–270.

Cleveland, David A., and Stephen C. Murray. 1997. "The World's Crop Genetic Resources and the Rights of Indigenous Farmers. *Current Anthropology* 38: 477–515.

Cleveland, David A., and Daniela Soleri. 2007. "Farmer Knowledge and Scientist Knowledge in Sustainable Agricultural Development." In *Local Science versus Global Science: Approaches to Indigenous Knowledge in International Development*, ed. Paul Sillitoe. Oxford: Berghahn Books.

Cleveland, David A.; Daniela Soleri; and Steven E. Smith. 2000. "A Biological Framework for Understanding Farmers' Plant Breeding." *Economic Botany* 54: 377–394.

Convention on Biological Diversity (CBD). 1992. Montreal. Available from http://www.cbd.int/convention/convention.shtml

Ellen, Roy. 1999. "Models of Subsistence and Ethnobiological Knowledge: Between Extraction and Cultivation in Southeast Asia." In *Folkbiology*, ed. Douglas L. Medin and Scott Atran. Cambridge, MA: MIT Press.

Food and Agriculture Organization of the United Nations (FAO). 1996. "Global Plan of Action Summary." Rome: FAO. Available from http://www.fao.org/focus/e/96/06/more/glopla-e.htm

Koch, P. L., and A. D. Barnosky. 2006. "Late Quaternary Extinctions: State of the Debate." *Annual Review of Ecology Evolution and Systematics* 37: 215–250.

Mann, Charles C. 2005. *1491: New Revelations of the Americas before Columbus*. New York: Random House.

Medin, Douglas L., and Scott Atran. 1999. "Introduction." In *Folkbiology*, ed. Douglas L. Medin and Scott Atran. Cambridge, MA: MIT Press.

Mkumbira, J.; L. Chiwona-Karltun; U. Lagercrantz; et al. 2003. "Classification of Cassava into 'Bitter' and 'Cool' in Malawi: From Farmers' Perception to Characterisation by Molecular Markers. *Euphytica* 132(1): 7–22.

Sillitoe, Paul. 1996. *A Place against Time: Land and Environment in the Papua New Guinea Highlands*. Amsterdam: Harwood Academic.

Soleri, Daniela; David A. Cleveland; Donald Eriacho; et al. 1994. "Gifts from the Creator: Intellectual Property Rights and Folk Crop Varieties." In *IPR for Indigenous Peoples: A Sourcebook*, ed. Tom Greaves. Oklahoma City, OK: Society for Applied Anthropology.

Soleri, Daniela; David A. Cleveland; Steven E. Smith; et al. 2002. "Understanding Farmers' Knowledge as the Basis for Collaboration with Plant Breeders: Methodological Development and Examples from Ongoing Research in Mexico, Syria, Cuba, and Nepal." In *Farmers, Scientists and Plant Breeding: Integrating Knowledge and Practice*, ed. David A. Cleveland and Daniela Soleri. Wallingford, UK: CAB International.

United Nations General Assembly. 2007. "General Assembly Adopts Declaration on Rights of Indigenous Peoples; 'Major Step Forward' toward Human Rights for All, Says President." Available from http://www.un.org/News/Press/docs/2007/ga10612.doc.htm

David A. Cleveland

TRAGEDY OF THE COMMONS

The word *tragedy* is used to refer generically to anything really bad. As Carol Rose (1986) notes though, the word has an older meaning, referring to literary works that depict a protagonist caught up in events inexorably leading to his or her doom. Some of that older meaning is implicit in the logic of what we now call the tragedy of the commons. The phrase was invented by Garrett Hardin, who credits the idea to W. F. Lloyd. (See also the work of H. Scott Gordon.)

THE LOGIC OF THE COMMONS

Suppose there is a plot of land. The land has a *carrying capacity*: a number of animals the land can sustain more or less indefinitely. (The concept of carrying capacity is somewhat problematic. While it points to something real, because there really are limits to what the land can support, such limits are not fixed. Carrying capacity is somewhat fluid, and a function of many variables. For example, whether Kruger National Park in South Africa can carry 15,000 elephants depends on whether we want to leave room for rhinos, which is not simply an ecological issue.) Suppose the parcel's carrying capacity is 100 animals. The land is jointly owned by ten shepherds, each of whom owns ten animals for a total flock of 100 animals. The land is thus at its carrying capacity. As things stand, each animal is worth, for example, one dollar to its owner, so that, at carrying capacity, 100 animals are worth $100. Crucially, although the ten shepherds treat their individual flocks as private property, they jointly treat the land as one large pasture, with no internal fences, so that each of their animals grazes freely.

Now suppose one shepherd adds an eleventh animal. We now have 101 animals altogether, and thus have exceeded the land's carrying capacity. There is not quite enough food per animal now; therefore they are a bit leaner, and the value per animal drops to 95 cents per head. The total stock of 101 animals is now worth $95.95, which is $4.95 less than the total stock was worth before, when it was within the land's carrying capacity.

Why would a shepherd add the extra animal, when it so clearly is a losing proposition? At the original carrying capacity, the individual flocks of ten were worth $10. Having added one more sheep, the shepherd now has eleven and each is worth 95 cents. That works out to $10.45, which means that the individual shepherd actually made a profit of forty-five cents by adding the extra animal, even though the value of the total stock went from $100 to $95.95.

Although the total cost to the group of adding the extra animal exceeded the total benefit, the individual shepherd receives 100 percent of the benefit while paying only 10 percent of the cost. The other nine shepherds own ninety percent of the animals, so they suffer ninety percent of the loss involved in the falling price per head. Individual shepherds, though, see only individual costs and benefits, and act accordingly. The logic of the commons has begun its seemingly inexorable grind toward its tragic fate.

The tragedy of the commons is one version of a more general problem of *externalities*. An externality, also called an "external" or "spillover" cost, is that portion of the cost of a decision borne by someone other than the decision-maker. We say cost is "internalized" when the arrangement is changed so that decision-makers now bear the entire cost of their decisions. One general purpose of property institutions is to internalize externalities, preventing people from shifting the cost of their activities onto others. Ideally, property regimes should evolve, internalizing externalities as they become significant—both "positive" externalities associated with productive effort and "negative" externalities associated with misuse and overuse of commonly held resources. A system is more likely to be economically and ecologically sustainable when overuse is costly not just to the community as a whole but also specifically to the individuals who decide to overuse.

PRIVATE PROPERTY AS A SOLUTION TO COMMONS PROBLEMS

In an unmanaged commons, individual shepherds are left to decide for themselves whether to step up the intensity of their resource use. They do not take full responsibility for the cost of their overuse, though, because the cost falls mainly on other members of the group of communal users. The payoff of overuse is negative for the group but positive for the individual who elects to overuse.

Is there nothing those shepherds can do? One option would be to cut their jointly owned territory into ten smaller parcels, so each shepherd owns a small parcel with its own individual carrying capacity. Under this new arrangement, instead of dispersing the environmental degradation over the entire commons, the damage is concentrated on the offender's own private land. Thus, in the hypothetical example, instead of dispersing damages worth $4.95 over a hundred animals and ten owners, the damage is concentrated within the individual shepherd's own parcel. To keep the example simple, suppose the parcel covers an area one-tenth the size of the original communal plot. Suppose also that when the damage is concentrated in one-tenth of the area, the resulting damage is ten times as great per square foot. In that case, the flock of ten, which had been worth $10, is now a starving flock of eleven, worth about $5.05. The value of each animal has been cut roughly in half, a painfully obvious mistake. Consequently, under a system of individual parcels, everyone learns in a hurry not to add an eleventh animal.

Private ownership gives an owner a right to exclude. By conferring a right to exclude, the system gives an owner the opportunity to conserve a resource. In giving such an opportunity, the system also provides an incentive, because whatever owners save, they save for themselves.

AN ALTERNATIVE SOLUTION: COMMUNAL MANAGEMENT

In a large range of cases, parcelization is a viable alternative to managing land as an unregulated commons. It is not always a viable alternative (before the invention of barbed wire, parcelizing western rangeland was not feasible), and it is not always the only viable alternative. Another option is for the shepherds to leave the territory in a common pool, and instead of each tending a small flock of ten sheep, ignoring the costs they impose on each other as they add more sheep, they could pool their flocks and become joint owners of a single large flock of one hundred.

Each shepherd now has an interest in all of the sheep. Under a communal arrangement, a shepherd considers not whether to add the eleventh sheep, but whether to add the 101st. Adding an extra sheep means that, for each shepherd, the result is not that the value of his flock goes from $10 to $10.45. Instead, the value goes from a 10 percent ownership stake in $100 to a 10 percent stake in $95.95. Therefore, under the communal arrangement, no one wants to add the extra sheep. Here, too, as in the case of switching to private parcels, an external cost has been internalized and each of the ten shepherds now has a self-interested reason to respect the land's carrying capacity.

Regardless of whether they cut their land into parcels, or pool their flocks of sheep, the fact remains that in the real world, a community of ten people has a good chance of being able to come together to devise and enforce rules governing the land's use that will enable them to avoid collective suicide. However, there is an additional issue to consider.

THE OPEN ACCESS COMMONS: A DIFFERENT SORT OF PROBLEM

Suppose the group has gone communal, pooling both their land and their livestock. So far, so good. Now, however, suppose that whatever rules the ten shepherds might devise to regulate the addition of extra animals, they are not able to stop an eleventh shepherd from entering the picture with yet another flock. With a fixed and known set of players, viable conventions tend to emerge, but if a community is not able to restrict the inflow of new users, then we have an *open access* commons, which makes the tragedy far more likely.

Sometimes, everything depends on whether it is possible to add the extra player rather than for existing players to add the extra animal. For a community to manage itself successfully, it must be able to control negative externalities within the community, but it is likewise critical that the community be able to restrict access, controlling the size and membership of the community of users. Robert Ellickson contrasts the unregulated or open-access commons with communes. A *commune* is a restricted-access commons. In a commune, property is owned by the group rather than by individual members. People as a group claim and exercise a right to exclude. Typically, communes draw a sharp distinction between members and non-members, and regulate access accordingly. Access to public property tends to be restricted by time of day or year. Some activities are permitted; others are not.

Some medieval commons lasted, non-tragically, for hundreds of years. Elinor Ostrom describes a Swiss commons whose written records date back to the thirteenth century. Cattle were privately owned but grazed in communal highlands in the summer. People grew private crops on individual plots in the valleys, intending to use part of their crops to sustain their cattle over the winter. The basic limitation on communal summer grazing was that owners could send only as many cattle to the highland meadows as their private land parcel could sustain over the winter, with fodder grown during summer.

Allowing individual owners freely to decide whether to add to their individual stock is above all what governors of a commons cannot do. To avoid tragedy, governors of a common pasture must manage the overall livestock population, based on their estimate of the pasture's overall carrying capacity. There are several ways to do this. Managers can allow a given owner to graze cattle on common land only in proportion to: (a) how much hay he produces, (b) what proportion of the land belongs to him, or (c) the number of shares he owns in the cooperative.

Ostrom says, "All of the Swiss institutions that used to govern commonly owned alpine meadows have one obvious similarity—the appropriators themselves make all major decisions about the use of the common property resource. . . . Thus, residents of Törbel and other Swiss villages who own communal land spend time governing themselves. Many of the rules they use, however, keep their monitoring and other transactions costs relatively low and reduce the potential for conflict" (p. 65). The lesson is that successful commons are flexible and under local control. Rules sometimes need to change in response to circumstances and local people know what needs changing locally.

In the Swiss commons, Ostrom says, no citizen could send more cows to the Alp than he could feed during the winter. As David Schmidtz and Elizabeth Willott (2003a) summarize, partners recognize an imperative to avoid the tragedy of the commons and in each case do so by taking the option of overgrazing out of the hands of individual partners. History indicates, though, that members of successful communes internalize the rewards that come with that collective responsibility. In particular, they reserve the right to exclude nonmembers. A successful commune does not run itself as an open-access commons.

Hardin himself viewed commons tragedies as problems for which there is no technical solution. To Hardin, the only solution, when there is a solution, is "mutual coercion, mutually agreed upon." Ensuing decades seem to have shown that Hardin was overly pessimistic. In particular, what Hardin deemed the ultimate commons tragedy, namely global human overpopulation, has not followed Hardin's predictions. The population is still increasing, but population growth rates have fallen everywhere, and in many countries, have fallen below zero. Part of the solution

EXCERPT FROM GARRETT HARDIN'S "THE TRAGEDY OF THE COMMONS"

Perhaps the simplest summary of this analysis of man's population problems is this: the commons, if justifiable at all, is justifiable only under conditions of low-population density. As the human population has increased, the commons has had to be abandoned in one aspect after another.

First we abandoned the commons in food gathering, enclosing farm land and restricting pastures and hunting and fishing areas. These restrictions are still not complete throughout the world.

Somewhat later we saw that the commons as a place for waste disposal would also have to be abandoned. Restrictions on the disposal of domestic sewage are widely accepted in the Western world; we are still struggling to close the commons to pollution by automobiles, factories, insecticide sprayers, fertilizing operations, and atomic energy installations. ...

Every new enclosure of the commons involves the infringement of somebody's personal liberty. ... When men mutually agreed to pass laws against robbing, mankind became more free, not less so. Individuals locked into the logic of the commons are free only to bring on universal ruin; once they see the necessity of mutual coercion, they become free to pursue other goals. I believe it was Hegel who said, "Freedom is the recognition of necessity.". ...

SOURCE: from Hardin, Garrett. 1968. "The Tragedy of the Commons." *Science* 162:1243-1248.

was birth control technology, which on a priori grounds Hardin believed women would never use owing to the biological "imperative to breed" (Willott 2002).

While Hardin was overly pessimistic, it would likewise be a mistake to be overly optimistic. Society's institutions have a history of evolving in response to commons problems. Property law evolves, as do easements, nuisance law, tort law, zoning law, and regulatory agencies, along with a matrix of conventions of neighborliness. However, such remedies have a history of being imperfect. (Some small solutions turn out to be small mistakes. Some big solutions turn out to be big mistakes. And even genuine solutions can do only so much.) Although some problems get solved, others are merely mitigated. As old problems are solved or mitigated, new ones emerge. Sometimes, new problems are caused by the very regulations people devise to solve old problems. In a world filled with producers, consumers, emerging technologies, people wanting to live in neighborhoods with public spaces, and recalcitrant difficulties with enclosing such public spaces as the atmosphere itself, there will always be commons problems.

SEE ALSO *Consumption; Europe: II. Western Europe; Private Property; Takings.*

BIBLIOGRAPHY

Ellickson. Robert C. 1993. "Property in Land." *Yale Law Journal* 102: 1315–1400.

Gordon, H. Scott. 1954. The Economic Theory of a Common-Property Resource: The Fishery." *Journal of Political Economy* 62: 124–42.

Hardin, Garrett. 1968. "The Tragedy of the Commons." *Science* 162: 1243–1248.

LLoyd. W. F. 1833. *Two Lectures on the Checks to Population.* Oxford: Oxford University Press.

Ostrom, Elinor. 1990. *Governing the Commons: the Evolution of Institutions for Collective Action.* Cambridge, UK: Cambridge University Press.

Rose, Carol. 1986. "The Comedy of the Commons: Custom, Commerce, and Inherently Public Property." *University of Chicago Law Review* 53: 711–87.

Schmidtz, David. 1994. "The Institution of Property." *Social Philosophy & Policy* 11: 42–62.

Schmidtz, David, and Elizabeth Willott. 2002. *Environmental Ethics: What Really Matters, What Really Works.* New York: Oxford University Press.

Schmidtz, David, and Elizabeth Willott. 2003a. "Reinventing the Commons: An African Case Study." *University of California at Davis Law Review* 36: 203–32.

Schmidtz, David, and Elizabeth Willott. 2003b. "The Tragedy of the Commons." In *Blackwell Companion to Applied Ethics*, ed. Raymond Frey and Christopher Wellman. Oxford: Blackwell, 662–73.

Willott, Elizabeth. 2002. "Recent Populations Trends." In *Environmental Ethics: What Really Matters, What Really Works*, ed. David Schmidtz and Elizabeth Willott. New York: Oxford University Press.

David Schmidtz
Elizabeth Willott

Early sections of this entry borrow from Schmidtz, and from Schmidtz and Willott (2002, 2003b). Later sections borrow from Schmidtz and Willott (2003a).

TRANSGENIC ANIMALS

Transgenic animals are animals whose genome has been changed by means of advanced biotechnology. There are several kinds of transgenic animals. An important distinction can be made between animals that have had their genome modified by having genes knocked out, or copied, and animals that have had genes not normally found in that species inserted into their genome. (The inserted genes can come from another species or can be artificial constructs.) Technically, only the second group of animals are transgenic. The term *transgenic*, however, is used widely to refer to all animals that have had their genome modified in some way.

THE SCIENCE

Most work on transgenic animals has been carried out on laboratory mice; rats have been the second most frequently modified animal. Other species that have been genetically modified include pigs, sheep, goats, cattle, fish, rabbits, and cats. The first and still widely used method of genetic modification is so-called pronuclear microinjection, in which DNA is injected into the pronucleus of an early embryo. However, this method is not very efficient, nor is it precise, and a number of other methods of gene transfer or gene knockout have been developed. One of those methods makes use of cloning technology. In this case the genetic modifications are made on individual cells from a cell line. Afterward a genetically modified cell is inserted into an enucleated egg and turned into an embryo through the use of a cloning technique. New viral vectors and sperm-mediated DNA transfer that bring the desired genetic material into predesignated areas of the genome are other methodologies that are being developed. These technologies are likely to make the production of transgenic animals technically more efficient (Robl et al. 2007).

THE APPLICATIONS

The vast majority of transgenic animals are used in basic research or biomedical research to study biological development and function or as disease models that mimic human diseases and are therefore useful in the study of medical conditions such as Parkinson's disease, cancer, and cystic fibrosis and the testing of new drugs. A well-known example is the OncoMouse. This mouse, which was produced by Harvard University and Dupont in the late 1980s, was genetically modified to carry an activated cancer gene that significantly increases its susceptibility to cancer. Tens of thousands of animal models, mainly mice, had been developed by the first decade of the twenty-first century.

Since the early 1990s researchers have attempted to develop a new biomedical application of transgenic animals: so-called bioreactors. These are animals, typically of farm animal species, with special traits that make them useful in pharmaceutical production. An example is the goat developed by GTC Biotherapeutics that produces recombinant human antithrombin, an anticlotting protein, in its milk that can be used as medicine for humans. This product was the first of its kind to reach the market (Choi 2006).

"Britney," a Transgenic Chicken. *Dr. Helen Sang, head of the Britney research team, is shown with a third generation chicken at the Roslin Institute in Edinburgh, 2000. The scientists began the project with plans to use the protein in the eggs to treat cancer. The Roslin Institute was also the creator of probably the most famous transgenic animal to date, Dolly the sheep.* © MC PHERSON, COLIN/CORBIS SYGMA.

There is perhaps a possibility of developing transgenic animals, typically pigs, for xenotransplantation. Thus, it is envisaged that complete organs from animals will be transplantable into humans. Here the aim of modification is to make the tissue of the animal immunologically compatible with the human body to prevent rejection. Progress has been made in this area (Yamada et al. 2005), but many problems remain.

In the agricultural area attempts are being made to produce animals with traits that will allow improved production, better animal health, and/or reduced environmental impact. The Enviropig developed by researchers at the University of Guelph in Canada is an example. It has been genetically modified to be able to digest the phosphorus in a cereal grain diet. This removes the need to add phytase enzymes to the pigs' diet and reduces the amount of phosphorus released into the environment. The Enviropig thus ideally reduces production costs and environmental impact at the same time (Forsberg et al. 2003). The number of agricultural applications of transgenic animals in the research and development phase has been limited by high production costs, technical barriers, and fears of rejection by the public.

Because it is in principle possible to transfer a gene from one living organism to another regardless of species barriers and it is becoming possible to produce artificial genes on demand, the theoretical applications of the technology are limitless. Several aspects of the technology, however, remain inefficient and imprecise because many basic aspects of the function and interconnectedness of genes are poorly understood. There is little doubt, though, that the technology will play a greater role in the future, especially in cases in which individual modified animals exhibit a trait that makes them valuable enough to cover the cost of their production, cases in which their contribution to scientific research cannot be obtained in other ways, and cases in which the use of the technology is socially acceptable. Thus, it is likely that transgenic animals will be utilized primarily for biomedical applications.

ETHICAL CONCERNS

The most common ethical concerns about transgenic animals can be divided into three main areas. Those areas and the most important issues within each of them are discussed below.

Humans and the Environment Concerns about potential risks to humans and the environment figure prominently in many discussions. Risks to humans most frequently are equated with risks to human health presented by food or medicinal products derived from transgenic animals. A very limited amount of research has been done in this area because few products have been developed. There is, however, a substantial literature on what risks should be taken into consideration when those products are evaluated. In the medical area it usually is suggested that risk assessments should follow the approach by which newly developed drugs are tested. In the food area risks arising from changes in amino acids leading to allergenicity, toxic effects, or changes in nutritional value are important parameters (National Research Council and Institute of Medicine of the National Academies 2004). At the beginning of the twenty-first century no food product from a transgenic animal had been developed far enough to necessitate serious attempts to conduct a risk assessment.

Risks relating to a specific use of transgenic animals can be found in the area of xenotransplantation. Here questions about the risk of transferring diseases from pig donors to human recipients have not been resolved. Especially important is the question whether there is a significant risk that the porcine endogenous retrovirus (PERV), which lies dormant in the pig genome, will become active after transfer to the human body. There is no doubt that this could cause serious health problems for humans, and the situation often is compared with the history of AIDS and severe acute respiratory syndrome (SARS); however, there is no agreement about how this risk should be evaluated (Moalic et al 2006; Martin et al. 2006; Levy et al. 2007).

Another risk to humans that is mentioned frequently relates to the possible socioeconomic effects of the technology, especially in agriculture, where it could accelerate the development of large-scale industrialized factory farming and deepen the divide between the developing world and richer countries. This risk is not specific to transgenic animals but relates to developments in technology in general.

There are also potential risks to the environment. The concern here is that transgenic animals might escape and breed with wild populations, thus spreading their genes in an uncontrollable environment. The most frequently cited example is transgenic fish, for example, salmon with genetic alterations that allow for faster growth. The concerns in this area can be about the indirect consequences this might have for humans (in this case economic losses for the fishing industry) or about direct concerns involving the animals and the wider ecosystem. Whether one is concerned about a particular application of the technology because it constitutes a risk to human interests or because it constitutes a risk to other species or the integrity of the ecosystem, the risk that transgenic animals will escape and evade human control and confinement is a socially important issue (PEW Initiative on Food and Biotechnology 2003).

There are also concerns about the use of transgenic animals constituting a step onto a slippery slope that will lead to unacceptable uses of biotechnology on humans. Although current uses of transgenic animals are intended primarily to gain basic scientific understanding of molecular biology and study human diseases, it is clear that the more skilled scientists become at applying biotechnology to animals, the easier it will be to apply the same technologies to humans. The factors that will prevent technologies from moving from the animal sphere to the human sphere are not the technical limitations but ethical objections, and people concerned about the slippery slope are worried that those objections eventually will be pushed aside by the appeal of the technical possibilities (Kass 1997).

Animal Welfare Transgenic animals have been used mainly in basic biological research and as disease models. Often the goal is to produce animals that underexpress or overexpress certain genes or express a mutated, disease-causing human gene. In all these cases normal body function in the organism is disrupted in some way. Modifications can involve any part of the animal genome, and the effects on the animal's phenotype range from those which are lethal to those which have no detectable effect on the health of the animal. It is therefore impossible to generalize about the effects of genetic modification on the welfare of animals.

The effects that occur can be divided into two main categories: intended and unintended. Welfare problems stemming from intended genetic change are hard to avoid because the point of inducing the change is to affect the animal. Thus, a mouse carrying the human Huntington's disease gene will inevitably suffer welfare problems as it develops the disease, including rapid progressive loss of neural control that leads to premature death. Unintended effects are connected with the inaccuracy of the technology and insufficient understanding of the function of different genes in different organisms. Both types of effects contribute to the unpredictable nature of genetic modification at the phenotypic level.

To deal properly with both intended and unintended effects on animal welfare, it is important to monitor the animals and, when severe effects occur, take action to alleviate or end their suffering. In laboratory animal science it should be considered part of good practice to find ways to conduct experiments that minimize the discomfort and suffering imposed on animals and to define so-called humane endpoints, that is, points at which animals have to be euthanized (Olsson and Sandøe 2004).

There is wide agreement about the need to limit the discomfort and suffering imposed on animals. However, from a philosophical perspective it may be questioned whether the focus should be only on preventing pain and other kinds of suffering in animals (and perhaps promoting positive experiences). It may be argued that animal welfare is also about the extent to which an animal is allowed to fulfill its species-specific potential regardless of its subjective experience.

Very often this broader perspective on animal welfare will point to an additional group of considerations that have to be taken into account in deliberations about animal welfare. Concern about an animal's opportunity to engage in certain kinds of behavior does not prevent one from caring about its subjective experiences. Nevertheless, the two kinds of considerations sometimes are difficult to reconcile in practice. Considerations within a narrow perspective in which the subjective experiences of the animal alone matter may be outweighed by considerations in the broader perspective (Gjerris, Olsson, and Sandøe 2005).

Some of those engaged in the ethical debate on transgenic animals argue that welfare is not all that matters in dealing with animals. They may defend the view that animal integrity also must be considered.

Integrity The term *integrity* means wholeness or fullness. In the literature two notions of animal integrity are prominent. The first is based on a biological understanding, and the second on a phenomenological understanding. The first stresses the genetic integrity of the animal and therefore focuses on the importance of not changing animal genomes to suit human purposes. The obvious objection to concerns about the violation of genetic integrity through gene technology is that the genome of an animal species is in constant flux because of naturally occurring evolutionary forces and through well-established breeding practices such as conventional selective breeding. A difference between genetic changes induced by natural selective forces and human-induced changes can be stated, but it is difficult to argue for a relevant difference between introducing genetic changes with modern biotechnology and introducing changes with older, conventional methods (Rollin 1996). This has led some to conclude that transgenic animals raise no new or additional ethical concerns. Others claim that this alone constitutes a reason to reexamine conventional breeding methods with a more critical eye (Gjerris and Sandøe 2006).

The second notion of integrity is based on the experience of the animal as an inviolable whole. Animal integrity can be understood as an inherent limit in the relationship between humans and nature, a "red line" that governs what is ethically acceptable for humans to do to animals. Integrity in this case derives from an experience and understanding of animals as beings that in and of themselves set up an ethical requirement of noninterference. This requirement may be violated only if the reasons are adequate from an ethical perspective. Integrity signifies a difference between the knowledge of the animal people have through their understanding of its usefulness to humans and the knowledge people have when they conceive of the animal independently of human needs. A cow is a producer of hide, milk, and meat; it holds no surprises when it is experienced from the perspective of human need. However, when it is experienced in a nonreductionist perspective, the cow amounts to more than that. Respecting the integrity of animals is thus the polar opposite of wholesale reification of an animal as a natural resource (Gjerris and Sandøe 2006).

SEE ALSO *Animal Cloning; Animal Ethics; Factory Farms; Farms; Genetically Modified Organisms and Biotechnology; Patenting Life.*

BIBLIOGRAPHY

Buehr, Mia; J. Peter Hjorth; A. K. Hansen; and P. Sandøe. 2004. "Genetically Modified Laboratory Animals—What Welfare Problems Do They Face?" *Journal of Applied Animal Welfare Science* 6: 319–338.

Choi, C. 2006. "Transgenic Market Heats Up." *The Scientist*, August 2006. Available from http://www.the-scientist.com

Cooper, D. E. 1997. "Intervention, Humility and Animal Integrity." In *Animal Biotechnology and Ethics*, ed. Alan J. Holland and Andrew Johnson. New York: Chapman and Hall.

Dahl, K.; P. Sandøe; P. F. Johnsen, et al. 2003. "Outline of a Risk Assessment: The Welfare of Future Xeno-Donor Pigs." *Animal Welfare* 12: 219–237.

Duncan, I. J. H., and D. Fraser. 1997. "Understanding Animal Welfare." In *Animal Welfare*, ed. Michael C. Appleby and Barry O. Hughes. Wallingford, UK: CAB International.

Faber, D. C.; J. A. Molina; C. L. Ohlrichs, et al. 2003. "Commercialization of Animal Biotechnology." *Theriogenology* 59: 125–138.

Forsberg, C. W.; J. P. Philips; S. P. Golovan, et al. 2003. "The Enviropig Physiology, Performance, and Contribution to Nutrient Management Advances in a Regulated Environment: The Leading Edge of Change in the Pork Industry." *Journal of Animal Science* 81: E68–E77.

Gjerris, Mickey; Anna Olsson; and Peter Sandøe. 2005. "Animal Biotechnology and Animal Welfare." In *Ethical Eye—Animal Welfare.* Strasbourg, France: Council of Europe.

Gjerris, M., and P. Sandøe. 2006. "Farm Animal Cloning: The Role of the Concept of Animal Integrity in Debating and Regulating the Technology." In *Ethics and the Politics of Food: Preprints of the 6th Congress of the European Society for Agricultural and Food Ethics,* ed. Mathias Kaiser and Marianne E. Lien. Wageningen, Netherlands: Wageningen Academic Publishers.

Kass, Leon R . 1997. "The Wisdom of Repugnance." *New Republic,* June 2, pp. 17–26.

Lassen, Jesper; Mickey Gjerris; and Peter Sandøe. 2006. "After Dolly—Ethical Limits to the Use of Biotechnology on Farm Animals." *Theriogenology* 65(5): 992–1004.

Lassen, J.; P. Sandøe; and B. Forkman. 2006. "Happy Pigs Are Dirty!—Conflicting Perspectives on Animal Welfare." *Livestock Science* 103(3): 221–230.

Levy, Marlon F.; Takele Argaw; Carolyn A. Wilson, et al. 2007. "No Evidence of PERV Infection in Healthcare Workers Exposed to Transgenic Porcine Liver Extracorporeal Support." *Xenotransplantation* 14(4): 309–315.

Martin, Stanley I.; Robert Wilkinson; and Jay A. Fishman. 2006. "Genomic Presence of Recombinant Porcine Endogenous Retrovirus in Transmitting Miniature Swine." *Virology Journal* 3: 91.

Moalic, Yann.; Yannick Blanchard; Hélène Félix; and André Jestin. 2006. "Porcine Endogenous Retrovirus Integration Sites in the Human Genome: Features in Common with Those of Murine Leukemia Virus." *Journal of Virology* 80(22): 10980–10988.

National Research Council of the National Academies, Committee on Defining Science-Based Concerns Associated with Products of Animal Biotechnology. 2002. *Animal Biotechnology: Science-Based Concerns.* Washington, DC: National Academies Press.

National Research Council and Institute of Medicine of the National Academies, Committee on Identifying and Assessing Unintended Effects of Genetically Engineered Foods on Human Health. 2004. *Safety of Genetically Engineered Foods: Approaches to Assessing Unintended Health Effects.* Washington, DC: National Academies Press.

Olsson, I. A. S., and P. Sandøe. 2004. "Ethical Decisions Concerning Animal Biotechnology: What Is the Role of Animal Welfare Science?" *Animal Welfare* 13: S139–S144.

Paterson, L.; P. DeSousa; W. Ritchie, et al. 2003. "Application of Reproductive Biotechnology in Animals: Implications and Potentials: Applications of Reproductive Cloning. *Animal Reproduction Science* 79(3–4): 137–143.

PEW Initiative on Food and Biotechnology. 2003. *Future Fish: Issues in Science and Regulation of Transgenic Fish.* Washington, DC: PEW Initiative on Food and Biotechnology. Available from pewagbiotech.org.

Robl J. M.; Z. Wang; P. Kasinathan; and Y. Kuroiwa. 2007. "Transgenic Animal Production and Animal Biotechnology." *Theriogenology* 67(1): 127–133.

Rollin, Bernard E. 1995. *Frankenstein Syndrome: Ethical and Social Issues in the Genetic Engineering of Animals.* Cambridge, UK, and New York: Cambridge University Press.

Rollin, B. E. 1996. "Bad Ethics, Good Ethics and the Genetic Engineering of Animals in Agriculture." *Journal of Animal Science* 74(3): 535–541.

U.S. Food and Drug Administration. 2006.: *Animal Cloning: A Draft Risk Assessment.* Rockville, MD: U.S. Food and Drug Administration.

Yamada Kazuhiko; Koji Yazawa; Akira Shimizu, et al. 2005. "Marked Prolongation of Porcine Renal Xenograft Survival in Baboons through the Use of Alpha1,3-Galactosyltransferase Gene-Knockout Donors and the Cotransplantation of Vascularized Thymic Tissue." *Nature Medicine* 11: 32–34.

Mickey Gjerris
Peter Sandøe

TRANSPORTATION

From the canal systems and railroad lines of nineteenth-century America to the current rapid growth of motorized transport in developing countries, the development and expansion of transportation systems has been a central feature of modern societies. In all its forms, transportation is the means by which human beings traverse the natural world, and is of great concern to environmental philosophy. The modes of transportation available to a population shape both the distribution of goods and services within that population and its manner of growth. Historically, advances in transportation have greatly increased the mobility of populations. For instance, in Roman times, individuals traveling on horses on Roman roads averaged around 10 miles per hour, and oxen, used for transporting many goods, averaged 2 miles per hour. Premodern sailing ships too rarely achieved average speeds of more than 9 knots (10 miles per hour). By the mid-nineteenth century, in contrast, it was not uncommon for steam-engine locomotives to move goods and passengers at speeds of over 50 miles per hour, and modern steam ships too could travel at speeds significantly faster than premodern sailing ships. By 2008, average travel times had decreased even further, with contemporary motor-vehicle travel routinely averaging 65 miles per hour and the airplanes in current use often traveling at speeds of over 500 miles per hour. Traveling by air, a contemporary individual can cross in a matter of hours distances that would have taken premodern predecessors months to traverse.

Developments in transportation have done more than just afford individuals greater mobility, though.

They have also determined the forms of human settlement. For instance, prior to modern refrigeration technologies, the transportation of perishable goods was inherently limited, and thus the areas from which food sources for urban populations could be derived was directly related to the distance that food goods could be safely transported. From early times, major population centers tended to be situated along waterways, which provided the most efficient means for transporting food and other goods. As various technological innovations in transportation occurred, developmental patterns followed suit. In the United States, many cities were founded as service hubs along the paths of the larger canals and railroad systems of the nineteenth century. As these cities grew, they in turn provided new destinations for workers and travelers in a synergistic process of expansion. In Europe and North America, the development of railroad systems in the nineteenth century had a particularly significant impact upon the movement of persons and goods. Creating an even greater change in the landscape, however, was the invention of the automobile and its rapid diffusion within Western societies in the twentieth century. With the advent of the automobile, virtually all locales became readily accessible to settlement and development. The ever increasing expansion of the roadway system and the relative affordability of automobile travel have allowed individual workers easily to commute long distances from home to work, and this ability eventually led to the suburbanization so prominent in the United States in the later half of the twentieth century. In the United States alone, there are currently over 4 million miles of public roads, and the total number of miles traveled by motor vehicles reached 2.7 trillion miles in 2000. The growth of airplane transportation has had a similar impact upon human movement, particularly at the global level, since the development of inexpensive and reliable airplane transport has allowed persons and goods to travel immense distances with little effort.

Within large urban centers, other modes of transportation have significantly affected passenger travel as well. Many modern cities developed subway or light-rail systems to move large numbers of persons, particularly commuters, more efficiently. Bicycles also provide a relatively inexpensive form of transportation for many urban dwellers, and in some countries, particularly China, bicycle traffic has constituted a major portion of intraurban transportation. While most modern technological developments have tended to encourage greater individual travel, in recent years some technological advances have had the opposite effect. Most important, the development of modern telecommunication networks and the Internet have allowed many workers to telecommute to work from their homes. Similarly, face-to-face business meetings, document transfers, and even medical diagnostics can all be done electronically in 2008, eliminating some of the need for real travel associated with such activities in the past.

The brief outline above illustrates how the history of social and economic development is tied to the history of transportation. The means of transportation available to people within a society crucially affects their access to various goods and their options concerning work and place of residence. In many ways the various technological advances in transportation that have taken place since the beginning of the modern era have undeniably provided many persons with a greater range of options concerning their choices for employment, residency, and leisure activities. This is no doubt why there exists in Western countries such a strong association between the idea of the automobile and notions of individual freedom and independence. Yet the benefits that increased mobility and greater access to goods provides for some members of society must be seen against the larger impact that various means of transportation have upon the greater good. In this sense, the ethical concerns about issues of transportation raised by environmental ethicists are best seen in terms of a discussion about the environmental costs of various modes of transportation. Unfortunately, these costs are often far less obvious to individual users of transportation than the direct benefits that such transportation provides them. Likewise, the historical and ongoing public-policy debates that deal with the environmental impact of various forms of transportation can most profitably be seen in terms of differences in how participants in those debates weight the various potential benefits and costs associated with transportation issues. Thus, to understand and evaluate such discussions and debates properly, one must first know the nature of the major environmental costs of transportation involved.

DIRECT ENVIRONMENTAL COSTS

The primary forms of direct environmental costs associated with modern forms of transportation stem from pollution. Pollution of one form or another has been associated with virtually all major forms of transportation utilized by human beings. Indeed, often the development of one form of transportation responds to problems associated with a form of pollution stemming from a previous mode of transportation, only to give rise to newer kinds of pollution. For instance, the use of horses as the primary mode of transportation in major urban centers in the nineteenth century produced such large quantities of solid waste, mainly manure, in those cities that it posed serious health risks, as well as raising quality-of-living issues, for residents. As such, the transition to the use of automobiles and rail systems in major cities represented the alleviation of one type of pollution and the problems associated with it. It was soon discovered, however, that the new forms of

Traffic Congestion in Beijing, China, 2007. *Along with the benefits provided by transportation, including increased mobilization and greater access to goods, there are a number of costs that must be considered as well, particularly negative effects to the environment. Automobiles are powered by internal-combustion engines, a major source of air pollution. Many nations have used legal measures, such as clean air acts and greater fuel-efficiency standards, in order to reduce the amount of harmful pollutants emitted by automobiles.* AP IMAGES.

transportation gave rise to new kinds of pollution with a fresh set of problems. This picture presents a cautionary reminder that we should carefully account for all forms of pollution associated with the development of a new transportation technology, lest we merely replace one set of environmental problems with another. Such considerations are particularly relevant to current discussions about the relative merits of various alternatives to the use of gasoline engines in automobiles, such as hydrogen-fuel-cell technologies, electric cars, and ethanol conversion.

Currently, the major direct environmental costs of transportation discussed by environmentalists center primarily around pollution generated by the use of vehicles powered by internal combustion engines, particularly automobiles and airplanes. Since these two forms of transportation account for most personal travel in the United States and for a large portion of commercial transportation as well, this focus is certainly appropriate. At the dawn of the twenty-first century, there were around 200 million automobiles in use in the United States alone, and well over 80 percent of Americans currently rely on personal automobiles or other motor vehicles, such as trucks and motorcycles, for their daily transportation needs. There has also been a significant increase in the number of airline miles traveled per capita. Indeed, in the United States, airline travel is the fastest growing segment of personal transportation. According to statistics of the U.S. Bureau of Transportation, air travel increased from 118 billion passenger-miles in 1975 to over 600 billion passenger-miles by 2000, and the number of aircraft in use by commercial carriers increased by more than 30 percent from 1990 to 2000. In 2008 there are nearly 30,000 commercially scheduled flights per day in the United States. This has resulted in a concurrent increase in the amount of pollutants produced by air transportation, which now accounts for the fastest growing source of carbon dioxide emissions, for instance. In 2000 airline traffic accounted for around 3.5 percent of all carbon dioxide emissions, and if current trends in air travel continue, that percentage is expected to grow to 15 percent by 2050. Also, transit by motor vehicles and airplanes involves significantly more pollution problems than major alternatives, such as rail travel. Air travel produces over five times, and automobile travel over three times, the total amount of pollutants per passenger-mile as rail travel. Finally, these forms of transport are also generally less fuel-efficient, domestic airline travel averaging around 3,800 British thermal units per passenger-mile, automobile travel averaging around 3,500 British thermal units per passenger-mile, and domestic rail travel averaging around 2,100 British thermal units per passenger-mile. This is not to say that other forms of transportation do not involve any significant pollution problems, but only to emphasize that the greatest environmental problems involving transportation stem from pollution related to our reliance on motor-vehicle and air transport.

The pollution problems stemming from heavy dependence on automobiles for transit in urban areas was brought strongly to the forefront of public attention during the early environmental movement and eventually contributed greatly to the passage of clean-air acts at both the federal and state levels. Automobiles produce pollution containing lead (many countries still allow the use of leaded gasoline), benzene, carbon monoxide, nitrogen oxides, ozone, volatile organic compounds, and other environmentally harmful chemicals. These pollutants have been linked to a host of environmental problems, including smog, acid rain, low-lying ozone, and toxic air pollution. Such problems have resulted in significant damage to the environment and human health. As a result, since the 1960s there have been a number of efforts in countries such as the United States and Great Britain to reduce the level of harmful pollutants emitted by automobiles. Some of these efforts have taken the form of legislation, such as clean-air acts, that restrict levels of harmful pollutants like carbon monoxide and nitrogen oxides or mandate fuel efficiency for vehicles, while others have stemmed from industry and consumer initiatives. As a result, the motorized vehicles of today emit significantly less pollutants than their counterparts of the 1960s and 1970s. However, automobile emissions still produce significant amounts of pollutants. The Environmental Protection Agency estimates that automobiles still account for about 75 percent of all carbon monoxide emissions in the United States. The greater number of vehicles on the roads in 2008 also mitigates to some extent the emission reductions that have been gained. Furthermore, while the fuel efficiency of motor vehicles is somewhat greater in 2008 than in past decades, the fuel efficiency of many vehicles, particularly popular sport utility vehicles, remains less than optimal. In the United States, the Corporate Average Fuel Economy standard, which represents the weighted average of the fuel economy of a manufacturer's fleet of passenger vehicles, was first introduced in Congressional legislation in 1975, but since 1985 remained unchanged at 27.5 miles per gallon for passenger vehicles until 2007. Furthermore, the category of light trucks, which sport utility vehicles were categorized under, was set at the much lower standard of 20.7 miles per gallon. In view of the increase in sport utility vehicles and light trucks in the 1990s, it is perhaps not surprising then that the total fuel economy for all passenger vehicles in the United States actually peaked in 1986 at 26.2 miles per gallon. In 2005 the standard for the light-truck category was changed in part to reflect environmental concerns, and for the year 2007 the standard for light trucks has been set at 22.2 miles per gallon. A number of other countries have set fuel-efficiency standards for manufacturers that are higher than the standards in the United States, including China, which introduced a new set of standards for fuel efficiency in 2004.

Since the turn of the twenty-first century, an even greater concern about the pollution associated with petroleum-based transportation is related to global climate change. Motor vehicles are major producers of the main greenhouse gases that have been linked to global warming. In particular, such vehicles are among the primary producers of carbon dioxide, the principle greenhouse gas that contributes to global warming. Global warming occurs when greenhouse gases such as carbon dioxide build up in the atmosphere and prevent heat from leaving the earth's atmosphere. It has been estimated that in Western nations, motor-vehicle travel accounts for around 20 to 30 percent of all carbon dioxide emissions. Though there remains some debate about the extent of global warming, the potential effects upon the global environment are significant, since human, animal, and plant habitats across the globe will be altered as a result. Rising sea levels, changes in weather patterns, and alterations in the distributions of animals and plants brought about by global warming could effect agricultural practices, displace populations, and contribute to the spread of disease. Thus, even though the extent and causes of global warming are not fully known, it seems clear that the amount of carbon dioxide emissions produced by the current transportation system in countries that rely significantly on personal-motor-vehicle transit poses at least a significant risk to the environment and future generations.

As with all forms of pollution, it is the cumulative effect of the pollution associated with transportation that is particularly problematic. While the pollution produced by any particular automobile ride is unlikely to have a seriously deleterious effect upon the environment, there are significant environmental risks and potential harms to humans, animals, and ecosystems associated with the large-scale utilization of automobiles. Thus, the costs of transportation to the environment outlined above must be evaluated in terms of the total system of transportation utilized within any given society. In the United States, that system clearly encourages the private use of automobiles as the primary mode of transportation. Since the mid-twentieth century, most governmental support and funding for transportation in the United States has gone into the construction of highways and roads, and only a small proportion has been provided for the creation and support of public transit systems, such as commuter trains or light-rail systems. Since the 1990s, some small initiatives have taken place to alleviate the impact of automobile traffic, such as the creation of high-occupancy-vehicle traffic lanes in some metropolitan areas, the expansion of light-rail systems in some urban districts, and the construction of bicycle lanes within many cities. Nonetheless, public policy on transportation in the United States still is focused on facilitating the use of automobiles, and despite the above mentioned efforts, the average commuting time has actually increased in recent years. For example, the number of drivers commuting more than sixty minutes to work by automobile rose by over 50 percent from 1990 to 2000, and the number of solo drivers commuting to work has increased by over 13 million from the 1990s.

INDIRECT ENVIRONMENTAL COSTS

While the pollution associated with contemporary forms of transportation might represent the largest and most direct impact upon the environment, many environmental ethicists have argued that it is equally important to understand the less direct effects associated with current transportation systems. First among these is the urban sprawl associated with the development of the roadway system in the United States and some other Western nations. The seemingly endless expansion of development afforded by the extension of roadway systems has raised a host of environmental concerns. Most obviously, such development makes more natural areas open to human development and thus allows the environmental degradation of those areas. The expansion of the roadway system and the accompanying development of areas along those roadways also require the use of vast amounts of resources, such as concrete and steel, and there are serious environmental impacts associated with the extraction and processing of these resources. By spreading populations out over greater distances, such sprawl creates inherent inefficiencies in the distribution of goods and services, which also entails a greater use of natural resources. In this vein, while cybercommuting might lessen the need for some workers to commute to work on a daily basis, it has the potential negative affect upon the environment of allowing individuals even greater opportunities to move into formally wild areas further away from urban centers. Finally, the reliance on petroleum inherent in a transportation system based primarily on personal motor vehicles exacerbates the well-known environmental costs related to the exploration, extraction, and refining of crude oil.

Our particular forms of transportation no doubt have other indirect environmental costs on the environment, such as those associated with invasive exotic species of plants and animals whose spread has been facilitated by various forms of modern transportation, the environmental costs related to disposing of toxic materials used in the construction of modern vehicles and vehicle-related products, and environmentally caused health problems brought about by the immense concentration of trucks and railways around major shipping ports. The point here is not to give an exhaustive list of all such costs, but merely to illustrate some of the more prominent ones and to show that a careful accounting of the environmental impact of our transportation systems needs to take in the full range of consequences that result from activities associated with the development and utilization of transportation if public policy for transportation is to be properly geared toward the overall good.

TRANSPORTATION IN A GLOBAL CONTEXT

In light of the extensive concern about the environmental effects that current modes of transportation are having within Western developed countries, it is not surprising that there is also a great deal of growing anxiety about the severe environmental impact of expanding these modes of transportation within developing nations. Within the evolving globalized economy, an ever greater amount of industry and commerce is shifting to formally less developed countries. As such countries industrialize and expand their economies, they also are taking on means of transportation that model those of the Western world, both because doing business in international markets requires having a sophisticated transportation infrastructure in place and because the creation of new wealth within these countries gives more people within their economies the ability to purchase hitherto unaffordable technologies like automobiles. The rate at which this expansion is taking place in recent years is exponential. For instance, automobile ownership in China more than tripled from the 1990s to the 2000s and has been accompanied by a corresponding and ongoing boom in the expansion of the highway system in that country. As countries such as China increasingly utilize motorized-vehicle transportation, they will incur the environmental costs that such transportation involves and can seriously aggravate global environmental problems. A particularly prominent example of this is the decline of Chinese bicycle culture. Until recent years, bicycle use dominated local travel in Chinese cities. However, with the increased use of automobiles in China, the dominance of the bicycle is beginning to fade in large urban centers. In

Beijing, for instance, bicycle travel accounted for over 60 percent of all trips within the city in 1995, whereas in 2006 that number had fallen to less than 40 percent. Other large Chinese cities have experienced a similar transition, and many Chinese cities are now even banning bicycle use on many roadways.

CHALLENGES TO DEVELOPING RESPONSIBLE TRANSPORTATION POLICY

As noted earlier, modern systems of transportation offer benefits and incur costs. At least two difficulties can make implementing transportation policies that properly balance these benefits and costs particularly troublesome. The first difficulty concerns the evaluation process itself. Transportation systems are exceedingly complex entities, and both the goods and services they engender and the problems associated with them are numerous and multifaceted. As mentioned previously, even the kinds of environmental harms associated with transportation come in a variety of forms, some of which are not readily apparent from the outset. It is no simple matter to account for all the various factors, good and bad, long-term and short-term, associated with transportation systems. Such a process also involves comparing and weighting goods and values of diverse kinds. For example, one needs to consider how values and goods such as those of environmental integrity and species welfare are to be gauged in relation to those such as individual freedom and economic growth.

The second problem is a pragmatic one. As many environmental philosophers have noted, one of the difficulties with getting people to take environmental problems seriously is the very scope of such problems. Environmental harms are often both indirect and long-term, and thus their costs are frequently ignored by the very people who bring them about. It is difficult enough to convince people to consider the long-term welfare of others in their decisions; it is even much more difficult when those considerations run contrary to the satisfaction of their immediate interests. The potential conflict between morality and self-interest becomes particularly thorny in such situations, especially because the harmful consequences are so far removed from view. This explains in part why it is difficult to get people voluntarily to change their choices of transportation, even when we can objectively evaluate the long-term consequences of the various options. As a result, it may be that societies can effect enough change in transportation to make a significant environmental difference only through public policies that involve strong incentives or strong restrictions. Because of the individualistic framework of much of modern society, restrictions on transportation choices are likely to be strongly opposed by many. Developing incentives that are comprehensive enough to alter transportation patterns and strong enough to sway individual choices will involve a degree of foresight and planning that has so far been lacking in the development of transportation policy. The task of developing effective transportation policies thus remains a difficult challenge for those concerned about the environmental impact of our transportation choices.

SEE ALSO *Automobiles; Built Environment; Energy; Pollution; Urban Environments.*

BIBLIOGRAPHY

Bureau of Transportation Statistics. "National Transportation Statistics." Available from http://www.bts.gov/publications/national_transportation_statistics/

Dower, Roger; Daryl Ditz; Paul Faeth; et al. 1997. *Frontiers of Sustainability: Environmentally Sound Agriculture, Forestry, Transportation, and Power Production*. Washington, DC: Island Press.

Duany, Andres; Elizabeth Plater-Zyberk; and Jeff Speck. 2001. *Suburban Nation: The Rise of Sprawl and the Decline of the American Dream*. New York: North Point Press.

Flink, James J. 1988. *The Automobile Age*. Cambridge, MA: MIT Press.

Goudie, Andrew. 2006. *The Human Impact on the Natural Environment*, 6th ed. Malden, MA: Blackwell Publishing.

Greenpeace. 1991. *The Environmental Impact of the Car*. Amsterdam: Greenpeace International.

Kunstler, William. 1994. *The Geography of Nowhere*. New York: Touchstone.

Meaton, Julia, and David Morrice. 1996. "The Ethics and Politics of Private Automobile Use." *Environmental Ethics* 18: 39–54.

Riley, Robert. 2004. *Alternative Cars in the 21st Century: A New Personal Transportation Paradigm*, 2nd ed. Warrendale, PA: SAE International.

U.S. Department of Transportation. 2001. *Transportation Statistics Annual Report 2000*. BTS01–02. Washington, DC: Bureau of Transportation Statistics. Available from http://www.bts.gov/publications/transportation_statistics_annual_report/2000/

Vig, Norman J., and Michael E. Kraft, eds. 2003. *Environmental Policy: New Directions for the Twenty-first Century*. Washington, DC: CQ Press.

Wenz, Peter S. 2001. *Environmental Ethics Today*. Oxford: Oxford University Press.

Zuckerman, Wolfgang. 1991. *End of the Road: The World Car Crisis and How We Can Solve It*. White River Junction, VT: Chelsea Green.

Daniel E. Palmer

U

UNITED KINGDOM

The United Kingdom (England, Scotland, Wales, and Northern Ireland) is a densely populated country with a complex history of human habitation and land use. Its cultural richness is due to centuries of invasions and settlement of peoples and, in the nineteenth century, the wealth generated by the industrial revolution, which occurred against a background of colonization and plunder of other nations' resources and labor. The population density and the complex historical tensions among the different countries, religious groups, and social classes that make up the United Kingdom (UK) have been instrumental in shaping the British psyche and the relationship to nature that underlies many of that nation's contemporary responses to environmental issues.

THE OPEN SPACE MOVEMENT

Robert Hunter (1844–1913) supplied the vision and rhetorical power needed to drive what was to become the open space movement. Hunter worked for the Commons Protection Society and used his legal training and painstaking research of historical documents to resist further enclosure of common land by landowners. That drive was intended not just to protect woodland and open moors from being built on or aggressively farmed but also to keep them open for ordinary people's recreational enjoyment.

Although ancient royal parks and forests provided open space for city dwellers close to them, other burgeoning industrialized areas had no such resources. The working and living conditions of factory workers and the poor air quality led to concern for the health of city dwellers. Time spent in nature was seen as morally uplifting as well as providing fresh air, and this led to an increasing number of wealthy benefactors and city corporations creating urban parks with open or very cheap access to provide healthy recreation for the masses.

The drive for open access to countryside has persisted to the present time. In 1884 the first attempt to pass a Freedom to Roam bill in Parliament failed, as did many subsequent attempts. In 1932 six people were jailed for leading a mass trespass of four to five hundred people on Kinder Scout in the Derbyshire Peak District, England, and that led to greater public support for the "right to roam." The Ramblers Association, whose origins lay in the local walking clubs that had sprung up in the 1880s, rekindled this form of civil disobedience by holding annual mass trespass events beginning in 1985. Those unlawful but peaceful events were instrumental in providing the pressure that resulted in the Countryside and Rights of Way Act 2000. That act led to a mapping exercise culminating in the Right to Roam 2005, which gives walkers access to mountains, moor, heath, down, and common land in England and Wales. Similarly, the Land Reform Act 2003, which was implemented fully in 2005, grants the statutory right of responsible access to almost all land and water in Scotland.

THE NATIONAL TRUST AND OTHER NGOS

Legislation such as the Right to Roam has opened up privately owned land, but one of the most effective means of social transformation in terms of giving people access to nature and conserving natural and cultural landscapes has been to buy land and have it designated for that purpose. The National Trust, established in 1895 by Robert Hunter, Octavia Hill, and Hardwicke Rawnsley, began doing that with the purchase of four and a half acres of coastline at Dinas Oleu, Wales. The National Trust now manages 250,000 hectares (617,761 acres) of countryside, moorland, beaches, and coastline in England, Wales, and Northern Ireland. It also owns and maintains historic houses, gardens, industrial monuments, churches, and pubs. Its focus is not just the conservation of nature but also the conservation of the historical fabric of the country and, perhaps most strikingly in the national parks of England and Wales, the conservation of the social fabric or ways of life that have shaped the cultural landscapes that characterize the UK. The National Trust for Scotland has 128 properties and a similar profile of aims and activities. Both organizations are nongovernmental charities funded by donations, legacies, and a broad membership (3.68 million combined in the first decade of the twenty-first century) and supported by volunteer labor. Although they are nongovernmental organizations (NGOs), these organizations are seen as part of the establishment: safe, nonthreatening, and often willing to compromise.

Other mainstream campaigning charity bodies have originated with visionary individuals or small groups, sometimes beginning by focusing on a small local issue in which they are out of step with public opinion. The Royal Society for the Protection of Birds began as a campaign to stop the use of grebe feathers on women's hats in 1889. It grew to have a million members, two hundred nature reserves (130,000 hectares [321,235 acres] in total), and a scope of action that is focused on protecting wildlife habitats. The larger mainstream campaigning bodies tend to work in the areas of informing the public, including educational programs for children; doing or funding research; lobbying Parliament; using the legal system to challenge damaging developments; and providing a focus for their memberships' concerns. Some focus more on lobbying and legal challenges; for example, the Campaign for the Protection of Rural England, along with its equivalent Welsh body, specializes in putting pressure on planning authorities. This organization began in 1926 and has maintained a sharp focus on resisting urban sprawl and ribbon developments: expansions of towns that eat into the countryside. Other campaigning bodies see themselves as umbrella organizations, such as Transport 2000 (and Transform Scotland), which began in 1973 and has maintained a focus on transport issues and worked to represent and inform government and other campaign groups about transport.

PUBLIC PARTICIPATION

Grassroots activity operates at a number of levels. There is mass membership in established organized campaign groups and the Green Party, where members of the public can pay for membership and see themselves as "doing their bit" for the environment. Practical volunteering is used widely; for example, the National Trust has 43,000 volunteers. More radical groups are characterized by many loose affiliations between different types of campaigns and multiple memberships by individuals, along with mass attendance at events.

An umbrella name predominant in the contemporary protest movement is Earth First! Both globally and nationally this has been a nonhierarchical rubric for other groups or individuals who use direct action to protest against destructive organizations or actions. Direct action under this name can consist in anything from a single individual stopping a shop operating for a morning by gluing his or her hands to its door to the organization of mass protests such as the 2007 climate camp to protest the addition of a new runway at Heathrow Airport. Contemporary direct action under any banner tends to consist in actions that are usually illegal but not dangerous to the public and often combine humor with a serious message. For example, critical mass cycling events involve large groups of cyclists using the same route during rush hour, clogging the road, but the cyclists are invited to wear fancy dress and create a colorful spectacle, adding an element of consciousness-raising to practical activities.

With road building protests there is more emphasis on stopping proposed projects. Direct action can involve damage to machinery or protesters placing themselves in danger to prevent tree felling or excavation. The Newberry bypass in Berkshire, England, was a new section of road intended to ease traffic congestion that ran through a designated area of outstanding natural beauty, a National Trust nature reserve, and a historic civil war battlefield. To defend those areas and protest road building in general, trees and specially dug tunnels were occupied by particularly devoted protesters (ecowarriors) and numerous camps were set up along the proposed route. Although all the camps were evacuated and work began on the bypass in 1996, the protesters received a significant amount of publicity, with the media highlighting a young man called Swampy (Daniel Hooper), the last to be evicted from the tunnels to protect Snelsmoor Common. The dedication and risk to their lives of those individuals over a long period as well as a march of five thousand people impressed the public. The cost to the construction company of delays and security, which was estimated at £23.7 million, as well as the costs of policing such protests, had an effect on plans for road building in the future.

Protests such as Newbury brought hard-core environmental campaigners who might travel from one protest to another together with local people whose concerns were more place-specific and conservation organizations

Climate Change Protesters, London, 2007. *Protestors demonstrated for the week-long Camp for Climate Action outside the BAA office near Heathrow Airport in London, England. The United Kingdom is home to many active grassroots environmental organizations, including well known groups such as Earth First! Despite the popular ethos of a love for nature, the UK deals with several environmental problems, particularly with litter.* CATE GILLON/GETTY IMAGES.

with specific interests. In this instance it also triggered a report from the World Wide Fund for Nature (WWF-UK) criticizing the government's watchdog for conservation, English Nature (EN). It appeared that if EN could not protect an area with the multiple designations that the Newberry site had from a stretch of road that was only 13.5 kilometers long and was predicted to be ineffective at reducing traffic elsewhere, it was powerless.

PROTESTS AGAINST GENETICALLY MODIFIED ORGANISMS

A campaign that brought together NGOs, campaigners, and the wider general public involved the introduction of genetically modified organisms (GMOs). Resistance to a new form of technology was the focus. Genetic modification (GM) was uniquely placed to bring together opposition from a number of fronts. The technology was seen as controlled and promulgated by multinational companies to the detriment of poor farmers globally, foodstuffs with GMOs were perceived by the public as dangerous, and the growing of GM crops was predicted to create "superweeds" that would have a negative impact on biodiversity.

Alongside campaigns by established groups such as Friends of the Earth, new groups emerged. One of the most effective groups at getting information to people and "outing" industry front groups was the Norfolk Genetic Information Network (now GMWatch). Another that has worked in a scientifically literate way at the interface of science and public information is the Institute for Science in Society (I-sis), a nonprofit organization founded in 1999 by Mae-Wan Ho and Peter Saunders.

Anti-GM, like many environmental campaigns, is not just about legislating against or preventing something but also about engaging the public in decision making and encouraging informed debate. Not only did the GM issue stimulate debate across the country, the wider general public got involved by not buying GM food, much to the surprise of major UK supermarkets, which had to advertise that they were "GM free." Resistance to GM was a wider European phenomenon, and legal battles still were being waged in the first decade of the twenty-first century between the European Union and the World Trade Organization about allowing the importation of GM foods. In Britain at that time, crop trials were destroyed by protesters' direct action of pulling up the plants, sometimes at the risk of imprisonment.

ENVIRONMENTAL PHILOSOPHY AND DEBATES

A number of themes characterize environmental debates in the UK, including the layperson's love of nature. In a country dominated by cultural landscapes and with no real wilderness this speaks to a love of countryside as a place of interaction between people and nature. In farming areas and in cities and towns previous human activity provides a richness that people love and work to conserve. However, the problem of litter, which may be worse than in many other places in Europe, suggests a lack of care at even the most basic level. As in other countries, commercial forces drive developers to build uninspiring houses on greenbelt land rather than well-designed ecodevelopments on brownfield inner-city sites, and town councils do nothing to prevent historically rich and socially cohesive towns from becoming what the New Economics Forum has called "clone towns" as chain stores move in. The Transition Towns movement addresses the related problems of global climate change and community disengagement and alienation at the local level.

Environmental philosophy in the UK has emphasized approaches that are scientifically and historically informed and the creative possibilities of human-nature collaboration. There is wide recognition of romantic figures such as Ruskin and Wordsworth as a resource one can go to for inspiration. Another underlying theme is the importance of transforming rather than denying the urban experience coupled with a politically aware concern for providing opportunities for everyone to experience nature and live in socially vibrant communities.

Robin Attfield (2003) developed a biocentric practice-consequentialist line of argument and has supported and represented it as well as broadening it to include global issues. Global justice issues and international relations also appear in the work of Nigel Dower. Environmental economics is represented by figures such as John O'Neill (2006), whose virtues ethic approach is interwoven with questions of social justice. Similarly, Alan Carter combines environmental and equality concerns and combines academic work on environmental questions with work supporting policy groups or campaign organizations.

Work on animal welfare is also evident (the term *speciesism* was coined by the psychologist Richard Ryder); perhaps the most nuanced work in this area is that of Mary Midgely (1983), who has mounted a defense of a benign form of speciesism while emphasizing shared nature between humans and nonhuman animals. A mainstream philosopher who has developed arguments with regard to animals (including a defense of fox hunting) and a somewhat nostalgic approach to the countryside is Roger Scruton. The nature of the place, the richness of its nature writing, and the inspirational science of Darwin have led Alan Holland to develop the fine-grained, culturally aware, and environmentally rich notion of a worthwhile life as an approach to ethics.

The beauty of some UK landscapes also can be seen as generating the resurgence of interest in the aesthetics of nature. Ronald Hepburn was responsible for the environmental turn in the study of aesthetics, which was taken up by Emily Brady, who has focused on both cultural and natural landscapes and developed an integrated aesthetic that brings together an experiential view and an ecologically informed view. The urban situation and the densely populated nature of the country have contributed to a shift in the work of Warwick Fox. His theory of responsive cohesion is crafted to cover not just interhuman ethics and the ethics of the natural environment but also the ethics of the built environment.

Clare Palmer, the founding editor of *Worldviews: Environment, Culture, Religion,* an emigrant to the United States, has done important critical work on process philosophy and environmental ethics as well as on animal ethics. Andrew Brennan, an emigrant to Australia, has argued for a pluralist approach to environmental ethics.

Environmental philosophers in the UK have to battle against the conservative forces of philosophy as well as the current Research Assessment Exercise, which has been perceived as privileging theoretical over practical philosophy in the distribution of academic funding. However, the relative academic freedom and openness to interdisciplinarity of the higher education sector has meant that it was the first place in the world to establish a dedicated master's course in environmental philosophy. Lancaster University's degree in values and environment, which now runs from the University of Central Lancashire, was set up in 1989, and the journal *Environmental Values* was established at Lancaster in 1992.

SEE ALSO *Animal Ethics; Civil Disobedience; Earth First!; Environmental Activism; Environmental Law; Environmental Philosophy: V. Contemporary Philosophy; Genetically Modified Organisms and Biotechnology; Green Politics in Germany; Land Ethic; Midgley, Mary; Nongovernmental Organizations; Ruskin, John; Space/Place; Speciesism; Urban Environments; Wordsworth, William.*

BIBLIOGRAPHY

Attfield, Robin. 2003. *Environmental Ethics: An Overview for the Twenty-First Century.* Cambridge, UK, and Malden, MA: Polity Press.

Brady, Emily. 2003. *Aesthetics of the Natural Environment.* Tuscaloosa: University of Alabama Press.

Fox, Warwick. 2006. *A Theory of General Ethics: Human Relationships, Nature, and the Built Environment.* Cambridge, MA: MIT Press.

Midgley, Mary. 1983. *Animals and Why They Matter.* Athens: University of Georgia Press.

Murphy, Graham. 2002. *Founders of the National Trust.* London: National Trust.

O'Neill, John. 2006. *Markets, Deliberation, and Environmental Value.* London: Routledge.

Sheail, John. 1998. *Nature Conservation in Britain: The Formative Years.* London: Stationary Office Books.

Isis Brook

UNITED NATIONS EDUCATIONAL, SCIENTIFIC AND CULTURAL ORGANIZATION

The United Nations Educational, Scientific and Cultural Organization (UNESCO) was founded in November 1945 and came into force in November 1946 after ratification by twenty countries. By 2007 UNESCO had 192 member states and 6 associate members. It disseminates and shares information and knowledge in the fields of education, science, culture, communication, and information. Furthermore, it also works as a laboratory of ideas and a standard setter in forging universal agreements on emerging ethical issues, such as those concerning contemporary environmental challenges. Environmental concerns currently occupy a prominent place on the organization's agenda and generate systematic reflection on how international policy can promote an ethical approach to the environment.

UNESCO has developed international programs that help reinforce the capacities of developing countries to improve the management of the earth's resources in the fields of the natural sciences, engineering, and technology. Priorities in this area are related to capacity building in the basic and engineering sciences for the use of water and associated ecosystems, including the oceans. To complement this effort, the organization promotes the proper use and maintenance of natural resources, preparedness for and mitigation of disasters, and the use of renewable sources of energy.

The Man and Biosphere Programme, launched in the early 1970s, is especially relevant to environmental ethics. It promotes interdisciplinary research to improve people's relationship with the environment, targeting the ecological, social, and economic dimensions of the loss of biodiversity. Its World Network of Biosphere Reserves works as a vehicle for knowledge sharing, research and monitoring, education and training, and participatory decision making. Designed to balance community needs with the conservation of natural resources, these biosphere reserves seek to provide a harmonious marriage of conservation and development. With more than 480 sites in over 100 countries, the World Network of Biosphere Reserves provides context-specific opportunities to combine scientific knowledge and governance practice in order to reduce loss of biodiversity; improve livelihoods; and enhance social, economic, and cultural conditions for sustaining the environment. Thus, these biosphere reserves contribute to the pursuit of the seventh Millennium Development Goal, to ensure environmental sustainability.

The biosphere reserves can also serve as learning and demonstration sites in the framework of the Decade of Education for Sustainable Development. UNESCO assumes that education lays the ground for environmentally and ethically informed, responsible, and sound decisions and actions. The educational priorities of UNESCO include technical and vocational education, science and technology education, and education for sustainable development. UNESCO values equity for all communities and the long-term stability of the economy and ecology.

According to UNESCO, education and training are primary agents for mobilizing communities toward sustainable development. They fulfill this role by increasing people's capacities, transforming their visions for themselves and their societies into reality, and enhancing linkages between cultural and biological diversity. Yet education must respect cultural landscapes and sacred sites, promote learning about the interactions of biological and cultural diversity, and maintain biosphere reserves and World Heritage sites.

UNESCO is actively pursuing the Millennium Development Goals, especially those aiming to halve the proportion of people living in extreme poverty in

CAPE HORN BIOSPHERE RESERVE

The Cape Horn Biosphere Reserve (CHBR) protects one of the world's most pristine ecoregions, the Magellanic subantarctic rain forests, at the southern end of the Americas (Rozzi et al. 2004). It includes the archipelagoes south of Tierra del Fuego, and the fjords, ice fields, and glaciers on Darwin Cordillera, just 1,000 kilometers north of the Antarctic Peninsula. With five million hectares of marine (three million ha) and terrestrial (two million ha) ecosystems, it is the largest biosphere reserve in southern South America. The CHBR has at least three attributes that are relevant to environmental philosophers.

First, its creation in 2005 resulted from a six-year collaborative effort between the regional government and an interdisciplinary team of ecologists, artists, and humanists led by the Chilean philosopher and ecologist Ricardo Rozzi. The creation of the CHBR involved ten guiding principles, which can be adapted for interdisciplinary research and conservation work in other regions:

1. interinstitutional cooperation,
2. a participatory approach,
3. an interdisciplinary integration of environmental philosophy, sciences, arts, and policy,
4. networking and international partnership,
5. communication through the media,
6. identification of flagship species,
7. "direct encounters" with human and nonhuman beings living in their habitats,
8. economic sustainability and ecotourism,
9. territorial planning and administrative sustainability,
10. "conceptual sustainability" based on continuous long-term *in situ* research (Rozzi et al. 2006).

Second, the Research, Education, and Conservation Center of the CHBR is in the Omora Ethnobotanical Park. In 1999 Omora Park inaugurated a program in field environmental philosophy and biocultural conservation that integrates comparative ethnoecology, ecotourism, and environmental ethics. In 2004 this program was further projected through a partnership with the Department of Philosophy and Religion Studies at the University of North Texas (University of North Texas 2008).

Third, the CHBR is located in a remote region of the Americas, only 1,000 kilometers north of Antarctica. With its location at the end of the continent, it is home to a uniquely rich biological and cultural diversity (Callicott et al. 2006). For example, more than 5 percent of the world's species of mosses and lichens are found here in less than 0.01 percent of the earth's land surface. This high diversity of small flora was critical in making the case to UNESCO representatives for the establishment of the CHBR. Lichens and mosses, although very small organisms in comparison to charismatic megaflora and megafauna, can play important roles in promoting conservation when their ecological and aesthetic values are understood by the general public and by policy makers (Rozzi et al. 2008). Such a shift in the "cognitive lenses," which allows a keener assessment and greater appreciation of biodiversity, has both scientific and philosophical value.

BIBLIOGRAPHY

Callicott, J. B., R. Rozzi, L. Delgado et al. 2006. "Biocomplexity and Biodiversity Hotspots: Three Case Studies from the Americas." *Philosophical Transactions of the Royal Society of London* 362: 321–333.

Rozzi, R., F. Massardo, and C. Anderson, eds. 2004. *The Cape Horn Biosphere Reserve: A Proposal for Conservation and Tourism to Achieve Sustainable Development at the Southern End of the Americas.* Punta Arenas, Chile: Ediciones de la Universidad de Magallanes.

Rozzi, R., F. Massardo, C. Anderson, K. Heidinger, and J. A. Silander, Jr. 2006. "Ten Principles for Biocultural Conservation at the Southern Tip of the Americas: The Approach of the Omora Ethnobotanical Park." *Ecology and Society* 11(1): 43. Available from http://www.ecologyandsociety.org/vol11/iss1/art43/

Rozzi, R., J. Armesto, B. Goffinet, W. Buck, et al. 2008. "Changing Biodiversity Conservation Lenses: Insights from the Sub-Antarctic Non-Vascular Flora of Southern South America." *Frontiers in Ecology and the Environment* 6: 131–137.

University of North Texas. 2008. "The Cape Horn Biosphere Reserve." http://www.chile.unt.edu/capehorn/index.htm

Ricardo Rozzi

developing countries by 2015, to achieve universal primary education in all countries by 2015, to eliminate gender disparity in primary and secondary education by 2005, to help countries implement a national strategy for sustainable development by 2005, and to reverse current trends in the loss of environmental resources by 2015.

The social and human sciences have a vital role to play in helping to understand and interpret the social, cultural, and economic environment. UNESCO's priorities in the social and human sciences are advancing human rights; fighting against all forms of discrimination, racism, xenophobia, and related intolerance; and promoting ethics in science and technology, with an emphasis on bioethics (United Nations 1997).

Different solutions and alternatives started to be developed in response to environmental problems. But especially in developing countries, incipient structures of environmental and educational management have been and are exposed or vulnerable to suffering the impact of various types of economic or political crisis in those countries or regions.

One of the most frequent effects of these crises has been and continues to be the elimination of budgets or the discontinuation of funds. These irregularities have sometimes prevented governments from complying with international agreements. This happens also with UNESCO's programs, which often arise as good intentions but have hitherto been difficult to materialize. All this results in a limited development of criteria to assess the effectiveness of management, cooperation, conservation, education and training, monitoring, and participation of integrating countries. For example, in the case of Latin America and the Caribbean, there is not enough information on whether the implementation is carried out by state agencies—federal, municipal, or mixed management—or by nongovernmental institutions. Although UNESCO's programs are seen with positive expectations, they will still have to face bureaucratic structures, lack of political commitment, or insufficient human resources and institutional capacity.

UNESCO has a Division of Philosophy and Ethics, the only outpost of philosophy in the United Nations system. Through its Universal Ethics Project, this division is leading a worldwide discussion on universal ethics. Emphasis has been placed on the ethical principles at stake in environmental sciences and policies. What is the moral value of the environment? What in nature is worth protecting, preserving, and respecting? What do we mean by global sustainability? How much should we protect the interests of future generations? What are the implications of the principle of justice, for policy decisions related to environmental issues?

Here we can also find difficulties to adequately attain such valuable objectives. But, in the case of philosophy, this is not only due to institutional budgetary limitations, but also to the very nature of the philosopher's task, often linked to purely theoretical work. The UNESCO's *Courier*, published in celebration of World Philosophy Day (November 15, 2007), gathered the critical opinion of various thinkers, aiming at a strengthening of the social responsibility of the philosopher, and they recognize that the philosophical impact on society is not what it could be. The question posed here is how they could make themselves useful.

Michel Onfray (France) advocates ceasing to lecture everybody and being comfortable to remain only in the realm of the word, and trying to produce philosophical effects on the existential level. Norwegian Jostein Gaarder stresses questions as what shift in consciousness we need, what is a sustainable wisdom, what qualities of life are important, and if the unsustainable consumerism is really the only possible model of life.

Gaarder relates this to his proposal of a universal declaration of human obligations, since it is no longer meaningful to talk about rights without simultaneously stressing the individual state's or person's obligations and responsibilities.

To promote sustainable consumption, UNESCO and the United Nations Environment Programme are working together to understand young people's consumer behavior and work with them in promoting more sustainable consumption patterns for the future. Launched in March 2000, the Youth Exchange Programme, for example, includes a training guide, a Web site, online discussions, and related workshops aimed at young people's consumption.

In 2007 UNESCO undertook several important initiatives. It launched the Biosphere Connections partnership to support conservation of biodiversity and sustainable development in conjunction with the airline coalition Star Alliance, the Man and Biosphere Programme, the Ramsar Convention on Wetlands, and the World Conservation Union. The same year it also launched the International Initiative in Defence of the Quality of the Night Sky as Mankind's Scientific, Cultural, and Environmental Right. This initiative maintains that future generations have the right to skies free of light pollution.

UNESCO's Communication and Information Programme will help to build a society based on the sharing of knowledge and incorporating the sociocultural and ethical dimensions of sustainable development. The world urgently requires global visions of sustainable development based on observance of human rights, mutual respect, and alleviation of poverty—goals which lie at the heart of UNESCO's mission and activities.

SEE ALSO *Biodiversity; Consumption; Environmental Education; Future Generations; Hunger; Sustainable Development; United Nations Environment Programme.*

BIBLIOGRAPHY

Bindé, Jérôme, ed. 2001. *Keys to the 21st Century*. Paris: UNESCO.

Boulières, François. 1993. *The Biosphere Conference 25 Years Later*. Paris: UNESCO. Available from http://unesdoc.unesco.org/images/0014/001471/147152eo.pdf

Diemer, Alwin; J. Hersch, P. Hountondji, et al. 1986. *Philosophical Foundations of Human Rights*. Paris: UNESCO.

Frodeman, Robert. 2006. "The Policy Turn in Environmental Philosophy." *Environmental Ethics* 28(1):10.

Have, Henk A. M. J. ten, ed. 2006. *Environmental Ethics and International Policy*. Paris: UNESCO.

Selgelid, Michael. 2005. "Universal Norms and Conflicting Values." *Developing World Bioethics* 5(3): 267–273.

UNESCO. 2000. *Solving the Puzzle: The Ecosystem Approach and Biosphere Reserves*. Paris: UNESCO. Available from http://unesdoc.unesco.org/images/0011/001197/119790eb.pdf

UNESCO. World Commission on the Ethics of Scientific Knowledge and Technology (COMEST). 2005. *The Precautionary Principle*. Paris: UNESCO. Available from http://unesdoc.unesco.org/images/0013/001395/139578e.pdf

United Nations. 1997. *Universal Declaration on the Human Genome and Human Rights*. Available from http://portal.unesco.org/en/ev.php-url_id=13177&url_do=do_topic&url_section=201.html

Alicia Irene Bugallo

UNITED NATIONS ENVIRONMENT PROGRAMME

The United Nations Environment Programme (UNEP) promotes international cooperation on the environment and serves as the focal point for environmental action and coordination within the United Nations system. Its origins can be traced back to the United Nations Conference on the Human Environment in Stockholm in June 1972. At that time it was acknowledged that a growing number of regional and environmental problems would require extensive cooperation among nations and action by international organizations in the common interest. Shortly after the Stockholm conference, the United Nations Twenty-Seventh General Assembly approved Resolution 2997 in December 1972 (United Nations 1972). In that resolution the United Nations established UNEP in the context of "the urgent need for a permanent institution within the United Nations system for the protection and improvement of the environment" (United Nations 1972, p. 43). The UNEP Governing Council is composed of fifty-eight members elected by the General Assembly for three-year terms. Seats are allocated on a regional basis.

ORIGINS AND STRUCTURE

In Resolution 2997 it was decided that UNEP should promote international cooperation on the environment; review the world environmental situation to ensure that emerging international issues would receive adequate consideration by governments; promote the acquisition, assessment, and exchange of environmental knowledge; and review the impact of environmental policies on developing countries. The mission of UNEP is "to provide leadership and encourage partnership in caring for the environment by inspiring, informing, and enabling nations and peoples to improve their quality of life without compromising that of future generations" (UNEP, "What UNEP Does," para. 2).

UNEP's headquarters are in Nairobi, Kenya. UNEP also supports six regional offices around the world and a growing number of liaison and outpost offices, collaborating centers, and convention secretariats. Based in Africa, it is positioned to have a firsthand understanding of environmental issues facing developing countries.

UNEP seeks partnerships in all sectors and describes its work as assessing global, regional, and national environmental conditions and trends; developing international and national environmental instruments; strengthening institutions for the wise management of the environment; the transfer of knowledge and technology for sustainable development; and encouraging new partnerships and mind-sets within civil society and the private sector. The implementation of this work is done through seven divisions: Early Warning and Assessment; Environmental Policy Implementation; Technology, Industry and Economics; Regional Cooperation; Environmental Law and Conventions; Global Environment Facility Coordination; and Communications and Public Information.

ACHIEVEMENTS AND PUBLICATIONS

Since its inception UNEP has been a principal actor in major international initiatives and events. Significant milestones include the Convention on International Trade in Endangered Species (1973), the Bonn Convention on Migratory Species (1979), the Vienna Convention for the Protection of the Ozone Layer (1985), the Montreal Protocol on Substances That Deplete the Ozone Layer (1987), the Intergovernmental Panel on Climate Change (IPCC) (1988), the Basel Convention on the Transboundary Movement of Hazardous Wastes (1989), the United Nations Conference on Environment and Development (1992), the Convention on Biological Diversity (1992), the Stockholm Convention on Persistent Organic Pollutants (2001), and the World Summit on Sustainable Development (2002).

UNEP promotes dialogue and cooperation among stakeholders and has assisted in the establishment of a number of notable demonstration projects, including the OzonAction program designed to help implement the Montreal Protocol, the International Coral Reef Action Network, and the Great Apes Survival Project.

In addition to a large number of books on a broad range of environmental topics, UNEP publishes annually the *UNEP Year Book*, formally known as the *GEO Year Book*; the magazine *Our Planet*; and the youth magazine *Tunza*.

UNEP's Medium-Term Strategy (2010–2013), approved in February 2008, sets out the next phase of the evolution of UNEP as it becomes more effective, efficient, and results-focused for the UNEP program of work. This focus will enable UNEP to better deliver on its mandate by building on its expertise and comparative advantage in a limited number of priority areas: climate change; disasters and conflicts; ecosystem management; environmental governance; harmful substances and hazardous waste; and resource efficiency–sustainable consumption and production.

UNEP will deliver on the six cross-cutting priorities by utilizing the capacity and expertise of UNEP divisions and regional offices and will actively reach out to government, other United Nations entities, international institutions, MEA secretariats, civil society, the private sector, and other relevant partners to support delivery of the MTS. The vision of UNEP for the medium-term future is to be "the leading global environmental authority that sets the global environmental agenda, that promotes the coherent implementation of the environmental dimension of sustainable development within the United Nations system and that serves as an authoritative advocate for the global environment" (UNEP Working Group on Medium-Term Strategy, 2010–2013, Consultation Paper No. 1).

ENVIRONMENTAL PHILOSOPHY

UNEP's values relating to environmental philosophy and ethics are stated in its founding documents and mission statement. The Declaration of the 1972 United Nations Conference on the Human Environment in Stockholm identified both natural and human-made elements of the environment as essential to human well-being and to the "enjoyment of basic human rights and the right to life itself" (UNEP 1972). In establishing UNEP, United Nations Resolution 2997 identified a need for measures designed to safeguard and enhance the environment "for the benefit of present and future generations of man [SIC]" (United Nations 1972, p. 43). UNEP's mission is "to provide leadership and encourage partnership in caring for the environment by inspiring, informing, and enabling nations and peoples to improve their quality of life without compromising that of future generations" (United National Environment Programme, "What UNEP Does"). These statements describe environmental values that direct UNEP activities. Historically, those activities primarily have been related to human rights, human rights to life, and concern for future generations.

On the 1997 World Environment Day in Korea the UNEP executive director chaired the Environment and Ethics Roundtable. The resulting document, *The Seoul Declaration on Environmental Ethics* (UNEP 1997), expanded UNEP's scope for ethical consideration. The declaration began by acknowledging that there is no choice but to redefine the values and principles that underlie human relationships with the earth. Without such fundamental changes, it suggested, further environmental degradation will lead to the collapse of the natural systems that support life. Again, there is a concern for safeguarding the rights of future generations. However, the document also rests on the assumption that the existence of all life, including human life, can be sustained only if the entire community of life on earth is sustained. *The Seoul Declaration* also stated that its framework of ideals, principles, and guidelines will evolve over time.

UNEP supported the development and publication of *Environmental Education, Ethics and Action: A Workbook to Get Started* (2006).

SEE ALSO *Convention on Biodiversity; Environmental Policy; Intergovernmental Panel on Climate Change.*

BIBLIOGRAPHY

Jickling, Bob; Heila Lotz-Sisitka; Rob O'Donoghue; and Akpeizi Ogbuigwe. 2006. *Environmental Education, Ethics, and Action: A Workbook to Get Started.* Nairobi: United Nations Environment Programme. Also published as *Educación Ambiental, Ética y Acción* (Spanish), *Éducation Écologique, Éthique et Agir* (French), *Educazione Ambientale, Etica e Azione* (Italian). Available from http://www.unep.org/Training/publications/index.asp

United Nations. 1972. United Nations General Assembly Resolution 2997 (XXVII). Available from http://daccessdds.un.org/doc

United Nations Environment Programme. Available from http://www.unep.org

United Nations Environment Programme. "What UNEP Does." Available from http://www.unep.org

United Nations Environment Programme. 1972. *Declaration of the United Nations Conference on the Human Environment.* Available from http://www.unep.org/Documents. multilingual/Default.asp?DocumentID=97&ArticleID=1503

United Nations Environment Programme. 1997. *Seoul Declaration on Environmental Ethics.* Available from http://www.nyo.unep.org/wed_eth.htm

United Nations Environment Programme Working Group on Medium-Term Strategy, 2010–2013. 2007. Consultation

Paper No. 1. Available from http://unep.org/civil_society/gcsf/1_MTS_Consultation_paper_1.pdf

Bob Jickling

UNIVERSITY-INDUSTRY RELATIONSHIPS

The ties between universities and private industry have been transformed from arms-length relationships between two distinct entities with different purposes into an interaction among overlapping institutional spheres in which each side has assumed some of its partner's traditional roles and characteristics. Government, at various levels, increasingly encourages university-industry interactions in order to foster public economic goals such as job creation and economic growth. As a result, university-industry relationships have become university-industry-government relationships, sometimes referred to as the "triple helix" (Etzkowitz 2008).

THE GROWTH OF SCIENTIFIC RESEARCH IN THE UNIVERSITY

The university, from its medieval origins until the late nineteenth century, was devoted to the production, preservation, and transmission of culture. Since 1900, however, it has become the source of new industries and private corporations while maintaining and expanding its traditional roles. In the late nineteenth century growing science-based electrical and chemical industries initiated relations with universities to serve their research needs and to supply them with personnel.

These university-industry relations were conducted across well-defined borders between the academic and commercial domains. The university's increasing involvement in science, however, engendered a more organized approach to managing research and its practical consequences. For instance, when researchers at the University of Toronto invented an insulin treatment for diabetes in 1922, the university found that it had to patent and license the technology in order to protect itself from potentially unethical manufacturers (Bliss 2007). The University of Wisconsin encountered a similar situation when a faculty member, Harry Steenbock, in 1924 patented the irradiation process based on his work on antirachitic vitamine (Apple 1989). Thus, even before universities realized that they could earn income from inventions made on campus, they were impelled to create mechanisms to insure an orderly process of technology transfer and protect their reputations.

Traditionally, university-industry relations denoted the provision of research support from a firm to a campus-based researcher. The contemporary form, however, involves the participation of academic scientists in the formation of private companies through the use of their academic research. The university takes on an entrepreneurial role, assisting in the founding of private firms and contributing to regional economic development. As the university acquires an industrial penumbra, industry takes on some of the values of the university, sharing as well as protecting knowledge.

THE FORGING OF UNIVERSITY-INDUSTRY RELATIONSHIPS

There are both formal and informal modes of university-industry relationships. The liaison and technology-transfer offices of the university constitute the formal mechanisms through which introductions are made in seminars for potential industrial partners, disclosure statements of inventions are collected, and patents licensed and contracts are negotiated. The activities of these formal programs to paint in broad brush-strokes are similar to the informal channels, with perhaps the key difference being that financial results from the formal channels are typically shared with the university (Etzkowitz and Webster 1995).

Informal relations with industry typically occur through contacts between professors and their former students and may lead to consulting and joint-research projects with a company. Informal relations may be viewed as an "underground economy" that is not counted as part of official academic work. Typically, both modes are present: the formal organizations and the informal ties through which social and intellectual capital moves from graduating students into firms and back again into the university.

Traditionally, university-industry relations denoted the provision of research support from a corporation to a campus-based researcher. Though offering far fewer restrictions than government support in many cases—and finding favor with the academic research staff—such funds represented a tiny proportion of academic research support. For instance, industrial R&D support to U.S. universities and colleges was 7.4 percent in 1959; 2.7 percent in 1969; 3.6 percent in 1979; 6.6 percent in 1989; 7.4 percent in 1999; 6.7 percent in 2001; and 5.1 percent in 2006 (NSF/SRS 2006). Most of these funds flowed through consulting relationships with faculty members who provided advice on campus to company visitors and at the industrial lab, conducted tests of materials and products in their laboratories, and occasionally carried out small research projects for a company (Etzkowitz 2002). From the early years of the research university in the late nineteenth century, university-industry relationships were largely established at the behest of industry to serve the needs of existing companies. Engineering schools reorganized themselves to serve the research needs of growing science-based electrical and chemical industries and to supply them with personnel. The linkages included cooperative programs that sent students to industry for part of their training, university professors undertaking research at the request of industry, and donations of money and equipment by industrial firms to support engineering education (Noble 1976). These relationships, however, declined in the 1930s because of the Depression, elevating foundations as important sources of sponsored research.

However, during the Depression a new series of relationship formats were being created at MIT—for example, the faculty-formed firm, an explicit role for the university in shaping regional economic development, the interdisciplinary center, and the invention of the venture capital firm (Etzkowitz 2002). The older forms of university-industry connections involved payment for services rendered, whether it was received directly in the form of consultation fees or indirectly as endowment gifts. The new formats of university-industry relationships are built upon the development of scientific-research capabilities and the creation of a series of boundary-spawning mechanisms like technology-transfer offices and spinoff firms. This institutional transformation is reflected in the enactment of the Bayh-Dole Act of 1980, which transferred ownership of intellectual property emanating from government-sponsored research to universities on the condition that they take steps to promote their utilization. Similar new legal frameworks followed in other countries, often supported by funding programs, to legitimate and foster university-industry interactions.

Contemporary university-industry relations arose from two distinct sources and an emerging third hybrid stream: basic research interests funded by research councils and similar bodies, industrial projects for which academic input is solicited and in a creative fusion, and a joint formulation of research programs with conjoint basic and applied goals and multiple funding sources. Basic research increasingly takes place in research groups that function as "quasi-firms" that have many of the attributes of a private corporation except for the profit motive.

Incubators provide a means to subsidize the infrastructure of company formation and a training mechanism to teach academics to operate a firm. Centers integrate disciplinary research groups into broader interdisciplinary collaborations and carry out "translational research" to bring research findings closer to utilization. Technology-transfer offices arrange intellectual property protection and negotiate the terms for movement of commercially promising research into private companies, including those founded by members of the university. Science parks provide a home for research units of corporations, which often emanate from universities, offering projects and collaborative opportunities to their academic counterparts.

As a "third mission" of contributing to economic and social development is integrated into the university, the dissemination of academic knowledge takes place through patents as well as publications. The hybrid forms of university-industry relationships involve the multiplication of resources through university and faculty participation in capital-formation projects such as real estate development in science parks and the formation of companies in incubator facilities. The objective is to multiply the value of intellectual property derived from academic research through the stock market, either directly or indirectly.

CONCLUSION

The university is undergoing a "second academic revolution," integrating teaching, research, and economic development. As the university engages in technology transfer, it becomes a source of new product development, which is, of course, a traditional industrial function. The growth of university-government relationship was intertwined with the formation of national identity in Germany in the early nineteenth century, with the so-called Humboltdian academic model integrating teaching and research—the first academic revolution (Jencks and Riesman 1969).

In the United States, university-government relationship transcended the emergency of World War II as academics realized, during the postwar period, that theoretical advances could arise from problem-oriented research and vice versa. As new arrangements are put in place, old formats remain in use, creating a complex interplay among organizations and roles with ensuing conflicts and confluences of interest. As the university acquires an industrial penumbra, industry takes on some of the values of the university, sharing as well as protecting knowledge. Governments assume a new role in innovation by encouraging university-industry interactions of various kinds.

The line between facts and values is in most instances blurred. But since the first expression of a plant gene in a different species of plant in 1983, ethical, environmental, and religious concerns have shaped debates over the developments in agricultural biotechnology and genetic engineering (Tokar 2001; Kleinman 2005). Environmental concerns have been at the forefront of the debates over genetically modified organisms (GMOs). The discovery in 1999 of deadly effects of pollen from genetically engineered corn on immature butterflies upped a notch the tenor of environmental consequences of genetic engineering (Tokar 2001). As a result, much of the public debate about GMO has centered on the risks of gene transfer, the role of big business, and instances of the connivance of university scientists.

Since the issues and implications of GMOs are global, such issues are best tackled within collaborative frameworks. This was illustrated by the farm-scale trials of genetically modified crops in the United Kingdom. As the exercise revealed, irrespective of the genetic design and constitution of the crop plants, ecological effects were expected to arise as a result of the conditions in which each crop was cultured (Ormerod et al. 2003, p. 940). It is plausible that any future ecological cost from genetic modification will reflect the diversifying opportunities, methods and conditions for crop growth as any other factor in the brave new GMO landscape (Ormerod et al. 2003).

The university-industry-government interaction is a global phenomenon. The increasing recognition of universities as actors in national and regional innovation systems is leading to the blurring of boundaries between corporations and the academy and their replacement with a web of ties.

SEE ALSO *Alternative Technology; Genetically Modified Organisms and Biotechnology; Technology.*

BIBLIOGRAPHY

Apple, Rima. 1989. "Patenting University Research: Harry Steenbock and the Wisconsin Alumni Research Foundation." *ISIS* 80(3): 374–394.

Bliss, Michael. 2007. *The Discovery of Insulin.* Chicago: University of Chicago Press.

Etzkowitz, Henry. 2002. *MIT and the Rise of Entrepreneurial Science.* London and New York: Routledge.

Etzkowitz, Henry. 2008. *The Triple Helix: University-Industry-Government Innovation in Action.* London and New York: Routledge.

Etzkowitz, Henry, and Andrew Webster. 1995. "Science as Intellectual Property." In *Handbook of Science and Technology Studies,* ed. Sheila Jasanoff, Gerald Markle, James Petersen, and Trevor Pinch. London: Sage.

Jencks, Christopher, and David Riesman, 1969. *The Academic Revolution.* New York: Doubleday.

Kleinman, Daniel Lee. 2005. *Science and Technology in Society: From Biotechnology to the Internet.* Oxford: Blackwell Publishing.

National Science Foundation/Division of Science Resources Statistics. 2006. "Survey of Research and Development Expenditures at Universities and Colleges, FY 2006." Available from http://www.nsf.gov/statistics/showsrvy.cfm?srvy_CatID=4&srvy_Seri=12

Noble, David. 1976. *America by Design: Science, Technology, and the Rise of Corporate Capitalism.* New York: Knopf.

Ormerod, Steve; E. J. P. Marshall; Gillian Kerby; and Steve P. Rushton. 2003. "Meeting The Ecological Challenges of Agricultural Change." *Journal of Applied Ecology* 40(6): 939–946.

Tokar, Brian, ed. 2001. *Redesigning Life? The Worldwide Challenge to Genetic Engineering.* London: Zed Books.

Henry Etzkowitz
James Dzisah

URBAN ENVIRONMENTS

Born in America, Australia, and Norway, environmental philosophy has its deepest roots in the tradition of wilderness advocacy and thus has tended to focus on the value of wild nature, the interests of wild creatures, and the integrity of unaltered ecosystems. Only recently have some environmental philosophers begun to turn their attention to urban environments, broadening the field and connecting it with branches of the environmental movement other than wilderness advocacy, including the environmental justice movement and the broader discourse on sustainability.

A BLIND SPOT

From its inception environmental ethics has focused mainly on the normative status of wilderness and wildlife. The many arguments about moral standing and intrinsic value have been aimed at finding a more appropriate balance between the domestic and the wild, mainly by valorizing the wild. One example is Paul W. Taylor's theory of respect for nature, which purports to be biocentric but provides inherent worth (intrinsic value) only for wild organisms, expressly excluding domestic ones. Landscapes that have not been touched by human hands are the normative reference against which all other landscapes are to be judged not only ethically but aesthetically, according to Allen Carlson's theory of positive environmental aesthetics.

The historical alliance of environmental ethics with the tradition of wilderness advocacy once all but precluded serious direct engagement with urban environments. There is thus what Andrew Light has called "an urban blind-spot" in environmental ethics: "[B]y and large, cities are considered sources of environmental disvalue: a landscape either to be mined for examples to be avoided or ignored altogether as a product of human intentions—an artifact rather than a part of nature and so outside the proper boundaries of the discipline" (Light 2001, p. 8).

A handful of environmental ethicists have attempted to remedy that blind spot, motivated by a sense that environmental ethics is incomplete if it automatically excludes from consideration the landscapes in which people live. Roger J. H. King, for example, found a "self-destructive logic" in an approach to environmental ethics that "presupposes that humans are at best interlopers on what should otherwise have been a nonhuman scene." The one-sided valorization of wilderness yields a "halt-and-withdraw" strategy that "holds nothing open for the future; the narrative line ends in tragedy for nature and for humans." (King 2000, pp. 115–116)

The literature on urban environments has expanded in two directions. One approach focuses on the role

Aerial View of Manhattan. *The island of Manhattan, a part of New York City seen here from the Empire State Building, is the epitome of an urban metropolis in the United States. Many believe that environmental ethics have ignored urban environments until only more recently, instead choosing to focus more on wilderness habitats. Some places, like New York's Central Park, provide ta mix of both environments: a lush natural habitat in the midst of one of the largest urban centers of the world.* IMAGE COPYRIGHT DONALD R. SCHWARTZ, 2008. USED UNDER LICENSE FROM SHUTTERSTOCK.COM.

urban environments can play in fostering concern for wild nature and putting that concern into practice. Light argued that in the short term at least, practicing "ecological restoration" in urban settings can be a means toward the goal of fostering "ecological citizenship" (Light 2001, p. 28). King maintained that cities should at least foster an "environmental conscience" and to that end proposed as a principle that "the built environment should make nature a visible, palpable presence in daily human experience" (King 2000, pp. 115, 130).

Another approach starts with the recognition that cities and suburbs are the environments of most direct concern to most people, with the implication that those environments should be understood and evaluated on their own account, not simply or exclusively held to the standard of wildness. In discussing long-term considerations, Light stated, "If environmental ethics is to fully embrace the urban, then it must describe the brown space of the city to be as important a locus of normative consideration as the green space" (Light 2001, p. 31). Alastair Gunn went so far as to redefine the natural environment in terms of the conditions for human thriving, stating that "to the extent that human beings are able to thrive in a given environment—to live long lives, to be physically and psychologically healthy, to fulfill a significant portion of their potential—then that environment is natural for humans" (Gunn 1998, p. 348). There is no reason in principle why a city or a suburb cannot be natural in this sense.

Taking both of these approaches into account, an urban turn in environmental ethics opens up a broad range of normative questions concerning human well-being, justice, sustainability, and political legitimacy (Kirkman, 2004).

GOOD PLACES

One set of normative considerations in urban environments concerns the conditions that contribute to or detract from human health and well-being. Those considerations need not be limited to matters of physical health, though the availability of food and clean water and protection from various kinds of risk are important. Also of interest, though, are features of the environment that contribute to or detract from mental health and personal development, including access to cultural and educational institutions, opportunities for recreation, and opportunities for economic activity. Taking all these elements together provides a basis for judging whether a particular urban setting constitutes a good place to live.

Attention to matters of human well-being in urban environments meshes with the concerns of environmentalism in a number of areas. First, attention to matters of human health in urban environments relates to the antitoxics branch of the environmental justice movement. The signal event for the rise of antitoxics activism was the discovery of toxic waste in the soil of Love Canal, a suburban neighborhood near Buffalo, New York, and the likely effects of that waste on the health of children who lived and played there.

Second, the experiential or aesthetic richness of landscapes has long been important to environmental advocates and their conservationist predecessors. An entirely technological landscape probably would be dreary and dispiriting; this attests to the need for some measure of scenic diversity, some space left open for wild nature in the urban fabric. Ironically, much of the initial appeal of suburban development was the location of houses in a parklike setting with carefully arranged bits of wildness close by. However, a major impetus for more recent critiques of suburbanization is the perception that the resulting landscape has degraded to a dreary monoculture of lawns interrupted only by a dreary monoculture of parking lots.

Third, in addition to the provision of green space, decisions about where and where not to build mesh with broader concerns about the relationship of the domestic to the wild and the need to be attentive to nonhuman nature. In places where human well-being is at stake, it is especially important to note that inattentive building can expose people to undue risks. Examples include development in flood-prone areas or even below sea level, schools or hospitals built on ground prone to shaking during earthquakes, and neighborhoods intruding into the habitat of large predators or into landscapes subject to frequent wildfires.

JUSTICE

Landscapes in and around metropolitan areas are diverse, affording different opportunities for and obstacles to well-being. The work of establishing a particular built environment incurs costs that may be externalized to other parts of the landscape. Economic and educational opportunity has shifted to suburbs, for example, leading to deterioration of the inner city. At the same time people in one area may have to live with pollution generated in the process of providing goods and services from which people in another area derive the greatest benefit.

These are matters of justice that may be distilled into two basic questions: Who has access to good places to live, work, and play? and Who bears the costs of establishing and maintaining those good places? The answers to those questions are intertwined with matters of distribution, consent, and compensation.

A concern for justice in urban environments connects to the broader environmental movement at two points. The first involves the rise of the environmental justice movement in the 1990s. At first considerations of race and class were brought into the antitoxics movement, and it was noted that poor and minority communities are exposed to disproportionate risks in their neighborhoods and that government agencies are not evenhanded in protecting communities from those risks.

Environmental-justice theorists have looked beyond the imposition of risk to patterns of exclusion and segregation in the urban landscape. The United States, for example, has a long history of excluding minority groups from the most desirable neighborhoods: A combination of federal laws, lending practices, and personal prejudice led to concentrations of poor and minority groups in deteriorating inner cities. Laws and, to some extent, lending practices have changed, but the legacy of past discrimination persists (Torres, Bullard, and Johnson 2000). In some cases that legacy is embodied in the physical infrastructure of the city, which was put in place with the explicit intent of physically blocking the expansion of minority communities into particular areas when that was both legal and acceptable public policy (Bayor 1996).

Matters of justice are at the root of environmental ethics in its traditional guise. The built environment is a rearrangement of the existing natural environment to suit particular human ends; intentionally or not, this has the effect of including and excluding other living things and changing the composition and form of the biotic community. Aside from the long-term impact of such changes on human well-being, people may owe some consideration to nonhuman living things and systems in their own right. From this perspective it is at least possible to pose the question of whether an injustice has been done, for example, in appropriating the habitat of an animal or plant population and altering it so that that population no longer can survive there.

SUSTAINABILITY

A further set of normative considerations in urban environments concerns the degree to which human life in those environments—and civilization itself—can be sustained in the future. As environmentalists have long maintained, the domestic environment is carved out of and remains dependent on its broader environmental context, which includes not only raw materials and sources of energy but also the relatively benign and predictable setting afforded by stable landforms, stable climate, healthy ecosystems, fertile soil, and clean air and water. Patterns of human activity in a domestic environment are sustainable to the extent that they tend to conserve finite resources and maintain the dynamics that underlie the benign and predictable setting on which they depend; they are unsustainable to the extent that they tend not to do that.

Urban form is deeply implicated in charges of unsustainability leveled against contemporary civilization.

Metropolitan growth in the United States has been shaped by nearly exclusive reliance on the automobile for local transportation. As a consequence, vast areas of the American landscape are all but uninhabitable without the automobile: The functions of economic and civic life are scattered across the landscape, connected to one another and to residential areas only by roads and highways. Automobiles powered by fossil fuels are a major factor in resource depletion, local pollution, and global climate change, in effect undermining the long-term sustainability of American-style metropolitan regions.

By contrast, moderately dense urban forms in which the functions of economic and civic life are intermixed with residential areas may be more sustainable, at least by some measures. For example, residential energy use is generally lower in relatively dense urban areas than in suburban or rural areas (Light 2001), and in that setting more efficient forms of transportation become feasible, including walking, cycling, and mass transit (Gillham 2002).

SEE ALSO *Built Environment; Ecological Restoration; Environmental Justice; Landscape Architecture, Design, and Preservation; Sustainability; Sustainable Architecture and Engineering; Taylor, Paul.*

BIBLIOGRAPHY

Bayor, Ronald H. 1996. *Race and the Shaping of Twentieth-Century Atlanta.* Chapel Hill: University of North Carolina Press.

Carlson, Allen. 2000. *Aesthetics and the Environment: The Appreciation of Nature, Art and Architecture.* London: Routledge.

Gillham, Oliver. 2002. *The Limitless City: A Primer on the Urban Sprawl Debate.* Washington, DC: Island Press.

Gunn, Alastair. 1998. "Rethinking Communities: Environmental Ethics in an Urbanized World." *Environmental Ethics* 20(4): 341–360.

King, Roger J. H. 2000. "Environmental Ethics and the Built Environment." *Environmental Ethics* 22: 115–131.

Kirkman, Robert. 2004. "The Ethics of Metropolitan Growth: A Framework." *Philosophy and Geography* 7(2): 201–218.

Light, Andrew. 2001. "The Urban Blind Spot in Environmental Ethics." *Environmental Politics* 10(1): 7–35.

Taylor, Paul. 1986. *Respect for Nature: A Theory of Environmental Ethics.* Princeton: Princeton University Press.

Torres, Angel O.; Robert D. Bullard; and Glenn S. Johnson. 2000. "Closed Doors: Persistent Barriers to Fair Housing." In *Sprawl City: Race, Politics, and Planning in Atlanta,* ed. Robert D. Bullard, Glenn S. Johnson, and Angel O. Torres. Washington, DC: Island Press.

Robert Kirkman

U.S. BUREAU OF LAND MANAGEMENT

The Bureau of Land Management (BLM) is responsible for managing more than 260 million acres of federal public lands, mostly scattered across the eleven western states. Created by Congress in 1946 through a merger of the General Land Office and the Grazing Service, the BLM is the largest land management agency in the Department of the Interior. Most of the BLM's acreage consists of arid and semiarid federal lands that went unclaimed during the disposal era (c. 1862–1891), when national policy was to transfer public lands into private ownership under the Homestead Act, General Mining Law, and other such laws. In addition, the BLM oversees 700 million acres of subsurface mineral resources underlying federal, state, private, and tribal lands, as well as nearly 3 million acres of timberland in western Oregon.

For its first thirty years, the BLM operated without a clear charter from Congress. In the then lightly populated West, the agency's principal focus on minerals and range resources evoked little concern among local residents, many of whom relied on mining and ranching for their economic well-being. But as the region's population expanded and urbanized and as public interest in recreation, wildlife, and wilderness grew, resource use and access conflicts became increasingly contentious, leading critics to complain that the BLM was too beholden to its traditional mineral and ranching constituencies.

In 1976 Congress responded by adopting the Federal Land Policy and Management Act (FLPMA), which is now the BLM's organic charter. The FLPMA directs the BLM to manage its lands under the "multiple-use" principle, which includes "recreation, range, timber, minerals, watershed, wildlife and fish, and natural scenic, scientific and historical values" but "without permanent impairment of the productivity of the land and the quality of the environment." To do so, the BLM must prepare comprehensive resource-management plans that effectively zone its lands for the various uses. In addition, the BLM is authorized to designate "areas of critical environmental concern" (ACECs) and was directed to inventory its lands for potential wilderness designation, a process that has placed 24 million acres in legally protected Wilderness Study Area (WSA) status. Though FLPMA includes wildlife as a delineated multiple use, it does not contain a biodiversity provision; rather, state wildlife agencies are primarily responsible for managing wildlife populations on the public lands.

Besides its FLPMA obligations, the BLM is governed by a diverse and sometimes conflicting array of resource-development and environmental laws. These laws range from the National Environmental Policy Act and the Endangered Species Act to the Mineral Leasing Act,

Grand Staircase-Escalante National Monument, Utah. *Grand Staircase-Escalante National Monument covers nearly 1.9 million acres of public land in the Unites States and serves as an outstanding source of scientific research and education. The U.S. Bureau of Land Management is responsible for managing public land, most of which consists of the arid and semi-arid land that went unclaimed in the latter half of the nineteenth century during westward expansion.* U.S. GEOLOGICAL SURVEY.

Taylor Grazing Act, and the Energy Policy Act of 2005. During the 1970s, once FLPMA and related environmental laws began taking hold, several western states and angry ranchers launched a political movement dubbed the Sagebrush Rebellion, which was designed to recapture control of the BLM lands. Predicated on the later-discredited theory that the states really owned the unreserved federal public lands, the Sagebrush Rebellion evaporated once Ronald Reagan was elected president in 1980 and installed James Watt as secretary of the interior to pursue an aggressive development agenda on the public lands.

During the Clinton presidency (1993–2001), however, the BLM radically changed its policy direction. Under the leadership of Secretary of the Interior Bruce Babbitt, the BLM pursued a vigorous reform agenda, not only revising its mining and rangeland regulations to incorporate new environmental standards, but also expanding the agency's wilderness review and protection authority. President Clinton created fourteen new BLM-managed national monuments (including the landmark Grand Staircase–Escalante National Monument in southern Utah) that were incorporated into a new National Landscape Conservation system, the first time that the BLM was invested with such extensive preservation responsibilities. On the BLM's Oregon and California timberlands, Secretary Babbitt instituted a new ecosystem management policy designed to promote biodiversity conservation.

But these shifts in BLM policy priorities have not held. In the aftermath of the 9/11 tragedy, the Bush administration pursued an aggressive energy-development agenda on BLM lands across the interior West. The result was an avalanche of new oil and gas leases and drilling projects that later encumber these lands. Furthermore, the BLM modified its mining responsibilities, loosened its rangeland-management regulations, and reduced its wilderness-protection obligations. Not all of these efforts succeeded, however, because the courts blocked several of these reforms. Nonetheless, the BLM remained into the early 2000s an Interior Department agency still in search of a clear identity, having been whipsawed between two very different competing visions of appropriate federal conservation policy.

SEE ALSO *Biodiversity; Environmental Law; Environmental Policy; Land Ethic; Wilderness.*

BIBLIOGRAPHY

Cawley, R. McGreggor. 1993. *Federal Land, Western Anger: The Sagebrush Rebellion and Environmental Politics.* Lawrence: University of Kansas Press.

Clarke, Jeanne Nienaber, and Daniel C. McCool. 1996. *Staking Out the Terrain: Power and Performance Among Natural Resource Agencies.* 2nd edition. Albany: State University of New York Press.

Clawson, Marion. 1971. *The Bureau of Land Management.* New York: Praeger.

Clawson, Marion. 1983. *The Federal Lands Revisited.* Baltimore: Johns Hopkins University Press.

Muhn, James, and Hanson R. Stuart. 1988. *Opportunity and Challenge: The Story of the BLM.* Washington, DC: Department of the Interior.

Public Land Law Review Commission. 1970. *One Third of the Nation's Land: A Report to the President and Congress.* Washington, DC: U.S. Government Printing Office.

Robert B. Keiter

U.S. DEPARTMENT OF AGRICULTURE

The act creating the U.S. Department of Agriculture (USDA) was signed by President Abraham Lincoln on May 15, 1862. In the same year Lincoln signed the Homestead Act and the Morrill Act, which created the land-grant-college system. These three acts had a significant bearing on the mission and methods of the department over the years.

The Morrill Act provided grants of federal land to states to establish colleges emphasizing agricultural and mechanic arts. The land-grant colleges developed strong research and educational ties to the agricultural communities in their states and to the USDA. Later, state-oriented systems of extension and secondary education were created to complement the university and USDA research efforts.

The USDA evolved from the U.S. Patent Office to become a research and education agency with a natural affinity to the state land-grant universities, in keeping with the mandate of the act that created the Department of Agriculture: "to acquire and diffuse . . . useful information on subjects connected with agriculture in the most general and comprehensive sense of that word, and to procure, propagate, and distribute among the people new and valuable seeds and plants" (Rasmussen 1975, Vol. 1, p. 614).

Agriculture in the United States accounts for over 933 million acres, 40 percent of America's land area. Virtually all cropland is privately owned, as is three-fifths of grassland pasture and range. Add the national forests and grasslands, which come under purview of the USDA Forest Service, and the USDA is directly involved in the land use of over 50 percent of the nation's land area. Its efforts in research, education, and cooperative extension service education indirectly affect resource use in other domains. For example, wetlands restoration affects the population size and diversity of wildlife, and wildlife can impact the income of farmers and farmland owners. Wildlife health and wellbeing is of concern to the USDA's Animal and Plant Health Inspection Service (see sidebar) and other agencies, notably the departments of the Interior and of Health and Human Services.

The USDA relates directly and indirectly to farmers; landowners; the food, fiber, and timber industries; recreationists; environmental interests; and others in rural America. The programs of the USDA are clustered under the management of seven undersecretaries: those for natural resources and environment; for farm and foreign agricultural services; for rural development; for food nutrition and consumer services; for food safety; for research, education, and economics; and for marketing and regulation. Under these clusters are many programs administered through its agencies, such as the following:

THE ANIMAL AND PLANT HEALTH INSPECTION SERVICE

The Department of Agriculture's Animal and Plant Health Inspection Service, through a system of regulations, permits, and controls, protects the health and welfare of the nation's agriculture and natural resources. Globalization and transportation has increased the exposure of America's people, animals, and plants to disease, bioterrorism, invasive or destructive species, and other threats. Since 1972, the animal- and plant-protection functions of the USDA have been consolidated under a single agency, the Animal and Plant Health Inspection Service, which monitors the condition of domestic and wild animals and plants, regulates international transportation of plants and animals, and evaluates and controls the welfare of animals in human care. (See the Web sites of the Animal and Plant Health Inspection Service and of the National Agricultural Library's Animal Welfare Information Center.)

The Farm Service Agency, through its 2,346 state and county offices, is the USDA's local interface with farmers and landowners through programs affecting production management, soil conservation, and agricultural markets and finance. The agency produces and stores aerial photography and land records; administers programs on crop production, storage, finance, and insurance; provides disaster assistance and insurance; and gives technical advice and support for conservation and environmental projects.

The Food and Nutrition Service claims, "No one should go hungry in America." It administers programs affecting about one in five people, from school children to the elderly. Key operations include supplemental food for women, infants, and children; the National School Lunch Program; Food Stamps for low-income persons; child and adult food care; food assistance for disaster relief; and related research and education.

The Food Safety and Inspection Service assures the safety, quality, and labeling of meat, poultry, and eggs in the nation's food supply through research, testing, and inspection. The federal-inspection program is allied with state programs and coordinates on standards with international organizations. Under the Organic Foods Production Act of 1990, the Food Safety and Inspection

Service is responsible for standards and labeling in the rapidly expanding organic meat and produce sector.

The Natural Resource Conservation Service (formerly the Soil Erosion Service, then the Soil Conservation Service) was formed during the drought and economic depression of the 1930s. The principal land-use-management agency of the USDA, it is a source of information and support for environmental improvement. The Natural Resource Conservation Service provides technical information and council through its Conservation Technical Assistance program. It administers resource programs such as the Wetlands Reserve Program and the National Resources Inventory, and provides technical support for resource programs in other agencies, such as the Conservation Reserve Program in the Farm Service Agency. (For an overview of conservation and other programs of the USDA, see its Web site.)

The USDA's knowledge base is rooted in the research programs of the Agricultural Research Service, the Economic Research Service, the National Agricultural Statistics Service, and the Cooperative State Research, Education, and Extension Service. Partnering with the research, education, and extension establishment in the USDA is the National Agricultural Library, with its agricultural and natural-resource collections and its information centers for alternative-farming systems and for animal welfare.

SEE ALSO *Agriculture; Animal Ethics; Environmental Education; Farms; Food; Food Safety; Forests; Hunger; Resource Management; Soils; U.S. Forest Service; Wetlands.*

BIBLIOGRAPHY

Baker, Gladys; Wayne Rasmussen; Vivian Wiser; and Jane Porter. 1963. *Century of Service: The First 100 Years of the United States Department of Agriculture*. Washington, DC: U.S. Department of Agriculture.

Rasmussen, Wayne. 1975. *Agriculture in the United States: A Documentary History*, 4 vols. New York: Random House.

Wiebe, Keith, and Noel Gollehon, eds. 2006. *Agricultural Resources and Environmental Indicators*, 2006 edition, EIB–16. Washington, DC: U.S. Department of Agriculture.

Gene Wunderlich

U.S. DEPARTMENT OF THE INTERIOR

Established in 1849 to oversee the young nation's internal affairs, the U.S. Department of the Interior (DOI) has evolved into the federal government's principal land-management and conservation agency. The DOI is responsible for nearly 450 million acres of federal public lands, most of which are apportioned between three agencies: the Bureau of Land Management (BLM), the National Park Service, and the U.S. Fish and Wildlife Service (FWS). In addition, the DOI houses several other agencies, including the Bureau of Reclamation, U.S. Geological Survey, Office of Surface Mining Reclamation and Enforcement, and the Minerals Management Service. Since its inception, the DOI has also handled federal relations with the nation's native inhabitants, through the Bureau of Indian Affairs (BIA). With oversight responsibility for these diverse agencies, the DOI does not have a single central mission but rather pursues an assortment of missions, some with a strong resource-development agenda and others with a much different preservationist agenda.

EARLY EVOLUTION OF THE DOI

In the beginning, the DOI had even more diffuse responsibilities. Congress created the DOI by consolidating several unrelated domestic bureaus into it to unburden the other cabinet agencies. Thus, Interior originally housed the General Land Office, Patent Office, Pensions Office, and Indian Affairs, and oversaw territorial governments and the District of Columbia jail system. Several of these bureaus were gradually incorporated into other newly established cabinet agencies, leaving Interior primarily responsible for the nation's publicly owned lands and resources. In 1905 Congress transferred the new forest reserves from Interior to the Department of Agriculture, thereby establishing the U.S. Forest Service and dividing oversight of federal public lands between these two cabinet departments. Despite periodic proposals to consolidate the four federal land-management agencies into a single Department of Natural Resources, Congress has consistently rejected this idea, leaving the bureaucratic landscape intact.

Within the DOI federal conservation policy has steadily evolved over the past 150 years. Acting through the General Land Office, DOI originally oversaw disposal of the nation's public lands into private ownership under such laws as the Homestead Act of 1862, the Pacific Railroad Act, and the General Mining Law. In 1872, however, Congress passed the Yellowstone Act and retained nearly 2 million acres of this unique landscape in public ownership as a "public park or pleasuring ground for the benefit and enjoyment of the people," thus opening a new preservationist chapter in federal land policy. (Runte 1987, p. 46). In 1891 Congress authorized creation of the forest reserves, which has led to the retention of more than 150 million acres of public forest lands in federal ownership. By the beginning of the twentieth century, under the leadership of President Theodore

Roosevelt and Gifford Pinchot, federal policy had shifted away from disposal of publicly owned lands and resources to retaining and managing them under the new banner of "conservation"—a Pinchot-inspired doctrine holding that the public interest was better served by government ownership and scientific management of the nation's natural resources for utilitarian goals.

This new conservation philosophy took hold, and the DOI was soon freighted with several new professional bureaus responsible for implementing these new policies. In 1902 Congress created the Bureau of Reclamation to provide water resources for the arid West through the construction of federally funded dams and aqueducts. In 1905 Congress established the U.S. Forest Service to manage the new national forest system, but placed it in the Department of Agriculture. In 1910 the Bureau of Mines was established to promote mine safety and mineral technology. In 1916 Congress passed the National Parks Organic Act to consolidate the existing national parks and monuments into a national park system and to create the National Park Service to manage them under a strong preservation mandate. During the dust-bowl era of the 1930s, Congress adopted the Taylor Grazing Act, giving the DOI regulatory oversight of livestock-grazing practices on the public-domain lands. In 1940 the U.S. Fish and Wildlife Service was created to manage the growing national wildlife refuge system. In 1946 Congress merged Interior's General Land Office and the Grazing Service to create the Bureau of Land Management to oversee the unreserved public lands, those federal lands that were not under the aegis of the other conservation agencies.

POST-WAR

Congressional legislation since World War II has further muddled the DOI mission and aggravated the latent tensions between its resource-development and preservation-oriented agencies. The 1964 Wilderness Act not only established the national-wilderness preservation system, but it also instructed the Park Service, Forest Service, and FWS to review their undeveloped lands for possible inclusion in the system. During the 1970s a plethora of new environmental laws, including the Clean Water Act, National Environmental Policy Act, and the Endangered Species Act, required all federal agencies to take into account the environmental consequences of their actions before undertaking any action. In 1976 the Federal Land Policy and Management Act gave the BLM a new organic charter that provided for multiple-use management through an interdisciplinary resource-planning process and new wilderness review and management responsibilities. In 1980 Congress passed the Alaska National Interest Lands Conservation Act, which transferred nearly 60 million acres of Alaskan public lands into national-park, wildlife-refuge, or wilderness status.

The result for the DOI agencies has been an increasingly overt tension between preservation and utilitarian management policies. This conflict is evident in such settings as the Arctic National Wildlife Reserve, which has long been coveted by oil companies for its petroleum potential while environmental interests have sought to protect it from industrial development. Another example of this tension can be found in the BLM. Under the leadership of Secretary of the Interior Bruce Babbitt (1993–2001), the BLM was given responsibility for fourteen new national monuments as part of a new National Landscape Conservation System. But Babbitt's successor, Secretary Gale Norton (2001–2006), moved oil and gas exploration across the interior West to the forefront of its agenda.

One effort to reconcile these divergent conservation philosophies has come in the form of new ecosystem-management policies designed to achieve ecologically and economically sustainable resource-management goals at the regional or landscape scale. The efficacy of these policies is still being tested. Meanwhile, the Interior agencies are pursuing their individual missions under a welter of sometimes conflicting laws and policies, leaving the DOI a pastiche of different agencies and often clashing conservation priorities.

SEE ALSO *Environmental Law; Environmental Policy; Pinchot, Gifford; U.S. Department of Agriculture; U.S. Fish and Wildlife Service; U.S. Forest Service.*

BIBLIOGRAPHY

Clawson, Marion. 1983. *The Federal Lands Revisited.* Baltimore, MD: Johns Hopkins University Press.

Public Land Law Review Commission. 1970. *One Third of the Nation's Land: A Report to the President and Congress.* Washington, DC: U.S. Government Printing Office.

Runte, Alfred. 1987. *National Parks: The American Experience.* Lincoln: University of Nebraska Press.

Utley, Robert M., and Barry Mackintosh. 1989. *The Department of Everything Else: Highlights of Interior History.* Washington, DC: Department of the Interior.

Robert B. Keiter

U.S. ENVIRONMENTAL PROTECTION AGENCY

The United State Environmental Protection Agency (EPA) was formed in 1970. As a dynamic, ever-changing, and powerful federal agency, the EPA often faces ethical choices on a range of issues, such as the mandatory cleanup of contaminated sites and assigning liability without culpability, assessing the varying impacts of environmental regulation by race and income, whether or how to regulate nanotechnology, how to pursue and define social preferences for sustainability, and how to navigate global environmental issues like climate change. The EPA must deal with a broad range of factors in its decision making: public involvement and participation, land use, cumulative environmental impacts, the institutional ethics of economic development, legal constraints, and risk assessment processes only partially based on science.

HISTORY

Prior to the founding of the EPA, other federal agencies handled parts of some environmental issues. Their authority was often unclear, inconsistent, and sometimes conflicting. Fifteen federal agencies or parts of federal agencies handled some aspects of some environmental problems. Some states were initiating their own environmental agencies, and often-conflicting sets of regulatory procedures were beginning to emerge. There was a growing recognition of the severity, national scope, and political urgency of environmental issues.

The EPA is a creature of many federal statutes and rules. In 1970 President Nixon signed Executive Order 1110.2 which began the process of forming this new, powerful federal agency. By 1979 at least twenty-seven new environmental laws had been passed. The first waves of national environmental regulations were solidified with the passage and implementation of the National Environmental Policy Act (NEPA), the Resource Recovery Act, and the Clean Air Act in the early 1970s.

The EPA is still a relatively young and very powerful federal agency. It was designed to provide independent and objective information about the environment and to develop rules and regulations to protect it. It was also designed to have a wide purview, covering environmental issues pertaining to air, water, and land. Because of the novelty of national environmental regulations, the EPA was charged with examining new policies, assisting Congress with legal advice, and mediating political controversies.

DEVELOPMENT OF THE AGENCY

The EPA's initial tasks were massive. With every new environmental law, the EPA had to develop and enforce rules and regulations. It also had to be ready to litigate to defend these new laws and to enforce them. Many new environmental laws, such as the Clean Air Act and Clean Water Act, contain provisions that allow citizens to sue the polluter and the EPA if the EPA fails to enforce the law. Environmental groups have successfully litigated against the EPA many times on issues that lie at the foundation of modern environmental policy. For example, litigation by environmental-advocacy organizations forced the EPA to develop clean air standards.

During its first twenty years the EPA reacted to environmental issues and controversies emerging from court cases and from Congress, setting and enforcing environmental regulations that were nationally uniform and legally defensible. The early years were also notable for the cadre of dedicated senior EPA staff working to improve the presence and acceptance of the EPA from industry, states, Congress, and the public generally.

The relationship of the EPA to the states is an evolving aspect of environmental federalism. Revenue allocations flow from EPA headquarters to the ten EPA regions. Each region has from three to five states and territories within its jurisdiction. From the EPA regional offices revenue is distributed to state environmental agencies. Because many states are historically hostile to environmental regulations, EPA revenues usually make up a substantial part of many state environmental-agency budgets. States are usually free to run their programs as long as they comply with the minimum EPA requirements, although exceptions abound. If states do not want to enforce or accept a particular national program, then EPA will run it for them. States usually prefer not to lose control of regulating the environment.

The dynamic of intergovernmental relations in environmental regulatory regimes in the United States is rapidly evolving to include land-use practices that affect the environment. Ecosystems and bioregions transcend political boundaries and require cooperative national action. Protecting and preserving the environment also requires grassroots implementation at the community land-use level, a challenge not yet met by the EPA.

CHANGING MANDATES AND MERCURY MISSIONS

As the EPA has grown and evolved, so have the number of environmental laws and regulations. Often they have emerged at the EPA as part of an agency mission. Many of the early activities of the EPA involved seemingly distinct concerns about air, water, land, and solid wastes. But the environment is, in reality, multifaceted and complexly integrated. Pollution can move through an ecological cycle of land, air, and water. Along the way, it can accumulate in parts of an ecosystem and do damage to the entire system. For example, metals like mercury may accumulate in the fat and in the nervous system of fish. People who rely on these fish for food are exposed to this mercury. According to a 1969 study, the amount of heavy-metals pollution in the environment was so high that it infected mothers' milk and babies' bones to such a degree that if mothers' milk were a commercial food item, the U.S. Food and Drug Administration would not have approved it.

The degree of danger from chemicals like DDT or metals like mercury often constitutes a separate set of controversies tied closely to concerns about liability for the damages from such exposures. There is some disparity in the degree of risk from various industrial pollutants that different cultures and communities are willing to accept, making uniform standards difficult to determine. Further, some of the cultures and communities that do not accept dangerous industrial facilities in their midst and/or require cleanup of such to residential standards of living are able to readily obtain legal remedy. Others, however, are forced into a series of "scientific" debates about the reasonability of their concern about risk. These communities end up with greater exposure to environmental risks, slower cleanup of hazardous waste sites, lower compensation and fines for environmental harms inflicted on them, and cleanups performed only to industrial, not residential, standards. These lower cleanup standards are less expensive and facilitate the sale of property. But meeting such minimal standards also keeps a higher amount of waste in those communities where such standards are imposed and makes them inappropriate for residential uses and uses for facilities for vulnerable populations like schools and hospitals

1970S: NEW LAW, RULES, AND REGULATIONS

The EPA's first public priorities were environmental protection and protecting public health, without unduly burdening the national economy. This set of priorities translated, in practice, to controlling pollution and dangerous chemicals. Standards for clean air and water were also developed and refined during the 1970s. The EPA was often sued by industries if such standards seemed too far reaching and the EPA was sued by environmentalists when it was seen to move too slowly in the development of standards or the enforcement of them.

1980S: CLEANUP

Polluting industries were not well regulated until the early 1970s, and even then regulations only partially covered some industries. Thus some egregious waste sites created before regulation was initiated or by some unregulated industries became of national concern. Mandatory cleanup laws were developed and enforced, often through the courts. As a result the Comprehensive Environmental Response, Compensation, and Liability Act (CERCLA), also known as Superfund, was enacted in 1980 to, among other reasons, identify potentially responsible parties (PRPs) for the waste-site cleanup. The law required the creator, shipper, or property owner to clean up the "Superfund" site—defined as one that was egregiously polluted. Once the PRPs are found, they must either do an adequate cleanup or pay for cleaning up the site. The EPA can force any one of a group of PRPs to pay because they are jointly and severally liable. That one then may sue the other PRPs if it has to clean up more than the share of the waste for which it was responsible. If there are many PRPs, and some declare bankruptcy or are otherwise unreachable, the secondary litigation can be extensive.

If the cleanup laws required proportional sharing of responsibility for the culpable parties, there would be much less resistance against the EPA's assigning liability for cleanup when there was no other culpability than property ownership. The reason for assigning responsibility to property owners who may not otherwise be culpable is simply that unabated and unmitigated pollution can spread via the air and water to schools, organic crops and fields, and whole communities, while the actual polluters may be hard to find, bankrupt, or corporately dissolved. There is tension between U.S. traditions of private-property rights and freedom on the one hand, and the rigorous and unforgiving requirements of pollution control and abatement on the other. As many communities became concerned about accumulating emissions, impacts on the environment and human health, and other risks from the pollution of the land, air, and water, waste sites became battlegrounds pitting neighbors against one another, neighborhoods against city hall, environmentalists against industrial corporations, and parents against commercial economic development. Liability for environmental cleanup costs have followed the owner of the land, so when a municipality forecloses on a land for failure to pay taxes, the municipality may be liable for the environmental cleanup costs. On the other hand, if the land remains abandoned, it does not generate tax revenue or get cleaned up. Thus there is a chance the pollution could migrate off the polluted property and that abandoned, polluted property will suppress the property values of surrounding properties.

In the 1980s the EPA forced the clean up of Love Canal, New York, which was built on top of a chemical company's waste dump and contaminated with carcinogenic chemicals. New York State declared it a health emergency in 1978. It was one of many events that provided the political impetus to the Superfund Act. In 1983 the EPA also relocated the community of Times Beach, Missouri, near St. Louis because of excessive dioxin contamination for using waste industrial lubricants to oil the communities' roads which was then spread over the soil by floodwaters.

In the early 1980s the EPA began to respond to the environmental justice movement. Distinct from any mainstream U.S. environmental organization, environmental justice groups were grassroots affairs, predominantly communities of color affected disproportionately by environmental harms. It had been business as usual in the United States to place toxic and noxious facilities near and around low-income and African-American communities. A study by the United Church of Christ performed in 1986 and revisited in 2006 (Bullard et al. 2007) showed that race, more than any other characteristic, indicated the likelihood of the placement of a commercial waste site, regulated or unregulated, near a community. The more African American a community was, and is, the more likely that it has one of these unwanted land uses, to a 99.9 percent certainty or one in ten thousand chance of randomness. With the rise of environmental regulation and citizen monitoring, knowledge of environmental impacts increased. With this knowledge came evidence of unequal access to legal remedy and unequal environmental results that had been previously unaddressed in U.S. environmental policy. The failure of the U.S. environmental movement to include the interests and values of oppressed people of color also contributed to the rise of environmental justice groups. Their motto, which came out of the First People of Color Environmental Leadership Summit, held in Washington, D.C., in 1991, is "We speak for ourselves." They are now formally represented by the Office of Environmental Justice, developed in the EPA during the first decade of the twenty-first century.

Lead was and is a high-priority pollutant for the EPA. Lead comes from vehicle emissions, industrial pollution, old paint in homes and on bridges, water pipes, and many other sources. It can irreversibly impair the nervous system of unborn children and children up to the age of five. It is associated with several types of cancers and other dangerous health risks. In 1985 the EPA set new limits on lead in gasoline because of air pollution. This has been one of the most successful programs implemented by the EPA. Lead emissions from motor vehicles were reduced, and lead levels in the U.S. population decreased. This measure greatly enhanced the EPA's credibility as a regulator of corporate activities in the interests of public health.

ENVIRONMENTAL INFORMATION AVAILABLE TO COMMUNITIES

Late in the 1980s the EPA began a push to make information about hazardous chemicals and pollution more available to the public. The primary concern then was with first responders to emergencies, such as fire and police personnel. With the Emergency Planning and Right to Know Act in 1986, industries that emitted over a certain threshold of chemicals from a short list of 300 had to report them annually. That list now includes over 650 chemicals. However, there are more than 80,000 chemicals used in U.S. commerce, and less than 2 percent of them have been checked for safety. Many states and some municipalities are now developing their own right-to-know laws. These often mimic the federal law. These laws have been criticized because the information is often self-reported by industry and because only when emissions exceed a certain threshold are industries required to report them. Not all industries are included. For example, universities and colleges emit significant chemicals without reporting requirements. Also, many claim that not all chemicals with significant environmental impacts are included. Despite these serious shortcomings, these laws have been a tremendous organizing tool for environmentalists and environmental justice advocates.

1990S: PREVENT THE POLLUTION

The 1990s saw environmental mandates shift to pollution prevention. The decade was inaugurated in 1990 with a major amendment to the Clean Air Act of 1963. Some 30 million tons of toxic chemicals were prevented from going into the air in the early years of its implementation. Food quality, second-hand smoke, and safe drinking water also dominated the environmental policy development at the EPA. Concern for environmental justice continued to grow in prominence, and EPA began to study issues of sustainability.

TWENTY-FIRST CENTURY: POLICIES IMPLEMENTED, GLOBAL CONCERNS LOOM

During the first decade of the twenty-first century many of the pollution-prevention, -control, and -regulation policies established over the last quarter of the previous century continued to be implemented. Like most federal agencies, the EPA responded to the 9-11 bombing of the World Trade Center towers in 2001. The EPA issued a strategic plan for homeland security in 2001 and updated it in 2004. However, because the United States did not endorse many important world environmental treaties such as Kyoto, it is difficult for the EPA to coordinate environmental policy with other countries. It is also difficult for international U.S. businesses which must comply with these global concerns in order to operate in other countries and compete internationally.

THE EPA TODAY

The strength of the EPA is its credibility here and abroad. Its information, monitoring, research, and power to control behavior through regulation in the face of unknown contingencies contribute to its credibility. Some of the vulnerabilities of the EPA stem from its lack of inclusion of all affected communities and from controversies that surround its enforcement policies. The primary enforcement theory at the EPA is compliance, not general or specific deterrence. For example, if the accused polluter simply confesses, the EPA applies a "reduction of gravity rule" to reduce the monetary penalty by 50 to 75 percent right off the top. Citizens are often excluded from EPA decision making, even when it affects them. For example, in a federal environmental-impact statement, citizens are not given notice of a preliminary environmental assessment of their neighborhood until after it has been determined, based on this preliminary assessment, whether there are significant environmental impacts that require a full environmental-impact assessment be done or not. Citizens who may care, may know, or may simply want to participate are not given an opportunity until the process is far advanced.

Because the EPA is tied tightly to the specifics of each piece of legislation setting environmental policy and mandating its responsibilities, it is difficult for it to strategically plan for future environmental contingencies. This uncertainty makes it difficult for it to work with new, non-legislated ways of approaching environmental controversies, issues, and problems, such as the rising social concern for sustainability. However, the EPA is adaptable to changing political environments. It is exploring supplemental environmental projects (SEPs), begun in the late 1990s, so that convicted polluters can mitigate the damage they have caused in a community. By exploring collaborative, multistakeholder SEPs—SEPs with more than two stakeholders, usually community, industry, environmental, and sometimes labor organizations—the EPA is moving beyond the mandated citizen participation of the 1970s and 1980s and into citizen involvement for long-term environmental planning.

SEE ALSO *Environmental Justice; Environmental Law; Pollution; Risk Assessment.*

BIBLIOGRAPHY

Bullard, Robert D., Pul Mohai, Robin Saha, and Beverely Wright. 2007. *Toxic Wastes and Race at Twenty: 1987–2007.* Atlanta: United Church of Christ.

Collin, Robert W. 2006. *The Environmental Protection Agency: Cleaning Up America's Act.* Westport, CT: Greenwood Press.

Collin, Robert W. 2008. *Battleground: Environment.* Westport, CT: Greenwood Press.

Mintz, Joel A. 1995. *Enforcement at the EPA: High Stakes and Hard Choices.* Austin: University of Texas Press.

Robert Collin

U.S. FISH AND WILDLIFE SERVICE

In the lands that later became the United States, early European settlers encountered an apparently unlimited supply of wildlife and other natural resources and often had utilitarian, negativistic, and dominionistic views of wildlife. They embarked on a process of wildlife slaughter and habitat destruction, including deforestation, on a mass scale. That process continued through much of the nineteenth century and caused the local or regional extirpation of many species, including most large predators, white-tailed deer, elk, turkey, waterfowl, and the American bison. The first conservation efforts in the United States occurred at the state level because wildlife was considered a public resource held in trust by the states for the benefit of all the people; this was known as the public trust doctrine. However, individual states struggled to protect wildlife species that crossed state and national boundaries or were in the possession of commercial hunters and trappers.

Vermont was one of the first states to face a landscape stripped of natural resources. Deforestation in that state influenced George Perkins Marsh's *Man and Nature* (1864), which suggested that societal collapse would follow environmental degradation, and led Gifford Pinchot, the first chief of the U.S. Forest Service, to craft an anthropocentric and utilitarian ethic for conservation in the United States. That ethic defined wise use of natural resources as generating "the greatest good for the greatest number for the longest time" (Pinchot 1947, pp. 325–326) and provided the context for the emergence of the U.S. Fish and Wildlife Service (USFWS).

EARLY HISTORY OF THE USFWS

The roots of the USFWS can be traced to 1871, when Congress created the U.S. Commission on Fish and Fisheries to protect rapidly disappearing food fish stocks. A parallel effort to study the food habits of migratory birds led to the establishment of the Office of Economic Ornithology in the Department of Agriculture in 1885. The U.S. Commission on Fish and Fisheries was renamed the Bureau of Biological Survey in 1905 and was given management responsibility for the U.S. wildlife refuge system that started with Theodore Roosevelt's establishment of the first Federal Bird Reservation on Pelican Island, Florida, in 1903. In 1939 the Fisheries and Biological Survey bureaus were transferred to the Department of the Interior.

In 1940 those bureaus were combined into the Fish and Wildlife Service, and in 1956 the Fish and Wildlife Act created the USFWS. During the early years the responsibilities of the USFWS reflected its utilitarian roots: It focused on enforcing the law, regulating trade, and conducting research. Those activities were intended to protect and manage game species and minimize conflicts between wildlife and agriculture. The utilitarian approach to conservation was largely responsible for the restoration of elk, white-tailed deer, the American bison, turkey, bear, and many other species throughout the United States.

CHANGES IN THE ROLE OF THE USFWS

The role of the USFWS changed drastically in the 1960s and early 1970s, when the environmental and animal rights movements began to flourish. During that period the environmental ethics guiding wildlife conservation diverged from the early anthropocentric and utilitarian focus articulated by Pinchot. Aldo Leopold's *Sand County Almanac* (1949) helped promote an ecocentric ethic—the land ethic—for wildlife conservation. The more ecocentric ethical views prevalent in the environmental movement played a role in the establishment of the Endangered Species Act (ESA) of 1973. The ESA prohibited the importation, exportation, taking, or possession of a registered endangered species and gave USFWS responsibility for listing of, recovery planning for, education about, and delisting of species.

The ESA protected animal and plant species without consideration of the economic, cultural, and social preferences of humans. The ecocentric approach to wildlife conservation taken by USFWS was responsible for some notable conservation successes, including the delisting of the American alligator, gray wolf, grizzly bear, and bald eagle. However, the tendency to subordinate human interests to the protection of listed endangered species led to conflicts between the USFWS and several rural communities. Conflicts surrounding the spotted owl in the Pacific Northwest and the coho salmon in the Klamath Basin in Oregon were among the most publicized. Those conflicts raised questions about how to integrate the interests of stakeholders at the local, regional, and national levels. In some cases the conflicts threatened wildlife management objectives by alienating private landowners. Critics of the ecocentric focus of the USFWS argued that fear of ESA-related restrictions was a disincentive for wildlife conservation on private land and might motivate landowners to get rid of threatened species before restrictions on property uses were mandated. Landowners in some areas did restrict access to their land in efforts to prevent the discovery of endangered species.

Fish and Wildlife Service Officer with Illegal Trade Items. *The responsibilities of the USFWS during their early years involved enforcing the law, regulating trade, and conducting research. Today, despite a period of a more ecocentric approach that emerged in the 1960s and 1970s, the USFWS continues to take an utilitarian approach to wildlife preservation and conservation.* **PHOTO BY CARL ZITSMAN/U.S. FISH AND WILDLIFE SERVICE.**

REESTABLISHMENT OF THE UTILITARIAN PERSPECTIVE

In the mid-1990s the ethical perspective of USFWS started moving back toward anthropocentric utilitarianism. Unlike the earlier shift to ecocentrism, this move was tied to presidential influence over the USFWS rather than to a national social movement. In the 1992 presidential race Bill Clinton promised to move the country beyond a false choice between environmental protection and economic growth. His administration pushed the Habitat Conservation Planning process to achieve that objective. Habitat Conservation Plans provided a loophole in the absolute rule of not harming endangered species. With an approved Habitat Conservation Plan, landowners, government entities, and corporations could kill individuals from a population of endangered species under the protection of incidental take permits. The 14 incidental take permits issued before the 1992 presidential race paled in comparison to the 425 approved as of July 2003.

The shift toward anthropocentric utilitarianism in USFWS operations expanded with the 1995 Safe Harbor Program and the 2007 Endangered Species Recovery Credits system. The Safe Harbor Program exempts landowners from further restrictions on their land when they agree to manage the land for endangered wildlife, and the

Credits system allows private and public entries to harm endangered species in exchange for purchasing conservation credits that require proper habitat management elsewhere.

This move back to anthropocentric utilitarianism has increased the ability of the ESA to accommodate human interests but could weaken protection for endangered species. Critics argue that sidestepping the exclusive focus on biotic integrity will allow human interests to displace the needs of other species.

SEE ALSO *Conservation; Environmental Law; Forests; Habitat Loss; Leopold, Aldo; Resource Management; Utilitarianism.*

BIBLIOGRAPHY

Carson, Rachel. 1962. *Silent Spring.* Boston: Houghton Mifflin.

Kellert, Stephen R., and Edward O. Wilson, eds. 1993. *The Biophilia Hypothesis.* Washington, DC: Island Press.

Leopold, Aldo. 1949. *A Sand County Almanac, and Sketches Here and There.* New York: Oxford University Press.

Marsh, George P. 1864. *Man and Nature; or Physical Geography as Modified by Human Action.* New York: C. Scribner.

Peterson, Tarla R., and Cristi C. Horton. 1995. "Rooted in the Soil: How Understanding the Perspectives of Landowners Can Enhance the Management of Environmental Disputes." *Quarterly Journal of Speech* 81(2): 139–166.

Pinchot, Gifford. 1947. *Breaking New Ground.* New York: Harcourt, Brace.

M. Nils Peterson
Aimee Rockhill
Christopher S. DePerno

U.S. FOOD AND DRUG ADMINISTRATION

The U.S. Food and Drug Administration (FDA) is responsible for protecting the public health by assuring the safety, efficacy, and security of human and veterinary drugs, the nation's food supply, medical devices, cosmetics, and several other products. The FDA is also responsible for advancing the public health by helping to speed innovations that make medicines and foods more effective, safe, and affordable (FDA Mission Statement). In the area of food safety, the FDA executes the Federal Food Drug and Cosmetic Act (FFDCA) (21 U.S.C§§ 301–399) by setting standards for food and food products, inspecting food production and distribution facilities, and ensuring proper labeling. In the area of animal health, the FDA is responsible for regulating the manufacture and distribution of food additives used in animal feed and of drugs that will be given to animals.

Under this diverse portfolio of products and activities, the FDA also has a mandate for environmental protection. The FDA is required under the National Environmental Policy Act (NEPA) (21 C.F.R. Part 25) to take environmental considerations into account in all final agency actions. For example, during the review of animal drugs under FFDCA, the FDA considers the excretion of drugs in animal waste and the effects of drug residues on the environment. NEPA requires that U.S. agencies include an environmental impact statement (EIS) with every major federal action that significantly affects the quality of the human environment. Environmental assessments (EAs) are prepared to help determine if an action will have a significant impact on the environment and whether an EIS is required.

The FDA is often faced with situations in which goals of human and animal health protection, speeding innovation, and environmental protection are at odds. Decisions under NEPA do not require that the action most beneficial to the environment be taken. For example, the FDA might be faced with a policy choice about whether to accept environmental harm from its actions to protect human or animal health or to make foods safer or more affordable. A contentious instance of this dilemma is the approval of the subtherapeutic use of antibiotics in feed to promote animal growth during agricultural production. Environmental and consumer groups have argued for a precautionary approach to limit or ban the use of antibiotics in animal feed. Concerns include the harmful effects of antibiotic residues on native species in the environment and the increased development of resistant microorganisms that cause disease.

The ethical principles of beneficence (doing good) and nonmaleficence (doing no harm) are prominent when the FDA makes decisions in the face of competing goals or interests. The distribution of risks and benefits to various stakeholder groups—the environment, animals, and humans—is an important consideration. Equity in decision making comes into play when the FDA considers this distribution. Integrity, autonomy, and justice are also prominent in how the agency makes decisions. For example, transparency in decision making, avoiding conflicts of interest in conducting safety studies, and giving consumers or users of products the right to know and choose based on good information are manifestations of these principles in the FDA's regulatory context.

An example of an emerging issue that spans environmental protection and the FDA's jurisdiction is the agency's proposal to regulate genetically engineered animals as new animal drugs (NADs) under the FFDCA. The claim by the agency is that the introduced and engineered gene is the "drug," because it alters the structure or function of the body of animals. The FDA has been considering regulatory approval of genetically engineered (GE) fish with growth-promoting genes since the late 1990s. This decision has been stalled in part because of continuing controversy over genetically engineered organisms in food and agriculture. Environmental and consumer groups and scientists are concerned about the environmental risks of GE fish, including adverse effects on native populations, relatives, or other species of predators or prey from the introduced fish if they were to escape or were intentionally released. Many in the aquaculture industry are excited about the potential of GE fish to resist disease, grow faster, or have fewer resource needs (thus improving the environment). As the regulatory agency with primary responsibility for GE animals, the FDA will need to carefully consider its authority in this area under NEPA.

The FDA will also be faced with considering ethical principles in its decision making about GE animals. Federal agencies like the FDA are asked to weigh the risks, benefits, and costs of their decisions (to use, in effect, a utilitarian ethical framework) to comply with the order of the Executive Office of the President (1993). However, procedural justice, autonomy, inherent objections on fundamental grounds ("playing god"), and the integrity of the regulatory approval system are also prominent in GE animal oversight, as reviewed by Thompson (2007).

SEE ALSO *Animal Ethics; Environmental Policy; Food Safety; Genetically Modified Organisms and Biotechnology; Risk Assessment; U.S. Department of Agriculture; U.S. Environmental Protection Agency.*

BIBLIOGRAPHY

Executive Office of the President. 1993. *Regulatory review and planning, Executive Order 12866. Federal Register* 58 (190): 51735–51744. Available from http://www.whitehouse.gov/news/releases/2007/01/20070118.html

Thompson, Paul. 2007. "Animal Health and Welfare." In *Food Biotechnology in Ethical Perspective.* 2nd edition, ed. Paul Thompson. Dordrecht, Netherlands: Springer, 121–146.

U.S. Food and Drug Administration. "Mission Statement." Available from www.fda.gov/opacom/morechoices/mission.html.

Jennifer Kuzma

U.S. FOREST SERVICE

The United States Forest Service (USFS) administers 155 national forests and twenty national grasslands—some 193 million acres of land—in forty-four states, Puerto Rico, and the Virgin Islands. The USFS was established in 1905 within the U.S. Department of Agriculture to administer the nation's newly created national forests and the related forestry programs of the federal government. The creation of the USFS was a key event of the Progressive-Era conservation movement as it gained definition and broad public and political support during the presidency of Theodore Roosevelt (1858–1919, president 1901–1909). Since then the USFS has played an important role in the evolution of conservation policy, science, and practice; it has been both a generator and barometer of changing environmental values and ethics. A major force in shaping American natural-resource management, it has had a broad influence on forestry and environmental policy internationally, at other levels of government, and in the private and nongovernmental sectors.

ORIGINS AND EVOLUTION OF THE USFS

By the mid-1800s, widespread clearing of the forests of the eastern United States for farming and for timber, fuel, and other forest products had begun to raise concerns among protoconservationists. In his classic *Man and Nature; or, Physical Geography as Modified by Human Action* (2006 [1864]), George Perkins Marsh critically examined deforestation and its impact on climate, soil conditions, hydrologic dynamics, watershed function, and forest plant and animal life. Correcting the "terrible evils" of deforestation, Marsh wrote, was among "the most obvious of the duties which this age owes to those that are to come after it" (p. 279). Through his writing and his advocacy Marsh influenced those who later created the USFS, echoing his mandate to "care for the moral and material interests of our own posterity" (p. 279).

In the three decades following the Civil War, the rapid development of the mid-continent's cities, towns, and farms, along with the advent of new sawing, milling, and transportation technologies, led to the near-complete removal of the white pine forests of the upper Great Lakes. As these forests were depleted, devastating slash fires often followed in the wake of the loggers and fed growing fears of a "timber famine."

As public attitudes and political sentiment began to shift, the federal government took its first concerted actions on behalf of forest conservation. In 1876 Congress appointed Franklin B. Hough (1822–1885) as a special agent in the Department of Agriculture, charged with assessing the state of the nation's forests. In 1881 Hough's responsibilities were expanded within a new Bureau (later Division) of Forestry, the precursor to the USFS. The Forest Reserve Act of 1891 empowered the president to withdraw forestlands from the nation's public domain and to designate them as forest reserves under the administration of the U.S. Department of Interior.

Over the following fifteen years 100 million acres of forest reserves were created, primarily in the western states and territories. The Organic Act of 1897 directed the Secretary of Interior "to make such rules and regulations and establish such service as will insure the objects for which forest reservations are created..." (U.S. Congress 1897). Those "objects" were "to protect and improve the forests for the purpose of securing a permanent supply of timber for the people and insuring conditions favorable to continuous water flow" (U.S. Congress 1897). Under the Transfer Act of 1905, the reserves were renamed national forests. The Transfer Act shifted responsibility for their administration to the Department of Agriculture, where the old Division of Forestry was reassigned and rechristened as the U.S. Forest Service.

These developments exposed latent philosophical tensions in the emerging national conservation movement. The change in nomenclature from *forest reserves* to *national forests*, with the implication that the nation's forest estate was not to be set aside but used, revealed the movement's clashing values. Utilitarian conservationists (often identified with Roosevelt's "chief forester," Gifford Pinchot) stressed the instrumental value of the nation's forest resources and the need for efficient, scientifically informed forest management. In its classic formulation the USFS administration of the national forests was to serve "the greatest good for the greatest number over the long run." It pursued this policy by applying principles of sustained-yield silviculture developed in the European forestry tradition. By contrast, preservation-minded conservationists (often identified with the author and naturalist John Muir) emphasized the intrinsic, aesthetic, and spiritual values of forests and pushed for their protection from the incursions of settlers, loggers, grazers, miners, and dam builders. The clash of these schools played out most famously in the extended conflict over a proposed dam (eventually approved in 1913) on the Tuolumne River in Yosemite National Park's Hetch Hetchy Valley.

The timber values of the national forests dominated USFS management goals and actions through the twentieth century. The service's views and values did not, however, go unchallenged. Through the 1920s and 1930s scientists, foresters, and conservationists, within and beyond the USFS, began to champion a wider variety of forestland assets: water, wilderness lands, wildlife, and recreational opportunities. Aldo Leopold, who joined the USFS in 1909 and spent the first half of his career working in the agency, became a leading voice for recognition of the full spectrum of forest values and for their effective integration into conservation planning. Leopold's appreciation of the implications of ecology for sustainable land management drew on his own early field experience in the national forests of the American Southwest. Leopold's land ethic, by emphasizing the integrity, diversity, and self-renewing capacity of ecological communities—what he termed "land health"—sought to bridge the divergent utilitarian and preservationist strains of conservation thought; all forest values, he held, depended on the healthy functioning of the forest as a whole.

POST–WORLD WAR II DEVELOPMENTS

The end of World War II brought new pressures to bear on the national forests and the USFS. Before the war exploitation of timber resources was concentrated in the nation's private forestlands. After the war the baby boom (and the corresponding construction boom) across the United States increased demand for timber from the national forests. Large-scale industrial forestry operations became the norm, most visibly in the form of expanded clear-cutting. In the USFS narrowly defined economic values overwhelmed ecological rationales for better integrated forest management.

At the same time the more urban and suburban postwar generation took to the national forests in growing numbers for recreation, which became an increasingly important use of forestlands. This shift in turn contributed to a widening public awareness of environmental values and the rise of the environmental movement. Thus, as material and recreational demands on the national forests intensified, so did the movement for protection of roadless wildlands within the forests, culminating in passage of the Wilderness Act in 1964.

In an attempt to reconcile these competing public demands—and by implication the varied interpretations of the "greatest good"—the USFS followed the mandate of the Multiple Use and Sustained Yield Act of 1960 (MUSY). MUSY directed that the national forests be "administered for outdoor recreation, range, timber, watershed, and wildlife and fish purposes" (U.S. Congress 1960). MUSY stated that all of these uses should be accorded equal importance and directed the USFS to "[achieve] and [maintain]...a high-level regular output of the renewable resources of the national forest without impairment of the land's productivity" (U.S. Congress 1960). MUSY's goals proved difficult to realize. In the absence of any robust set of common values or shared commitment to land health, multiple competing interests continued to strive for predominance in national forest management, and the USFS remained a magnet for political controversy.

By the late 1980s an array of intractable forest management issues were playing out on the national stage: forest fragmentation, with its the attendant impacts on biological diversity; delineation and protection of the remaining roadless areas in the national forest system;

The Northern Spotted Owl. *In 1986 the U.S. Forest Service initiated steps to protect the spotted owl from extinction by limiting timber sales in mature portions of National Forests, the species' natural habitat. This sparked a lengthy and heated controversy between environmentalists and the timber industry when measures to protect the spotted owl may have led to layoffs and decreased business for loggers and mill workers. In 2008 the spotted owl continued to be classified as a threatened species.* PHOTO BY JOHN AND KAREN HOLLINGSWORTH/U.S. FISH AND WILDLIFE SERVICE.

the impacts of heavy grazing on forests, grasslands, watersheds, and riparian communities; disruption of historic fire regimes because of a century of fire suppression and postwar development at the urban/wildland interface; intensified recreational demands on forests; and local timber economies that were faltering because of mechanization, dislocation, and the impacts of the globalizing economy. Caught between the intensified scrutiny of the environmental movement (and especially the deep ecology–inspired Earth First! Movement) on the one hand and the property rights–focused wise-use movement on the other, the USFS (and other public land agencies) found itself torn between the opposing forces of forest utility and forest protection. The struggle in the late 1980s and early 1990s over the fate of the northern spotted owl and the logging economy in the national forests of the Pacific Northwest was emblematic of the broader effort to overcome conservation's philosophical divide. Out of this complex set of circumstances, foresters and other resource managers began to revisit their philosophical foundations and to search for a new management paradigm. In the USFS this trend was reflected by the formation, in 1989, of the Forest Service Employees for Environmental Ethics (FSEEE). This self-criticism within the forest service, and within professional forestry more generally, was evidenced in other land-management agencies and in the natural-resource management professions. One result was the emergence of *ecosystem management* as a potentially more integrated and unified approach to land stewardship. Embracing a broader range of intrinsic and instrumental values, recognizing the dynamic nature and scale-dependent processes of ecosystems, and acknowledging the active and mutual influence of human and natural communities, ecosystem management (and its variants) remained as of the early 2000s a work in progress, subject to varied interpretations and vulnerable to changing political forces. It has, at minimum, ushered land ethics back toward the center of discussions about national forest management, USFS decision making, the public interest, and "the greatest good."

As the USFS enters its second century, it continues to play a leading role, nationally and internationally, in shaping the ethos of the conservation/environmental movement. Even as the demand for forest products ("green"-certified or otherwise) continues, national forests are increasingly appreciated for nontimber values and services: as water sources, carbon sinks, and biodiversity repositories; as anchors for whole and sustainable landscapes; as open space and settings for healthy human communities; and as classrooms for an increasingly land-detached public. The USFS faces the challenge of managing the national forests in response to shifting demographics, new scientific knowledge, and uncertain economics, even while facing continuing threats from invasive species, habitat fragmentation, accelerating climate change, and other widespread forces of environmental change.

SEE ALSO *Conservation; Ecology: III. Ecosystems; Environmental Law; Forests; Hetch Hetchy; Marsh, George Perkins; Muir, John; Pinchot, Gifford; Preservation; Resource Management; U.S. Department of the Interior; Utilitarianism; Wilderness Act of 1964.*

BIBLIOGRAPHY

Clary, David A. 1986. *Timber and the Forest Service.* Lawrence: University Press of Kansas.

Hays, Samuel P. 2007. *War in the Woods: The Rise of Ecological Forestry in America*. Pittsburgh, PA: University of Pittsburgh Press.

Hirt, Paul W. 1994. *A Conspiracy of Optimism: Management of the National Forests since World War Two*. Lincoln: University of Nebraska Press.

Langston, Nancy. 1995. *Forest Dreams, Forest Nightmares: The Paradox of Old Growth in the Inland West*. Seattle: University of Washington Press.

Marsh, George P. 2006 [1864]. *Man and Nature; or, Physical Geography as Modified by Human Action*. Ann Arbor: Scholarly Publishing Office, University of Michigan Library.

Miller, Char. 2001. *Gifford Pinchot and the Making of Modern Environmentalism*. Washington, DC: Island Press.

Pyne, Stephen J. 1997. *Fire in America: A Cultural History of Wildland and Rural Fire*. Seattle: University of Washington Press.

Steen, Harold K., and Christine Guth. 2004. *The U.S. Forest Service: A Centennial History*. Seattle: University of Washington Press.

U.S. Congress. 1897. *Organic Act of 1897*. 55th Congress, 1st session. Available from http://www.cfr.washington.edu/classes.common/comweb/Case%20Studies/usa/yellowstone/yellowstone/Project1/Yellowstone_Docs/ORGANIC%20ACT%20OF%201897.pdf

U.S. Congress. 1960. *Multiple Use and Sustained Yield Act of 1960 (MUSY)*. Public Law 86-517. Approved June 12, 1960. Available from http://www.fs.fed.us/emc/nfma/includes/musya60.pdf

U.S. Forest Service. 2005. *The Greatest Good: A Forest Service Centennial Film*. Washington, DC: USFS.

Williams, Michael. 1989. *Americans and Their Forests: A Historical Geography*. Cambridge, UK: Cambridge University Press.

Yaffee, Steven L. 1994. *The Wisdom of the Spotted Owl: Policy Lessons for a New Century*. Washington, DC: Island Press.

Curt Meine

U.S. NATIONAL PARK SERVICE

The design of the U.S. national parks and the management philosophy of the National Park Service (NPS) have grown out of the mainstream principles and practices of American landscape design. The agency's two main missions are (a) to protect natural scenery, cultural features, and wildlife of the parklands and (b) to make them accessible for public enjoyment. The tensions between these two goals make up much of the history of the NPS and have spawned many changes and posed many challenges in the management of the National Park System.

FOUNDING VISION OF NATIONAL PARKS' DESIGN AND MANAGEMENT

The three men who most heavily influenced the initial design and management philosophy of the national parks were George Perkins Marsh (1810–1882), Andrew Jackson Downing (1815–1852), and Frederic Law Olmsted Sr. (1822–1903). Marsh advised the nation to set aside lands as public parks and preserves and to protect its natural resources (1864). Downing translated the idea of wilderness into design concepts. He was, according to NPS historian Linda Flint McClelland, "intensely aware of the tremendous influence that primeval nature, with its dramatically changing landforms, variations of light and shadow, sounds of moving water, and enveloping vegetation could exert on the human senses" (1997, p. 20). He introduced the picturesque style to the United States, adapting English landscape garden techniques to heighten the observer's experience of nature. Olmsted elaborated this tradition, creating systems designed to promote the circulation of human traffic amid a series of pictorially composed views. Olmsted's recommendations—which became the guiding principles of the NPS—articulated the model for national parks: Undiminished nature should be made accessible, accommodating development while subordinating it to the environment and scenic values. Accordingly, park designers adopted naturalistic approaches to landscape preservation and development, integrating roads and structures into their native surroundings and minimally disrupting topography and vegetation to provide recreational access and to educate the public about their environment's natural and cultural legacies.

ORIGINS, PUBLIC EXPECTATIONS, AND PARK MANAGEMENT

The first national park in the history of the United States—and the world—was Yellowstone National Park, created in 1872. The year 1890 saw the creation of Yosemite, Sequoia, and General Grant national parks as the idea of protecting unique natural environments became a solidly entrenched feature of U.S. policy. The park system expanded over the ensuing quarter century, culminating in the founding of the National Park Service in 1916.

J. Baird Callicott has argued that humans go to the wilderness in a spiritual quest to engage the natural world and counter the alienation fostered by modern urban civilization (1998). Fittingly, then, the law that founded the NPS, the National Park Service Organic Act of 1916, expressly set down the service's dual purposes: "to conserve the scenery and the natural and historic objects and the wildlife" and "to provide for the enjoyment of same," but so as to "leave them unimpaired for the

Everglades National Park. *A Great Egret (*Adrea alba*) stands among cypress trees in Everglades National Park in south Florida. The park, formally established in 1947, encompasses 2,354 square miles of mangrove swamps, pinelands, pond apple and cypress forests, and saw grass prairie. Hundreds of thousands of tourisits visit it each year.* NPS.

enjoyment of future generations." (National Park Service 2008b). Reflecting the then-fashionable aesthetic of the sublime and picturesque, America's national parks evolved as sanctified places, neither simply religious nor aesthetic (Frow 1997).

The National Park Service began to develop a transportation system in the 1920s that focused primarily on road construction to facilitate automobile usage. The early transportation planning by the Bureau of Public Roads in the 1920s sought to design a system of "touring" roads that connected many of the largest western parks (Hartzog 1988). As automobile usage increased, the deficiencies of the early road design became more apparent, bringing still more road design and construction. Attempting to solve automobile congestion, the NPS upgraded existing roads and infrastructure, thereby stimulating renewed cycles of increased use and ecosystem deterioration. Park visitation rose dramatically during the ensuring years: from 358,000 in 1916 to 287 million in 1999, dropping to 266 million in 2003 and rising to 275 million in 2007 (National Park Service 2008a).

FROM PRESERVATION AND CONSERVATION TO RESTORATION AND RESILIENCE

The tension between preservation of the parklands and the depredations caused by visitors' use has played out against the tradition of American Progressivism's conservation policy. As Worster (1979) has noted, in contrast to preservation, conservation holds that (a) the resources of public lands, including national parks, are abundant or renewable and thus meant to be consumed and (b) rationalized long-term practices can manage continuing production to maximize prosperity for the whole nation. An implication of conservation philosophy is that even once-ravaged parts of the natural environment can be maintained and restored (as accomplished, for example, between 1900 and 1925 with once-decimated deer herds. As sanctuaries the national parks now play a major role in restoring native species, protecting their genetic integrity, and affording the public the chance to become acquainted—or reacquainted—with them. Invasive species are removed, and the negative effects of erosion,

WOLVES RETURN TO YELLOWSTONE

Attempts to manage wildlife respond to changes in policy, practice, natural processes, and scientific knowledge. When the national parks were established, a variety of large mammals were present (gray wolves, grizzly bears, deer, and elk), although hunting and trapping almost exterminated many once-abundant creatures (deer had been killed by hunters until almost none remained by the 1880s). Changes in policy and actions based on limited observations often led to dramatic swings in the populations of individual species, oscillating from severely diminished to far in excess of a habitat's carrying capacity, with considerable variation from region to region. For example, the Progressives' view that natural resources needed to be efficiently managed included increasing "crops" such as cattle, sheep, and deer by eliminating predators. Serious predator extermination began in the late 1800s and proceeded through the early 1900s (Worster 1979). By the 1940s gray wolves were rare in the Yellowstone National Park region, with no evidence of any population (save for a random stray) by the 1970s. It appeared that the elimination of large carnivores led to the recovery of the population of deer, elk, bison, and moose (as well as to fewer lost livestock) (US–Parks 2008).

As the numbers of grazing animals increased, montane riparian willow and aspen stands became overgrazed. Because the change along rivers and streams was pronounced, the adverse effects rippled to beavers, fish, birds, and then organisms in the networks of interdependency through adjacent grasslands.

These adverse conditions prompted a turnabout in policy: It seemed a good idea to reduce the elk, bison, moose, and mule deer to remove the direct threat to vegetation and the rest of the ecosystem. There were many options—each loaded with socially charged and contested assumptions and implications—ranging from increased hunting to covert shooting by government personnel to attempts to redistribute the large animals. In 1987 the U.S. Fish and Wildlife Northern Rocky Mountain Wolf Recovery Plan experimented with the reintroduction of gray wolves; in 1991 they were restored to Yellowstone and central Montana.

The reintroduced gray wolves have thrived to the point that they have been removed from the endangered species list. There certainly are now fewer elk, and they have changed their habitat use. It is not clear, however, to what extent this is due to wolves, other predators, or changes in the weather (National Park Service 2008, US–Parks 2008). More research and interpretation need to be done to understand the impact of the wolves, not only on the entire ecosystem but on the "behaviors and life-history traits that confer resilience to environmental disturbances at various temporal and spatial scales . . . at different hierarchical levels" (Weaver et al. 1996, p. 964).

At the same time that native species have been preserved or restored, the tourist population in the parks has boomed. Since 1995 more than 100,000 visitors have swarmed over the park to observe the once-absent species. Hence the NPS still faces the perennial dilemma: developing management policies and practices that can realize its two elusive and often conflicting goals: preserving both wildlife and public access.

BIBLIOGRAPHY

National Park Service. 2008. "Wolf Restoration in Yellowstone Successful Beyond Expectations." Washington, DC: U.S. Government Printing Office.

US–Parks. 2008. "Yellowstone National Park: Wolf Restoration." Available from http://www.us-parks.com.yellowstone/wolf_restoration.html

Weaver, John; Paul Pacquet; and Leonard Ruggerio. 1996. "Resilience and Conservation of Large Carnivores in the Rocky Mountains." *Conservation Biology* 10(4): 964–976.

Worster, David. 1979. *Nature's Economy*. New York: Anchor.

flooding, and other disturbances caused by human activities such as "road construction, visitor impact, and facility maintenance" are remediated (Majerus 2000, p. 77; National Park Service 2001). Such projects are undertaken in diverse environments: grassland prairies, hardwood forests, wetlands, streams, lakes, and marine systems such as coral reefs.

In the process of restoration, managers have to decide what to restore (to what state) and what to remove. Some critics have argued that NPS policies have to generate a more sophisticated idea of checks and balances among species to achieve stabilized populations and environments. They contend that management practices need to deal with the cases in which an isolated species was restored only to have the ecosystem unpredictably collapse in new ways. Some argue that complexity theory may provide the best scientific basis for management policy: "Ecosystem restoration is invariably difficult and disturbed;... ecosystems often exhibit classic traits of complex systems... [and] responses to perturbations, including restoration efforts, can be highly nonlinear and lead to management surprises" (Gross 2004).

As Walker and Salt explain, with complexity theory comes another concept that dramatically changes goals and practices: "Resilience [is] the capacity of a system to absorb disturbance and still retain its basic formation and structure.... Resilience thinking presents an approach to managing natural resources that embraces human and natural systems as complex systems continually adapting through cycles of change" (2006, p. 10). They argue that, because current research usually simulates the future by extrapolating from the past (using conventional data about average conditions and incremental growth but setting aside major events), it provides an inadequate, even misleading, basis for sustainable resource management (Walker and Salt 2006, Peterson et al. 2003). There is a growing consensus that new modes of research and practice are needed (National Park Service 2000).

In the view of many policy analysts, studies and actions concerning complex, dynamic ecosystems need to respond to whole communities and events for longer periods and over wider areas (including regional variations and differences in elevation). Complexity studies often revise previous interpretations, such as the impact of pavement on stream communities: "Thresholds appear to exist, which may be catastrophic," but are not immediately apparent, because paved roads and parking lots "most likely do not cause degradation directly. Instead, [they] are a surrogate for a wide range of other variables, such as [altered water flow], increased stream temperature, enhanced erosion, habitat degradation, and so on" (Hilderbrand et al. 2001, p. 401). Sprawl reduces and fragments habitat and erodes existing buffer zones, in turn harming wildlife and plants (Trombulak 2000). The NPS has attempted to address these problems, temporarily closing areas to protect sensitive species, capping commercial development, improving wildlife corridors, and funding alternative modes of transportation, but traditional NPS management practices favoring development and some recreational activities continue to dominate park policy, perpetuating many of the threats to the ecological viability of national parks.

To be properly understood and operationalized the key concepts need to be more precisely defined and the environmental factors more thoroughly researched. For example, "habitat types are rated for development potential (e.g., camping areas and trails) on the basis of their resistance... (the ability of a habitat to tolerate human impacts, such as trampling, without undergoing major changes in community composition and structure) and their resilience (the ability of vegetation on a habitat to recover once it has been destroyed or severely disrupted)"; but there is no universal rule of relationship (NPS 2007, p. 4). Counterintuitively, sometimes resilience and resistance vary inversely to each other, as in the forest communities of Mount Rainier National Park or with the fauna's low resistance but high resiliency to nonnative fish introductions in Yosemite National Park lakes (Hilderbrand et al. 2001).

Complexity theory is pushing management practices away from the goal of restoration and toward resilience, resulting in a tension between these two newer approaches in both scientific research and management policy, even as the older differences between preservation and conservation persist. Although the various tasks facing the NPS do have some positive connections, they also pull in notably different directions—certainly the research and practices vary for each, thus multiplying management challenges to generate alternative scenarios that can offer satisfactory visitor experiences that will not damage the parks' ecological integrity.

SEE ALSO *Ecological Restoration; Endangered Species Act; Land Ethic; Preservation; Wilderness.*

BIBLIOGRAPHY

Congress of the United States. 1916. "Organic Act of 1916: (U.S.C., title 16, sec 1.)." Washington, DC: U.S. Government Printing Office.

Frow, John. 1997. *Time and Culture.* Oxford, UK: Clarendon Press.

Gross, John. 2004. "Management and Restoration of Dynamic Ecosystems." Ecological Society of America. Available from http://www.abstracts.co.allenpress.com

Knapp, Roland A., Kathleen R. Matthews, and Orlando Sarnelle. 2001. "Resistance and Resilience of Alpine Lake Fauna to Fish Introductions." *Ecological Monographs* 71(3): 401–421.

Knapp, Roland, et al. 2005. "Fauna of Yosemite National Park Lakes Has Low Resistance but High Resilience to Fish Introductions." *Ecological Applications* 15(3): 835–847.

Majerus, Mark. 2000. "Restoration with Native Indigenous Plants in Yellowstone and Glacier National Parks." In *Billings Land Reclamation Symposium.* Bridger, MT: USDA-NRCS Plant Materials Center. Available from http://www.ott.wrcc.osmre.gov/library/proceed/recsym/blrs.pdf

McClelland, Linda Flint. 1997. *Building the National Parks.* Baltimore: The Johns Hopkins Press.

National Park Service. 2000. *Ecosystem Management in National Parks.* Washington, DC: U.S. Government Printing Office.

National Park Service. 2001. "Natural Resource Management." In *Management Polices.* Washington, DC: U.S. Government Printing Office.

National Park Service. 2007. "Management Interpretations of the Habitat Types." In *The Forest Communities of Mount Rainier National Park.* Washington, DC: U.S. Government Printing Office.

National Park Service. 2008a. *NPS Stats.* Available from http://nature.nps.gov/stats

National Park Service. 2008b. National Park Service Act. Available from http://www.nps.gov/legacy/organic-act.htm

Peterson, G. D., et al. 2003. "Assessing Future Ecosystem Services: A Case Study of the Northern Highlands Lake District, Wisconsin." *Conservation Ecology* 7(3): 1.

Trombulak, Stephen C., and C. Frissell. 2000. "Review of Ecological Effects of Roads on Terrestrial and Aquatic Communities." *Conservation Biology* 14: 19–29.

Walker, Brian, and David Salt. 2006. *Resilience Thinking.* Washington, DC: Island Press.

Robert Mugerauer
J. Watson

UTILITARIANISM

Utilitarianism is a popular ethical theory that has greatly influenced the development of law, economic theory, and many areas of public policy, including technological and environmental planning. Along with rights theory, it may be said to form the ethical basis of modern liberal democracy. This entry explores the possibility of basing environmental ethics on utilitarianism.

According to the most famous advocate of utilitarianism, John Stuart Mill, "The creed which accepts as the foundation of morals, Utility, or the Greatest Happiness Principle, holds that actions are right in proportion as they tend to promote happiness, wrong as they tend to produce the reverse of happiness. By happiness is intended pleasure, and the absence of pain; by unhappiness, pain, and the privation of pleasure" (2001 [1863], chap. 2).

John Stuart Mill, 1884. *Mill (1806–1873) is considered one of the greatest advocates of the ethical theory of utilitarianism. Mill attempted to defend the controversial theory against its critics. For example, his idea of "rule utilitarianism" qualified the belief that an act is good or bad based only on the amount of happiness it produces, by suggesting that an actor must consider whether an experience shows an action to produce positive consequences for it to be "right."* THE LIBRARY OF CONGRESS.

Utilitarianism is the most widely accepted consequentialist theory. Consequentialism is the approach to ethics that considers that the rightness of an act is determined by the good achieved by that act for those affected, and by nothing else. Thus no act is inherently right or wrong, and the rightness or wrongness of an act is independent of such factors as the motive or character of the agent. The British jurist Jeremy Bentham (1789) is usually considered to be the founder of utilitarianism. He, Mill, and Henry Sidgwick (the third great classical utilitarian) believed that nothing is good in itself except happiness, which they identified with pleasure; they are thus known as hedonistic utilitarians, from the Greek word for pleasure. Everything else is valuable only instrumentally, as a means to producing pleasure. Thus utilitarians reject claims about the intrinsic value of environmental objects, species, and ecosystems, which are commonly believed to be central to environmental ethics.

More recently, utility has been seen as the satisfaction of interests or preferences. Its proponents, notably the celebrated Australian philosopher Peter Singer (1993), regard it as a complete theory of ethics, that is, as capable of identifying the right thing to do in every conceivable situation.

SOME PROBLEMS

From the beginning, utilitarianism was controversial, for several reasons. First, many people were offended by the determinedly secular nature of the theory. It was not so much that it questioned the widely accepted idea that morality was a function of divine command, but that it left no room for it or any other moral authority.

Second, since the amount of happiness is all that counts, ethically it does not matter whose happiness is affected. Insofar as institutions such as slavery and racist and sexist practices diminish the happiness of their victims (a reasonable thing to suppose), slavery and racism and sexism must be abolished. Thus utilitarianism is profoundly democratic and egalitarian, unlike the society of the day.

Third, conservative moralists were shocked by utilitarianism's implicit acceptance of behavior that was widely regarded as sinful. If pleasure is the only good, then any activity that gives pleasure must be good, and the more pleasure it gives, the better it is. The only moral or legal constraint that utilitarians accept is what Mill called the "Harm Principle": An act is wrong only if it causes avoidable nontrivial harm to others.

Fourth, if only pleasure is valuable in itself, then no source of pleasure is any better or worse than any other. As Bentham put it, "Quantity of pleasure being equal, pushpin [a children's game played with pins and a hat] is as good as poetry." Not only that, the pleasure of torture afforded a sadist is not bad in itself; torture is abominable only because it inflicts great pain on the sadist's victim. Worse still, the ideal life could be achieved by plugging people into a computer-generated reality-simulation system (as in the film *The Matrix*) or a pleasure machine (which would constantly stimulate the pleasure center of the brain).

Finally, Bentham's *act utilitarianism* seems to require that in each situation the agent should predict the consequences of an action and calculate the effects on happiness, but, it was argued, this would cripple decision making because it is not possible to make any such calculation.

Classical utilitarians ignored or dismissed the first three criticisms, which are less relevant at the beginning of the twenty-first century. Mill attempted to deal with the fourth by arguing that as well as degrees of *quantity* there are also degrees of *quality* of pleasure, and that those who have experienced both lower and higher pleasures will acknowledge that, say, philosophical speculation is superior to mud wrestling. Most commentators have considered this argument elitist and unconvincing. Logically, it appears to involve a contradiction, for pleasure per se is acknowledged to be the only good, but if some pleasures are qualitatively better than others, then some other standard of good appears to be implicitly invoked.

Mill addressed the fifth concern by proposing what has become known as *rule utilitarianism*: The right act is the one that follows a rule that experience has shown to have positive consequences. This gets around the problem that act utilitarianism requires the agent successfully to figure out the consequences of an action, thus requiring what is often referred to as moral luck. This solution, however, is bought at the expense of also requiring the agent to ignore special features of a situation. Experience suggests that adherence to the rule "Do not murder" has good consequences overall, but surely a utilitarian would accept that it would have been ethical to take out the Nazi high command in the early days of World War II, thus perhaps saving many lives and much misery. Rule utilitarianism, it is said, thus collapses into act utilitarianism.

ANIMALS AND THE ENVIRONMENT

Utilitarianism is usually associated with anthropocentrism, for instance in the work of William Baxter (1974), who famously claimed that penguins are valuable only because people happen to enjoy watching them walk about on rocks. However, Bentham argued that since pain and suffering are inherently bad, it is as wrong to inflict suffering on any sentient beings (animals capable of feeling pleasure or pain) as it is to inflict it on humans. This idea was largely ignored until it was taken up by Singer (1975). Singer argued that since it is wrong to discriminate on the grounds of sex (sexism) and race (racism), it is equally wrong to discriminate on the grounds of species membership (speciesism). Suffering is bad, regardless of the species of the suffering organism. Many of our uses of animals cause suffering, and hence Singer advocates that we should cease animal farming, research and testing involving animals, and hunting, fishing, and other recreational use of animals.

Singer's utilitarian position has both positive and negative implications for environmental protection. On the one hand, protection of the interests of animals often happens to coincide with protection of the environment. For instance, industrial agriculture, as exemplified by giant pig-raising operations, may be condemned because it causes the animals to suffer, has effects injurious to humans, and destroys natural values. On the other hand, the interests of individual animals may happen not to coincide with the protection of the environment. Advocates for animals and advocates for the environment can and do find themselves in opposition, as in the case of invasive introduced species. The eastern brush-tail possum, an Australian marsupial that was introduced to New Zealand in the mid-nineteenth century, has now reached plague proportions, with a population of up to 70 million in a country with an area of some 269,000 square kilometers, the size of Colorado. These possums destroy entire areas of native forest, and the only way to protect the forest is to kill millions of the animals, which, even with the most humane methods, inevitably involves the suffering of many animals. In Singer's view, killing possums to protect forests cannot be justified, because trees, not being sentient, do not have interests.

A number of other problems may be briefly mentioned. First is the problem of predation in nature. This undoubtedly causes considerable suffering. For instance, African wild dogs in pursuit of their prey rip chunks of flesh out of the latter. Some chimpanzees reportedly captured a living bonobo, ripped the arm from the creature, and ate it while the bonobo screamed in agony. Domestic cats are not the only animals that play with their food while it is still alive. Killer whales toss around their favorite prey, seal pups, in a similar manner. We could reduce this suffering by painlessly eliminating predators, though of course we would then have to painlessly cull prey animals to keep them from overpopulating. Alternatively, we could painlessly kill grazing animals and feed them to predators. Indeed, utilitarianism would seem to require that we engage in these or similar activities.

Second, if we could increase total utility by transforming natural areas for the benefit of sentient beings, this might be permissible or even obligatory. Mark Sagoff (1984) argues that wilderness areas and national parks could be converted into game parks, where animals could be fed and receive veterinary care. This would also solve the predation problem.

Third, because utilitarianism focuses entirely on the interests of sentient individual organisms, it does not attach any value to species or ecosystems, except insofar as these can be reduced to the aggregated value of individuals. Thus, for a utilitarian, it is no worse to kill a member of a common species, such as a minke whale or a black bear, than to kill a member of an endangered species, such as a blue whale or a polar bear.

For these and other reasons, Sagoff (1984) has argued that animal liberation and environmental ethics are fundamentally incompatible. Similarly, utilitarianism cannot recognize value in naturally occurring inanimate objects, such as rock formations and caves, or in the built environment and works of art, except insofar as sentient beings happen to enjoy or prefer them. Even then, if something is valued only instrumentally, there is no reason why it cannot be substituted for by something providing equal or greater happiness. Thus if the Board of Trustees of Britain's National Gallery were dedicated utilitarians, they might decide secretly to sell one of the Gallery's more celebrated paintings—Leonardo da Vinci's *Madonna of the Rocks*, say—to a wealthy collector, replacing it with a perfect copy and using the proceeds to add to its collection.

PUBLIC POLICY

Utilitarianism has been very influential in shaping public policy, in particular through risk-cost-benefit analysis. Such analysis operates by identifying all the risk, costs, and benefits of a proposed action, assigning a dollar value to each, and summing them. One then accepts only proposals with a favorable risk-cost-benefit balance (or, in the case of competing proposals, the one with the most favorable balance). Risk-cost-benefit analysis is widely used to evaluate and rank projects such as dams and new highways, regulations of emissions and food additives, social programs such as free mammograms for women of varying ages, environmental actions such as endangered-species management plans, and so on. The best known philosophical advocate of risk-cost-benefit analysis is Kristin Schrader-Frechette (2001).

Bentham and Singer notwithstanding, risk-cost-benefit analysis is anthropocentric. This is because the methods used in such analysis—such as calculations based on the economic value of a life predicted to be saved by highway improvements or industrial health and safety standards, and willingness to pay—can be applied only to humans. This does not mean that endangered species, forests, and ecosystems are not valuable, but it does mean that their value effectively depends on human preferences.

Despite the theoretical defects that many philosophers see in utilitarianism, in practical terms it may turn out to be an effective basis for environmental policy. This is because its egalitarianism implies intergenerational equity, and therefore an obligation to future generations to leave the earth in at least as good a condition as we inherited it in. True, we do not have detailed knowledge of what will make future generations happy or what their preferences will be. However, we can be reasonably certain that these will not include exposure to a toxic environment, drowned coastlines, desertification, and mass species extinction. Since we do not wish these things on ourselves or our direct descendents, we should not wish them on future generations as a whole. This has profound implications for our consumerist lifestyles.

SEE ALSO *Animal Ethics; Consumption; Cost-Benefit Analysis; Environmental Policy; Future Generations; Natural Law Theory; Risk Assessment; Shrader-Frechette, Kristin; Singer, Peter; Speciesism.*

BIBLIOGRAPHY

Baxter, William F. 1974. *People or Penguins: The Case for Optimal Pollution.* New York: Columbia University Press.

Bentham, Jeremy. 1996 [1789]. *An Introduction to the Principles of Morals and Legislation.* New York: Oxford University Press.

Mill, John Stuart. 2001 [1863]. *Utilitarianism,* 2nd edition. Indianapolis, IN: Hackett Publishing.

Sagoff, Mark. 1984. "Animal Liberation and Environmental Ethics: Bad Marriage, Quick Divorce." *Osgoode Hall Law Journal* 22: 297–307.

Schrader-Frechette, Kristin. 2001. "A Defense of Risk-Cost-Benefit Analysis." In *Environmental Ethics: Readings in Theory and Application*, ed. Louis P. Pojman. Belmont, CA: Wadsworth Publishing.

Singer, Peter. 1975. *Animal Liberation*. New York: New York Review.

Singer, Peter. 1993. *Practical Ethics*, 2nd edition. Cambridge, UK: Cambridge University Press.

Alastair S. Gunn

V

VEGANISM

SEE *Vegetarianism.*

VEGETARIANISM

In the United States, between 2 percent and 5 percent of the population "classify themselves as *vegetarians*; of that number perhaps five percent are strict vegans" (Koerner 2007). Although vegetarians renounce animal flesh, they consume animal fluids (milk and milk-derivates such as cheese, yogurt, butter, and ice cream) and/or eggs.

TYPES OF VEGETARIANISM

Thus the vegetarian tribe is divided into "lacto-ovo" vegetarians; "lacto-vegetarians," who eat dairy but no eggs; and "ovo-vegetarians," who eat eggs but no dairy. Some describing themselves as vegetarians eat fish ("pesco-vegetarians") or chicken ("pollo-vegetarians") or both ("pesco-pollo vegetarians").

There are many reasons for choosing a vegetarian diet: personal distaste for meat; a personal health, especially avoidance of saturated animal fats (a concern that is similar to the avoidance of saturated vegetable fats); or ethics. Although eating yogurt and/or egg whites, chicken and/or fish may be either preferable to or more healthful than eating beef or pork, from an ethical point of view, many strict vegetarians consider these eaters of animal products and/or animals such as fish and chickens to be carnivores whose claim to vegetarianism depends on equating "meat" with "red meat."

Vegetarianism itself has been criticized as ethically inadequate and inconsistent by a more radical group known as "vegans." The same ethical considerations that lead vegetarians to renounce meat-eating lead vegans to repudiate dairy, cheese, eggs, and honey; clothing items such as fur, leather, wool, and silk; and animal-tested products, including shampoo, cosmetics, and, drugs. Vegans believe that vegetarians only partially— and therefore inconsistently—break from a cruel, violent, and ecocidal system of food production. Milk cows and birds in battery cages—no less than veal calves—are confined for "lacto-ovo" consumption; and dairy and egg farms pollute the air and water.

The vegan pioneer Donald Watson (1910–2005) disparaged vegetarianism as "but a half-way house between flesh eating and a truly humane, civilised diet" (1944, p. 1).

As with vegetarianism, there are subcategories of vegans, including fruitarians, raw-food vegans, and freegans (who practice a minimal consumption, "dumpster-diving" lifestyle).

HISTORY OF VEGETARIANISM

Vegetarianism (which will be used here to include veganism) has a long and rich history that is as old as European and American cultures (see Berry 1998, Iacobbo and Iacobbo 2004, Spencer 2002, Walters and Portmess 1999 and 2001, Spencer 2004, Tristram 2007). As a health-promoting diet and an ethic rooted in compassion for all living beings (*ahimsa*), vegetarianism emerged more than 3,000 years ago as a philosophy and practice of the ancient South Asian religions: Hinduism, Jainism, and Buddhism. The vegetarianism of the Greek philosopher Pythagoras (ca. 496–552 BCE) and animal-protectionist ethics spread throughout the ancient world and resurfaced in the seventeenth century (indeed, until the mid-nineteenth century, those who abstained from meat were called "Pythagoreans"). At the dawn of modernity, vegetarianism became increasingly influential throughout European society; radicals deployed its nonviolent and egalitarian outlook as a critical weapon against class rule and what they viewed as European barbarism, and prominent medical figures espoused it as ideal for health as well as morality (Stuart 2007).

VEGETARIANISM AND THE ENVIRONMENT

As the twentieth century unfolded, however, the influence of vegetarianism in the United States began to wane as the livestock industry became increasingly powerful and meat became an affordable staple for working-class families (Rifkin 1992). In a culture trained in the mindset that meat promotes strength and vegetarianism fosters weakness, a dramatic revival, growth, and broadening of vegetarianism began in 1971, with the publication of Francis Moore Lappé's book *Diet for a Small Planet.* In this and subsequent books (1977, 1998, 2003), Lappé described a corporate-controlled, industrialized, factory-farmed system of animal agriculture that is inefficient, wasteful, cruel, and destructive to every facet of the environment. The global livestock industry is, she argued, a vehicle of European-American imperialism that displaces millions of people from the land, destroys the livelihood of independent farmers, exacerbates poverty and inequality, and aggravates world hunger by diverting resources into producing feed rather than food. To this destructive, unethical, unjust, and unsustainable system of agriculture, Lappé contrasted a vegetarian mode of farming that produces maximum output with minimum input; that promotes health, rights, justice, and democracy; and that is environmentally sound and sustainable.

Lappé's work—along with Peter Singer's *Animal Liberation* (2001 [1975]), Singer's and Jim Mason's *Animal Factories* (1990 [1980]), and John Robbins's *Diet for a New America* (1998 [1987])—vividly portrays the human, animal, and environmental costs of the global meat culture, and this systemic outlook inspired the *vegetarian environmental* movement. This movement has fused issues of health, animal rights, social justice, world hunger, violence, globalization, and environmental concerns into a holistic theory unrivaled in depth, comprehensiveness, and awareness of the multidimensional crisis—health, moral, social, and environmental—facing humanity. Pursuing the lead of these theorists, a number of books have documented the central role of the livestock industry in the devastation of the social and natural worlds (see Mason and Singer 1990 [1980], Jacobs 1992, Rifkin 1992, Hill 1996, Robbins 2001, Lyman 2001, and Jacobson 2006).

By 2000 growing alarm over the human, animal, and environmental toll of the global meat, dairy, and egg industries percolated into scientific sectors, international government bodies, and—in a slow and hesitant way—some mainstream environmental groups such as the Sierra Club. Throughout 400 pages, a landmark 2006 United Nations report, "Livestock's Long Shadow," identified the livestock industry "as one of the top two or three most significant contributors to the most serious environmental problems, at every scale from local to global" (Steinfeld et. al. 2006). The data compiled in this report and countless thousands of corroborating studies leave little room for doubt that the livestock industry is a significant threat to the planet. The number of farmed animals (including fish) in the world has quadrupled in the last fifty years, putting a great strain on air, land, and water. Livestock uses 70 percent of all agricultural land and 30 percent of the earth's entire land surface (Steinfeld et. al. 2006). Crops grown for animal feed rather than human food consume 87 percent of the nation's fresh water, 90 percent of the soy crop, 80 percent of its corn, and 50 percent of all grains (Vesterby and Krupa 1997, Pimentel 1997). Compared to a vegetarian diet, meat production demands seven times more land (Leckie 2007), eight times more fossil fuel energy (Pimenel 1997), and ten times as many crops (Pimentel 1997; Robbins 1998 and 2001; Horrigan et. al. 2002). In this system of carnivorous consumption, 41 million tons of plant protein for cows returns 7 million tons of protein for humans (Pimentel 1997).

The livestock industry is a major cause of air pollution, soil erosion, desertification, water pollution, and acid rain. U.S. farms generate 130 times as much excrement as the nation's entire human population (Worldwatch Institute 1998). Factory farm effluvia—a slurry of manure, pesticides, antibiotics, hormones, and fertilizers—poison water supplies, decimate fish populations, degrade coral reefs, and have contributed to the emergence of more than 150 oxygen-starved "dead zones" in the oceans (Larsen 2004).

Moreover, 70 percent of the Amazon rain forest has been slashed and burned to graze cattle, and much of the remainder has been converted to fields for growing feed. In addition to being a principal cause of forest destruction and species extinction, the livestock industry is a significant factor in global climate change (Steinfeld et. al. 2006). Meat, dairy, and egg industries emit 18 percent of all carbon dioxide, 37 percent of the methane gas (a heat-trapping gas that is twenty times stronger than carbon dioxide), and 65 percent of nitrous oxide gases (300 times more potent than carbon dioxide). The UN report concluded that the livestock industry produces more greenhouse gases than all the world's transportation systems (Steinfeld et al. 2006).

CONFLICTS BETWEEN VEGETARIANS AND ENVIRONMENTALISTS

These facts suggest the possible importance of vegetarianism and animal rights for the environmental movement and the urgency of finding common ground for a triangular alliance. Yet rather than uniting in the war to prevent massive species extinction, catastrophic ecological breakdown, and irreversible climate change, vegetarian and environmental camps remain divided by deep differences in philosophy and lifestyle (Motavalli 2002, Sapontzis 2004). Both camps break with anthropocentrism (the belief that only human beings deserve ethical consideration). Environmentalists advocate a new holistic "ecological consciousness" and "land ethic" but often neglect to advocate concern for individual sentient or conscious nonhuman beings. Whereas vegetarians now identify themselves as ipso facto environmentalists, certainly not all environmentalists embrace vegetarianism. At stake are competing views on animal rights, whether or not hunting and meat-eating are ethical and compatible with environmental values, and how to balance the values of individuals and ecosystems.

Ethical vegetarians shift the criterion for having rights from rationality to the far broader characteristics of consciousness or sentience. For Singer (1975) and Regan (1983) a necessary and sufficient condition of meriting moral consideration is the capacity to experience pleasure and pain, although both place a moral premium on advanced levels of consciousness and intelligence. If it is a fundamental moral axiom that it is wrong to cause injury, suffering, or death to another individual unless there is a compelling reason to do so, ethical vegetarians argue that—except in very rare cases such as self-defense—we never have adequate reason to harm animals. This is true not only for exploiting animals for "sport," "entertainment," and fur, but also for killing them for food, because humans cannot only live but flourish on a purely plant-based diet.

Many environmentalists opposed to industrial agriculture agree that factory farming is cruel and unethical but nonetheless assert that raising animals on small "family" farms without intensive confinement and manipulation is acceptable and good. Their justifications for raising animals for slaughter include the argument that such animals would not live at all if not bred for food, that they live a satisfying and worthy life on nonindustrial farms, and that killing and consuming other animals is a natural fact of life. This position turns on a "welfare" rather than "rights" position (see Regan 2004); it contends that the moral wrong lies in causing animals severe or unnecessary suffering (such as on factory farms) rather than in exploiting them for human purposes. On the welfare view, slaughtering animals for food is ethical if done "humanely"—a concept ethical vegetarians dismiss as Orwellian doublespeak, arguing that there is nothing humane about any kind of killing and taking a being's life against its will.

Whereas vegetarians view hunting as unnecessary and therefore unjustifiable killing, some environmentalists support hunting as a means of affirming our evolved human place in the biotic community. And some argue that hunting has positive ecological benefits by stabilizing "game" (prey) populations such as deer that would otherwise overpopulate (Lott 2007, Miniter 2007). Vegetarians respond that hunting in fact is the prime cause of deer overpopulation and argue that hunters' predilection for killing large, healthy males over weaker individuals and females disrupts ecological and evolutionary dynamics (Pickover 2005). Unlike the animal-rights ethic that defends the rights of sentient individuals as inviolable, environmental ethics is often holistic, valuing ecosystems and species populations over individuals. Whereas many environmentalists champion Aldo Leopold's "land ethic" (1970) as the most comprehensive embrace of the biotic community (Callicott 1993), the animal-rights philosopher Tom Regan (1983) has denounced it as "environmental fascism" that sacrifices the individual to the good of the whole. Other environmental ethicists have worked to reconcile these contrasting positions (Jamieson 1997).

Although, some environmentalists agree with vegetarians that factory farming is cruel, they also support obtaining meat from noncommercial wild sources through sustainable hunting and fishing, just as they might argue that small-scale, organic farming is humane and beneficial for the environment (Pollan 2007). Some land spaces—such as high-elevation, short-grass prairies and steppes—are unsuited for cultivation but can support cows and sheep; thus it might be argued that their highest use is meat production (*Science Daily* 2007). Rebutting vegetarians who boast the ecological virtues of a plant-based diet, environmentalists point out that a frugal organic farmer who consumes modest amounts of meat from his or her own livestock can leave a lighter "ecological footprint" than a vegetarian who drives a Hummer, is a frequent flyer, and buys produce from global rather than local sources.

Vegetarians counter that such environmentalists have not explained how their vision of a global network of small farms can satisfy the competitive need for profits

(Collin 2003), much less the surging demand for meat—especially in the world's most populous nations, China and India, and a burgeoning overall population projected to double to 12 billion by 2050 (Worldwatch Institute 1998, Steinfeld et. al. 2006, Freston 2007). Moreover, vegetarians argue, environmentalists' uncritical praise for organic farming is naïve and romantic because organic-farm products are satisfying high-end consumer demand and becoming just another form of mass production and large-scale killing of animals (Cienfuegos 2004, Davis 2007, Lucas 2007, PETA n.d.)

CONCLUSION

Vegetarianism is not a panacea for ever-worsening social and environmental crises, but it could be a significant part of major changes that people—especially those in the developed world—can make to avert ecological disaster. These changes include reducing consumption and shifting from industrial to local agriculture, from chemically intensive to genuinely organic farming, and from fossil-fuel to alternative energy. The shift from a meat-based to a vegetarian diet would not only benefit the environment but also save billions of animals from suffering in factory farms and slaughterhouses, keep small-scale farmers from being displaced from their land, and protect billions of people against the suffering of diseases of excess (in the developed world) and of lack (in the undeveloped world).

SEE ALSO *Animal Ethics; Biocentrism; Factory Farms; Farms; Food; Global Climate Change; Hunger; Population; Regan, Tom; Singer, Peter.*

BIBLIOGRAPHY

Agence-France Presse. October 25, 2007. "Save the Planet? It's Now or Never, Warns Landmark UN Report." Available from http://afp.google.com/article/ALeqM5ituweJvQetTICry Y4W-NlA86Cbsg

Berry, Rynn. 1998. *Food for the Gods: Vegetarianism and the World's Religions*. New York: Pythagorean.

Callicott, J. Baird. 2004. "The Conceptual Foundations of the Land Ethic." In *Environmental; Philosophy: From Animal Rights to Radical Ecology*, eds. Michael E. Zimmerman. 4th edition. Englewood Cliffs, NJ: Prentice Hall.

Cienfuegos, Paul. May 31, 2004. "The Organic Foods Movement—Led by Heinz Corporation or We the People?" Available from http://www.commondreams.org/views04/0531-11.htm

Colin, Molly. July 14, 2003. "Elite Meat Shoppers Sold on Organic Produce Find Its Main-Course Counterpart—Certified Beef, Poultry, and Pork—To Be Elusive." *Christian Science Monitor*. Available from http://www. csmonitor.com/2003/0714/p13s02-wmcn.html

Davis, Joyzelle. October 20, 2007. "Huge Dairy Doesn't Fit Organic Image: Aurora Operation Foes Say Farm Pays Lip Service to Ideal." *RockyMountainNews.com*. Available from http://www.rockymountainnews.com/drmn/other_business/article/0,2777,DRMN_23916_5727225,00.html

"Diet with a Little Meat Uses Less Land than Many Vegetarian Diets," October 10, 2007. *Science Daily*. Available from http://www.sciencedaily.com/releases/2007/10/071008130203.htm

Freston, Kathy. January 18, 2007. "Vegetarian Is the New Prius." *Huffington Post*. Available from http://www.huffingtonpost.com/kathy-freston/vegetarian-is-the-new-pri_b_39014.html

Hill, John Lawrence. 1996. *The Case for Vegetarianism: Philosophy for a Small Planet*. New York: Rowman & Littlefield.

Horrigan, Leo; Robert S. Lawrence; and Polly Walker. 2002. "How Sustainable Agriculture Can Address the Environmental and Human Health Harms of Industrial Agriculture." *Environmental Health Perspectives* 110 (5). Available from http://www.ehponline.org/members/2002/110p445-456horrigan/horrigan-full.html#sust

Iacobbo, Karen, and Michael Iacobbo. 2004. *Vegetarian America: A History*. Westport, CT: Praeger.

Jacobs, Jynn. 1992. *The Waste of the West: Public Lands Ranching*. Tucson, AZ: Lynn Jacobs.

Jacobson, Michael F. 2006. *Six Arguments for a Greener Diet: How a Plant-based Diet Could Save Your Health and the Environment*. Washington, DC: Center for Science in the Public Interest.

Jamieson, Dale. 2002. "Animal Liberation Is an Environmental Ethic." In *Morality's Progress*. New York: Oxford University Press.

Koerner, Brendan. October 23, 2007. "Vegans v. Vegetarians: What Kind of Diet Is Best for the Environment?" *Slate*. Available from http://www.slate.com/id/2176420/

Lappé, Frances Moore. 1971. *Diet for a Small Planet*. New York: Ballantine Books.

Lappé, Frances Moore, and Joseph Collins. 1977. *Food First: Beyond the Myth of Scarcity*. New York: Ballantine.

Lappé, Francis Moore, et al. 1998. *World Hunger: Twelve Myths*. New York: Grove Press.

Lappé, Francis Moore, and Anna Lappé. 2003. *Hope's Edge: The Next Diet for a Small Planet*. New York: Tarcher.

Larsen, Janet. June 16, 2004. "Dead Zones Increasing in World's Coastal Waters," Earth Policy Institute. Available from http://www.earth-policy.org/Updates/Update41.htm

Leckie, Stephen. February 1, 2007. "Meat Production's Environmental Toll: Wilderness Destruction, Soil Erosion, Energy Waste, and Pollution." *Toronto Vegetarian Association Newsletter*. Available from http://veg.ca/content/view/133/111/

Leopold, Aldo. 1970. *A Sand County Almanac*. New York: Ballantine.

Lott, John R. Jr. October 19, 2007. "Get Your Hunt On: It's Good for the Animals," *National Review Online*. Available from http://article.nationalreview.com/?q=NGYzZjdjNz JiYWQwMWFkYmMyNTA1MzY1ZjFlMDAyODQ=

Lucas, Joanna. November 13, 2007. "Coming Home," Peaceful Prairie Sanctuary blog. Available from http://peacefulprairie.blogspot.com/2007/11/coming-home_13.html

Lyman, Howard. 2001. *The Mad Cowboy: Plain Truth from the Cattle Rancher Who Won't Eat Meat*. New York: Scribner.

Mason, Jim. 1993. *An Unnatural Order: Uncovering the Roots of Our Domination of Nature and Each Other.* New York: Simon and Schuster.

Mason, Jim, and Peter Singer. 1990. *Animal Factories: What Agribusiness Is Doing to the Family Farm, the Environment, and Your Health.* New York: Three Rivers Press.

Miniter, Frank. 2007. *The Politically Incorrect Guide to Hunting.* Washington, DC: Regnery.

Motavalli, Jim. 2002. "Across the Great Divide: Environmentalists and Animal Rights Activists Battle Over Vegetarianism." *E Magazine* 13(1). Available from http://www.emagazine.com/view/?145&src=

PETA, "Free-Range and Organic Meat, Eggs, and Dairy Products: Conning Consumers?" Available from http://www.peta.org/mc/factsheet_display.asp?ID=96

Rifkin, Jeremy. 1992. *Beyond Beef: The Rise and Fall of the Cattle Culture.* New York: Dutton.

Pimentel, David. August 7, 1997. "Eight Meaty Facts about Animal Food." *Cornell University Science News.* Available from http://www.news.cornell.edu/releases/Aug97/livestock.hrs.html

Pollan, Michael. 2007. *The Omnivore's Dilemma: A Natural History of Four Meals.* New York: Penguin.

Pickover, Michel. 2005. *Animal Rights in South Africa.* Wetton, Cape Town: Double Story Books.

Robbins, John. 1998. *Diet for a New America: How Your Food Choices Affect Your Health, Happiness, and the Future of Life on Earth.* 2nd edition. Tiburon, CA: HJ Kramer.

Robbins, John. 2001. *The Food Revolution: How Your Diet Can Help Save Your Life and Our World.* San Francisco: Conari Press.

Regan, Tom. 1983. *The Case for Animal Rights.* Berkeley: University of California Press.

Regan, Tom. 2004. *Empty Cages: Facing the Challenge of Animal Rights.* Lanham, MD: Rowman & Littlefield.

Sapontzis, Steve F., ed. 2004. *Food for Thought: The Debate over Eating Meat.* Amherst, NY: Prometheus Books.

Singer, Peter. 2001. *Animal Liberation.* New York: Harper Perennial.

Spencer, Colin. 2004. *Vegetarianism: A History.* New York: Four Walls Eight Windows.

Stuart, Tristram. 2007. *The Bloodless Revolution: A Cultural History of Vegetarianism from 1600 to Modern Times.* New York: W. W. Norton.

Steinfeld, H. et al. November 2006. "Livestock's Long Shadow: Environmental Issues and Options," Livestock, Environment and Development Centre. Available from http://www.virtualcentre.org/en/frame.htm

Varner, Gary E. 2002. "Can Animal Rights Activists Be Environmentalists?" In *Environmental Ethics: An Anthology,* ed. Holmes Rolston and Andrew Light. Malden, MA: Blackwell.

Varner, Gary E. 2002. *In Nature's Interests? Interests, Animal Rights, and Environmental Ethics.* New York: Oxford University Press.

Vesterby, Marlow, and Kenneth Krupa. 1997. "Major Uses of Land in the United States, 1997," *Statistical Bulletin* (973). Available from http://www.ers.usda.gov/publications/sb973/sb973.pdf

Walters, Kerry, and Lisa Portmess, eds. 1999. *Ethical Vegetarianism: From Pythagoras to Peter Singer.* Albany: State University of New York Press.

Watson, Donald. November 1944. Article in *The Vegan News* 1 (1): Available from http://www.ukveggie.com/vegan_news/vegan_news_1.pdf

Worldwatch Institute. July 2, 1998. "United States Leads World Meat Stampede." Available from https://www.worldwatch.org/node/1626.

Steven Best

VIRTUE ETHICS

Virtue ethics is the branch of ethics that focuses on issues of character, excellence, and human flourishing. Some philosophers believe that virtue ethics is the correct approach to ethics in general; other philosophers see it as a necessary supplement to Kantian and utilitarian approaches that focus on the treatment of others (Swanton 2003). In recent years more philosophers have come to believe that virtue ethics is a valuable approach to environmental issues, although they have different reasons for that belief.

THE VALUE OF VIRTUE ETHICS

Some philosophers see standard attempts to prove the intrinsic value or moral considerability of nonhuman beings as inconclusive or failed. In embracing environmental virtue ethics they seek a more solid grounding for environmentalism, arguing that protecting the environment is necessary to promote human well-being or flourishing (Hursthouse 2007). Other philosophers find some intrinsic value claims convincing but believe that appeals to human flourishing provide further arguments for environmental protection. They attempt to supplement appeals to altruism with appeals to people's enlightened self-interest (O'Neill 1993). Yet other philosophers see an urgent need to specify the kinds of character traits that are needed to live environmentally sustainable lives. For this group environmental virtue ethics is the proper framework in which to discuss the creation of sustainable societies (Newton 2003).

INFLUENTIAL WORKS

Four books have been particularly influential in the development of environmental virtue ethics. John O'Neill's *Ecology, Policy, and Politics* (1993) was an early attempt to ground environmentalist claims in human flourishing. Rejecting accounts of human well-being based on subjective states or mere preference satisfaction, O'Neill developed an Aristotelian conception of well-being in terms of objective goods such as health, friendship, knowledge, and the ability to develop people's natural capacities. Nature must be protected both to preserve the basic resources people need to live and to protect opportunities for people to develop their higher capabilities, such as scientific knowledge, artistic creation, and personal connection to the natural world. Like many environmental virtue ethicists, O'Neill attempted to rein in an overly economic view of the world. He questioned the wisdom of cost-benefit analysis, asserting that it provides "policy without [political] debate" (O'Neill 1993, p. 78), and argued that people must limit the power of markets through regulations that will uphold strong conceptions of the common good.

Louke van Wensveen's *Dirty Virtues* (2000) showed philosophers that popular and scholarly environmental discourse already contained discussions of virtue and character (an appendix lists 189 virtues and 174 vices mentioned in the previous three decades of environmental literature). She found this environmental virtue discourse to be productive and dynamic both in its discussion of new ecological virtues such as attunement and earthiness and in its reinterpretation of established virtues such as frugality and gratitude in an ecological context. Wensveen's analysis attempted to bring order to that discourse by developing criteria for what should count as genuine ecological virtue. Her procedure was a mixture of the old and the new; to Aristotelian criteria focused on the social sustainability of various human qualities she added the newer criteria of psychological and ecological sustainability.

Philip Cafaro's *Thoreau's Living Ethics* (2004) argues that Henry David Thoreau's life and writings provide a coherent and inspiring environmental virtue ethic that links attentiveness to nature to human excellence and flourishing. Thoreau thus points the way toward a comprehensive, life-affirming environmental ethics in which the traditional "thou shalt nots" of environmentalism are complemented by a description of positive ideals of character. By recognizing nature's value, people enrich their lives. By restraining physical consumption, people are more likely to lead healthy and enjoyable lives and allow future generations to do the same thing. By devoting themselves to pursuits higher than moneymaking, they act in their enlightened self-interest, with great benefits for the many other species with which they share the earth.

Cafaro also argued that Thoreau's evolutionary experimentalism—in which human nature and hence human virtue may change over time, and our particular virtue judgments are always assumed fallible and hence must be tested in life—provides a more plausible foundation for environmental virtue ethics than do the unchanging human nature and timeless objectivity to which many contemporary virtue ethicists subscribe (following Aristotle).

Ronald Sandler's *Character and Environment* (2007) is a theoretically rigorous defense of environmental virtue ethics that contains detailed positions on issues such as the proper criteria for judging right action and whether virtue ethics approaches are necessarily anthropocentric. It provides a naturalistic account of the proper characterization of virtue that is grounded in a broad sense of human flourishing but leaves open the possibility that virtues may be virtues because they recognize or promote other goods, such as the intrinsic value of nonhuman beings. Surveying the many accounts of environmental virtue, Sandler developed a typology that does justice to the many ways in which character traits may be environmental virtues. His typology includes virtues of sustainability (temperance and frugality), communion with nature (attentiveness and wonder) and respect for nature (care, compassion, and ecological sensitivity), stewardship (honesty and diligence), and environmental activism (cooperativeness, perseverance).

CRITICISMS AND DEFENSES

One criticism of environmental virtue ethics is that its account of environmental virtues is undertheorized and ad hoc. In light of Wensveen's list of 189 environmental virtues, it is not clear how to make sense of such a plenitude. Environmentalists can identify their favorite virtues and exhort like-minded people to cultivate them, but what imposes discipline on discussions of particular virtues? What allows philosophers to put those accounts together into a cohesive whole, whether in an account of general human virtue or in terms of particular lives well lived? Because environmental virtue ethicists see the need for new virtues and the reinterpretation of old ones, they cannot rely on accepted accounts of virtue from general ethical theory; in addition, general virtue theorists disagree about what makes a virtue a virtue.

Environmental virtue ethicists have tried to answer this criticism by developing specific criteria for what makes a virtue a virtue. Wensveen (2000) argued that genuine virtues must contribute to ecological sustainability because environmental degradation undermines the ability to live and pursue virtue. Most proponents of environmental virtue ethics attempt to ground their general accounts of virtue in a substantive account of human flourishing; the virtues are those qualities which constitute or contribute to that flourishing. Their accounts of this flourishing differ, however, and more work needs to be done to generate an account of human flourishing that also upholds diversity and possibility. Moving beyond exhortation, the hope is to converge on the personal qualities that make people good people who are likely to live well and protect nature.

Reflecting a criticism of virtue ethics in general, some argue that environmental virtue ethics is not sufficiently action-guiding. They claim that virtue ethics tells people what sort of character to cultivate but cannot do what an ethical philosophy must do: tell people how to act in particular situations. One response to this criticism is that it is not the job of ethics to provide rules and that the failure of deontology and utilitarianism to supply compelling rules for dealing with hard ethical cases shows that such a goal is impossible to reach (Hursthouse 2007). Cultivating the virtues, especially practical wisdom, will put people in the best position to make good moral decisions in hard cases and act on them. Another response is to develop "v-rules" that specify how people with the requisite virtues would respond to particular ethical challenges (Sandler 2007). Thus, it is possible to say something about how a temperate person might respond to the blandishments of modern commercial society or how someone with the virtues of attentiveness and ecological sensitivity might decide which forms of recreation are acceptable in natural areas.

Perhaps the strongest answer to the objection that virtue ethics is not sufficiently action-guiding is to remove the focus from virtue and place it on the notion of flourishing. Peter Wenz (2005) argued that there is evidence that the excesses of modern consumerism harm human health, distract people from more enjoyable activities, and impede human flourishing in other ways; thus, there are good reasons to rein them in. Cafaro argues that the pursuits of the naturalist lead to greater health, happiness, knowledge, and wisdom; thus, there are good reasons to explore and protect nature (Cafaro 2001). Once the definition of human flourishing is broadened to include the flourishing of all living things, people have even stronger guidance in limiting environmentally destructive practices and protecting nature.

The most persistent criticism of environmental virtue ethics states that it is anthropocentric because it stresses the contribution of environmental protection to people's well-being and focuses excessively on human character. This focus on people leads to the same human selfishness that created contemporary environmental problems. Holmes Rolston makes this criticism in "Environmental Virtue Ethics: Half the Truth but Dangerous as a Whole" (Sandler and Cafaro 2005): If people protect nature solely to help themselves live better lives, they are not acting from correct motives. Even worse, they will sacrifice nature if it is in their own selfish interests to do so.

In response, a few environmental virtue ethicists embrace anthropocentrism. Yes we are anthropocentric, they concede, but any reasonable ethics necessarily focuses on human needs and wants. They claim that those considerations provide powerful reasons for environmental protection that will actually motivate people to protect nature.

Most environmental virtue ethicists, however, make a place in their philosophies for the intrinsic value or moral considerability of nonhuman nature. Some say that they ground ethics in human and nonhuman flourishing; the virtues are those qualities which further both (Cafaro 2004). Others say that people can best get from the "is" of nature's intrinsic value to the "ought" that compels people to protect that value, by bringing in issues of human flourishing (O'Neill 1993). Still others argue that a pluralistic account of moral considerability in which different kinds of entities are considered from the perspective of the appropriate virtues is more plausible and accurate than monistic accounts of intrinsic value (Sandler 2007). All these approaches emphasize that human flourishing and the flourishing of nature are necessarily intertwined; the same actions and personal characteristics allow people to be good neighbors and citizens and good environmentalists.

APPLICATIONS

Ethical approaches should not just be judged by how well proponents respond to criticisms, or defend the weakest links in their arguments. Perhaps more important is how well they help philosophers explore new areas and illuminate old ones. Some of the best work in environmental virtue ethics has focused on giving detailed "thick" accounts of particular environmental virtues. This includes Geoffrey Frasz on "Benevolence as an Environmental Virtue" (Sandler and Cafaro 2005) and on friendship ("What Is Environmental Virtue Ethics That We Should Be Mindful of It?" Cafaro 2001); and Louke van Wensveen on temperance ("Attunement: An Ecological Spin on the Virtue of Temperance," Cafaro 2001) and on the Seven Deadly Sins (Wensveen 2000). It includes Thomas Hill, Jr. on humility ("Ideals of Human Excellence and Preserving Natural Environments," Sandler and Cafaro 2005), Philip Cafaro on wisdom ("The Naturalist's Virtues," Cafaro 2001), and Rosalind Hursthouse (2007) on wonder. This work can deepen people's understanding of environmentalism and provide a better sense of what it would mean to be benevolent, temperate, properly humble, and wise people.

Lisa Newton's textbook *Ethics and Sustainability* (2003) provides another example of applied philosophy built on a base of environmental virtue ethics. After giving an account of environmental virtue grounded in Aristotle and Aldo Leopold, among others, Newton works out the kinds of technological, social, and economic policies needed to create sustainable, flourishing societies. She also discusses the sorts of lives people will have to lead to make those societies work. The book culminates in an account of simplicity as the virtue most in demand at this point in societal evolution. Newton's work shows that environmental virtue ethics can help specify the sorts of lives people need to live to do justice to nature and live good lives within it. It thus helps people think comprehensively and practically about how to make sustainable, just, generous human societies a reality.

SEE ALSO *Cost-Benefit Analysis; Sustainability; Thoreau, Henry David; Utilitarianism.*

BIBLIOGRAPHY

Cafaro, Philip, ed. 2001. *Environmental Virtue Ethics*. Special issue of *Philosophy in the Contemporary World*, Vol. 8, No. 2.

Cafaro, Philip. 2004. *Thoreau's Living Ethics: Walden and the Pursuit of Virtue*. Athens: University of Georgia Press.

Hursthouse, Rosalind. 2007. "Environmental Virtue Ethics." In *Working Virtue: Virtue Ethics and Contemporary Moral Problems*, ed. Rebecca L. Walker and Philip J. Ivanhoe. Oxford: Clarendon Press.

Newton, Lisa H. 2003. *Ethics and Sustainability: Sustainable Development and the Moral Life*. Upper Saddle River, NJ: Prentice Hall.

O'Neill, John. 1993. *Ecology, Policy, and Politics: Human Well-Being and the Natural World*. London and New York: Routledge.

Rolston, Holmes. 2005. "Environmental Virtue Ethics: Half the Truth but Dangerous as a Whole." In *Environmental Virtue Ethics*, ed. Ronald Sandler and Philip Cafaro. Lanham, MD: Rowman & Littlefield.

Sandler, Ronald L. 2007. *Character and Environment: A Virtue-Oriented Approach to Environmental Ethics*. New York: Columbia University Press.

Sandler, Ronald, and Philip Cafaro, eds. 2005. *Environmental Virtue Ethics*. Lanham, MD: Rowman & Littlefield.

Swanton, Christine. 2003. *Virtue Ethics: A Pluralistic View*. New York: Oxford University Press.

Wensveen, Louke van. 2000. *Dirty Virtues: The Emergence of Ecological Virtue Ethics*. Amherst, NY: Humanity Books.

Wenz, Peter. 2005. "Synergistic Environmental Virtues." In *Environmental Virtue Ethics*, ed. Philip Cafaro and Ronald Sandler. Lanham, MD: Rowman & Littlefield.

Philip Cafaro

VON HUMBOLDT, ALEXANDER

1769–1859

Alexander von Humboldt, who was a polymath, an explorer, and one of the founders of modern geography, was born in Berlin on September 14, 1769, the year in which James Cook sailed the *Endeavour*, and died in that city on May 6, 1859, the year of publication of Charles Darwin's *Origin of Species*. A Prussian, he received training in biology, geology, and political economy at Göttingen and Frankfurt. After a period in the Prussian civil service, he sailed with Aimé Bonpland for South America in 1799 and travelled across the llanos, the Andes, and Mexico. On that journey (Von Humboldt and Bonpland 1881 [1852]) he determined heights and positions by using a barometer and a chronometer, developed ideas about the geography of plants (Von Humboldt and Bonpland 1977 [1805]), and made an innovative attempt to relate the political and economic conditions of New Spain to its physical geography and to natural and human resources (Von Humboldt 1988 [1811]). Later in his life he visited the Urals and the Altai.

In *Cosmos*, which was published between 1845 and 1862, von Humboldt stressed relationships between phenomena and attempted to explain the links between phenomena in different climatic zones. He believed that the earth is an integrated organic whole. He also had a vision of man as a part of nature:

> The general picture of nature which I have endeavoured to delineate would be incomplete if I did not venture to trace a few of the most marked features of the human race, considered with reference to physical gradations—to the geographical distribution of contemporaneous types, to the influence exercised upon man by the forces of nature and the reciprocal, although weaker, action which he in his turn exercises on these natural forces. Dependant, although in a lesser degree than plants and animals, on the soil, and on the meteorological processes of the atmosphere with which he is surrounded, escaping more readily from the control of natural forces, by activity of his mind, and the advance of intellectual cultivation, no less than, by his wonderful capacity of adapting himself to all climates—man everywhere becomes most essentially associated with terrestrial life. (Von Humboldt 1850–1859, vol. 1, pp. 360–361)

Von Humboldt, as D. N. Livingstone pointed out, "constantly sought for the universal behind the particular, for underlying patterns and unities that tied nature together in such a beautiful, functioning system" (1992, p. 135).

Cosmos included a survey of the current theories about the structure of the universe, a sketch of the historical progress of science and of geographical exploration and discovery, an exposition of the content of the disparate sciences, and a discussion of nature poetry, aesthetics, and landscape painting.

Von Humboldt was influenced by many other thinkers—Kant, Rousseau, J. G. Forster, Agassiz, Ritter,

Goethe, von Liebig, and Schiller, among others—and had a strong influence on scientists such as Darwin and Alfred Russel Wallace. His holistic views were fundamental for the development of geography (Mathewson 2006), but he also demonstrated the importance of detailed and reliable measurements, established many of the basic tenets of plant geography, and was a founder of global and regional climatology, a term he probably coined (Dickinson 1969). In his essay on Mexico he demonstrated that he was a regionalist, illustrating the interrelated conditions that give its character to an area. This was a striking contrast to the encyclopedic compilations of many previous geographers, topographers, and explorers.

Von Humboldt's work spawned what has been termed Humboldtian science:

> To signify a scientific style that conducted observations with the latest instruments, corrected measurements for errors, and linked these to mathematical laws; constructed maps of isolines connecting points with the same average values; identified large, even global, units of investigation; and used nature rather than the laboratory as a site of investigation. The term as applied to nineteenth century science has since acquired other connotations, including connecting different types of large-scale phenomena, demonstrating their independencies, seeking a universal science of nature, and using large-scale international organisational structures to execute local readings as part of a global effort. (Olesko 2003, p. 384)

Von Humboldt was the last truly universal man. His funeral was one of the most imposing state funerals in the history of Berlin, and his name is given to over a thousand places in the world.

SEE ALSO *Darwin, Charles; Environmental Philosophy: IV. Nineteenth-Century Philosophy; Regionalism.*

BIBLIOGRAPHY

WORKS BY ALEXANDER VON HUMBOLDT

Von Humboldt, Alexander. 1850–1859. *Cosmos: A Sketch of a Physical Description of the Universe,* trans. E. C. Otté. New York: Harper & Brothers.

Von Humboldt, Alexander. 1988 [1811]. *Political Essay on the Kingdom of New Spain,* trans John Black. Norman: University of Oklahoma Press.

Von Humboldt, Alexander, and Aimé Bonpland. 1881 [1852]. *Personal Narrative of Travels to the Equinoctial Regions of America during the Years 1799–1804.* London: G. Bell & Sons.

Von Humboldt, Alexander, and Aimé Bonpland. 1977 [1805]. *Essai sur la Géographie des Plantes.* New York: Arno Press.

WORKS ABOUT ALEXANDER VON HUMBOLDT

Dickinson, Robert E. 1969. *The Makers of Modern Geography.* London: Routledge & K. Paul.

Livingstone, D. N. 1992. *The Geographical Tradition: Episodes in the History of a Contested Enterprise.* Oxford, UK and Cambridge, MA: Blackwell.

Mathewson, Kent. 2006. "Alexander von Humboldt's Image and Influence in North American Geography, 1804–2004." *Geographical Review* 96: 416–438.

Olesko, K. 2003. "Humboldtian Science." In *The Oxford Companion to the History of Modern Science,* ed. J. L. Heilbron. Oxford, UK and New York: Oxford University Press.

Andrew S. Goudie

WAR

War, armed conflict with the intent to kill and destroy, has been pursued between groups, societies, and nations throughout human history. Such activity, with all of its associated devastation, continues into the twenty-first century, despite a widespread international attempt in 1928 to renounce for all time at least war between nations. Its brutality also continues into the 2000s, despite a lengthy series of intergovernmental attempts—especially in 1899, 1907, 1929, 1949, and 1977—at constraining the more vicious and pernicious aspects of international war. Its frequency notwithstanding, few comparable attempts have been made over the years to constrain internal (noninternational, civil, insurgency) war.

The environment—that is, the world at large, encompassing the atmosphere, lithosphere, hydrosphere, and biosphere, upon which all humankind depends for its well-being and very survival—is undergoing ever more serious deterioration by the civil sector of society. It is thus particularly troubling that damage to the environment is also an inevitable consequence of warfare. Such wartime damage can be incidental or intentional. It occurs within the theater of military operations and beyond it. Moreover, environmental damage occurs not only during wartime, but also as a result of preparing for war and as a result of some postwar recovery efforts. Perversely, it must be noted that war can also lead to a number of beneficial environmental outcomes.

This entry covers both the environmental consequences of war and the environmental causes of war. It further examines the ethical development and implications of wartime environmental protection, whether as an anthropocentric extension of social justice or perhaps in part as an ecocentric concern in its own right, and briefly also religious norms related to war and the environment.

INCIDENTAL ENVIRONMENTAL IMPACT ON THE BATTLEFIELD

Incidental environmental damage to the battlefield is an inevitable outcome of war. Such incidental damage can result from the profligate use of high-explosive munitions against enemy personnel and matériel. Another common source of incidental damage is the use of tanks and other heavy off-road vehicles. These environmental insults are especially disruptive of local habitats and the wildlife depending on them. Battle-related actions can also result in less persistent local soil, water, and air pollution.

Conversely, a theater of military operations often precludes the usual peacetime exploitation (often over-exploitation) of a rural area by hunting, logging, fishing, grazing, or farming. This permits population recoveries of the local flora and fauna, the most notable case being in the demilitarized zone between North and South Korea.

INTENTIONAL ENVIRONMENTAL IMPACT ON THE BATTLEFIELD

Intentional environmental impact in a theater of military operations can take various forms, including especially forest clearing, oil releases, area denial (e.g., land mines), and on occasion even the use of chemical, biological, or nuclear weapons.

Intentional forest destruction is carried out during wartime primarily to deny the enemy cover and concealment.

Agent Orange. *The aftermath of a herbicidal attack with agent orange by the United States during the Vietnam conflict of 1961–1975 against a lush mangrove forest for the immediate purpose of denying the enemy cover and concealment is depicted here. It was U.S. actions of this sort that awoke a widespread revulsion against massive military disruption of the environment and, more generally, contributed significantly to ushering in the newly emerging environmental ethic in the world at large.* PHOTO TAKEN BY ARTHUR H. WESTING IN GIA DINH PROVINCE, SOUTH VIETNAM, AUGUST 15, 1970.

This has been accomplished by spraying the area with herbicides (including the notorious agent orange), by using heavy tractors equipped with special forest clearing blades (so-called Rome plows), by saturation bombing, and, at propitious times and places, by setting self-propagating wild fires. Depending upon the site and severity of attack, recovery from any such assault can take years to decades. Indeed, it was U.S. forces' sustained use of all of these tactics during the Vietnam War of 1961–1975 that alerted the world to the levels of environmental destruction that can be associated with war.

Intentional oil releases are possible in certain theaters of military operation for purposes of enemy harassment or area denial as well as for punitive purposes. This can be variously accomplished by sabotaging oil wells; by breaching pipelines, collection centers, and storage tanks; and by attacking oil tankers—in each instance with or without igniting the escaping oil. The most spectacular example of such action was by Iraqi forces during the Persian Gulf War of 1991. Much rural surface area, groundwater, and Persian Gulf seawater and shoreline became severely contaminated with liquid oil, and the torched escaping oil heavily contaminated the regional atmosphere with dense smoke (soot plus combustion gases), with adverse health effects on wildlife, livestock, and humans.

Area denial with land mines (both antivehicle and antipersonnel) or with remotely delivered scatterable mines and cluster-bomb submunitions is a common military practice for purposes of hindering, slowing down, or channeling the movements of enemy forces, as well as of sapping their morale. Such area denial is a pernicious military tactic because the area remains perilous for years, even decades, after its battlefield use, thereby hindering subsequent forestry, herding, farming, or other development. Postwar rehabilitation efforts are a daunting task that is technically difficult, time-consuming, expensive, exceedingly dangerous, often environmentally disruptive, and never fully successful. On the other side of the ledger, after a war, previously overgrazed or otherwise overexploited minefields can undergo considerable habitat recovery, both floral and faunal. For example, such recovery occurred following World War II in the heavily mined rangelands in Libya and elsewhere in North Africa. A somewhat comparable maritime example was the recovery during World War II of the drastically depleted North Sea fishery, where a combination of naval actions and sea mines made it long impossible for fishing boats to enter.

Chemical, biological, and nuclear weapons are weapons of mass destruction that lead to intentional environmental disruption. Antipersonnel chemical agents were used extensively (and openly) by both Allied and Axis forces during World War I, but not before or since on such a grand scale. Antiplant chemical agents are covered above. Biological agents have been used essentially only once in modern times, on a relatively minor scale (and secretly) by Japanese forces during World War II. Nuclear weapons have also been used only once, by U.S. forces during World War II over Hiroshima and Nagasaki. The potential for long-term environmental disruption by any of these so-called unconventional weapons is truly enormous.

INTENTIONAL RELEASE OF DANGEROUS FORCES

Under certain conditions, it is possible for a belligerent to manipulate some component of the natural or built environment so as to result in the release of pent-up energy, so-called dangerous forces. This sort of hostile action—often referred to as *environmental warfare*—becomes especially tempting when the hostile manipulation involves a relatively modest expenditure of effort (i.e., of triggering energy) leading to the release of a substantially greater amount of directed destructive energy. Environmental modifications of particular concern here involve the pent-up forces contained in fresh-water impoundments, nuclear power stations, certain industrial facilities, and, to a lesser extent, forest and other wild lands.

Water impoundments formed by the many hundreds of major dams that have been constructed throughout the world contain enormous quantities of water. Many such dams could be breached with relative ease either through direct attack or sabotage, and the release of the impounded reservoir would cause immense levels of death and destruction. Hostile actions of this nature have been spectacularly successful in various wars, for example, the Sino-Japanese War of 1937–1945, World War II, and the Korean War of 1950–1952.

Nuclear facilities have become essentially permanent additions to the human environment. Almost two hundred clusters of nuclear power stations, plus a number of additional nuclear reprocessing plants and nuclear waste repositories, are distributed in over thirty nations. These sites are amenable to direct attack or sabotage, with the possible attendant release into the surrounding area, measurable in thousands to millions of hectares, of iodine-131, cesium-137, strontium-90, and other radioactive elements. What has been learned from the aftermath of U.S., British, and French nuclear testing on several Pacific islands in the 1940s and 1950s and from the Soviet Chernobyl accident of April 1986 is that the contaminated zone essentially defies rehabilitation. Thus, both the Pacific islands and the region surrounding Chernobyl remain uninhabitable despite massive attempts at decontamination.

Certain industrial facilities, if attacked, can release a cloud of toxic chemicals spreading over hundreds of hectares and proving variously detrimental to plants, animals, and humans. Providing ample evidence of this source of wartime danger are the horrifying releases into the air in Seveso, Italy, of dioxin from a factory explosion in July 1976, and similarly in Bhopal, India, of methyl isocyanate in December 1984, as well as the catastrophic explosion in Mexico of a huge liquefied-natural-gas storage facility in Mexico City in November 1984.

Wild fires can be started and become self-propagating over large areas under special habitat and weather conditions in forest, grassland (prairie), and tundra ecosystems. Initiated by napalm or other incendiary munitions (and sometimes preconditioned by herbicidal attack), such fires have the potential to cause extensive damage to wildlife, livestock, and humans, and under certain site conditions lead to severe soil erosion and loss of nutrients (so-called *nutrient dumping*). Grassland recovery is measurable in years, forest recovery in decades, and tundra recovery in many, many decades.

WARTIME ENVIRONMENTAL IMPACT BEYOND THE BATTLEFIELD

Environmental disruptions caused by military activities during wartime but beyond the theater of military operations can be more or less severe, and are generally excused as part and parcel of the necessities of war. These disruptions derive from the construction of base camps, fortifications, and lines of communication, and also often from armed forces' heavy exploitation of timber, food, and feed. Preemptive self-inflicted scorched-earth tactics in anticipation of an enemy advance can be utterly devastating, as exemplified during World War II by such actions carried out by Soviet forces in western Russia and also by German forces in northern Norway. Finally, persons displaced from a war zone and concentrated into refugee camps can cause severe environmental damage to the surrounding flora and fauna, as seen during and following a number of early-twenty-first-century internal wars in Africa.

PREWAR AND POSTWAR ENVIRONMENTAL IMPACTS

Whether at peace or war, most nations continually maintain armed forces. Such armed forces are kept for various reasons, especially (1) to deter an attack from outside the country's borders or, failing that, to defend against such an attack; (2) to threaten an attack on another nation in support of some foreign-policy objective or, failing that, to carry out such an attack; and (3) to deter or quell internal uprisings. In fact, a majority of the wars since the 1980s have fallen into the third of these categories.

There are manifold environmental ramifications of nations' maintaining armed forces for any or all of the three reasons mentioned above. Environmental disruption can result from any of the following:

- Establishing military fortifications and other military facilities
- Equipping and supplying armed forces with weapons and other military needs
- Disposing of those items once they become obsolete or otherwise unwanted
- Training armed forces and testing the weapons they use
- Routine deployment of armed forces nationally, in other nations, and in areas beyond any national jurisdiction.

Among benefits, the often large exclusion zones surrounding nuclear and other sensitive military facilities, especially in the industrialized nations, often become important de facto nature reserves and wildlife sanctuaries. Moreover, during peacetime, a number of nations (India being a prime example) employ a portion of their armed forces to police their nature reserves, controlling wildlife and timber poachers and other harmful trespassers.

ENVIRONMENTAL CAUSES OF WAR

Over the past nine millennia or more, war has been widely recognized as a successful societal strategy for acquiring needed or desired land and other natural resources. So it should come as no surprise that environmental scarcities, especially in the nonindustrialized world, lead to human violence—and indeed already have in such diverse places as Haiti, Pakistan, the Philippines, and South Africa.

The growing numbers of people, especially in the nonindustrialized world—together with their ever expanding demands for such increasingly scarce natural resources as cropland, forestland, freshwater supplies, and ocean fish—have increasingly significant social consequences. In many such impoverished nations, the social consequences are certain to cause further internal unrest, population movements, and even occasional insurgencies. And in many instances the adversities of climate change exacerbate this intractable problem of growing environmental scarcities. Indeed, among military planners, climate change is already identified as a threat multiplier, perhaps even promoting terrorism, within many of the more fragile and unstable regions of the world.

SOCIETAL CONSTRAINTS ON ENVIRONMENTAL DAMAGE IN WARFARE

Social values and environmental values have been of philosophical and religious concern to both Western and non-Western cultures since ancient times. But only since the late twentieth century have social and environmental concerns begun to intersect and become mutually reinforcing, for example, in the emerging ethical norms related to the notion of sustainable development. Modern tangible manifestations first of social values and subsequently of environmental values did not arise until the after World War II. And (as indicated earlier) the widespread emergence of recognizable ethical concerns over wartime environmental damage had to await the depredations of the Vietnam War of 1961–1975.

It was the human atrocities associated with World War II that fostered and crystallized worldwide ethical concerns and gave rise to meaningful expression of those concerns. Such concerns were first tangibly expressed through passage of the 1948 Convention on the Prevention and Punishment of the Crime of Genocide (UN Treaty Series No. 1021), soon followed by adoption of the 1948 Universal Declaration of Human Rights (United Nations General Assembly, resolution 217 A [III]). Those two early postwar manifestos would not have received the overwhelming governmental support they each enjoyed had they not been expressing the newly burgeoning cultural norms (ethical values) they were codifying.

Neither of these two instruments recognized environmental concerns. In fact, widely shared environmental norms did not begin to surface for another two decades or so, and gained momentum only over a period of some further years. Thus, expressions of emerging, widely shared core environmental values did not surface until adoption of the 1972 Stockholm Declaration on the Human Environment (UN document No. A/conf.48/14/rev.1), to be followed a decade later by the almost universal adoption of the 1982 World Charter for Nature (UN resolution No. 37/7), and that in turn by the 1992

Rio Declaration on Environment and Development (UN document No. A/conf.151/5/rev.1). Unsurprisingly, all three of these instruments were based essentially on anthropocentric concerns. But interestingly, all three variously proclaimed the need to avoid wartime damage to the environment.

Ethical concerns related to war—aimed not at its abolition but rather at civilizing its conduct—began to be seriously expressed as early as 1899 and 1907 via the then almost universal adoption of the core international treaties comprising what came to be known as the law of war, although later often referred to as international humanitarian law, together with the related and overlapping body of international arms control law. Largely through the tireless efforts of the International Committee of the Red Cross, the law of war was updated and enlarged at various times since, especially in 1929, 1949, and 1977. Most of the law of war concerns international war, and again not surprisingly, it derives almost entirely from anthropocentric concerns.

Three major bodies of international law have the potential for expressing (revealing) widely held cultural norms or ethical values relating to the environmental impact of war. Two of these—international environmental law and international human rights law—fail to do so. The third, the law of war, does contain a number of provisions that incidentally offer environmental benefit, among them the clearly expressed nearly universal anathema against the use of chemical and biological weapons and the expanding revulsion against the use of antipersonnel land mines. (Curiously, as of 2008, there had not clearly surfaced a similar sentiment that would renounce the use of nuclear weapons, doubtlessly the most socially and environmentally destructive weapon of war.) Of comparable environmental benefit, substantial support based on social concerns has been demonstrated for constraining attacks upon certain targets, including agricultural areas, dams, and nuclear electrical generating facilities, the latter two because they would release the dangerous forces noted earlier. Similarly, it has long been widely accepted that an occupying power is enjoined from the nonusufructuary exploitation of forests and agricultural works in enemy territory.

The most interesting expression of a cultural norm concerning war and the environment was the addition to the law of war of the 1977 Protocol on International Armed Conflicts (UN Treaty Series No. 17512, articles 35.3 and 55.1), prohibiting means of warfare that may be expected to cause widespread, long-term, and severe damage to the natural environment. The sentiment expressed there was an unveiled international rebuke of U.S. actions during the Vietnam War and seemed to derive in part from ecocentric concerns.

Religious norms have had hardly any influence on societies' cultural norms, whether military or environmental. Indeed, the influence seems to flow largely in the other direction. By way of example, the Christian religion of Western Civilization and the Buddhist religion of Central and East Asian Civilizations are strongly pacifist in their origins and essence. Indeed, one of the central tenets of Christianity is, "Blessed are the peacemakers; for they shall be called the sons of God" (Matthew 5:9). And the very first of the five precepts of Buddhism is the rule to abstain from taking life, including animal life. Nonetheless, the associated societies have long records of military activity unsurpassed in magnitude and ferocity. And secular proenvironmental norms are now being ever more widely embraced despite the antienvironmental tenets found in both Eastern and Western (Genesis 1:28) teachings to be fruitful and fill the earth, to subdue it, and to have dominion over all other living things on earth.

SEE ALSO *Chernobyl; Ecological Restoration; Ecotage and Ecoterrorism; Environmental Law; Fire; Forests; Pollution; Rio Declaration; Water.*

BIBLIOGRAPHY

Ehrlich, Anne H., and John W. Birks, eds. 1990. *Hidden Dangers: Environmental Costs of Preparing for War*. San Francisco: Sierra Books.

Goldblat, Jozef. 2002. *Arms Control: The New Guide to Negotiations and Agreements*, 2nd edition. London: Sage.

Homer-Dixon, Thomas F. 1999. *Environment, Scarcity, and Violence*. Princeton, NJ: Princeton University Press.

Renner, Michael. 1997. "Environmental and Health Effects of Weapons Production, Testing, and Maintenance." In *War and Public Health*, ed. Barry S. Levy and Victor W. Sidel. New York: Oxford University Press.

Roberts, Adam, and Richard Guelff, eds. 1989. *Documents on the Laws of War*, 2nd edition. Oxford: Clarendon Press.

Stone, Christopher D. 1974. *Should Trees Have Standing? Toward Legal Rights for Natural Objects*. Los Altos, CA: William Kaufmann.

Westing, Arthur H. 1980. *Warfare in a Fragile World: Military Impact on the Human Environment*. London: Taylor and Francis.

Westing, Arthur H. 1987. "Ecological Dimension of Nuclear War." *Environmental Conservation* 14(4): 295–306.

Westing, Arthur H., ed. 1988. *Cultural Norms, War, and the Environment*. Oxford: Oxford University Press.

Westing, Arthur H. 1989. "Herbicides in Warfare: The Case of Indochina." In *Ecotoxicology and Climate*, ed. Philippe Bourdeau; John A. Haines; Werner Klein; and C. R. Krishna Murti. Chichester, UK: John Wiley.

Westing, Arthur H., ed. 1990. *Environmental Hazards of War: Releasing Dangerous Forces in an Industrialized World*. London: Sage.

Westing, Arthur H. 1996. "Core Values for Sustainable Development." *Environmental Conservation* 23(3): 218–225.

Westing, Arthur H. 1997. "Environmental Protection from Wartime Damage: The Role of International Law." In *Conflict and the Environment*, ed. Nils Petter Gleditsch. Dordrecht, Netherlands: Kluwer Academic.

Westing, Arthur H. 2003. "Environmental Dimension of the Gulf War of 1991." In *Security and Environment in the Mediterranean*, ed. Hans Günter Brauch; P. H. Liotta; Antonio Marquina, et al. Berlin: Springer Verlag.

White, Lynn, Jr. 1967. "Historical Roots of Our Ecologic Crisis." *Science* 155(3767): 1203–1207.

Arthur H. Westing

WARREN, KAREN J.
1947–

Karen J. Warren was born on Long Island, New York, on September 10, 1947. She received a bachelor of arts degree from the University of Minnesota in 1970. The University of Massachusetts at Amherst awarded her a master's degree in 1974 and a doctorate in 1978 for one of the earliest dissertations on environmental ethics. She was the second Ecofeminist Scholar-in-Residence at Murdoch University, Australia, in 1995; held the Women's Chair in Humanistic Studies at Marquette University in 2004; and as of 2008 is a professor of philosophy at Macalester College in St. Paul, Minnesota, where she has taught since 1985. An international expert in environmental ethics, feminism, and peace studies, she is known for her work in connecting those fields. Committed to philosophy as democratic practice, she has taught prison inmates and developed award-winning environmental curricula for schoolchildren.

Warren's publications and philosophical practices demonstrate the potential of ecofeminism to restructure human relationships with other humans and with nonhuman others; create life-affirming, intentional interspecies communities; and rethink philosophy as a forum for public thinking and action. Her editorial work in introducing ecofeminism to North American audiences has influenced feminist scholars worldwide. She shares the social ecologist's insight that environmental issues are connected with social justice, in contrast to approaches, such as wilderness-oriented ethics, that tend to overlook the relationship between environmental degradation and human suffering. Warren edited the section "Ecofeminism and Social Justice" in *Environmental Philosophy: From Animal Rights to Radical Ecology* (2005). That collection, widely used as a university text, appeared in a fourth edition. Her *Gendering the History of Western Philosophy* (2008), a more directly feminist project, is a fifteen-chapter book that pairs male and female philosophers through the centuries, with introductions by Warren and commentaries by other feminists.

Warren's best-known work as of 2008 is "The Power and Promise of Ecological Feminism," first printed in *Environmental Ethics* in 1990, widely reprinted, translated into five languages, and revised in 2005. She argues that there are important connections between the domination of women and the exploitation of nature; that is, women and nature are objects of the same oppressive patriarchal conceptual framework. She defines key features of an oppressive conceptual framework: oppositional value dualisms, hierarchical thinking, and a logic of domination that assumes that alleged superiority justifies the subordination of alleged inferiors. Patriarchy structures a man-woman dualism hierarchically to establish male privilege and uses that privilege to validate women's oppression.

Man-nature dualism is inseparable historically from gender oppression, and so feminism and environmental ethics are linked. However, ecofeminism unifies through solidarity, not sameness, and so it can celebrate difference and oppose all forms of domination. Warren articulates the boundary conditions of ecofeminism: It (1) does not promote social domination but is (2) contextualist, (3) pluralistic, and (4) inclusive; (5) it is theory in process that (6) takes objectivity as a false promise and opts instead for critical analysis of which biases are better than others, (7) makes a place for traditionally neglected values (e.g., care and friendship), moral emotions, and emotional intelligence, and (8) reconceives ethics and the meaning of being human.

A distinctive feature of ecofeminism is attitude. Using her experience in mountain climbing, Warren advocates caring over conquering to show the limitations of the conquering and the potential for success, growth, and development of caring. She defends first-person narration as ecofeminist methodology: It gives voice to felt sensitivity and experience, expresses values overlooked in mainstream ethics, develops a stance not imposed but emergent from diverse individual voices, and suggests what might count as an ethical solution in a particular situation.

Warren's first single-authored book is *Ecofeminist Philosophy: A Western Perspective on What It Is and Why It Matters* (2000), a philosophical work for nonspecialists that integrates previous research with original material to define ecofeminism, explore its potential, and defend it against typical criticisms. Key metaphors are a quilt and a fruit bowl. A quilt's borders define its limits, but different quilters can contribute to its design. Similarly, theory has necessary conditions, but diverse thinkers can add specifics. Ecofeminism is thus inclusive theory in process and always is revisable, like a quilt whose individual patches can be replaced or repaired without abandoning the larger design. The fruit bowl is used to explain ecofeminist ethics. Fruit is selected from the bowl to fit specific situations, for example, bananas when one is making banana bread, but none is inherently better. Similarly, no ethical principle has intrinsic superiority, though monist exclusionary principles that disallow emotional intelligence are prohibited. Methodologically, Warren revisits her arguments for taking empirical data seriously in a significant and original challenge to the traditional pursuit of universal truth through abstract, detached philosophical reasoning.

Warren's feminism has been criticized for political naiveté. She has claimed, for instance, that feminism can be defined as "the movement to end all forms of oppression," though a completely non-oppressive context is impossible. Nonetheless, Warren draws attention to women's lived experience of disproportionate harm in consequence of environmental degradation and shows that ecofeminism neither favors a privileged class of white Western women nor reinscribes essentialism. Rather, ecofeminism is a conceptually mature possibility of novel ethical relations between humans and between humans and nonhuman others.

SEE ALSO *Animal Ethics; Ecological Feminism; Queer Theory.*

BIBLIOGRAPHY

WORKS BY KAREN J. WARREN

Warren, Karen J., ed. 1991. *Hypatia* 6(1): Special Issue: *Ecological Feminism.* Revised and republished as *Ecological Feminist Philosophies* (Bloomington: Indiana University Press, 1996).

Warren, Karen J., ed. 1994. *Ecological Feminism.* London and New York: Routledge.

Warren, Karen J., and Duane L. Cady, eds. 1996. *Bringing Peace Home: Feminism, Violence, and Nature.* Bloomington: Indiana University Press.

Warren, Karen J., ed. 1997. *Ecofeminism: Women, Culture, Nature.* Bloomington: Indiana University Press.

Warren, Karen J. 2000. *Ecofeminist Philosophy: A Western Perspective on What It Is and Why It Matters.* Lanham, MD: Rowman & Littlefield.

Warren, Karen J. 2005. "The Power and Promise of Ecological Feminism, Revisited." In *Environmental Philosophy: From Animal Rights to Radical Ecology*, 4th edition, ed. Michael E. Zimmerman; J. Baird Callicott; George Sessions; et al. Upper Saddle River, NJ: Prentice Hall.

WORKS ABOUT KAREN J. WARREN

Frodeman, Robert, 1992. "Radical Environmentalism and the Political Roots of Postmodernism: Differences That Make a Difference." *Environmental Ethics* 14(4): 307–319. Reprinted in *Postmodernist Environmental Ethics*, ed. Max Oelschlaeger (Albany: State University of New York Press, 1995).

Trish Glazebrook

WASTE MANAGEMENT

Waste can be defined as unwanted material, or as material that the holder discards or intends to discard. The distinction between wastes and resources depends on one's willingness and technical ability to reuse artifacts and materials. One person's waste can be another person's resources.

Industrial mass manufacture and modern packaging have led to a dramatic and still ongoing increase in the volume and variety of the waste produced by households, industrial concerns, and other workplaces. Such waste includes various categories of hazardous waste, such as anthropotoxic, ecotoxic, infectious, and radioactive waste.

Waste management is often discussed in terms of a *waste hierarchy* that lists the major treatment methods in order of decreasing priority. One common variant of the hierarchy has six steps:

1. *Prevent* the creation of future waste.
2. *Minimize* the volumes and the harmful properties of future waste.
3. *Reuse* the artifact.
4. *Recycle* the material in the artifact.
5. Incinerate the waste, and use the heat to *recover energy*.
6. *Dispose* of the waste.

The waste hierarchy is not always applicable to hazardous waste. In the Third World, recycling often takes the form of scavenging. The waste pickers are poor, marginalized people, including children, with no or inadequate protection against toxic or infectious waste. When recycling involves serious occupational risks, it may be wrong to prefer recycling to final disposal.

WASTE DISASTERS

Public discussions on waste disposal have largely focused on a few major environmental disasters and scandals, among them the following: In 1978 a local resident in Love Canal, close to Niagara Falls, New York, suspected that her children's health problems were related to leakage of chemical waste in the neighborhood. It emerged that the entire settlement was built on top of an old landfill containing more than 20,000 tons of chemical waste. Leakage of several toxic chemicals was confirmed, a federal emergency was declared, and eventually more than 800 families were evacuated and relocated. The Love Canal events stimulated the creation of the Superfund Act, adopted by the U.S. Congress in 1980, which provides funds for cleanup of contaminated areas.

In 1982 the U.S. Environmental Protection Agency closed down Times Beach, a town in Missouri. More than

Computer Waste, Guiyu, China. *A worker with a hammer smashes a junk computer to pieces to obtain the usable parts. Environmental groups have contacted Guiyu, a cluster of five villages southwest of Shanghai, with a cautionary tale for poor countries that accept high-tech waste, drawing attention to the toxic chemicals released when the locals melt the computer parts to extract precious metals such as gold and platinum.* AP IMAGES.

2,000 inhabitants were relocated, and all buildings were demolished. The agency had discovered potentially dangerous levels of dioxin, originating in waste oil that had been sprayed on streets and parking lots to control dust. In hindsight, it has been questioned whether the actual risks were severe enough to justify the drastic measures taken by the Environmental Protection Agency at Times Beach. However, it should be remembered that these decisions were based on more uncertain information about the health effects of dioxins than what was later available.

In 1986 the cargo ship *Khian Sea* sailed with around 14,000 tons of toxic incinerator ash from Philadelphia to the Bahamas. After being denied permission to unload there, the ship sailed around the world for sixteen months, changing its name twice, in a vain search for a country where it could get rid of its freight. Eventually, most of the waste was dumped into the ocean.

In the early 2000s the exportation of e-waste, waste from electronic appliances, from industrial to developing countries became a major issue. One of the best-known examples is the Guiyu area in China's Guangdong Province, which receives large quantities of electronic waste, in particular from the United States. Here and in numerous places in the Third World, unprotected workers disassemble computers by hand and sort the material for recycling.

GEOGRAPHICAL JUSTICE

A large part of the ethical discussion of waste management has focused on questions of geographical justice. Empirical evidence shows that the disadvantages associated with waste treatment and disposal fall mostly on underprivileged persons, who benefit little or none from the products that gave rise to the waste. Internationally, rich countries figure as the major exporters of waste, and poor countries and regions as the major importers. Internally in the United States, studies have shown that communities receiving hazardous wastes from other communities are economically underprivileged and have high proportions of ethnic minorities. This phenomenon has been called *environmental racism.*

Companies and public authorities searching for places to site waste-treatment plants and dumps have often found it difficult to convince local populations and their elected representatives. Local resistance to waste facilities has been named the NIMBY phenomenon (not in my backyard). The allegation is that locals obstruct the creation of a facility that would contribute to the common good. The NIMBY phenomenon has sometimes been interpreted as an expression of irrational risk aversion of an uneducated public.

The NIMBY attitude is not necessarily irrational or unethical. Individuals need not be wrong in refusing to accept disadvantages for themselves (waste dumps in their neighborhood) to achieve advantages for others (removal of waste from other areas). In other contexts, taking a risk to help others is often considered supererogatory, virtuous beyond the call of duty. When members of a local community are required to accept a risk exposure not imposed on others, they may legitimately ask why they and not others are selected for this sacrifice.

Attempts are sometimes made to solve siting problems by compensation, for example, by offering monetary or other advantages to homeowners in exchange for accepting a waste facility in the vicinity. Such compensation is often controversial. On the one hand, it may be claimed that those who (have to) accept a disadvantage of this kind should not do it for free. On the other hand, the practice can be interpreted as buying people off to make them accept dangers that they would not otherwise accept. The line between fair compensation and corruptive practices is not easy to draw. One important factor in this demarcation is the magnitude of the dangers compensated for. It is much more controversial to buy acceptance of a large risk than to buy acceptance of a small risk.

The exportation of waste has often been criticized. In particular, protests have been waged against the transportation of waste from industrialized to developing countries. The Basel Convention, which entered into force in 1992, prohibits such exportation for final disposal, but allows it for recycling. A 1995 amendment, the Basel Ban Amendment, prohibits all exports of hazardous waste from industrialized to developing countries. This amendment was as of 2008 not in force (because of too few ratifications). The European Union has nevertheless adopted it and made it binding on all its member states. Other industrial nations, including the United States, oppose it and do not implement it.

INTERGENERATIONAL JUSTICE

Leakage from toxic waste can make neighborhoods uninhabitable and waters unpotable for many generations. Therefore, waste management involves issues of intergenerational justice.

Discussions of intergenerational justice in waste management have focused mostly on spent nuclear fuel. This deadly material remains lethal for tens of thousands of years. (Owing to radioactive decay, some dangerous isotopes disappear in relatively short periods, but isotopes with long half-lives remain for tens of thousands of years or longer.) The waste-management solution preferred by governments and the nuclear industry is deep geological disposal, which involves placing the waste in final, sealed repositories in stable geological formations usually around 500 to 1,000 meters below ground. Then humans and the environment are protected by a combination of highly durable containers and a geological environment so selected that any leakage will move very slowly. Critics claim that there is nevertheless insufficient guarantee against future groundwater contamination, which would threaten the health of coming generations.

The planned Yucca Mountain Repository in Nevada was as of 2008 projected to receive spent fuel from American nuclear plants from 2017. However, crucial regulatory, legal, and political decisions remained to be made. The project is based on extensive scientific and engineering studies showing that the risks associated with the repository are very small. However, some of these studies have been subject to controversy. The repository has also met with strong local opposition, and Native Americans have claimed that it infringes on their holy lands.

A major alternative solution to sealed repositories is monitored retrievable storage. Such storage would give future generations control over the waste. They would have to monitor it, unless they chose at some stage to transfer it to permanent storage. This solution has the comparative advantage of not binding future generations to possible imperfections in our current technology, the disadvantage of leaving a problem for them to solve, and also the disadvantage of keeping nuclear material easily retrievable for conversion into weapons-grade material for nuclear warheads. Hence, the choice of principles for managing nuclear waste involves complex ethical issues concerning what risks and what responsibilities we can and should hand over to coming generations.

PRODUCER RESPONSIBILITY

According to the polluter-pays principle, the polluting party should pay for the damage done to the environment. According to the principle of extended producer responsibility, the manufacturer's responsibility covers the entire life cycle of the product and the waste it gives rise to. This means that the producer has to plan and pay for recycling or other measures needed for responsible waste management. Deposit systems for bottles exemplify application of extended producer responsibility.

Extended producer responsibility, combined with the principles for waste management enshrined in the waste hierarchy, lead to far-reaching demands on producers. The highest levels of the waste hierarchy, prevention and minimization, are often in conflict with producers' interests in expanding business and maximizing sales of their products. Waste management is an area with many potential conflicts between business interests and the environment. Hence, it is also an area in great need of ethical discussion.

SEE ALSO *Environmental Law; Future Generations; Industrial Ecology; Intergenerational Justice; Pollution; Risk Assessment; Sustainability; U.S. Environmental Protection Agency; Water.*

BIBLIOGRAPHY

Basel Action Network. Available at http://www.ban.org

Hadjilambrinos, Constantine. 2000. "An Egalitarian Response to Utilitarian Analysis of Long-Lived Pollution: The Case of High-Level Radioactive Waste." *Environmental Ethics* 22(1): 43–62.

Hermansson, Hélène. 2007. "The Ethics of NIMBY Conflicts." *Ethical Theory and Moral Practice* 10(1): 23–34.

Luloff, A. E.; S. L. Albrecht; and L. Bourke. 1998. "NIMBY and the Hazardous and Toxic Waste Siting Dilemma: The Need for Concept Clarification." *Society and Natural Resources* 11: 81–89.

McKay, Ruth Burnice. 2000. "Consequential Utilitarianism: Addressing Ethical Deficiencies in the Municipal Landfill Siting Process." *Journal of Business Ethics* 26: 289–306.

Peterson, Martin, and Sven Ove Hansson. 2004. "On the Application of Rights-Based Moral Theories to Siting Controversies." *Journal of Risk Research* 7: 269–275.

Shrader-Frechette, Kristin. 2000. "Duties to Future Generations, Proxy Consent, Intra- and Intergenerational Equity: The Case of Nuclear Waste." *Risk Analysis* 20: 771–778.

Wilson, David C.; Costas Velis; and Chris Cheeseman. 2006. "Role of Informal Sector Recycling in Waste Management in Developing Countries." *Habitat International* 30: 797–808.

Sven Ove Hansson

WATER

Earth is the blue planet, the water planet. Nearly three-quarters of the earth's surface is covered by water, mostly in oceans. Water gives rhythm and pulse to life. Moving through all living entities—our bodies, the land, the atmosphere, and our cultures—water connects, transports, and transforms.

That water is life has become a truism. But the fact is, the earth has a biosphere because it has a hydrosphere. Owing to its vital place in life, water has turned into a culture and language unto itself. There are countless sayings, imageries, and references to water. Many fluid phenomena—the fluidity of globalization, liquid capital, and streams of consciousness—have been tied to a water metaphor as flexible and mysterious as the molecule itself.

Ubiquitous as water may be, freshwater is a finite resource. Most water is saline; only 2.5 percent is fresh. More than two-thirds of all freshwater is locked in ice at the polar regions or in glaciers in distant mountainous areas, a little less than one-third is groundwater, and only 0.3 percent is surface water (rivers, lakes, and reservoirs).

Water is in constant motion all around us: in the atmosphere, on the earth's surface, and in its depths. It constantly alternates among three physical states: gas, liquid, and solid. Evaporated or transpired by plants, it rises up into the sky and falls to the earth again as rain or snow, where it finds its way back underground or into lakes, rivers, and oceans. The hydrological cycle then repeats itself. The amount of water on the earth is basically the same as it was 4 billion years ago, and it has been recycled ever since.

Water has the rare characteristic of being less dense as a solid than as a liquid. Consequently, ice floats. Without this crucial property, rivers and lakes in the higher latitudes would freeze from the bottom up, with only a melted puddle on top during the warm season. Water freezing and thawing is capable of breaking granite. Water seeps into cracks, expands and contracts, and with a slow steady force turns even the hardest stone into soil. Water also has a great capacity to absorb heat. Ocean currents play a large role in the earth's climate patterns, tempering climates in many coastal regions.

Water has a pervasive presence in every aspect of our daily lives. We drink it, bathe in it, and cook with it. Our food consists in considerable part of water. Spinach equals milk in being approximately 90 percent water; tomatoes 95 percent; and beef, seemingly so solid, 60 percent. We ourselves are composed mainly of water (two-thirds of our body weight), and we need about 2 to 3 liters a day to live. We can survive for a month without food, but only 5 to 7 days without water.

CULTURAL, RELIGIOUS, AND PHILOSOPHICAL LEGACIES OF WATER

Because of its vital role, water has always been intrinsically linked with culture. Virtually all civilizations developed around water: Tribes settled on the shores or banks of water bodies, and cities originated at the confluence of rivers. The first complex societies—societies as diverse as ancient China, India, Mesopotamia, Mexico, and Peru—were irrigation-based cultures with ingenious water-management structures, a centralized government, and extensive division of labor. The Romans were also skilled water engineers, building innumerable aqueducts for public water supply using various techniques of hydraulic architecture. Karl A. Wittfogel called them "hydraulic civilizations."

Through the ages, poetry, music, and religion have found a deep well of meaning in water and crosscurrents of meaning in a vast pool of reflections and emotions about water. Narcissus epitomized the rise of self-consciousness in his self-reflection on the surface of a pond. The cosmogonic power of water has been a major theme in many ancient accounts of origin. These poetic sources are precursors of the later scientifically developed theory of evolution, which confirmed that all life forms emerged phylogenetically and ontogenetically out of water.

In the *Enuma Elish*, the Mesopotamian-Babylonian creation epic of the third millennium B.C.E., the primordial waters are Apsu (meaning sweet-water "ocean," "deep abyss," or "outermost limit") and Tiamat (meaning "primeval waters," the one who "is too deep to fathom," the salt sea). Apsu and Tiamat are brought under control by gods (their offspring) to create the topography of earth and sky.

We see a similar structure in Genesis (compiled between approximately 1400 and 400 B.C.E.), the first book of the Old Testament. Clearly influenced by the *Enuma Elish*, its opening lines read, "In the beginning … darkness was upon the face of the deep; and the Spirit of God was moving over the face of the waters." The Hebrew word *Tehom*, meaning "deep [waters]," is etymologically related to *Tiamat*.

The Greek philosopher Thales of Miletus (c. 624–545 B.C.E.) considered water an originating and guiding principle (*archē*). There were similarities between Thales' philosophy and the ancient legends and myths, specifically the ones about Oceanus and his consort Tethys, who was both sister and wife of Oceanus and whose name has etymological ties to *Tiamat* and *Tehom*. Thales, however, broke with the traditional belief that the gods organized, shaped, and controlled the cosmos. Hence, for Aristotle, Thales was no longer a theologian, like the old poets, but the founder of natural philosophy, investigating the basic principles of matter and theoretically moving toward a scientific treatment of natural phenomena.

Nearly a century later Heraclitus of Ephesus (c. 535–475 B.C.E.) found in the *movement* of water a guiding principle: "All things flow" (*Panta rhei*), and "One cannot step in the same river twice." Heraclitus, less concerned with finding unity in a material substrate, was a protophenomenologist, interested in the everyday experience of change: "Cold things warm up, the hot cools off, wet becomes dry, dry becomes wet." Heraclitus emphasized that opposites are equally capable of transforming and are permanently changing into one another—a constant flux governed by Logos, the first proposed law of nature in Western natural philosophy.

Similarly, the ancient Chinese text *Tao Te Ching* of Lao Tzu (sixth century B.C.E.) masterfully invokes the paradoxical powers of water: "Nothing in the world is as soft and yielding as water. Yet for dissolving the hard and inflexible, nothing can surpass it" (chap. 78). Water is powerful yet unassuming. Thus, "The supreme good is like water, which nourishes all things without trying to. It is content with the low places that people disdain. Thus it is like the Tao" (chap. 8). Water is exemplary for leadership in that as the highest power, it accepts the lowest place: "All streams flow to the sea because it is lower than they are. Humility gives it its power. ... If you want to govern the people, you must place yourself below them. If you want to lead the people, you must learn how to follow them" (chap. 66).

Humility is also an important value in the Bible, often thematically combined with God's punishment of human hubris. The Flood is the most famous biblical water story, the ultimate expression of God's wrath. In the deluge, only Noah, his family, and the pairs of animals on his arc are saved. At that point the imagery moves from water as an agent of punishment to water as a cleansing agent, and there appears a rainbow, the symbol of God's covenant with his people, in which humans have the first right to the goods of his creation, but in return have to take care of the earth.

In Islam there is a similar ethic of stewardship: The blessings of water come with human responsibility for its proper use. All life forms, including plants and animals, should be supported according to their needs. The Koran explicitly states that the supply of water is fixed and should not be wasted. Water is architecturally revered most majestically and aesthetically in Alhambra, the fourteenth-century Muslim palace in southern Spain. From every room can be heard fountains with running water, which have a cooling effect. A stately pool, a symbol of power, reflects the building at its entrance.

The revered place of water appears in a wide variety of religious rituals and cultural practices, such as baptism and pilgrimages to sacred wells. The spring in the Grotto of Lourdes (in southern France), where Mary is said to have appeared to fourteen-year-old Saint Bernadette in 1858, soon became the world's leading pilgrimage site even though the holy water was "not exactly inviting," in the words of Emile Zola. The healing powers attributed to the water that wells up from the grotto attracts 5 million pilgrims a year from all over the world. People drink it, bathe in it, and buy plastic Mary figures filled with it. It is probably no coincidence that the Virgin Mary is portrayed as dressed in blue and white, water's colors, which contributes to Mary's image as the life-giving mother of Christ.

Water is a recurrent theme in literature. The French philosopher of science Gaston Bachelard (1884–1962), in *Water and Dreams*, saw water as a prime source for the imagination. He connected water's symbolic power to purify to its material power to clean and rinse and ultimately rejuvenate, as expressed metaphorically in the notion of the Fountain of Youth. According to the philosopher of technology Ivan Illich (1926–2002), this symbolic force of water has been compromised in modern life. For him, water in its engineered form reduced to H_2O hardly speaks to the imagination. Chemically purified and piped, water as a living element has been separated from our daily consciousness.

Water has always played an important role in the arts. Famous Dutch seventeenth-century landscape painting was born out of seascapes (a fact that resonates with the trajectories of early myths of origin). In sculpture, architecture, multimedia, and landscape art, contemporary artists such as Tadao Ando, William Pye, Roni Horn, Basia Irland, and Herbert Dreiseitl—to name just a few—all explore innovative ways with water, implicitly articulating the larger cultural interest in the element. William Shakespeare in *King Henry VIII* invokes water to convey the fleeting nature of virtue: "Men's evil manners live in brass; their virtues we write in water." Virtues are seen as ephemeral, as fluid, like water, while vices are like brass—here to last.

LEGACIES OF MODERNITY

The late nineteenth century and entire twentieth century saw a massive harnessing of water through extensive damming, draining, and diverting. Progress was measured, in part, in terms of the control, mainly through large-scale water-engineering projects. The leading water-development paradigm was one of economic growth driven by a utilitarian ethic. Any drop reaching the ocean instead of being used for agriculture, industry, or hydropower was considered a waste. From the late nineteenth century, an unprecedented boom of large-scale water projects ensued, followed a century later by a rising tide

Delivering Water under the Hot Desert Sun. *A group of Kenyan Oromos (an ethnic division of Ethiopia) travel through the desert to deliver water to Oromo Liberation Front (OLF) troops in 2006. The walk takes a full day, and they each carry at least 30 kilos of water on their back. Although three-quarters of the earth is covered with water, freshwater for drinking is a finite resource, and in many places such as Africa this scarcity is already displaying its effects. In addition, the quality of water too is declining, as toxins and other industrial waste pollute the supply.* GETTY IMAGES.

of problems varying from silting reservoirs to oxygen-depleted dead zones and heightened coastal vulnerability because the massive destruction of wetlands has taken away a buffer zone of natural barriers against hurricanes and cyclones.

In this whole process, water consumption has sky-rocketed. A growing world population has led to greater needs for food and industrial production and an explosive rate of urbanization. Consequentially, the amount of water available for humans and other species keeps declining. About 70 percent of global water use goes to agriculture, and 22 percent to industry, while domestic and municipal use accounts for a mere 8 percent. The high percentage of water for agriculture is partially due to low water-use efficiency, further aggravated by archaic water laws and irrigation subsidies, which take away incentives to use water more efficiently. Moreover, meeting the Millennium Development Goal on hunger entails doubling food production by 2050, which means more irrigation.

An important closeted use of water has been exposed through John Anthony Allan's notion of *virtual water*, also referred to as embedded water, embodied water, hidden water, or water footprint. It is the water used in the production of a good or service. Allan called it "virtual" because once, for example, an irrigated crop of wheat is grown, the water used to grow it is no longer actually contained, molecule for molecule, in the wheat. It thus takes 2,000 liters of water to make a cotton T-shirt, 2,400 liters for a hamburger, and 400,000 liters for a car.

Also declining is the quality of water, polluted by heavy metals and other industrial toxic wastes, by microbial pathogens and excessive nutrients from untreated sewage and intensive use of agricultural fertilizers. Whereas once dilution was the solution to pollution, such practice has reached its saturation point. According to Jeremy Jackson, a leading oceanographer, the oceans are at a tipping point; that is, environmental damage any moment now could pass an unknown point of no return, at which the ocean's resilience, its capacity to bounce back into a healthy state, is exhausted. With the bulk of the human population (60% of people worldwide) already living along or near coasts, and with ongoing growth (by 2025 coastal populations are expected to reach 6 billion people), coastal and ocean ecosystems are on a fast track to devastation. Worldwide fish stocks are already in steep decline. Sea-level rise caused by climate warming will have tremendous consequences for millions of people.

Some foresee that, owing to population growth, increasing environmental degradation, and global climate change, by the end of the twenty-first century water will be the single dominant factor in world natural-resource politics. Freshwater is a crucial limiting factor for health, food security, economic growth, biodiversity, and environmental sustainability. The total volume of water on the earth may be sufficient to accommodate our needs on a sustainable basis, but, as has been argued by various U.N. agencies, nongovernmental organizations, and other water organizations, creating sustainable conditions requires a serious political commitment. Much water is wasted, polluted, drained, or misallocated.

Uneven distribution of water is one of the most poignant problems. People in developing countries use on average 10 liters of water per person each day, while in Europe the average is 135 liters, and in the United States, 570 liters. Roughly 1.2 billion people, one-fifth of the world's population in 2008, lack access to potable water. And approximately 2.4 billion people, two-fifths of the population, do not have adequate sanitation services. Water-related diseases cause 80 percent of all illnesses and deaths in the developing world. In addition, many African and Asian women and girls spend hours a day walking to get water, an activity that severely reduces their participation in other productive activities, including education.

Increasingly, water scarcity is seen in terms of a crisis of management. Moreover, the water crisis is linked to a development crisis and an energy crisis. Its solution is a matter not only of engineering know-how, which we have in abundance, but also of political will. To build a secure and sustainable future for a huge and still growing world population, considerations of equity may become more crucial than concerns of economic efficiency and invisible-hand allocation of water by the free market. Emerging is a paradigm shift from large-scale, utilitarian, supply-side management of water to an environmentally sound and equitably just allocation of water.

TOWARD AN INTEGRATED SENSE OF WATER

Since the early 1990s an approach known as integrated water-resource management has gained ground. Integrated management explicitly advocates a holistic approach, managing water at the basin or watershed level and integrating land and water; upstream and downstream; surface water, groundwater, and coastal water. A more transparent and participatory style of management replaces sector-by-sector and top-down management. In multiple-stakeholder processes, hydrological and engineering expertise is complemented with ecological concerns, while also tending to urban, agricultural, industrial, and recreational interests. Water connects them all and calls for a water ethics that integrates local problems of water quality and quantity in transregional and global political contexts.

Peter Gleick, a leading water-policy expert and director of the Pacific Institute, speaks of a "soft path" that complements the twentieth-century large-scale centralized infrastructure with "lower cost community-scale systems, decentralized and open decision-making, water markets and equitable pricing, application of efficient technology, and environmental protection." Conservation is one of the main strategies to keep more water in the system. Sandra Postel, another well-known water expert and director of the Global Water Policy Project, calls conservation our "last oasis."

A variety of changes in technology and legal structures have been explored, and these, together with an increased awareness of the need for water conservation and proper economic incentives, have resulted in more efficient water use. There has been a resurgence of traditional technologies, such as rainwater harvesting (the systematic capturing of rainfall or floodwater) and small-scale run-of-river irrigation systems. Such technologies are often more cost-effective and less disruptive to the social and environmental functioning of local communities. Various new water-conservation techniques have been implemented, such as highly efficient time-released drip irrigation systems. Two commonly mentioned contemporary technologies to increase available freshwater are desalination and water reuse. Desalination, however, is energy- and capital-intensive and generates vast amounts of wastewater, twice as saline as seawater. Oil-rich and water-poor countries like Saudi Arabia get most of their water through desalination and account for almost one-fourth of the world's desalinized water. Reuse is more widespread. Countries as different as Japan and Dubai make extensive use of grey water (nonsewage wash water used in the home) for landscape irrigation. Water gets polluted, but it can be cleaned, one of the great assets of water. More and more municipalities consider treated sewage water as an option for providing water.

New laws also have beneficial effects. Two excellent examples of such legal change are the implementation in the United States in the 1970s of the Clean Water Act and the Safe Drinking Water Act, which require industries to clean up their wastewater. These laws made many industries more water-efficient, because it turned out to be cheaper not to produce the waste than to clean it up. As of 2008 it takes 5 to 6 tons of water to make a ton of steel, for example, whereas it used to take 200 tons.

Water privatization may be seen as another tool of efficiency. While most water-provision and sanitation systems are publicly owned and operated (globally, 90%), there is a tendency toward private-sector participation in these very basic municipal services. The proponents of privatization argue that it will improve the quality of service, reduce costs, and mobilize more financial investment and technical expertise. Opponents counter that privatization leads to poorer service and higher costs because competitive water markets are hard to arrange (owing to the nature of the service) and profits are valued over service. There have been intense political campaigns against privatization in Ghana (2003), Uruguay (2004), and, most famously, Bolivia, the site of the Cochabamba Water Riots of 2000. Some countries (e.g., the Netherlands, 2004) have enacted laws banning the privatization of public water supply.

Progressive water pricing is often mentioned as a regulative approach to conservation that encourages more responsible water use by means of an incentive. In such a system, a daily minimum amount of water is sold at an affordable price. As a customer's use increases, the price per unit also increases in a stepwise manner. This is the reverse of the usual approach of markets, in which high-use customers are charged less per unit than low-use customers. From a progressive-water-pricing perspective, agricultural water stands out as seriously underpriced.

In reaction to conventional antiprogressive water pricing and water privatization, a growing movement has emerged for implementing a U.N.-mandated human right to water, which would entitle all people to sufficient, safe, accessible, and affordable water. This right would trump contractual rights to water and property rights in water.

There are awareness campaigns about the value of water, water-education programs in schools and municipalities, venues to involve various stakeholders in water management and in citizen-based projects of wetland, riparian, and coastal restoration. Environmental restoration requires amending our relation to the land as much as restoring the land itself. Watershed restoration involves a broad spectrum of human-water-land relations spanning ethics, aesthetics, politics, and participatory activities and encompasses modes of knowledge as diverse as science, engineering, elders' experiences, storytelling, and children's imagination. Celebrations such as World Water Day and local river festivals provide playful ways to reconnect with water and to enhance motivation to learn about water quality and quantity.

Almost 2,500 years later, we are relearning the old Daoist and Heraclitean aquatic wisdom: In water's humility lies its power; in its constant flow lies its stability. All living entities—including the earth's biosphere—depend on the ongoing cycling of water. On a planetary level, we are all downstream. Our future and the future of the planet are written in water.

SEE ALSO *Aquifers; Conservation; Consumption; Daoism; Global Climate Change; Islam; Nongovernmental Organizations; Population; Sustainability; Watersheds; Wetlands.*

BIBLIOGRAPHY

Barlow, Maude. 2007. *Blue Covenant: The Global Water Crisis and the Coming Battle for the Right to Water.* New York: New Press.

Barlow, Maude, and Tony Clarke. 2002. *Blue Gold: The Fight to Stop the Corporate Theft of the World's Water.* New York: New Press.

Blatter, Joachim, and Helen Ingram. 2001. *Reflections on Water: New Approaches to Transboundary Conflicts and Cooperation.* Cambridge, MA: MIT Press.

Cech, Tom. 2005. *Principles of Water Resources: History, Development, Management, and Policy.* 2nd ed. Hoboken, NJ: John Wiley and Sons.

Conca, Ken. 2006. *Governing Water: Contentious Transnational Politics and Global Institution Building.* Cambridge, MA: MIT Press.

De Villiers, Marq. 2000. *Water: The Fate of Our Most Precious Resource.* Boston: Houghton Mifflin.

Donahue, John M., and Barbara Rose Johnston, eds. 1998. *Water, Culture, and Power: Local Struggles in a Global Context.* Washington, DC: Island Press.

Gleick, Peter, ed. 1993. *Water in Crisis: A Guide to the World's Fresh Water Resources.* New York: Oxford University Press.

Gleick, Peter, et al. 2006. *The World's Water, 2006–2007: The Biennial Report on Freshwater Resources.* Washington, DC: Island Press.

Glennon, Robert. 2002. *Water Follies: Groundwater Pumping and the Fate of America's Fresh Waters.* Washington, DC: Island Press.

Hinrichsen, Don. 1998. *Coastal Waters of the World: Trends, Threats, and Strategies.* Washington, DC: Island Press.

Iyer, Ramaswamy R. 2007. *Towards Water Wisdom: Limits, Justice, Harmony.* New Dehli, India: Sage Publications.

Latour, Bruno, and Peter Weibel. 2005. *Making Things Public: Atmospheres of Democracy.* Cambridge, MA: MIT Press.

Postel, Sandra. 1992. *Last Oasis: Facing Water Scarcity.* New York: W. W. Norton.

Postel, Sandra. 1999. *Pillar of Sand: Can the Irrigation Miracle Last?* New York: W. W. Norton.

Reisner, Marc. 1993. *Cadillac Desert: The American West and Its Disappearing Water.* New York: Penguin Books.

Strang, Veronica. 2004. *The Meaning of Water.* Oxford: Berg.

Swyngedouw, Erik. 2004. *Social Power and the Urbanization of Water: Flows of Power.* New York: Oxford University Press.

United Nations. World Water Assessment Programme. 2003. *Water for People, Water for Life.* New York: UNESCO and Berghahn Books.

Whiteford, Linda, and Scott Whiteford, eds. 2005. *Globalization, Water, and Health: Resource Management in Times of Scarcity.* Santa Fe, NM: School of American Research Press.

Irene J. Klaver

WATERSHEDS

Watersheds are land areas in which all the ground and surface water flows to the same place, creating a systemic link among land, water, and everything that lives on the land. Shaped by biological and geological forces, watersheds give rise to a variety of ethical issues involving political boundaries, population growth, urbanization, externalities, income inequality, changing knowledge, and future generations.

Shaped by nature rather than by political forces, watersheds often cross political boundaries. For instance, rivers often have been perceived as convenient boundary markers. Consequently, because watersheds are situated on both sides of rivers, there is often no single political entity to provide oversight of watershed matters. Rivers that provide water to more than one nation-state include the Danube, Jordan, Ganges, Indus, Nile, and Colorado rivers. Within nations watersheds cross a multitude of provincial, prefectural, and state and local boundaries. Effective management requires the cooperation of diverse political entities. Because the power of a jurisdiction does not always correspond to its demand for water, there is a need for cooperation and ethical behavior in caring for and distributing water.

ALLOCATION

The press of population growth on the resources of some watersheds has been so great that ethical issues have been created that involve not only allocation of water among humans but also the fulfillment of the water needs of a watershed's land, plants, and animals. Although water allocation has been a topic of global conferences since the 1970s, the conferences have been slow to recognize the ethical issues implied in allocation decisions within watersheds. Before the 1990s some conferences treated water solely as a marketable commodity without any discussion of ethical issues. However, subsequently the ethical issues of providing a basic amount of water as a human right and meeting the needs of the entire ecosystem of a watershed were acknowledged. In 2002 the United Nations Committee on Economic, Social, and Cultural Rights called for the combined basic physical and cultural and social needs for water to be met without regard to costs. The Earth and Johannesburg summits of 1992 and 2002 gave explicit recognition to the need of the entire ecosystem for water.

URBANIZATION AND EXTERNALITIES

The growing urbanization of human populations creates an ethical challenge when cities draw on water from outside their boundaries, often from rural watersheds. In essence cities are drawing on nature's services in distant watersheds. This raises two ethical issues. The first is the ethical obligation to leave enough water to meet the water needs of the supplying watershed. The second is compensation of rural areas for the use of the natural services. For instance, New York, Beijing, and Los Angeles draw on water far from their boundaries. Los Angeles and Beijing have caused depletion of distant water supplies, and New York used eminent domain to flood villages to create space for its reservoirs. The procedure used in New York, though it created conflict, was fairly open and included compensation for the actions taken by the city; in contrast, the Los Angeles story is famous for deception and chicanery.

The interconnected and open nature of watersheds has created a class of ethical issues through what economists call *externalities*: the impact of decisions on third parties not responsible for those decisions. One example stems from the generally open access to watershed waters and land. Because watercourses are ubiquitous and impossible to patrol, the decisions of individuals and corporations to refrain from polluting them depend on an ethical determination rather than fear of a legal penalty. There is even more open access to the air that affects watersheds. Fuels selected by power companies can create acid rain in distant watersheds.

Private ownership of property provides another example. For instance, much U.S. forest land is privately owned, but little is under a management plan to maintain diversity in the types and ages of trees needed for watershed protection. Economic incentives favor cutting down the biggest trees, which often offer the most protection. Although state governments may be able to make legal arguments in favor of taking over the privately held forest land in a watershed by eminent domain to protect the water supply, the local communities would be likely to consider this approach unethical. Because of such community pressures, New York City signed an agreement with communities in its distant rural watershed not to take land by eminent domain and instead seek land through voluntary sales and has created incentives to induce private forest land owners to put their land under conservation management.

Inequality of income among watershed residents and users creates two types of ethical questions. The first concerns the way society provides and finances watershed services; the second concerns the way wealthy persons and companies observe laws. Often high-cost water protection and distribution infrastructure systems, even if built by governments, bring water to the middle and upper classes, leaving lower-income persons with relatively more expensive water sources or without safe water. Even when water is provided universally within a watershed, the taxes used to finance the infrastructure are often regressive, resulting in lower-income persons paying a higher percent on their incomes than higher-income persons. Also, although taxes and fines may be set to encourage conservation and prevent pollution, persons with a high income and wealthy companies can afford to pay them. Additionally, companies can pass the cost of fines to the public in the form of higher prices, which can also, depending on the consumption pattern of the product, affect lower-income groups more than upper-income groups.

Because knowledge, weather, and technology are always changing, there is an ongoing ethical need to devise new laws as the situation demands and redress the harms that have been done. For instance, the water in most river basin watersheds is insufficient to meet current needs, and global climate change is exacerbating the problem. Existing water agreements may be unrealistic. For instance, the Colorado River Basin agreement was based on atypical, historically high water flows. Around the world many others rivers face demands that exceed their potential supply, including the Yangtze in China, the Nile in Egypt, and the Litani and Jordan rivers in the Levant. However, many international water treaties do not address allocation issues. Further, serious pollution of water supplies may have occurred as a result of the use of technologies whose toxicity was not well understood. An example is the polluting of the Hudson River in New York with cancer-inducing polychlorinated biphenyls (PCBs) by the General Electric Corporation. The responsibility to remediate situations such as these faces companies and governments around the world.

Future generations depend on the ethical choices made by the contemporary generation to provide them with adequate water supplies. Many contend that free markets cannot ensure such protection because people tend to value near-term benefits and market criteria cannot capture the holistic nature and needs of watersheds. They argue that preserving watershed services for future generations requires decision making that honors the need for safety and precaution, based on an understanding of how the web of relationships in a watershed works.

POSSIBLE SOLUTIONS

Political boundaries and urbanization problems can be addressed in watershed treaties and collaborations. Treaties on the sharing of water across boundaries, although hammered out over time, can represent a set of rules rather than an ongoing process. Collaboration, however, is a process in which the diverse stakeholders in a watershed make a commitment to work with one another to evolve workable compromises to protect and share the water in a watershed. Collaborations that involve distant partners or technological complexities are likely to require regulations to protect water.

Collaboration Collaboration tools are being developed to help stakeholders understand the implications of choices before they make them. For instance, a model system for water allocation that was developed for Middle Eastern countries not only allows consideration of diverse options but allows variation in the input of social values as well as financial costs. Both treaties and collaborations can address inequalities of need through compensation programs that are in essence payments for the ecosystem services of a watershed. There has been growing recognition that residents of rural areas should be compensated for the provision of nature's services from their land to cities.

Water as a Human Right The need to prioritize watershed services has generated a variety of approaches to providing basic water allocation as a human right. It is argued that this right should become embodied in international law to help build protection of basic water rights into water treaties and collaborations. Within single watersheds with poor communities ladders of use have been recommended. The ladder gives first priority to enhancing open sources, then public taps, then communal sources, and finally individual taps. Another technique is the use of virtual water: the importation of water-intensive crops such as wheat from water-rich to water-poor watersheds to allow the water-poor areas to retain their water for basic needs. For example, Egypt saves water by importing wheat. However, the dependence on virtual water would create an ethical demand for the continued provision of those crops. The impact of the financing of the imported crop on the distribution of income is also an ethical issue.

Education Education and access to information are vital tools for addressing the externality problem created by open access to watersheds. For instance, many governments fund education about the value of and ways to use a watershed for both children and adults through schools and other community facilities. An example of an information tool is the U.S. toxic-release inventory law, which requires companies to publish releases of certain toxic substances into the environment. It was discovered that the companies reduced their emissions to avoid adverse publicity. Also, the information has been used for legal suits by citizens against polluters.

Financial, Taxation, and Subsidy Policies Ethical issues created by inequality of income distribution and private ownership of land within watersheds can be addressed through progressive financial, taxation, and subsidy policies. For instance, rising unit prices can be charged for escalating use of water for inessential purposes such as private swimming pools. The United States provides an example of subsidies helping to reduce inequality. Because U.S. national taxes tend to be progressive and local taxes tend to be regressive, national water infrastructure grants were especially beneficial to poorer members of watersheds. In the United States tax deductions for privately owned land put under conservation easements address environmental and economic needs simultaneously.

Restoration and financial compensation by polluters will not prevent the loss of many lives (human, animal, and plant) and enduring harm to landscapes and habitat. This is why many people recommend the use of the precautionary principle and advocate cradle-to-grave clean technologies so that people can keep track of what they bring to and create in watersheds, honoring the privilege of using those resources by leaving a watershed and its inhabitants and constituent parts at least as well off as they originally were.

SEE ALSO *Environmental Education; Future Generations; Habitat Loss; Pollution; Population; Rivers; Urban Environments; Water.*

BIBLIOGRAPHY

Beach, Heather, et al. 2000. "Organizational and Economic Theory." In *Transboundary Freshwater Dispute Resolution: Theory, Practice, and Annotated References*, ed. Heather Beach, Jesse Hammer, J. Joseph Hewitt et al. Tokyo and New York: United Nations University Press.

Bouwer, H. 2002. "Integrated Water Management for the 21st Century: Problems and Solutions." *Journal of Irrigation and Drainage Engineering* 128(4): 193.

CleanUpGE.org. 2007. "Truth about Hudson River PCBs: A Counter to GE's Claims." Available from http://www.cleanupge.org/pcbs.html

Costanza, Robert; John H. Cumberland; Herman Daly et al. 1997. *An Introduction to Ecological Economics*. Boca Raton, FL: St. Lucie Press.

Daily, Gretchen C., ed. 1997. *Nature's Services: Societal Dependence on Natural Ecosystems*. Washington, DC: Island Press.

Fisher, Franklin M., and Annette Huber-Lee. 2006. "Economics, Water Management, and Conflict Resolution in the Middle East and Beyond." *Environment* 48(3): 26–41.

Galusha, Diane. 1999. *Liquid Assets: A History of New York City's Water System*. Fleishmanns, NY: Purple Mountain Press.

Giltmier, James W. 1996. "A Federal Commitment to Forest Conservation on Private Lands: The Story of State and Private Forestry in the USDA Forest Service." Pinchot Institute for Conservation. Available from http://fs.jorge.com/archives/indexNational.html

Gleick, Peter. 2007. "The Human Right to Water." Pacific Institute. Available from http://www.pacinst.org/reports/human_right_may_07.pdf

Goodstein, Eban S. 2002. *Economics and the Environment*, 3rd edition. New York: Wiley.

Hoffman, Joan. 2005. "Economic Stratification and Management of Water Quality: A Case Study of the New York City Catskill/Delaware Watershed." *Environmental Values* 14: 44–470.

Reisner, Marc. 1993. *Cadillac Desert: The American West and Its Disappearing Water*. New York: Penguin Books.

Riverkeeper.org. 2007. "Clean Up GE PCBs." Available from http://www.riverkeeper.org/campaign.php/ge_pcbs

Sabatier, Paula W.; Will Focht; Mark Lubell, et al. 2005. *Swimming Upstream: Collaborative Approaches to Watershed Management*. Cambridge, MA: MIT Press.

Satterwaithe, David, and Gordon McGranahan. 2007. "Providing Clean Water and Sanitation." In *State of the World: Our Urban Future*, ed. Molly O'Meara et al. Washington, DC: World Watch Institute.

United Nations Development Programme. 2006. "Managing Transboundary Waters." In *The Human Development Report 2006*. Available from http://www.undp.org/hdr2006

U.S. Environmental Protection Agency. 2007. "What Is a Watershed?" Available from http://www.epa.gov/owow/watershed/whatis.html

Ward, F. A.; J. F. Booker; and A. M. Michelsen. 2006. "Integrated Economic, Hydrologic, and Institutional Analysis of Policy Responses to Mitigate Drought Impacts in Rio Grande Basin." *Journal of Water Resources Planning and Management* 132(6): 488–502.

World Water Council. 2007. "Virtual Water in Brief." Available from http://www.worldwatercouncil.org/index.php?id=866

Joan Hoffman

WETLANDS

Wetlands are found on every continent except Antarctica, and at every latitude from the tropics to the tundra. About 6 percent of the land surface of the Earth is wetlands, and in the boreal regions this is 11 percent. Wetland ecosystems on global scales cover more than 1,280 million hectares, an area 33 percent larger than the United States.

Individual wetlands are often relatively transient features of a landscape. They may have differing water levels depending on seasonal changes and climate. They may fill with debris. Water-loving plants invade the margins of a lake; as detritus collects, marsh-loving plants replace them; afterward the bog fills and shrubs and trees can enter. Meanwhile wetlands will be generated elsewhere on the landscape.

Though sometimes transient, wetlands considered as a biological phenomenon are often long-lived and constantly present throughout Earth's evolutionary past, and there is no cause for wetlands not to persist indefinitely in the future. Like the forest and the sea, mountains and rivers, lakes and islands, wetlands are a form of landscape that Earth has regularly been producing over the epochs of natural history.

Wetlands vary in size from small soggy areas and pools to vast tracts covering many thousands of hectares, both freshwater and brackish. Wetlands are dominated by water, but there is a wide water gradient from drier uplands to deep water, and this often differs depending on rainfall from season to season and year to year. As a result, what is recognized as wetlands varies, depending somewhat on whether the focus is plants adapted to flooding and saturated soil or wildlife so adapted, or the water table relative to ground surface, or the wet/dry season of the year.

Wetlands may be the most threatened of all landscape types. The world has lost half its wetlands since 1900. The United States has lost over half its wetlands. Most wetlands in Europe have been drained or filled for development. Africa, South America, and Asia are continually developing vast wetland areas for food and fiber. The Millennium Ecosystem Assessment, sponsored by the United Nations, has an assessment: *Ecosystems and Human Well-Being: Wetlands and Water*. The authors express concern: "The degradation and loss of wetlands is more rapid than that of other ecosystems. Similarly, the status of both freshwater and coastal wetland species is deteriorating faster than those of other ecosystems" (Millennium Ecosystem Assessment 2000, p. ii).

Since 1989, the U.S. Environmental Protection Agency has had a policy of "no net loss" of wetlands, resulting in a quagmire of controversies about what counts as wetlands (a golf course lake with swampy shoreline?), wetland remediation, loss (of acres or function?) and gain (when does "wet land" become "wetland"?). With wetland legislation, this has resulted in legal quarrels about what is a jurisdictional wetland, what permits may be required to dredge or fill them, or when a mitigation is successful—often destroying a wetland in one place and creating a substitute one elsewhere. A frequent judgment is that recreated wetlands are no match for the originals (Kaiser 2001). A National Research Council report was quite blunt: "The goal of no net loss of wetlands is not being met for wetland functions by the mitigation process, despite progress in the last 20 years" (National Research Council 2001, p. 2).

On international scales, the Convention on Wetlands of International Importance (commonly called the Ramsar Convention, from its first adoption at Ramsar, Iran, in 1971) has become increasingly important, with, since 1987, a permanent secretariat headquartered at the International Union for the Conservation of Nature and Natural Resources in Switzerland. By the turn of the twenty-first century, 117 nations had joined the convention and there were over a thousand wetland sites (Frazier 1999).

The wide varieties of wetlands and the varieties of peoples experiencing them over many centuries have generated a rich terminology: bogs; marshes; mires; muskegs; aapa peatlands, palsa bogs; fens; swamps; wetland moors; wetland prairies; tidal salt marshes; mangrove wetlands; river floodplains, deltas; wetland alluvial fans. Scientists have attempted more precision in classifying schemes (Cowardin et al. 1979; Gore 1983; National Research Council 1995; Mitsch and Gosselink 2000). The National Research Council report cautioned that certain types of wetlands (fens and bogs) are much more difficult to restore than others; some may be impossible to restore.

Human encounters with wetlands have distinctive dimensions (Vileisis 1997). Wetlands have been much misunderstood environments, perhaps the most misunderstood (Miller 1989). The contemporary term *wetlands* has been chosen partly to avoid classical terms such as *swamp, bog, mire*, which have negative connotations. A "pleasant mire" is almost a contradiction in terms. Swamps are damp, marshy, overgrown, rank, dismal, gloomy. They are uninviting places where one has to contend with insects while trying to keep from falling into the treacherous mud. Wetlands are often believed to be wastelands, best to be filled, drained, and converted into a useful resource.

Traditionally, there was an element of truth in such dislike. *Miasma,* from a Greek word for pollution, was poisonous air rising from the rotting bogs. *Malaria* means "bad air," and the disease was more often caught by those who lived near wetlands, breathing this bad air. That the disease was carried by a protist in mosquitoes, breeding in stagnant or slow moving waters, was unknown until the 1890s.

Wetlands have economic uses and provide ecological services (Maltby 1986; Gore 1983; Richardson 1994; Mitsch and Gosselink 2000, Chapter 16). They provide habitats for fishing and for hunting waterfowl; they may contain timber that can be harvested. Rice, a staple grain for over half the world, is grown on managed agricultural wetlands. Peatlands provide fuel and energy. Coastal marshes are critical to the marine fishing, shellfish, and shrimp industries. Eighty percent of the commercial catch off the southeast U.S. coast is linked to salt marshes. There may be important biogeochemical transformations involving phosphates, nitrogen

Wetland Habitat in Wisconsin. *The term "wetland" refers to various transient features of a landscape, but essentially includes any area dominated by water. Wetlands may be one of the most threatened ecosystems, and because of their important economic and biological uses, recent efforts have been made in an attempt to preserve these landscapes. However, as in the United States, legal measures to retain a net amount of wetlands produces additional problems and controversy.* **PHOTO BY RYAN HAGERTY/U.S. FISH AND WILDLIFE SERVICE.**

compounds, sulfur compounds, and carbon. Wetlands serve for water storage and as filters for wastewater treatment. Wetlands provide flood control. In one notable example, damage in New Orleans from hurricane Katrina (2005) was much worse than it might have been, as a result of wetland losses. The four decades of wetlands loss before the turn of the twenty-first century increased storm surges about three feet (Stokstad 2005). The Millennium Ecosystem Assessment analysis also laments: "Many water resource developments undertaken to increase access to water have not given adequate consideration to harmful trade-offs with other services provided by wetlands" (Millennium Ecosystem Assessment 2000, p. ii).

Scientific understanding of wetlands has led to their better appreciation. The first misperception to be set aside is that wetlands are wastelands biologically. Wetlands can be high in biodiversity and biomass productivity, especially if open to hydrologic and nutrient fluxes. "Wetlands ... are among the most fertile and productive ecosystems in the world" (Maltby 1986, p. 9). There is often less diversity in northern wetlands than in those in warmer regimes, but the fewer species there may be present in enormous numbers. In wetlands, obtaining nutrients and oxygen can be problematic, sometimes resulting in ingenious solutions to these stresses.

Wetlands offer unusual experiences of natural history (Rolston 2000). The slowed processes of decay underwater keep the evidences of former life close to the surface. The black ooze is a mixture of silt and partly decayed plants and animals that have gradually piled up on the bottom. The waterlogged remains are oxygen-starved, compared to terrestrial sites, and decay slowly: waterlogged logs, a soggy thatch of dead plants, or peat. The retreat of the glaciers left a Finnish landscape of lakes, scoured hollows, kettles, bogs, and mires. Many Finnish mires are thought to have existed continuously for 8,000 years (Ruuhijärvi 1983, p. 48). Studies of the pollen preserved therein record life in these wetlands, ongoing with vigor for eight millennia.

Wetland plants can tolerate extremes of moisture, nutrients, and oxygen in the soil. Insectivorous plants have adapted to the nitrogen-deficient soils of bogs by reversing the usual trophic pyramids, in which insects typically eat plants. Here plants eat insects, as with the sundews, pitcher plants, Venus flytraps, or bladderworts. Marshes and mires are remarkable places for adapted fit, complex networks connected in biotic community, as we learn when we try to recreate them.

Bird life in wetlands can be abundant, with distinctive adaptations to life in the wetlands. The ducks include mallards, pintails, shovelers, buffleheads, teal. There are also the blackbirds, coots, grebes, cormorants, pelicans, mergansers, gallinules, jacanas, herons, marsh hawks, cranes (Burt 2007). Perhaps the most celebrated of the northland waterbirds are the loons, ancient, deep-diving birds with their striking call.

In sum, wetlands are "biological supermarkets" for the diversity of life (Mitsch and Gosselink 2000, p. 4). Wetlands are especially demanding environments to understand and to conserve. Wetlands are challenging scientifically, economically, politically, socially, philosophically, ethically. Conserving them, we respect life; and, as the Millennium Ecosystem Report insists, the integrity of wetlands is closely linked to human well-being.

SEE ALSO *Biodiversity; Hurricane Katrina; Millennium Ecosystem Assessment; Rivers; U.S. Environmental Protection Agency; Water.*

BIBLIOGRAPHY

Burt, William. 2007. *Marshes: The Disappearing Edens.* New Haven, CT: Yale University Press.

Callicott, J. Baird, 2003. "Wetland Gloom, Wetland Glory." *Philosophy and Geography* 6: 33–45.

Cowardin, Lewis M.; Virginia Carter; Francis C. Golet; and Edward T. LaRoe. 1979. *Classification of Wetlands and Deepwater Habitats of the United States.* Washington, DC: Fish and Wildlife Service, U.S. Department of the Interior. Available from http://www.fws.gov

Frazier, Scott. 1999. *Ramsar Sites Overview: A Synopsis of the World's Wetlands of International Importance.* Wageningen, Netherlands: Wetlands International.

Gore, A. J. P., ed. 1983. *Ecosystems of the World,* Vol. 4, *Mires—Swamp, Bog, Fen, and Moor.* Amsterdam: Elsevier Scientific.

Kaiser, Joselyn. 2001. "Recreated Wetlands No Match for Original." *Science* 293(5527): 25.

Maltby, Edward. 1986. *Waterlogged Wealth: Why Waste the World's Wet Places?* London: Earthscan.

Millennium Ecosystem Assessment. 2000. *Ecosystems and Human Well-Being: Wetlands and Water.* Washington, DC: World Resources Institute. Available from http://www.millenniumassessment.org/documents/document.358.aspx.pdf

Miller, David C. 1989. *Dark Eden: The Swamp in Nineteenth-Century American Culture.* Cambridge, UK: Cambridge University Press.

Mitsch, William J. 2006. *Wetland Creation, Restoration, and Conservation: The State of the Science.* Boston: Elsevier.

Mitsch, William J., and James G. Gosselink. 2000. *Wetlands,* 3rd edition. New York: Wiley.

National Research Council, Committee on Characterization of Wetlands. 1995. *Wetlands: Characteristics and Boundaries.* Washington, DC: National Academy Press.

National Research Council, Committee on Mitigating Wetland Losses. 2001. *Compensating for Wetland Losses under the Clean Water Act.* Washington, DC: National Academy Press.

Richardson, Curtis J. 1994. "Ecological Functions and Human Values in Wetlands: A Framework for Assessing Forestry Impacts." *Wetlands* 14: 1–9.

Rolston, Holmes III. 2000. "Aesthetics in the Swamps." *Perspectives in Biology and Medicine* 43(4): 584–597.

Ruuhijärvi, R. 1983. "The Finnish Mire Types and Their Regional Distribution." In Vol. 4B of *Ecosystems of the World,* ed. A. J. P. Gore. Amsterdam: Elsevier Scientific.

Stokstad, Erik, 2005. "After Katrina: Louisana's Wetlands Struggle for Survival," *Science* 310(5752): 1264–1266.

Vileisis, Ann, 1997. *Discovering the Unknown Landscape: A History of America's Wetlands.* Washington, DC: Island Press.

Holmes Rolston III

WHITE, LYNN, JR.

1907–1987

A leading historian of medieval technology, Lynn White Jr. was born on April 29, 1907, in San Francisco and died March 30, 1987. In his seminal paper "The Historical Roots of Our Ecologic Crisis" (1967), he argued that the Judeo-Christian worldview underlies the distinctively Western tendency to exploit nature through technology, and only a revision of that worldview can resolve our ecological problems. White's view that ecological problems are caused by Judeo-Christian values has been much debated by environmental philosophers, theologians, and others.

White maintains that while nineteenth- and twentieth-century attempts to dominate nature arose from the coalescence of science and technology in the industrial revolution, the attitudes underlying these practices are a millennium older, deriving from the conversion of Western Europe to Christianity. (Orthodox Christianity is exempted from these charges, although White's interpretations of Judeo-Christian scriptures make it implicitly subject to them as well.) Christianity, according to White, is the most anthropocentric religion of the world, because Christianity teaches that God desires humanity to exploit nature in its own interest, with indifference to other creatures. In the early middle ages, these exploitative attitudes were apparent in the new technology of deep plowing ("Formerly man had been part of nature; now he was the exploiter of nature" [White 1967, p. 1205]) and in new calendars representing man modifying nature ("Man and nature are two things, and man is master" [White 1967, p. 1205]). Environmental problems cannot be solved unless the West changes religious values by turning to Zen Buddhism or, failing that, to Saint Francis's belief in the equality of all creatures.

Curiously, White elsewhere expressed admiration for Western medieval technology as a humane and liberating force (see 1962), besides stressing the long-term significance of such technology. Indeed, his staccato remarks about the changed relation of man and nature seem disproportionate to the alterations he actually describes. Did previous civilizations, including the Romans, really behave as "part of nature"? Did deep plowing really betoken universal exploitation of nature or signify an end in the West to love and respect for land? Did symbols of farming in Frankish calendars exhibit a significantly greater sense of mastery than the works of (say) Hesiod or Vitruvius (fl. first century BCE)? Historical debates about White's thesis have also concerned methodology; Elspeth Whitney (1993), for example, questions the assumption that religious beliefs could drive technological change, rather than expressing changes driven by economic and social forces.

White's article also generated debates among biblical scholars, historians of ideas, and ecologically concerned theologians on the interpretation of the Old and New Testaments and the attitudes of Judaism and Christianity toward nature. John Passmore, for instance, rejects an anthropocentric interpretation of the Hebrew Bible and Judaism. Certainly passages such as Psalm 104 express God's concern for nonhuman creatures, while books such as Leviticus and Deuteronomy convey the idea that the land is not human property but God's, and is held subject to a range of ethical conditions, including concern for other creatures. Clarence Glacken interprets humanity's role in the Bible as that of steward of the natural kingdom, as do many ecologically concerned theologians (such as Paul Santmire) when expounding the teachings of Christianity.

Certainly, Old Testament passages affirming "man's dominion" (Genesis 1, Psalm 8) raise theological problems. But the Hebrew term *rada* may convey nothing more than governorship of a kind answerable to God (as James Barr and Michael Northcott argue). *Dominion*, with its Latin-derived overtones of mastery, is misleading enough; reinterpreting *rada* as meaning domination (disproportionate power) or domineering (oppressive exercise of power) is a distortion. Interpretations stressing responsible stewardship are thus preferable to White's reading of these texts as conveying monarchical and unconditional mastery. Not even the "in the image of God" teaching of Genesis (1:26–27) warrants such an interpretation.

EXCERPT FROM LYNN WHITE'S "THE HISTORICAL ROOTS OF OUR ECOLOGIC CRISIS"

. . . Christianity inherited from Judaism not only a concept of time as nonrepetitive and linear but also a striking story of creation. By gradual stages a loving and all-powerful God had created light and darkness, the heavenly bodies, the earth and all its plants, animals, birds, and fishes. Finally, God had created Adam and, as an afterthought, Eve, to keep man from being lonely. Man named all the animals, thus establishing his dominance over them. God planned all of this explicitly for man's benefit and rule: no item in the physical creation had any purpose save to serve man's purposes. And, although man's body is made of clay, he is not simply part of nature: he is made in God's image.

Especially in its Western form, Christianity is the most anthropocentric religion the world has seen. As early as the second century both Tertullian and Saint Irenaeus of Lyons were insisting that when God shaped Adam he was foreshadowing the image of the Incarnate Christ, the Second Adam. Man shares, in great measure, God's transcendence of nature. Christianity, in absolute contrast to ancient paganism and Asia's religions (except, perhaps, Zoroastrianism), not only established a dualism of man and nature but also insisted that it is God's will that man exploit nature for his proper ends.

SOURCE: (from White, Lynn, Jr. 1967. "The Historical Roots of Our Ecologic Crisis." *Science* 155: 1203–1207.)

Jesus's teaching about sparrows and lilies (Matthew 6) embodies a nonanthropocentric view of nature, and other New Testament passages suggest that the whole of nature has a place in God's plan of salvation (Romans 8, Colossians 1). Christianity has sometimes been interpreted as legitimizing human domination of nature yet has just as often offered prayers for beasts and encouraged humanity to adorn the world so as to complete God's creation. As for the widespread medieval use of Christian language to endorse technology, perhaps this just reflects, as Whitney has suggested, "an age in which a religious perspective permeated every dimension of human life" (2005, p. 1736).

As Santmire shows, Christian writers' stances on the environment have been ambivalent. Some—such as Origen (185?–254?), Thomas Aquinas (1224/5–1274), and John Calvin (1509–1564)—adopted metaphysical forms of anthropocentrism, while others, such as Augustine (354–430) and René Descartes (1596–1650), without embracing the kind of anthropocentrism depicted by White, endorsed despotic interpretations. Yet many others—such as Irenaeus (c. 120/40–c. 200/3), Basil (c. 329–379), Ambrose (339–397), Cuthbert (635?–687), and Saint Francis (1181/2–1226)—adopted much gentler stances. Matthew Hale (1609–1676) explicitly viewed humanity as the steward of nature, and has had many modern successors.

The main historical significance of White's provocative paper has consisted not in his arguments but in the challenge it presented to theologians and philosophers worldwide to reappraise their attitudes toward the environment and the links of their thought with religion and culture. White expressly declares that reforming human behavior toward nature must be preceded by a reformed conception of the relationship between humanity and nature. Yet White's influence may unduly ascribe our environmental problems to culturally shared attitudes and values (Whitney 1993, pp. 168–169) and make their solution turn on a reform of worldview. Effectively solving such problems may require greater focus on reforming institutions, such as economic and political structures, rather than on reforming worldviews.

SEE ALSO *Buddhism; Christianity; Descartes, Rene; Ecotheology; Environmental Philosophy: II. Medieval Philosophy; Judaism; Passmore, John Arthur; St. Francis of Assisi; Stewardship; Technology.*

BIBLIOGRAPHY

Attfield, Robin. 1991. *The Ethics of Environmental Concern*, 2nd edition. Athens: University of Georgia Press.

Barr, James. 1972. "Man and Nature: The Ecological Controversy in the Old Testament." *Bulletin of the John Rylands Library* 55: 9–32.

Berry, R. J., ed. 2006. *Environmental Stewardship: Critical Perspectives, Past and Present.* London: T & T Clark International.

Glacken, Clarence J. 1967. *Traces on the Rhodian Shore: Nature and Culture in Western Thought from Ancient Times to the End of the Eighteenth Century.* Berkeley: University of California Press.

Hargrove, Eugene C., ed. 1986. *Religion and Environmental Crisis.* Athens: University of Georgia Press, 1986.

Northcott, Michael. 2006. "Soil, Stewardship, and Spirit in the Age of Chemical Agriculture." In *Environmental Stewardship*, ed. R. J. Berry. London: T. & T. Clark International, 213–219.

Passmore, John. 1974. *Man's Responsibility for Nature.* London: Duckworth.

Santmire H. Paul. 1985. *The Travail of Nature: The Ambiguous Ecological Promise of Christian Theology.* Philadelphia: Fortress Press.

White, Lynn, Jr. 1962. *Medieval Technology and Social Change.* Oxford: Clarendon Press.

White, Lynn, Jr. 1967. "The Historical Roots of Our Ecologic Crisis." *Science* 155: 1203–1207.

White, Lynn, Jr. 1978. *Medieval Technology and Social Change: Collected Essays.* Berkeley: University of California Press.

Whitney, Elspeth. 1993. "Lynn White, Ecotheology and History." *Environmental Ethics* 15(2): 151–169.

Whitney, Elspeth. 2005. "Lynn White (1907–1987), Thesis of." In *The Encyclopedia of Nature and Religion*, ed. Bron R. Taylor et al. London: Continuum Press. 2: 1735–1737.

Robin Attfield

WHITEHEAD, ALFRED NORTH

SEE *Process Philosophy.*

WILDERNESS

Throughout history wilderness has had many definitions and connotations. In the Book of Joel in the Bible the wilderness is characterized as a "desolate" place to put behind one and set in contrast to the "Garden of Eden" before one. For American transcendentalists such as John Muir wilderness was to be preserved because it was viewed as the handiwork of God. For Muir's father, Daniel, in contrast, wilderness was to be destroyed because it was the foothold of the Devil. Dictionary definitions range from the Middle English denotation of "a place of wild beasts," to a place "uncultivated," to a place "undisturbed" or "uninhabited" by human beings. Aldo Leopold referred to wilderness as "the raw material out of which man has hammered the artifact called civilization" and therefore considered it "a resource which can shrink but cannot grow" (1949, pp. 188, 199). For the historian Frederick Jackson Turner experience with the wilderness frontier shaped and has continued to influence the American character.

ORIGINS OF WILDERNESS PRESERVATION

Though wilderness is a variously defined and ancient concept, many environmentalists consider the preservation of wilderness one of the most important goals of environmentalism. The focus on wilderness preservation is perhaps

Hiking in the Arctic National Wildlife Refuge, Alaska. *The Arctic National Wildlife Refuge encompasses 19.2 million acres of Arctic and subartic habitat, 8 million acres of which are designated wilderness. Established in 1960, the refuge became a place of debate in the 1980s and onward because of the presence of (projectedly) billions of barrels of crude oil underneath its surface.* U.S. FISH AND WILDLIFE SERVICE.

the most characteristic component and contribution of North American environmentalism. The movement to preserve areas of wilderness in the United States, for instance, goes back to the early 1900s and can be seen as a reaction against a certain level of civilizing transformation and the despoliation of a presumably pristine landscape, whether the battle to save Hetch Hetchy Valley in the Yosemite from damming in the early 1900s or the early twenty-first-century battle over oil drilling in the Arctic National Wildlife Refuge.

Wilderness preservation has been the cause célèbre of a number of environmental groups. In late 1934 and early 1935 a group of American environmental leaders that included Benton MacKaye, Robert Marshall, Aldo Leopold, and Robert Sterling Yard founded the Wilderness Society, whose purpose is "to save from invasion ... that extremely minor fraction of outdoor America which yet remains free from mechanical sights and sounds and smell." Other environmental groups, including the Wildlands Project and the Rewilding Institute, also focus their efforts on the preservation of wilderness, and groups such as the Sierra Club (founded by John Muir in 1892) see wilderness preservation as a significant dimension of their work. The uniqueness of North American ideas about wilderness is pointed out by the fact that although the preservation of "protected areas" has become a component of conservation efforts in other parts of the world, the term *wilderness* seldom is evoked in those places; when it is used, American ideas about the meaning of wilderness almost always are cited.

In the late 1980s and early 1990s a number of environmental historians and philosophers introduced a series of criticisms of the concept of wilderness. Those criticisms prompted defensive responses from other environmental scholars and activists. This "great new wilderness debate" continues to rage. The concept is "alleged" by its critics "to be ethnocentric, androcentic, phallogocentric, unscientific, unphilosophic, impolitic, outmoded, even genocidal" (Callicott and Nelson 1998, p. 2).

THE RECEIVED WILDERNESS IDEA

The legitimacy of such charges depends on the existence of a "received wilderness idea," or a conception of wilderness that is socially crafted and that infiltrates the collective consciousness of a specific community in an essentially uniform fashion. Thus, those who offer a critique of the concept of wilderness do so with the assumption that wilderness is a social construction and that even from an environmental standpoint it is flawed, counterproductive, and even dangerous. Many of those who consider themselves defenders of wilderness deny that there is a received wilderness idea, asserting instead a wilderness realism or the idea that wilderness has a reality beyond that which people socially construct for it. Others defend the concept of wilderness by agreeing that it is a social construction but argue that it is not the social construction that critics believe it to be.

Wilderness constructivists point out that the concept of wilderness has been defined variously and incommensurably over millennia, that the word *wilderness* does not appear in all languages, and that current ideas and laws about wilderness preservation reflect previous definitions of and justifications for wilderness preservation while ignoring others. Hence, when the most important piece of wilderness legislation, the Wilderness Act of 1964, was enacted in the United States, a wilderness area was defined in a way that reflected—and now codified as law—the dominant received wilderness idea: "in contrast with those areas where man and his own works dominate the landscape ... an area where earth and its community of life are untrammeled by man, where man is a visitor who does not remain."

CRITICISMS BY WILDERNESS CONSTRUCTIVISTS

Constructivist critics suggest that even though variations in definition occur in characterizations of wilderness, all the definitions set wilderness in opposition to humans and human civilization ("visitors who do not remain"), make the presumed lack of human influence the measure of "untrammeled," and generally consider wilderness areas to be places "in contrast" to human works. Further, critics argue that this characterization has a series of significant shortcomings.

First, the received wilderness idea is said to be ethnocentric to the point of being genocidal when transported outside North America. Wilderness critics suggest that the concept as constructed in its North American context is so specific to that context that it has negative human implications when exported to other countries. They point out that if one considers long-term human habitation as anathema to wilderness, if a country desires to create a wilderness area and there are aboriginal people living in that area, those people must be deported. To the degree that the identity of a people is embedded in its landscape, the forced removal of that people from that landscape may constitute a form of cultural genocide. Wilderness defenders sometimes deny this implication and sometimes admit it and openly opt for the preservation of the nonhuman over the preservation of specific human communities. Some historical work (Spence 1999, Burnham 2000) suggests that evicting native peoples from their homelands to establish national parks and other "protected areas" occurred in the United States in a number of instances.

Second, the received wilderness idea is said to be inappropriately andro- or phallogocentric. Former U.S. president Theodore Roosevelt touted the importance of wilderness experience because it "promoted that lacking vigorous manliness," and the nature writer Sigurd Olson championed wilderness travel because it provided "that virile, masculine type of experience men need today" (quoted in Callicott and Nelson 1998). This hypermasculine image of wilderness is thought by some to be offensive and exclusionary.

Third, constructivists argue that wilderness (set in opposition to humanity) is viewed as the highest manifestation of nature. In light of the fact that the received wilderness idea evolved parallel to and was influenced directly by the development of the science of ecology and the fact that people often look to ecology to deliver the clearest images of the ontology of nature, ideas about wilderness and the preservation of wilderness reflect those early ecological paradigms. Whether portrayed as a collection of superorganisms by Frederick Clements or likened to a functioning economy by Charles Elton, the background reality of nature—and therefore wilderness—was thought to be harmonious and balanced, static and unchanging unless spoiled by human impact. Since the 1980s and drawing on ecological thought going back to the 1920s, however, the reigning ecological image of nature has been one of disturbance, flux, change, and discord. This background image of an ever-changing nature contrasts sharply and incommensurably with received ideas of wilderness as primeval, a place frozen in time, land as it was before human conquest. This rethought image of nature and wilderness affects people's corresponding assumptions about how they should interact with wilderness or what constitutes harm to wilderness. If wilderness is protected properly only when it remains static, any impact that alters wilderness also harms wilderness. However, if nature and wilderness are inherently dynamic, the idea of an untouched and unchanged wilderness as a properly treated wilderness has to be revised.

Fourth, critics of the received idea suggest that this view is unphilosophical and impolitic. The image of wilderness as land at the far end of the spectrum between the natural and the unnatural perpetuates a metaphysical bifurcation between humans and nature. Such an image also creates and enforces a value dualism in light of the positive value that wilderness advocates assign to wilderness and the corresponding and consistent negative value they are obligated to assign to humans and human activities. Wilderness critics point out that these dualisms are malignant. For instance, from them flows the inevitable condemnation of human interactions with nature, including not only oil spills and species eradication but also acts of ecological restoration. Moreover, because of the logic of these dualisms, wilderness advocacy has been criticized for being elitist and exclusionary in that non-wilderness areas are treated as places of lesser value, and the people who love them are by implication environmentalists of lesser worth.

Fifth, wilderness critics suggest that the received wilderness idea interferes with the acquisition of an inclusive environmental ethic, especially the land ethic of Aldo Leopold. That is seemingly the environmental ethic of many environmental activists, natural resource managers, and the environmentally literate public, and it promises to deliver direct moral standing to the nonhuman world. If Leopold is correct and an appropriate moral relationship between humans and land depends on people viewing themselves as fully and properly part of an inclusive biotic community, anything that conceptually separates people from land, even people's images of wilderness, stands in the way of an inclusive environmental ethic. According to this line of thought, to the degree that people's ideas about wilderness conceptually separate humans from nature, wilderness destroys the ability to extend direct moral consideration to nature.

RESPONSES TO THE CRITICS

In response to these criticisms, wilderness defenders deny that wilderness is a received concept and attempt to refute these criticisms one by one or admit that wilderness is a received idea but suggest that wilderness critics have portrayed it in the wrong light or that it can and should be reconceived. Some wilderness critics agree that the concept of wilderness should be and can be reworked, whereas others have suggested that the idea is too burdened with past associations and should be jettisoned in favor of some other term and corresponding "protected area" strategy. J. Baird Callicott (Callicott and Nelson 1998, Nelson and Callicott 2008), for example, suggests that *biodiversity reserve* is a preferable label in that it clearly indicates that such set-asides are protected first and foremost for the good of biodiversity, not for the good of human recreational, scientific, or aesthetic interests.

Ideas about the nature of wilderness and prehuman conditions can affect ecological sciences in different ways. The effect of a prehuman or nonhuman landscape is not apparent in "pure" ecology or in ecological descriptions and modeling aimed at understanding how a specific system works; examples of this would be the answers to questions such as: Why are there so many species in an ecosystem? and Why do predator-prey systems seem more stable than theory suggests they should be? However, that effect is present in more normative "applied" ecology, in which the goal is to predict the future with an eye toward guiding people's actions. Various forms of applied ecology—from restoration to wildlife management—seem to evoke a kind of naturalism (equating the good with the natural) in which a wilderness or pre- or nonhuman condition is seen as good or desirable and constitutes the proper target of conservation efforts, whereas the humanized is thought of as a bad or undesirable state of affairs. The alternative interpretation in this context appears to be uncritically anthropocentric.

SEE ALSO *Bible; Callicott, J. Baird; Land Ethic; Leopold, Aldo; Muir, John; North America; Preservation; Roosevelt, Theodore; Sierra Club; Social Constructivism; Wilderness Act of 1964.*

BIBLIOGRAPHY

Burnham, Philip. 2000. *Indian Country, God's Country: Native Americans and the National Parks.* Washington, DC: Island Press.

Callicott, J. Baird, and Michael P. Nelson, eds. 1998. *The Great New Wilderness Debate.* Athens: University of Georgia Press.

Cronon, William. 1996. *Uncommon Ground: Toward Reinventing Nature.* New York: W. W. Norton.

Leopold, Aldo. 1949. *A Sand County Almanac, and Sketches Here and There.* New York: Oxford University Press.

Nash, Roderick. 2001. *Wilderness and the American Mind,* 4th edition. New Haven, CT: Yale University Press.

Nelson, Michael P., and J. Baird Callicott, eds. 2008. *The Wilderness Debate Rages On: Continuing the Great New Wilderness Debate.* Athens: University of Georgia Press.

Oelschlaeger, Max. 1991. *The Idea of Wilderness: From Prehistory to the Age of Ecology.* New Haven, CT: Yale University Press.

Spence, Mark David. 1999. *Dispossessing the Wilderness: Indian Removal and the Making of the National Parks.* New York: Oxford University Press.

Michael P. Nelson

WILDERNESS ACT OF 1964

The Wilderness Act of 1964 was signed into law by President Lyndon Johnson on September 3 of that year. It is one of the most significant pieces of environmental legislation because of its continental scope and the idea of wilderness it contains. It was sent to President Johnson after eight years of rewriting and negotiations in Congress.

PURPOSE OF THE ACT AND DEFINITION OF WILDERNESS

The overall purpose of the act is to establish a National Wilderness Preservation System that identifies, protects, and administers areas designated as wilderness for the future recreational use of the American people. Proponents of the act believed that in its absence expanding population and development would leave no lands in their natural condition.

Some portions of the lands that came to be designated as wilderness under the act already were protected by various overlapping designations. Some of the lands were controlled by the federal agencies with jurisdiction, including the Forest Service, the National Parks Service, and the Bureau of Land Management. This was the source of an important controversy regarding the act. Conservationists and other protectors of wilderness believed that those agencies were open to manipulation by business interests and political groups. Proponents of the act cited cases of logging, mining, and dam building within lands already under some form of federal protection. They feared that wilderness areas might disappear through administrative neglect. Federal agencies, with the exception of U.S. Fish and Wildlife Service, saw the act as a challenge to their authority and expertise. Further, the Forest Service and other agencies were committed to a multiple-use doctrine by which all lands under their control should be used for more than wilderness recreation, whether logging or mining or hydroelectric power.

This set of issues is reflected in the concessions to existing federal agencies in the act: They would be allowed to continue to manage and have jurisdiction over the newly designated wilderness areas and further, the secretary of agriculture would have ten years after the act is signed to review and submit primitive areas as wilderness. Once areas are designated as wilderness by the act, their uses are limited with some exceptions. Deciding whether a specific primitive or roadless area is wilderness is still an agency decision, to be confirmed by the president, who then advises Congress.

OBJECTIONS

In addition to the initial opposition by some federal agencies, logging, mining, hydropower, motorized recreation, and tourist lodging interests opposed the passage of the act, arguing that it was not necessary, would impede economic development, was designed to benefit only an elite group, and undermined the wisdom of the multiple-use doctrine.

Those objections resulted in some important qualifications in the act. For instance, mining and prospecting as allowed by U.S. mining law were to continue with some restriction until 1983. Also, grazing and the use of motorboats and planes were allowed in areas where they already had been established. Those concessions did not eliminate opposition to the act but made it sufficiently palatable for senators and representatives in western states and congressional districts to vote for it.

Several people and groups played key roles in passing the act. Howard Zahniser became the leader of the Wilderness Society and the editor of its magazine in 1945. In the years from 1956 to 1964 Zahniser produced several dozen drafts of the act, continuously lobbied, and wrote and spoke for its passage. Although that eight-year battle culminated in the act becoming law, Zahniser did not live to see it; he died of a heart attack shortly before the signing.

Many other conservationists were involved in drafting and lobbying for the act. The Minnesota native Sigurd Olson, an acquaintance of Zahniser, advised Senator Hubert H. Humphrey to support the act after Zahniser solicited Humphrey's help. Largely on the basis of Olson's recommendation, Humphrey introduced the act in the Senate in 1956. Humphrey proved to be a crucial proponent of the act through the years of hearings, compromises, and rewriting. Besides Olson, David Brower worked on editing early versions of the act.

EFFECTS ON ENVIRONMENTAL ETHICS

The act has figured into environmental and philosophical thinking in several ways insofar as it provides a definition of wilderness that continues to draw sharp criticism. That definition says in part that wilderness is primeval and uninhabited and untrammeled by humans and also that it should provide opportunities for solitude and primitive recreation. One kind of criticism focuses on the relative vagueness of the terms used and the possibility of misunderstanding that may result. Other criticisms revolve around the inherent conflict involved in trying to satisfy both of those conditions for wilderness. For instance, it seems as if increases in the numbers of people wanting primitive recreation will have an adverse effect on the natural, primitive, and untrammeled character of wilderness areas, to say nothing of solitude. A final conceptual difficulty is that the definition of wilderness endorsed by the act insists on a sharp division between wilderness and humans that is subject to serious challenge because, for example, it seems to ignore the historical presence of aboriginal peoples in many of those areas and, when imitated in other countries, can result in the eviction and dispossession of indigenous peoples.

The act set aside some 9 million acres of land as wilderness, a number that has grown to over 100 million acres. Controversy was expected to continue and even intensify in light of increasing demands for resources.

SEE ALSO *Environmental Law; Environmental Policy; Forests; Mining: I. Overview; U.S. Forest Service; Wilderness.*

BIBLIOGRAPHY

Allin, Craig W. 1982. *The Politics of Wilderness Preservation.* Westport, CT: Greenwood Press.

Backes, David. 1997. *A Wilderness Within: The Life of Sigurd F. Olson.* Minneapolis: University of Minnesota Press.

Callicott, J. Baird, and Michael P. Nelson, eds. 1998. *The Great New Wilderness Debate.* Athens: University of Georgia Press.

Frome, Michael. 1997. *Battle for the Wilderness,* rev. edition. New York: Praeger.

Harvey, Mark. 2005. *Wilderness Forever: Howard Zahniser and the Path to the Wilderness Act.* Seattle: University of Washington Press.

Nelson, Michael P., and J. Baird Callicott, eds. 2008. *The Wilderness Debate Rages On.* Athens: University of Georgia Press.

Scott, Doug. 2004. *The Enduring Wilderness.* Golden, CO: Fulcrum.

Woods, Mark. 1998. "Federal Wilderness Preservation in the United States: The Preservation of Wilderness?" In *The Great New Wilderness Debate,* ed. J. Baird Callicott and Michael P. Nelson. Athens: University of Georgia Press.

Charles J. List

WILSON, EDWARD O.
1929–

Born on June 10, 1929, in Birmingham, Alabama, Edward Osborne Wilson, one of the twentieth century's exemplary individuals, started out as an ecological scientist and transitioned to applied conservationist and environmental ethicist. Early in his life, Wilson began to observe some of the smallest and most diverse organisms: insects, especially ants. "In 1942, when I was 13 years old," he remembers, "I was studying ants for a Boy Scout project…, and so I discovered a nest of red fire ants" (2006, p. 71), which was the first known colony of the invasive exotic species *Solenopsis invicta* in the United States. In his autobiography *Naturalist* (1994), Wilson recalls how his childhood years in the U.S. South imbued him with a curiosity for all aspects of natural history and provided the foundation of his scientific career. This boyhood fascination with ants eventually led to a Ph.D. and faculty career at Harvard University where he explored diverse topics and scales, ranging from chemical ecology to biogeography, from taxonomy to the evolution of social interactions. For *The Ants* (coauthored with Bert Hölldobler), he was awarded a Pulitzer Prize for nonfiction in 1990.

In the 1960s, ant-collecting expeditions to Pacific islands and the New World Tropics grounded Wilson's scientific work in basic natural history, which emphasized descriptions of species and their interactions. Then, however, he used firsthand experiences in observing small organisms as illustrations for addressing major ecological and evolutionary questions, in the process positing the taxon cycle (Wilson 1959, 1961) and island biogeography theory (MacArthur and Wilson 1967). Both hypotheses were major conceptual advances in explaining patterns of species richness from biogeographic and demographic principles. This body of work, resting on theoretical principles, was also rigorously shown in the field with whole island experiments in the Florida Keys. It has since proved to be not only a powerful current of thought in general ecological theory, but also in conservation, as the idea of island patches of suitable habitats for particular species can be used to manage vulnerable populations across the landscape matrix.

Wilson's investigation of ants also led to insights regarding their communication and colony organization. His seminal research in chemical ecology showed that ants use pheromone cues to coordinate their complex group superstructure. Such results eventually led to a path of inquiry regarding the biological basis of social interactions in general. With *Sociobiology: The New Synthesis* (1975), Wilson sought to explain behavioral traits within strictly biological confines—an idea that proved controversial, but also founded a new discipline and earned him his first Pulitzer Prize for nonfiction in 1978 for *On Human Nature.*

As a writer, Edward O. Wilson was able to transcend his academic discipline and link it with society by effectively communicating scientific understanding to a wider audience, often employing ground-breaking neologisms. For example, the term *biophilia* (humans' innate attraction to living systems) was coined by Erich Fromm in *The Heart of Man* (1964), but Wilson's homonymous book generated dynamic discussion about the concept in 1984. Likewise, in an academic symposium moderated by Wilson, Walter G. Rosen coined the term *biodiversity* in 1985, but Wilson's 1988 book of the same name introduced the term into our collective vocabulary and imagination and from there helped to coalesce the environmental movement and governmental policy around concern over the modern crisis of mass extinctions of species. In 2006 Wilson himself invented the term *Eremozoic Era* as a provocative depiction of the "Age of Loneliness" that will succeed the sixth mass extinction if humanity does not undertake immediate actions to protect the planet's biodiversity.

To perpetuate his legacy, the Edward O. Wilson Biodiversity Foundation was launched in 2007 to "preserve biological diversity in the living environment by inventing and implementing business and education strategies in the science of conservation." Its approach parallels Wilson's own development. The foundation attempts to maintain Wilson's traditional emphasis on the need to understand all biodiversity with a natural history-oriented program consisting of hands-on education and citizen science. Yet the organization also uses the experience of its cofounders and capital derived from the biotechnology industry to put forward a technology-based, capitalistic model for future research and conservation. This market-based approach to conservation and development—involving, for example, tapping genetic resources and creating win-win cooperative agreements with industry—coincides with Wilson's own strong belief in the power of science and technology to solve problems. This approach is ironic to some, as many conservation problems are in fact the result of science and technology.

As a scientist, Wilson has earned the highest professional recognition (e.g., the National Medal of Science, the Craaford Prize, and the Tyler Award), but unlike most academics he has also attained great social relevance. In both arenas, his popular and scientific work has significantly contributed to shedding light on the beauty and value of the diversity of insects and other inconspicuous organisms found in the living systems around us. Prestige, however, does not confer immunity from disapproval. By placing such a high value on technological and scientific solutions, Wilson has also been the focus of criticism. In particular, his defense of sociobiology created a storm of rancorous debate, leading to a confrontation in which activists dumped a pitcher of water on Wilson's head at a conference. The current orientation of his foundation may also prove problematic for some conservationists. Nonetheless, Edward O. Wilson established himself as one of the most influential thinkers of his day precisely by working in the natural sciences and simultaneously at the interface of science and society.

SEE ALSO *Biodiversity; Biophilia; Conservation; Conservation Biology; Extinction.*

BIBLIOGRAPHY

Hölldobler, Bert, and Edward O. Wilson 1990. *The Ants.* Cambridge, MA: Harvard University Press.

MacArthur, Robert H., and Edward O. Wilson. 1967. *The Theory of Island Biogeography.* Princeton, NJ: Princeton University Press.

Wilson, Edward O. 1959. "Adaptive Shift and Dispersal in a Tropical Ant Fauna." *Evolution* 13: 122–144.

Wilson, Edward O. 1961. "The Nature of the Taxon Cycle in the Melanesian Ant Fauna." *American Naturalist* 95: 169–193.

Wilson, Edward O. 1975. *Sociobiology: The New Synthesis.* Cambridge, MA: Harvard University Press.

Wilson, Edward O. 1978. *On Human Nature.* Cambridge, MA: Harvard University Press.

Wilson, Edward O. 1984. *Biophilia.* Cambridge, MA: Harvard University Press.

Wilson, Edward O. 1988. *Biodiversity.* Cambridge, MA: Harvard University Press.

Wilson, Edward O. 1994. *Naturalist.* Washington, DC: Island Press.

Wilson, Edward O. 2006. *The Creation: An Appeal to Save Life on Earth.* New York: W. W. Norton.

Christopher B. Anderson
Ricardo Rozzi

WORDSWORTH, WILLIAM
1770–1850

William Wordsworth was born on April 7, 1770, in Cockermouth in Cumberland's Lake District and is considered one of England's finest nature poets. His poetry and critical works manifest a complex understanding of the relationship among the natural environment, language, and human passions. That complexity is apparent in his best-known critical essay, the Preface to *Lyrical Ballads* (Wordsworth and Coleridge 1969 [1800, 1802]), which was written to justify the experimental style of those poems. The Preface indicts the artificial diction of earlier poets and calls for a new way of describing and relating to the natural world that will honor the qualities most fundamental to human life: imagination and emotion. Wordsworth did not advocate a simple return to nature. Instead, he reminded his readers that the source of what is best in human manners, feelings, language, and community resides in a positive relationship to the natural world. Wordsworth died on April 23 in Rydal Mount, Westmorland, England.

POETRY AND THE BOND WITH NATURE

The bond of humanity with nature, Wordsworth suggested—though as indestructible as the deep, universal emotions that structure people's inner lives and influence their actions and relationships—was being attenuated in an increasingly urban Great Britain. The radical industrialization of English life and the tumultuous events connected with the French Revolution of 1789, he suggested, were "acting with a combined force to blunt the discriminating powers of the mind" (Wordsworth and Coleridge 1969, p. 160). That situation can be considered the predecessor of the contemporary information age, with its taste for news, novelty, and speed. The more city people's cravings for "extraordinary incident" (Wordsworth and Coleridge 1969, p. 160) are satisfied by tabloid-style accounts and the more their desire for racy or mawkish entertainment is satisfied by pandering artists, the more they tend toward a paradoxical, narcotized state of "savage torpor" (Wordsworth and Coleridge 1969, p. 160). Essentially, Wordsworth's complaint is about the hollowing out of emotion and experience until no feeling, event, or utterance seems authentic: Cheap spectacle replaces genuinely artistic representation, and quantity replaces quality in all areas of life.

In response to a development destructive to individual identity and genuine community, Wordsworth outlined a redemptive poetics. Because the corruption of language and sentiment lay at the heart of the problem, he considered it vital to have recourse to devise a better model in both areas: life in the English countryside. The Preface states that the main object of *Lyrical Ballads* was "to chuse incidents and situations from common life, and to relate or describe them, throughout, as far as was possible, in a selection of language really used by men" (Wordsworth and Coleridge 1969, p. 156). Readers are told, "Low and rustic life was generally chosen, because in that condition, the essential passions of the heart find a better soil in which they can attain their maturity, are less under restraint, and speak a plainer and more emphatic language ..." (Wordsworth and Coleridge 1969, p. 156). Country people are less distanced from their vital passions than are urban dwellers; they express what they feel more readily to other members of the community and have not lost the emotional connection to nature that has been forfeited by the average Londoner. They also have not lost the ability to see what is in front of them in a meadow or a wood, in all its particularity as well as in its universal dimension among "the beautiful and permanent forms of nature" (Wordsworth and Coleridge 1969, p. 156).

What the pastoral citizen has achieved over a lifetime, the poet can accomplish by an initially more self-conscious effort involving careful selection from the chosen language model and by methods of composition best characterized as meditative. The latter quality underlies Wordsworth's famous definition: "Poetry is the spontaneous overflow of powerful feelings: it takes its origin from emotion recollected in tranquillity: the emotion is contemplated till by a species of reaction the tranquillity gradually disappears, and an emotion, kindred to that which was before the subject of contemplation, is gradually produced ..." (Wordsworth and Coleridge 1969, p. 173). By this process, the lyricist will compose poems that help ordinary people recover the passion and sense of community from which city life bars them. The poet who can do this, Wordsworth declared, will have attained the status of a priest or prophet and will be "the rock of defence of human nature; an upholder and preserver, carrying every where with him relationship and love" (Wordsworth and Coleridge 1969, p. 168).

NATURE AND HUMANITY

In keeping with that ideal and in terms of descriptive quality and aesthetic appeal, Wordsworth's poetry is not timid about suffusing nature with human reference and spiritual significance. His reference in "I Wandered Lonely as a Cloud" to common meadow flowers transfigured into an almost biblical "host of golden daffodils" suggests this, as does his attention to what he calls in Book 11 of the 1805 *Prelude* "spots of time": moments in the presence of nature that form highly charged "renovating" memories to which people may repair in times of trouble. (The ode "Tintern Abbey" involves strong memories connected to a beautiful natural scene.) Many of Wordsworth's poems are not simply about nature; they are about the joys and travails of human beings in the presence of or in painful separation from nature. The *Prelude* traces the development of his own artistic and spiritual qualities even if those qualities owe almost everything to the natural environment in which he grew up and with which he sought to keep communion.

Wordsworth was the British poet laureate from 1843 to 1850, and his poetry and poetic theory are central to British romanticism. They also figure in the work of later authors, among them John Ruskin. Wordsworth's close attention to nature's particularities, insistence that the language of poetry should be the language of ordinary uncorrupted humanity, and impassioned argument that literary art can help people achieve both individual happiness and community have been used as points of departure by authors from his own time through the present.

SEE ALSO *Romantic Poetry, English; Romanticism; Ruskin, John; United Kingdom.*

BIBLIOGRAPHY

Gill, Stephen. 1989. *William Wordsworth: A Life*. Oxford: Clarendon Press.

Gill, Stephen. 2003. *The Cambridge Companion to Wordsworth.* Cambridge, UK, and New York: Cambridge University Press.

Roe, Nicholas. 2002. *The Politics of Nature: William Wordsworth and Some Contemporaries,* 2nd edition. Houndmills, Basingstoke, Hampshire, UK, and New York: Palgrave.

Wordsworth, William, and Samuel Taylor Coleridge. 1969. *Lyrical Ballads, 1798 [by] Wordsworth and Coleridge,* ed. W. J. B. Owen, 2nd edition. London and Oxford: Oxford University Press.

Wu, Duncan. 2002. *Wordsworth: An Inner Life*. Oxford and Malden, MA: Blackwell.

Alfred J. Drake

WORLD TRADE ORGANIZATION

The product of a decades-long global movement to ease trade barriers among nations, the World Trade Organization (WTO) was founded on January 1, 1995, to provide an international negotiating forum, enforcement agency, and dispute-resolution mechanism for countries seeking to benefit from liberalized trade policies. As of June 2008, 152 nations were members of the WTO.

The underlying principle of all WTO regulations is the most-favored-nation policy: Each member nation is obliged to grant to all members the most favorable trade terms that it grants to any other member, and members are required to give equal treatment to overseas and domestic suppliers of goods and services. Although the WTO has achieved substantial reductions in tariffs and other trade barriers since its founding, it has also generated storms of controversy and protest among various constituencies—chiefly trade unions, student groups, and environmental organizations—that contend that the WTO's basic orientation favors the commercial interests of transnational corporations at the expense of labor and environmental standards and that its rules and governance procedures constitute a threat to the institutions of civil society. Many environmental organizations share the basic view expressed by Greenpeace International: "The World Trade

Organisation (WTO) promotes free trade for the gain of private interests, over and above our health and the environment" (Greenpeace International 2008).

ORIGINS AND STRUCTURE

The WTO is the successor to the General Agreement on Tariffs and Trade (GATT). GATT was founded in 1947 to foster a worldwide reduction in tariffs, mainly on manufactured goods. The final Uruguay round of negotiations under GATT resulted in a pact that aimed to achieve an eventual one-third reduction in global tariffs and a gradual elimination of other hurdles to open trade among nations, especially in agricultural goods. This agreement also took unprecedented steps toward liberalizing international commerce in investments and services and bolstering international intellectual-property rights. These initiatives were carried over into the WTO, which was created to provide stronger enforcement and dispute-resolution capabilities for the agreements forged by GATT and to further lower hurdles to global trade.

The WTO is headquartered in Geneva, Switzerland. Its highest governing body is the Ministerial Conference, which is composed of all the members of the organization. A General Council carries out the decisions of the conference and is responsible for daily administrative tasks. The director-general, appointed by the Ministerial Conference, is the group's chief executive officer. Substantive trade negotiations take place in various rounds launched by the Ministerial Conference. The Doha Round was launched in 2001 in Doha, Qatar but stalled over disagreements between the industrialized countries and the nations of the third world, largely over agricultural subsidies in the European Union and the United States that many third-world governments believe force their countries to import cheaper agribusiness foodstuffs that were once grown locally, thereby placing their farmers at a competitive disadvantage and threatening the livelihoods of millions of farm families.

CONFLICTS AND CONTROVERSIES

The free-trade agenda of the WTO began to generate gales of protest and controversy in the late 1990s. The watershed event in this wave of anti-WTO sentiment was the series of mass demonstrations and street disruptions that greeted the WTO Ministerial Conference in Seattle, which began on November 30, 1999. These protests arose as part of a worldwide surge of "antiglobalization" sentiment in the opening decade of the twenty-first century. Many of the leaders of this movement deny that its intent is antiglobalist. They insist, rather, that they are opposed to what they term "corporate globalization"—the erosion of the sovereign powers of nation-states and the institutions of civil society in the face of institutions like the WTO that, they believe, reflect the interests of private transnational corporations rather than the public interests of the poor, the labor movement, and the environment.

Although some of this antiglobalist ire has been directed against other instruments of international commerce, such as the World Bank and the International Monetary Fund, the WTO has borne the brunt of the protests because of its aggressive agenda of dismantling barriers to transnational corporate investment. There are three main areas of controversy: governance and dispute-resolution procedures, labor standards, and environmental policies.

Governance and Dispute-Resolution Procedures The WTO prides itself on its democratic practice of according one vote to each country, regardless of its size or wealth. Yet, purportedly in the interests of avoiding conflict, the WTO has never taken a formal vote on trade proposals, preferring instead to rely on negotiation and consensus decision making to minimize conflicts and polarization. Critics have noted, however, that much of the real agenda setting and substantive negotiation in the WTO takes place among informal groups, the so-called miniministerials, consisting mostly of the most powerful members, so the interests of developing countries tend to be underrepresented in the final agreements. Even the WTO acknowledges, "It would be wrong to suggest that every country has the same bargaining power" (WTO 2008a).

The WTO's dispute-resolution procedures have also aroused considerable controversy. The member nations agree to avoid unilateral action and to rely instead on the WTO's multilateral system. If the contending parties cannot resolve their issues through mediation and negotiation, the complainant can request the convening of a special panel, the decisions of which are binding on all parties. These panels consist of three officials appointed by the WTO secretariat. Critics have raised a number of objections to these tribunals: The panelists are unelected and are under no obligation to disclose conflicts of interest; only the official national trade representatives of the contending parties can attend the proceedings—all other government officials are prohibited from attending; all proceedings, documents, and transcripts are secret; and the press is barred from the tribunals. Whether the closed, secretive nature of these panels vitiates the WTO's claims to be democratic and open to public concerns remains a topic of vigorous debate.

A matter of broader concern is the WTO's requirement that member nations' future laws conform to WTO rules, which some critics assert is bound to have a chilling effect on the law making of sovereign states. In this view, transnational corporations looking for overseas investment opportunities will seek out the countries with the fewest impediments to low-cost production, including labor and environmental regulations, thus setting off an economic race to the bottom in workers' rights and environmental protection.

Labor and Workers' Rights On the labor front the WTO asserts that "all WTO member governments are committed to a narrower set of internationally recognized 'core standards' for labor," but it also acknowledges that there is "no work on this subject in the WTO's Councils and Committees," that "it is not easy for [member nations] to agree" on labor standards, that "the question of international enforcement is a minefield," and that "WTO agreements do not deal with labor standards as such" (WTO 2008b).

The pressure to minimize labor standards in WTO rules comes chiefly from two sources: the transnational corporations seeking low-wage production opportunities abroad and third-world governments desiring such investments. Both parties have characterized the quest for rigorous labor standards in trade agreements as a form of protectionism. In the words of the WTO's 1996 Singapore ministerial declaration on core labor standards, "We reject the use of labour standards for protectionist purposes, and agree that the comparative advantage of countries, particularly low-wage developing countries, must in no way be put into question" (WTO 2008b). Despite the WTO's pledge to work with the International Labor Organization on this issue, there was little progress as of 2008. Labor groups throughout the world regard this charge of "protectionism" as a guise for promoting flagrant exploitation in the interests of large corporations. Labor-movement representatives still decry the shift in investments from the highly unionized, high-wage countries of the developed world to third-world export-processing zones, areas specially designated for export-oriented manufacturing that are often exempt from national labor codes (Jauch 2002).

The WTO and the Environment The WTO states, "Sustainable development and protection and preservation of the environment are fundamental goals of the WTO." It further asserts, "The Doha Agenda includes specific negotiations on trade and environment and some tasks assigned to the regular Trade and Environment Committee" (WTO 2008c).

Critics claim that the reality belies this reassuring rhetoric. They note, for example, that WTO panels have consistently ruled that trade regulations cannot discriminate among products on the basis of their method of production or harvesting, so that, for example, preferring sustainably harvested wood to clear-cut timber from tropical forests would be considered a restraint of trade. Likewise, sustainably caught fish or products resulting from humane labor conditions or humane treatment of animals cannot be accorded preferential treatment under WTO rules. Critics further charge that the WTO's tariff policies on raw materials encourage the depletion of dwindling natural resources and put pressure on third-world countries to focus on extracting and exporting raw materials, often in partnership with giant transnational mining companies, rather than on sustainable industrial development.

The record of WTO rulings in environment-related disputes has not reassured critics or lent credibility to the WTO's claims of environmental sensitivity. The original General Agreement on Tariffs and Trade contained a provision, Article XX, that claimed to allow exceptions to trade rules based on domestic concerns about health, safety, and the environment. Nevertheless, nearly all the decisions of the WTO panel have ruled in favor of challenges to environmental protections. Of the nine cases that bear on the environment, three have aroused the most controversy:

- In 1997 Venezuela and Brazil challenged a U.S. policy that limited gasoline imports based on the chemical makeup of the product. The WTO's ruling in favor of the plaintiffs led the United States to comply with the ruling by weakening provisions of the Clean Air Act, downgrading cleanliness rules to allow imports that would result in a 5- to 7-percent increase in nitrous oxide emissions.
- The United States once banned shrimp imports from countries whose boats did not use turtle-excluder devices. In 1998 India, Malaysia, Pakistan, and Thailand challenged this U.S. ban on their shrimp as a restraint of trade. The WTO found in favor of the plaintiffs, obliging the United States to weaken that provision of the Endangered Species Act. It still restricts imports of shrimp to those brought in by ships with turtle-excluder nets, but with no requirement that those ships be verified as the ones that actually caught the fish.
- After years of Mexican threats of a trade challenge to a U.S. policy requiring that domestically sold tuna have a "dolphin-safe" label, the Bush administration in 2002 softened U.S. regulations to permit "dolphin-safe" labeling for tuna harvested with techniques that had previously been deemed dangerous to dolphins.

Another major area of environmental concern is the WTO's posture on trade in genetically modified organisms. In response to a formal complaint from three major growers of genetically modified crops—the United States, Canada, and Argentina—the WTO ruled in

May 2006 that the European Union's delays in coming up with a systematic regulatory policy regarding genetically modified organisms were a violation of WTO rules. The European Union policy in place in 2008 does require that genetically modified products be labeled, segregated, and traceable. The WTO has ruled, however, that outright national bans on specific genetically modified crops are illegal restraints of trade if the country's regulations are based on criteria stricter than those contained in the WTO's Sanitary and Phytosanitary Measures, which many environmental groups consider to be inadequate.

THE WTO AND POLITICAL AND SOCIAL PHILOSOPHY

The opponents in the controversies that swirl around the WTO fall broadly into two camps. The WTO's champions are, for the most part, proponents of a free-market liberalism—often dubbed neoliberalism in contemporary discourse—that regards the unfettered operation of free markets as the most efficient way to create wealth and to address social and economic ills. The WTO's critics—a disparate coalition that includes labor unions, antiglobalist student activists, displaced third-world farmers, and environmental organizations—typically regard the unregulated free market as a recipe for increasing inequality and exploitation of the environment and other humans. The WTO's supporters tend to be suspicious of the public sphere and the state, which they view as a brake on individual initiative and creativity. WTO opponents regard the public sphere as the domain of democracy, where citizens can deliberate about which policies best serve their interests as a cooperative community, not merely as an array of discrete, self-interested individuals. These WTO critics point to the irony that the antistatist free-traders seem all too ready to accept the dictates of a secretive, powerful, privately run de facto world government that enforces the interests of corporations while these same "neoliberals" resent and resist the incursions of governmental bodies that represent a broader public interest. This clash of the social and the individual, the public and the private, has been a long-standing theme in world politics and social theory. The WTO has become the latest, and in some respects the most prominent, lightning rod for these clashing views of how humans should relate to one another and their environment.

SEE ALSO *Genetically Modified Organisms and Biotechnology; Greenpeace; Sustainability.*

BIBLIOGRAPHY

Bernasconi-Osterwalder, Nathalie, et al. 2006. *Environment and Trade: A Guide to WTO Jurisprudence.* London: Earthscan.

Greenpeace International. 2008. "Encourage Sustainable Trade." http://www.greenpeace.org/international/campaigns/trade-and-the-environment

Hoekman, Bernard M., and Michel M. Kostecki. 1995. *The Political Economy of the World Trading System.* Oxford, UK: Oxford University Press.

Jackson, John. 1997. *The World Trading System: Law and Policy of International Economic Relations.* Cambridge, MA: MIT Press.

Jauch, Herbert. 2002. "Export Processing Zones and the Quest for Sustainable Development: A Southern African Perspective." *Environment and Urbanization* 14(1): 101–113. Available from http://www.gpn.org/research/namibia/jauch.pdf

Moellendorf, Darrel. 2005. "The World Trade Organization and Egalitarian Justice." *Metaphilosophy* 36(1, 2): 145–162.

Singer, Peter. 2002. *One World: The Ethics of Globalization.* New Haven, CT: Yale University Press.

Stiglitz, Joseph E. 2003. *Globalization and Its Discontents.* New York: W. W. Norton.

Wallach, Lori, and Michelle Sforza. 2000. *The WTO: Five Years of Reasons to Resist Corporate Globalization.* New York: Seven Stories Press.

Wallach; Lori; Patrick Woodall; and Ralph Nader. 2004. *Whose Trade Organization? A Comprehensive Guide to the World Trade Organization.* 2nd edition. New York: New Press.

World Trade Organization (WTO). 2008a. "The WTO Is NOT Undemocratic." Available from http://www.wto.org/english/thewto_e/whatis_e/10mis_e/10m10_e.htm

World Trade Organization (WTO). 2008b. "Labour Standards: Consensus, Coherence and Controversy." Available from http://www.wto.org/English/thewto_e/whatis_e/tif_e/bey5_e.htm

World Trade Organization (WTO). 2008c. "Trade and Environment." Available from http://www.wto.org/english/tratop_e/envir_e/envir_e.htm

William Kaufman

WORLD WIDE FUND FOR NATURE

Environmental and ecological degradation occurs worldwide, and international organizations have become major players in nature and environmental conservation. One of the leading organizations in this area is the World Wide Fund for Nature (WWF), whose founding was brought about by, among others, Sir Julian Huxley, then director-general of the United Nations Educational, Scientific and Cultural Organization (UNESCO). After visiting eastern Africa in 1960, he was shocked by the destruction of wildlife habitats and the rate of hunting and published his findings in the British newspaper *The Observer.* One of his readers, the businessman Victor Stolan, suggested establishing an international organization for fund-raising. Huxley interested conservationists and naturalists such as Max Nicholson, director of Britain's Nature Conservancy, and Peter Scott, vice president of the International Union for the Conservation of Nature (IUCN), in the idea of establishing a fund-raising organization to support scientifically sound conservation projects of existing nongovernmental organizations (NGOs).

ESTABLISHMENT AND ACTIVITIES OF THE WWF

In September 1961 the new organization was established as the World Wildlife Fund with headquarters in Switzerland, later to be known as WWF International. Within six weeks independent WWF offices were set up in a number of other countries. According to the Deed of Foundation, the organization was created to ensure the "conservation of world fauna, flora, forests, landscape, water, soils and other natural resources by the acquisition and management of land, research and investigation, education at all levels, information and publicity, coordination of efforts, cooperation with other interested parties and all other appropriate means."

Although the Deed of Foundation expressed a broad agenda, the first twenty years of the WWF were characterized by the establishment of wildlife reserves and the protection of impressive animal species, often in developing countries. For example, whales, elephants, rhinoceroses, and tigers played an important role in public campaigns. Through contacts in higher social circles and appeals to the public, WWF raised large amounts of money in the subsequent years.

However, by the end of the 1970s WWF realized that it was not addressing the fact that entire ecosystems were under threat from socioeconomic and political developments. In collaboration with the IUCN and the United Nations Environment Programme (UNEP), the *World Conservation Strategy* was published in 1980, stressing the interrelationship between conservation and socioeconomic development. That document introduced the concept of sustainable development, which was elaborated further in 1987 in *Our Common Future* by the World Commission on Environment and Development. In line with those documents, IUCN, UNEP, and WWF published *Caring for the Earth: A Strategy for Sustainable Living* in 1991, a document that explored strategic perspectives on sustainable development.

Because of its broadening scope, in 1985 WWF decided to change its name to the World Wide Fund for Nature, although the U.S. and Canadian offices retained the old name. Beginning in the early 1990s, a new WWF mission statement expressed the changing perspectives: "To stop the degradation of the planet's natural environment and to build a future in which humans live in harmony with nature." Preserving biodiversity, ensuring the use of renewable natural resources, and reducing pollution and wasteful consumption became the main aim. Because it was recognized that many threatened ecosystems had been the homes of indigenous peoples for thousands of years, traditional knowledge and a respect for indigenous traditions became part of the conservation goals of WWF. That led to alliances with NGOs for indigenous people and to conservation guidelines announced by the IUCN, WPCA (World Commission on Protected Areas), and WWF in 1999.

To keep track of the state of the natural habitat and human pressure on the planet, in 1998 the WWF began to publish the biennial *Living Planet Report,* in which the living planet index, covering population trends in over 1,300 species, and the ecological footprint, measuring the area of the planet needed for sustainable use, were explored. Those reports showed that the planet was in an alarming state. The ecological footprint had tripled since 1960, and the living planet index had decreased by about one-third since 1970.

According to WWF, recognition of this worldwide environmental crisis called for larger geographical scales of conservation. To that end, the term *ecoregion* was coined. David M. Olson and Eric Dinerstein defined an ecoregion as "a large unit of land or water containing a geographically distinct assemblage of species, natural communities, and environmental conditions" (Olson and Dinerstein 1998, p. 502). Recognizing that resources had to be used effectively and efficiently, WWF announced the Global 200, a set of over 200 ecoregions prioritized for conservation (Olson and Dinerstein 1998).

In the same period WWF intensified its focus on economic issues. The organization took part in the establishment of the Forest Steward Council, an independent agency that certifies sustainable timber production. WWF also established the Marine Steward Council (MSC) in cooperation with Unilever, one of the world's largest producers of seafood. This alliance points to another development within WWF: cooperation with multinational companies to develop standards and practices for corporate responsibility in regard to social and conservation issues.

CRITICISMS AND RESPONSES

This globalizing aspect of the work of WWF has led to criticism. In 2004 the anthropologist Mac Chapin criticized WWF and other large international nature protection organizations for their domination of the conservation agenda, neglect of indigenous people in spite of agreements and documents, and increasing financial dependency on multinationals and governmental agencies. In their response in 2005 the WWF representatives

Carter S. Roberts and Chris Hails pointed out that they shared Chapin's broader concerns "in spite of exaggerations and inaccuracies" (Roberts and Hails 2005, p. 7).

Chapin's article and the many responses to it made clear that the relationship between the rights and interests of indigenous people and conservation in a globalizing world is complex, implying often painstaking decisions. In the first decade of the twenty-first century WWF had offices in over fifty countries and was active in more than a hundred, with nearly five million individual members and an annual budget of around $500 million. Established as an organization that primarily focused on animal species with great public appeal, it gradually came to include ecological, environmental, developmental, and economic issues in its mission. WWF became a powerful player in international civil society, bringing states, companies, and NGOs together. Balancing those different forces has been one of its main challenges.

SEE ALSO *Conservation; Consumption; Nongovernmental Organizations; Sustainability; Sustainable Development; United Nations Educational, Scientific and Cultural Organization; United Nations Environment Programme; Wilderness.*

BIBLIOGRAPHY

Chapin, Mac. 2004. "A Challenge to Conservationists." *World Watch* 17(6): 17–31.

Hails, Chris. 2007. "The Evolution of Approaches to Conserving the World's Natural Heritage: The Experience of WWF." *International Journal of Heritage Studies* 13(4–5): 365–379.

International Union for the Conservation of Nature, United Nations Environment Program, and World Wide Fund for Nature. 1980. *World Conservation Strategy: Living Resource Conservation for Sustainable Development.* Gland, Switzerland: IUCN/UNEP/WWF.

International Union for the Conservation of Nature, United Nations Environment Program, and World Wide Fund for Nature. 1991. *Caring for the Earth: A Strategy for Sustainable Living.* Gland, Switzerland: IUCN/UNEP/WWF.

International Union for the Conservation of Nature, World Commission on Protected Areas, and World Wide Fund for Nature. 1999. *Principles and Guidelines on Indigenous and Traditional Peoples and Protected* Areas. Available from http://www.iucn.org

Olson, David M., and Eric Dinerstein. 1998. "The Global 200: A Representation Approach to Conserving the Earth's Most Biologically Valuable Ecoregions." *Conservation Biology* 12(3): 502–515.

Roberts, Carter S., and Chris Hails. 2005. "From the World Wildlife Fund (WWF)." *World Watch* 18(1): 6–7.

Van Koppen, C. S. A., and William T. Markham, eds. 2007. *Protecting Nature: Organizations and Networks in Europe and the USA.* Cheltenham, UK, and Northampton, MA: Edward Elgar.

World Commission on Environment and Development. 1987. *Our Common Future.* Oxford and New York: Oxford University Press.

World Wide Fund for Nature. 2006. *Living Planet Report 2006.* Gland, Switzerland: WWF. Available from http://worldwildlife.org/news/livingplanet/pdfs/living_planet_report.pdf .

World Wild Fund for Nature. http://www.panda.org

Jac. A. A. Swart

WRIGHT, FRANK LLOYD

1867–1959

Frank Lloyd Wright, who was born in Richland Center, Wisconsin, on June 8, 1867, is recognized as one of the most influential modern architects, the preeminent American architect of the late nineteenth and early twentieth centuries, and, along with Louis Sullivan, the inspiration for the Prairie School of architecture. After studying engineering at the University of Wisconsin, Wright spent his formative years as an architect under the tutelage of Joseph Lyman Silsbee and Sullivan. Sullivan's architectural maxim "form follows function" prompted Wright to develop a distinct approach called *organic architecture* that was driven by the maxim "form and function are one." That maxim was expressed in Wright's written works, especially "The Art and Craft of the Machine" (1992 [1901]), *An Organic Architecture* (1970 [1939]), *The Future of Architecture* (1953), *The Natural House* (1954), and *The Living City* (1963 [1958]), and in many of his over 500 completed designs, especially at Taliesin in Wisconsin, Falling Water in Pennsylvania, and Taliesin West in Arizona. Wright's significance for environmental ethics and philosophy lies in the commitment of his architectural ethos to organicism and openness through the use of the horizontal line.

Wright believed that architectural principles should be modeled after forms found in nature. Because forms have a quality of life that is the same as their function, they are formal declarations of function. The continuity Wright saw between architectural and natural forms drove his organic architecture, at the heart of which was the building that declares its purpose (function) without pretense and is enmeshed in and indeed is one with its urban or rural site.

Wright's organic architecture centered on projects that exemplified harmony with nature. To realize that harmony, architecture had to break down the nature-culture dualism that informed the creation of conceptually and architecturally situating human-made objects that stand in opposition to nature. Wright's strategy for breaking down that dualism included a focus on the horizontal over the vertical. When preference is given to the horizontal, more specifically the horizontal line (for

Fallingwater, Bear Run, Pennsylvania. *One of Frank Lloyd Wright's most recognizable works of architecture, Fallingwater was designed in 1936 as a summer home for Pittsburgh department store owner Edgar Kaufmann. The home is toured by over 70 thousand people each year.* AP IMAGES.

example, in Robie House in Illinois), buildings stand a better chance of expressing the openness and freedom of nature because deference to the horizontal leads to structures close to and guided by the forms and lines of the earth, starting with the horizon.

If one evaluates Wright's architectural ethos in terms of specific buildings or projects, one is likely to focus on Wright's success in developing a set of principles useful for guiding green architects as they work to bring about unity within buildings and in the way buildings fit into their sites in an unobtrusive and ecologically sensitive manner (Rogers 2004). On this level, a question can be asked about Wright's ethos that looks similar to a question that often comes up in relation to ecological design generally and biomimicry more specifically, namely is it justified to base norms and principles that are to guide human conduct on the perceived norms in and of nonhuman nature?

However, troubling philosophical issues arise with respect to the use of Wright's approach as an ecologically informed guide for planning and sustainability. Wright's commitment to openness through the horizontal line led him to develop a response to what he saw as the failures of urbanism. According to Wright, the artificial, constricted, and centralized industrial city demanded a response. In what can be seen as a precursor to the type of antiurbanism found in the work of later environmental ethicists, most notably Murray Bookchin (1974), Wright called for a decentralized and horizontal approach to human settlement. In his effort to make human settlements look more like natural organisms, Wright arrived at his "Broadacre City" concept, a precursor to later suburban development. Insofar as basic elements of the concept can be viewed as being related to environmental problems faced in the twenty-first century, Broadacre City delimits what is considered part of the ecological equation. For example, Wright's allowance of one acre per person attempts to honor a commitment to openness and freedom but ends up encouraging populations to spread out over vast distances and as a result contribute to a need for massive transportation infrastructure. Site and immediate surroundings are chosen over and at the expense of the ecological region, as well as wider ecological cycles and processes. It is paradoxical that Wright's ethos, which accomplishes so much in terms of ecological insight on the micro level, ultimately stands opposed to an ecological approach to architecture that recognizes the vital role of human-made structures in ecosystem operations on the macro level.

SEE ALSO *Bookchin, Murray; Space/Place; Sustainable Architecture and Engineering; Urban Environments.*

BIBLIOGRAPHY

WORKS BY FRANK LLOYD WRIGHT

Wright, Frank Lloyd. 1953. *The Future of Architecture.* New York: Horizon Press.

Wright, Frank Lloyd. 1954. *The Natural House.* New York: Horizon Press.

Wright, Frank Lloyd. 1963 [1958]. *The Living City.* New York: New American Library.

Wright, Frank Lloyd. 1970 [1939]. *An Organic Architecture: The Architecture of Democracy.* London: Lund Humphries.

Wright, Frank Lloyd. 1992 [1901]. "The Art and Craft of the Machine." In *Frank Lloyd Wright Collected Writings*, ed. Bruce Brooks Pfeiffer, Vol 1. New York: Rizzoli International.

WORKS ABOUT FRANK LLOYD WRIGHT

Bookchin, Murray. 1974. *The Limits of the City.* New York: Harper & Row.

Gill, Brendan. 1987. *Many Masks: A Life of Frank Lloyd Wright.* New York: Putnam.

Rogers, W. Kim. 2004. "Frank Lloyd Wright's 'Organic Architecture': An Ecological Approach in Theory and Practice." *Analecta Husserliana* 83: 381–390.

Secrest, Meryle. 1998. *Frank Lloyd Wright: A Biography.* Chicago: University of Chicago Press.

James W. Sheppard

Appendices

Text Acknowledgments

COPYRIGHTED EXCERPTS IN THE *ENCYCLOPEDIA OF ENVIRONMENTAL ETHICS AND PHILOSOPHY* WERE REPRODUCED FROM THE FOLLOWING SOURCES:

Jeffers, Robinson. From "The Answer," in ***The Selected Poetry of Robinson Jeffers,*** 1728–1938, Volume 2, edited by Tim Hunt. Random House, 1938. Copyright 1927, 1928, 1938 by Robinson Jeffers. Copyright renewed 1955, 1956, 1966 by Robinson Jeffers. Used with the permission of Stanford University Press, **www.sup.org.**—Byron, (Lord). From "Darkness," in ***Lord Byron: The Major Works.*** Edited by Jerome J. McGann. Oxford University Press, 2000. Introduction, edited text, and editorial matter copyright © Jerome J. McGann, 1986. Reproduced by permission of Oxford University Press.—Snyder, Gary. From ***Axe Handles.*** North Point Press, 1983. Copyright © 2005 by Gary Snyder. Reprinted by permission of the publisher.—Snyder, Gary. From ***Riprap and Cold Mountain Poems.*** North Point Press, 1990. Copyright © 2003 by Gary Snyder. Reprinted by permission of the publisher.—Snyder, Gary. From ***Left Out in the Rain: New Poems, 1947-1985.*** North Pointe Press, 1986. Copyright © 2005 by Gary Snyder. Reprinted by permission of the publisher.—"NSS Conservation Policy," ***The National Speleological Society,*** 2008. Copyright © The National Speleological Society 1995–2008. All rights reserved. Reproduced by permission.—Leopold, Aldo. From ***A Sand County Almanac: And Sketches Here And There.*** Oxford University Press, 1968 reprint (original 1949). Copyright © 1949 by Oxford University Press, Inc. Copyright renewed 1976 by Oxford University Press. Reproduced by permission of Oxford University Press.—Carson, Rachel. From ***Silent Spring.*** Houghton Mifflin Company, 1962. Copyright © 1962 by Rachel L. Carson. Copyright renewed 1990 by Roger Christie. All rights reserved. Reproduced by permission of Houghton Mifflin Harcourt Publishing Company.—***Science,*** v. 155, March 10, 1967. Copyright © 1967 by AAAS. Reprinted with permission from AAAS.—Shepard, Paul. From "Ecology and Man–a Viewpoint," in ***The Subversive Science.*** Edited by Shepard, P., and D. McKinley. Houghton Mifflin Company, 1969. Copyright © 1969 by Paul Shepard and Daniel McKinley. All rights reserved. Reproduced by permission of the Estate of the Author.—***Science,*** v. 162, December 13, 1968. Copyright © 1968 by AAAS. Reprinted with permission from AAAS.—Stone, Christopher D. From ***Southern California Law Review.*** 1972. Copyright © 1972, *Southern California Law Review.* Reproduced by permission.—Sylvan (Routley), Richard. From "Is There a Need for a New, an Environmental Ethic?" in ***Environmental Philosophy,*** **2nd ed.** Edited by Michael

E. Zimmerman, J. Baird Callicott, et al. Prentice Hall, 1998. Copyright © 1998, 1983 by Prentice-Hall, Inc. Reproduced by permission of the Estate of the author.—***Ethics,*** v. 85, January 1975. Copyright © 1975 University of Chicago Press. All rights reserved. Reproduced by permission.—***Inquiry Magazine (Oslo),*** v. 16, 1973 for "The Shallow and the Deep, Long-range Ecology Movements: A Summary" by Arne Naess. Reproduced by permission of the author.—***Earth Charter Initiative, 2000.*** Copyright © 2000. Reprinted with the permission of Earth Charter International Secretariat.

Glossary

Acoustic ecology: The interdisciplinary field that studies the auditory relationship between living organisms and their environments. In urban and suburban settings acoustic ecologists investigate the sonic effects of constant sound and noise pollution. In oceans, national parks, and wilderness areas there is heated debate over the presence of jet skis, airplane overflights, off-road vehicles, and military monitoring systems that rely on high-intensity underwater sonar blasts. Acoustic ecologists generally advocate the value of listening, the quality of the soundscape, and more consciously choosing the sounds humans create.

Androcentrism: Literally "male-centered," the conscious or unconscious practice of emphasizing the male viewpoint and male interests over the female viewpoint and female interests. Feminist theory commonly criticizes androcentrism on the grounds that it assumes a universal human nature corresponding to male attitudes while female attitudes are considered deviant.

Animal law: The branch of law concerning all interactions with animals from the perspective of traditional statutory and case law. The subjects of animal law include wildlife, animals kept in captivity, companion animals, and animals used for various entertainment, research, and food purposes. In the United States animal laws exist at local, state, and national levels.

Anthropocentrism: Literally "man-centered," the conscious or unconscious practice of emphasizing the human viewpoint and human interests over nonhuman viewpoints and interests. Anthropocentrism is frequently justified on the grounds that, because humans are the most significant fact of the universe, they are the most important. See also *chauvinism, human.*

Anthropocosmism: An alternative to the dichotomy between anthropocentrism and nonanthropocentrism that seeks to encapsulate both humanity and the natural world without placing greater value on a particular center—anthropocentric, biocentric, ecocentric—and thereby excluding or marginalizing something of peripheral value. From the anthropocosmic perspective, the values of humans ("anthropoi") and the world ("cosmos") are not in opposition, but are intimately interwoven.

Autecology: The study of the relationship between an individual organism or an individual species and its physical environment. Contrast to *synecology.*

Behaviorism: A psychological theory and method of investigation that seeks to explain behavior solely in terms of observable and quantifiable responses to environmental stimuli. Behaviorism ignores conscious experience and subjective phenomena like desires, motives, and emotions.

Biocentric egalitarianism: See *egalitarianism.*

Biocide: Any chemical agent capable of destroying living organisms. Although biocides are commonly associated with pesticides (herbicides, fungicides, insecticides), antimicrobials (antibacterials, antivirals, antiparasites) are biocides as well. See also *persistent organic pollutant (POP).*

Bioethics: The branch of ethics that investigates issues surrounding health care and the biological sciences. These include access to the allocation of limited resources (organs, treatment); the authority of the patient, the physician, and others; and the scope and limits of confidentiality. While bioethics has traditionally focused on abortion, euthanasia, surrogacy, in vitro fertilization, and organ transplants, it is increasingly concerned with drug research, embryonic stem cell research, genetic engineering (cloning, screening, gene therapy), recent work in synthetic biology (biotechnology), and a wide spectrum of environmental concerns.

Biome: The largest, most comprehensive ecological community adapted to a climatic type. While biomes are often thought of in terms of vegetative communities, animals, fungi, and microbes also constitute biomes. The terrestrial biomes are tundra, desert, chaparral, coniferous (boreal) forest, temperate forest, temperate grassland, tropical rainforest, and tropical savanna and scrub. The two basic aquatic biomes, freshwater and marine, include reefs, marshes, swamps, estuaries, shoreline, flowing waters, the continental shelf, lakes and ponds, and the open ocean.

Biomimicry (bionics): Literally "life imitation," a design strategy that takes nonhuman natural process as models to solve human problems. Velcro was inspired by burrs; glue-free tape by the Gecko lizard's sticky feet; and, to achieve passive air conditioning, Eastgate Center in Harare, Zimbabwe, was designed after the termite mound. Some of the underlying assumptions of the biomimetic approach are that "nature knows best;" natural things and processes work because they have been "field-tested" by evolution; and engineering projects based upon biomimetic principles will be conducive to life, or at least not be detrimental to it.

Biopiracy (bioprospecting): The unauthorized and uncompensated appropriation of biological material (plants, animals, genetic cell lines) or indigenous knowledge for the purposes of commercial development. Insofar as it promises to provide just compensation to aggrieved parties, the concept of biopiracy is attractive to indigenous rights advocates. Yet the concept has been criticized on the grounds that laws against biopiracy cannot provide adequate protection for all indigenous people—not every indigenous territory contains useful biological material—and it is unclear whether biological materials can and should be owned as a matter of natural right.

Bioregion: A territory defined in ecological units (watershed or ecosystem) with similar flora, fauna, and environmental conditions as opposed to a territory defined in political or administrative terms.

Bioregionalism: A loose-knit movement formed in the late 1970s, in response to the modern environmental crisis, which advocates transforming human societies to mirror naturally occurring bioregions. The hope is that if people live in accordance with and in awareness of these ecological units, they will live in a sustainable fashion.

Carbon sequestration: The process of removing and storing atmospheric carbon in forests, oceans, or fossil fuel reservoirs. Efforts are underway to develop technologies that can capture CO_2 emissions and sequester them under pressure underground where they cannot be reemitted into the atmosphere.

Carrying capacity: The maximum population size that a given environment can sustain without degradation.

Charismatic megafauna: See *flagship species.*

Chauvinism, human: The exaggerated belief in and devotion to the supremacy of the human species over all other species. See also *anthropocentrism.*

Coevolution: The biological process, sometimes called an evolutionary arms race, whereby two or more species evolve in response to one another. Such a process includes predator-prey relationships (robins and earthworms) and various mutualistic and parasitic relationships (legumes and nitrogen-fixing bacteria, wasps and caterpillars).

Communalism: An umbrella term centered on notions of the community, shared living, public ownership, and so on.

Consequentialism: Any ethical approach that understands the moral rightness of an action as determined solely by its results. Roughly, if the consequences of an act are good, the act is right; if they are bad, the act is wrong. For the consequentialist, the character and motives of an actor, as well as any formal characteristics of an act itself, are irrelevant to its moral status. The two chief consequentialist theories are ethical egoism and utilitarianism.

Constructivism, ethical: In metaethics, the antirealist view that moral facts exist insofar as they are the result of an actual or hypothetical constructive process. Hence, prior to the constructive process, there are no moral standards or moral facts.

Contractarianism (contractualism): Any theory that justifies moral principles or political arrangements by appealing to a voluntarily accepted social contract that is committed to under ideal conditions—no ignorance of relevant facts and no personal biases, power inequalities, or malicious ambitions.

Cornucopianism: The pro-technology view that material resources are less important than resources of the mind. Necessity, the mother of invention, will see to it that once a problem becomes sufficiently acute—be it pollution, resource depletion, or global climate change—creative minds, motivated by the promise of wealth, will find appropriate solutions. Such an approach implies that there is little reason to conserve resources as they will be replaced by alternatives, that toxic waste is an opportunity to develop waste disposal technologies, and so forth.

Cradle-to-cradle design: A burgeoning design and production approach that seeks not only to reduce but to eliminate all waste from the manufacturing process. Much like a food chain, the goal is for all products and byproducts to feed into production systems thereby wasting nothing. The cradle-to-cradle approach is generally pro-technology and pro-consumerist: sustainability is not about consuming less, but about consuming differently.

Cybernetics: The study of the control and communication processes in living organisms and artificial systems, especially the comparison between the two. A key feature of cybernetics is a system's use of feedback to steer it toward a goal: when the feedback causes changes in the system itself, it appears to be self-organizing.

Cyborg: Short for "cybernetic organism," a self-regulating organism, part-human and part-machine. The proportion of human qualities to machine qualities in cyborgs ranges from humans with synthetic implants (artificial hearts and pacemakers) to futuristic visions of sentient computers that can think thoughts and feel emotions.

Deconstruction: The view that language is a closed system of meanings with little direct reference to actual objects or relations among objects and thus that "reality" is "socially constructed"; deconstruction is the critical process of revealing and undermining the contingent way that the world is organized conceptually, including such binary categories as male/female, mind/body, and nature/culture.

Deontology: Literally the "science of duty," a general approach in ethics that sees the morality of an action in terms of motives for acting as well as the rightness and wrongness of the act itself regardless of consequences.

Distributive justice: Principles about the proper distribution of benefits (power, wealth, privilege) and burdens (taxes, environmental harms) among members of a society. Common bases for distribution are desert, needs, and entitlement.

Divine Command Theory: The theory that morality is grounded in God's will and that ethical principles are simply commandments of God. For many, God is required to make morality strong and firm. The fear is that without an ultimate guarantor of right and wrong, there will be only moral nihilism. The character Ivan Karamazov in Dostoevsky's *The Brothers Karamazov* famously sums up this fear: "If God doesn't exist, everything is permissible?"

Dualism, human/nature: The conscious or unconscious view that humanity and nature are fundamentally distinct, independent, and mutually exclusive. Some interpret the emphasis on the separation of humanity and nature as encouraging the belief that nature is to be controlled or dominated, or else simply a domain of facts to be studied—all hallmark attitudes of Renaissance humanism and contemporary Western science.

Dualism, substance: The view that the mind and the body are fundamentally different substances each with an independent existence and having a unique set of properties. Substance dualism is often used to support the view that, despite changes to the body, up to the point of death, the mental or spiritual substance abides unaffected.

Ecocentrism: In contrast to an *anthropocentrism*, an emphasis on the ecological point of view, frequently crediting ecological units of nature (rivers, species, communities, populations, ecosystems) with rights. Through metaphors like "the web of life," the ecocentric approach tends to view ecological interactions holistically.

Ecofascism: Primarily an inflammatory criticism of ecocentrism asserting that granting rights to ecological units of nature leads to the inappropriate sacrifice of the rights of the individual. As the term is an insult, the connection between ecocentrism and fascist societies—Nazi Germany in particular—is rarely explored. Rather, because supporters of German fascism exalted nature and the "Land" and the Nazi party tapped into this sentiment, or because Hitler was allegedly a vegetarian and animal lover who supported organic farming, any environmental cause or ecocentric approach is bound to be fascist.

Ecological anthropology: The study of how human culture mediates human-environment interactions. Not only does culture influence the character of human interactions with the environment (land, climate, species), but the environment shapes culture (beliefs, traditions, organizations). The principal subdisciplines of contemporary ecological anthropology include cultural ecology, historical ecology, political ecology, spiritual ecology, and environmental anthropology.

Ecological footprint: A spatial metaphor to communicate the total amount of resources required to produce and dispose of the goods and services of a particular lifestyle. An ecological footprint's size is determined by calculating the amount of land needed for food production, housing, transportation, consumer goods, services, and so forth.

Egalitarianism: The doctrine or belief in the equality of humankind—morally, politically, economically, and socially. *Biocentric egalitarianism* extends this notion to confer intrinsic value on all living entities, human and nonhuman, sentient and non-sentient.

Embodied knowledge: Knowledge derived from the subjective experience of one's own body as opposed to knowledge derived from objective, scientific understanding.

Emissions trading: A market-based response to pollution that allows polluters to select cost-effective solutions to achieve specific emissions goals. The open-market trading system allows a polluter to earn emission rights by reducing its emissions to levels below an established standard. In a cap-and-trade system, the government determines an acceptable level of pollution (the "cap") and then issues permits to pollute. Companies producing fewer emissions than allowed can sell or trade their excess capacity to others who might otherwise exceed the cap and incur a penalty.

Endangered species: Any species at risk of extinction throughout all or a significant portion of its range if existing pressures persist. Such pressures include pollution, habitat destruction, invasive species, and unsustainable exploitation. A related term, "threatened species," refers to any species liable to become endangered in the near future.

Enlightenment: Also known as the "Age of Reason," a European intellectual movement of the seventeenth and eighteenth centuries that emphasized the use of reason to critically inspect longstanding doctrines, traditions, and authorities of previous generations.

Environmental determinism: The largely discredited view that the physical environment, as opposed to social circumstances, wholly determines cultural characteristics. Environmental determinism has been associated with outdated racist theories that identify persons from tropical climates as inferior to those from northern climates: tropical climates purportedly cause laziness and promiscuity while variability of weather in northern latitudes purportedly determines a strong work ethic.

Environmental literature: A diffuse, quickly evolving literary genre that includes *nature writing,* oral storytelling, and ecological fiction, drama, and poetry. It is not necessarily concerned with conservationist causes nor is it simply trying to solve environmental problems. Rather, environmental literature is about fundamental human attitudes about the nonhuman natural world and our experience of it. While such explorations may be critical, they may also be celebratory, inquisitive, or downright humorous.

Environmental racism: Deliberate or unintentional racial discrimination in environmental policy making, enforcement of environmental regulations, the targeting of communities for the disposal of toxic wastes and siting of polluting industries, and the distribution of environmental amenities (parks, vistas, open spaces). *Environmental justice* is the movement to end environmental racism.

Ethnobotany: The study of the relationship between people and plants including plant lore, agricultural customs, and the use of plants in medicines and artifacts (houses, storage, modes of transportation).

Ethnocentrism: The conscious or unconscious practice of emphasizing the viewpoint of one's ethnic group and interests over the viewpoint of other cultures. Such bias can lead to the evaluation of other cultures in terms of one's own and perhaps even the belief that one's culture is superior to others.

Eutrophication: The process of nutrient enrichment (nitrogen and phosphorus), increased production of organic matter, and succeeding ecosystem degradation in a water body. In the United States nutrient enrichment is often the result of excess synthetic fertilizers on commercial farmland that gets into riverine systems and eventually into the ocean.

Existentialism: A broad philosophical movement emphasizing subjective choice over objective description, lived experience over abstract reasoning, individuality over mass culture, freedom over determinism, and authenticity over inauthenticity. Insofar as humans create value and meaning in an otherwise meaningless universe, existentialism may also stress one's emotional reaction to such a realization: sadness, dread, or feelings of absurdity about life.

Extinct in the wild: A species whose only living members are in captivity or live as a naturalized population outside of their traditional range. Examples of species extinct in the wild include the Barbary Lion, Spix's Macaw, the Hawaiian Crow, and the Wyoming Toad.

Extirpation: See *local extinction.*

Fact/value distinction: The assumption that facts and values are mutually exclusive. Facts represent what is while values describe what ought to be. The fact/value distinction has been criticized on the grounds that it is impossible to precisely distinguish facts from values as the two necessarily interpermeate.

Flagship species: Any species having broad public appeal and that can be used to promote conservation efforts. The hope is that, in protecting flagship species, entire biological communities and their associated ecosystems will also be protected. *Charismatic megafauna* are large animals belonging to flagship species including lions and tigers, bison and elephants, gorillas and orangutans, pandas and polar bears, and whales and sea turtles.

Fluorocarbon: See *perfluorinated compounds (PFCs).*

Fossil fuels: Any hydrocarbon deposit such as petroleum, coal, or natural gas that, when burned with air directly, produces heat or indirectly produces energy. All fossil fuels are the result of geologic processes acting on the fossilized remains of plants and animals that lived millions of years ago.

Gene bank: A facility that preserves genetic material for the ex situ conservation of individuals (seeds), tissues, and reproductive cells of plants or animals.

Genome, human: The complete set of genetic information or hereditary material in the chromosomes of each cell of a human being.

Green architecture: See *sustainable architecture.*

Green revolution: The intensification of industrial agriculture in the 1960s in the developing world. The green revolution led to a dramatic increase of crop yields per unit area of farmland. Practices supporting such an increase include turning large tracts of land over to monocultures of high-yield cereal crops (rice and wheat); the heavy use of irrigation, pesticides, and synthetic fertilizers; and sowing and harvesting on a single piece of land multiple times per season or per year. The green revolution has been criticized for its long-term deleterious ecological and social effects.

Green space: Open, undeveloped space frequently containing forests, gardens, or grass within or adjacent to a built-up area. Green spaces are often designated for parks, trails, gardens, preserves, playgrounds, or habitat restoration.

Greenhouse gas: Any gas contributing to the greenhouse effect by absorbing some of the outgoing terrestrial infrared radiation and re-emitting it back to the earth's surface. The most important greenhouse gases, in order of relative abundance, are water vapor, carbon dioxide (CO_2), methane (CH_4), nitrous oxide (N_2O), and the halocarbon gases.

Human ecology: The study of the interactions between humans, their communities, and the environment including responses to and effects on the environment.

Humanism: A nineteenth-century term for the values and ideals of the European Renaissance. Broadly speaking humanism centers on and assigns a positive role to the human individual as the ultimate source of value, and as capable of developing the moral, physical, spiritual, and rational faculties. The movement was originally sparked by the rediscovery of Greek and Roman literature in the twelfth century.

Hydrography: The study, measurement, and description of large bodies of water. The central goal of hydrography is to provide for safe navigation and protection of the marine environment through the creation of accurate nautical charts and related publications.

Idealism: The metaphysical theory that the external world is not physical but is in fact mental or psychical. The epistemological claim that reality is best explained in terms of psychic phenomena like minds, spirits, or ideas instead of in terms of matter. The ethical device of positing an action, person, political organization, or state of being that is free of all the imperfections that characterize actual actions, persons, political organizations, and states of being.

Individualism/holism debate: The debate over whether the individual or the whole is primary: ethically, semantically, ontologically, or epistemologically. For instance, the ontological individualist claims that individual elements are independent and self-existent whereas for the holist they are interdependent and constituted by the qualities of others. The ontological individualist claims, and the holist denies, that all complex phenomena are reducible; for the holist some qualities only emerge at the level of the whole.

Island biogeography: The study of the geographical distribution of plants and animals on islands or in isolated locales ("habitat islands"). According to the equilibrium theory of island biogeography, species richness on an island stems from a balance between the number of species added by immigration and the number lost to extirpation. First proposed by Edward O. Wilson and Robert H. MacArthur in 1963, the theory is increasingly important to conservation efforts—the creation of habitat islands represents one of the chief threats to biodiversity.

Kantianism: Any philosophical approach or theory that follows the work of Immanuel Kant and, perhaps, shares his interest in elevating philosophy to the level of science. In ethics, Kantianism refers to an emphasis on moral duty, the universalizability of moral commands, and the idea that humans, as rational beings, are ends unto themselves, never simply a means to an end.

Kyoto protocol: First drafted in 1997, the international agreement on global climate change to reduce the emissions by signatory nations of six greenhouse gases to levels below those in 1990. While the United States. has staunchly refused to ratify the protocol, it nevertheless came into effect in 2005 with Russia's ratification of it. By 2004 only the United Kingdom (closely followed by Germany) had met its reduction target while the United States, Canada, Italy, and Japan had all increased their greenhouse gas emissions.

Local extinction: The disappearance or elimination of a population or species in a particular locale but not from its entire range (global extinction). Also known as *extirpation.*

Materialism: The belief that only matter exists. Mind or consciousness is (somehow) reducible to matter, its properties, or interactions.

Mechanism: The view that all phenomena can be explained using the principles by which machines are explained: classical physics and chemistry, and mechanical science.

The seventeenth-century mechanists—Galileo, Boyle, Descartes, Hobbes, and Newton, among others—advanced the theory of mechanism to eliminate all non-observable and mathematically untreatable explanations, in particular vitalistic and teleological explanations. While mechanism has been extremely successful in terms of epistemological clarity and technical application, it has been criticized for unduly reducing the richness and complexity of life to a few basic entities and forces. Opposite of *organicism*.

Mendelian inheritance: As opposed to extrachromosomal inheritance, the pattern of gene inheritance, first described by Gregor Johann Mendel, in which one copy of a gene (allele) is inherited from each parent by an offspring.

Metaphysics: The philosophical study of the most general or abstract characteristics of reality: identity, existence, substance; permanence and change, time and space, cause and effect, difference and sameness, unity and variety, mind and matter, and so forth.

Microcosm/macrocosm: Literally "small-order"/"large-order," the very old and very common belief, found in both Western and Eastern cultures, that parts of the human body correspond to aspects of the universe, that every part mirrors the whole, or that the microcosm corresponds to the macrocosm. Variations on this theme include the analogies between the human being and society, and between society and the universe. Since the Renaissance, the microcosm/macrocosm analogy has largely been displaced by a mechanistic, ontologically materialistic model of the universe.

Naturalism: A broad concept including the notions that (a) the universe is wholly natural, not derived from or constituted by any nonnatural component (supernatural or transcendental); (b) empirical science in principle can explain all phenomena; (c) humans are no different in kind from the rest of the universe; and (d) values do not have a supernatural origin or sanction, but are either humanly constructed or grounded in natural phenomena.

Naturalistic fallacy: First identified by G. E. Moore, the practice of identifying goodness with a natural quality like pleasantness or beauty. For Moore moral goodness is a primitive, unanalyzable, and nonnatural concept that cannot be equated with or explained in terms of natural properties.

Nature writing: Literary nonfiction that intertwines careful, often scientifically oriented, personal observation of the natural world with spiritual, philosophical, and perhaps even political reflections. Nature writers are especially concerned with exploring epistemologies of place, encouraging an appreciation of sensual experience, and tracing the relationship between humans and the "more-than-human world." If nature writing can be said to have an overarching goal, it is to nudge Western culture toward a more sustainable relationship with the world. Nature writing is a subset of *environmental literature*.

Neo-colonialism: The control and management of a weaker nation by a stronger one through economic and cultural measures like trade agreements, the operations of transnational corporations, and particular business models.

Neo-Darwinism: The synthesis of Darwin's theory of evolution by natural selection with Gregor Mendel's theory of genetic inheritance by chance mutation and recombination. Also called the "modern synthesis" and "neo-Darwinianism."

North/South divide: A geographic division thought to reflect the socioeconomic and political divisions between the developed wealthy "northern" countries (Japan, Europe, United States) and the developing or least developed "southern" countries (southern Asia, Africa, Central and South America). As more and more countries in the South industrialize (Mexico and South Korea, for instance), the usefulness of the term will further diminish.

Old growth forest: The North American term for a late successional forest that has not been significantly altered by human activity and that contains numerous mature, dying, and fallen trees.

Organicism: In philosophy, any theory describing the universe as the analog of a living organism, especially in terms of development and organization. Opposite of *mechanism*.

Perennial polyculture: A biomimetic (see *biomimicry*) approach to agriculture in which numerous plantings are intercropped for three or more seasons in rows or mixed arrangements. Perennial polyculture is one response to the degradation and loss of topsoil in industrial monocultural farming with its heavy emphasis on annual crops, frequent plowing, and inputs of synthetic fertilizer and pesticides. Perennial polyculture emphasizes renewable natural resources and the self-regeneration of local ecosystems.

Perfluorinated compounds (PFCs): Also known as "fluorocarbons," human-made chemicals composed of only carbon and fluorine. PFCs are widely used in manufacturing, particularly in food packaging and Teflon products like nonstick cookware. They are a *persistent organic pollutant* and are found in the vast majority of people living in the industrialized world today. While PFCs do not harm the ozone layer, in a gaseous state they are a powerful greenhouse gas.

Persistent Organic Pollutant (POP): A toxic chemical substance persisting in the environment for an unreasonable amount of time, usually decades or more. POPs also bioaccumulate in food chains and can travel long distances. DDT, PCBs, PFCs, Heptachlor, and Furans are examples of POPs.

Population viability analysis (PVA): Analysis predicting the likelihood that a population or species will persist or go extinct in an environment over a certain period of time. Such analysis includes estimating the minimum size required for a population to persist and the effects on endangered species of habitat loss, fragmentation, and deterioration.

Positivism: Developed by Auguste Comte, the philosophy that there exist three stages of thought—the theological, the metaphysical, and the positive—each having an economic and cultural correlate. The last and highest stage is "positive" insofar as human thought is limited only to facts, to what is positively given, while scrupulously avoiding any *a priori* speculation. Positivism is sanguine about science's ability to explain all phenomena as well as the advantages of a thoroughly scientific society.

Post-materialism: A cultural shift of focus from material affluence and physical security to quality-of-life concerns, including civil liberties, minority rights, and environmental protection. It is uncertain whether the United States has ever fully experienced a post-material revolution.

Postmodernism: In philosophy, the view that universal statements about value, progress, or historical causation are impossible because all knowledge is shaped by the conceptual framework of the knower. Everything we perceive and interpret is necessarily influenced by specific circumstances, including, for instance, sex, gender, class, culture, biology, and historical era. Thus, knowledge can never be universal, but must always be partial, situated, and embodied. Postmodernism has led to a proliferation of philosophical theories critical of grand narratives, most notably post-structuralism. See also *relativism, ethical.*

Precautionary principle: The principle that an action should not be undertaken until its effects are adequately understood and deemed safe. The precautionary principle requires that new technology be considered guilty until proven innocent.

Preservationism: Any movement that seeks to protect natural areas, historical sites, or endangered species from loss or danger due to human intervention.

Primatology: The interdisciplinary study of nonhuman primates. Today many primatologists work with conservation groups to preserve the habitats and populations of the species they study.

Progressivism: A period of economic and social reform in the United States that occurred roughly between 1900 and 1920. Progressives were composed of Democrats, Republicans, and nonpartisans, all of whom were distressed by the concentration of political and economic power, which they believed was contrary to equality and democracy. Progressives called for the government to be more active in reform and to eliminate inequities created by the rapid industrialization of America. Much of the Progressive agenda passed into law during the presidencies of Theodore Roosevelt, William Howard Taft, and Woodrow Wilson.

Reductionism (reductivism): Either the explanation of all fields of knowledge in terms of a single, simpler science (physics) or the metaphysical belief that all things, regardless of outward appearance, are really just one kind of thing. While reductionism was popular in the early days of analytic philosophy, the failure of philosophers to come up with satisfactory inter-theoretic reductions along with the recognition of emergent or supervenient properties has undermined enthusiasm about reductionism as a philosophical program. Nonetheless, as a general methodological tendency, reductionism still holds sway in most academic disciplines today.

Relativism, ethical: The doctrine that moral values are relative to particular persons or cultures and cannot be assessed apart from these criteria. If values are relative, no overarching standards exist according to which individuals and cultures can be judged—what is right in one place may be wrong in another simply because the only criteria for distinguishing right from wrong are the moral values of the society or individual. See also *postmodernism.*

Relativity theory: First proposed by Albert Einstein, the theory that time, space, and mass, rather than being absolutes, are relative to the observer and the observed.

Rights: A certain type of relationship between two parties, the rights holder and the rights observer. For the holder, a right is a permission to act; for the observer, it is an obligation to respect that permission. While rights of noninterference (negative rights) are widely accepted in the United States, rights that obligate others to take positive steps in helping a person exercise his or her right (positive rights)—such as Affirmative Action or basic health care—are controversial insofar as they may allow for an endless expansion of rights. Perhaps due to the ascendancy of individualism in Western society, rights talk is the commonest form of moral discourse today.

Scholasticism: The methods and teachings of academic philosophers and theologians of the Middle Ages beginning with St. Augustine in the fifth century and lasting up to the mid-seventeenth century with the birth of Renaissance humanism. A primary goal of scholasticism was to reconcile faith with reason, Christian theology with the Greek philosophy of Aristotle (and to a lesser extent Plato). With the coming of the Renaissance, scholasticism's methods of deductive logic and dialectical reasoning were fiercely criticized and replaced by the methods of modern science as first articulated by Galileo Galilei, Francis Bacon, René Descartes, and Thomas Hobbes, among others.

Speciation: The biological process by which new species arise from pre-existing species. The main type of species development is allopatric speciation or development owing to the physical separation of populations over a geographic distance. A new species emerges when it can no longer interbreed with the population from which it was isolated. At this point each species embarks upon an independent evolutionary trajectory.

Supernaturalism: Belief in a domain of existence over and above the natural or material domain. Belief in ultimate forces or agencies that transcend the universe but somehow influence natural events in it.

Superorganism: A collection of individual yet interdependent organisms that behave as if they were a single organism. Examples of superorganisms include coral, insect colonies, and the Portuguese man-of-war. Frederick Clements argued that what later came to be called ecosystems were also superorganisms.

Synecology: The study of entire communities of organisms and interactions within them. Contrast to *autecology*.

Theocentrism: Literally "God-centered." In contrast to anthropocentrism, the belief that God's values, including the God-ordained goodness of creation, trump human values, including the way humans value nature.

Transcendentalism: A literary and philosophical movement of nineteenth-century American philosophers and writers who postulated the existence of a nonphysical, spiritual reality knowable only through intuition, the highest form of knowledge. They also emphasized humanity's essential goodness, God's immanence in nature, and the essential unity of creation; along with the values of individualism, self-reliance, and the rejection of authority. The central Transcendentalists were Ralph Waldo Emerson, Henry David Thoreau, Margaret Fuller, and Amos Bronson Alcott. The movement has influenced scores of American writers including Nathaniel Hawthorne, Herman Melville, and Walt Whitman.

List of Primary Sources

THE LAND ETHIC

SOURCE Leopold, Aldo. 1968 (1949). In *A Sand County Almanac and Sketches Here and There.* Oxford, U.K.: Oxford University Press. Copyright renewed 1976. Reproduced by permission of Oxford University Press.

INTRODUCTION *Aldo Leopold (1887–1948) worked for nineteen years in the U.S. Forest Service beginning in 1909. His writings have been extremely influential in conservation and in the development of the field of environmental ethics. Leopold begins the essay "The Land Ethic" by discussing the development of ethics, which historically has steadily extended outward to encompass more and more subjects. Ethics evolved as a means of social cohesion and developed correlatively to the development of societies or communities—from small bands of hunter-gatherers to the global village. Leopold argues for the necessity of taking the next step in the sequence of social-ethical evolution—an ethic dealing with our relationship to the land and nonhuman life. Ethics acknowledges the membership of an individual in a community, and Leopold urges ecologically broadening our understanding of community to include the land itself. A shift is required from humanity's role of conqueror over nature to "plain member and citizen of it." Embracing this shift would prevent approaching land-use issues purely out of self-interest. Bringing about such an ethic would require a change in conscience, leading individuals to feel a sense of personal obligation to the land. To explain this, Leopold characterizes the land pyramid to show that the land is something toward which we can and must act ethically, rather than basing our actions on what is economically expedient.*

When god-like Odysseus returned from the wars in Troy, he hanged all on one rope a dozen slave-girls of his household whom he suspected of misbehavior during his absence.

This hanging involved no question of propriety. The girls were property. The disposal of property was then, as now, a matter of expediency, not of right and wrong.

Concepts of right and wrong were not lacking from Odysseus' Greece: witness the fidelity of his wife through the long years before at last his black-prowed galleys clove the wine-dark seas for home. The ethical structure of that day covered wives, but had not yet been extended to human chattels. During the three thousand years which have since elapsed, ethical criteria have been extended to many fields of conduct, with corresponding shrinkages in those judged by expediency only.

THE ETHICAL SEQUENCE

This extension of ethics, so far studied only by philosophers, is actually a process in ecological evolution. Its sequences may be described in ecological as well as in philosophical terms. An ethic, ecologically, is a limitation on freedom of action in the struggle for existence. An ethic, philosophically, is a differentiation of social from anti-social conduct. These are two definitions of one thing. The thing has its origin in the tendency of interdependent individuals or groups to evolve modes of co-operation. The ecologist calls these symbioses. Politics and economics are advanced symbioses in which the original free-for-all competition has been replaced, in part, by co-operative mechanisms with an ethical content.

The complexity of co-operative mechanisms has increased with population density, and with the efficiency of tools. It was simpler, for example, to define the anti-social uses of sticks and stones in the days of the mastodons than of bullets and billboards in the age of motors.

The first ethics dealt with the relation between individuals; the Mosaic Decalogue is an example. Later accretions dealt with the relation between the individual and society. The Golden Rule tries to integrate the individual to society; democracy to integrate social organization to the individual.

There is as yet no ethic dealing with man's relation to land and to the animals and plants which grow upon it. Land, like Odysseus' slave-girls, is still property. The land-relation is still strictly economic, entailing privileges but not obligations.

The extension of ethics to this third element in human environment is, if I read the evidence correctly, an evolutionary possibility and an ecological necessity. It is the third step in a sequence. The first two have already been taken. Individual thinkers since the days of Ezekiel and Isaiah have asserted that the despoliation of land is not only inexpedient but wrong. Society, however, has not yet affirmed their belief. I regard the present conservation movement as the embryo of such an affirmation.

An ethic may be regarded as a mode of guidance for meeting ecological situations so new or intricate, or involving such deferred reactions, that the path of social expediency is not discernible to the average individual. Animal instincts are modes of guidance for the individual in meeting such situations. Ethics are possibly a kind of community instinct in-the-making.

THE COMMUNITY CONCEPT

All ethics so far evolved rest upon a single premise: that the individual is a member of a community of interdependent parts. His instincts prompt him to compete for his place in that community, but his ethics prompt him also to co-operate (perhaps in order that there may be a place to compete for).

The land ethic simply enlarges the boundaries of the community to include soils, waters, plants, and animals, or collectively: the land.

This sounds simple: do we not already sing our love for and obligation to the land of the free and the home of the brave? Yes, but just what and whom do we love? Certainly not the soil, which we are sending helter-skelter downriver. Certainly not the waters, which we assume have no function except to turn turbines, float barges, and carry off sewage. Certainly not the plants, of which we exterminate whole communities without batting an eye. Certainly not the animals, of which we have already extirpated many of the largest and most beautiful species. A land ethic of course cannot prevent the alteration, management, and use of these 'resources,' but it does affirm their right to continued existence, and, at least in spots, their continued existence in a natural state.

In short, a land ethic changes the role of Homo sapiens from conqueror of the land-community to plain member and citizen of it. It implies respect for his fellow-members, and also respect for the community as such.

In human history, we have learned (I hope) that the conqueror role is eventually self-defeating. Why? Because it is implicit in such a role that the conqueror knows, *ex cathedra*, just what makes the community clock tick, and just what and who is valuable, and what and who is worthless, in community life. It always turns out that he knows neither, and this is why his conquests eventually defeat themselves.

In the biotic community, a parallel situation exists. Abraham knew exactly what the land was for: it was to drip milk and honey into Abraham's mouth. At the present moment, the assurance with which we regard this assumption is inverse to the degree of our education.

The ordinary citizen today assumes that science knows what makes the community clock tick; the scientist is equally sure that he does not. He knows that the biotic mechanism is so complex that its workings may never be fully understood.

That man is, in fact, only a member of a biotic team is shown by an ecological interpretation of history. Many historical events, hitherto explained solely in terms of human enterprise, were actually biotic interactions between people and land. The characteristics of the land determined the facts quite as potently as the characteristics of the men who lived on it.

Consider, for example, the settlement of the Mississippi valley. In the years following the Revolution, three groups were contending for its control: the native Indian, the French and English traders, and the American settlers. Historians wonder what would have happened if the English at Detroit had thrown a little more weight into the Indian side of those tipsy scales which decided the outcome of the colonial migration into the cane-lands of Kentucky. It is time now to ponder the fact that the cane-lands, when subjected to the particular mixture of forces represented by the cow, plow, fire, and axe of the pioneer, became bluegrass. What if the plant succession inherent in this dark and bloody ground had, under the impact of these forces, given us some worthless sedge, shrub, or weed? Would Boone and Kenton have held out? Would there have been any overflow into Ohio, Indiana, Illinois, and Missouri? Any Louisiana Purchase?

Any transcontinental union of new states? Any Civil war?

Kentucky was one sentence in the drama of history. We are commonly told what the human actors in this drama tried to do, but we are seldom told that their success, or the lack of it, hung in large degree on the reaction of particular soils to the impact of the particular forces exerted by their occupancy. In the case of Kentucky, we do not even know where the bluegrass came from—whether it is a native species, or a stowaway from Europe.

Contrast the cane-lands with what hindsight tells us about the Southwest, where the pioneers were equally brave, resourceful, and persevering. The impact of occupancy here brought no bluegrass, or other plant fitted to withstand the bumps and buffetings of hard use. This region, when grazed by livestock, reverted through a series of more and more worthless grasses, shrubs, and weeds to a condition of unstable equilibrium. Each recession of plant types bred erosion; each increment to erosion bred a further recession of plants. The result today is a progressive and mutual deterioration, not only of plants and soils, but of the animal community subsisting thereon. The early settlers did not expect this: on the ciénegas of New Mexico some even cut ditches to hasten it. So subtle has been its progress that few residents of the region are aware of it. It is quite invisible to the tourist who finds this wrecked landscape colorful and charming (as indeed it is, but it bears scant resemblance to what it was in 1848).

This same landscape was 'developed' once before, but with quite different results. The Pueblo Indians settled the Southwest in pre-Columbian times, but they happened not to be equipped with range livestock. Their civilization expired, but not because their land expired.

In India, regions devoid of any sod-forming grass have been settled, apparently without wrecking the land, by the simple expedient of carrying the grass to the cow, rather than vice versa. (Was this the result of some deep wisdom, or was it just good luck? I do not know.)

In short, the plant succession steered the course of history; the pioneer simply demonstrated, for good or ill, what successions inhered in the land. Is history taught in this spirit? It will be, once the concept of land as a community really penetrates our intellectual life.

THE ECOLOGICAL CONSCIENCE

Conservation is a state of harmony between men and land. Despite nearly a century of propaganda, conservation still proceeds at a snail's pace; progress still consists largely of letterhead pieties and convention oratory. On the back forty we still slip two steps backward for each forward stride.

The usual answer to this dilemma is 'more conservation education.' No one will debate this, but is it certain that only the *volume* of education needs stepping up? Is something lacking in the *content* as well?

It is difficult to give a fair summary of its content in brief form, but, as I understand it, the content is substantially this: obey the law, vote right, join some organizations, and practice what conservation is profitable on your own land; the government will do the rest.

Is not this formula too easy to accomplish anything worth-while? It defines no right or wrong, assigns no obligation, calls for no sacrifice, implies no change in the current philosophy of values. In respect of land-use, it urges only enlightened self-interest. Just how far will such education take us? An example will perhaps yield a partial answer.

By 1930 it had become clear to all except the ecologically blind that southwestern Wisconsin's topsoil was slipping seaward. In 1933 the farmers were told that if they would adopt certain remedial practices for five years, the public would donate CCC labor to install them, plus the necessary machinery and materials. The offer was widely accepted, but the practices were widely forgotten when the five-year contract period was up. The farmers continued only those practices that yielded an immediate and visible economic gain for themselves.

This led to the idea that maybe farmers would learn more quickly if they themselves wrote the rules. Accordingly the Wisconsin Legislature in 1937 passed the Soil Conservation District Law. This said to farmers, in effect: *We, the public, will furnish you free technical service and loan you specialized machinery, if you will write your own rules for land-use. Each county may write its own rules, and these will have the force of law.* Nearly all the counties promptly organized to accept the proffered help, but after a decade of operation, *no county has yet written a single rule.* There has been visible progress in such practices as strip-cropping, pasture renovation, and soil liming, but none in fencing woodlots against grazing, and none in excluding plow and cow from steep slopes. The farmers, in short, have elected those remedial practices which were profitable anyhow, and ignored those which were profitable to the community, but not clearly profitable to themselves.

When one asks why no rules have been written, one is told that the community is not yet ready to support them; education must precede rules. But the education actually in progress makes no mention of obligations to land over and above those dictated by self-interest. The net result is that we have more education but less soil, fewer healthy woods, and as many floods as in 1937.

The puzzling aspect of such situations is that the existence of obligations over and above self-interest is taken for granted in such rural community enterprises as the betterment of roads, schools, churches, and baseball teams. Their existence is not taken for granted, nor as yet seriously discussed, in bettering the behavior of the water that falls on the land, or in the preserving of the beauty or diversity of the farm landscape. Land-use ethics are still governed wholly by economic self-interest, just as social ethics were a century ago.

To sum up: we asked the farmer to do what he conveniently could to save his soil, and he has done just that and only that. The farmer who clears the woods off a 75 per cent slope, turns his cows into the clearing, and dumps its rainfall, rocks, and soil into the community creek, is still (if otherwise decent) a respected member of society. If he puts lime on his fields and plants his crops on contour, he is still entitled to all the privileges and emoluments of his Soil Conservation District. The District is a beautiful piece of social machinery, but it is coughing along on two cylinders because we have been too timid, and too anxious for quick success, to tell the farmer the true magnitude of his obligations. Obligations have no meaning without conscience, and the problem we face is the extension of the social conscience from people to land.

No important change in ethics was ever accomplished without an internal change in our intellectual emphasis, loyalties, affections, and convictions. The proof that conservation has not yet touched these foundations of conduct lies in the fact that philosophy and religion have not yet heard of it. In our attempt to make conservation easy, we have made it trivial.

SUBSTITUTES FOR A LAND ETHIC

When the logic of history hungers for bread and we hand out a stone, we are at pains to explain how much the stone resembles bread. I now describe some of the stones which serve in lieu of a land ethic.

One basic weakness in a conservation system based wholly on economic motives is that most members of the land community have no economic value. Wildflowers and songbirds are examples. Of the 22,000 higher plants and animals native to Wisconsin, it is doubtful whether more than 5 per cent can be sold, fed, eaten, or otherwise put to economic use. Yet these creatures are members of the biotic community, and if (as I believe) its stability depends on its integrity, they are entitled to continuance.

When one of these non-economic categories is threatened, and if we happen to love it, we invent subterfuges to give it economic importance. At the beginning of the century songbirds were supposed to be disappearing. Ornithologists jumped to the rescue with some distinctly shaky evidence to the effect that insects would eat us up if birds failed to control them. The evidence had to be economic in order to be valid.

It is painful to read these circumlocutions today. We have no land ethic yet, but we have at least drawn nearer the point of admitting that birds should continue as a matter of biotic right, regardless of the presence or absence of economic advantage to us.

A parallel situation exists in respect of predatory mammals, raptorial birds, and fish-eating birds. Time was when biologists somewhat overworked the evidence that these creatures preserve the health of game by killing weaklings, or that they control rodents for the farmer, or that they prey only on 'worthless' species. Here again, the evidence had to be economic in order to be valid. It is only in recent years that we hear the more honest argument that predators are members of the community, and that no special interest has the right to exterminate them for the sake of a benefit, real or fancied, to itself. Unfortunately this enlightened view is still in the talk stage. In the field the extermination of predators goes merrily on: witness the impending erasure of the timber wolf by fiat of Congress, the Conservation Bureaus, and many state legislatures.

Some species of trees have been 'read out of the party' by economics-minded foresters because they grow too slowly, or have too low a sale value to pay as timber crops: white cedar, tamarack, cypress, beech, and hemlock are examples. In Europe, where forestry is ecologically more advanced, the non-commercial tree species are recognized as members of the native forest community, to be preserved as such, within reason. Moreover some (like beech) have been found to have a valuable function in building up soil fertility. The interdependence of the forest and its constituent tree species, ground flora, and fauna is taken for granted.

Lack of economic value is sometimes a character not only of species or groups, but of entire biotic communities: marshes, bogs, dunes, and 'deserts' are examples. Our formula in such cases is to relegate their conservation to government as refuges, monuments, or parks. The difficulty is that these communities are usually interspersed with more valuable private lands; the government cannot possibly own or control such scattered parcels. The net effect is that we have relegated some of them to ultimate extinction over large areas. If the private owner were ecologically minded, he would be proud to be the custodian of a reasonable proportion of such areas, which add diversity and beauty to his farm and to his community.

In some instances, the assumed lack of profit in these 'waste' areas has proved to be wrong, but only after most of them had been done away with. The present scramble to reflood muskrat marshes is a case in point.

There is a clear tendency in American conservation to relegate to government all necessary jobs that private landowners fail to perform. Government ownership, operation, subsidy, or regulation is now widely prevalent in forestry, range management, soil and watershed management, park and wilderness conservation, fisheries management, and migratory bird management, with more to come. Most of this growth in governmental conservation is proper and logical, some of it is inevitable. That I imply no disapproval of it is implicit in the fact that I have spent most of my life working for it. Nevertheless the question arises: What is the ultimate magnitude of the enterprise? Will the tax base carry its eventual ramifications? At what point will governmental conservation, like the mastodon, become handicapped by its own dimensions? The answer, if there is any, seems to be in a land ethic, or some other force which assigns more obligation to the private landowner.

Industrial landowners and users, especially lumbermen and stockmen, are inclined to wail long and loudly about the extension of government ownership and regulation to land, but (with notable exceptions) they show little disposition to develop the only visible alternative: the voluntary practice of conservation on their own lands.

When the private landowner is asked to perform some unprofitable act for the good of the community, he today assents only with outstretched palm. If the act costs him cash this is fair and proper, but when it costs only forethought, open-mindedness, or time, the issue is at least debatable. The overwhelming growth of land-use subsidies in recent years must be ascribed, in large part, to the government's own agencies for conservation education: the land bureaus, the agricultural colleges, and the extension services. As far as I can detect, no ethical obligation toward land is taught in these institutions.

To sum up: a system of conservation based solely on economic self-interest is hopelessly lopsided. It tends to ignore, and thus eventually to eliminate, many elements in the land community that lack commercial value, but that are (as far as we know) essential to its healthy functioning. It assumes, falsely, I think, that the economic parts of the biotic clock will function without the uneconomic parts. It tends to relegate to government many functions eventually too large, too complex, or too widely dispersed to be performed by government.

An ethical obligation on the part of the private owner is the only visible remedy for these situations.

THE LAND PYRAMID

An ethic to supplement and guide the economic relation to land presupposes the existence of some mental image of land as a biotic mechanism. We can be ethical only in relation to something we can see, feel, understand, love, or otherwise have faith in.

The image commonly employed in conservation education is 'the balance of nature.' For reasons too lengthy to detail here, this figure of speech fails to describe accurately what little we know about the land mechanism. A much truer image is the one employed in ecology: the biotic pyramid. I shall first sketch the pyramid as a symbol of land, and later develop some of its implications in terms of land-use.

Plants absorb energy from the sun. This energy flows through a circuit called the biota, which may be represented by a pyramid consisting of layers. The bottom layer is the soil. A plant layer rests on the soil, an insect layer on the plants, a bird and rodent layer on the insects, and so on up through various animal groups to the apex layer, which consists of the larger carnivores.

The species of a layer are alike not in where they came from, or in what they look like, but rather in what they eat. Each successive layer depends on those below it for food and often for other services, and each in turn furnishes food and services to those above. Proceeding upward, each successive layer decreases in numerical abundance. Thus, for every carnivore there are hundreds of his prey, thousands of their prey, millions of insects, uncountable plants. The pyramidal form of the system reflects this numerical progression from apex to base. Man shares an intermediate layer with the bears, raccoons, and squirrels which eat both meat and vegetables.

The lines of dependency for food and other services are called food chains. Thus soil-oak-deer-Indian is a chain that has now been largely converted to soil-corn-cow-farmer. Each species, including ourselves, is a link in many chains. The deer eats a hundred plants other than oak, and the cow a hundred plants other than corn. Both, then, are links in a hundred chains. The pyramid is a tangle of chains so complex as to seem disorderly, yet the stability of the system proves it to be a highly organized structure. Its functioning depends on the co-operation and competition of its diverse parts.

In the beginning, the pyramid of life was low and squat; the food chains short and simple. Evolution has added layer after layer, link after link. Man is one of thousands of accretions to the height and complexity of the pyramid. Science has given us many doubts, but it has given us at least one certainty: the trend of evolution is to elaborate and diversify the biota.

Land, then, is not merely soil; it is a fountain of energy flowing through a circuit of soils, plants, and animals. Food chains are the living channels which conduct energy upward; death and decay return it to the soil. The circuit is not closed; some energy is dissipated in decay, some is added by absorption from the air, some is stored in soils, peats, and long-lived forests; but it is a sustained circuit, like a slowly augmented revolving fund of life. There is always a net loss by downhill wash, but this is normally small and offset by the decay of rocks. It is deposited in the ocean and, in the course of geological time, raised to form new lands and new pyramids.

The velocity and character of the upward flow of energy depend on the complex structure of the plant and animal community, much as the upward flow of sap in a tree depends on its complex cellular organization. Without this complexity, normal circulation would presumably not occur. Structure means the characteristic numbers, as well as the characteristic kinds and functions, of the component species. This interdependence between the complex structure of the land and its smooth functioning as an energy unit is one of its basic attributes.

When a change occurs in one part of the circuit, many other parts must adjust themselves to it. Change does not necessarily obstruct or divert the flow of energy; evolution is a long series of self-induced changes, the net result of which has been to elaborate the flow mechanism and to lengthen the circuit. Evolutionary changes, however, are usually slow and local. Man's invention of tools has enable him to make changes of unprecedented violence, rapidity, and scope.

One change is in the composition of floras and faunas. The larger predators are lopped off the apex of the pyramid; food chains, for the first time in history, become shorter rather than longer. Domesticated species from other lands are substituted for wild ones, and wild ones are moved to new habitats. In this world-wide pooling of faunas an floras, some species get out of bounds as pests and disease,s others are extinguished. Such effects are seldom intended or foreseen; they represent unpredicted and often untraceable readjustments in the structure. Agricultural science is large a race between the emergence of new pests and the emergence of new techniques for their control.

Another change touches the flow of energy through plants and animals and its return to the soil. Fertility is the ability of soil to receive, store, and release energy. Agriculture, by overdrafts on the soil, or by too radical a substitution of domestic for native species in the superstructure, may derange the channels of flow or deplete storage. Soils depleted of their storage, or of the organic matter which anchors it, wash away faster than they form. This is erosion.

Waters, like soil, are part of the energy circuit. Industry, by polluting waters or obstructing them with dams, may exclude the plants and animals necessary to keep energy in circulation.

Transportation brings about another basic change: the plants or animals grown in one region are now consumed and returned to the soil in another. Transportation taps the energy stored in rocks, and in the air, and uses it elsewhere; thus we fertilize the garden with nitrogen gleaned by the guano birds from the fishes of seas on the other side of the Equator. Thus the formerly localized and self-contained circuits are pooled on a world-wide scale.

The process of altering the pyramid for human occupation releases stored energy, and this often gives rise, during the pioneering period, to a deceptive exuberance of plant and animal life, both wild and tame. These releases of biotic capital tend to becloud or postpone the penalties of violence.

* * *

This thumbnail sketch of land as an energy circuit conveys three basic ideas:

(1) That land is not merely soil.

(2) That the native plants and animals kept the energy circuit open; others may or may not.

(3) That man-made changes are of a different order than evolutionary changes, and have effects more comprehensive than is intended or foreseen.

These ideas, collectively, raise two basic issues: Can the land adjust itself to the new order? Can the desired alterations be accomplished with less violence?

Biotas seem to differ in their capacity to sustain violent conversion. Western Europe, for example, carries a far different pyramid than Caesar found there. Some large animals are lost; swampy forests have become meadows or plowlands; many new plants and animals are introduced, some of which escaped as pests; the remaining natives are greatly changed in distribution and abundance. Yet the soil is still there and, with the help of imported nutrients, still fertile; the waters flow normally; the new structure seems to function and to persist. There is no visible stoppage or derangement of the circuit.

Western Europe, then, has a resistant biota. Its inner processes are tough, elastic, resistant to strain. No matter how violent the alterations, the pyramid, so far, has developed some new modus vivendi which preserves its habitability for man, and for most of the other natives.

Japan seems to present another instance of radical conversion without disorganization.

Most other civilized regions, and some as yet barely touched by civilization, display various stages of disorganization, varying from initial symptoms to advanced wastage. In Asia Minor and North Africa diagnosis is confused by climatic changes, which may have been either the cause or the effect of advanced wastage. In the United States the degree of disorganization varies locally; it is worst in the Southwest, the Ozarks, and parts of the South, and least in New England and the Northwest. Better land-uses may still arrest it in the less advanced regions. In parts of Mexico, South America,

South Africa, and Australia a violent and accelerating wastage is in progress, but I cannot assess the prospects.

This almost world-wide display of disorganization in the land seems to be similar to disease in an animal, except that it never culminates in complete disorganization or death. The land recovers, but at some reduced level of complexity, and with a reduced carrying capacity for people, plants, and animals. Many biotas currently regarded as 'lands of opportunity' are in fact already subsisting on exploitative agriculture, i.e. they have already exceeded their sustained carrying capacity. Most of South America is overpopulated in this sense.

In arid regions we attempt to offset the process of wastage by reclamation, but it is only too evident that the prospective longevity of reclamation projects is often short. In our own West, the best of them may not last a century.

The combined evidence of history and ecology seems to support one general deduction: the less violent the manmade changes, the greater the probability of successful readjustment in the pyramid. Violence, in turn, varies with human population density; a dense population requires a more violent conversion. In this respect, North America has a better chance for permanence than Europe, if she can contrive to limit her density.

This deduction runs counter to our current philosophy, which assumes that because a small increase in density enriched human life, that an indefinite increase will enrich it indefinitely. Ecology knows of no density relationship that holds for indefinitely wide limits. All gains from density are subject to a law of diminishing returns.

Whatever may be the equation for men and land, it is improbable that we as yet know all its terms. Recent discoveries in mineral and vitamin nutrition reveal unsuspected dependencies in the up-circuit: incredibly minute quantities of certain substances determine the value of soils to plants, of plants to animals. What of the down-circuit? What of the vanishing species, the preservation of which we now regard as an esthetic luxury? They helped build the soil; in what unsuspected ways may they be essential to its maintenance? Professor Weaver proposes that we use prairie flowers to reflocculate the wasting soils of the dust bowl; who knows for what purpose cranes and condors, otters and grizzlies may some day be used?

LAND HEALTH AND THE A–B CLEAVAGE

A land ethic, then, reflects the existence of an ecological conscience, and this in turn reflects a conviction of individual responsibility for the health of the land. Health is the capacity of the land for self-renewal. Conservation is our effort to understand and preserve this capacity.

Conservationists are notorious for their dissensions. Superficially these seem to add up to mere confusion, but a more careful scrutiny reveals a single plane of cleavage common to many specialized fields. In each field one group (A) regards the land as soil, and its function as commodity-production; another group (B) regards the land as a biota, and its function as something broader. How much broader is admittedly in a state of doubt and confusion.

In my own field, forestry, group A is quite content to grow trees like cabbages, with cellulose as the basic forest commodity. It feels no inhibition against violence; its ideology is agronomic. Group B, on the other hand, sees forestry as fundamentally different from agronomy because it employs natural species, and manages a natural environment rather than creating an artificial one. Group B prefers natural reproduction on principle. It worries on biotic as well as economic grounds about the loss of species like chestnut, and the threatened loss of the white pines. It worries about whole series of secondary forest functions: wildlife, recreation, watersheds, wilderness areas. To my mind, Group B feels the stirrings of an ecological conscience.

In the wildlife field, a parallel cleavage exists. For Group A the basic commodities are sport and meat; the yardsticks of production are ciphers of take in pheasants and trout. Artificial propagation is acceptable as a permanent as well as a temporary recourse—if its unit costs permit. Group B, on the other hand, worries about a whole series of biotic side-issues. What is the cost in predators of producing a game crop? Should we have further recourse to exotics? How can management restore the shrinking species, like prairie grouse, already hopeless as shootable game? How can management restore the threatened rarities, like trumpeter swan and whooping crane? Can management principles be extended to wildflowers? Here again it is clear to me that we have the same A–B cleavage as in forestry.

In the larger field of agriculture I am less competent to speak, but there seem to be somewhat parallel cleavages. Scientific agriculture was actively developing before ecology was born, hence a slower penetration of ecological concepts might be expected. Moreover the farmer, by the very nature of his techniques, must modify the biota more radically than the forester or the wildlife manager. Nevertheless, there are many discontents in agriculture which seem to add up to a new vision of 'biotic farming.'

Perhaps the most important of these is the new evidence that poundage or tonnage is no measure of the food-value of farm crops; the products of fertile soil may be qualitatively as well as quantitatively superior. We can bolster poundage from depleted soils by pouring on imported fertility, but we are not necessarily bolstering food-value. The possible ultimate ramifications of this idea are so immense that I must leave their exposition to abler pens.

The discontent that labels itself 'organic farming,' while bearing some of the earmarks of a cult, is nevertheless biotic in its direction, particularly in its insistence on the importance of soil flora and fauna.

The ecological fundamentals of agriculture are just as poorly known to the public as in other fields of land-use. For example, few educated people realize that the marvelous advances in technique made during recent decades are improvements in the pump, rather than the well. Acre for acre, they have barely sufficed to offset the sinking level of fertility.

In all of these cleavages, we see repeated the same basic paradoxes: man the conqueror versus man the biotic citizen; science the sharpener of his sword versus science the search-light on his universe; land the slave and servant versus land the collective organism. Robinson's injunction to Tristram may well be applied, at this juncture, to *Homo sapiens* as a species in geological time:

> Whether you will or not
> You are a King, Tristram, for you are one
> Of the time-tested few that leave the world,
> When they are gone, not the same place it was.
> Mark what you leave.

THE OUTLOOK

It is inconceivable to me that an ethical relation to land can exist without love, respect, and admiration for land, and a high regard for its value. By value, I of course mean something far broader than mere economic value; I mean value in the philosophical sense.

Perhaps the most serious obstacle impeding the evolution of a land ethic is the fact that our educational and economic system is headed away from, rather than toward, an intense consciousness of land. Your true modern is separated from the land by many middlemen, and by innumerable physical gadgets. He has no vital relation to it; to him it is the space between cities on which crops grow. Turn him loose for a day on the land, and if the spot does not happen to be a golf links or a 'scenic' area, he is bored stiff. If crops could be raised by hydroponics instead of farming, it would suit him very well. Synthetic substitutes for wood, leather, wool, and other natural land products suit him better than the originals. In short, land is something he has 'outgrown.'

Almost equally serious as an obstacle to a land ethic is the attitude of the farmer for whom the land is still an adversary, or a taskmaster that keeps him in slavery. Theoretically, the mechanization of farming ought to cut the farmer's chains, but whether it really does is debatable.

One of the requisites for an ecological comprehension of land is an understanding of ecology, and this is by no means co-extensive with 'education'; in fact, much higher education seems deliberately to avoid ecological concepts. An understanding of ecology does not necessarily originate in courses bearing ecological labels; it is quite as likely to be labeled geography, botany, agronomy, history, or economics. This is as it should be, but whatever the label, ecological training is scarce.

The case for a land ethic would appear hopeless but for the minority which is in obvious revolt against these 'modern' trends.

The 'key-log' which must be moved to release the evolutionary process for an ethic is simply this: quit thinking about decent land-use as solely an economic problem. Examine each question in terms of what is ethically and esthetically right, as well as what is economically expedient. A thing is right when it tends to preserve the integrity, stability, and beauty of the biotic community. It is wrong when it tends otherwise.

It of course goes without saying that economic feasibility limits the tether of what can or cannot be done for land. It always has and it always will. The fallacy the economic determinists have tied around our collective neck, and which we now need to cast off, is the belief that economics determines *all* land-use. This is simply not true. An innumerable host of actions and attitudes, comprising perhaps the bulk of all land relations, is determined by the land-users' tastes and predilections, rather than by his purse. The bulk of all land relations hinges on investments of time, forethought, skill, and faith rather than on investments of cash. As a land-user thinketh, so is he.

I have purposely presented the land ethic as a product of social evolution because nothing so important as an ethic is ever 'written.' Only the most superficial student of history supposes that Moses 'wrote' the Decalogue; it evolved in the minds of a thinking community, and Moses wrote a tentative summary of it for a 'seminar.' I say tentative because evolution never stops.

The evolution of a land ethic is an intellectual as well an emotional process. Conservation is paved with good intentions which prove to be futile, or even dangerous, because they are devoid of critical understanding either of the land, or of economic land-use. I think it is a truism that as the ethical frontier advances from the individual to the community, its intellectual content increases.

The mechanism of operation is the same for any ethic: social approbation for right actions: social disapproval for wrong actions.

By and large, our present problem is one of attitudes and implements. We are remodeling the Alhambra with a steam-shovel, and we are proud of our yardage. We shall hardly relinquish the shovel, which after all has many good points, but we are in need of gentler and more objective criteria for its successful use.

A FABLE FOR TOMORROW

SOURCE Carson, Rachel. 1962. *Silent Spring.* Boston: Houghton Mifflin. Copyright renewed 1990 by Roger Christie. Reproduced by permission of Houghton Mifflin.

INTRODUCTION *Rachel Carson (1907–1964) was a marine biologist for the U.S. Bureau of Fisheries who later devoted herself full time to nature writing. Her concern about pesticides, in relation to conservation and environmental issues, led to the writing of* Silent Spring, *which is often cited as the book that launched the modern environmental movement. In "A Fable for Tomorrow," the book's first chapter, Carson traces the effects of the pesticide DDT on an imaginary town. She evokes a town once full of life, color, and voices, where human and nonhuman life mysteriously becomes sick and dies. An uncharacteristic stillness descends upon this town, the cause of which can be traced back to a "white granular powder." Who or what is responsible? It is the people of this town who have brought this unnatural stillness and death upon themselves. Carson closes with the claim that, while this is an imaginary town, these events were happening in real places around the country. Her book is an attempt to explain how and why.*

There was once a town in the heart of America where all life seemed to live in harmony with its surroundings. The town lay in the midst of a checkerboard of prosperous farms, with fields of grain and hillsides of orchards where, in spring, white clouds of bloom drifted above the green fields. In autumn, oak and maple and birch set up a blaze of color that flamed and flickered across a backdrop of pines. Then foxes barked in the hills and deer silently crossed the fields, half hidden in the mists of the fall mornings.

Along the roads, laurel, viburnum and alder, great ferns and wildflowers delighted the traveler's eye through much of the year. Even in winter the roadsides were places of beauty, where countless birds came to feed on the berries and on the seed heads of the dried weeds rising above the snow. The countryside was, in fact, famous for the abundance and variety of its bird life, and when the flood of migrants was pouring through in spring and fall people traveled from great distances to observe them. Others came to fish the streams, fish flowed clear and cold out of the hills and contained shady pools where trout lay. So it had been from the days many years ago when the first settlers raised their houses, sank their wells, and built their barns.

Then a strange blight crept over the area and everything began to change. Some evil spell had settled on the community: mysterious maladies swept the flocks of chickens; the cattle and sheep sickened and died. Everywhere was a shadow of death. The farmers spoke of much illness among their families. In the town the doctors had become more and more puzzled by new kinds of sickness appearing among their patients. There had been several sudden and unexplained deaths, not only among adults but even among children, who would be stricken suddenly while at play and die within a few hours.

There was a strange stillness. The birds, for example—where had they gone? Many people spoke of them, puzzled and disturbed. The feeding stations in the backyards were deserted. The few birds seen anywhere were moribund; they trembled violently and could not fly. It was spring without voices. On the mornings that had once throbbed with the dawn chorus of robins, catbirds, doves, jays, wrens, and scores of other bird voices there was now no sound; only silence lay over the fields and woods and marsh.

On the farms the hens brooded, but no chicks hatched. The farmers complained that they were unable to raise any pigs—the litters were small and the young survived only a few days. The apple trees were coming into bloom but no bees droned among the blossoms, so there was no pollination and there would be no fruit.

The roadsides, once so attractive, were now lined with browned and withered vegetation as though swept by fire. These, too, were silent, deserted by all living things. Even the streams were now lifeless. Anglers no longer visited them, for all the fish had died.

In the gutters under the eaves and between the shingles of the roofs, a white granular powder still showed a few patches; some weeks before it had fallen like snow upon the roofs and the lawns, the fields and streams.

No witchcraft, no enemy action had silenced the rebirth of new life in this stricken world. The people had done it themselves.

This town does not actually exist, but it might easily have a thousand counterparts in America or elsewhere in the world. I know of no community that has experienced all the misfortunes I describe. Yet every one of these disasters has actually happened somewhere, and many real communities have already suffered a substantial number of them. A grim specter has crept upon us almost unnoticed, and this imagined tragedy may easily become a stark reality we all shall know.

What has already silenced the voices of spring in countless towns in America? This book is an attempt to explain.

THE HISTORICAL ROOTS OF OUR ECOLOGIC CRISIS

SOURCE White Jr., Lynn. 1967. *Science.* 155 (March 10): 1203–1207. Reproduced by permission of AAAS.

INTRODUCTION *Lynn White Jr. (1907–1987) was professor of medieval history at the University of California, Los Angeles; he also taught at Princeton and Stanford and served as president of Mills College in Oakland, California, for fifteen years. "The Historical Roots of Our Ecologic Crisis" is a seminal piece in environmental ethics that often serves as the starting point for addressing environmental issues as they relate to philosophy. White argues that while all living things modify their environments, the ability of humanity to transform our environment radically changed when we married science to technology, which is the proximate cause of the current ecological crisis. White argues that science and technology are distinctively Western in provenance, born when Europe was dominated by the Christian worldview. He characterizes Christianity as anthropocentric and argues that it desanctifies nature, leaving it vulnerable to exploitation. Christianity thus bears significant blame for the ecological crisis. White does allow for an alternative Christian tradition that focuses on St. Francis of Assisi. Science and technology cannot solve our environmental problems, and indeed will only exacerbate them unless we address the fact that the crisis is largely conceptual in nature. Ultimately, White famously claims, we must either "find a new religion or rethink our old one." More generally, he claims that what we do in regard to the natural environment depends on what we think about it and about ourselves in relationship to it.*

A conversation with Aldous Huxley not infrequently put one at the receiving end of an unforgettable monologue. About a year before his lamented death he was discoursing on a favorite topic: Man's unnatural treatment of nature and its sad results. To illustrate his point he told how, during the previous summer, he had returned to a little valley in England where he had spent many happy months as a child. Once it had been composed of delightful grassy glades; now it was becoming overgrown with unsightly brush because the rabbits that formerly kept such growth under control had largely succumbed to a disease, myxomatosis, that was deliberately introduced by the local farmers to reduce the rabbits' destruction of crops. Being something of a Philistine, I could be silent no longer, even in the interests of great rhetoric. I interrupted to point out that the rabbit itself had been brought as a domestic animal to England in 1176, presumably to improve the protein diet of the peasantry.

All forms of life modify their contexts. The most spectacular and benign instance is doubtless the coral polyp. By serving its own ends, it has created a vast undersea world favorable to thousands of other kinds of animals and plants. Ever since man became a numerous species he has affected his environment notably. The hypothesis that his fire-drive method of hunting created the world's great grasslands and helped to exterminate the monster mammals of the Pleistocene from much of the globe is plausible, if not proved. For 6 millennia at least, the banks of the lower Nile have been a human artifact rather than the swampy African jungle which nature, apart from man, would have made it. The Aswan Dam, flooding 5000 square miles, is only the latest stage in a long process. In many regions terracing or irrigation, overgrazing, the cutting of forests by Romans to build ships to fight Carthaginians or by Crusaders to solve the logistics problems of their expeditions, have profoundly changed some ecologies. Observation that the French landscape falls into two basic types, the open fields of the north and the bocage of the south and west, inspired Marc Bloch to undertake his classic study of medieval agricultural methods. Quite unintentionally, changes in human ways often affect nonhuman nature. It has been noted, for example, that the advent of the automobile eliminated huge flocks of sparrows that once fed on the horse manure littering every street.

The history of ecologic change is still so rudimentary that we know little about what really happened, or what the results were. The extinction of the European aurochs as late as 1627 would seem to have been a simple case of overenthusiastic hunting. On more intricate matters it often is impossible to find solid information. For a thousand years or more the Frisians and Hollanders have been pushing back the North Sea, and the process is culminating in our own time in the reclamation of the Zuider Zee. What, if any, species of animals, birds, fish, shore life, or plants have died out in the process? In their epic combat with Neptune have the Netherlanders overlooked ecological values in such a way that the quality of human life in the Netherlands has suffered? I cannot discover that the questions have ever been asked, much less answered.

People, then, have often been a dynamic element in their own environment, but in the present state of historical scholarship we usually do not know exactly when, where, or with what effects man-induced changes came. As we enter the last third of the 20th century, however, concern for the problem of ecologic backlash is mounting feverishly. Natural science, conceived as the effort to understand the nature of things, had flourished in several eras and among several peoples. Similarly there had been an age-old accumulation of technological skills, sometimes growing rapidly, sometimes slowly. But it was not until about four generations ago that Western Europe and North America arranged a marriage between science and technology, a union of the theoretical and the empirical approaches to our natural environment. The emergence in widespread practice of the Baconian creed that scientific knowledge means technological power over nature can scarcely be dated before about 1850, save in the chemical industries, where it is anticipated in the 18th century. Its acceptance as a normal pattern of action may mark the greatest event in human history since the invention of agriculture, and perhaps in nonhuman terrestrial history as well.

Almost at once the new situation forced the crystallization of the novel concept of ecology; indeed, the word ecology first appeared in the English language in 1873. Today, less than a century later, the impact of our race upon the environment has so increased in force that it has changed in essence. When the first cannons were fired, in the early 14th century, they affected ecology by sending workers scrambling to the forests and mountains for more potash, sulphur, iron ore, and charcoal, with some resulting erosion and deforestation. Hydrogen bombs are of a different order: a war fought with them might alter the genetics of all life on this planet. By 1285 London had a smog problem arising from the burning of soft coal, but our present combustion of fossil fuels threatens to change the chemistry of the globe's atmosphere as a whole, with consequences which we are only beginning to guess. With the population explosion, the carcinoma of planless urbanism, the now geological deposits of sewage and garbage, surely no creature other than man has ever managed to foul its nest in such short order.

There are many calls to action, but specific proposals, however worthy as individual items, seem too partial, palliative, negative: ban the bomb, tear down the billboards, give the Hindus contraceptives and tell them to eat their sacred cows. The simplest solution to any suspect change is, of course, to stop it, or better yet, to revert to a romanticized past: make those ugly gasoline stations look like Anne Hathaway's cottage or (in the Far West) like ghost-town saloons. The "wilderness area" mentality invariably advocates deep-freezing an ecology, whether San Gimignano or the High Sierra, as it was before the first Kleenex was dropped. But neither atavism nor prettification will cope with the ecologic crisis of our time.

What shall we do? No one yet knows. Unless we think about fundamentals, our specific measures may produce new backlashes more serious than those they are designed to remedy.

As a beginning we should try to clarify our thinking by looking, in some historical depth, at the presuppositions that underlie modern technology and science. Science was traditionally aristocratic, speculative, intellectual in intent; technology was lower-class, empirical, action-oriented. The quite sudden fusion of these two, towards the middle of the 19th century, is surely related to the slightly prior and contemporary democratic revolutions which, by reducing social barriers, tended to assert a functional unity of brain and hand. Our ecologic crisis is the product of an emerging, entirely novel, democratic culture. The issue is whether a democratized world can survive its own implications. Presumably we cannot unless we rethink our axioms.

THE WESTERN TRADITIONS OF TECHNOLOGY AND SCIENCE

One thing is so certain that it seems stupid to verbalize it: both modern technology and modern science are distinctively Occidental. Our technology has absorbed elements from all over the world, notably from China; yet everywhere today, whether in Japan or in Nigeria, successful technology is Western. Our science is the heir to all the sciences of the past, perhaps to the work of the great Islamic scientists of the Middle Ages, who so often outdid the ancient Greeks in skill and perspicacity: al-Razi in medicine, for example; or ibn-al-Haytham in optics; or Omar Khayyam in mathematics. Indeed, not a few works of such geniuses seem to have vanished in the original Arabic and to survive only in medieval Latin translations that helped to lay the foundations for later Western developments. Today, around the globe, all significant science is Western in style and method, whatever the pigmentation or language of the scientists.

A second pair of facts is less well recognized because they result from quite recent historical scholarship. The leadership of the West, both in technology and in science, is far older than the so-called Scientific Revolution of the 17th century or the so-called Industrial Revolution of the 18th century. These terms are in fact outmoded and obscure the true nature of what they try to describe—significant stages in two long and separate developments. By A.D. 1000 at the latest—and perhaps, feebly, as much as 200 years earlier—the West began to apply water power to industrial processes other than milling grain. This was followed in the late 12th century by the harnessing of wind power. From simple beginnings, but with remarkable consistency of style, the West rapidly expanded its skills in the development of power machinery, labor-saving devices, and automation. Those who doubt should contemplate that most monumental achievement in the history of automation: the weight-driven mechanical clock, which appeared in two forms in the early 14th century. Not in craftsmanship but in basic technological capacity, the Latin West of the later Middle Ages far outstripped its elaborate, sophisticated, and esthetically magnificent sister cultures, Byzantium and Islam. In 1444 a great Greek ecclesiastic, Bessarion, who had gone to Italy, wrote a letter to a prince in Greece. He is amazed by the superiority of Western ships, arms, textiles, glass. But above all he is astonished by the spectacle of waterwheels sawing timbers and pumping the bellows of blast furnaces. Clearly, he had seen nothing of the sort in the Near East.

By the end of the 15th century the technological superiority of Europe was such that its small, mutually hostile nations could spill out over all the rest of the world, conquering, looting, and colonizing. The symbol of this technological superiority is the fact that Portugal, one of the weakest states of the Occident, was able to become, and to remain for a century, mistress of the East Indies. And we must remember that the technology of Vasco da Gama and Albuquerque was built by pure empiricism, drawing remarkably little support or inspiration from science.

In the present-day vernacular understanding, modern science is supposed to have begun in 1543, when both Copernicus and Vesalius published their great works. It is no derogation of their accomplishments, however, to point out that such structures as the Fabrica and the De revolutionibus do not appear overnight. The distinctive Western tradition of science, in fact, began in the late 11th century with a massive movement of translation of Arabic and Greek scientific works into Latin. A few notable books—Theophrastus, for example—escaped the West's avid new appetite for science, but within less than 200 years effectively the entire corpus of Greek and Muslim science was available in Latin, and was being eagerly read and criticized in the new European universities. Out of criticism arose new observation, speculation, and increasing distrust of ancient authorities. By the late 13th century Europe had seized global scientific leadership from the faltering hands of Islam. It would be as absurd to deny the profound originality of Newton, Galileo, or Copernicus as to deny that of the 14th century scholastic scientists like Buridan or Oresme on whose work they built. Before the 11th century, science scarcely existed in the Latin West, even in Roman times. From the 11th century onward, the scientific sector of Occidental culture has increased in a steady crescendo.

Since both our technological and our scientific movements got their start, acquired their character, and achieved world dominance in the Middle Ages, it would seem that we cannot understand their nature or their present impact upon ecology without examining fundamental medieval assumptions and developments.

MEDIEVAL VIEW OF MAN AND NATURE

Until recently, agriculture has been the chief occupation even in "advanced" societies; hence, any change in methods of tillage has much importance. Early plows, drawn by two oxen, did not normally turn the sod but merely scratched it. Thus, cross-plowing was needed and fields tended to be squarish. In the fairly light soils and semi-arid climates of the Near East and Mediterranean, this worked well. But such a plow was inappropriate to the wet climate and often sticky soils of northern Europe. By the latter part of the 7th century after Christ, however, following obscure beginnings, certain northern peasants were using an entirely new kind of plow, equipped with a vertical knife to cut the line of the furrow, a horizontal share to slice under the sod, and a moldboard to turn it over. The friction of this plow with the soil was so great that it normally required not two but eight oxen. It attacked the land with such violence that cross-plowing was not needed, and fields tended to be shaped in long strips.

In the days of the scratch-plow, fields were distributed generally in units capable of supporting a single family. Subsistence farming was the presupposition. But no peasant owned eight oxen: to use the new and more efficient plow, peasants pooled their oxen to form large plow-teams, originally receiving (it would appear) plowed strips in proportion to their contribution. Thus, distribution of land was based no longer on the needs of a family but, rather, on the capacity of a power machine to till the earth. Man's relation to the soil was profoundly changed. Formerly man had been part of nature; now he was the exploiter of nature. Nowhere else in the world did farmers develop any analogous agricultural implement. Is it coincidence that modern technology, with its ruthlessness toward nature, has so largely been produced by descendants of these peasants of northern Europe?

This same exploitive attitude appears slightly before A.D. 830 in Western illustrated calendars. In older calendars the months were shown as passive personifications. The new Frankish calendars, which set the style for the Middle Ages, are very different: they show men coercing the world around them—plowing, harvesting, chopping trees, butchering pigs. Man and nature are two things, and man is master.

These novelties seem to be in harmony with larger intellectual patterns. What people do about their ecology depends on what they think about themselves in relation to things around them. Human ecology is deeply conditioned by beliefs about our nature and destiny—that is, by religion. To Western eyes this is very evident in, say, India or Ceylon. It is equally true of ourselves and of our medieval ancestors.

The victory of Christianity over paganism was the greatest psychic revolution in the history of our culture. It has become fashionable today to say that, for better or worse, we live in the "post-Christian age." Certainly the forms of our thinking and language have largely ceased to be Christian, but to my eye the substance often remains amazingly akin to that of the past. Our daily habits of action, for example, are dominated by an implicit faith in perpetual progress which was unknown either to Greco-Roman antiquity or to the Orient. It is rooted in, and is indefensible apart from, Judeo-Christian theology. The fact that Communists share it merely helps to show what can be demonstrated on many other grounds: that Marxism, like Islam, is a Judeo-Christian heresy. We continue today to live, as we have lived for about 1700 years, very largely in a context of Christian axioms.

What did Christianity tell people about their relations with the environment? While many of the world's mythologies provide stories of creation, Greco-Roman mythology was singularly incoherent in this respect. Like Aristotle, the intellectuals of the ancient West denied that the visible world had a beginning. Indeed, the idea of a beginning was impossible in the framework of their cyclical notion of time. In sharp contrast, Christianity inherited from Judaism not only a concept of time as nonrepetitive and linear but also a striking story of creation. By gradual stages a loving and all-powerful God had created light and darkness, the heavenly bodies, the earth and all its plants, animals, birds, and fishes. Finally, God had created Adam and, as an afterthought, Eve to keep man from being lonely. Man named all the animals, thus establishing his dominance over them. God planned all of this explicitly for man's benefit and rule: no item in the physical creation had any purpose save to serve man's purposes. And, although man's body is made of clay, he is not simply part of nature: he is made in God's image.

Especially in its Western form, Christianity is the most anthropocentric religion the world has seen. As early as the 2nd century both Tertullian and Saint Irenaeus of Lyons were insisting that when God shaped Adam he was foreshadowing the image of the incarnate Christ, the Second Adam. Man shares, in great measure, God's transcendence of nature. Christianity, in absolute contrast to ancient paganism and Asia's religions (except, perhaps, Zorastrianism), not only established a dualism of man and nature but also insisted that it is God's will that man exploit nature for his proper ends.

At the level of the common people this worked out in an interesting way. In Antiquity every tree, every spring, every stream, every hill had its own genius loci, its guardian spirit. These spirits were accessible to men, but were very unlike men; centaurs, fauns, and mermaids show their ambivalence. Before one cut a tree, mined a mountain, or dammed a brook, it was important to placate the spirit in charge of that particular situation, and to keep it placated. By destroying pagan animism, Christianity made it possible to exploit nature in a mood of indifference to the feelings of natural objects.

It is often said that for animism the Church substituted the cult of saints. True; but the cult of saints is functionally quite different from animism. The saint is not in natural objects; he may have special shrines, but his citizenship is in heaven. Moreover, a saint is entirely a man; he can be approached in human terms. In addition to saints, Christianity of course also had angels and demons inherited from Judaism and perhaps, at one remove, from Zorastrianism. But these were all as mobile as the saints themselves. The spirits in natural objects, which formerly had protected nature from man, evaporated. Man's effective monopoly on spirit in this world was confirmed, and the old inhibitions to the exploitation of nature crumbled.

When one speaks in such sweeping terms, a note of caution is in order. Christianity is a complex faith, and its consequences differ in differing contexts. What I have said may well apply to the medieval West, where in fact technology made spectacular advances. But the Greek East, a highly civilized realm of equal Christian devotion, seems to have produced no marked technological innovation after the late 7th century, when Greek fire was invented. The key to the contrast may perhaps be found in a difference in the tonality of piety and thought which students of comparative theology find between the Greek and the Latin Churches. The Greeks believed that sin was intellectual blindness, and that salvation was found in illumination, orthodoxy—that is, clear thinking. The Latins, on the other hand, felt that sin was moral evil, and that salvation was to be found in right conduct. Eastern theology has been intellectualist. Western theology has been voluntarist. The Greek saint contemplates; the Western saint acts. The implications of Christianity for the conquest of nature would emerge more easily in the Western atmosphere.

The Christian dogma of creation, which is found in the first clause of all the Creeds, has another meaning for our comprehension of today's ecologic crisis. By revelation, God had given man the Bible, the Book of Scripture. But since God had made nature, nature also must reveal the divine mentality. The religious study of nature for the better understanding of God was known as natural theology. In the early Church, and always in the Greek East, nature was conceived primarily as a symbolic system through which God speaks to men: the ant is a sermon to sluggards; rising flames are the symbol of the soul's aspiration. The view of nature was essentially artistic rather than scientific. While Byzantium preserved and copied great numbers of ancient Greek scientific texts, science as we conceive it could scarcely flourish in such an ambience.

However, in the Latin West by the early 13th century natural theology was following a very different bent. It was ceasing to be the decoding of the physical symbols of

God's communication with man and was becoming the effort to understand God's mind by discovering how his creation operates. The rainbow was no longer simply a symbol of hope first sent to Noah after the Deluge: Robert Grosseteste, Friar Roger Bacon, and Theodoric of Freiberg produced startlingly sophisticated work on the optics of the rainbow, but they did it as a venture in religious understanding. From the 13th century onward, up to and including Leitnitz and Newton, every major scientist, in effect, explained his motivations in religious terms. Indeed, if Galileo had not been so expert an amateur theologian he would have got into far less trouble: the professionals resented his intrusion. And Newton seems to have regarded himself more as a theologian than as a scientist. It was not until the late 18th century that the hypothesis of God became unnecessary to many scientists.

It is often hard for the historian to judge, when men explain why they are doing what they want to do, whether they are offering real reasons or merely culturally acceptable reasons. The consistency with which scientists during the long formative centuries of Western science said that the task and the reward of the scientist was "to think God's thoughts after him" leads one to believe that this was their real motivation. If so, then modern Western science was cast in a matrix of Christian theology. The dynamism of religious devotion shaped by the Judeo-Christian dogma of creation, gave it impetus.

AN ALTERNATIVE CHRISTIAN VIEW

We would seem to be headed toward conclusions unpalatable to many Christians. Since both science and technology are blessed words in our contemporary vocabulary, some may be happy at the notions, first, that viewed historically, modern science is an extrapolation of natural theology and, second, that modern technology is at least partly to be explained as an Occidental, voluntarist realization of the Christian dogma of man's transcendence of, and rightful master over, nature. But, as we now recognize, somewhat over a century ago science and technology—hitherto quite separate activities—joined to give mankind powers which, to judge by many of the ecologic effects, are out of control. If so, Christianity bears a huge burden of guilt.

I personally doubt that disastrous ecologic backlash can be avoided simply by applying to our problems more science and more technology. Our science and technology have grown out of Christian attitudes toward man's relation to nature which are almost universally held not only by Christians and neo-Christians but also by those who fondly regard themselves as post-Christians. Despite Copernicus, all the cosmos rotates around our little globe. Despite Darwin, we are not, in our hearts, part of the natural process. We are superior to nature, contemptuous of it, willing to use it for our slightest whim. The newly elected Governor of California, like myself a churchman but less troubled than I, spoke for the Christian tradition when he said (as is alleged), "when you've seen one redwood tree, you've seen them all." To a Christian a tree can be no more than a physical fact. The whole concept of the sacred grove is alien to Christianity and to the ethos of the West. For nearly 2 millennia Christian missionaries have been chopping down sacred groves, which are idolatrous because they assume spirit in nature.

What we do about ecology depends on our ideas of the man-nature relationship. More science and more technology are not going to get us out of the present ecologic crisis until we find a new religion, or rethink our old one. The beatniks, who are the basic revolutionaries of our time, show a sound instinct in their affinity for Zen Buddhism, which conceives of the man-nature relationship as very nearly the mirror image of the Christian view. Zen, however, is as deeply conditioned by Asian history as Christianity is by the experience of the West, and I am dubious of its viability among us.

Possibly we should ponder the greatest radical in Christian history since Christ: Saint Francis of Assisi. The prime miracle of Saint Francis is the fact that he did not end at the stake, as many of his left-wing followers did. He was so clearly heretical that a General of the Franciscan Order, Saint Bonavlentura, a great and perceptive Christian, tried to suppress the early accounts of Franciscanism. The key to an understanding of Francis is his belief in the virtue of humility—not merely for the individual but for man as a species. Francis tried to depose man from his monarchy over creation and set up a democracy of all God's creatures. With him the ant is no longer simply a homily for the lazy, flames a sign of the thrust of the soul toward union with God; now they are Brother Ant and Sister Fire, praising the Creator in their own ways as Brother Man does in his.

Later commentators have said that Francis preached to the birds as a rebuke to men who would not listen. The records do not read so: he urged the little birds to praise God, and in spiritual ecstasy they flapped their wings and chirped rejoicing. Legends of saints, especially the Irish saints, had long told of their dealings with animals but always, I believe, to show their human dominance over creatures. With Francis it is different. The land around Gubbio in the Apennines was ravaged by a fierce wolf. Saint Francis, says the legend, talked to the wolf and persuaded him of the error of his ways. The wolf repented, died in the odor of sanctity, and was buried in consecrated ground.

What Sir Steven Ruciman calls "the Franciscan doctrine of the animal soul" was quickly stamped out. Quite possibly it was in part inspired, consciously or unconsciously, by the belief in reincarnation held by the Cathar heretics who at that time teemed in Italy and southern

France, and who presumably had got it originally from India. It is significant that at just the same moment, about 1200, traces of metempsychosis are found also in western Judaism, in the Provencal Cabbala. But Francis held neither to transmigration of souls nor to pantheism. His view of nature and of man rested on a unique sort of pan-psychism of all things animate and inaminate, designed for the glorification of their transcendent Creator, who, in the ultimate gesture of cosmic humility, assumed flesh, lay helpless in a manger, and hung dying on a scaffold.

I am not suggesting that many contemporary Americans who are concerned about our ecologic crisis will be either able or willing to counsel with wolves or exhort birds. However, the present increasing disruption of the global environment is the product of a dynamic technology and science which were originating in the Western medieval world against which Saint Francis was rebelling in so original a way. Their growth cannot be understood historically apart from distinctive attitudes toward nature which are deeply grounded in Christian dogma. The fact that most people do not think of these attitudes as Christian is irrelevant. No new set of basic values has been accepted in our society to displace those of Christianity. Hence we shall continue to have a worsening ecologic crisis until we reject the Christian axiom that nature has no reason for existence save to serve man.

The greatest spiritual revolutionary in Western history, Saint Francis, proposed what he thought was an alternative Christian view of nature and man's relation to it; he tried to substitute the idea of the equality of all creatures, including man, for the idea of man's limitless rule of creation. He failed. Both our present science and our present technology are so tinctured with orthodox Christian arrogance toward nature that no solution for our ecologic crisis can be expected from them alone. Since the roots of our trouble are so largely religious, the remedy must also be essentially religious, whether we call it that or not. We must rethink and refeel our nature and destiny. The profoundly religious, but heretical, sense of the primitive Franciscans for the spiritual autonomy of all parts of nature may point a direction. I propose Francis as a patron saint for ecologists.

THE TRAGEDY OF THE COMMONS

SOURCE Hardin, Garrett. 1968. *Science* 162 (December 13): 1243–1248. Reproduced by permission of AAAS.

INTRODUCTION *Garrett Hardin (1915–2003) was for fifteen years professor of human ecology at the University of California, Santa Barbara. He begins his hugely influential "The Tragedy of the Commons" by drawing attention to the fact that there are some problems for which a technical solution does not exist. The "population problem" falls into this class of problems; failure to acknowledge this fact will lead to increased human suffering in a world of finite resources. Hardin argues that to adequately address the urgent population problem, we must reject Adam Smith's notion of the "invisible hand." He provides the now-famous example of herdsmen adding cattle to a pasture open to all, to disprove the operative assumption in Smith that an individual, pursuing his own good, will produce a collective good. In this imagined commons, a rational agent is compelled to seek his own benefit at the expense of the collective good, thus bringing ruin to all. Hardin argues that population growth (as well as a host of other problems) is a problem of the commons, and that we must abandon the idea of the unregulated freedom to breed if we wish to preserve other freedoms.*

At the end of a thoughtful article on the future of nuclear war, Wiesner and York[1] concluded that: "Both sides in the arms race are . . . confronted by the dilemma of steadily increasing military power and steadily decreasing national security. *It is our considered professional judgment that this dilemma has no technical solution.* If the great powers continue to look for solutions in the area of science and technology only, the result will be to worsen the situation."

I would like to focus your attention not on the subject of the article (national security in a nuclear world) but on the kind of conclusion they reached, namely that there is no technical solution to the problem. An implicit and almost universal assumption of discussions published in professional and semipopular scientific journals is that the problem under discussion has a technical solution. A technical solution may be defined as one that requires a change only in the techniques of the natural sciences, demanding little or nothing in the way of change in human values or ideas of morality.

In our day (though not in earlier times) technical solutions are always welcome. Because of previous failures in prophecy, it takes courage to assert that a desired technical solution is not possible. Wiesner and York exhibited this courage; publishing in a science journal, they insisted that the solution to the problem was not to be found in the natural sciences. They cautiously qualified their statement with the phrase, "It is our considered professional judgment. . . ." Whether they were right or not is not the concern of the present article. Rather, the concern here is with the important concept of a class of human problems which can be called "no technical solution problems," and, more specifically, with the identification and discussion of one of these.

It is easy to show that the class is not a null class. Recall the game of tick-tack-toe. Consider the problem,

"How can I win the game of tick-tack-toe?" It is well known that I cannot, if I assume (in keeping with the conventions of game theory) that my opponent understands the game perfectly. Put another way, there is no "technical solution" to the problem. I can win only by giving a radical meaning to the word "win." I can hit my opponent over the head; or I can drug him; or I can falsify the records. Every way in which I "win" involves, in some sense, an abandonment of the game, as we intuitively understand it. (I can also, of course, openly abandon the game—refuse to play it. This is what most adults do.)

The class of "No technical solution problems" has members. My thesis is that the "population problem," as conventionally conceived, is a member of this class. How it is conventionally conceived needs some comment. It is fair to say that most people who anguish over the population problem are trying to find a way to avoid the evils of overpopulation without relinquishing any of the privileges they now enjoy. They think that farming the seas or developing new strains of wheat will solve the problem—technologically. I try to show here that the solution they seek cannot be found. The population problem cannot be solved in a technical way, any more than can the problem of winning the game of tick-tack-toe.

WHAT SHALL WE MAXIMIZE?

Population, as Malthus said, naturally tends to grow "geometrically," or, as we would now say, exponentially. In a finite world this means that the per capita share of the world's goods must steadily decrease. Is ours a finite world?

A fair defense can be put forward for the view that the world is infinite; or that we do not know that it is not. But, in terms of the practical problems that we must face in the next few generations with the foreseeable technology, it is clear that we will greatly increase human misery if we do not, during the immediate future, assume that the world available to the terrestrial human population is finite. "Space" is no escape.[2]

A finite world can support only a finite population; therefore, population growth must eventually equal zero. (The case of perpetual wide fluctuations above and below zero is a trivial variant that need not be discussed.) When this condition is met, what will be the situation of mankind? Specifically, can Bentham's goal of "the greatest good for the greatest number" be realized?

No—for two reasons, each sufficient by itself. The first is a theoretical one. It is not mathematically possible to maximize for two (or more) variables at the same time. This was clearly stated by von Neumann and Morgenstern[3], but the principle is implicit in the theory of partial differential equations, dating back at least to D'Alembert (1717–1783).

The second reason springs directly from biological facts. To live, any organism must have a source of energy (for example, food). This energy is utilized for two purposes: mere maintenance and work. For man, maintenance of life requires about 1600 kilocalories a day ("maintenance calories"). Anything that he does over and above merely staying alive will be defined as work, and is supported by "work calories" which he takes in. Work calories are used not only for what we call work in common speech; they are also required for all forms of enjoyment, from swimming and automobile racing to playing music and writing poetry. If our goal is to maximize population it is obvious what we must do: We must make the work calories per person approach as close to zero as possible. No gourmet meals, no vacations, no sports, no music, no literature, no art.... I think that everyone will grant, without argument or proof, that maximizing population does not maximize goods. Bentham's goal is impossible.

In reaching this conclusion I have made the usual assumption that it is the acquisition of energy that is the problem. The appearance of atomic energy has led some to question this assumption. However, given an infinite source of energy, population growth still produces an inescapable problem. The problem of the acquisition of energy is replaced by the problem of its dissipation, as J. H. Fremlin has so wittily shown.[4] The arithmetic signs in the analysis are, as it were, reversed: but Bentham's goal is still unobtainable.

The optimum population is, then, less than the maximum. The difficulty of defining the optimum is enormous; so far as I know, no one has seriously tackled this problem. Reaching an acceptable and stable solution will surely require more than one generation of hard analytical work—and much persuasion.

We want the maximum good per person; but what is good? To one person it is wilderness, to another it is ski lodges for thousands. To one it is estuaries to nourish ducks for hunters to shoot; to another it is factory land. Comparing one good with another is, we usually say, impossible because goods are incommensurable. Incommensurables cannot be compared.

Theoretically this may be true; but in real life incommensurables are commensurable. Only a criterion of judgment and a system of weighting are needed. In nature the criterion is survival. Is it better for a species to be small and hideable, or large and powerful? Natural selection commensurates the incommensurables. The compromise achieved depends on a natural weighting of the values of the variables.

Man must imitate this process. There is no doubt that in fact he already does, but unconsciously. It is when the hidden decisions are made explicit that the arguments begin. The problem for the years ahead is to work out an acceptable theory of weighting. Synergistic effects, nonlinear variation, and difficulties in discounting the future make the intellectual problem difficult, but not (in principle) insoluble.

Has any cultural group solved this practical problem at the present time, even on an intuitive level? One simple fact proves that none has: there is no prosperous population in the world today that has, and has had for some time, a growth rate of zero. Any people that has intuitively identified its optimum point will soon reach it, after which its growth rate becomes and remains zero.

Of course, a positive growth rate might be taken as evidence that a population is below its optimum. However, by any reasonable standards, the most rapidly growing populations on earth today are (in general) the most miserable. This association (which need not be invariable) casts doubt on the optimistic assumption that the positive growth rate of a population is evidence that it has yet to reach its optimum.

We can make little progress in working toward optimum population size until we explicitly exorcise the spirit of Adam Smith in the field of practical demography. In economic affairs, *The Wealth of Nations* (1776) popularized the "invisible hand," the idea that an individual who "intends only his own gain," is, as it were, "led by an invisible hand to promote . . . the public interest".[5] Adam Smith did not assert that this was invariably true, and perhaps neither did any of his followers. But he contributed to a dominant tendency of thought that has ever since interfered with positive action based on rational analysis, namely, the tendency to assume that decisions reached individually will, in fact, be the best decisions for an entire society. If this assumption is correct it justifies the continuance of our present policy of laissez-faire in reproduction. If it is correct we can assume that men will control their individual fecundity so as to produce the optimum population. If the assumption is not correct, we need to reexamine our individual freedoms to see which ones are defensible.

TRAGEDY OF FREEDOM IN A COMMONS

The rebuttal to the invisible hand in population control is to be found in a scenario first sketched in a little-known pamphlet[6] in 1833 by a mathematical amateur named William Forster Lloyd (1794–1852). We may well call it "the tragedy of the commons," using the word "tragedy" as the philosopher Whitehead used it[7]: "The essence of dramatic tragedy is not unhappiness. It resides in the solemnity of the remorseless working of things." He then goes on to say, "This inevitableness of destiny can only be illustrated in terms of human life by incidents which in fact involve unhappiness. For it is only by them that the futility of escape can be made evident in the drama."

The tragedy of the commons develops in this way. Picture a pasture open to all. It is to be expected that each herdsman will try to keep as many cattle as possible on the commons. Such an arrangement may work reasonably satisfactorily for centuries because tribal wars, poaching, and disease keep the numbers of both man and beast well below the carrying capacity of the land. Finally, however, comes the day of reckoning, that is, the day when the long-desired goal of social stability becomes a reality. At this point, the inherent logic of the commons remorselessly generates tragedy.

As a rational being, each herdsman seeks to maximize his gain. Explicitly or implicitly, more or less consciously, he asks, "What is the utility to *me* of adding one more animal to my herd?" This utility has one negative and one positive component.

1. The positive component is a function of the increment of one animal. Since the herdsman receives all the proceeds from the sale of the additional animal, the positive utility is nearly +1.
2. The negative component is a function of the additional overgrazing created by one more animal. Since, however, the effects of overgrazing are shared by all the herdsmen, the negative utility for any particular decision-making herdsman is only a fraction of −1.

Adding together the component partial utilities, the rational herdsman concludes that the only sensible course for him to pursue is to add another animal to his herd. And another; and another. . . . But this is the conclusion reached by each and every rational herdsman sharing a commons. Therein is the tragedy. Each man is locked into a system that compels him to increase his herd without limit—in a world that is limited. Ruin is the destination toward which all men rush, each pursuing his own best interest in a society that believes in the freedom of the commons. Freedom in a commons brings ruin to all.

Some would say that this is a platitude. Would that it were! In a sense, it was learned thousands of years ago, but natural selection favors the forces of psychological denial.[8] The individual benefits as an individual from his ability to deny the truth even though society as a whole, of which he is a part, suffers.

Education can counteract the natural tendency to do the wrong thing, but the inexorable succession of generations requires that the basis for this knowledge be constantly refreshed.

A simple incident that occurred a few years ago in Leominster, Massachusetts, shows how perishable the knowledge is. During the Christmas shopping season the parking meters downtown were covered with plastic bags that bore tags reading: "Do not open until after Christmas. Free parking courtesy of the mayor and city council." In other words, facing the prospect of an increased demand for already scarce space, the city fathers reinstituted the system of the commons. (Cynically, we suspect that they gained more votes than they lost by this retrogressive act.)

In an approximate way, the logic of the commons has been understood for a long time, perhaps since the discovery of agriculture or the invention of private property in real estate. But it is understood mostly only in special cases which are not sufficiently generalized. Even at this late date, cattlemen leasing national land on the western ranges demonstrate no more than an ambivalent understanding, in constantly pressuring federal authorities to increase the head count to the point where overgrazing produces erosion and weed-dominance. Likewise, the oceans of the world continue to suffer from the survival of the philosophy of the commons. Maritime nations still respond automatically to the shibboleth of the "freedom of the seas." Professing to believe in the "inexhaustible resources of the oceans," they bring species after species of fish and whales closer to extinction.[9]

The National Parks present another instance of the working out of the tragedy of the commons. At present, they are open to all, without limit. The parks themselves are limited in extent—there is only one Yosemite Valley—whereas population seems to grow without limit. The values that visitors seek in the parks are steadily eroded. Plainly, we must soon cease to treat the parks as commons or they will be of no value to anyone.

What shall we do? We have several options. We might sell them off as private property. We might keep them as public property, but allocate the right to enter them. The allocation might be on the basis of wealth, by the use of an auction system. It might be on the basis of merit, as defined by some agreed-upon standards. It might be by lottery. Or it might be on a first-come, first-served basis, administered to long queues. These, I think, are all the reasonable possibilities. They are all objectionable. But we must choose—or acquiesce in the destruction of the commons that we call our National Parks.

POLLUTION

In a reverse way, the tragedy of the commons reappears in problems of pollution. Here it is not a question of taking something out of the commons, but of putting something in—sewage, or chemical, radioactive, and heat wastes into water; noxious and dangerous fumes into the air, and distracting and unpleasant advertising signs into the line of sight. The calculations of utility are much the same as before. The rational man finds that his share of the cost of the wastes he discharges into the commons is less than the cost of purifying his wastes before releasing them. Since this is true for everyone, we are locked into a system of "fouling our own nest," so long as we behave only as independent, rational, free-enterprisers.

The tragedy of the commons as a food basket is averted by private property, or something formally like it. But the air and waters surrounding us cannot readily be fenced, and so the tragedy of the commons as a cesspool must be prevented by different means, by coercive laws or taxing devices that make it cheaper for the polluter to treat his pollutants than to discharge them untreated. We have not progressed as far with the solution of this problem as we have with the first. Indeed, our particular concept of private property, which deters us from exhausting the positive resources of the earth, favors pollution. The owner of a factory on the bank of a stream—whose property extends to the middle of the stream, often has difficulty seeing why it is not his natural right to muddy the waters flowing past his door. The law, always behind the times, requires elaborate stitching and fitting to adapt it to this newly perceived aspect of the commons.

The pollution problem is a consequence of population. It did not much matter how a lonely American frontiersman disposed of his waste. "Flowing water purifies itself every 10 miles," my grandfather used to say, and the myth was near enough to the truth when he was a boy, for there were not too many people. But as population became denser, the natural chemical and biological recycling processes became overloaded, calling for a redefinition of property rights.

HOW TO LEGISLATE TEMPERANCE?

Analysis of the pollution problem as a function of population density uncovers a not generally recognized principle of morality, namely: *the morality of an act is a function of the state of the system at the time it is performed.*[10] Using the commons as a cesspool does not harm the general public under frontier conditions, because there is no public, the same behavior in a metropolis is unbearable. A hundred and fifty years ago a plainsman could kill an American bison, cut out only the tongue for his dinner, and discard the rest of the animal. He was not in any important sense being wasteful. Today, with only a few thousand bison left, we would be appalled at such behavior.

In passing, it is worth noting that the morality of an act cannot be determined from a photograph. One does not know whether a man killing an elephant or setting fire to the grassland is harming others until one knows the total system in which his act appears. "One picture is worth a thousand words," said an ancient Chinese; but it may take 10,000 words to validate it. It is as tempting to ecologists as it is to reformers in general to try to persuade others by way of the photographic shortcut. But the essence of an argument cannot be photographed: it must be presented rationally—in words.

That morality is system-sensitive escaped the attention of most codifiers of ethics in the past. "Thou shalt not . . ." is the form of traditional ethical directives which make no allowance for particular circumstances. The laws of our society follow the pattern of ancient ethics, and therefore are poorly suited to governing a complex, crowded, changeable world. Our epicyclic solution is to augment statutory law with administrative law. Since it is practically impossible to spell out all the conditions under which it is safe to burn trash in the back yard or to run an automobile without smog-control, by law we delegate the details to bureaus. The result is administrative law, which is rightly feared for an ancient reason—*Quis custodiet ipsos custodes?*—"Who shall watch the watchers themselves?" John Adams said that we must have a "government of laws and not men." Bureau administrators, trying to evaluate the morality of acts in the total system, are singularly liable to corruption, producing a government by men, not laws.

Prohibition is easy to legislate (though not necessarily to enforce); but how do we legislate temperance? Experience indicates that it can be accomplished best through the mediation of administrative law. We limit possibilities unnecessarily if we suppose that the sentiment of *Quis custodiet* denies us the use of administrative law. We should rather retain the phrase as a perpetual reminder of fearful dangers we cannot avoid. The great challenge facing us now is to invent the corrective feedbacks that are needed to keep custodians honest. We must find ways to legitimate the needed authority of both the custodians and the corrective feedbacks.

FREEDOM TO BREED IS INTOLERABLE

The tragedy of the commons is involved in population problems in another way. In a world governed solely by the principle of "dog eat dog"—if indeed there ever was such a world—how many children a family had would not be a matter of public concern. Parents who bred too exuberantly would leave fewer descendants, not more, because they would be unable to care adequately for their children. David Lack and others have found that such a negative feedback demonstrably controls the fecundity of birds.[11] But men are not birds, and have not acted like them for millenniums, at least.

If each human family were dependent only on its own resources; if the children of improvident parents starved to death; if, thus, overbreeding brought its own "punishment" to the germ line—*then* there would be no public interest in controlling the breeding of families. But our society is deeply committed to the welfare state[12], and hence is confronted with another aspect of the tragedy of the commons.

In a welfare state, how shall we deal with the family, the religion, the race, or the class (or indeed any distinguishable and cohesive group) that adopts overbreeding as a policy to secure its own aggrandizement?[13] To couple the concept of freedom to breed with the belief that everyone born has an equal right to the commons is to lock the world into a tragic course of action.

Unfortunately this is just the course of action that is being pursued by the United Nations. In late 1967, some 30 nations agreed to the following[14]:

> The Universal Declaration of Human Rights describes the family as the natural and fundamental unit of society. It follows that any choice and decision with regard to the size of the family must irrevocably rest with the family itself, and cannot be made by anyone else.

It is painful to have to deny categorically the validity of this right; denying it, one feels as uncomfortable as a resident of Salem, Massachusetts, who denied the reality of witches in the 17th century. At the present time, in liberal quarters, something like a taboo acts to inhibit criticism of the United Nations. There is a feeling that the United Nations is "our last and best hope," that we shouldn't find fault with it; we shouldn't play into the hands of the archconservatives. However, let us not forget what Robert Louis Stevenson said: "The truth that is suppressed by friends is the readiest weapon of the enemy." If we love the truth we must openly deny the validity of the Universal Declaration of Human Rights, even though it is promoted by the United Nations. We should also join with Kingsley Davis[15] in attempting to get Planned Parenthood-World Population to see the error of its ways in embracing the same tragic ideal.

CONSCIENCE IS SELF-ELIMINATING

It is a mistake to think that we can control the breeding of mankind in the long run by an appeal to conscience. Charles Galton Darwin made this point when he spoke on the centennial of the publication of his grandfather's great book. The argument is straightforward and Darwinian.

People vary. Confronted with appeals to limit breeding, some people will undoubtedly respond to the plea more than others. Those who have more children will produce a larger fraction of the next generation than those with more susceptible consciences. The difference will be accentuated, generation by generation.

In C. G. Darwin's words: "It may well be that it would take hundreds of generations for the progenitive instinct to develop in this way, but if it should do so, nature would have taken her revenge, and the variety *Homo contracipiens* would become extinct and would be replaced by the variety *Homo progenitivus*".[16]

The argument assumes that conscience or the desire for children (no matter which) is hereditary—but hereditary only in the most general formal sense. The result will be the same whether the attitude is transmitted through germ cells, or exosomatically, to use A. J. Lotka's term. (If one denies the latter possibility as well as the former, then what's the point of education?) The argument has here been stated in the context of the population problem, but it applies equally well to any instance in which society appeals to an individual exploiting a commons to restrain himself for the general good—by means of his conscience. To make such an appeal is to set up a selective system that works toward the elimination of conscience from the race.

PATHOGENIC EFFECTS OF CONSCIENCE

The long-term disadvantage of an appeal to conscience should be enough to condemn it; but it has serious short-term disadvantages as well. If we ask a man who is exploiting a commons to desist "in the name of conscience," what are we saying to him? What does he hear?—not only at the moment but also in the wee small hours of the night when, half asleep, he remembers not merely the words we used but also the nonverbal communication cues we gave him unawares? Sooner or later, consciously or subconsciously, he senses that he has received two communications, and that they are contradictory: (i) (intended communication) "If you don't do as we ask, we will openly condemn you for not acting like a responsible citizen"; (ii) (the unintended communication) "If you do behave as we ask, we will secretly condemn you for a simpleton who can be shamed into standing aside while the rest of us exploit the commons."

Everyman then is caught in what Bateson has called a "double bind." Bateson and his co-workers have made a plausible case for viewing the double bind as an important causative factor in the genesis of schizophrenia.[17] The double bind may not always be so damaging, but it always endangers the mental health of anyone to whom it is applied. "A bad conscience," said Nietzsche, "is a kind of illness."

To conjure up a conscience in others is tempting to anyone who wishes to extend his control beyond the legal limits. Leaders at the highest level succumb to this temptation. Has any President during the past generation failed to call on labor unions to moderate voluntarily their demands for higher wages, or to steel companies to honor voluntary guidelines on prices? I can recall none. The rhetoric used on such occasions is designed to produce feelings of guilt in noncooperators.

For centuries it was assumed without proof that guilt was a valuable, perhaps even an indispensable, ingredient of the civilized life. Now, in this post-Freudian world, we doubt it.

Paul Goodman speaks from the modern point of view when he says: "No good has ever come from feeling guilty, neither intelligence, policy, nor compassion. The guilty do not pay attention to the object but only to themselves, and not even to their own interests, which might make sense, but to their anxieties".[18]

One does not have to be a professional psychiatrist to see the consequences of anxiety. We in the Western world are just emerging from a dreadful two-centuries-long Dark Ages of Eros that was sustained partly by prohibition laws, but perhaps more effectively by the anxiety-generating mechanisms of education. Alex Comfort has told the story well in *The Anxiety Makers*[19]; it is not a pretty one.

Since proof is difficult, we may even concede that the results of anxiety may sometimes, from certain points of view, be desirable. The larger question we should ask is whether, as a matter of policy, we should ever encourage the use of a technique the tendency (if not the intention) of which is psychologically pathogenic. We hear much talk these days of responsible parenthood; the coupled words are incorporated into the titles of some organizations devoted to birth control. Some people have proposed massive propaganda campaigns to instill responsibility into the nation's (or the world's) breeders. But what is the meaning of the word responsibility in this context? Is it not merely a synonym for the word conscience? When we use the word responsibility in the absence of substantial sanctions are we not trying to browbeat a free man in a commons into acting against his own interest? Responsibility is a verbal counterfeit for a substantial *quid pro quo.* It is an attempt to get something for nothing.

If the word responsibility is to be used at all, I suggest that it be in the sense Charles Frankel uses it.[20] "Responsibility," says this philosopher, "is the product of definite social arrangements." Notice that Frankel calls for social arrangements—not propaganda.

MUTUAL COERCION MUTUALLY AGREED UPON

The social arrangements that produce responsibility are arrangements that create coercion, of some sort. Consider bank-robbing. The man who takes money from a bank acts as if the bank were a commons. How do we prevent such action? Certainly not by trying to control his behavior solely by a verbal appeal to his sense of responsibility. Rather than rely on propaganda we follow Frankel's lead and insist that a bank is not a commons; we seek the definite social arrangements that will keep it from becoming a commons. That we thereby infringe on the freedom of would-be robbers we neither deny nor regret.

The morality of bank-robbing is particularly easy to understand because we accept complete prohibition of this activity. We are willing to say "Thou shalt not rob banks," without providing for exceptions. But temperance also can be created by coercion. Taxing is a good coercive device. To keep downtown shoppers temperate in their use of parking space we introduce parking meters for short periods, and traffic fines for longer ones. We need not actually forbid a citizen to park as long as he wants to; we need merely make it increasingly expensive for him to do so. Not prohibition, but carefully biased options are what we offer him. A Madison Avenue man might call this persuasion; I prefer the greater candor of the word coercion.

Coercion is a dirty word to most liberals now, but it need not forever be so. As with the four-letter words, its dirtiness can be cleansed away by exposure to the light, by saying it over and over without apology or embarrassment. To many, the word coercion implies arbitrary decisions of distant and irresponsible bureaucrats; but this is not a necessary part of its meaning. The only kind of coercion I recommend is mutual coercion, mutually agreed upon by the majority of the people affected.

To say that we mutually agree to coercion is not to say that we are required to enjoy it, or even to pretend we enjoy it. Who enjoys taxes? We all grumble about them. But we accept compulsory taxes because we recognize that voluntary taxes would favor the conscienceless. We institute and (grumblingly) support taxes and other coercive devices to escape the horror of the commons.

An alternative to the commons need not be perfectly just to be preferable. With real estate and other material goods, the alternative we have chosen is the institution of private property coupled with legal inheritance. Is this system perfectly just? As a genetically trained biologist I deny that it is. It seems to me that, if there are to be differences in individual inheritance, legal possession should be perfectly correlated with biological inheritance—that those who are biologically more fit to be the custodians of property and power should legally inherit more. But genetic recombination continually makes a mockery of the doctrine of "like father, like son" implicit in our laws of legal inheritance. An idiot can inherit millions, and a trust fund can keep his estate intact. We must admit that our legal system of private property plus inheritance is unjust—but we put up with it because we are not convinced, at the moment, that anyone has invented a better system. The alternative of the commons is too horrifying to contemplate. Injustice is preferable to total ruin.

It is one of the peculiarities of the warfare between reform and the status quo that it is thoughtlessly governed by a double standard. Whenever a reform measure is proposed it is often defeated when its opponents triumphantly discover a flaw in it. As Kingsley Davis has pointed out[21], worshipers of the status quo sometimes imply that no reform is possible without unanimous agreement, an implication contrary to historical fact. As nearly as I can make out, automatic rejection of proposed reforms is based on one of two unconscious assumptions: (i) that the status quo is perfect; or (ii) that the choice we face is between reform and no action; if the proposed reform is imperfect, we presumably should take no action at all, while we wait for a perfect proposal.

But we can never do nothing. That which we have done for thousands of years is also action. It also produces evils. Once we are aware that the status quo is action, we can then compare its discoverable advantages and disadvantages with the predicted advantages and disadvantages of the proposed reform, discounting as best we can for our lack of experience. On the basis of such a comparison, we can make a rational decision which will not involve the unworkable assumption that only perfect systems are tolerable.

RECOGNITION OF NECESSITY

Perhaps the simplest summary of this analysis of man's population problems is this: the commons, if justifiable at all, is justifiable only under conditions of low-population density. As the human population has increased, the commons has had to be abandoned in one aspect after another.

First we abandoned the commons in food gathering, enclosing farm land and restricting pastures and hunting and fishing areas. These restrictions are still not complete throughout the world.

Somewhat later we saw that the commons as a place for waste disposal would also have to be abandoned. Restrictions on the disposal of domestic sewage are widely accepted in the Western world; we are still struggling to close the commons to pollution by automobiles, factories, insecticide sprayers, fertilizing operations, and atomic energy installations.

In a still more embryonic state is our recognition of the evils of the commons in matters of pleasure. There is almost no restriction on the propagation of sound waves in the public medium. The shopping public is assaulted with mindless music, without its consent. Our government is paying out billions of dollars to create a supersonic transport which would disturb 50,000 people for every one person who is whisked from coast to coast 3 hours faster. Advertisers muddy the airwaves of radio and television and pollute the view of travelers. We are a long way from outlawing the commons in matters of pleasure. Is this because our Puritan inheritance makes us view pleasure as something of a sin, and pain (that is, the pollution of advertising) as the sign of virtue?

Every new enclosure of the commons involves the infringement of somebody's personal liberty. Infringements made in the distant past are accepted because no contemporary complains of a loss. It is the newly proposed infringements that we vigorously oppose; cries of "rights" and "freedom" fill the air. But what does "freedom" mean? When men mutually agreed to pass laws against robbing, mankind became more free, not less so. Individuals locked into the logic of the commons are free only to bring on universal ruin once they see the necessity of mutual coercion, they become free to pursue other goals. I believe it was Hegel who said, "Freedom is the recognition of necessity."

The most important aspect of necessity that we must now recognize, is the necessity of abandoning the commons in breeding. No technical solution can rescue us from the misery of overpopulation. Freedom to breed will bring ruin to all. At the moment, to avoid hard decisions many of us are tempted to propagandize for conscience and responsible parenthood. The temptation must be resisted, because an appeal to independently acting consciences selects for the disappearance of all conscience in the long run, and an increase in anxiety in the short.

The only way we can preserve and nurture other and more precious freedoms is by relinquishing the freedom to breed, and that very soon. "Freedom is the recognition of necessity"—and it is the role of education to reveal to all the necessity of abandoning the freedom to breed. Only so, can we put an end to this aspect of the tragedy of the commons.

ENDNOTES

1. J. B. Wiesner and H. F. York, *Sci. Amer.* 211 (No. 4), 27 (1964).

2. G. Hardin, *J. Hered.* 50, 68 (1959); S. von Hoernor, *Science* **137**, 18 (1962).

3. J. von Neumann and O. Morgenstern, *Theory of Games and Economic Behavior* (Princeton Univ. Press, Princeton, N.J., 1947), p. 11.

4. J. H. Fremlin, *New Sci.*, No. 415 (1964), p. 285.

5. A. Smith, *The Wealth of Nations* (Modern Library, New York, 1937), p. 423.

6. W. F. Lloyd, *Two Lectures on the Checks to Population* (Oxford Univ. Press, Oxford, England, 1833), reprinted (in part) in *Population, Evolution, and Birth Control*, G. Hardin, Ed. (Freeman, San Francisco, 1964), p. 37.

7. A. N. Whitehead, *Science and the Modern World* (Mentor, New York, 1948), p. 17.

8. G. Hardin, Ed. *Population, Evolution, and Birth Control* (Freeman, San Francisco, 1964), p. 56.

9. S. McVay, *Sci. Amer.* 216 (No. 8), 13 (1966).

10. J. Fletcher, *Situation Ethics* (Westminster, Philadelphia, 1966).

11. D. Lack, *The Natural Regulation of Animal Numbers* (Clarendon Press, Oxford, 1954).

12. H. Girvetz, *From Wealth to Welfare* (Stanford Univ. Press, Stanford, Calif., 1950).

13. G. Hardin, *Perspec. Biol. Med.* 6, 366 (1963).

14. U Thant, *Int. Planned Parenthood News*, No. 168 (February 1968), p. 3.

15. K. Davis, *Science* 158, 730 (1967).

16. S. Tax, Ed., *Evolution after Darwin* (Univ. of Chicago Press, Chicago, 1960), vol. 2, p. 469.

17. G. Bateson, D. D. Jackson, J. Haley, J. Weakland, *Behav. Sci.* 1, 251 (1956).

18. P. Goodman, *New York Rev. of Books* 10(8), 22 (23 May 1968).

19. A. Comfort, *The Anxiety Makers* (Nelson, London, 1967).

20. C. Frankel, *The Case for Modern Man* (Harper, New York, 1955), p. 203.

21. J. D. Roslansky, *Genetics and the Future of Man* (Appleton-Century-Crofts, New York, 1966), p. 177.

ECOLOGY AND MAN—A VIEWPOINT

SOURCE Shepard, Paul. 1969. "Ecology and Man—a Viewpoint." *The Subversive Science.* Paul Shepard and Daniel McKinley, eds. Boston: Houghton Mifflin. Reproduced by permission of the estate of the author.

INTRODUCTION *Paul Shepard (1925–1996) was Avery Professor of natural philosophy and human ecology at Pitzer College and Claremont Graduate School, where he taught for twenty-one years. In the introduction to his influential collection of essays,* The Subversive Science: Essays Toward an Ecology of Man *(co-edited with Daniel McKinley), Shepard begins by criticizing the "web of life" metaphor in ecology for failing to accurately reflect the complexity of the natural world. Shepard urges that we revive a more ancient way of perceiving humanity in the world. To do so entails redefining the self as something not distinct from nature, but extended into nature, where we overcome the dualism that has dominated much recent human history. For Shepard, humanity and the environment are interdependent; the complexity of humanity is directly linked to the complexity of natural systems.*

Shepard insists that this fact be highlighted in a time where the diversity and complexity of systems are diminishing. Part of the role of ecology, as Shepard describes it, is to realize that the "ecological crisis" is partly a matter of shifting values, and ecology must strive to communicate our interconnectedness with, and likeness to, nonhuman nature.

Ecology is sometimes characterized as the study of a natural "web of life." It would follow that man is somewhere in the web or that he in fact manipulates its strands, exemplifying what Thomas Huxley called "man's place in nature." But the image of a web is too meager and simple for the reality. A web is flat and finished and has the mortal frailty of the individual spider. Although elastic, it has insufficient depth. However solid to the touch of the spider, for us it fails to denote the *eikos*—the habitation—and to suggest the enduring integration of the primitive Greek domicile with its sacred hearth, bonding the earth to all aspects of society.

Ecology deals with organisms in an environment and with the processes that link organism and place. But ecology as such cannot be studied, only organisms, earth, air, and sea can be studied. It is not a discipline: there is no body of thought and technique which frames an ecology of man.[1] It must be therefore a scope or a way of seeing. Such a *perspective* on the human situation is very old and has been part of philosophy and art for thousands of years. It badly needs attention and revival.

Man is in the world and his ecology is the nature of that *inness*. He is in the world as in a room, and in transience, as in the belly of a tiger or in love. What does he do there in nature? What does nature do there *in him*? What is the nature of the transaction? Biology tells us that the transaction is always circular, always a mutual feedback. Human ecology cannot be limited strictly to biological concepts, but it cannot ignore them. It cannot even transcend them. It emerges from biological reality and grows from the fact of interconnection as a general principle of life. It must take a long view of human life and nature as they form a mesh or pattern going beyond historical time and beyond the conceptual bounds of other humane studies. As a natural history of what it means to be human, ecology might proceed the same way one would define a stomach, for example, by attention to its nervous and circulatory connections as well as its entrance, exit, and muscular walls.

Many educated people today believe that only what is unique to the individual is important or creative, and turn away from talk of populations and species as they would from talk of the masses. I once knew a director of a wealthy conservation foundation who had misgivings about the approach of ecology to urgent environmental problems in America because its concepts of communities and systems seemed to discount the individual. Communities to him suggested only followers, gray masses without the tradition of the individual. He looked instead—or in reaction—to the profit motive and capitalistic formulas, in terms of efficiency, investment, and production. It seemed to me that he had missed a singular opportunity. He had shied from the very aspect of the world now beginning to interest industry, business, and technology as the biological basis of their—and our—affluence, and which his foundation could have shown to be the ultimate basis of all economics.

Individual man *has* his particular integrity, to be sure. Oak trees, even mountains, have selves or integrities too (a poor word for my meaning, but it will have to do). To our knowledge, those other forms are not troubled by seeing themselves in more than one way, as man is. In one aspect the self is an arrangement of organs, feelings, and thoughts—a "me"—surrounded by a hard body boundary: skin, clothes, and insular habits. This idea needs no defense. It is conferred on us by the whole history of our civilization. Its virtue is verified by our affluence. The alternative is a self as a center of organization, constantly drawing on and influencing the surroundings, whose skin and behavior are soft zones contacting the world instead of excluding it. Both views are real and their reciprocity significant. We need them both to have a healthy social and human maturity.

The second view—that of relatedness of the self—has been given short shrift. Attitudes toward ourselves do not change easily. The conventional image of a man, like that of the heraldic lion, is iconographic; its outlines are stylized to fit the fixed curves of our vision. We are hidden from ourselves by habits of perception. Because we learn to talk at the same time we learn to think, our language, for example, encourages us to see ourselves—or a plant or animal—as an isolated sack, a thing, a contained self. Ecological thinking, on the other hand, requires a kind of vision across boundaries. The epidermis of the skin is ecologically like a pond surface or a forest soil, not a shell so much as a delicate interpenetration. It reveals the self enobled and extended rather than threatened as part of the landscape and the ecosystem, because the beauty and complexity of nature are continuous with ourselves.

And so ecology as applied to man faces the task of renewing a balanced view where now there is man-centeredness, even pathology of isolation and fear. It implies that we must find room in "our" world for all plants and animals, even for their otherness and their opposition. It further implies exploration and openness across an inner boundary—an ego boundary—and appreciative understanding of the animal in ourselves which our heritage of Platonism, Christian morbidity, duality, and mechanism have long held repellant and degrading. The older counter-currents—relics of pagan myth, the universal application of Christian compassion, philosophical naturalism, nature romanticism and pantheism—have been swept away, leaving only odd bits of wreckage. Now we find ourselves in a deteriorating environment which breeds aggressiveness and hostility toward ourselves and our world.

How simple our relationship to nature would be if we only had to choose between protecting our natural home and destroying it. Most of our efforts to provide for the natural in our philosophy have failed—run aground on their own determination to work out a peace at arm's length. Our harsh reaction against the peaceable kingdom of sentimental romanticism was evoked partly by the tone of its dulcet facade but also by the disillusion to which it led. Natural dependence and contingency suggest togetherness and emotional surrender to mass behavior and other lowest common denominators. The environmentalists matching culture and geography provoke outrage for their over-simple theories of cause and effect, against the sciences which sponsor them and even against a natural world in which the theories may or may not be true. Our historical disappointment in the nature of nature has created a cold climate for ecologists who assert once again that we are limited and obligated. Somehow they must manage in spite of the chill to reach the centers of humanism and technology, to convey there a sense of our place in a universal vascular system without depriving us of our self-esteem and confidence.

Their message is not, after all, all bad news. Our natural affiliations define and illumine freedom instead of denying it. They demonstrate it better than any dialectic. Being more enduring than we individuals, ecological patterns—spatial distributions, symbioses, the streams of energy and matter and communication—create among individuals the tensions and polarities so different from dichotomy and separateness. The responses, or what theologians call "the sensibilities" of creatures (including ourselves) to such arrangements grow in part from a healthy union of the two kinds of self already mentioned, one emphasizing integrity, the other relatedness. But it goes beyond that to something better known to 12th century Europeans or Paleolithic hunters than to ourselves. If nature is not a prison and earth a shoddy way-station, we must find the faith and force to affirm its metabolism as our own—or rather, our own as part of it. To do so means nothing less than a shift in our whole frame of reference and our attitude towards life itself, a wider perception of the landscape as a creative, harmonious being where relationships of things are as real as the things. Without losing our sense of a great human destiny and without intellectual surrender, we must affirm that the world is a being, a part of our own body.[2]

Such a being may be called an ecosystem or simply a forest or landscape. Its members are engaged in a kind of choreography of materials and energy and information, the creation of order and organization. (Analogy to corporate organization here is misleading, for the distinction between social (one species) and ecological (many species) is fundamental). The pond is an example. Its ecology includes all events: the conversion of sunlight to food and the food-chains within and around it, man drinking, bathing, fishing, plowing the slopes of the watershed, drawing a picture of it, and formulating theories about the world based on what he sees in the pond. He and all the other organisms at and in the pond act upon one another, engage the earth and atmosphere, and are linked to other ponds by a network of connections like the threads of protoplasm connecting cells in living tissues.

The elegance of such systems and delicacy of equilibrium are the outcome of a long evolution of interdependence. Even society, mind and culture are parts of that evolution. There is an essential relationship between them and the natural habitat: that is, between the emergence of higher primates and flowering plants, pollinating insects, seeds, humus, and arboreal life. It is unlikely that a manlike creature could arise by any other means than a long arboreal sojourn following and followed by a time of terrestriality. The fruit's complex construction and the mammalian brain are twin offspring of the maturing earth, impossible, even meaningless, without the deepening soil and the mutual development of savannas and their faunas in the last geological epoch. Internal complexity, as the mind of a primate, is an extension of natural complexity, measured by the variety of plants and animals and the variety of nerve cells—organic extensions of each other.

The exuberance of kinds as the setting in which a good mind could evolve (to deal with a complex world) was not only a past condition. Man did not arrive in the world as though disembarking from a train in the city. He continues to arrive, somewhat like the birth of art, a train in Roger Fry's definition, passing through many stations, none of which is wholly left behind. This idea of natural complexity as a counterpart to human intricacy is central to an ecology of man. The creation of order, of which man is an example, is realized also in the number of species and habitats, an abundance of landscapes lush and poor. Even deserts and tundras increase the planetary opulence. Curiously, only man and possibly a few birds can appreciate this opulence, being the world's travelers. Reduction of this variegation would, by extension then, be an amputation of man. To convert all "wastes"—all deserts, estuaries, tundras, ice-fields, marshes, steppes and moors—into cultivated fields and cities would impoverish rather than enrich life esthetically as well as ecologically. By esthetically, I do not mean that weasel term connoting the pleasure of baubles. We have diverted ourselves with litterbug campaigns and greenbelts in the name of esthetics while the fabric of our very environment is unravelling. In the name of conservation, too, such things are done, so that conservation becomes ambiguous. Nature is a fundamental "resource" to be sustained for our own well-being. But it loses in the translation into usable energy and commodities. Ecology may testify as often against our uses of the world, even against conservation techniques of control and management for sustained yield, as it does for them. Although ecology may be treated as a science, its greater and overriding wisdom is universal.

That wisdom can be approached mathematically, chemically, or it can be danced or told as a myth. It has been embodied in widely scattered economically different cultures. It is manifest, for example, among pre-Classical Greeks, in Navajo religion and social orientation, in Romantic poetry of the 18th and 19th centuries, in Chinese landscape painting of the 11th century, in current Whiteheadian philosophy, in Zen Buddhism, in the world view of the cult of the Cretan Great Mother, in the ceremonials of Bushman hunters, and in the medieval Christian metaphysics of light. What is common among all of them is a deep sense of engagement with the landscape, with profound connections to surroundings and to natural processes central to all life.

It is difficult in our language even to describe that sense. English becomes imprecise or mystical—and therefore suspicious—as it struggles with "process" thought. Its noun and verb organization shapes a divided world of static doers separate from the doing. It belongs to an idiom of social hierarchy in which all nature is made to mimic man. The living world is perceived in that idiom as an upright ladder, a "great chain of being," an image which seems at first ecological but is basically rigid, linear, condescending, lacking humility and love of otherness.

We are all familiar from childhood with its classifications of everything on a scale from the lowest to the highest: inanimate matter/vegetative life/lower animals/higher animals/men/angels/gods. It ranks animals themselves in categories of increasing good: the vicious and lowly parasites, pathogens and predators/the filthy decay and scavenging organisms/indifferent wild or merely useless forms/good tame creatures/and virtuous beasts domesticated for human service. It shadows the great man-centered political scheme upon the world, derived from the ordered ascendency from parishioners to clerics to bishops to cardinals to popes, or in a secular form from criminals to proletarians to aldermen to mayors to senators to presidents.

And so is nature pigeonholed. The sardonic phrase, "the place of nature in man's world," offers, tongue-in-cheek, a clever footing for confronting a world made in man's image and conforming to words. It satirizes the prevailing philosophy of anti-nature and human omniscience. It is possible because of an attitude which—like ecology—has ancient roots, but whose modern form was shaped when Aquinas reconciled Aristotelian homocentrism with Judeo-Christian dogma. In a later setting of machine technology, puritanical capitalism, and an urban ethos it carves its own version of reality into the landscape like a schoolboy initialing a tree. For such a philosophy nothing in nature has inherent merit. As one professor recently put it, "The only reason anything is done on this earth is for people. Did the rivers, winds, animals, rocks, or dust ever consider my wishes or needs? Surely, we do all our acts in an earthly environment, but I have never had a tree, valley, mountain, or flower thank me for preserving it."[3] This view carries great force, epitomized in history by Bacon, Descartes, Hegel, Hobbes, and Marx.

Some other post-Renaissance thinkers are wrongly accused of undermining our assurance of natural order. The theories of the heliocentric solar system, of biological evolution, and of the unconscious mind are held to have deprived the universe of the beneficence and purpose to which man was a special heir and to have evoked feelings of separation, of antipathy towards a meaningless existence in a neutral cosmos. Modern despair, the arts of anxiety, the politics of pathological individualism and predatory socialism were not, however, the results of Copernicus, Darwin and Freud. If man was not the center of the universe, was not created by a single stroke of Providence, and is not ruled solely by rational intelligence, it does not follow therefore that nature is defective where we thought it perfect. The astronomer, biologist and psychiatrist each achieved for mankind corrections in sensibility. Each showed the interpenetration of human life and the universe to be richer and more mysterious than had been thought.

Darwin's theory of evolution has been crucial to ecology. Indeed, it might have helped rather than aggravated the growing sense of human alienation had its interpreters emphasized predation and competition less (and, for this reason, one is tempted to add, had Thomas Huxley, Herbert Spencer, Samuel Butler and G. B. Shaw had less to say about it). Its bases of universal kinship and common bonds of function, experience and value among organisms were obscured by pre-existing ideas of animal depravity. Evolutionary theory was exploited to justify the worst in men and was misused in defense of social and economic injustice. Nor was it better used by humanitarians. They opposed the degradation of men in the service of industrial progress, the slaughter of American Indians, and child labor, because each treated men "like animals." That is to say, men were not animals, and the temper of social reform was to find good only in attributes separating men from animals. Kindness both towards and among animals was still a rare idea in the 19th century, so that using men as animals could mean only cruelty.

Since Thomas Huxley's day the non-animal forces have developed a more subtle dictum to the effect that, "Man may be an animal, but he is more than an animal, too!" The *more* is really what is important. This appealing aphorism is a kind of anesthetic. The truth is that we are ignorant of what it is like or what it means to be any other kind of creature than we are. If we are unable to truly define the animal's experience of life or "being an animal" how can we isolate our animal part?

The rejection of animality is a rejection of nature as a whole. As a teacher, I see students develop in their humanities studies a proper distrust of science and technology. What concerns me is that the stigma spreads to the natural world itself. C. P. Snow's "Two Cultures," setting the sciences against the humanities, can be misunderstood as placing nature against art. The idea that the current destruction of people and environment is scientific and would be corrected by more communication with the arts neglects the hatred for this world carried by our whole culture. Yet science as it is now taught does not promote a respect for nature. Western civilization breeds no more ecology in Western science than in Western philosophy. Snow's two cultures cannot explain the antithesis that splits the world, nor is the division ideological, economic or political in the strict sense. The antidote he proposes is roughly equivalent to a liberal education, the traditional prescription for making broad and well-rounded men. Unfortunately, there is little even in the liberal education of ecology-and-man. Nature is usually synonymous with either natural resources or scenery, the great stereotypes in the minds of middle class, college-educated Americans.

One might suppose that the study of biology would mitigate the humanistic—largely literary—confusion between materialism and a concern for nature. But biology made the mistake at the end of the 17th century of adopting a *modus operandi* or life style from physics, in which the question why was not to be asked, only the question how. Biology succumbed to its own image as an esoteric prologue to technics and encouraged the whole society to mistrust naturalists. When scholars realized what the sciences were about it is not surprising that they threw out the babies with the bathwater: the information content and naturalistic lore with the rest of it. This is the setting in which academia and intellectual America undertook the single-minded pursuit of human uniqueness, and uncovered a great mass of pseudo distinctions such as language, tradition, culture, love, consciousness, history and awe of the supernatural. Only men were found to be capable of escape from predictability, determinism, environmental control, instincts and other mechanisms which "imprison" other life. Even biologists, such as Julian Huxley, announced that the purpose of the world was to produce man, whose social evolution excused him forever from biological evolution. Such a view incorporated three important presumptions: that nature is a power structure shaped after human political hierarchies; that man has a monopoly of immortal souls; and omnipotence will come through technology. It seems to me that all of these foster a failure of responsible behavior in what Paul Sears calls "the living landscape" except within the limits of immediate self-interest.

What ecology must communicate to the humanities—indeed, as a humanity—is that such an image of the world and the society so conceived are incomplete. There is overwhelming evidence of likeness, from molecular to mental, between men and animals. But the dispersal of this information is not necessarily a solution. The Two Culture idea that the problem is an information bottleneck is only partly true; advances in biochemistry, genetics, ethology, paleoanthropology, comparative physiology and psychobiology are not self-evidently unifying. They need a unifying principle not found in any of them, a wisdom in the sense that Walter B. Cannon used the word in his book *Wisdom of the Body*,[4] about the community of self-regulating systems within the organism. If the ecological extension of that perspective is correct, societies and ecosystems as well as cells have a physiology, and insight into it is built into organisms, including man. What was intuitively apparent last year—whether aesthetically or romantically—is a find of this year's inductive analysis. It seems apparent to me that there is an ecological instinct which probes deeper and more comprehensively than science, and which anticipates every scientific confirmation of the natural history of man.

It is not surprising, therefore, to find substantial ecological insight in art. Of course there is nothing wrong with a poem or dance which is ecologically neutral; its merit may have nothing to do with the transaction of man and nature. It is my impression, however, that students of the arts no longer feel that the subject of a work of art—what it "represents"—is without importance, as was said about 40 years ago. But there are poems and dances as there are prayers and laws attending to ecology. Some are more than mere comments on it. Such creations become part of all life. Essays on nature are an element of a functional or feedback system influencing men's reactions to their environment, messages projected by men to themselves through some act of design, the manipulation of paints or written words. They are natural objects, like bird nests. The essay is as real a part of the community—in both the one-species sociological and many-species ecological senses—as are the songs of choirs or crickets. An essay is an Orphic sound, words that make knowing possible, for it was Orpheus as Adam who named and thus made intelligible all creatures.

What is the conflict of Two Cultures if it is not between science and art or between national ideologies? The distinction rather divides science and art within themselves. An example within science was the controversy over the atmospheric testing of nuclear bombs and the effect of radioactive fallout from the explosions. Opposing views were widely published and personified when Linus Pauling, a biochemist, and Edward Teller, a physicist, disagreed. Teller, one of the "fathers" of the bomb, pictured the fallout as a small factor in a world-wide struggle, the possible damage to life in tiny fractions of a percent, and even noted that evolutionary progress comes from mutations. Pauling, an expert on the hereditary material, knowing that most mutations are detrimental, argued that a large absolute number of people might be injured, as well as other life in the world's biosphere.

The humanness of ecology is that the dilemma of our emerging world ecological crises (over-population, environmental pollution, etc.) is at least in part a matter of values and ideas. It does not divide men as much by their trades as by the complex of personality and experience shaping their feelings towards other people and the world at large. I have mentioned the disillusion generated by the collapse of unsound nature philosophies. The anti-nature position today is often associated with the focusing of general fears and hostilities on the natural world. It can be seen in the behavior of control-obsessed engineers, corporation people selling consumption itself, academic superhumanists and media professionals fixated on political and economic crisis; neurotics working out psychic problems in the realm of power over men or nature, artistic symbol-manipulators disgusted by anything organic. It includes many normal, earnest people who are unconsciously defending themselves or their families against a vaguely threatening universe. The dangerous eruption of humanity in a deteriorating environment does not show itself as such in the daily experience of most people, but is felt as general tension and anxiety. We feel the pressure of events not as direct causes but more like omens. A kind of madness arises from the prevailing nature-conquering, nature-hating and self- and world-denial. Although in many ways most Americans live comfortable, satiated lives, there is a nameless frustration born of an increasing nullity. The aseptic home and society are progressively cut off from direct organic sources of health and increasingly isolated from the means of altering the course of events. Success, where its price is the misuse of landscapes, the deterioration of air and water and the loss of wild things, becomes a pointless glut, experience one-sided, time on our hands an unlocalized ache.

The unrest can be exploited to perpetuate itself. One familiar prescription for our sick society and its loss of environmental equilibrium is an increase in the intangible Good Things: more Culture, more Security and more Escape from pressures and tempo. The "search for identity" is not only a social but an ecological problem having to do with a sense of place and time in the context of all life. The pain of that search can be cleverly manipulated to keep the *status quo* by urging that what we need is only improved forms and more energetic expressions of what now occupy us: engrossment with ideological struggle and military power, with productivity and consumption as public and private goals, with commerce and urban growth, with amusements, with fixation on one's navel, with those tokens of escape or success already belabored by so many idealists and social critics so ineffectually.

To come back to those Good Things: the need for culture, security and escape are just near enough to the truth to take us in. But the real cultural deficiency is the absence of a true *cultus* with its significant ceremony, relevant mythical cosmos, and artifacts. The real failure in security is the disappearance from our personal lives of the small human group as the functional unit of society and the web of other creatures, domestic and wild, which are part of our humanity. As for escape, the idea of simple remission and avoidance fails to provide for the value of solitude, to integrate leisure and natural encounter. Instead of these, what are foisted on the puzzled and troubled soul as Culture, Security and Escape are more art museums, more psychiatry, and more automobiles.

The ideological status of ecology is that of a resistance movement. Its Rachel Carsons and Aldo Leopolds are subversive (as Sears recently called ecology itself[5]). They challenge the public or private right to pollute the environment, to systematically destroy predatory animals, to spread chemical pesticides indiscriminately, to meddle chemically with food and water, to appropriate without hindrance space and surface for technological and military ends; they oppose the uninhibited growth of human populations, some forms of "aid" to "underdeveloped" peoples, the needless addition of radioactivity to the landscape, the extinction of species of plants and animals, the domestication of all wild places, large-scale manipulation of the atmosphere or the sea, and most other purely engineering solutions to problems of and intrusions into the organic world.

If naturalists seem always to be *against* something it is because they feel a responsibility to share their understanding, and their opposition constitutes a defense of the natural systems to which man is committed as an organic being. Sometimes naturalists propose projects too, but the project approach is itself partly the fault, the need for projects a consequence of linear, compartmental thinking, of machine-like units to be controlled and manipulated. If the ecological crisis were merely a matter of alternative techniques, the issue would belong among the technicians and developers (where most schools and departments of conservation have put it).

Truly ecological thinking need not be incompatible with our place and time. It does have an element of humility which is foreign to our thought, which moves us to silent wonder and glad affirmation. But it offers an essential factor, like a necessary vitamin, to all our engineering and social planning, to our poetry and our understanding. There is only one ecology, not a human ecology on one hand and another for the subhuman. No one school or theory or project or agency controls it. For us it means seeing the world mosaic from the human vantage without being man-fanatic. We must use it to confront the great philosophical problems of man—transience, meaning, and limitation—without fear. Affirmation of its own organic essence will be the ultimate test of the human mind.

ENDNOTES

1. There is a branch of sociology called Human Ecology, but it is mostly about urban geography.

2. See Alan Watts, "The World is Your Body," in *The Book on the Taboo Against Knowing Who You Are*. New York: Pantheon Books, 1966.

3. Clare A. Gunn in *Landscape Architecture*, July 1966, p. 260.

4. New York: W. W. Norton, 1932.

5. Paul B. Sears, "Ecology—a subversive subject," *BioScience*, 14(7):11, July 1964.

SHOULD TREES HAVE STANDING?—TOWARD LEGAL RIGHTS FOR NATURAL OBJECTS

SOURCE *Stone, Christopher D. 1972. *Southern California Law Review* 45: 450–457. Reproduced by permission.

INTRODUCTION *In 2008 Christopher D. Stone (1937–) was J. Thomas McCarthy Trustee Chair in law at the University of Southern California. "Should Trees Have Standing?" is a highly influential essay that weaves together environmental philosophy and law. Stone begins by stating that a parallel phenomenon to moral extensionism is observable in the history of the law—the class of rights bearers has grown through history. Once unthinkable, inanimate things such as trusts and corporations are now possessors of rights. Stone argues for giving legal rights to natural objects, and the natural environment as a whole. He qualifies what it might mean for the environment to be a holder of rights, advocating a "guardianship approach" whereby an appointed guardian represents the interests of natural objects themselves, not as their degradation impacts human interests, and where monetary remedies would run to the benefit of the natural objects directly. Stone also argues that a change in our "environmental consciousness" is necessary, but may not be sufficient to address environmental problems. Nonetheless, he proposes that the law may be instrumental in creating a "new theory or myth" for humanity's relationship to nature, and that the courts, by conferring rights on nature, might help bring about a necessary shift in popular consciousness.*

INTRODUCTION: THE UNTHINKABLE

In *Descent of Man*, Darwin observes that the history of man's moral development has been a continual extension in the objects of his "social instincts and sympathies." Originally each man had regard only for himself and those of a very narrow circle about him; later, he came to regard more and more "not only the welfare, but the happiness of all his fellowmen"; then "his sympathies became more tender and widely diffused, extending to men of all races, to the imbecile, maimed, and other useless members of society, and finally to the lower animals...."[1]

The history of the law suggests a parallel development. Perhaps there never was a pure Hobbesian state of nature, in which no "rights" existed except in the vacant sense of each man's "right to self-defense." But it is not unlikely that so far as the earliest "families" (including extended kinship groups and clans) were concerned, everyone outside the family was suspect, alien, rightless.[2] And even within the family, persons we presently regard as the natural holders of at least some rights had none. Take, for example, children. We know something of the early rights-status of children from the widespread practice of infanticide—especially of the deformed and female.[3] (Senicide,[4] as among the North American Indians, was the corresponding rightlessness of the aged).[5] Maine tells us that as late as the Patria Potestas of the Romans, the father had *jus vitae necisque*—the power of life and death—over his children. A fortiori, Maine writes, he had power of "uncontrolled corporal chastisement; he can modify their personal condition at pleasure; he can give a wife to his son; he can give his daughter in marriage; he can divorce his children of either sex; he can transfer them to another family by adoption; and he can sell them." The child was less than a person: an object, a thing.[6]

The legal rights of children have long since been recognized in principle, and are still expanding in practice. Witness, just within recent time, *In re Gault*,[7] guaranteeing basic constitutional protections to juvenile defendants, and the Voting Rights Act of 1970.[8] We have been making persons of children although they were not, in law, always so. And we have done the same, albeit imperfectly some would say, with prisoners,[9] aliens, women (especially of the married variety), the insane,[10] Blacks, foetuses,[11] and Indians.

Nor is it only matter in human form that has come to be recognized as the possessor of rights. The world of the lawyer is peopled with inanimate right-holders: trusts, corporations, joint ventures, municipalities, Subchapter R partnerships,[12] and nation-states, to mention just a few. Ships, still referred to by courts in the feminine gender, have long had an independent jural life, often with striking consequences.[13] We have become so accustomed to the idea of a corporation having "its" own rights, and being a "person" and "citizen" for so many statutory and constitutional purposes, that we forget how jarring the notion was to early jurists. "That invisible, intangible and artificial being, that mere legal entity" Chief Justice Marshall wrote of the corporation in *Bank of the United States v. Deveaux*[14]—could a suit be brought in its name? Ten years later, in the *Dartmouth College* case,[15] he was still refusing to let pass unnoticed the wonder of an entity "existing only in contemplation of law."[16] Yet, long before Marshall worried over the personifying of the modern corporation, the best medieval legal scholars had spent hundreds of years struggling with the notion of the legal nature of those great public "corporate bodies," the Church and the State. How could they exist in law, as entities transcending the living Pope and King? It was clear how a king could bind *himself*—on his honor—by a treaty. But when the king died, what was it that was burdened with the obligations of, and claimed the rights under, the treaty *his* tangible hand had signed? The medieval mind saw (what we have lost our capacity to see)[17] how *unthinkable* it was, and worked out the most elaborate conceits and fallacies to serve as anthropomorphic flesh for the Universal Church and the Universal Empire.[18]

It is this note of the *unthinkable* that I want to dwell upon for a moment. Throughout legal history, each successive extension of rights to some new entity has been, theretofore, a bit unthinkable. We are inclined to suppose the rightlessness of rightless "things" to be a decree of Nature, not a legal convention acting in support of some status quo. It is thus that we defer considering the choices involved in all their moral, social, and economic dimensions. And so the United States Supreme Court could straight-facedly tell us in *Dred Scott* that Blacks had been denied the rights of citizenship "as a subordinate and inferior class of beings, who had been subjugated by the dominant race...."[19] In the nineteenth century, the highest court in California explained that Chinese had not the right to testify against white men in criminal matters because they were "a race of people whom nature has marked as inferior, and who are incapable of progress or intellectual development beyond a certain point ... between whom and ourselves nature has placed an impassable difference.[20] The popular conception of the Jew in the 13th Century contributed to a law which treated them as "men *ferae naturae*, protected by a quasi-forest law. Like the roe and the deer, they form an order apart."[21] Recall, too, that it was not so long ago that the foetus was "like the roe and the deer." In an early suit attempting to establish a wrongful death action on behalf of a negligently killed foetus (now widely accepted practice), Holmes, then on the Massachusetts Supreme Court, seems to have thought it simply inconceivable "that a man might owe a civil duty and incur a conditional prospective liability in tort to one not yet in being."[22] The first woman in Wisconsin who thought she might have a right to practice law was told that she did not, in the following terms:

> The law of nature destines and qualifies the female sex for the bearing and nurture of the children of our race and for the custody of the homes of the world .... [A]ll life-long callings of women, inconsistent with these radical and sacred duties of their sex, as is the profession of the law, are departures from the order of nature; and when voluntary, treason against it .... The peculiar qualities of womanhood, its gentle graces, its quick sensibility, its tender susceptibility, its purity, its delicacy, its emotional impulses, its subordination of hard reason to sympathetic feeling, are surely not qualifications for forensic strife. Nature has tempered woman as little for the juridical conflicts of the court room, as for the physical conflicts of the battle field ....[23]

The fact is, that each time there is a movement to confer rights onto some new "entity," the proposal is bound to sound odd or frightening or laughable.[23a] This is partly because until the rightless thing receives its rights, we cannot see it as anything but a *thing* for the use of "us"—those who are holding rights at the time.[24] In this vein, what is striking about the Wisconsin case above is that the court, for all its talk about women, so clearly was never able to see women as they are (and might become). All it could see was the popular "idealized" version of *an object it needed.* Such is the way the slave South looked upon the Black.[25] There is something of a seamless web involved: there will be resistance to giving the thing "rights" until it can be seen and valued for itself; yet, it is hard to see it and value it for itself until we can bring ourselves to give it "rights"—which is almost inevitably going to sound inconceivable to a large group of people.

The reason for this little discourse on the unthinkable, the reader must know by now, if only from the title of the paper. I am quite seriously proposing that we give legal rights to forests, oceans, rivers and other so-called "natural objects" in the environment—indeed, to the natural environment as a whole.[26]

As strange as such a notion may sound, it is neither fanciful nor devoid of operational content. In fact, I do not think it would be a misdescription of recent developments in the law to say that we are already on the verge of assigning some such rights, although we have not faced up to what we are doing in those particular terms.[27] We should do so now, and begin to explore the implications such a notion would hold.

TOWARD RIGHTS FOR THE ENVIRONMENT

Now, to say that the natural environment should have rights is not to say anything as silly as that no one should be allowed to cut down a tree. We say human beings have rights, but—at least as of the time of this writing—they can be executed.[28] Corporations have rights, but they cannot plead the fifth amendment[29]; *In re Gault* gave 15-year-olds certain rights in juvenile proceedings, but it did not give them the right to vote. Thus, to say that the environment should have rights is not to say that it should have every right we can imagine, or even the same body of rights as human beings have. Nor is it to say that everything in the environment should have the same rights as every other thing in the environment.

What the granting of rights does involve has two sides to it. The first involves what might be called the legal-operational aspects; the second, the psychic and socio-psychic aspects. I shall deal with these aspects in turn.

THE LEGAL-OPERATIONAL ASPECTS

What it Means to be a Holder of Legal Rights

There is, so far as I know, no generally accepted standard for how one ought to use the term "legal rights." Let me indicate how I shall be using it in this piece.

First and most obviously, if the term is to have any content at all, an entity cannot be said to hold a legal right unless and until *some public authoritative body* is prepared to give *some amount of review* to actions that are colorably inconsistent with that "right." For example, if a student can be expelled from a university and cannot get any public official, even a judge or administrative agent at the lowest level, either (i) to require the university to justify its actions (if only to the extent of filling out an affidavit alleging that the expulsion "was not wholly arbitrary and capricious") or (ii) to compel the university to accord the student some procedural safeguards (a hearing, right to counsel, right to have notice of charges), then the minimum requirements for saying that the student has a legal right to his education do not exist.[30]

But for a thing to be *a holder of legal rights*, something more is needed than that some authoritative body will review the actions and processes of those who threaten it. As I shall use the term, "holder of legal rights," each of three additional criteria must be satisfied. All three, one will observe, go towards making a thing *count* jurally—to have a legally recognized worth and dignity in its own right, and not merely to serve as a means to benefit "us" (whoever the contemporary group of rights-holders may be). They are, first, that the thing can institute legal actions *at its behest*; second, that in determining the granting of legal relief, the court must take *injury to it* into account; and, third, that relief must run to the *benefit* of it.

To illustrate, even as between two societies that condone slavery there is a fundamental difference between S_1, in which a master can (if he chooses), go to court and collect reduced chattel value damages from someone who has beaten his slave, and S_2, in which the slave can institute the proceedings *himself*, for *his* own recovery, damages being measured by, say, *his* pain and suffering. Notice that neither society is so structured as to leave wholly unprotected the slave's interests in not being beaten. But in S_2 as opposed to S_1 there are three operationally significant advantages that the slave has, and these make the slave in S_2, albeit a slave, a holder of rights. Or, again, compare two societies, S_1, in which pre-natal injury to a live-born child gives a right of action against the tortfeasor at the mother's instance, for the mother's benefit, on the basis of the mother's mental anguish, and S_2, which gives the child a suit in its own name (through a guardian *ad litem*) for its own recovery, for damages to it.

When I say, then, that at common law "natural objects" are not holders of legal rights, I am not simply remarking what we would all accept as obvious. I mean to emphasize three specific legal-operational advantages that the environment lacks, leaving it in the position of the slave and the foetus in S_1, rather than the slave and foetus of S_2.

The Rightlessness of Natural Objects at Common Law

Consider, for example, the common law's posture toward the pollution of a stream. True, courts have always been able, in some circumstances, to issue orders that will stop the pollution—just as the legal system in S_1 is so structured as incidentally to discourage beating slaves and being reckless around pregnant women. But the stream itself is fundamentally rightless, with implications that deserve careful reconsideration.

The first sense in which the stream is not a rights-holder has to do with standing. The stream itself has none. So far as the common law is concerned, there is in general no way to challenge the polluter's actions save at the behest of a lower riparian—another human being—able to show an invasion of *his* rights. This conception of the riparian as the holder of the right to bring suit has more than theoretical interest. The lower riparians may simply not care about the pollution. They themselves may be polluting, and not wish to stir up legal waters. They may be economically dependent on their polluting neighbor.[31] And, of course, when they discount the value of winning by the costs of bringing suit and the chances of success, the action may not seem worth undertaking. Consider, for example, that while the polluter might be injuring 100 downstream riparians $10,000 a year *in the aggregate*, each riparian separately might be suffering injury only to the extent of $100—possibly not enough for any one of them to want to press suit by himself, or even to go to the trouble and cost of securing co-plaintiffs to make it worth everyone's while. This hesitance will be especially likely when the potential plaintiffs consider the burdens the law puts in their way:[32] proving, *e.g.*, specific damages, the "unreasonableness" of defendant's use of the water, the fact that practicable means of abatement exist, and overcoming difficulties raised by issues such as joint causality, right to pollute by prescription, and so forth. Even in states which, like California, sought to overcome these difficulties by empowering the attorney-general to sue for abatement of pollution in limited instances, the power has been sparingly invoked and, when invoked, narrowly construed by the courts.[33]

The second sense in which the common law denies "rights" to natural objects has to do with the way in which the merits are decided in those cases in which someone is competent and willing to establish standing. At its more primitive levels, the system protected the "rights" of the property owning human with minimal weighing of any values: "*Cujus est solum, ejus est usque ad coelum et ad infernos.*"[34] Today we have come more and more to make balances—but only such as will adjust the economic best interests of identifiable humans. For example, continuing with the case of streams, there are commentators who speak of a "general rule" that "a riparian owner is legally entitled to have the stream flow by his land with its quality unimpaired" and observe that "an upper owner has, prima facie, no right to pollute the water."[35] Such a doctrine, if strictly invoked, would protect the stream absolutely whenever a suit was brought; but obviously, to look around us, the law does not work that way. Almost everywhere there are doctrinal qualifications on riparian "rights" to an unpolluted stream.[36] Although these rules vary from jurisdiction to jurisdiction, and upon whether one is suing for an equitable injunction or for damages, what they all have in common is some sort of balancing. Whether under language of "reasonable use," "reasonable methods of use," "balance of convenience" or "the public interest doctrine,"[37] what the courts are balancing, with varying degrees of directness, are the economic hardships on the upper riparian (or dependent community) of abating the pollution vis-à-vis the economic hardships of continued pollution on the lower riparians. What does not weigh in the balance is the damage to the stream, its fish and turtles and "lower" life. So long as the natural environment itself is rightless, these are not matters for judicial cognizance. Thus, we find the highest court of Pennsylvania refusing to stop a coal company from discharging polluted mine water into a tributary of the Lackawana River because a plaintiff's "grievance is for a mere personal inconvenience; and ... mere private personal inconveniences ... must yield to the necessities of a great public industry, which although in the hands of a private corporation, subserves a great public interest."[38] The stream itself is lost sight of in "a quantitative compromise between *two* conflicting interests."[39]

The third way in which the common law makes natural objects rightless has to do with who is regarded as the beneficiary of a favorable judgment. Here, too, it makes a considerable difference that it is not the natural object that counts in its own right. To illustrate this point, let me begin by observing that it makes perfectly good sense to speak of, and ascertain, the legal damage to a natural object, if only in the sense of "making it whole" with respect to the most obvious factors.[40] The costs of making a forest whole, for example, would include the costs of reseeding, repairing watersheds, restocking wildlife—the sorts of costs the Forest Service undergoes after a fire. Making a polluted stream whole would include the costs of restocking with fish, water-fowl, and other animal and vegetable life, dredging, washing out impurities, establishing natural and/or artificial aerating agents, and so forth. Now, what is important to note is that, under our present system, even if a plaintiff riparian wins a water pollution suit for damages, no money goes to the benefit of the stream itself to repair *its* damages.[41] This omission has the further effect that, at most, the law confronts a polluter with what it takes to make the plaintiff riparians whole; this may be far less than the damages to the stream,[42] but not so much as to force the polluter to desist. For example, it is easy to imagine a polluter whose activities damage a stream to the extent of $10,000 annually, although the aggregate damage to all the riparian plaintiffs who come into the suit is only $3000. If $3000 is less than the cost to the polluter of shutting down, or making the requisite technological changes, he might prefer to pay off the damages (*i.e.*, the legally cognizable damages) and continue to pollute the stream. Similarly, even if the jurisdiction issues an injunction at the plaintiffs' behest (rather than to order payment of damages), there is nothing to stop the plaintiffs from "selling out" the stream, *i.e.*, agreeing to dissolve or not enforce the injunction at some price (in the example above, somewhere between plaintiffs' damages—$3000—and defendant's next best economic alternative). Indeed, I take it this is exactly what Learned Hand had in mind in an opinion in which, after issuing an anti-pollution injunction, he suggests that the defendant "make its peace with the plaintiff as best it can."[43] What is meant is a peace between *them*, and not amongst them and the river.

I ought to make clear at this point that the common law as it affects streams and rivers, which I have been using as an example so far, is not exactly the same as the law affecting other environmental objects. Indeed, one would be hard pressed to say that there was a "typical" environmental object, so far as its treatment at the hands of the law is concerned. There are some differences in the law applicable to all the various resources that are held in common: rivers, lakes, oceans, dunes, air, streams (surface and subterranean), beaches, and so forth.[44] And there is an even greater difference as between these traditional communal resources on the one hand, and natural objects on traditionally private land, *e.g.*, the pond on the farmer's field, or the stand of trees on the suburbanite's lawn.

On the other hand, although there be these differences which would make it fatuous to generalize about a law of the natural environment, most of these differences simply underscore the points made in the instance of rivers and streams. None of the natural objects, whether held in common or situated on private land, has any of the three criteria of a rights-holder. They have no standing in their own right; their unique damages do not count in determining outcome; and they are not the beneficiaries of awards. In such fashion, these objects have traditionally been regarded by the common law, and even by all but the most recent legislation, as objects for man to conquer and master and use—in such a way as the law once looked upon "man's" relationships to African Negroes. Even where special measures have been taken to conserve them, as by seasons on game and limits on timber cutting, the dominant motive has been to conserve them *for us*—for the greatest good of the greatest number of human beings. Conservationists, so far as I am aware, are generally reluctant to maintain otherwise.[45] As the name implies, they want to conserve and guarantee *our* consumption and *our* enjoyment of these other living things. In their own right, natural objects have counted for little, in law as in popular movements.

As I mentioned at the outset, however, the rightlessness of the natural environment can and should change; it already shows some signs of doing so.

Toward Having Standing in its Own Right

It is not inevitable, nor is it wise, that natural objects should have no rights to seek redress in their own behalf. It is no answer to say that streams and forests cannot have standing because streams and forests cannot speak. Corporations cannot speak either; nor can states, estates, infants, incompetents, muncipalities or universities. Lawyers speak for them, as they customarily do for the ordinary citizen with legal problems. One ought, I think, to handle the legal problems of natural objects as one does the problems of legal incompetents—human beings who have become vegetable. If a human being shows signs of becoming senile and has affairs that he is de jure incompetent to manage, those concerned with his well being make such a showing to the court, and someone is designated by the court with the authority to manage the incompetent's affairs. The guardian[46] (or "conservator"[47] or "committee"[48]—the terminology varies) then represents the incompetent in his legal affairs. Courts make similar appointments when a corporation has become "incompetent"—they appoint a trustee in bankruptcy or reorganization to oversee its affairs and speak for it in court when that becomes necessary.

On a parity of reasoning, we should have a system in which, when a friend of a natural object perceives it to be endangered, he can apply to a court for the creation of a guardianship.[49] Perhaps we already have the machinery to do so. California law, for example, defines an incompetent as "any person, whether insane or not, who by reason of old age, disease, weakness of mind, or other cause, is unable, unassisted, properly to manage and take care of himself or his property, and by reason thereof is likely to be deceived or imposed upon by artful or designing persons."[50] Of course, to urge a court that an endangered river is "a person" under this provision will call for lawyers as bold and imaginative as those who convinced the Supreme Court that a railroad corporation was a "person" under the fourteenth amendment, a constitutional provision theretofore generally thought of as designed to secure the rights of freedmen.[51] (As this article was going to press, Professor Byrn of Fordham petitioned the New York Supreme Court to appoint him legal guardian for an unrelated foetus scheduled for abortion so as to enable him to bring a class action on behalf of all foetuses similarly situated in New York City's 18 municipal hospitals. Judge Holtzman granted the petition of guardianship.[52]) If such an argument based on present statutes should fail, special environmental legislation could be enacted along traditional guardianship lines. Such provisions could provide for guardianship both in the instance of public natural objects and also, perhaps with slightly different standards, in the instance of natural objects on "private" land.[53]

The potential "friends" that such a statutory scheme would require will hardly be lacking. The Sierra Club, Environmental Defense Fund, Friends of the Earth, Natural Resources Defense Counsel, and the Izaak Walton League are just some of the many groups which have manifested unflagging dedication to the environment and which are becoming increasingly capable of marshalling the requisite technical experts and lawyers. If, for example, the Environmental Defense Fund should have reason to believe that some company's strip mining operations might be irreparably destroying the ecological balance of large tracts of land, it could, under this procedure, apply to the court in which the lands were situated to be appointed guardian.[54] As guardian, it might be given rights of inspection (or visitation) to determine and bring to the court's attention a fuller finding on the land's condition. If there were indications that under the substantive law some redress might be available on the land's behalf, then the guardian would be entitled to raise the land's rights in the land's name, *i.e.*, without having to make the roundabout and often unavailing demonstration, discussed below, that the "rights" of the club's members were being invaded. Guardians would also be looked to for a host of other protective tasks, *e.g.*, monitoring effluents (and/or monitoring the monitors), and representing their "wards" at legislative and administrative hearings on such matters as the setting of state water quality standards. Procedures exist, and can be strengthened, to move a court for the removal and substitution of guardians, for conflicts of interest or for other reasons,[55] as well as for the termination of the guardianship.[56]

In point of fact, there is a movement in the law toward giving the environment the benefits of standing, although not in a manner as satisfactory as the guardianship approach. What I am referring to is the marked liberalization of traditional standing requirements in recent cases in which environmental action groups have challenged federal government action. *Scenic Hudson Preservation Conference v. FPC*[57] is a good example of this development. There, the Federal Power Commission had granted New York's Consolidated Edison a license to construct a hydroelectric project on the Hudson River at Storm King Mountain. The grant of license had been opposed by conservation interests on the grounds that the transmission lines would be unsightly, fish would be destroyed, and nature trails would be inundated. Two of these conservation groups, united under the name Scenic Hudson Preservation Conference, petitioned the Second Circuit to set aside the grant. Despite the claim that Scenic Hudson had no standing because it had not made the traditional claim "of any personal economic injury resulting from the Commission's actions,"[58] the petitions were heard, and the case sent back to the Commission. On the standing point, the court noted that Section 313(b) of the Federal Power Act gave a right of instituting review to any party "aggrieved by an order issued by the Commission"[59]; it thereupon read "aggrieved by" as not limited to those alleging the traditional personal economic injury, but as broad enough to include "those who by their activities and conduct have exhibited a special interest" in "the aesthetic, conservational, and recreational aspects of power development...."[60] A similar reasoning has swayed other circuits to allow proposed actions by the Federal Power Commission, the Department of Interior, and the Department of Health, Education and Welfare to be challenged by environmental action groups on the basis of, *e.g.*, recreational and esthetic interests of members, in lieu of direct economic injury.[61] Only the Ninth Circuit has balked, and one of these cases, involving the Sierra Club's attempt to challenge a Walt Disney development in the Sequoia National Forest, is at the time of this writing awaiting decision by the United States Supreme Court.[62]

Even if the Supreme Court should reverse the Ninth Circuit in the Walt Disney-Sequoia National Forest matter, thereby encouraging the circuits to continue their trend toward liberalized standing in this area, there are significant reasons to press for the guardianship approach notwithstanding. For one thing, the cases of this sort have extended standing on the basis of interpretations of specific federal statutes—the Federal Power Commission Act,[63] the Administrative Procedure Act,[64] the Federal Insecticide, Fungicide and Rodenticide Act,[65] and others. Such a basis supports environmental suits only where acts of federal agencies are involved; and even there, perhaps, only when there is some special statutory language, such as "aggrieved by" in the Federal Power Act, on which the action groups can rely. Witness, for example, *Bass Angler Sportsman Society v. United States Steel Corp.*[66] There, plaintiffs sued 175 corporate defendants located throughout Alabama, relying on 33 U.S.C. § 407 (1970), which provides:

> It shall not be lawful to throw, discharge, or deposit ... any refuse matter ... into any navigable water of the United States, or into any tributary of any navigable water from which the same shall float or be washed into such navigable water...[67]

Another section of the Act provides that one-half the fines shall be paid to the person or persons giving information which shall lead to a conviction.[68] Relying on this latter provision, the plaintiff designated his action a *qui tam* action[69] and sought to enforce the Act by injunction and fine. The District Court ruled that, in the absence of express language to the contrary, no one outside the Department of Justice had standing to sue under a criminal act and refused to reach the question of whether violations were occurring.[70]

Unlike the liberalized standing approach, the guardianship approach would secure an effective voice for the environment even where federal administrative action and public-lands and waters were not involved. It would also allay one of the fears courts—such as the Ninth Circuit—have about the extended standing concept: if any ad hoc group can spring up overnight, invoke some "right" as universally claimable as the esthetic and recreational interests of its members and thereby get into court, how can a flood of litigation be prevented?[71] If an ad hoc committee loses a suit brought *sub nom.* Committee to Preserve our Trees, what happens when its very same members reorganize two years later and sue *sub nom.* the Massapequa Sylvan Protection League? Is the new group bound by res judicata? Class action law may be capable of ameliorating some of the more obvious problems. But even so, court economy might be better served by simply designating the guardian de jure representative of the natural object, with rights of discretionary intervention by others, but with the understanding that the natural object is "bound" by an adverse judgment.[72] The guardian concept, too, would provide the endangered natural object with what the trustee in bankruptcy provides the endangered corporation: a continuous supervision over a period of time, with a consequent deeper understanding of a broad range of the ward's problems, not just the problems present in one particular piece of litigation. It would thus assure the courts that the plaintiff has the expertise and genuine adversity in pressing a claim which are the prerequisites of a true "case or controversy."

The guardianship approach, however, is apt to raise two objections, neither of which seems to me to have much force. The first is that a committee or guardian could not judge the needs of the river or forest in its charge; indeed, the very concept of "needs," it might be said, could be used here only in the most metaphorical way. The second objection is that such a system would not be much different from what we now have: is not the Department of Interior already such a guardian for public lands, and do not most states have legislation empowering their attorneys general to seek relief—in a sort of *parens patriae* way—for such injuries as a guardian might concern himself with?

As for the first objection, natural objects *can* communicate their wants (needs) to us, and in ways that are not terribly ambiguous. I am sure I can judge with more certainty and meaningfulness whether and when my lawn wants (needs) water, than the Attorney General can judge whether and when the United States wants (needs) to take an appeal from an adverse judgment by a lower court. The lawn tells me that it wants water by a certain dryness of the blades and soil—immediately obvious to the touch—the appearance of bald spots, yellowing, and a lack of springiness after being walked on; how does "the United States" communicate to the Attorney General? For similar reasons, the guardian-attorney for a smog-endangered stand of pines could venture with more confidence that his client wants the smog stopped, than the directors of a corporation can assert that "the corporation" wants dividends declared. We make decisions on behalf of, and in the purported interests of, others every day; these "others" are often creatures whose wants are far less verifiable, and even far more metaphysical in conception, than the wants of rivers, trees, and land.[73]

As for the second objection, one can indeed find evidence that the Department of Interior was conceived as a sort of guardian of the public lands.[74] But there are two points to keep in mind. First, insofar as the Department already is an adequate guardian it is only with respect to the federal public lands as per Article IV, section 3 of the Constitution.[75] Its guardianship includes neither local public lands nor private lands. Second, to judge from the environmentalist literature and from the cases environmental action groups have been bringing, the Department is itself one of the bogeys of the environmental movement. (One thinks of the uneasy peace between the Indians and the Bureau of Indian Affairs.) Whether the various charges be right or wrong, one cannot help but observe that the Department has been charged with several institutional goals (never an easy burden), and is currently looked to for action by quite a variety of interest groups, only one of which is the environmentalists. In this context, a guardian outside the institution becomes especially valuable. Besides, what a person wants, fully to secure his rights, is the ability to retain independent counsel even when, and perhaps especially when, the government is acting "for him" in a beneficent way. I have no reason to doubt, for example, that the Social Security System is being managed "for me"; but I would not want to abdicate my right to challenge its actions as they affect me, should the need arise.[76] I would not ask more trust of national forests, vis-à-vis the Department of Interior. The same considerations apply in the instance of local agencies, such as regional water pollution boards, whose members' expertise in pollution matters is often all too credible.[77]

The objection regarding the availability of attorneys-general as protectors of the environment within the existing structure is somewhat the same. Their statutory powers are limited and sometimes unclear. As political creatures, they must exercise the discretion they have with an eye toward advancing and reconciling a broad variety of important social goals, from preserving morality to increasing their jurisdiction's tax base. The present state of our environment, and the history of cautious application and development of environmental protection laws long on the books,[78] testifies that the burdens of an attorney-general's broad responsibility have apparently not left much manpower for the protection of nature. (*Cf. Bass Anglers*, above.) No doubt, strengthening interest in the environment will increase the zest of public attorneys even where, as will often be the case, well-represented corporate polluters are the quarry. Indeed, the United States Attorney General has stepped up anti-pollution activity, and ought to be further encouraged in this direction.[79] The statutory powers of the attorneys-general should be enlarged, and they should be armed with criminal penalties made at least commensurate with the likely economic benefits of violating the law.[80] On the other hand, one cannot ignore the fact that there is increased pressure on public law-enforcement offices to give more attention to a host of other problems, from crime "on the streets" (why don't we say "in the rivers"?) to consumerism and school bussing. If the environment is not to get lost in the shuffle, we would do well, I think, to adopt the guardianship approach as an additional safeguard, conceptualizing major natural objects as holders of their own rights, raisable by the court-appointed guardian.

Toward Recognition of its Own Injuries

As far as adjudicating the merits of a controversy is concerned, there is also a good case to be made for taking into account harm to the environment—in its own right. As indicated above, the traditional way of deciding whether to issue injunctions in law suits affecting the environment, at least where communal property is involved, has been to strike some sort of balance regarding the economic hardships *on human beings*. Even recently, Mr. Justice Douglas, our jurist most closely associated with conservation sympathies in his private life, was deciding the propriety of a new dam on the basis of, among other things, anticipated lost profits from fish catches, some $12,000,000 annually.[81] Although he decided to delay the project pending further findings, the reasoning seems unnecessarily incomplete and compromising. Why should the environment be of importance only indirectly, as lost profits to someone else? Why not throw into the balance the cost *to the environment*?

The argument for "personifying" the environment, from the point of damage calculations, can best be demonstrated from the welfare economics position. Every well-working legal-economic system should be so structured as to confront each of us with the full costs that our activities are imposing on society.[82] Ideally, a paper-mill, in deciding what to produce—and where, and by what methods—ought to be forced to take into account not only the lumber, acid and labor that its production "takes" from other uses in the society, but also what costs alternative production plans will impose on society through pollution. The legal system, through the law of contracts and the criminal law, for example, makes the mill confront the costs of the first group of demands. When, for example, the company's purchasing agent orders 1000 drums of acid from the *Z* Company, the *Z* Company can bind the mill to pay for them, and thereby reimburse the society for what the mill is removing from alternative uses.

Unfortunately, so far as the pollution costs are concerned, the allocative ideal begins to break down, because the traditional legal institutions have a more difficult time "catching" and confronting us with the full social costs of our activities. In the lakeside mill example, major riparian interests might bring an action, forcing a court to weigh *their* aggregate losses against the costs to the mill of installing the anti-pollution device. But many other interests—and I am speaking for the moment of recognized homocentric interests—are too fragmented and perhaps "too remote" causally to warrant securing representation and pressing for recovery: the people who own summer homes and motels, the man who sells fishing tackle and bait, the man who rents rowboats. There is no reason not to allow the lake to prove damages to them as the prima facie measure of damages to it. *By doing so, we in effect make the natural object, through its guardian, a jural entity competent to gather up these fragmented and otherwise unrepresented damage claims, and press them before the court even where, for legal or practical reasons, they are not going to be pressed by traditional class action plaintiffs.*[83] Indeed, one way—the homocentric way—to view what I am proposing so far, is to view the guardian of the natural object as the guardian of unborn generations, as well as of the otherwise unrepresented, but distantly injured, contemporary humans.[84] By making the lake itself the focus of these damages, and "incorporating" it so to speak, the legal system can effectively take proof upon, and confront the mill with, a larger and more representative measure of the damages its pollution causes.

So far, I do not suppose that my economist friends (unremittent human chauvanists, every one of them!) will have any large quarrel in principle with the concept. Many will view it as a *trompe l'oeil* that comes down, at best, to effectuate the goals of the paragon class action, or the paragon water pollution control district. Where we are apt to part company is here—I propose going beyond gathering up the loose ends of what most people would presently recognize as economically valid damages. The guardian would urge before the court injuries not presently cognizable—the death of eagles and inedible crabs, the suffering of sea lions, the loss from the face of the earth of species of commercially valueless birds, the disappearance of a wilderness area. One might, of course, speak of the damages involved as "damages" to us humans, and indeed, the widespread growth of environmental groups shows that human beings do feel these losses. But they are not, at present, economically measurable losses: how can they have a monetary value for the guardian to prove in court?

The answer for me is simple. Wherever it carves out "property" rights, the legal system is engaged in the process of *creating* monetary worth. One's literary works would have minimal monetary value if anyone could copy them at will. Their economic value to the author is a product of the law of copyright; the person who copies a copyrighted book has to bear a cost to the copyright-holder because the law says he must. Similarly, it is through the law of torts that we have made a "right" of—and guaranteed an economically meaningful value to—privacy. (The value we place on gold—a yellow inanimate dirt—is not simply a function of supply and demand—wilderness areas are scarce and pretty too—, but results from the actions of the legal systems of the world, which have institutionalized that value; they have even done a remarkable job of stabilizing the price). I am proposing we do the same with eagles and wilderness areas as we do with copyrighted works, patented inventions, and privacy: *make* the violation of rights in them to be a cost by declaring the "pirating" of them to be the invasion of a property interest.[85] If we do so, the net social costs the polluter would be confronted with would include not only the extended homocentric costs of his pollution (explained above) but also costs to the environment *per se.*

How, though, would these costs be calculated? When we protect an invention, we can at least speak of a fair market value for it, by reference to which damages can be computed. But the lost environmental "values" of which we are now speaking are by definition over and above those that the market is prepared to bid for: they are priceless.

One possible measure of damages, suggested earlier, would be the cost of making the environment whole, just as, when a man is injured in an automobile accident, we impose upon the responsible party the injured man's medical expenses. Comparable expenses to a polluted river would be the costs of dredging, restocking with fish, and so forth. It is on the basis of such costs as these, I assume, that we get the figure of $1 billion as the cost of saving Lake Erie.[86] As an ideal, I think this is a good guide applicable in many environmental situations. It is by no means free from difficulties, however.

One problem with computing damages on the basis of making the environment whole is that, if understood most literally, it is tantamount to asking for a "freeze" on environmental quality, even at the costs (and there will be costs) of preserving "useless" objects.[87] Such a "freeze" is not inconceivable to me as a general goal, especially considering that, even by the most immediately discernible homocentric interests, in so many areas we ought to be cleaning up and not merely preserving the environmental status quo. In fact, there is presently strong sentiment in the Congress for a total elimination of all river pollutants by 1985,[88] notwithstanding that such a decision would impose quite large direct and indirect costs on us all. Here one is inclined to recall the instructions of Judge Hays, in remanding Consolidated Edison's Storm King application to the Federal Power Commission in *Scenic Hudson*:

> The Commission's renewed proceedings must include as a basic concern the preservation of natural beauty and of natural historic shrines, keeping in mind that, in our affluent society, the cost of a project is only one of several factors to be considered.[89]

Nevertheless, whatever the merits of such a goal in principle, there are many cases in which the social price tag of putting it into effect are going to seem too high to accept. Consider, for example, an oceanside nuclear generator that could produce low cost electricity for a million homes at a savings of $1 a year per home, spare us the air pollution that comes of burning fossil fuels, but which through a slight heating effect threatened to kill off a rare species of temperature-sensitive sea urchins; suppose further that technological improvements adequate to reduce the temperature to present environmental quality would expend the entire one million dollars in anticipated fuel savings. Are we prepared to tax ourselves $1,000,000 a year on behalf of the sea urchins? In comparable problems under the present law of damages, we work out practicable compromises by abandoning restoration costs and calling upon fair market value. For example, if an automobile is so severely damaged that the cost of bringing the car to its original state by repair is greater than the fair market value, we would allow the responsible tortfeasor to pay the fair market value only. Or if a human being suffers the loss of an arm (as we might conceive of the ocean having irreparably lost the sea urchins), we can fall back on the capitalization of reduced earning power (and pain and suffering) to measure the damages. But what is the fair market value of sea urchins? How can we capitalize their loss to the ocean, independent of any commercial value they may have to someone else?

One answer is that the problem can sometimes be sidestepped quite satisfactorily. In the sea urchin example, one compromise solution would be to impose on the nuclear generator the costs of making the ocean whole somewhere else, in some other way, *e.g.*, reestablishing a sea urchin colony elsewhere, or making a somehow comparable contribution.[90] In the debate over the laying of the trans-Alaskan pipeline, the builders are apparently prepared to meet conservationists' objections half-way by re-establishing wildlife away from the pipeline, so far as is feasible.[91]

But even if damage calculations have to be made, one ought to recognize that the measurement of damages is rarely a simple report of economic facts about "the market," whether we are valuing the loss of a foot, a foetus, or a work of fine art. Decisions of this sort are always hard, but not impossible. We have increasingly taken (human) pain and suffering into account in reckoning damages, not because we think we can ascertain them as objective "facts" about the universe, but because, even in view of all the room for disagreement, we come up with a better society by making rude estimates of them than by ignoring them.[92] We can make such estimates in regard to environmental losses fully aware that what we are really doing is making implicit normative judgments (as with pain and suffering)—laying down rules as to what the society is going to "value" rather than reporting market evaluations. In making such normative estimates decision-makers would not go wrong if they estimated on the "high side," putting the burden of trimming the figure down on the immediate human interests present. All burdens of proof should reflect common experience; our experience in environmental matters has been a continual discovery that our acts have caused more long-range damage than we were able to appreciate at the outset.

To what extent the decision-maker should factor in costs such as the pain and suffering of animals and other sentient natural objects, I cannot say; although I am prepared to do so in principle.[93] Given the conjectural nature of the "estimates" in all events, and the roughness of the "balance of conveniences" procedure where that is involved, the practice would be of more interest from the socio-psychic point of view, discussed below, than from the legal-operational.

Toward Being a Beneficiary in its Own Right

As suggested above, one reason for making the environment itself the beneficiary of a judgment is to prevent it from being "sold out" in a negotiation among private litigants who agree not to enforce rights that have been established among themselves.[94] Protection from this will be advanced by making the natural object a party to an injunctive settlement. Even more importantly, we should make it a beneficiary of money awards. If, in making the balance requisite to issuing an injunction, a court decides *not* to enjoin a lake polluter who is causing injury to the extent of $50,000 annually, then the owners and the lake ought both to be awarded damages. The natural object's portion could be put into a trust fund to be administered by the object's guardian, as per the guardianship recommendation set forth above. So far as the damages are proved, as suggested in the previous section, by allowing the natural object to cumulate damages to others as prima facie evidence of damages to it, there will, of course, be problems of distribution. But even if the object is simply construed as representing a class of plaintiffs under the applicable civil rules,[95] there is often likely to be a sizeable amount of recovery attributable to members of the class who will not put in a claim for distribution (because their pro rata share would be so small, or because of their interest in the environment). Not only should damages go into these funds, but where criminal fines are applied (as against water polluters) it seems to me that the monies (less prosecutorial expenses, perhaps) ought sensibly to go to the fund rather than to the general treasuries. Guardians fees, including legal fees, would then come out of this fund. More importantly, the fund would be available to preserve the natural object as close as possible to its condition at the time the environment was made a rights-holder.[96]

The idea of assessing damages as best we can and placing them in a trust fund is far more realistic than a hope that a total "freeze" can be put on the environmental status quo. Nature is a continuous theatre in which things and species (eventually man) are destined to enter and exit.[97] In the meantime, co-existence of man and his environment means that *each* is going to have to compromise for the better of both. Some pollution of streams, for example, will probably be inevitable for some time. Instead of setting an unrealizable goal of enjoining absolutely the discharge of all such pollutants, the trust fund concept would (a) help assure that pollution would occur only in those instances where the social need for the pollutant's product (via his present method of production) was so high as to enable the polluter to cover *all* homocentric costs, plus some estimated costs to the environment *per se*, and (b) would be a corpus for preserving monies, if necessary, while the technology developed to a point where repairing the damaged portion of the environment was feasible. Such a fund might even finance the requisite research and development.

(Incidentally, if "rights" are to be granted to the environment, then for many of the same reasons it might bear "liabilities" as well—as inanimate objects did anciently.[98] Rivers drown people, and flood over and destroy crops; forests burn, setting fire to contiguous communities. Where trust funds had been established, they could be available for the satisfaction of judgments against the environment, making it bear the costs of some of the harms it imposes on other right holders. In effect, we would be narrowing the claim of Acts of God. The ontological problem would be troublesome here, however; when the Nile overflows, is it the "responsibility" of the river? the mountains? the snow? the hydrologic cycle?[99])

Toward Rights in Substance

So far we have been looking at the characteristics of being a *holder of rights*, and exploring some of the implications that making the environment a holder of rights would entail. Natural objects would have standing in their own right, through a guardian; damage to and through them would be ascertained and considered as an independent factor; and they would be the beneficiaries of legal awards. But these considerations only give us the skeleton of what a meaningful rights-holding would involve. To flesh out the "rights" of the environment demands that we provide it with a significant body of rights for it to invoke when it gets to court.

In this regard, the lawyer is constantly aware that a right is not, as the layman may think, some strange substance that one either has or has not. One's life, one's right to vote, one's property, can all be taken away. But those who would infringe on them must go through certain procedures to do so; these procedures are a measure of what we value as a society. Some of the most important questions of "right" thus turn into questions of degree: how much review, and of which sort, will which agencies of state accord us when we claim our "right" is being infringed?

We do not have an absolute right either to our lives or to our driver's licenses. But we have a greater right to our lives because, if even the state wants to deprive us of that "right," there are authoritative bodies that will demand that the state make a very strong showing before it does so, and it will have to justify its actions before a grand jury, petit jury (convincing them "beyond a reasonable doubt"), sentencing jury, and, most likely, levels of appellate courts. The carving out of students "rights" to their education is being made up of this sort of procedural fabric. No one, I think, is maintaining that in no circumstances ought a student to be expelled from school. The battle for student "rights" involves shifting the answers to questions like: before a student is expelled, does he have to be given a hearing; does he have to have prior notice of the hearing, and notice of charges; may he bring counsel, (need the state provide counsel if he cannot?); need there be a transcript; need the school carry the burden of proving the charges; may he confront witnesses; if he is expelled, can he get review by a civil court; if he can get such review, need the school show its actions were "reasonable," or merely "not unreasonable," and so forth?[100]

In this vein, to bring the environment into the society as a rights-holder would not stand it on a better footing than the rest of us mere mortals, who every day suffer injuries that are *damnum absque injuria*. What the environment must look for is that its interests be taken into account in subtler, more procedural ways.

The National Environmental Policy Act is a splendid example of this sort of rights-making through the elaboration of procedural safeguards. Among its many provisions, it establishes that every federal agency must:

> (C) include in every recommendation or report on proposals for legislation and other major Federal actions significantly affecting the quality of the human environment, a detailed statement by the responsible official on—
>
> (i) the environmental impact of the proposed action,
>
> (ii) any adverse environmental effects which cannot be avoided should the proposal be implemented,
>
> (iii) alternatives to the proposed action,
>
> (iv) the relationship between local short-term uses of man's environment and the maintenance and enhancement of long-term productivity, and
>
> (v) any irreversible and irretrievable commitments of resources which would be involved in the proposed action should it be implemented.

Prior to making any detailed statement, the responsible Federal official shall consult with and obtain the comments of any Federal agency which has jurisdiction by law or special expertise with respect to any environmental impact involved. Copies of such statement and the comments and views of the appropriate Federal, State, and local agencies, which are authorized to develop and enforce environmental standards, shall be made available to the President, the Council on Environmental Quality and to the public as provided by section 552 of title 5, United States Code, and shall accompany the proposal through the existing agency review processes;

> (D) study, develop, and describe appropriate alternatives to recommended courses of action in any proposal which involves unresolved conflicts concerning alternative uses of available resources;
>
> (E) recognize the worldwide and long-range character of environmental problems and, where consistent with the foreign policy of the United States, lend appropriate support to initiatives, resolutions, and programs designed to maximize international cooperation in anticipating and preventing a decline in the quality of mankind's environment;
>
> (F) make available to States, counties, municipalities, institutions, and individuals, advice and information useful in restoring, maintaining, and enhancing the quality of the environment ....[101]

These procedural protections have already begun paying off in the courts. For example, it was on the basis of the Federal Power Commission's failure to make adequate inquiry into "alternatives" (as per subsection (iii)) in *Scenic Hudson*, and the Atomic Energy Commission's failure to make adequate findings, apparently as per subsections (i) and (ii), in connection with the Amchitka Island underground test explosion,[102] that Federal Courts delayed the implementation of environment-threatening schemes.

Although this sort of control (remanding a cause to an agency for further findings) may seem to the layman ineffectual, or only a stalling of the inevitable, the lawyer and the systems analyst know that these demands for further findings can make a difference. It may encourage the institution whose actions threaten the environment to really *think about* what it is doing, and that is neither an ineffectual nor a small feat. Indeed, I would extend the principle beyond federal agencies. Much of the environment is threatened not by them, but by private corporations. Surely the constitutional power would not be lacking to mandate that all private corporations whose actions may have significant adverse affect on the environment make findings of the sort now mandated for federal agencies. Further, there should be requirements that these findings and reports be channeled to the Board of Directors; if the directors are not charged with the knowledge of what their corporation is doing to the environment, it will be all too easy for lower level management to prevent such reports from getting to a policy-making level. We might make it grounds for a guardian to enjoin a private corporation's actions if such procedures had not been carried out.

The rights of the environment could be enlarged by borrowing yet another page from the Environmental Protection Act and mandating comparable provisions for "private governments." The Act sets up within the Executive Office of the President a Council on Environmental Quality "to be conscious of and responsive to the scientific, economic, social, esthetic, and cultural needs of the Nation; and to formulate and recommend national policies to promote the improvement of the quality of the environment."[103] The Council is to become a focal point, within our biggest "corporation"—the State—to gather and evaluate environmental information which it is to pass on to our chief executive officer, the President. Rather than being ineffectual, this may be a highly sophisticated way of steering organizational behavior. Corporations—especially recidivist polluters and land despoilers—should have to establish comparable internal reorganization, *e.g.*, to set up a Vice-President for Ecological Affairs. The author is not offering this suggestion as a cure-all, by any means, but I do not doubt that this sort of control over internal corporate organization would be an effective supplement to the traditional mechanisms of civil suits, licensing, administrative agencies, and fines.[104]

Similarly, courts, in making rulings that may affect the environment, should be compelled to make findings with respect to environmental harm—showing how they calculated it and how heavily it was weighed—even in matters outside the present Environmental Protection Act. This would have at least two important consequences. First, it would shift somewhat the focus of court-room testimony and concern; second, the appellate courts, through their review and reversals for "insufficient findings," would give content to, and build up a body of, environmental rights, much as content and body has been given, over the years, to terms like "Due Process of Law."

Beyond these procedural safeguards, would there be any rights of the environment that might be deemed "absolute," at least to the extent of, say, Free Speech? Here, the doctrine of irreparable injury comes to mind. There has long been equitable support for an attorney-general's enjoining injury to communal property if he can prove it to be "irreparable." In other words, while repairable damage to the environment might be balanced and weighed, irreparable damage could be enjoined absolutely. There are several reasons why this doctrine has not been used effectively (witness Lake Erie).[105] Undoubtedly, political pressures (in the broadest sense) have had an influence. So, too, has the failure of all of us to understand just how delicate the environmental balance is; this failure has made us unaware of how early "irreparable" injury might be occurring, and, if aware, unable to prove it in court. But most important I think, is that the doctrine simply is not practical as a rule of universal application. For one thing, there are too many cases like the sea urchin example above, where the marginal costs of abating the damage seem too clearly to exceed the marginal benefits, even if the damage to the environment itself is liberally estimated. For another, there is a large problem in how one defines "irreparable." Certainly the great bulk of the environment in civilized parts of the world has been injured "irreparably" in the sense of "irreversably"; we are not likely to return it to its medieval quality. Despite the scientific ring to the term, judgments concerning "irreparable injury" are going to have to subsume questions both of degree of damage and of value—to all of "us" including the environment, *i.e.*, to "spaceship earth"—of the damaged object. Thus, if we are going to revitalize the "irreparable damages" doctrine, and expect it to be taken seriously, we have to recognize that what will be said to constitute "irreparable damage" to the ionosphere, because of its importance to all life, or to the Grand Canyon, because of its uniqueness, is going to rest upon normative judgments that ought to be made explicit.

This suggests that some (relatively) absolute rights be defined for the environment by setting up a constitutional list of "preferred objects," just as some of our Justices feel there are "preferred rights" where humans are concerned.[106] Any threatened injury to these most jealously-to-be-protected objects should be reviewed with the highest level of scrutiny at all levels of government, including our "counter-majoritarian" branch, the court system. Their "Constitutional rights" should be implemented, legislatively and administratively, by, *e.g.*, the setting of environmental quality standards.

I do not doubt that other senses in which the environment might have rights will come to mind, and, as I explain more fully below, would be more apt to come to mind if only we should speak in terms of their having rights, albeit vaguely at first. "Rights" might well lie in unanticipated areas. It would seem, for example, that Chief Justice Warren was only stating the obvious when he observed in *Reynolds v. Sims* that "legislators represent people, not trees or acres." Yet, could not a case be made for a system of apportionment which *did* take into account the wildlife of an area?[107] It strikes me as a poor idea that Alaska should have no more congressmen than Rhode Island primarily *because there are in Alaska all those trees and acres, those waterfalls and forests.*[108] I am not saying anything as silly as that we ought to overrule *Baker v. Carr* and retreat from one man-one vote to a system of one man-or-tree one vote. Nor am I even taking the position that we ought to count each acre, as we once counted each slave, as three-fifths of a man. But I am suggesting that there is nothing unthinkable about, and there might on balance even be a prevailing case to be made for, an electoral apportionment that made some systematic effort to allow for the representative "rights" of non-human life. And if a case can be made for that, which I offer here mainly for purpose of illustration, I suspect that a society that grew concerned enough about the environment to make it a holder of rights would be able to find quite a number of "rights" to have waiting for it when it got to court.

Do We Really Have to Put it that Way?

At this point, one might well ask whether much of what has been written could not have been expressed without introducing the notion of trees, rivers, and so forth "having rights." One could simply and straightforwardly say, for example, that (R_1) "the class of persons competent to challenge the pollution of rivers ought to be extended beyond that of persons who can show an immediate adverse economic impact on themselves," and that (R_2), "judges, in weighing competing claims to a wilderness area, ought to think beyond the economic and even esthetic impact on man, and put into the balance a concern for the threatened environment as such." And it is true, indeed, that to say trees and rivers have "rights" is not in itself a stroke of any operational significance—no more than to say "people have rights." To solve any concrete case, one is always forced to more precise and particularized statements, in which the word "right" might just as well be dropped from the elocution.

But this is not the same as to suggest that introducing the notion of the "rights" of trees and rivers would accomplish nothing beyond the introduction of a set of particular rules like (R_1) and (R_2), above. I think it is quite misleading to say that "*A* has a right to ..." can be fully explicated in terms of a certain set of specific legal rules, and the manner in which conclusions are drawn from them in a legal system. That is only part of the truth. Introducing the notion of something having a "right" (simply *speaking* that way), brings into the legal system a flexibility and open-endedness that no series of specifically stated legal rules like R_1, R_2, R_3, ... R_n can capture. Part of the reason is that "right" (and other so-called "legal terms" like "infant," "corporation," "reasonable time") have meaning—vague but forceful—in the ordinary language, and the force of these meanings, inevitably infused with our thought, becomes part of the context against which the "legal language" of our contemporary "legal rules" is interpreted.[109] Consider, for example, the "rules" that govern the question, on whom, and at what stages of litigation, is the burden of proof going to lie? Professor Krier has demonstrated how terribly significant these decisions are in the trial of environmental cases, and yet, also, how much discretion judges have under them.[110] In the case of such vague rules, it is *context*—senses of direction, of value and purpose—that determines how the rules will be understood, every bit as much as their supposed "plain meaning." In a system which spoke of the environment "having legal rights," judges would, I suspect, be inclined to interpret rules such as those of burden of proof far more liberally from the point of the environment. There is, too, the fact that the vocabulary and expressions that are available to us influence and even steer our thought. Consider the effect that was had by introducing into the law terms like "motive," "intent," and "due process." These terms work a subtle shift into the rhetoric of explanation available to judges; with them, new ways of thinking and new insights come to be explored and developed.[111] In such fashion, judges who could unabashedly refer to the "legal rights of the environment" would be encouraged to develop a viable body of law—in part simply through the availability and force of the expression. Besides, such a manner of speaking by courts would contribute to popular notions, and a society that spoke of the "legal rights of the environment" would be inclined to legislate more environment-protecting rules by formal enactment.

If my sense of these influences is correct, then a society in which it is stated, however vaguely, that "rivers have legal rights" would evolve a different legal system than one which did not employ that expression, even if the two of them had, at the start, the very same "legal rules" in other respects.

THE PSYCHIC AND SOCIO-PSYCHIC ASPECTS

There are, as we have seen, a number of developments in the law that may reflect a shift from the view that nature exists *for men*. These range from increasingly favorable procedural rulings for environmental action groups—as regards standing and burden of proof requirements, for example—to the enactment of comprehensive legislation such as the National Environmental Policy Act and the thoughtful Michigan Environmental Protection Act of 1970. Of such developments one may say, however, that it is not the environment *per se* that we are prepared to take into account, but that man's increased awareness of possible long range effects on himself militate in the direction of stopping environmental harm in its incipiency. And this is part of the truth, of course. Even the far-reaching National Environmental Policy Act, in its preambulatory "Declaration of National Environmental Policy," comes out both for "restoring and maintaining environmental quality *to the overall welfare and development of man*" as well as for creating and maintaining "conditions under which *man and nature can exist in productive harmony*."[112] Because the health and well-being of mankind depend upon the health of the environment, these goals will often be so mutually supportive that one can avoid deciding whether our rationale is to advance "us" or a new "us" that includes the environment. For example, consider the Federal Insecticide, Fungicide, and Rodenticide Act (FIFRA) which insists that, *e.g.*, pesticides, include a warning "adequate to prevent injury to living man and other vertebrate animals, vegetation, and useful invertebrate animals."[113] Such a provision undoubtedly reflects the sensible notion that the protection of humans is best accomplished by preventing dangerous accumulations in the food chain. Its enactment does not necessarily augur far-reaching changes in, nor even call into question, fundamental matters of consciousness.

But the time is already upon us when we may have to consider subordinating some human claims to those of the environment *per se*. Consider, for example, the disputes over protecting wilderness areas from development that would make them accessible to greater numbers of people. I myself feel disingenuous rationalizing the environmental protectionist's position in terms of a utilitarian calculus, even one that takes future generations into account, and plays fast and loose with its definition of "good." Those who favor development have the stronger argument—they at least hold the protectionist to a standstill—from the point of advancing the greatest good of the greatest number of people. And the same is true regarding arguments to preserve useless species of animals, as in the sea urchin hypothetical. One *can* say that we never know what is going to prove useful at some future time. In order to protect ourselves, therefore, we ought to be conservative now in our treatment of nature. I agree. But when conservationists argue this way to the exclusion of other arguments, or find themselves speaking in terms of "recreational interests" so continuously as to play up to, and reinforce, homocentrist perspectives, there is something sad about the spectacle. One feels that the arguments lack even their proponent's convictions. I expect they want to say something less egotistic and more emphatic but the prevailing and sanctioned modes of explanation in our society are not quite ready for it. In this vein, there must have been abolitonists who put their case in terms of getting more work out of the Blacks. Holdsworth says of the early English Jew that while he was "regarded as a species of res nullius ... [H]e was valuable for his acquisitive capacity; and for that reason the crown took him under its protection."[114] (Even today, businessmen are put in the position of insisting that their decent but probably profitless acts will "help our company's reputation and be good for profits."[115])

For my part, I would prefer a frank avowal that even making adjustments for esthetic improvemens, what I am proposing is going to cost "us," *i.e.*, reduce our standard of living as measured in terms of our present values.

Yet, this frankness breeds a frank response—one which I hear from my colleagues and which must occur to many a reader. Insofar as the proposal is not just an elaborate legal fiction, but really comes down in the last analysis to a compromise of *our* interests for *theirs*, why should we adopt it? "What is in it for 'us'?"

This is a question I am prepared to answer, but only after permitting myself some observations about how *odd* the question is. It asks for me to justify my position in the very anthropocentric hedonist terms that I am proposing we modify. One is inclined to respond by a counter: "couldn't you (as a white) raise the same questions about compromising your preferred rights-status with Blacks?"; or "couldn't you (as a man) raise the same question about compromising your preferred rights-status with women?" Such counters, unfortunately, seem no more responsive than the question itself. (They have a nagging ring of "yours too" about them.) What the exchange actually points up is a fundamental problem regarding the nature of philosophical argument. Recall that Socrates, whom we remember as an opponent of hedonistic thought, confutes Thrasymachus by arguing that immorality makes one miserably unhappy! Kant, whose moral philosophy was based upon the categorical imperative ("Woe to him who creeps through the serpent windings of Utilitarianism"[116]) finds himself justifying, *e.g.*, promise keeping and truth telling, on the most prudential—one might almost say, commercial—grounds.[117] This "philosophic irony" (as Professor Engel calls it) may owe to there being something unique about ethical argument.[118] "Ethics cannot be put into words", Wittgenstein puts it; such matters "make themselves "manifest."[119] On the other hand, perhaps the truth is that in any argument which aims at persuading a human being to action (on ethical or any other bases), "logic" is only an instrument for illuminating positions, at best, and in the last analysis it is psycho-logical appeals to the listener's self-interest that hold sway, however "principled" the rhetoric may be.

With this reservation as to the peculiar task of the argument that follows, let me stress that the strongest case can be made from the perspective of human advantage for conferring rights on the environment. Scientists have been warning of the crises the earth and all humans on it face if we do not change our ways—radically—and these crises make the lost "recreational use" of rivers seem absolutely trivial. The earth's very atmosphere is threatened with frightening possibilities: absorption of sunlight, upon which the entire life cycle depends, may be diminished; the oceans may warm (increasing the "greenhouse effect" of the atmosphere), melting the polar ice caps, and destroying our great coastal cities; the portion of the atmosphere that shields us from dangerous radiation may be destroyed. Testifying before Congress, sea explorer Jacques Cousteau predicted that the oceans (to which we dreamily look to feed our booming populations) are headed toward their own death: "The cycle of life is intricately tied up with the cycle of water ... the water system has to remain alive if we are to remain alive on earth."[120] We are depleting our energy and our food sources at a rate that takes little account of the needs even of humans now living.

These problems will not be solved easily; they very likely can be solved, if at all, only through a willingness to suspend the rate of increase in the standard of living (by present values) of the earth's "advanced" nations, and by stabilizing the total human population. For some of us this will involve forfeiting material comforts; for others it will involve abandoning the hope someday to obtain comforts long envied. For all of us it will involve giving up the right to have as many offspring as we might wish. Such a program is not impossible of realization, however. Many of our so-called "material comforts" are not only in excess of, but are probably in opposition to, basic biological needs. Further, the "costs" to the advanced nations is not as large as would appear from Gross National Product figures. G.N.P. reflects social gain (of a sort) without discounting for the social *cost* of that gain, *e.g.*, the losses through depletion of resources, pollution, and so forth. As has well been shown, as societies become more and more "advanced," their real marginal gains become less and less for each additional dollar of G.N.P.[121] Thus, to give up "human progress" would not be as costly as might appear on first blush.

Nonetheless, such far-reaching social changes are going to involve us in a serious reconsideration of our consciousness towards the environment. I say this knowing full well that there is something more than a trifle obscure in the claim: is popular consciousness a meaningful notion, to begin with? If so, what is our present consciousness regarding the environment? Has it been causally responsible for our material state of affairs? Ought we to shift our consciousness (and if so, to what exactly, and on what grounds)? How, if at all, would a shift in consciousness be translated into tangible institutional reform? Not one of these questions can be answered to everyone's satisfactions, certainly not to the author's.

It is commonly being said today, for example, that our present state of affairs—at least in the West—can be traced to the view that Nature is the dominion of Man, and that this attitude, in turn, derives from our religious traditions.

> Whatever the origins, the text is quite clear in Judaism, was absorbed all but unchanged into Christianity, and was inflated in Humanism to become the implicit attitude of Western man to Nature and the environment. Man is exclusively divine, all other creatures and things occupy lower and generally inconsequential stature; man is given dominion over all creatures and things; he is enjoined to subdue the earth.... This environment was created by the man who believes that the cosmos is a pyramid erected to support man on its pinnacle, that reality exists only because man can perceive it, that God is made in the image of man, and that the world consists solely of a dialogue between men. Surely this is an infantalism which is unendurable. It is a residue from a past of inconsequence when a few puny men cried of their supremacy to an unhearing and uncaring world. One longs for a psychiatrist who can assure man that his deep seated cultural inferiority is no longer necessary or appropriate.... It is not really necessary to destroy nature in order to gain God's favor or even his undivided attention.[122]

Surely this is forcibly put, but it is not entirely convincing as an explanation for how we got to where we are. For one thing, so far as intellectual influences are to be held responsible for our present state of affairs, one might as fairly turn on Darwin as the Bible. It was, after all,

Darwin's views—in part through the prism of Spencer—that gave moral approbation to struggle, conquest, and domination; indeed, by emphasizing man's development as a product of chance happenings, Darwin also had the effect—intended or not—of reducing our awareness of the mutual interdependency of everything in Nature. And besides, as Professor Murphy points out, the spiritual beliefs of the Chinese and Indians "in the unity between man and nature had no greater effect than the contrary beliefs in Europe in producing a balance between man and his environment"; he claims that in China, *tao* notwithstanding, "ruthless deforestation has been continuous.[123] I am under the impression, too, that notwithstanding the vaunted "harmony" between the American Plains Indians and Nature, once they had equipped themselves with rifles their pursuit of the buffalo expanded to fill the technological potential.[124] The fact is, that "consciousness" explanations pass too quickly over the less negative but simpler view of the situation: there are an increasing number of humans, with increasing wants, and there has been an increasing technology to satisfy them at "cost" to the rest of nature. Thus, we ought not to place too much hope that a changed environmental consciousness will in and of itself reverse present trends. Furthermore, societies have long since passed the point where a change in human consciousness on any matter will rescue us from our problems. More then ever before we are in the hands of institutions. These institutions are not "mere legal fictions" moreover—they have wills, minds, purposes, and inertias that are in very important ways their own, *i.e.*, that can transcend and survive changes in the consciousnesses of the individual humans who supposedly comprise them, and whom they supposedly serve. (It is more and more the individual human being, with his consciousness, that is the legal fiction.[125])

For these reasons, it is far too pat to suppose that a western "environmental consciousness" is solely or even primarily responsible for our environmental crisis. On the other hand, it is not so extravagant to claim that it has dulled our resentment and our determination to respond. For this reason, whether we will be able to bring about the requisite institutional and population growth changes depends in part upon effecting a radical shift in our feelings about "our" place in the rest of Nature.

A radical new conception of man's relationship to the rest of nature would not only be a step towards solving the material planetary problems; there are strong reasons for such a changed consciousness from the point of making us far better humans. If we only stop for a moment and look at the underlying human qualities that our present attitudes toward property and nature draw upon and reinforce, we have to be struck by how stultifying of our own personal growth and satisfaction they can become when they take rein of us. Hegel, in "justifying" private property, unwittingly reflects the tone and quality of some of the needs that are played upon:

> A person has as his substantive end the right of putting his will into any and every thing and thereby making it his, because it has no such end in itself and derives its destiny and soul from his will. This is the absolute right of appropriation which man has over all "things."[126]

What is it within us that gives us this need not just to satisfy basic biological wants, but to extend our wills over things, to object-ify them, to make them ours, to manipulate them, to keep them at a psychic distance? Can it all be explained on "rational" bases? Should we not be suspect of such needs within us, cautious as to why we wish to gratify them? When I first read that passage of Hegel, I immediately thought not only of the emotional contrast with Spinoza, but of the passage in Carson McCullers' *A Tree, A Rock, A Cloud*, in which an old derelict has collared a twelve year old boy in a streetcar cafe. The old man asks whether the boy knows "how love should be begun?"

> The old man leaned closer and whispered:
>
> "A tree. A rock. A cloud."
>
> . . .
>
> "The weather was like this in Portland," he said. "At the time my science was begun. I meditated and I started very cautious. I would pick up something from the street and take it home with me. I bought a goldfish and I concentrated on the goldfish and I loved it. I graduated from one thing to another. Day by day I was getting this technique.. . .
>
> . . .
>
> . . . "For six years now I have gone around by myself and built up my science. And now I am a master. Son. I can love anything. No longer do I have to think about it even. I see a street full of people and a beautiful light comes in me. I watch a bird in the sky. Or I meet a traveler on the road. Everything, Son. And anybody. All stranger and all loved! Do you realize what a science like mine can mean?"[127]

To be able to get away from the view that Nature is a collection of useful senseless objects is, as McCullers' "madman" suggests, deeply involved in the development of our abilities to love—or, if that is putting it too strongly, to be able to reach a heightened awareness of our own, and others' capacities in their mutual interplay. To do so, we have to give up some psychic investment in our sense of separateness and specialness in the universe. And this, in turn, is hard giving indeed, because it involves us in a flight backwards, into earlier stages of civilization and childhood in which we had to trust (and perhaps fear) our environment, for we had not then the power to master it. Yet, in doing so, we—as persons—gradually free ourselves of needs for supportive illusions. Is not this one of the triumphs for "us" of our giving legal rights to (or acknowledging the legal rights of) the Blacks and women?[128]

Changes in this sort of consciousness are already developing, for the betterment of the planet and us. There is now federal legislation which "establishes by law"[129]

> the humane ethic that animals should be accorded the basic creature comforts of adequate housing, ample food and water, reasonable handling, decent sanitation, sufficient ventilation, shelter from extremes of weather and temperature, and adequate veterinary care including the appropriate use of pain-killing drugs. . . .[130]

The Vietnam war has contributed to this movement, as it has to others. Five years ago a Los Angeles mother turned out a poster which read "War is not Healthy for children and other living things."[131] It caught on tremendously—at first, I suspect, because it sounded like another clever protest against the war, *i.e.*, another angle. But as people say such things, and think about them, the possibilities of what they have stumbled upon become manifest—in its suit against the Secretary of Agriculture to cancel the registration of D.D.T., Environmental Defense Fund alleged "biological injury to man and other living things."[132] A few years ago the pollution of streams was thought of only as a problem of smelly, unsightly, unpotable water *i.e.*, to us. Now we are beginning to discover that pollution is a process that destroys wondrously subtle balances of life within the water, and as between the water and its banks. This heightened awareness enlarges our sense of the dangers to us. But it also enlarges our empathy. We are not only developing the scientific capacity, but we are cultivating the personal capacities *within us* to recognize more and more the ways in which nature—like the woman, the Black, the Indian and the Alien—is like us (and we will also become more able realistically to define, confront, live with and admire the ways in which we are all different).[133]

The time may be on hand when these sentiments, and the early stirrings of the law, can be coalesced into a radical new theory or myth—felt as well as intellectualized—of man's relationships to the rest of nature. I do not mean "myth" in a demeaning sense of the term, but in the sense in which, at different times in history, our social "facts" and relationships have been comprehended and integrated by reference to the "myths" that we are co-signers of a social contract, that the Pope is God's agent, and that all men are created equal. Pantheism, Shinto and Tao all have myths to offer. But they are all, each in its own fashion, quaint, primitive and archaic. What is needed is a myth that can fit our growing body of knowledge of geophysics, biology and the cosmos. In this vein, I do not think it too remote that we may come to regard the Earth, as some have suggested, as one organism, of which Mankind is a functional part—the mind, perhaps: different from the rest of nature, but different as a man's brain is from his lungs.

> Ever since the first Geophysical Year, international scientific studies have shown irrefutably that the Earth as a whole is an organized system of most closely interrelated and indeed interdependent activities. It is, in the broadest sense of the term, an "organism." The so-called life-kingdoms and the many vegetable and animal species are dependent upon each other for survival in a balanced condition of planet-wide existence; and they depend on their environment, conditioned by oceanic and atmospheric currents, and even more by the protective action of the ionosphere and many other factors which have definite rhythms of operation. Mankind is part of this organic planetary whole; and there can be no truly new global society, and perhaps in the present state of affairs no society at all, as long as man will not recognize, accept and enjoy the fact that mankind has a definite function to perform within this planetary organism of which it is an active part.
>
> In order to give a constructive meaning to the activities of human societies all over the globe, these activities—physical and mental—should be understood and given basic value with reference to the wholesome functioning of the entire Earth, and we may add of the entire solar system. This cannot be done (1) if man insists on considering himself an alien Soul compelled to incarnate on this sorrowful planet, and (2) if we can see in the planet, Earth, nothing but a mass of material substances moved by mechanical laws, and in "life" nothing but a chance combination of molecular aggregations.
>
> . . . As I see it, the Earth is only one organized "field" of activities—and so is the *human person*—but these activities take place at various levels, in different "spheres" of being and realms of consciousness. The lithosphere is not the biosphere, and the latter not the . . . ionosphere. The Earth is not *only* a material mass. Consciousness is not only "human"; it exists at animal and vegetable levels, and most likely must be latent, or operating in some form, in the molecule and the atom; and all these diverse and in a sense hierarchical modes of activity and consciousness should be seen integrated in and perhaps transcended by an all-encompassing and "eonic" planetary Consciousness.
>
> . . . .
>
> Mankind's function within the Earth-organism is to extract from the activities of all other operative systems within this organism the type of consciousness which we call "reflective" or "self"-consciousness—or, we may also say to *mentalize*

and give meaning, value, and "name" to all that takes place anywhere within the Earth-field....

This "mentalization" process operates through what we call culture. To each region of, and living condition in the total field of the Earth-organism a definite type of culture inherently corresponds. Each region is the "womb" out of which a specific type of human mentality and culture can and sooner or later will emerge. All these cultures—past, present and future—and their complex interrelationships and interactions are the collective builders of the Mind of humanity; and this means of the *conscious Mind of the Earth.*[134]

As radical as such a consciousness may sound today, all the dominant changes we see about us point in its direction. Consider just the impact of space travel, of world-wide mass media, of increasing scientific discoveries about the interrelatedness of all life processes. Is it any wonder that the term "spaceship earth" has so captured the popular imagination? The problems we have to confront are increasingly the world-wide crises of a global organism: not pollution of a stream, but pollution of the atmosphere and of the ocean. Increasingly, the death that occupies each human's imagination is not his own, but that of the entire life cycle of the planet earth, to which each of us is as but a cell to a body.

To shift from such a lofty fancy as the planetarization of consciousness to the operation of our municipal legal system is to come down to earth hard. Before the forces that are at work, our highest court is but a frail and feeble—a distinctly human—institution. Yet, the Court may be at its best not in its work of handing down decrees, but at the very task that is called for: of summoning up from the human spirit the kindest and most generous and worthy ideas that abound there, giving them shape and reality and legitimacy.[135] Witness the School Desegregation Cases which, more importantly than to integrate the schools (assuming they did), awakened us to moral needs which, when made visible, could not be denied. And so here, too, in the case of the environment, the Supreme Court may find itself in a position to award "rights" in a way that will contribute to a change in popular consciousness. It would be a modest move, to be sure, but one in furtherance of a large goal: the future of the planet as we know it.

How far we are from such a state of affairs, where the law treats "environmental objects" as holders of legal rights, I cannot say. But there is certainly intriguing language in one of Justice Black's last dissents, regarding the Texas Highway Department's plan to run a six-lane expressway through a San Antonio Park. [136] Complaining of the Court's refusal to stay the plan, Black observed that "after today's decision, the people of San Antonio and the birds and animals that make their home in the park will share their quiet retreat with an ugly, smelly stream of traffic.... Trees, shrubs, and flowers will be mown down."[137] Elsewhere he speaks of the "burial of public parks," of segments of a highway which "devour parkland," and of the park's heartland.[138] Was he, at the end of his great career, on the verge of saying—just saying—that "nature has 'rights' on its own account"? Would it be so hard to do?

* Professor of Law, University of Southern California. A.B. 1959, Harvard; LL.B. 1962, Yale. Chairman, Committee on Law and the Humanities, Association of American Law Schools. The author wishes to express his appreciation for the financial support of the National Endowment for the Humanities.

1. C. Darwin, Descent of Man 119, 120–21 (2d ed. 1874). *See also* R. Waelder, Progress and Revolution 39 *et seq.* (1967).

2. *See* Darwin, *supra* note 1, at 113–14:

> ... No tribe could hold together if murder, robbery, treachery, etc., were common; consequently such crimes within the limits of the same tribe "are branded with everlasting infamy"; but excite no such sentiment beyond these limits. A North-American Indian is well pleased with himself, and is honored by others, when he scalps a man of another tribe; and a Dyak cuts off the head of an unoffending person, and dries it as a trophy ... It has been recorded that an Indian Thug conscientiously regretted that he had not robbed and strangled as many travelers as did his father before him. In a rude state of civilization the robbery of strangers is, indeed, generally considered as honorable.

See also Service, *Forms of Kinship* in Man in Adaptation 112 (Y. Cohen ed. 1968).

3. *See* Darwin, *supra* note 1, at 113. *See also* E. Westermarck, 1 The Origin and Development of the Moral Ideas 406–12 (1912).

The practice of allowing sickly children to die has not been entirely abandoned, apparently, even at our most distinguished hospitals. *See Hospital Let Retarded Baby Die, Film Shows*, L. A. Times, Oct. 17, 1971, § A, at 9, col. 1.

4. There does not appear to be a word "gericide" or "geronticide" to designate the killing of the aged. "Senicide" is as close as the Oxford English Dictionary comes, although, as it indicates, the word is rare. 9 Oxford English Dictionary 454 (1933).

5. *See* Darwin, *supra* note 1, at 386–93. Westermarck, *supra* note 3, at 387–89, observes that where the killing of the aged and infirm is practiced, it is often supported by humanitarian justification; this, however, is a far cry from saying that the killing is *requested* by the victim as his right.

6. H. Maine, Ancient Law 153 (Pollock ed. 1930). Maine claimed that these powers of the father extended to all regions of private law, although not to the Jus Publicum, under which a son, notwithstanding his subjection in private life, might vote alongside his father *Id.* at 152. Westermarck, *supra* note 3, at 393–94, was skeptical that the arbitrary power of the father over the children extended as late as into early Roman law.

7. 387 U.S. 1 (1967).

8. 42 U.S.C. §§ 1973 *et seq.* (1970).

9. *See* Landman v. Royster, 40 U.S.L.W. 2256 (E.D. Va., Oct. 30, 1971) (eighth amendment and due process clause of the fourteenth amendment require federal injunctive relief, including compelling the drafting of new prison rules, for Virginia prisoners against prison conduct prohibited by vague rules or no rules, without disciplinary proceedings embodying rudiments of procedural due process, and by various penalties that constitute cruel and unusual punishment). *See* Note, *Courts, Corrections and the Eighth Amendment: Encouraging Prison Reform by Releasing Inmates*, 44 S. CAL. L. REV. 1060 (1971).

10. *But see* T. SZASZ, LAW, LIBERTY AND PSYCHIATRY (1963).

11. *See* notes 22, 52 and accompanying text *infra.* The trend toward liberalized abortion can be seen either as a legislative tendency back in the direction of rightlessness for the foetus—or toward increasing rights of women. This inconsistency is not unique in the law of course; it is simply support for Hohfeld's scheme that the "jural opposite" of someone's right is someone else's "no-right." W. HOHFELD, FUNDAMENTAL LEGAL CONCEPTIONS (1923).

Consider in this regard a New York case in which a settlor *S* established a trust on behalf of a number of named beneficiaries and "lives in being." Desiring to amend the deed of trust, the grantor took steps pursuant to statute to obtain "the written consent of all persons beneficially interested in [the] trust." At the time the grantor was pregnant and the trustee Chase Bank advised it would not recognize the proposed amendment because the child *en ventre sa mere* might be deemed a person beneficially interested in the trust. The court allowed the amendment to stand, holding that birth rather than conception is the controlling factor in ascertaining whether a person is beneficially interested in the trust which the grantor seeks to amend. *In re* Peabody, 5 N.Y.2d 541, 158 N.E.2d 841 (1959).

The California Supreme Court has recently refused to allow the deliberate killing of a foetus (in a non-abortion situation) to support a murder prosecution. The court ruled foetuses not to be denoted by the words "human being" within the statute defining murder. Keeler v. Superior Court, 2 Cal. 3d 619, 87 Cal. Rptr. 481, 470 P.2d 617 (1970). But see note 52 and accompanying text *infra.*

Some jurisdictions have statutes defining a crime of "feticide"—deliberately causing the death of an unborn child. The absence of such a specific feticide provision in the California case was one basis for the ruling in *Keeler. See* 2 Cal. 3d at 633 n.16, 87 Cal. Rptr. at 489 n.16, 470 P.2d at 625 n.16.

12. INT. REV. CODE of 1954, § 1361 (repealed by Pub. L. No. 89-389, effective Jan. 1, 1969).

13. For example, *see* United States v. Cargo of the Brig Malek Adhel, 43 U.S. (2 How.) 210 (1844). There, a ship had been seized and used by pirates. All this was done without the knowledge or consent of the owners of the ship. After the ship had been captured, the United States condemned and sold the "offending vessel." The owners objected. In denying release to the owners, Justice Story cited Chief Justice Marshall from an earlier case: "This is not a proceeding against the owner; it is a proceeding against the vessel for an offense committed by the vessel; which is not the less an offense ... because it was committed without the authority and against the will of the owner." 43 U.S. at 234, quoting from United States v. Schooner Little Charles, 26 F. Cas. 979 (No. 15,612) (C.C.D. Va. 1818).

14. 9 U.S. (5 Cranch) 61, 86 (1809).

15. Trustees of Darmouth College v. Woodward, 17 U.S. (4 Wheat.) 518 (1819).

16. *Id.* at 636.

17. Consider, for example, that the claim of the United States to the naval station at Guantanamo Bay, at $2000-a-year rental, is based upon a treaty signed in 1903 by José Montes for the President of Cuba and a minister representing Theodore Roosevelt; it was subsequently ratified by two-thirds of a Senate no member of which is living today. Lease [from Cuba] of Certain Areas for Naval or Coaling Stations, July 2, 1903, T.S. No. 426; C. BEVANS, 6 TREATIES AND OTHER INTERNATIONAL AGREEMENTS OF THE UNITED STATES 1776–1949, at 1120 (U.S. Dep't of State Pub. 8549, 1971).

18. O. GIERKE, POLITICAL THEORIES OF THE MIDDLE AGE (Maitland transl. 1927), especially at 22–30. The reader may be tempted to suggest that the "corporate" examples in the text are distinguishable from environmental objects in that the former are comprised by and serve humans. On the contrary, I think that the more we learn about the sociology of the firm—and the realpolitik of our society—the more we discover the ultimate reality of these institutions, and the increasingly legal fictiveness of the individual human being. *See* note 125 and accompanying text *infra.*

19. Dred Scott v. Sandford, 60 U.S. (19 How.) 396, 404–05 (1856). In Bailey v. Poindexter's Ex'r, 56 Va. (14 Gratt.) 132, 142–43 (1858) a provision in a will that testator's slaves could choose between emancipation and public sale was held void on the ground that slaves have no legal capacity to choose:

> These decisions are legal conclusions flowing naturally and necessarily from the one clear, simple, fundamental idea of chattel slavery. That fundamental idea is, that, in the eye of the law, so far certainly as civil rights and relations are concerned, the slave is not a person, but a thing. The investiture of a chattel with civil rights or legal capacity is indeed a legal solecism and absurdity. The attribution of legal personality to a chattel slave,—legal conscience, legal intellect, legal freedom, or liberty and power of free choice and action, and corresponding legal obligations growing out of such qualities, faculties and action—implies a palpable contradiction in terms.

20. People v. Hall, 4 Cal. 399, 405 (1854). The statute there under interpretation provided that "no Black or Mulatto person, or Indian shall be allowed to give evidence in favor of, or against a white man," but was silent as to Chinese. The "policy" analysis by which the court brings Chinese under "Black ... or Indian" is a fascinating illustration of the relationship between a "policy" decision and a "just" decision, especially in light of the exchange between Hart, *Positivism and the Separation of Law and Morals*, 71 HARV. L. REV. 593 (1958) and Fuller, *Positivism and Fidelity to Law—A Reply to Professor Hart, id.* at 630.

21. Schechter, *The Rightlessness of Mediaeval English Jewry*, 45 JEWISH Q. REV. 121, 135 (1954) quoting from M. BATESON, MEDIEVAL ENGLAND 139 (1904). Schechter also quotes Henry de

Bracton to the effect that "a Jew cannot have anything of his own, because whatever he acquires he acquires not for himself but for the king.. . ." *Id.* at 128.

22. Dietrich v. Inhabitants of Northampton, 138 Mass. 14, 16 (1884).

23. *In re* Goddell, 39 Wisc. 232, 245 (1875). The court continued with the following "clincher":

> And when counsel was arguing for this lady that the word, person, in sec. 32. ch. 119 [respecting those qualified to practice law], necessarily includes females, her presence made it impossible to suggest to him as *reductio ad absurdum* of his position, that the same construction of the same word ... would subject woman to prosecution for the paternity of a bastard, and ... prosecution for rape.

Id. at 246.

The relationship between our attitudes toward woman, on the one hand, and, on the other, the more central concern of this article—land—is captured in an unguarded aside of our colleague, Curt Berger. "... after all, land, like woman, was meant to be possessed.. . ." LAND OWNERSHIP AND USE 139 (1968).

23a. Recently, a group of prison inmates in Suffolk County tamed a mouse that they discovered, giving him the name Morris. Discovering Morris, a jailer flushed him down the toilet. The prisoners brought a proceeding against the Warden complaining, *inter alia*, that Morris was subjected to discriminatory discharge and was otherwise unequally treated. The action was unsuccessful, on grounds that the inmates themselves were "guilty of imprisoning Morris without a charge, without a trial, and without bail," and that other mice at the prison were not treated more favorably. "As to the true victim the Court can only offer again the sympathy first proffered to his ancestors by Robert Burns.. . ." The Judge proceeded to quote from Burns' "To a Mouse." Morabito v. Cyrta, 9 CRIM. L. REP. 2472 (N.Y. Sup. Ct. Suffolk Co. Aug. 26, 1971).

The whole matter seems humorous, of course. But what we need to know more of is the function of humor in the unfolding of a culture, and the ways in which it is involved with the social growing pains to which it is testimony. Why do people make jokes about the Women's Liberation Movement? Is it not on account of—rather than in spite of—the underlying validity of the protests, and the uneasy awareness that a recognition of them is inevitable? A. Koestler rightly begins his study of the human mind, ACT OF CREATION (1964), with an analysis of humor, entitled "The Logic of Laughter." And *cf.* Freud, *Jokes and the Unconscious*, 8 STANDARD EDITION OF THE COMPLETE PSYCHOLOGICAL WORKS OF SIGMUND FREUD (J. Strachey transl. 1905). (Query too: what is the relationship between the conferring of proper *names, e.g.*, Morris, and the conferring of social and legal *rights?*)

24. Thus it was that the Founding Fathers could speak of the inalienable rights of all men, and yet maintain a society that was, by modern standards, without the most basic rights for Blacks, Indians, children and women. There was no hypocrisy; emotionally, no one *felt* that these other things were men.

25. The second thought streaming from ... the older South [is] the sincere and passionate belief that somewhere between men and cattle, God created a *tertium quid*, and called it a Negro—a clownish, simple creature, at times even lovable within its limitations, but straitly foreordained to walk within the Veil.

W. E. B. DUBOIS, THE SOULS OF BLACK FOLK 89 (1924).

26. In this article I essentially limit myself to a discussion of non-animal but natural objects. I trust that the reader will be able to discern where the analysis is appropriate to advancing our understanding of what would be involved in giving "rights" to other objects not presently endowed with rights—for example, not only animals (some of which already have rights in some senses) but also humanoids, computers, and so forth. *Cf.* the National Register for Historic Places, 16 U.S.C. § 470 (1970), discussed in Ely v. Velde, 321 F. Supp. 1088 (E.D. Va. 1971).

As the reader will discover, there are large problems involved in defining the boundaries of the "natural object." For example, from time to time one will wish to speak of that portion of a river that runs through a recognized jurisdiction; at other times, one may be concerned with the entire river, or the hydrologic cycle—or the whole of nature. One's ontological choices will have a strong influence on the shape of the legal system, and the choices involved are not easy. *See* notes 49, 73 and accompanying text *infra.*

On the other hand, the problems of selecting an appropriate ontology are problems of all language—not merely of the language of legal concepts, but of ordinary language as well. Consider, for example, the concept of a "person" in legal *or* in everyday speech. Is each *person* a fixed bundle of relationships, persisting unaltered through time? Do our molecules and cells not change at every moment? Our hypostatizations always have a pragmatic quality to them. *See* D. HUME, *Of Personal Identity*, in TREATISE OF HUMAN NATURE bk. 1, pt. IV, § VI, in THE PHILOSOPHICAL WORKS OF DAVID HUME 310–18, 324 (1854); T. MURTI, THE CENTRAL PHILOSOPHY OF BUDDHISM 70–73 (1955). In LOVES BODY 146–47 (1966) Norman O. Brown observes:

> The existence of the "let's pretend" boundary does not prevent the continuance of the real traffic across it. Projection and introjection, the process whereby the self as distinct from the other is constituted, is not past history, an event in childhood, but a present process of continuous creation. The dualism of self and external world is built up by a constant process of reciprocal exchange between the two. The self as a stable substance enduring through time, an identity, is maintained by constantly absorbing good parts (or people) from the outside world and expelling bad parts from the inner world. "There is a continual 'unconscious' wandering of other personalities into ourselves."
>
> Every person, then, is many persons; a multitude made into one person; a corporate body; incorporated, a corporation. A "corporation sole"; every man a parson-person. The unity of the person is as real, or unreal, as the unity of the corporation.

See generally, W. BISHIN & C. STONE, LAW, LANGUAGE AND ETHICS Ch. 5 (1972).

In different legal systems at different times, there have been many shifts in the entity deemed "responsible" for harmful acts: an entire clan was held responsible for a crime before the notion of individual responsibility emerged; in some societies the offending hand, rather than an entire body, may be "responsible." Even today, we treat father and son as separate jural entities for some purposes, but as a single jural entity for others. I do not see why, in principle, the task of working out a legal ontology of natural objects (and "qualities." *e.g.*, climatic warmth) should be any more unmanageable. Perhaps someday all mankind shall be, for some purposes, one jurally recognized "natural object."

27. The statement in text is not quite true; *cf.* Murphy, *Has Nature Any Right to Life?*, 22 HAST. L.J. 467 (1971). An Irish court, passing upon the validity of a testamentary trust to the benefit of someone's dogs, observed in dictum that "'lives' means lives of human beings, not of animals or trees in California." Kelly v. Dillon, 1932 Ir. R. 255, 261. (The intended gift over on the death of the last surviving dog was held void for remoteness, the court refusing "to enter into the question of a dog's expectation of life," although prepared to observe that "in point of fact neighbor's [sic] dogs and cats are unpleasantly long-lived...." *Id.* at 260–61).

28. Four cases dealing with the Constitutionality of the death penalty under the eighth and fourteenth amendments are pending before the United States Supreme Court. Branch v. Texas, 447 S.W.2d 932 (Tex. 1969), *cert. granted*, 91 S. Ct. 2287 (1970); Aikens v. California, 70 Cal. 2d 369, 74 Cal. Rptr. 882, 450 P.2d 258 (1969). *cert. granted*, 91 S. Ct. 2280 (1970); Furman v. Georgia, 225 Ga. 253, 167 S.E.2d 628 (1969), *cert. granted*, 91 S. Ct. 2282 (1970); Jackson v. Georgia, 225 Ga. 790, 171 S.E.2d 501 (1969), *cert. granted*, 91 S. Ct. 2287 (1970).

29. *See* George Campbell Painting Corp. v. Reid, 392 U.S. 286 (1968); Oklahoma Press Pub. Co. v. Walling, 327 U.S. 186 (1946); Baltimore & O.R.R. v. ICC, 221 U.S. 612 (1911); Wilson v. United States, 221 U.S. 361 (1911); Hale v. Henkel, 201 U.S. 43 (1906).

30. *See* Dixon v. Alabama State Bd. of Educ., 294 F.2d 150 (5th Cir.), *cert. denied*, 368 U.S. 930 (1961).

31. For example, *see* People *ex rel.* Ricks Water Co. v. Elk River Mill & Lumber Co., 107 Cal. 221, 40 Pac. 531 (1895) (refusing to enjoin pollution by a upper riparian at the instance of the Attorney General on the grounds that the lower riparian owners, most of whom were dependent on the lumbering business of the polluting mill, did not complain).

32. The law in a suit for injunctive relief is commonly easier on the plaintiff than in a suit for damages. *See* J. GOULD, LAW OF WATERS § 206 (1883).

33. However, in 1970 California amended its Water Quality Act to make it easier for the Attorney General to obtain relief, *e.g.*, one must no longer allege irreparable injury in a suit for an injunction. CAL. WATER CODE § 13350(b) (West 1971).

34. To whomsoever the soil belongs, he owns also to the sky and to the depths. *See* W. BLACKSTONE, 2 COMMENTARIES *18.

At early common law, the owner of land could use all that was found under his land "at his free will and pleasure" without regard to any "inconvenience to his neighbour." Acton v. Blundell, 12 Meeson & Welsburg 324, 354, 152 Eng. Rep. 1223, 1235 (1843). "He [the landowner] may waste or despoil the land as he pleases...." R. MECARRY & H. WADE, THE LAW OF REAL PROPERTY 70 (3d ed. 1966). *See* R. POWELL, 5 THE LAW OF REAL PROPERTY 725 (1971).

35. *See* Note, *Statutory Treatment of Industrial Stream Pollution*, 24 GEO. WASH. L. REV. 302, 306 (1955); H. FARNHAM, 2 LAW OF WATERS AND WATER RIGHTS § 461 (1904); GOULD, *supra* note 32, at § 204.

36. For example, courts have upheld a right to pollute by prescription. Mississippi Mills Co. v. Smith, 69 Miss. 299, 11 So. 26 (1882), and by easement, Luama v. Bunker Hill & Sullivan Mining & Concentrating Co., 41 F.2d 358 (9th Cir. 1930).

37. *See* Red River Roller Mills v. Wright, 30 Minn. 249, 15 N.W. 167 (1883) (enjoyment of stream by riparian may be modified or abrogated by reasonable use of stream by others); Townsend v. Bell, 167 N.Y. 462, 60 N.E. 757 (1901) (riparian owner not entitled to maintain action for pollution of stream by factory where he could not show use of water was unreasonable); Smith v. Staso Milling Co., 18 F.2d 736 (2d Cir. 1927) (in suit for injunction, right on which injured lower riparian stands is a quantitative compromise between two conflicting interests); Clifton Iron Co. v. Dye, 87 Ala. 468, 6 So. 192 (1889) (in determining whether to grant injunction to lower riparian, court must weigh interest of public as against injury to one or the other party). *See also* Montgomery Limestone Co. v. Bearder, 256 Ala. 269, 54 So. 2d 571 (1951).

38. Pennsylvania Coal Co. v. Sanderson, 113 Pa. 126, 149, 6 A. 453, 459 (1886).

39. Hand, J. in Smith v. Staso Milling Co., 18 F.2d 736, 738 (2d Cir. 1927) (emphasis added). *See also* Harrisonville v. Dickey Clay Co., 289 U.S. 334 (1933) (Brandeis, J.).

40. Measuring plantiff's damages by "making him whole" has several limitations; these and the matter of measuring damages in this area generally are discussed more fully at notes 83–93 and accompanying text *infra.*

41. Here, again, an analogy to corporation law might be profitable. Suppose that in the instance of negligent corporate management by the directors, there were no institution of the stockholder derivative suit to force the directors to make *the corporation* whole, and the only actions provided for were direct actions by stockholders to collect for damages *to themselves qua* stockholders. Theoretically and practically, the damages might come out differently in the two cases, and not merely because the creditors' looses are not aggregated in the stockholders' direct actions.

42. And even far less than the damages to all human economic interests derivately through the stream; *see* text accompanying notes 83–84, 120 *infra.*

43. Smith v. Staso, 18 F.2d, 736, 738 (2d Cir. 1927).

44. Some of these public properties are subject to the "public trust doctrine," which, while ill-defined, might be developed in such fashion as to achieve fairly broad-ranging environmental protection. *See* Gould v. Greylock Reservation Comm'n, 350 Mass. 410, 215 N.E.2d 114 (1966), discussed in Sax, *The Public Trust Doctrine in Natural Resource Law: Effective Judicial Intervention*, 68 MICH. L. REV. 471, 492–509 (1970).

45. By contrast, for example, with humane societies.

46. *See, e.g.*, Cal. Prob. Code §§ 1460–62 (West Supp. 1971).

47. CAL PROB. CODE § 1751 (West Supp. 1971) provides for the appointment of a "conservator."

48. In New York the Supreme Court and county courts outside New York City have jurisdiction to appoint a committee of the person and/or a committee of the property for a person "incompetent to manage himself or his affairs." N.Y. MENTAL HYGIENE LAW § 100 (McKinney 1971).

49. This is a situation in which the ontological problems discussed in note 26 *supra* become acute. One can conceive a situation in which a guardian would be appointed by a county court with respect to a stream, bring a suit against alleged polluters, and lose. Suppose now that a federal court were to appoint a guardian with respect to the larger river system of which the stream were a part, and that the federally appointed guardian subsequently were to bring suit against the same defendants in state court, now on behalf of the river, rather than the stream. (Is it possible to bring a still subsequent suit, if the one above fails, on behalf of the entire hydrologic cycle, by a guardian appointed by an international court?)

While such problems are difficult, they are not impossible to solve. For one thing, pre-trial hearings and rights of intervention can go far toward their amelioration. Further, courts have been dealing with the matter of potentially inconsistent judgments for years, as when one state appears on the verge of handing down a divorce decree inconsistent with the judgment of another state's courts. Kempson v. Kempson, 58 N.J. Eg. 94, 43 A. 97 (Ch. Ct. 1899). Courts could, and of course would, retain some natural objects in the res nullius classification to help stave off the problem. Then, too, where (as is always the case) several "objects" are interrelated, several guardians could all be involved, with procedures for removal to the appropriate court—probably that of the guadian of the most encompassing "ward" to be acutely threatened. And in some cases subsequent suit by the guardian of more encompassing ward, not guilty of laches, might be appropriate. The problems are at least no more complex than the corresponding problems that the law has dealt with for years in the class action area.

50. CAL. PROB. CODE § 1460 (West Supp. 1971). The N.Y. MENTAL HYGIENE LAW (McKinney 1971) provides for jurisdiction "over the custody of a person and his property if he is incompetent to manage himself or his affairs by reason of age, drunkenness, mental illness or other cause... "

51. Santa Clara County v. Southern Pac. R.R., 118 U.S. 394 (1886). Justice Black would have denied corporations the rights of "persons" under the fourteenth amendment. *See* Connecticut Gen. Life Ins. Co. v. Johnson, 303 U.S. 77, 87 (1938) (Black, J. dissenting): "Corporations have neither race nor color."

52. *In re* Byrn, L. A. Times, Dec. 5, 1971, § 1, at 16, col. 1. A preliminary injunction was subsequently granted, and defendant's cross-motion to vacate the guardianship was denied. Civ. 13113/71 (Sup. Ct. Queens Co., Jan. 4, 1972) (Smith, J.). Appeals are pending. Granting a guardianship in these circumstances would seem to be a more radical advance in the law than granting a guardianship over communal natural objects like lakes. In the former case there is a traditionally recognized guardian for the object—the mother—and her decision has been in favor of aborting the foetus.

53. The laws regarding the various communal resources had to develop along their own lines, not only because so many different persons' "rights" to consumption and usage were continually and contemporaneously involved, but also because no one had to bear the costs of his consumption of public resources in the way in which the owner of resources on private land has to bear the costs of what he does. For example, if the landowner strips his land of trees, and puts nothing in their stead, he confronts the costs of what he has done in the form of reduced value of his land: but the river polluter's actions are costless, so far as he is concerned—except insofar as the legal system can somehow force him to internalize them. The result has been that the private landowner's power over natural objects on his land is far less restrained by law (as opposed to economics) than his power over the public resources that he can get his hands on. If this state of affairs is to be changed, the standard for interceding in the interests of natural objects on traditionally recognized "private" land might well parallel the rules that guide courts in the matter of people's children whose upbringing (or lack thereof) poses social threat. The courts can, for example, make a child "dependent of the court" where the child's "home is an unfit place for him by reason of neglect, cruelty, or depravity of either of his parents... " CAL. WELF. & INST. CODE § 600(b) (West 1966). *See also id* at § 601: any child "who from any cause is in danger of leading an idle, dissolute, lewd, or immoral life [may be adjudged] a ward of the court."

54. *See* note 53 *supra.* The present way of handling such problems on "private" property is to try to enact legislation of general application under the police power, *see* Pennsylvania Coal Co. v. Mahon, 260 U.S. 393 (1922), rather than to institute civil litigation which, though a piecemeal process, can be tailored to individual situations.

55. CAL. PROB. CODE § 1580 (West Supp. 1971) lists specific causes for which a guardian may, after notice and a hearing, be removed.

Despite these protections, the problem of overseeing the guardian is particularly acute where, as here, there are no immediately identifiable human beneficiaries whose self-interests will encourage them to keep a close watch on the guardian. To ameliorate this problem, a page might well be borrowed from the law of ordinary charitable trusts, which are commonly placed under the supervision of the Attorney General. *See* CAL. PROB. CODE §§ 9505, 10207 (West 1955).

56. *See* CAL. PROB. CODE § 1472, 1590 (West 1956 and Supp. 1971).

57. 354 F.2d 608 (2d Cir. 1965), *cert. denied*, Consolidated Edison Co. v. Scenic Hudson Preservation Conf., 384 U.S. 941 (1966).

58. 354 F.2d 608, 615 (2d Cir. 1965).

59. Act of Aug. 26, 1935, ch. 687, Title II, § 213, 49 Stat. 860 (*codified in* 16 U.S.C. § 8251(b) (1970).

60. 354 F.2d 608, 616 (2d Cir. 1965). The court might have felt that because the New York-New Jersey Trial Conference, one of the two conservation groups that organized Scenic Hudson, had some 17 miles of trailways in the area of Storm King Mountain, it therefore had sufficient economic interest to establish standing; Judge Hays' opinion does not seem to so rely, however.

61. Road Review League v. Boyd, 270 F. Supp. 650 (S.D.N.Y. 1967). Plaintiffs who included the Town of Bedford and the Road Review League, a non-profit association concerned with community problems, brought an action to review and set

aside a determination of the Federal Highway Administrator concerning the alignment of an interstate highway. Plaintiffs claimed that the proposed road would have an adverse effect upon local wildlife sanctuaries, pollute a local lake, and be inconsistent with local needs and planning. Plaintiffs relied upon the section of the Administrative Procedure Act, 5 U.S.C. § 702 (1970), which entitles persons "aggrieved by agency action within the meaning of a relevant statute" to obtain judicial review. The court held that plaintiffs had standing to obtain judicial review of proposed alignment of the road:

> I see no reason why the word "aggrieved" should have different meaning in the Administrative Procedure Act from the meaning given it under the Federal Power Act.... The "relevant statute," i.e., the Federal Highways Act, contains language which seems even stronger than that of the Federal Power Act, as far as local and conservation interests are concerned.

Id. at 661.

In Citizens Comm. for the Hudson Valley v. Volpe, 425 F.2d 97 (2d Cir. 1970), plaintiffs were held to have standing to challenge the construction of a dike and causeway adjacent to the Hudson Valley. The Sierra Club and the Village of Tarrytown based their challenge upon the provisions of the Rivers and Harbors Act of 1899. While the Rivers and Harbors Act does not provide for judicial review as does the Federal Power Act, the court stated that the plaintiffs were "aggrieved" under the Department of Transportation Act, the Hudson River Basin Compact Act, and a regulation under which the Corps of Engineers issued a permit, all of which contain broad provisions mentioning recreational and environmental resources and the need to preserve the same. Citing the *Road Review League* decision, the court held that as "aggrieved" parties under the Administrative Procedure Act, plaintiffs similarly had standing. Other decisions in which the court's grant of standing was based upon the Administrative Procedure Act include: West Virginia Highlands Conservancy v. Island Creek Coal Co., 441 F.2d 231 (4th Cir. 1971); Environmental Defense Fund, Inc. v. Hardin, 428 F.2d 1093 (D.C. Cir. 1970); Allen v. Hickel, 424 F.2d 944 (D.C. Cir. 1970); Brooks v. Volpe, 329 F. Supp. 118 (W.D. Wash. 1971); Delaware v. Pennsylvania N.Y. Cent. Transp. Co., 323 F. Supp. 487 (D. Del. 1971); Izaak Walton League of America v. St. Clair, 313 F. Supp. 1312 (D. Minn. 1970); Pennsylvania Environmental Council, Inc. v. Bartlett, 315 F. Supp. 238 (M.D. Pa. 1970).

62. Sierra Club v. Hickel, 433 F.2d 24 (9th Cir. 1970), *cert. granted sub nom.* Sierra Club v. Morton, 401 U.S. 907 (1971) (No. 70-34). The Sierra Club, a non-profit California corporation concerned with environmental protection, claimed that its interest in the conservation and sound management of natural parks would be adversely affected by an Interior permit allowing Walt Disney to construct the Mineral King Resort in Sequoia National Forest. The court held that because of the Sierra Club's failure to assert a direct legal interest, that organization lacked standing to sue. The court stated that the Sierra Club had claimed an interest only in the sense that the proposed course of action was displeasing to its members. The court purported to distinguish *Scenic Hudson* on the grounds that the plaintiff's claim of standing there was aided by the "aggrieved party" language of the Federal Power Act.

63. 16 U.S.C. §§ 791(a) *et seq.* (1970). *See* note 59 and accompanying text *supra.*

64. 5 U.S.C. §§ 551 *et seq.* (1970). Decisions relying upon 5 U.S.C. § 702 are listed in note 56 *supra.*

65. 7 U.S.C. §§ 135 *et seq.* (1970). Section 135b(d) affords a right of judicial review to anyone "adversely affected" by an order under the Act. *See* Environmental Defense Fund, Inc. v. Hardin, 428 F.2d 1093, 1096 (D.C. Cir. 1970).

66. 324 F. Supp. 412 (N.D., M.D. & S.D. Ala. 1970), *aff'd mem., sub nom.* Bass Anglers Sportsman Soc'y of America, Inc. v. Koppers Co., 447 F.2d 1304 (5th Cir. 1971).

67. Section 13 of Rivers and Harbors Appropriation Act of 1899.

68. 33 U.S.C. § 411 (1970) reads:

> Every person and every corporation that shall violate, or that shall knowingly aid, abet, authorize, or instigate a violation of the provisions of sections 407, 408, and 409 of the title shall ... be punished by a fine ... or by imprisonment ... in the discretion of the court, one-half of said fine to be paid to the person or persons giving information which shall lead to conviction.

69. This is from the latin, "who brings the action as well for the King as for himself," referring to an action brought by a citizen for the state as well as for himself.

> 70. These sections create a criminal liability. No civil action lies to enforce it; criminal statutes can only be enforced by the government. A qui tam action lies only when expressly or impliedly authorized by statute to enforce a penalty by civil action, not a criminal fine.

324 F. Supp. 412, 415–16 (ND., M.D. & S.D. Ala. 1970). Other *qui tam* actions brought by the Bass Angler Sportsman Society have been similarly unsuccessful. *See* Bass Anglers Sportsman Soc'y of America v. Scholze Tannery, 329 F. Supp. 339 (E.D. Tenn. 1971); Bass Anglers Sportsman's Soc'y of America v. United States Plywood-Champion Papers, Inc., 324 F. Supp. 302 (S.D. Tex. 1971).

71. Concern over an anticipated flood of litigation initiated by environmental organizations is evident in Judge Trask's opinion in Alameda Conservation Ass'n v. California, 437 F.2d 1087 (9th Cir.), *cert. denied,* Leslie Salt Co. v. Alameda Conservation Ass'n, 402 U.S. 908 (1971), where a non-profit corporation having as a primary purpose protection of the public's interest in San Francisco Bay was denied standing to seek an injunction prohibiting a land exchange that would allegedly destroy wildlife, fisheries and the Bay's unique flushing characteristics:

> Standing is not established by suit initiated by this association simply because it has as one of its purposes the protection of the "public interest" in the waters of the San Francisco Bay. However well intentioned the members may be, they may not by uniting create for themselves a super-administrative agency or a *parens patriae* official status with the capability of over-seeing and of challenging the action of the appointed and elected officials of the

> state government. Although recent decisions have considerably broadened the concept of standing, we do not find that they go this far. [Citation.]
>
> Were it otherwise the various clubs, political, economic and social now or yet to be organized, could wreak havoc with the administration of government, both federal and state. There are other forums where their voices and their views may be effectively presented, but to have standing to submit a "case or controversy" to a federal court, something more must be shown.

437 F.2d at 1090.

72. *See* note 49 *supra.*

73. Here, too, we are dogged by the ontological problem discussed in note 26 *supra.* It is easier to say that the smog-endangered stand of pines "wants" the smog stopped (assuming that to be a jurally significant entity) then it is to venture that the mountain, or the planet earth, or the cosmos, is concerned about whether the pines stand or fall. The more encompassing the entity of concern, the less certain we can be in venturing judgments as to the "wants" of any particular substance, quality, or species within the universe. Does the cosmos care if we humans persist or not? "Heaven and earth … regard all things as insignificant, as though they were playthings made of straw." LAO-TZU, TAO TEH KING 13 (D. Goddard transl. 1919).

74. *See* Knight v. United States Land Ass'n, 142 U.S. 161, 181 (1891).

75. Clause 2 gives Congress the power "to dispose of and make all needful Rules and Regulations respecting the Territory or other Property belonging to the United States."

76. *See* Flemming v. Nestor, 363 U.S. 603 (1960).

77. *See* the L. A. Times editorial *Water: Public vs. Polluters* criticizing:

> … the ridiculous built-in conflict of interests on Regional Water Quality Control Board. By law, five of the seven seats are given to spokesmen for industrial, governmental, agricultural or utility users. Only one representative of the public at large is authorized, along with a delegate from fish and game interests.

Feb. 12, 1969, Part II, at 8, cols. 1–2.

78. The Federal Refuse Act is over 70 years old. Refuse Act of 1899, 33 U.S.C. § 407 (1970).

79. *See* Hall, *Refuse Act of 1899 and the Permit Program,* 1 NAT'L RES. DEFENSE COUNCIL NEWSLETTER 1 (1971).

80. To be effective as a deterrent, the sanction ought to be high enough to bring about an internal reorganization of the corporate structure which minimizes the chances of future violations. Because the corporation is not necessarily a profit-maximizing "rationally economic man," there is no reason to believe that setting the fine as high as—but no higher than—anticipated profits from the violation of the law, will bring the illegal behavior to an end.

81. Udall v. FPC, 387 U.S. 428, 437 n.6 (1967). *See also* Holmes, J. in New Jersey v. New York, 283 U.S. 336, 342 (1931): "A river is more than an amenity, it is a treasure. It offers a necessity of life that must be rationed among those who have power over it."

82. To simplify the description, I am using here an ordinary language sense of causality, *i.e.*, assuming that the pollution causes harm to the river. As Professor Coase has pointed out in *The Problem of Social Cost,* 3 J. LAW & ECON. 1 (1960), harm-causing can be viewed as a reciprocal problem, *i.e.*, in the terms of the text, the mill wants to harm the river, and the river—if we assume it "wants" to maintain its present environmental quality—"wants" to harm the mill. Coase rightly points out that at least in theory (if we had the data) we ought to be comparing the alternative social product of different social arrangements, and not simply imposing full costs on the party who would popularly be identified as the harm-causer.

83. I am assuming that one of the considerations that goes into a judgment of "remoteness" is a desire to discourage burdensome amounts of petty litigation. This is one of the reasons why a court would be inclined to say—to use the example in the text—that the man who sells fishing tackle and bait has not been "proximately" injured by the polluter. Using proximate cause in this manner, the courts can protect themselves from a flood of litigation. But once the guardian were in court anyway, this consideration would not obtain as strongly, and courts might be more inclined to allow proof on the damages to remotely injured humans (although the proof itself is an added burden of sorts).

84. *Cf.* Golding, *Ethical Issues in Biological Engineering,* 15 U.C.L.A.L. REV. 443, 451–63 (1968).

85. Of course, in the instance of copyright and patent protection, the creation of the "property right" can be more directly justified on homocentric grounds.

86. *See* Schrag, *Life on a Dying Lake,* in THE POLITICS OF NEGLECT 167, at 173 (R. Meek & J. Straayer eds. 1971).

87. One ought to observe, too, that in terms of real effect on marginal welfare, the poor quite possibly will bear the brunt of the compromises. They may lack the wherewithal to get out to the countryside—and probably want an increase in material goods more acutely than those who now have riches.

88. On November 2, 1971, the Senate, by a vote of 86–0, passed and sent to the House the proposed Federal Water Pollution Control Act Amendments of 1971, 117 CONG. REC. S17464 (daily ed. Nov. 2, 1971). Sections 101(a) and (a)(1) of the bill declare it to be "national policy that, consistent with the provisions of this Act—(1) the discharge of pollutants into the navigable waters be eliminated by 1985." S.2770, 92d Cong., 1st Sess., 117 CONG. REC. S17464 (daily ed. Nov. 2, 1971).

89. 354 F.2d 608, 624 (2d Cir. 1965).

90. Again, there is a problem involving what we conceive to be the injured entity. *See* notes 26, 73 *supra.*

91. N.Y. Times, Jan. 14, 1971, § 1, col. 2, and at 74, col. 7.

92. Courts have not been reluctant to award damages for the destruction of heirlooms, literary manuscripts or other property having no ascertainable market value. In Willard v. Valley Gas Fuel Co., 171 Cal. 9, 151 Pac. 286 (1915), it was held that the measure of damages for the negligent destruction of a rare old book written by one of plaintiff's ancestors was the amount which would compensate the owner for all detriment including

sentimental loss proximately caused by such destruction. The court, at 171 Cal. 15, 151 Pac. 289, quoted approvingly from Southern Express Co. v. Owens, 146 Ala. 412, 426, 41 S. 752, 755 (1906):

> Ordinarily, where property has a market value that can be shown, such value is the criterion by which actual damages for its destruction or loss may be fixed. But it may be that property destroyed or lost has no market value. In such state of the case, while it may be that no rule which will be absolutely certain to do justice between the parties can be laid down, it does not follow from this, nor is it the law, that the plaintiff must be turned out of court with nominal damages merely. Where the article or thing is so unusual in its character that market value cannot be predicated of it, its value, or plaintiff's damages, must be ascertained in some other rational way and from such elements as are attainable.

Similarly, courts award damages in wrongful death actions despite the impossibility of precisely appraising the damages in such cases. In affirming a judgment in favor of the administrator of the estate of a child killed by defendant's automobile, the Oregon Supreme Court, in Lane v. Hatfield, 173 Or. 79, 88–89, 143 P.2d 230, 234 (1943), acknowledged the speculative nature of the measure of damages:

> No one knows or can know when, if at all, a seven year old girl will attain her majority, for her marriage may take place before she has become twenty-one years of age.... Moreover, there is much uncertainty with respect to the length of time anyone may live. A similar uncertainty veils the future of a minor's earning capacity or habit of saving. Illness or a non-fatal accident may reduce an otherwise valuable and lucrative life to a burden and liability.
>
> The rule, that the measure of recovery by a personal representative for the wrongful death of his decedent is the value of the life of such decedent, if he had not come to such an untimely end, has been termed vague, uncertain and speculative if not, conjectural. It is, however, the best that judicial wisdom has been able to formulate.

93. It is not easy to dismiss the idea of "lower" life having consciousness and feeling pain, especially since it is so difficult to know what these terms mean even as applied to humans. *See* Austin, *Other Minds*, in *Logic and Language* 342 (S. Flew ed. 1965); Schopenhauer, *On the Will in Nature*, in TWO ESSAYS BY ARTHUR SCHOPENHAUER 193, 281–304 (1889). Some experiments on plant sensitivity—of varying degrees of extravagance in their claims—include Lawrence, *Plants Have Feelings, Too ...*, ORGANIC GARDENING & FARMING 64 (April 1971); Woodlief, Royster & Huang, *Effect of Random Noise on Plant Growth*, 46 J. ACOUSTICAL SOC. AM. 481 (1969); Backster, *Evidence of a Primary Perception in Plant Life*, 10 INT'L J. PARAPSYCHOLOGY 250 (1968).

94. *See* note 39 *supra*, and Coase, note 82 *supra*.

95. *See* FED. R. CIV. P. 23 and note 49 *supra*.

96. This is an ideal, of course—like the ideal that no human being ought to interfere with any other human being. *See* Dyke, *Freedom, Consent and the Costs of Interaction*, and Stone, *Comment*, in IS LAW DEAD? 134–67 (E. Rostow ed. 1971). Some damages would inevitably be *damnum absque injuria*. *See* note 93 *supra*.

97. The inevitability of some form of evolution is not inconsistent with the establishment of a legal system that attempts to interfere with or ameliorate the process: is the same not true of the human law we now have, *e.g.*, the laws against murder?

98. Holmes, *Early Forms of Liability*, in THE COMMON LAW (1881), discusses the liability of animals and inanimate objects in early Greek, early Roman and some later law. Alfred's Laws (A.D. 871–901) provided, for example, that a tree by which a man was killed should "be given to the kindred, and let them have it off the land within 30 nights." *Id.* at 19. In Edward I's time, if a man fell from a tree the tree was deodand. *Id.* at 24. Perhaps the liability of non-human matter is, in the history of things, part of a paranoid, defensive phase in man's development; as humans become more abundant, both from the point of material wealth and iternally, they may be willing to allow an advance to the stage where non-human matter has rights.

99. *See* note 26 *supra*. In the event that a person built his house near the edge of a river that flooded, would "assumption of the risk" be available on the river's behalf?

100. *See* Dixon v. Alabama State Bd. of Educ., 294 F.2d 150 (5th Cir.), *cert. denied*, 368 U.S. 930 (1961); Comment, *Private Government on the Campus—Judicial Review of University Expulsions*, 72 YALE L.J. 1362 (1963).

101. National Environmental Policy Act. 92 U.S.C. § 4332 (1970).

102. *See* Committee for Nuclear Responsibility Inc. v. Schlesinger, 40 U.S.L.W. 3214 (Nov. 5, 1971) (Douglas, J. dissent to denial of application for injunction in aid of jurisdiction).

103. 42 U.S.C. § 4342 (1970).

104. As an indication of what lower-level management is apt to do, *see* Ehrenreich & Ehrenreich, *Conscience of a Steel Worker*, 213 THE NATION 268 (1971). One steel company's "major concession [toward obedience to the 1899 Refuse Act, note 78 *supra*] was to order the workers to confine oil dumping to the night shift. 'During the day the Coast Guard patrols. But at night, the water's black, the oil's black; no one can tell.'" An effective corporation law would assure that the internal information channels within a corporation were capable of forcing such matters to the attention of high-level officials. Even then, there is no guarantee that the law will be obeyed—but we may have improved the odds.

105. In the case of Lake Erie, in addition to the considerations that follow in the text, there were possibly additional factors such as that no one polluter's acts could be characterized as inflicting irreparable injury.

106. *See* for example Justice Reed's opinion for the Court in Kovacs v. Cooper, 336 U.S. 77 (1949) (*but see* Mr. Justice Frankfurter's concurring opinion, 336 U.S. at 89–96), and United States v. Carolene Products, 304 U.S. 144, 152 n.4 (1938).

107. Note that in the discussion that follows I am referring to legislative apportionment, not voting proper.

108. In point of fact, there is no reason to suppose that an increase of Congressmen for Alaska would be a benefit to the environment; the reality of the political situation might just as likely result in the election of additional Congressmen with closer ties to oil companies and other developers.

109. *See* Simpson, *The Analysis of Legal Concepts*, 80 LAW Q. REV. 535 (1964).

110. Krier, *Environmental Litigation and the Burden of Proof*, in LAW AND THE ENVIRONMENT 105 (M. Baldwin & J. Page eds. 1970). *See* Texas East Trans. Corp. v. Wildlife Preserves, 48 N.J. 261, 225 A.2d 130 (1966). There, where a corporation set up to maintain a wildlife preserve resisted condemnation for the construction of plaintiff's pipe line, the court ruled that "... the *quantum* of proof required of this defendant to show arbitrariness against it would not be as substantial as that to be assumed by the ordinary property owner who devotes his land to conventional uses." 225 A.2d at 137.

111. See Stone, *Existential Humanism and the Law*, in EXISTENTIAL HUMANISTIC PSYCHOLOGY 151 (T. Greening ed. 1971).

112. National Environmental Policy Act, 42 U.S.C. §§ 4321–47 (1970).

113. *See* note 65 *supra.*

114. W. HOLDSWORTH, HISTORY OF ENGLISH LAW 45 (5th ed. 1931).

115. Note that it is in no small way the law that imposes this manner of speech on businessmen. *See* Dodge v. Ford Motor Co., 204 Mich. 459, 499–505, 170 N.W. 668, 682–83 (1919) (holding that Henry Ford, as dominant stockholder in Ford Motor Co., could not withhold dividends in the interests of operating the company "as a semi-elecmosynary institution and not as a business institution").

116. I. KANT, PHILOSOPHY OF LAW 195 (Hastle Transl. 1887).

117. I. KANT, *The Metaphysics of Morality*, in THE PHILOSOPHY OF KANT § 1 at 230–31 (J. Watson transl. 1908).

118. Engel, *Reasons, Morals and Philosophic Irony*, in LANGUAGE AND ILLUMINATION 60 (1969).

119. L. WITTGENSTEIN, TRACTATUS LOGICO-PHILOSOPHICUS §§ 6.421, 6.522 (D. Pears & B. McGuinness transl. 1961).

120. Cousteau, *The Oceans: No Time to Lose*, L.A. Times. Oct. 24, 1971, § (opinion), at 1, col. 4.

121. *See* J. HARTE & R. SOCOLOW, PATIENT EARTH (1971).

122. McHarg, *Values, Process and Form*, in THE FITNESS OF MAN'S ENVIRONMENT 213–14 (1968).

123. Murphy, *supra* note 27, at 477.

124. On the other hand, the statement in text, and the previous one of Professor Murphy, may be a bit severe. One could as easily claim that Christianity has had no influence on overt human behavior in light of the killings that have been carried out by professed Christians, often in its name. *Feng shui* has, on all accounts I am familiar with, influenced the development of land in China *See* Freedman, *Geomancy*, 1968 PROCEEDINGS OF THE ROYAL ANTHROPOLOGICAL INSTITUTE OF GREAT BRITAIN AND IRELAND 5; March, *An Appreciation of Chinese Geomancy*, 27 J. ASIAN STUDIES 253 (1968).

125. The legal system does the best it can to maintain the illusion of the reality of the individual human being. Consider, for example, how many constitutional cases, brought in the name of some handy individual, represent a power struggle between institutions—the NAACP and a school board, the Catholic Church and a school board, the ACLU and the Army, and so forth. Are the individual human plaintiffs the real moving causes of these cases—or an afterthought?

When we recognize that our problems are increasingly institutional, we would see that the solution, if there is one, must involve coming to grips with how the "corporate" (in the broadest sense) entity is directed, and we must alter our views of "property" in the fashion that is needed to regulate organizations successfully. For example, instead of ineffectual, after-the-fact criminal fines we should have more preventive in-plant inspections, notwithstanding the protests of "invasion of [corporate] privacy."

In-plant inspection of production facilities and records is presently allowed only in a narrow range of areas, *e.g.*, in federal law, under the Federal Food, Drug, and Cosmetic Act, 21 U.S.C. § 374 *et seq.* (1970), and provisions for meat inspection, 21 U.S.C. § 608 (1970). Similarly, under local building codes we do not wait for a building to collapse before authoritative sources inquire into the materials and procedures that are being used in the construction; inspectors typically come on site to check the progress at every critical stage. A sensible preventive legal system calls for extending the ambit of industries covered by comparable "privacy invading" systems of inspection.

126. G. HEGEL, HEGEL'S PHILOSOPHY OF RIGHT 41 (T. Knox transl. 1945).

127. C. MCCULLERS, THE BALLAD OF THE SAD CAFE AND OTHER STORIES 150–51 (1958).

128. Consider what Schopenhauer was writing "Of Women," about the time the Wisconsin Supreme Court was explaining why women were unfit to practice law, note 23 *supra:*

> You need only look at the way in which she is formed, to see that woman is not meant to undergo great labour, whether of the mind or of the body. She pays the debt of life not by what she does, but by what she suffers; by the pains of childbearing and care for the child, and by submission to her husband, to whom she should be a patient and cheering companion. The keenest sorrows and joys are not for her, nor is she called upon to display a great deal of strength. The current of her life should be more gentle, peaceful and trivial than man's, without being essentially happier or unhappier.
>
> Women are directly fitted for acting as the nurses and teachers of our early childhood by the fact that they are themselves childish, frivolous and short-sighted; in a word, they are big children all their life long—a kind of intermediate stage between the child and the full-grown man, who is man in the strict sense of the word....
>
> However many disadvantages all this may involve, there is at least this to be said in its favour: that the woman lives more in the present than the man, and that, if the present is at all tolerable, she enjoys it more eagerly. This is the source of that cheerfulness

> which is peculiar to woman, fitting her to amuse man in his hours of recreation, and, in case of need, to console him when he is borne down by the weight of his cares.
>
> ....
>
> ... [I]t will be found that the fundamental fault of the female character is that it has *no sense of justice.* This is mainly due to the fact, already mentioned, that women are defective in the powers of reasoning and deliberation; but it is also traceable to the position which Nature has assigned to them as the weaker sex. They are dependent, not upon strength, but upon craft; and hence their instinctive capacity for cunning, and their ineradicable tendency to say what is not true. ***For as lions are provided with claws and teeth, and elephants and boars with tusks, bulls with horns, and the cuttle fish with its cloud of inky fluid, so Nature has equipped woman, for her defense and protection, with the arts of dissimulation; and all the power which Nature has conferred upon man in the shape of physical strength and reason, has been bestowed upon women in this form. Hence, dissimulation is innate in woman, and almost as much a quality of the stupid as of the clever....

A. SCHOPENHAUER, *On Women*, in STUDIES IN PESSIMISM 105–10 (T. B. Saunders transl. 1893).

If a man should write such insensitive drivel today, we would suspect him of being morally and emotionally blind. Will the future judge us otherwise, for venting rather than examining the needs that impel us to treat the environment as a senseless object—to blast to pieces some small atoll to find out whether an atomic weapon works?

129. Of course, the phase one looks toward is a time in which such sentiments need not be prescribed *by law.*

130. The "Purpose of the Legislation" in H.R. Rep. No. 91-1651, 91st Cong., 2d Sess. to the "[Animal] Welfare Act of 1970," 3 U.S. CODE CONG. & ADMIN. NEWS 5103, 5104 (1970). Some of the West Publishing Co. typesetters may not be quite ready for this yet; they printed out the title as "Annual Welfare Act of 1970."

131. *See* MCCALL'S May, 1971, at 44.

132. Environmental Defense Fund, Inc. v. Hardin, 428 F.2d 1093, 1096 (D.C. Cir. 1970). Plaintiffs would thus seem to have urged a broader than literal reading of the statute, 7 U.S.C. § 135(z) (2) (d) (1970), which refers to "... living man and other vertebrate animals, vegetation, and useful invertebrate animals."

E.D.F. was joined as petitioners by the National Audubon Society, the Sierra Club, and the West Michigan Environmental Action Council, 428 F.2d at 1094–95 n.5.

133. In the case of the bestowal of rights on other humans, the action also helps the recipient to discover new personal depths and possibilities—new dignity—within himself. I do not want to make much of the possibility that this effect would be relevant in the case of bestowing rights on the environment. But I would not dismiss it out of hand, either. How, after all, do we judge that a man is, say, "flourishing with a new sense of pride and dignity?" What we mean by such statements, and the nature of the evidence upon which we rely in support of them, is quite complex. *See* Austin, note 93 *supra.* A tree treated in a "rightful" manner would respond in a manner that, when described, would sound much like the response of a person accorded "new dignity." *See also* note 93 *supra.*

134. D. RUDHYAR, DIRECTIVES FOR NEW LIFE 21–23 (1971).

135. *See* Stone, note 111 *supra.*

136. San Antonio Conservation Soc'y v. Texas Highway Dep't, *cert. denied,* 400 U.S. 968 (1970) (Black, J. dissenting to denial of certiorari).

137. *Id.* at 969.

138. *Id.* at 971.

IS THERE A NEED FOR A NEW, AN ENVIRONMENTAL, ETHIC?

SOURCE Sylvan (Routley), Richard. 2001 (1973). In *Environmental Philosophy,* 2nd edition, 17–25. Michael E. Zimmerman, J. Baird Callicott, et al. Saddle River, NJ: Prentice-Hall. Reprinted with permission of the author.

INTRODUCTION *Richard Sylvan (formerly Routley) (1935–1996) taught at the University of Sydney, University of New England, and Monash University. For twenty-five years he was a fellow in the Research School of Social Sciences at the Australian National University. "Is There a Need for a New, an Environmental, Ethic?" was originally published in the 1973* Proceedings of the XV World Congress of Philosophy. *Sylvan begins by questioning whether we need a new ethic regulating human interaction with the natural environment. The dominant ethical systems do address environmental concerns, but only in an unsatisfactory, anthropocentric manner. The question is, do we need to extend or modify our ethics to include the natural world, or create a new one? Sylvan identifies three approaches to the environment in prevailing Western ethics—dominance, stewardship, and cooperation. He argues that all three are inconsistent with an adequate environmental ethic, and thus that we need a new—a distinctly environmental—ethic. Sylvan demonstrates his point by identifying a core principle of Western ethical systems—*basic (human) chauvinism*— and shows how it is inconsistent with a true environmental ethic. He then provides a series of counter-examples to this principle, such as the famous* last man *example, which supposes a sole survivor of a major collapse who would be free, according to the dominant Western ethical paradigm, to exterminate all life around him. Sylvan argues that our intuition is that he would still be acting unethically to do so—thus pointing up the need for a new, an environmental ethic.*

1

It is increasingly said that civilization, Western civilization at least, stands in need of a new ethic (and derivatively of a new economics) setting out people's relations to the natural environment, in Leopold's words "an ethic dealing with man's relation to land and to the animals and plants which grow upon it."[1] It is not of course that old and prevailing ethics do not deal with man's relation to nature; they do, and on the prevailing view man is free to deal with nature as he pleases, i.e., his relations with nature, insofar at least as they do not affect others, are not subject to moral censure. Thus assertions such as "Crusoe ought not to be mutilating those trees" are significant and morally determinate but, inasmuch at least as Crusoe's actions do not interfere with others, they are false or do not hold—and trees are not, in a good sense, moral objects.[2] It is to this, to the values and evaluations of the prevailing ethics, that Leopold and others in fact take exception. Leopold regards as subject to moral criticism, as wrong, behaviour that on prevailing views is morally permissible. But it is not, as Leopold seems to think, that such behavior is beyond the scope of the prevailing ethics and that an extension of traditional morality is required to cover such cases, to fill a moral void. If Leopold is right in his criticism of prevailing conduct what is required is a change in the ethics, in attitudes, values and evaluations. For as matters stand, as he himself explains, men do not feel morally ashamed if they interfere with a wilderness, if they maltreat the land, extract from it whatever it will yield, and then move on; and such conduct is not taken to interfere with and does not rouse the moral indignation of others. "A farmer who clears the woods off a 75% slope, turns his cows into the clearing, and dumps its rainfall, rocks, and soil into the community creek, is still (if otherwise decent) a respected member of society."[3] Under what we shall call an *environmental ethic* such traditionally permissible conduct would be accounted morally wrong, and the farmer subject to proper moral criticism.

Let us grant such evaluations for the purpose of the argument. What is not so clear is that a new ethic is required even for such radical judgments. For one thing it is none too clear what is going to count as a new ethic, much as it is often unclear whether a new development in physics counts as a new physics or just as a modification or extension of the old. For, notoriously, ethics are not clearly articulated or at all well worked out, so that the application of identity criteria for ethics may remain obscure.[4] Furthermore we tend to cluster a family of ethical systems which do not differ on core or fundamental principles together as one ethic; e.g. the Christian ethic, which is an umbrella notion covering a cluster of differing and even competing systems. In fact then there are two other possibilities, apart from a new environmental ethic, which might cater for the evaluations, namely that of an extension or modification of the prevailing ethics or that of the development of principles that are already encompassed or latent within the prevailing ethic. The second possibility, that environmental evaluations can be incorporated within (and ecological problems solved within) the framework of prevailing Western ethics, is open because there isn't a single ethical system uniquely assumed in Western civilization: on many issues, and especially on controversial issues such as infanticide, women's rights, and drugs, there are competing sets of principles. Talk of a new ethic and prevailing ethics tends to suggest a sort of monolithic structure, a uniformity, that prevailing ethics, and even a single ethic, need not have.

Indeed Passmore has mapped out three important traditions in Western ethical views concerning man's relation to nature; a dominant tradition, the despotic position, with man as despot (or tyrant), and two lesser traditions, the stewardship position, with man as custodian, and the co-operative position with man as perfecter.[5] Nor are these the only traditions; primitivism is another, and both romanticism and mysticism have influenced Western views.

The dominant Western view is simply inconsistent with an environmental ethic; for according to it nature is the dominion of man and he is free to deal with it as he pleases (since—at least on the mainstream Stoic-Augustine view—it exists only for his sake), whereas on an environmental ethic man is not so free to do as he pleases. But it is not quite so obvious that an environmental ethic cannot be coupled with one of the lesser traditions. Part of the problem is that the lesser traditions are by no means adequately characterized anywhere, especially when the religious backdrop is removed, e.g. *who* is man steward for and responsible to? However both traditions are inconsistent with an environmental ethic because they imply policies of complete interference, whereas on an environmental ethic some worthwhile parts of the earth's surface should be preserved from substantial human interference, whether of the "improving" sort or not. Both traditions would in fact prefer to see the earth's land surfaces reshaped along the lines of the tame and comfortable north-European small farm and village landscape. According to the co-operative position man's proper role is to develop, cultivate and perfect nature—all nature eventually—by bringing out its potentialities, the test of perfection being primarily usefulness for human purposes; while on the stewardship view man's role, like that of a farm manager, is to make nature productive by his efforts though not by means that will deliberately degrade its resources. Although these positions both depart from the dominant position in a way which enables the incorporation of some evaluations of an environmental ethic, e.g. some of those concerning the irresponsible farmer, they do not go far enough: for in the present situation of expanding populations confined to finite natural areas, they will lead to, and enjoin, the perfecting, farming and utilizing of all natural areas. Indeed these lesser traditions lead to, what a thoroughgoing environmental ethic would reject, a principle of total use, implying that every natural area should be cultivated or otherwise used for human ends, "humanized."[6]

As the important Western traditions exclude an environmental ethic, it would appear that such an ethic, not primitive, mystical or romantic, would be new all right. The matter is not so straightforward; for the dominant ethic has been substantially qualified by the rider that one is not always entitled to do as one pleases where this physically interferes with others. Maybe some such proviso was implicit all along (despite evidence to the contrary), and it was simply assumed that doing what one pleased with natural items would not affect others (the non-interference assumption). Be this as it may, the *modified* dominant position appears, at least for many thinkers, to have supplanted the dominant position; and the modified position can undoubtedly go much further towards an environmental ethic. For example, the farmer's polluting of a community stream may be ruled immoral on the grounds that it physically interferes with others who use or would use the streams. Likewise business enterprises which destroy the natural environment for no satisfactory returns or which cause pollution deleterious to the health of future humans, can be criticized on the sort of welfare basis (e.g. that of Barkley and Seckler) that blends with the modified position; and so on.[7] The position may even serve to restrict the sort of family size one is entitled to have since in a finite situation excessive population levels will interfere with future people. Nonetheless neither the modified dominant position nor its Western variants, obtained by combining it with the lesser traditions, is adequate as an environmental ethic, as I shall try to show. A new ethic is wanted.

2

As we noticed (an) *ethic* is ambiguous, as between a specific ethical system, a *specific* ethic, and a more generic notion, a super ethic, under which specific ethics cluster.[8] An ethical system S is, near enough, a propositional system (i.e. a structured set of propositions) or theory which includes (like individuals of a theory) a set of values and (like postulates of a theory) a set of general evaluative judgments concerning conduct, typically of what is obligatory, permissible and wrong, of what are rights, what is valued, and so forth. A general or lawlike proposition of a system is a principle; and certainly if systems S_1 and S_2 contain different principles, then they are different systems. It follows that any environmental ethic differs from the important traditional ethics outlined. Moreover if environmental ethics differ from Western ethical systems on some core principle embedded in Western systems, then these systems differ from the Western super ethic (assuming, what seems to be so, that it can be uniquely characterized)—in which case if an environmental ethic is needed then a new ethic is wanted. It suffices then to locate a core principle and to provide environmental counter examples to it.

It is commonly assumed that there are what amount to core principles of Western ethical systems, principles that will accordingly belong to the super ethic. The fairness principle inscribed in the Golden Rule provides one example. Directly relevant here, as a good stab at a core principle, is the commonly formulated liberal principle of the modified dominance position. A recent formulation runs as follows:

> "The liberal philosophy of the Western world holds that one should be able to do what lie wishes, providing (1) that he does not harm others and (2) that he is not likely to harm himself irreparably."[9]

Let us call this principle *basic (human) chauvinism*—because under it humans, or people, come first and everything else a bad last—though sometimes the principle is hailed as a *freedom* principle because it gives permission to perform a wide range of actions (including actions which mess up the environment and natural things) providing they do not harm others. In fact it tends to cunningly shift the onus of proof to others. It is worth remarking that *harming others* in the restriction is narrower than a restriction to the (usual) interests of others; it is not enough that it is in my interests, because I detest you, that you stop breathing; you are free to breathe, for the time being anyway, because it does not harm me. There remains a problem however as to exactly what counts as harm or interference. Moreover the width of the principle is so far obscure because "other" may be filled out in significantly different ways: it makes a difference to the extent, and privilege, of the chauvinism whether "other" expands to "other human"—which is too restrictive—or to "other person" or to "other sentient being"; and it makes a difference to the adequacy of the principle, and inversely to its economic applicability, to which class of others it is intended to apply, whether to future as well as to present others, whether to remote future others or only to non-discountable future others and whether to possible others. The latter would make the principle completely unworkable, and it is generally assumed that it applies at most to present and future others.

It is taken for granted in designing counter examples to basic chauvinist principles, that a semantic analysis of permissibility and obligation statements stretches out over ideal situations (which may be incomplete or even inconsistent), so that what is permissible holds in some ideal situation, what is obligatory in every ideal situation, and what is wrong is excluded in every ideal situation. But the main point to grasp for the counter examples that follow, is that ethical principles if correct are universal and are assessed over the class of ideal situations.

> (i) The *last man* example. The last man (or person) surviving the collapse of the world system lays about him, eliminating, as far as he can, every living thing, animal or plant (but painlessly if you like, as at the best abattoirs). What he does is quite permissible according to basic chauvinism, but on environmental grounds what he does is wrong. Moreover one does not have to be committed to esoteric values to regard Mr. Last Man as behaving badly (the reason being perhaps that radical thinking and values have shifted in an environmental direction in advance of corresponding shifts in the formulation of fundamental evaluative principles).

> (ii) The *last people* example. The last man example can be broadened to the last people example. We can assume that they know they are the last people, e.g. because they are aware that radiation effects have blocked any chance of reproduction. One considers the last people in order to rule out the possibility that what these people do harms or somehow physically interferes with later people. Otherwise one could as well consider science fiction cases where people arrive at a new planet and destroy its ecosystems, whether with good intentions such as perfecting the planet for their ends and making it more fruitful or, forgetting the lesser traditions, just for the hell of it.

Let us assume that the last people are very numerous. They humanely exterminate every wild animal and they eliminate the fish of the seas, they put all arable land under intensive cultivation, and all remaining forests disappear in favour of quarries or plantations, and so on. They may give various familiar reasons for this, e.g. they believe it is the way to salvation or to perfection, or they are simply satisfying reasonable needs, or even that it is needed to keep the last people employed or occupied so that they do not worry too much about their impending extinction. On an environmental ethic the last people have behaved badly; they have simplified and largely destroyed all the natural ecosystems, and with their demise the world will soon be an ugly and largely wrecked place. But this conduct may conform with the basic chauvinist principle, and as well with the principles enjoined by the lesser traditions. Indeed the main point of elaborating this example is because, as the last man example reveals, basic chauvinism may conflict with stewardship or co-operation principles. The conflict may be removed it seems by conjoining a further proviso to the basic principle, the effect (3) that he does not willfully destroy natural resources. But as the last people do not destroy resources willfully, but perhaps "for the best of reasons," the variant is still environmentally inadequate.

> (iii) The *great entrepreneur* example. The last man example can be adjusted so as to not fall foul of clause (3). The last man is an industrialist; he runs a giant complex of automated factories and farms which he proceeds to extend. He produces automobiles among other things, from renewable and recyclable resources of course, only he dumps and recycles these shortly after manufacture and sale to a dummy buyer instead of putting them on the road for a short time as we do. Of course he has the best of reasons for his activity, e.g. he is increasing gross world product, or he is improving output to fulfill some plan, and he will be increasing his own and general welfare since he much prefers increased output and productivity. The entrepreneur's behavior is on the Western ethic quite permissible; indeed his conduct is commonly thought to be quite fine and may even meet Pareto optimality requirements given prevailing notions of being "better off."

Just as we can extend the last man example to a class of last people, so we can extend this example to the *industrial society* example: the society looks rather like ours.

> (iv) The *vanishing species* example. Consider the blue whale, a mixed good on the economic picture. The blue whale is on the verge of extinction because of his qualities as a private good, as a source of valuable oil and meat. The catching and marketing of blue whales does not harm the whalers; it does not harm or physically interfere with others in any good sense, though it may upset them and they may be prepared to compensate the whalers if they desist; nor need whale hunting be willful destruction. (Slightly different examples which eliminate the hunting aspect of the blue whale example are provided by cases where a species is eliminated or threatened through destruction of its habitat by man's activity or the activities of animals he has introduced, e.g. many plains-dwelling Australian marsupials and the Arabian oryx.) The behavior of the whalers in eliminating this magnificent species of whale is accordingly quite permissible—at least according to basic chauvinism. But on an environmental ethic it is not. However, the free-market mechanism will not cease allocating whales to commercial uses, as a satisfactory environmental economics would; instead the market model will grind inexorably along the private demand curve until the blue whale population is no longer viable—if that point has not already been passed.[10]

In sum, the class of permissible actions that rebound on the environment is more narrowly circumscribed on an environmental ethic than it is in the Western super ethic. But aren't environmentalists going too far in claiming that these people, those of the examples and respected industrialists, fishermen and farmers are behaving, when engaging in environmentally degrading activities of the sort described, in a morally impermissible way? No, what these people do is to a greater or lesser extent evil, and hence in serious cases morally impermissible. For example, insofar as the killing or forced displacement of primitive peoples who stand in the way of an industrial development is morally indefensible and impermissible, so also is the slaughter of the last remaining blue whales for private profit. But how to reformulate basic chauvinism as a satisfactory freedom principle is a more difficult matter. A tentative, but none too adequate beginning might be made by extending (2) to include harm to or interference with others who would be so affected by the action in question were they placed in the environment and (3) to exclude speciecide. It may be preferable, in view of the way the freedom principle sets the onus of proof, simply to scrap it altogether, and instead to specify classes of rights and permissible conduct, as in a bill of rights.

3

A radical change in a theory sometimes forces changes in the meta-theory; e.g. a logic which rejects the Reference Theory in a thoroughgoing way requires a modification of the usual meta-theory which also accepts the Reference Theory and indeed which is tailored to cater only for logics which do conform. A somewhat similar phenomenon seems to occur in the case of a meta-ethic adequate for an environmental ethic. Quite apart from introducing several environmentally important notions, such as *conservation, pollution, growth* and *preservation*, for meta-ethical analysis, an environmental ethic compels re-examination and modified analyses of such characteristic actions as *natural right, ground* of right, and of the relations of obligation and permissibility to rights; it may well require re-assessment of traditional analyses of such notions as *value* and *right*, especially where these are based on chauvinist assumptions; and it forces the rejection of many of the more prominent meta-ethical positions. These points are illustrated by a very brief examination of accounts of *natural right* and then by a sketch of the species bias of some major positions.[11]

Hart accepts, subject to defeating conditions which are here irrelevant, the classical doctrine of natural rights according to which, among other things, "any adult human ... capable of choice is at liberty to do (i.e. is under no obligation to abstain from) any action which is not one coercing or restraining or designed to injure other persons."[12] But this sufficient condition for a human natural right depends on accepting the very human chauvinist principle an environmental ethic rejects, since if a person has a natural right he has a right; so too the *definition* of a natural right adopted by classical theorists and accepted with minor qualifications by Hart presupposes the same defective principle. Accordingly an environmental ethic would have to amend the classical notion of a natural right, a far from straightforward matter now that human rights with respect to animals and the natural environment are, like those with respect to slaves not all that long ago, undergoing major re-evaluation.

An environmental ethic does not commit one to the view that natural objects such as trees have rights (though such a view is occasionally held, e.g. by pantheists. But pantheism is false since artefacts are not alive). For moral prohibitions forbidding certain actions with respect to an object do not award that object a correlative right. That it would be wrong to mutilate a given tree or piece of property does not entail that the tree or piece of property has a correlative right not to be mutilated (without seriously stretching the notion of a right). Environmental views can stick with mainstream theses according to which rights are coupled with corresponding responsibilities and so with bearing obligations, and with corresponding interests and concern; i.e. at least, whatever has a right also has responsibilities and therefore obligations, and whatever has a right has interests. Thus although any person may have a right by no means every living thing can (significantly) have rights, and arguably most sentient objects other than persons cannot have rights. But persons can relate morally, through obligations, prohibitions and so forth, to practically anything at all.

The species bias of certain ethical and economic positions which aim to make principles of conduct or reasonable economic behavior calculable is easily brought out. These positions typically employ a single criterion p, such as preference or happiness, as a *summum bonnum*; characteristically each individual of some base class, almost always humans, but perhaps including future humans, is supposed to have an ordinal p ranking of the states in question (e.g. of affairs, of the economy); then some principle is supplied to determine a collective p ranking of these states in terms of individual p rankings, and what is best or ought to be done is determined either directly, as in act-utilitarianism under the Greatest Happiness principle, or indirectly, as in rule-utilitarianism, in terms of some optimization principle applied to the collective ranking. The species bias is transparent from the selection of the base class. And even if the base class is extended to embrace persons, or even some animals (at the cost, like that of including remotely future humans, of losing testability), the positions are open to familiar criticism, namely that the whole of the base class may be prejudiced in a way which leads to unjust principles. For example if every member of the base class detests dingoes, on the basis of mistaken data as to dingoes' behavior, then by the Pareto ranking test the collective ranking will rank states where dingoes are exterminated very highly, from which it will generally be concluded that dingoes ought to be exterminated (the evaluation of most Australian farmers anyway). Likewise it would just be a happy accident, it seems, if collective demand (horizontally ed from individual demand) for a state of the economy with blue whales as a mixed good, were to succeed in outweighing private whaling demands; for if no one in the base class happened to know that blue whales exist or cared a jot that thesummy do then "rational" economic decision-making would do nothing to prevent their extinction. Whether the blue whale survives should not have to depend on what humans know or what they see on television. Human interests and preferences are far too parochial to provide a satisfactory basis for deciding on what is environmentally desirable.

These ethical and economic theories are not alone in their species chauvinism; much the same applies to most going meta-ethical theories which, unlike intuitionistic theories, try to offer some rationale for their basic principles. For instance, on social contract positions obligations are a matter of mutual agreements between individuals of the base class; on a social justice picture rights and obligations spring from the application of symmetrical fairness principles to members of the base class, usually a rather special class of persons, while on a Kantian position which has some vague obligations somehow arise from respect for members of the base class persons. In each case if members of the base class happen to be ill-disposed to items outside the base class then that is too bad for them: that is (rough) justice.

NOTES

1. Aldo Leopold, *A Sand County Almanac with Essays on Conservation from Round River* (New York: Ballantine, 1966), p. 238.

2. A view occasionally tempered by the idea that trees house spirits.

3. Leopold, *Sand County*, p. 245.

4. To the consternation no doubt of Quineans. But the fact is that we can talk perfectly well about inchoate and fragmentary systems the identity of which may be indeterminate.

5. John Passmore, *Man's Responsibility for Nature: Ecological Problems and Western Traditions* (New York: Scribner's, 1974).

6. If "use" is extended, somewhat illicitly, to include use for preservation, this total use principle is rendered innocuous at least as regards its actual effects. Note that the total use principle is tied to the resource view of nature.

7. P. W. Barkley and D. W. Seckler, *Economic Growth and Environmental Decay: The Solution Becomes the Problem* (New York: Harcourt, Brace, Jovanovich, 1972).

8. A *meta-ethic* is, as usual, a theory about ethics, super ethics, their features and fundamental notions.

9. Barkley and Seckler, *Economic Growth and Environmental Decay*, p. 58. A related principle is that (modified) free enterprise can operate within similar limits.

10. For the tragedy of the commons type reasons well explained in Barkley and Seckler, *Economic Growth and Environmental Decay.*

11. Some of these points are developed by those protesting about human maltreatment of animals; see especially the essays collected in S. and R. Godlovitch and J. Harris, eds., *Animals, Men and Morals: An Enquiry into the Maltreatment of Non-humans* (New York: Grove Press, 1971).

12. H. L. A. Hart, "Are There any Natural Rights?" reprinted in A. Quinton, ed., *Political Philosophy* (London: Oxford University Press, 1967).

THE SHALLOW AND THE DEEP

SOURCE Naess, Arne. 1973. *Inquiry Magazine (Oslo)* 16: 95–100. Reproduced by permission of the author.

INTRODUCTION *Arne Naess (1912–) was for thirty years professor of philosophy at the University of Olso. He continues to practice philosophy and environmental activism from his home in Norway. Naess is generally considered to be the founder of Deep Ecology. In "The Shallow and the Deep," Naess claims there are two movements in ecology: the shallow ecology movement, which focuses on environmental concerns such as pollution solely in regards to their impact on human welfare, particularly humans in developed countries; and the deep ecology movement, which includes concerns about pollution, but only as one of seven points, with six others to be considered, according to Naess' summary. He makes the further claim that an "ethics of responsibility" requires that environmentalists join the deep ecological movement, not the shallow. These principles of Deep Ecology are suggested by ecology, says Naess, but not derived from ecology; they are philosophical principles, or rather, as Naess says, they are "ecophilosophical." Naess states that these principles provide the framework for the creation of an "ecosophical" worldview.*

The emergence of ecologists from their former relative obscurity marks a turning point in our scientific communities. But their message is twisted and misused. A shallow, but presently rather powerful movement, and a deep, but less influential movement, compete for our attention. I shall make an effort to characterize the two.

I. THE SHALLOW ECOLOGY MOVEMENT:

Fight against pollution and resource depletion.

Central objective: the health and affluence of people in the developed countries.

II. THE DEEP ECOLOGY MOVEMENT:

1. Rejection of the man-in-environment image in favor the relational, total-field image. Organisms as knots in the biospherical net or field of intrinsic relations. An intrinsic relation between two things A and B is such that the relation belongs to the definitions of basic constitutions of A and B, so that without the relation, A and B are no longer the same things. The total-field dissolves not on the man-in-environment concept, but every compact thing-in-milieu concept-except when talking at a superficial or preliminary level of communication.

2. Biospherical egalitarianism-in principle. The "in principle" clause is inserted because any realistic praxis necessitates some killing, exploitation, and suppression. The ecological field-worker acquires a deep-seated respect, or even veneration, for ways and forms of life. He reaches an understanding from within, a kind of understanding that others reserve for fellow men and for a narrow section of ways and forms of life. To the ecological field-worker, the equal right to live and blossom is an intuitively clear and obvious value axiom. Its restriction to humans is an anthropocentrism with detrimental effects upon the life quality of humans themselves. The quality depends in part upon the deep pleasure and satisfaction we receive from close partnership with other forms of life. The attempt to ignore our dependence and to establish a master-slave role has contributed to the alienation of man from himself. Ecological egalitarianism implies the reinterpretation of the future-research variable, "level of crowding," so that general mammalian crowding and loss of life-equality is taken seriously, not only human crowding. (Research on the high requirements of free space of certain mammals has, incidentally, suggested that theorists of human urbanism have largely underestimated human life-space requirements. Behavioral crowding symptoms, such as neuroses, aggressiveness, loss of traditions, are largely the same among mammals.)

3. Principles of diversity and of symbiosis. Diversity enhances the potentialities of survival, the chances of new modes of life, the richness of forms. And the so-called struggle for life, and survival of the fittest, should be interpreted in the sense of the ability to coexist and cooperate in complex relationships, rather than the ability to kill, exploit, and suppress. "Live and let live" is a more powerful ecological principle than "Either you or me." The latter tends to reduce the multiplicity of kinds of forms of life, and also to create destruction within the communities of the same species. Ecologically inspired attitudes therefore favor diversity of human ways of life, of cultures, of occupations, of economies. They support the fight against economic and cultural, as much as military, invasion and domination, and they are opposed to the annihilation of seals and whales as much as to that of human tribes and cultures.

4. Anti-class posture. Diversity of human ways of life is in part due to (intended or unintended) exploitation and suppression on the part of certain groups. The exploiter lives differently from the exploited, but both are adversely affected in their potentialities of self-realization. The principle of diversity does not cover differences due merely to certain attitudes or behaviors forcibly blocked or restrained. The principles of ecological egalitarianism and of symbiosis support the same anti-class posture. The ecological attitude favors the extension of all three principles to any group conflicts, including those of today between developing and developed nations. The three principles also favor extreme caution toward any over-all plans for the future, except those consistent with wide and widening classless diversity.

5. Fight against pollution and resource depletion. In this fight ecologists have found powerful supporters, but sometimes to the detriment of their total stand. This happens when attention is focused on pollution and resource depletion rather than on the other points, or when projects are implemented which reduce pollution but increase evils of other kinds. Thus, if prices of life necessities increase because of the installation of anti-pollution devices, class differences increase too. An ethics of responsibility implies that ecologists do not serve the shallow, but the deep ecological movement. That is, not only point five, but all seven points must be considered together.

Ecologists are irreplaceable informants in any society, whatever their political color. If well organized, they have the power to reject jobs in which they submit themselves to institutions or to planners with limited ecological objectives. As it is now, ecologists sometimes serve masters who deliberately ignore the wider perspectives.

6. Complexity, not complication. The theory of ecosystems contains an important distinction between what is complicated without any Gestalt or unifying principles—we may think of finding our way through a chaotic city—and what is complex. A multiplicity of more or less lawful, interacting factors may operate together to form a unity, a system. We make a shoe or use a map or integrate a variety of activities into a workaday pattern. Organisms, ways of life, and interactions in the biosphere in general, exhibit complexity of such an astoundingly high level as to color the general outlook of ecologists. Such complexity makes thinking in terms of vast systems inevitable. It also makes for a keen, steady perception of the profound human

ignorance of biospherical relationships and therefore of the effect of disturbances.

Applied to humans, the complexity-not-complication principle favors division of labor, not fragmentation of labor. It favors integrated actions in which the whole person is active, not mere reactions. It favors complex economies, an integrated variety of means of living. (Combinations of industrial and agricultural activity, of intellectual and manual work, of specialized and nonspecialized occupations, of urban and non-urban activity, of work in city and recreation in nature with recreation in city and work in nature...)

It favors soft technique and "soft future-research," less prognosis, more clarification of possibilities. More sensitivity toward continuity and live traditions, and more importantly, towards our state of ignorance.

The implementation of ecologically responsible policies requires in this century an exponential growth of technical skill and invention—but in new directions, directions which today are not consistently and liberally supported by the research policy organs of our nation states.

7. Local autonomy and decentralization. The vulnerability of a form of life is roughly proportional to the weight of influences from afar, from outside the local region in which that form has obtained an ecological equilibrium. This lends support to our efforts to strengthen local self-government and material and mental self-sufficiency. But these efforts presuppose an impetus towards decentralization. Pollution problems, including those of thermal pollution and recirculation of materials, also lead us in this direction, because increased local autonomy, if we are able to keep other factors constant, reduces energy consumption. (Compare an approximately self-sufficient locality with one requiring the importation of foodstuff, materials for house construction, fuel and skilled labor from other continents. The former may use only five percent of the energy used by the latter.)

Local autonomy is strengthened by a reduction in the number of links in the hierarchical chains of decision. (For example a chain consisting of a local board, municipal council, highest sub-national decision-maker, a state-wide institution in a state federation, a federal national government institution, a coalition of nations, and of institutions, e.g., E. E. C. top levels, and a global institution, can be reduced to one made up of a local board, nation-wide institution, and global institution.) Even if a decision follows majority rule at each step, many local interests may be dropped along the line, if it is too long.

Summing up then, it should, first of all, be borne in mind that the norms and tendencies of the Deep Ecology movement are not derived from ecology by logic or induction. Ecological knowledge and the life style of the ecological field-worker have suggested, inspired, and fortified the perspectives of the Deep Ecology movement. Many of the formulations in the above seven-point survey are rather vague generalizations, only tenable if made more precise in certain directions. But all over the world the inspiration from ecology has shown remarkable convergences. The survey does not pretend to be more than one of the possible condensed codifications of these convergences.

Secondly, it should be fully appreciated that the significant tenets of the Deep Ecology movement are clearly and forcefully normative. They express a value priority system only in part based on results (or lack of results, cf. point six) of scientific research. Today, ecologists try to influence policy-making bodies largely through threats, through predictions concerning pollutants and resource depletion, knowing that policy-makers accept at least certain minimum norms concerning health. But it is clear that there is a vast number of people in all countries, and even a considerable number of people in power, who accept as valid the wider norms and values characteristic of the Deep Ecology movement. There are political potentials in this movement which should not be overlooked and which have little to do with pollution and resource depletion. In plotting possible futures, the norms should be freely used and elaborated. Thirdly, insofar as ecology movements deserve our attention, they are ecophilosophical rather than ecological. Ecology is a limited science which makes use of scientific methods. Philosophy is the most general forum of debate on fundamentals, descriptive as well as prescriptive, and political philosophy is one of its subsections. By an ecosophy I mean a philosophy of ecological harmony or equilibrium. A philosophy is a kind of sophia wisdom, is openly normative, it contains both norms, rules, postulates, value priority announcements and hypotheses concerning the state of affairs in our universe. Wisdom is policy wisdom, prescription, not only scientific description and prediction.

The details of an ecosophy will show many variations due to significant differences concerning not only "facts" of pollution, resources, population, etc., but also value priorities. Today, however, the seven points listed provide one unified framework for ecosophical systems.

In general systems theory, systems are mostly conceived in terms of causally or functionally interacting or interrelated items. An ecosophy, however, is more like a system of the kind constructed by Aristotle or Spinoza. It is expressed verbally as a set of sentences with a variety of functions, descriptive and prescriptive. The basic relation is that between subsets of premises and subsets of conclusions, that is, the relation of derivability. The relevant notions of derivability may be classed according to rigor, with logical and mathematical deducations topping the list, but also according to how much is implicitly taken for granted. An exposition of an ecosophy must necessarily be only moderately precise considering the vast scope of relevant ecological and normative (social, political, ethical) material. At the moment, ecosophy might profitably use models of systems, rough approximations of global systematizations. It is the global character, not preciseness in detail, which distinguishes an ecosophy. It articulates and integrates the efforts of an ideal ecological team, a team comprising not only scientists from an extreme variety of disciplines, but also students of politics and active policy-makers.

Under the name of ecologism, various deviations from the deep movement have been championed-primarily with a one-sided stress on pollution and resource depletion, but also with a neglect of the great differences between underand over-developed countries in favor of a vague global approach. The global approach is essential, but regional differences must largely determine policies in the coming years.

IS THERE AN ECOLOGICAL ETHIC?

SOURCE Rolston III, Holmes. 1975. *Ethics* 85(2): 93–109. Reproduced with permission.

INTRODUCTION *Holmes Rolston III (1932–) taught philosophy for nearly fifty years at Colorado State University, retiring in 2008 as University Distinguished Professor. His writings have been central to the development of environmental ethics. In "Is There an Ecological Ethic?" Rolston questions first whether an ecological ethic would only be* about *the environment, or whether it would rather be formed* by *the environment, that is, by ecology. Interested in this relationship between science and morality, facts and values, Rolston states that traditionally facts fall under the domain of the sciences while values fall under ethics; but in the case of an environmental ethic, such distinctions prove complex if not artificial. Rolston distinguishes between two possible types of an environmental ethic. The first he claims is primarily anthropological and only secondarily ecological, in that humans become interested in an ecological ethic for the purpose of protecting their own interests and only secondarily the interests of nature. The second blurs these distinctions uniting the ultimate good on humanity* and *the environment in the realization that maximizing the one is the same as maximizing the other. In the latter sense, which is primarily ecological, humanity may construct value, but only in obedience to ecosystemic principles; we do not derive, but rather we discover, moral imperatives in nature. Facts and values are found simultaneously existing in nature.*

The Ecological Conscience[1] is the arresting title of a representative environmental anthology. The puzzlement lies neither in the noun nor in the by now familiar modifier, but in their operation on each other. We are comfortable with a Christian or humanist ethic, but the moral noun does not regularly take a scientific adjective: a biological conscience, a geological conscience. In a celebrated survey, *The Subversive Science*,[2] where ecology reaches into our ultimate commitments, Paul Sears entitles an essay "The Steady State: Physical Law and Moral Choice." To see how odd, ethically and scientifically, is the conjunction, replace homeostasis with gravity or entropy.

The sense of anomaly will dissipate, though moral urgency may remain, if an environmental ethic proves to be only an ethic—utilitarian, hedonist, or whatever—*about* the environment, brought to it, informed concerning it, but not in principle ecologically formed or reformed. This would be like medical ethics, which is applied to but not derived from medical science. But we are sometimes promised more, a derivation in which the newest bioscience shapes (not to say, subverts) the ethic, a resurgent naturalistic ethics. "We must learn that nature includes an intrinsic value system," writes Ian McHarg.[3] A *Daedalus* collection is introduced with the same conviction: Environmental science "is the building of the structure of concepts and natural laws that will enable man to understand his place in nature. Such understanding must be one basis of the moral values that guide each human generation in exercising its stewardship over the earth. For this purpose ecology—the science of interactions among living things and their environments—is central."[4] We shall presently inquire into the claim that an ecological ultimacy lies in "The Balance of Nature: A Ground for Values." Just what sort of traffic is there here between science and morality?

The boundary between science and ethics is precise if we accept a pair of current (though not unargued) philosophical categories: the distinction between descriptive and prescriptive law. The former, in the indicative, marks the realm of science and history. The latter, including always an imperative, marks the realm of ethics. The route from one to the other, if any, is perhaps the most intransigent issue in moral philosophy, and he who so moves will be accused of the naturalistic fallacy. No set of statements of fact by themselves entails any evaluative statement, except as some additional evaluative premise has been introduced. With careful analysis this evaluation will reappear, the ethics will separate out from the science. We shall press this logic on ecological ethics. Environmental science describes what is the case. An ethic prescribes what ought to be. But an environmental ethic? If our categories hold, perhaps we have a muddle. Or perhaps a paradox that yields light on the linkage between facts and values.

We find representative spokesman for ecological morality not of a single mind. But the multiple species can, we suggest, be classified in two genera, following two concepts that are offered as moral sources. (*A*) Prominent in, or underlying, those whom we hear first is the connection of homeostasis with morality. This issues largely in what we term an ethic that is secondarily ecological. (*B*) Beyond this, surpassing though not necessarily gainsaying it, is the discovery of a moral ought inherent in recognition of the holistic character of the ecosystem, issuing in an ethic that is primarily ecological.

But first, consider an analogue. When advised that we ought to obey the laws of health, we analyze the injunction. The laws of health are nonmoral and operate inescapably on us. But, circumscribed by them, we have certain options: to employ them to our health, or to neglect them ("break them") to our hurt. Antecedent to the laws of health, the moral ought reappears in some such form as, "You ought not to harm yourself." Similarly the laws of psychology, economics, history, the social sciences, and indeed all applied sciences describe what is (has been, or may be) the case; but in confrontation with human agency, they prescribe what the agent must do if he is to attain a desired end. They yield a technical ought related to an if-clause at the agent's option. So far they are nonmoral; they become moral only as a moral principle binds the agent to some end. This, in turn, is transmitted through natural law to a proximate moral ought. Let us map this as follows:

Technical Ought	*Natural Law*	*Antecedent If-Option*
You ought not to break the laws of health	for the laws of health describe the conditions of welfare	if you wish not to harm yourself.
Proximate Moral Ought	*Natural Law*	*Antecedent Moral Ought*
You ought not to break the laws of health	for the laws of health describe the conditions of welfare	and you ought not to harm yourself.

Allow for the moment that (in the absence of overriding considerations) prudence is a moral virtue. How far can ecological ethics transpose to an analogous format?

A

Perhaps the paramount law in ecological theory is that of homeostasis. In material, our planetary ecosystem is essentially closed, and life proceeds by recycling transformations. In energy, the system is open, with balanced solar input and output, the cycling being in energy subsystems of aggradation and degradation. Homeostasis, it should be noted, is at once an achievement and a tendency. Systems recycle, and there is energy balance; yet the systems are not static, but dynamic, as the forces that yield equilibrium are in flux, seeking equilibrium yet veering from it to bring counter-forces into play. This perpetual stir, tending to and deviating from equilibrium, drives the evolutionary process.

1. How does this translate morally? Let us consider first a guarded translation. In "The Steady State: Physical Law and Moral Choice," Paul Sears writes: "Probably men will always differ as to what constitutes the good life. They need not differ as to what is necessary for the long survival of man on earth. Assuming that this is our wish, the conditions are clear enough. As living beings we must come to terms with the environment about us, learning to get along with the liberal budget at our disposal, promoting rather than disrupting those great cycles of nature—of water movement, energy flow, and material transformation that have made life itself possible. As a physical goal, we must seek to attain what I have called a steady state."[5] The title of the article indicates that this is a moral "must." To assess this argument, begin with the following:

Technical Ought	*Ecological Law*	*Antecedent If-Option*
You ought to recycle	for the life-supporting ecosystem recycles or perishes	if you wish to preserve human life.

When we replace the if-option by an antecedent moral ought, we convert the technical ought to a proximate moral ought. Thus the "must" in the citation is initially one of physical necessity describing our circumscription by ecological law, and subsequently it is one of moral necessity when this law is conjoined with the life-promoting ought.

Proximate Moral Ought	*Ecological Law*	*Antecedent Moral Ought*
You ought to recycle	for the life-supporting ecosystem recycles or perishes	and you ought to preserve human life.

The antecedent ought Sears takes, fairly enough, to be common to many if not all our moral systems. Notice the sense in which we can break ecological law. Spelling the conditions of stability and instability, homeostatic laws operate on us willy-nilly, but within a necessary obedience we have options, some of which represent enlightened obedience. To break an ecological law, means then, to disregard its implications in regard to an antecedent moral ought.

Thus far ecological morality is informed about the environment, conforming to it, but is not yet an ethic in which environmental science affects principles.

Antecedent to ecological input, there is a classical ethical principle, "promoting human life," which, when ecologically tutored, better understands life's circulations, whether in homeostasis, or in DDT, or strontium 90. Values do not (have to) lie in the world but may be imposed on it, as man prudentially manages the world.

2. Much attention has focused on a 1968 address, "The Tragedy of the Commons," given by Garrett Hardin to the American Association for the Advancement of Science. Hardin's argument, recently expanded to book length, proposes an ecologically based "fundamental extension in morality."[6] While complex in its ramifications and deserving of detailed analysis, the essential ethic is simple, built on the model of a village commons. Used by the villagers to graze cattle, the commons is close to its carrying capacity. Any villager who does not increase his livestock will be disadvantaged in the market. Following self-interest, each increases his herd; and the commons is destroyed. Extended to the planet, seen as a homeostatic system of finite resources the model's implication of impending tragedy is obvious. (The propriety of the extrapolation is arguable, but not at issue here.) The prescription of an ecological morality is "mutual coercion, mutually agreed on" in which we limit freedom to grow in order to stabilize the ecosystem to the mutual benefit of all.

To distill the ethics here is not difficult. We begin as before, with ecological law that yields options, which translate morally only with the addition of the life-promoting obligation.

Technical Ought	*Ecological Law*	*Antecedent If-Option*
We ought to stabilize the ecosystem thru mutually imposed limited growth	for the life-supporting ecosystem stabilizes at a finite carrying capacity or is destroyed	if we wish mutually to preserve human life.
Proximate Moral Ought	*Ecological Law*	*Antecedent Moral Ought*
We ought to stabilize the ecosystem thru mutually imposed self-limited growth	for the life-supporting ecosystem stabilizes at a finite carrying capacity or is destroyed	and we ought mutually to preserve human life.

To clarify the problem of mutual preservation, Hardin uses an essentially Hobbesian scheme. Every man is an ego set over against the community, acting in his own self-interest. But to check his neighbor's aggrandizement, he compromises and enters a social contract where, now acting in enlightened self-interest, he limits his own freedom to grow in return for a limitation of the encroaching freedom of his competitors. The result is surprisingly atomistic and anthropocentric, recalling the post-Darwinian biological model, lacking significant place for the mutal interdependence and symbiotic cooperation so prominent in recent ecology. In any event, it is clear enough that Hardin's environmental ethic is only a classical ethic applied in the matrix of ecological limitations.

Typically, ecological morality generated by population pressure resolves itself into a particular case of this kind, as for instance in the analysis of Paul Ehrlich in *The Population Bomb*. This is an ethic of scarcity, but morality since its inception has been conceived in scarcity.

3. Let us pass to a more venturesome translation of homeostasis into moral prescription, that of Thomas B. Colwell, Jr. "The balance of Nature provides an objective normative model which can be utilized as the ground of human value.... Nor does the balance of Nature serve as the source of all our values. It is only the *ground* of whatever other values we may develop. But these other values must be consistent with it. The balance of Nature is, in other words, a kind of ultimate value.... It is a *natural* norm, not a product of human convention or supernatural authority. It says in effect to man: 'This much at least you must do, this much you must be responsible for. You must at least develop and utilize energy systems which recycle their products back into Nature.'... Human values are founded in objectively determinable ecological relations with Nature. The ends which we propose must be such as to be compatible with the ecosystems of Nature."[7]

Morality and homeostasis are clearly blended here, but it is not so clear how we relate or disentangle them. Much is embedded in the meanings of "ground of human value," "ultimate value," the mixed moral and physical "must," and the identification of a moral norm with a natural limit. Let us mark out first a purely technical ought, followed by an antecedent moral ought which may convert to a proximate moral ought.

Technical Ought	*Ecological Law*	*Antecedent If-Option*
You ought to recycle	for the value-supporting ecosystem recycles or perishes	if you wish to preserve the ground of human value.
Proximate Moral Ought	*Ecological Law*	*Antecedent Moral Ought*
You ought to recycle	for the value-supporting ecosystem recycles or perishes	and you ought to preserve the ground of human value.

The simplest reading of Colwell is to hold, despite his exaggerated terms, that the "ground of human value" means only the limiting condition, itself value free, within which values are to be constructed. Homeostasis is not "an ultimate value," only a precondition of the value enterprise, necessary but not sufficient for value. But then it is misleading to say that "human values have a root base in ecological relationships." For homeostasis, like scarce resources, or the cycling seasons, or soil characteristics, or the conservation of matter-energy, is a natural given, the stage on which the value-drama is played.

If, seeking to manage my finances wisely, I ask, "How shall I spend my money?" and you counsel, "You ought to balance your budget," the advice is sound enough, yet only preparatory to serious discussion of economic values. The balanced budget is necessary but not sufficient for value, a ground of value only in an enabling, not a fundamental sense; certainly not what we would ordinarily call an ultimate value. It is true, of course, that the means to any end can, in contexts of desperation and urgency, stand in short focus as ultimate values. Air, food, water, health, if we are deprived of them, become at once our concern. Call them ultimate values if you wish, but the ultimacy is instrumental, not intrinsic. We should think him immature whose principal goal was just to breathe, to eat, to drink, to be healthy—merely this and nothing more. We would judge a society stagnant whose ultimate goal was but to recycle. To say that the balance of nature is a ground for human values is not to draw any ethics from ecology, as may first appear, but only to recognize the necessary medium of ethical activity.

Thus far, ecological ethics reduces rather straightforwardly to the classical ethical query now advised of certain ecological boundaries. The stir is, to put it so, about the boundedness, not the morality. The ultimate science may well herald limits to growth; it challenges certain presumptions about rising standards of living, capitalism, progress, development, and so on; convictions that, though deeply entrenched parameters of human value, are issues of what is, can, or will be the case, not of what ought to be. This realization of limits, dramatically shift ethical application though it may, can hardly be said to reform our ethical roots, for the reason that its scope remains (when optimistic) a maximizing of human values or (when pessimistic) human survival. All goods are human goods, with nature an accessory. There is no endorsement of any natural rightness, only the acceptance of the natural given. It is ecological secondarily, but primarily anthropological.

B

The claim that morality is a derivative of the holistic character of the ecosystem proves more radical, for the ecological perspective penetrates not only the secondary but also the primary qualities of the ethic. It is ecological in substance, not merely in accident; it is ecological per se, not just consequentially.

Return, for instance, to Colwell. He seems to mean more than the minimal interpretation just given him. The mood is that the ecological circumscription of value is not itself amoral or premoral, neatly articulated from morality. Construct values though man may, he operates in an environmental context where he must ground his values in ecosystemic obedience. This "must" is ecologically descriptive: certain laws in fact circumscribe him and embrace his value enterprises. And it is also morally prescriptive: given options within parameters of necessary obedience, he morally ought to promote homeostasis. But here, advancing on the preceding argument, the claim seems to be that following ecological nature is not merely a prudential means to moral and valuational ends independent of nature but is an end in itself; or, more accurately, it is within man's relatedness to his environment that all man's values are grounded and supported. In that construction of values, man doubtless exceeds any environmental prescription, but nevertheless his values remain environmental reciprocals. They complement a homeostatic world. His valuations, like his other perceptions and knowings, are interactionary, drawn from environmental transactions, not merely brought to it. In this environmental encounter, he finds homeostasis a key to all values—the precondition of values, if you will—but one which, for all that, informs and shapes his other values by making them relational, corporate, environmental. But we are passing over to moral endorsement of the ecosystemic character, and to a tenor of argument that others make clearer.

Perhaps the most provocative such affirmation is in a deservedly seminal essay, "The Land Ethic," by Aldo Leopold. He concludes, "A thing is right when it tends to preserve the integrity, stability, and beauty of the biotic community. It is wrong when it tends otherwise."[8] Leopold writes in search of a morality of land use that escapes economic expediency. He too enjoins, proximately, recycling, but it is clear that his claim transcends the immediate context to teach that we morally ought to preserve the excellences of the ecosystem (or, more freely as we shall interpret him, to maximize the integrity, beauty, and stability of the ecosystem). He is seeking, as he says, to advance the ethical frontier from the merely interpersonal to the region of man in transaction with his environment.

Here the environmental perspective enters not simply at the level of the proximate ought which, environmentally informed and preceded by homocentrist moral principles, prescribes protection of the ecosystem. It acts at a higher level, as itself an antecedent ought, from which proximate oughts, such as the one earlier considered, about recycling, may be derived.

Proximate Moral Ought	*Ecological Law*	*Antecedent Moral Ought*
You ought to recycle	for recycling preserves the ecosystem	and you ought to preserve the integrity of the ecosystem.

Note how the antecedent parallels upper-level axioms in other systems (e.g., "You ought to maximize human good," or "You ought not to harm yourself or others," or "Love your neighbor as yourself"). Earlier, homeostatic connectedness did not really alter the moral focus; but here, in a shift of paradigms, the values hitherto reserved for man are reallocated to man in the environment.

Doubtless even Leopold's antecedent ought depends on a yet prior ought that one promote beauty and integrity, wherever he finds it. But this, like the injunction that one ought to promote the good, or that one ought to keep his promises, is so high level as to be, if not definitional or analytic, so general as to be virtually unarguable and therefore without any real theoretical content. Substantive values emerge only as something empirical is specified as the locus of value. In Leopold's case we have a feedback from ecological science which, prior to any effect on proximate moral oughts, informs the antecedent ought. There is a valuational element intrinsically related to the concepts utilized in ecological description. That is, the character of what is right in some basic sense, not just in application, is stated postecologically. Doubtless too, the natural course we choose to preserve is filtered through our concepts of beauty, stability, and integrity, concepts whose origins are not wholly clear and which are perhaps nonnatural. But, perspectival though this invariably is, what counts as beauty and integrity is not just brought to and imposed on the ecosystem but is discovered there. Let us map this as follows:

Proximate Moral Ought	*Ecological Law*	*Antecedent Moral Ought*	*Ecosystemic Evaluation*
You ought to recycle	for recycling preserves the integral ecosystem	and you ought to preserve the integrity of the ecosystem	for the integral ecosystem has value.

Our antecedent ought is not eco-free. Though preceding ecological law in the sense that, given this ought, one can transmit it via certain ecological laws to arrive at proximate oughts, it is itself a result of an ecosystemic appraisal.

This evaluation is not scientific description; hence not ecology per se, but metaecology. No amount of research can verify that the right is the optimum biotic community. Yet ecological description generates this evaluation of nature, endorsing the systemic rightness. The transition from "is" to "good" and thence to "ought" occurs here; we leave science to enter the domain of evaluation, from which an ethic follows. The injunction to recycle is technical, made under circumscription by ecological necessity and made moral only by the presence of an antecedent. The injunction to maximize the ecosystemic excellence is also ecologically *derived* but is an evaluative transition which is not made under necessity.

Our account initially suggests that ecological description is logically (if not chronologically) prior to the ecosystemic evaluation, the former generating the latter. But the connection of description with evaluation is more complex, for the description and evaluation to some extent arise together, and it is often difficult to say which is prior and which is subordinate. Ecological description finds unity, harmony, interdependence, stability, etc., and these are valuationally endorsed, yet they are found, to some extent, because we search with a disposition to value order, harmony, stability, unity. Still, the ecological description does not merely confirm these values, it informs them; and we find that the character, the empirical content, of order, harmony, stability is drawn from, no less than brought to, nature. In post-Darwinian nature, for instance, we looked for these values in vain, while with ecological description we now find them; yet the earlier data are not denied, only redescribed or set in a larger ecological context, and somewhere enroute our notions of harmony, stability, etc., have shifted too and we see beauty now where we could not see it before. What is ethically puzzling, and exciting, in the marriage and mutual transformation of ecological description and evaluation is that here an "ought" is not so much *derived* from an "is" as discovered simultaneously with it. As we progress from descriptions of fauna and flora, of cycles and pyramids, of stability and dynamism, on to intricacy, planetary opulence and interdependence, to unity and harmony with oppositions in counterpoint and synthesis, arriving at length at beauty and goodness, it is difficult to say where the natural facts leave off and where the natural values appear. For some observers at least, the sharp is/ought dichotomy is gone; the values seem to be there as soon as the facts are fully in, and both alike are properties of the system.

While it is frequently held that the basic criterion of the obligatory is the nonmoral value that is produced or sustained, there is novelty in what is taken as the nonmoral good—the ecosystem. Our ethical heritage largely attaches values and rights to persons, and if nonpersonal realms enter, they enter only as tributary to the personal. What is proposed here is a broadening of value, so that nature will cease to be merely "property" and become a commonwealth. The logic by which goodness is discovered or appreciated is notoriously evasive, and we can only reach it suggestively. "Ethics cannot be put into words," said Wittgenstein, such things "*make themselves manifest.*"[9] We have a parallel, retrospectively, in the checkered advance of the ethical frontier recognizing intrinsic goodness, and accompanying rights, outside the self. If we now universalize "person," consider how slowly the circle has been enlarged fully to include aliens, strangers, infants, children, Negroes, Jews, slaves, women, Indians, prisoners, the elderly, the insane, the deformed, and even now we ponder the status of fetuses.

Ecological ethics queries whether we ought again to universalize, recognizing the intrinsic value of every ecobiotic component.

Are there, first, existing ethical sentiments that are subecological, that is, which anticipate the ecological conscience, and on which we might build? Second, is the ecological evaluation authentic, or perhaps only a remodeled traditional humanist ethic? Lastly, what are the implications of maximizing the ecosystem, and what concept of nature warrants such evaluation?

1. Presumably the evaluation of a biotic community will rest partly on the worth of its elements, if not independently, then in matrix. We have a long-standing, if (in the West) rather philosophically neglected, tradition that grants some moral ought to the prevention of needless animal suffering: "A righteous man has regard for the life of his beasts" (Proverbs 12.10). Consider what we oddly call "humane" societies or laws against cockfighting, bear baiting, and (in our nation) bullfighting, and (in most states) steer busting. We prohibit a child's torture of a cat; we prosecute the rancher who carelessly lets horses starve. Even the hunter pursues a wounded deer. That one ought to prevent needless cruelty has no obvious ecological foundation, much less a natural one, but the initial point is that animals are so far endowed with a value that conveys something like rights, or at least obligates us.

More revelatory is the increasingly common claim that one ought not to destroy life, or species, needlessly, regardless of suffering. We prevent the wanton slaughter of eagles, whether they suffer or not. Even the zealous varmint hunter seems to need the rationalization that crows rob the cornfield. He must malign the coyote and wolf to slay them enthusiastically. He cannot kill just for fun. We abhor the oilspills that devastate birdlife. The Sierra Club defends the preservation of grizzlies or whooping cranes on many counts as means to larger ends—as useful components of the ecosystem, for scientific study, or for our children's enjoyment. (We shall return to the integrated character of such argument.) But sufficiently pressed, the defense is that one ought not destroy a life form of beauty. Since ecosystems regularly eliminate species, this may be a nonecological ought. Yet it is not clearly so, for part of a species' evaluation arises as it is seen in environmental matrix. Meanwhile, we admit they should continue to exist, "as a matter of biotic right."[10]

This caliber of argument can be greatly extended. A reason given for the preservation of Cades Cove in the Great Smoky Mountains National Park is the variety of rare salamanders there. Certain butterflies occur rarely in isolated hummocks in the African grasslands. Formerly, unscrupulous collectors would collect a few hundred then burn out the hummock to destroy the species, and thereby drive up the price of their collections. I find myself persuaded that they morally ought not do this. Nor will the reason resolve into the evil of greed, but it remains the needless destruction of even a butterfly species. At scattered occurrences of rare ferns in Tennessee I refused to collect, not simply to leave them for others to enjoy, but morally unwilling to imperil a species. Such species are a fortiori environmentally pressed, yet they remain, and even prosper, in selected environmental niches, and their dispatch by human whim seems of a different order from their elimination by natural selection—something like the difference between murder and death by natural causes.

This respect enlarges to the landscape. We preserve certain features of natural beauty—the Grand Canyon, or Rainbow Bridge, or the Everglades. Though it seems odd to accord them "rights" (for proposals to confer rights on some new entity always sound linguistically odd), we go so far as to say that, judged to be places of beauty or wonder, they ought to be preserved. Is this only as a means to an end, that we and others may enjoy them? The answer is complex. At least some argue that, as with persons, they are somehow violated, even prostituted, if treated merely as means; we enjoy them very largely for what they are in themselves. To select some landscapes is not to judge the omitted ones valueless. They may be sacrificed to higher values, or perhaps selected environments are judged sufficiently representative of more abundant ones. That we do preserve any landscape indicates our discovery of value there, with its accompanying ought. Nor are such environments only the hospitable ones. We are increasingly drawn to the beauty of wilderness, desert, tundra, the arctic, and the sea. Planetary forces ever reshape landscapes, of course, and former environments are now extinct; nevertheless, we find in extant landscapes an order of beauty that we are unwilling to destroy.

2. Do we perhaps have, even in this proposed primary ecological ethic, some eco-free ought? If Leopold's preserving the ecosystem is merely ancillary to human interests, the veiled antecedent ought is still that we ought to maximize human good. Were we so to maximize the ecosystem we should have a corporate anthropological egoism, "human chauvinism," not a planetary altruism. The optimum ecosystem would be but a prudential means to human welfare, and our antecedent ought would no longer be primarily ecological, but as before, simply a familiar one, incidentally ecological in its prudence.

Even when richly appreciative of nature's values, much ecological moralizing does in fact mix the biosystemic welfare with an appeal to human interests.

Reminiscent of Leopold, René Dubos suggests extending the Decalogue with an eleventh commandment, "Thou shalt strive for environmental quality." The justification may have a "resources" cast. We preserve wilderness and the maximally diverse ecosystem for reasons scientific and aesthetic. Natural museums serve as laboratories. Useless species may later be found useful. Diversity insures stability, especially if we err and our monocultures trigger environmental upset. Wild beauty adds a spiritual quality to life. "Were it only for selfish reasons, therefore, we must maintain variety and harmony in nature.... Wilderness is not a luxury; it is a necessity for the protection of humanized nature and for the preservation of mental health."[11]

But the "were it only ..." indicates that such reasons, if sufficient, are not ultimate. Deeper, nonselfish reasons respect "qualities inherent" in fauna, flora, landscape, "so as to foster their development." Haunting Western civilization is "the criminal conceit that nature is to be considered primarily as a source of raw materials and energy for human purposes," "the crude belief that man is the only value to be considered in managing the world and that the rest of nature can be thoughtlessly sacrificed to his welfare and whims." While holding that man is the creature who humanizes nature, the ecological conscience is sensitive to other worth. Indeed, somewhat paradoxically, it is only as man grants an intrinsic integrity to nature that he discovers his truest interests. "An enlightened anthropocentrism acknowledges that, in the long run, the world's good always coincides with man's own most meaningful good. Man can manipulate nature to his best interests only if he first loves her for her own sake."[12]

This coincidence of human and ecosystemic interests, frequent in environmental thought, is ethically confusing but fertile. To reduce ecological concern merely to human interests does not really exhaust the moral temper here, and only as we appreciate this will we see the ethical perspective significantly altered. That alteration centers in the dissolution of any firm boundary between man and the world. Ecology does not know an encapsulated ego over against his environment. Listen, for instance, to Paul Shepard: "Ecological thinking, on the other hand, requires a kind of vision across boundaries. The epidermis of the skin is ecologically like a pond surface or a forest soil, not a shell so much as a delicate interpenetration. It reveals the self ennobled and extended, rather than threatened, as part of the landscape, because the beauty and complexity of nature are continuous with ourselves."[13] Man's vascular system includes arteries, veins, rivers, oceans, and air currents. Cleaning a dump is not different in kind from filling a tooth. The self metabolically, if metaphorically, interpenetrates the ecosystem. The world is my body.

This mood frustrates and ultimately invalidates the effort to understand all ecological ethics as disguised human self-interest, for now, with the self expanded into the system, their interests merge. One may, from a limited perspective, maximize the systemic good to maximize human good, but one can hardly say that the former is only a means to the latter, since they both amount to the same thing differently described. We are acquainted with egoism, *égoïsme à deux, trois, quatres*, with familial and tribal egoism. But here is an *égoïsme à la système*, as the very etymology of "ecology" witnesses: the earth is one's household. In this planetary confraternity, there is a confluence of egoism and altruism. Or should we say that egoism is transformed into ecoism? To advocate the interests of the system as a means of promoting the interests of man (in an appeal to industry and to congressmen) is to operate with a limited understanding. If we wish, for rhetorical or pragmatic reasons, we may begin with maximizing human good. But when ecologically tutored, we see that this can be redescribed as maximizing the ecosystem. Our classical ought has been transformed, stretched, coextensively with an ecosystemic ought.

To illustrate, ponder the observation that biotic-environmental complexity is integrally related to the richness of human life. That the stability and integrity of an ecosystem is a function of its variety and diversity is a fairly well-established point; and it is frequently observed that complex life forms evolve only in complex environments. The long evolution of man, accordingly, has been possible only under the stimulation of many environments—marine, arboreal, savannah, tropical, temperate, even arctic. Even when man lives at a distance from some of these, they remain tributary to his life support. Without oceans, forests, and grasslands, human life would be imperiled. Thus man's complex life is a product of and is underlain by environmental complexity.

This complexity is not simply biological but also mental and cultural. For maximum noetic development, man requires an environmental exuberance. So Shepard eloquently introduces the "universal wisdom" of *The Subversive Science*:

> Internal complexity, as the mind of a primate, is an extension of natural complexity, measured by the variety of plants and animals and the variety of nerve cells—organic extensions of each other. The exuberance of kinds as the setting in which a good mind could evolve (to deal with a complex world) was not only a past condition. Man did not arrive in the world as though disembarking from a train in the city. He continues to arrive.... This idea of natural complexity as a counterpart to human intricacy is central to an ecology of

> man. The creation of order, of which man is an example, is realized also in the number of species and habitats, an abundance of landscapes lush and poor. Even deserts and tundras increase the planetary opulence.... Reduction of this variegation would, by extension then, be an amputation of man. To convert all "wastes"—all deserts, estuaries, tundras, ice-fields, marshes, steppes and moors—into cultivated fields and cities would impoverish rather than enrich life esthetically as well as ecologically.[14]

Mountains have both physical and psychic impact. Remove eagles from the sky and we will suffer a spiritual loss. For every landscape, there is an inscape; mental and environmental horizons reciprocate.

This supports, but only by curiously transforming, the preservation of the ecosystem in human self-interest, for the "self" has been so extended as to be ecosystemically redefined. The human welfare which we find in the enriched ecosystem is no longer recognizable as that of anthropocentrism. Man judges the ecosystem as "good" or "bad" not in short anthropocentric focus, but with enlarged perspective where the integrity of other species enriches him. The moral posture here recalls more familiar (if frequently unsettled) ethical themes: that self-interest and benevolence are not necessarily incompatible, especially where one derives personal fulfillment from the welfare of others; that treating the object of ethical concern as an end in itself is uplifting; that one's own integrity is enhanced by recognition of other integrities.

3. This environmental ethic is subject both to limits and to development, and a fair appraisal ought to recognize both. As a partial ethical source, it does not displace functioning social-personal codes, but brings into the scope of ethical transaction a realm once regarded as intrinsically valueless and governed largely by expediency. The new ethical parameter is not absolute but relative to classical criteria. Such extension will amplify conflicts of value, for human goods must now coexist with environmental goods. In operational detail this will require a new casuistry. Mutually supportive though the human and the ecosystemic interests may be, conflicts between individuals and parties, the rights of the component members of the ecosystem, the gap between the real and the ideal, will provide abundant quandaries.

Further, interpreting charitably, we are not asked to idolize the whole except as it is understood as a cosmos in which the corporate vision surrounds and limits, but does not suppress the individual. The focus does not only enlarge from man to other ecosystemic members, but from individuals of whatever kind to the system. Values are sometimes personalized; here the community holds values. This is not, of course, without precedent, for we now grant values to states, nations, churches, trusts, corporations, and communities. And they hold these values because of their structure in which individuals are beneficiaries. It is similar with the ecosystem, only more so; for when we recall its diffusion of the boundary between the individual and the ecosystem, we cannot say whether value in the system or in the individual is logically prior.

Leopold and Shepard do not mean to deep freeze the present ecosystem. Despite their preservationist vocabulary, their care for the biosystemic welfare allows for "alteration, management, and use."[15] We are not committed to this as the best possible ecosystem; it may well be that the role of man—at once "citizen" and "king"—is to govern what has hitherto been the partial success of the evolutionary process. Though we revere the earth, we may yet "humanize" it, a point made forcefully by Réné Dubos.[16] This permits interference with and rearrangement of nature's spontaneous course. It enjoins domestication, for part of the natural richness is its potential in human life support. We recognize man's creativity, development, openness, and dynamism.

Species regularly enter and exit nature's theater; perhaps natural selection currently tests species for their capacity to coexist with man. Orogenic and erosional forces have produced perpetual environmental flux; man may well transform his environment. But this should complement the beauty, integrity, and stability of the planetary biosystem, not do violence to it. There ought to be some rational showing that the alteration is enriching; that values are sacrified for greater ones. For this reason the right is not that which maintains the ecosystemic status quo, but that which preserves its beauty, stability, and integrity.

What ought to be does not invariably coincide with what is; nevertheless, here is a mood that, recalling etymology again, we can best describe as man's being "at home" in his world. He accepts, cherishes his good earth. Purely scientific descriptions of an ecosystem may warrant the term "stability," neutrally used; they facilitate the estimate of its beauty and integrity. Added, though, is a response of the ecologist to his discoveries, an evocation of altering consciousness. We see integrity and beauty we missed before, partly through new realization of fact—interdependence, environmental fitness, hydrologic cycles, population rhythms, and feedback loops—and partly through transformed concepts of what counts as beauty and integrity, for world and concept mutually transform each other.

Though the full range of that shifting concept of nature and the ecological description which underlies it are beyond our scope, we can suggest their central axis. After Darwin (through misunderstanding him, perhaps), the world of design collapsed, and nature, for all its law, seemed random, accidental, chaotic, blind, crude, an "odious scene of violence."[17] Environmental science has been resurveying the post-Darwinian natural jungle and has increasingly set its conflicts within a dynamic web of life. Nature's savagery is much less wanton and clumsy than formerly supposed, and we are invited to see the ecosystem not merely in awe, but in "love, respect, and admiration."[18] Ecological thinking "moves us to silent wonder and glad affirmation."[19] Oppositions remain in ecological models, but in counterpoint. The system resists the very life it supports; indeed it is by resistance not less than environmental conductivity that life is stimulated. The integrity of species and individual is a function of a field where fullness lies in interlocking predation and symbiosis, construction and destruction, aggradation and degradation. The planet that Darrow characterized, in the post-Darwinian heyday, as a miserable little "wart"[20] in the universe, eminently unsuited to life, especially human life, is now a sheltered oasis in space. Its harmony is often strange, and it is not surprising that in our immaturity we mistook it, yet it is an intricate and delicate harmony nevertheless.

Man, an insider, is not spared environmental pressures, yet, in the full ecosystemic context, his integrity is supported by and rises from transaction with his world and therefore requires a corresponding dignity in his world partner. Of late, the world has ceased to threaten, save as we violate it. How starkly this gainsays the alienation that characterizes modern literature, seeing nature as basically rudderless, antipathetical, in need of monitoring and repair. More typically modern man, for all his technological prowess, has found himself distanced from nature, increasingly competent and decreasingly confident, at once distinguished and aggrandized, yet afloat on and adrift in an indifferent, if not a hostile universe. His world is at best a huge filling station; at worst a prison, or "nothingness." Not so for ecological man; confronting his world with deference to a community of value in which he shares, he is at home again. The new mood is epitomized, somewhat surprisingly, in reaction to space exploration, prompted by vivid photography of earth and by astronaut's nostalgia, generating both a new love for Spaceship Earth and a resolution to focus on reconciliation with it.

We shall surely not vindicate the natural sequence in every detail as being productive of ecosystemic health, and therefore we cannot simplify our ethic to an unreflective acceptance of what naturally is the case. We do not live in Eden, yet the trend is there, as ecological advance increasingly finds in the natural given stability, beauty, and integrity, and we are henceforth as willing to open our concepts to reformation by the world as to prejudge the natural order. The question of evolution as it governs our concept of nature is technically a separate one. We must judge the worth of the extant ecosystem independently of its origins. To do otherwise would be to slip into the genetic fallacy. A person has rights for what he is, regardless of his ancestry; and it may well be that an ignoble evolutionary process has issued in a present ecosystem in which we rightly rejoice. No one familiar with paleontology is likely to claim that the evolutionary sequence moves unfailingly and without loss toward an optimally beautiful and stable ecosystem. Yet many ecological mechanisms are also evolutionary, and the ecological reappraisal suggests as a next stage an evolutionary redescription, in which we think again whether evolutionary history, for all its groping, struggle, mutation, natural selection, randomness, and statistical movement, does not yield direction enough to ponder that nature has been enriching the ecosystem. The fossil record is all of ruins. We survey it first with a certain horror; but then out of the ruins emerges this integral ecosystem. He who can be persuaded of this latter truth will have an even more powerful ecological ethic, for the injunction to maximize the ecosystemic excellences will be an invitation to get in gear with the way the universe is operating. Linking his right to nature's processes, he will have, at length, an authentic naturalistic ethic.

The perils of transposing from a new science to a world view, patent in the history of scientific thought, are surpassed only by the perils of omitting to do so. Granted that we yet lack a clear account of the logic by which we get our values, it seems undeniable that we shape them in significant measure in accord with our notion of the kind of universe that we live in. Science has in centuries before us upset those values by reappraising the character of the universe. One has but to name Copernicus and Newton, in addition to our observation that we have lately lived in the shadow of Darwin. The ecological revolution may be of a similar order; it is undeniably at work reilluminating the world.

Darwin, though, often proves more fertile than his interpreters. When, in *The Descent of Man*, he traces the natural history of man's noblest attribute, the moral sense, he observes that "the standard of his morality rises higher and higher." Initially each attended his self-interest. The growth of conscience has been a continual expansion of the objects of his "social instincts and sympathies," first to family and tribe; then he "regarded more and more, not only the welfare, but the happiness of all his fellow-men;" then "his sympathies became more tender and widely diffused, extending to men of all races, to the imbecile, maimed, and other useless members of society, and finally to the lower animals.. ."[21] After the fauna, can we add the flora, the landscape, the seascape, the ecosystem? There would be something magnificent about an evolution of conscience that circumscribed the whole. If so, Leopold lies in the horizon of Darwin's vision. Much of the search for an ecological morality will, perhaps in necessary pragmatism, remain secondary, "conservative," where the ground is better charted, and where we mix ethics, science, and human interests under our logical control. But we judge the ethical frontier to be beyond, a primary revaluing where, in ethical creativity, conscience must evolve. The topography is largely uncharted; to cross it will require the daring, and caution, of a community of scientists and ethicists who can together map both the ecosystem and the ethical grammar appropriate for it.

Perhaps the cash value is the same whether our ethic is ecological in secondary or primary senses; yet in the latter I find appeal enough that it has my vote to be so if it can. To the one, man may be driven while he still fears the world that surrounds him. To the other, he can only be drawn in love.

1. Robert Disch, ed., *The Ecological Conscience: Values for Survival* (Englewood Cliffs, N.J.: Prentice-Hall, Inc., 1970).

2. Paul Shepard and Daniel McKinley, eds., *The Subversive Science* (Boston: Houghton Mifflin Co., 1969).

3. Ian L. McHarg, "Values, Process, and Form," in Disch, p. 21.

4. Roger Revelle and Hans H. Landsberg, eds., *America's Changing Environment* (Boston: Beacon Press, 1970), p. xxii.

5. Shepard and McKinley, p. 401.

6. Garrett Hardin, "The Tragedy of the Commons," *Science* 162 (1968): 1243–48.

7. Thomas B. Colwell, Jr., "The Balance of Nature: A Ground for Human Values," *Main Currents in Modern Thought 26* (1969): 50.

8. Aldo Leopold, "The Land Ethic," in *A Sand County Almanac* (New York: Oxford University Press, 1949), pp. 201–26.

9. Ludwig Wittgenstein, *Tractatus Logico-Philosophicus*, trans. D. F. Pears and B. F. McGuiness (London: Routledge & Kegan Paul, 1969), 6:421, 522.

10. Leopold, p. 211.

11. Réné Dubos, *A God Within* (New York: Charles Scribner's Sons, 1972) pp. 166–67.

12. Ibid., pp. 40–41, 45.

13. Shepard, p. 2.

14. Ibid., pp. 4–5.

15. Leopold, p. 204.

16. Dubos, chap. 8.

17. John Stuart Mill, "Nature," in *Collected Works* (Toronto: University of Toronto Press, 1969), 10:398. The phrase characterizes Mill's estimate of nature.

18. Leopold, p. 223.

19. Shepard, p. 10.

20. Clarence Darrow, *The Story of My Life* (New York: Charles Scribner's Sons, 1932), p. 417.

21. Charles Darwin, *The Descent of Man*, new ed. (New York: D. Appleton & Co., 1895), pp. 124–25.

THE EARTH CHARTER

SOURCE 2000. *Earth Charter Commission, 2000.* Reprinted with the permission of Earth Charter International Secretariat.

INTRODUCTION *The origins of the Earth Charter lie in the 1987 World Commission on Environment and Development's call for a new set of principles to encourage and guide sustainable development. From that point until its completion and launch in 2000, it was developed and supported by a diverse group of individuals and institutions. The charter claims that humanity is at a critical point in its history and is in need of fundamental changes to our "values, institutions, and ways of living." It attempts to outline a code of ethics to guide us into a future just and sustainable world. The Earth Charter consists of a total of sixteen principles, organized under the four headings of Respect and Care for the Community of Life; Ecological Integrity; Social and Economic Justice; and Democracy, Nonviolence, and Peace. It claims that these principles are interdependent, and that the global community must realize its own internal interdependence with the planet. The Earth Charter expresses the hope that it will serve as a foundation for a "new beginning," inspiring new policies toward the achievement of a "sustainable global society founded on respect for nature, universal human rights, economic justice, and a culture of peace."*

PREAMBLE

We stand at a critical moment in Earth's history, a time when humanity must choose its future. As the world becomes increasingly interdependent and fragile, the future at once holds great peril and great promise. To move forward we must recognize that in the midst of a magnificent diversity of cultures and life forms we are one human family and one Earth community with a common destiny. We must join together to bring forth a sustainable global society founded on respect for nature, universal human rights, economic justice, and a culture of peace. Towards this end, it is imperative that we, the peoples of Earth, declare our responsibility to one another, to the greater community of life, and to future generations.

Earth, Our Home

Humanity is part of a vast evolving universe. Earth, our home, is alive with a unique community of life. The forces of nature make existence a demanding and uncertain adventure, but Earth has provided the conditions essential to life's evolution. The resilience of the community of life and the well-being of humanity depend upon preserving a healthy biosphere with all its ecological systems, a rich variety of plants and animals, fertile soils, pure waters, and clean air. The global environment with its finite resources is a common concern of all peoples. The protection of Earth's vitality, diversity, and beauty is a sacred trust.

The Global Situation

The dominant patterns of production and consumption are causing environmental devastation, the depletion of resources, and a massive extinction of species. Communities are being undermined. The benefits of development are not shared equitably and the gap between rich and poor is widening. Injustice, poverty, ignorance, and violent conflict are widespread and the cause of great suffering. An unprecedented rise in human population has overburdened ecological and social systems. The foundations of global security are threatened. These trends are perilous—but not inevitable.

The Challenges Ahead

The choice is ours: form a global partnership to care for Earth and one another or risk the destruction of ourselves and the diversity of life. Fundamental changes are needed in our values, institutions, and ways of living. We must realize that when basic needs have been met, human development is primarily about being more, not having more. We have the knowledge and technology to provide for all and to reduce our impacts on the environment. The emergence of a global civil society is creating new opportunities to build a democratic and humane world. Our environmental, economic, political, social, and spiritual challenges are interconnected, and together we can forge inclusive solutions.

Universal Responsibility

To realize these aspirations, we must decide to live with a sense of universal responsibility, identifying ourselves with the whole Earth community as well as our local communities. We are at once citizens of different nations and of one world in which the local and global are linked. Everyone shares responsibility for the present and future well-being of the human family and the larger living world. The spirit of human solidarity and kinship with all life is strengthened when we live with reverence for the mystery of being, gratitude for the gift of life, and humility regarding the human place in nature.

We urgently need a shared vision of basic values to provide an ethical foundation for the emerging world community. Therefore, together in hope we affirm the following interdependent principles for a sustainable way of life as a common standard by which the conduct of all individuals, organizations, businesses, governments, and transnational institutions is to be guided and assessed.

PRINCIPLES

I. RESPECT AND CARE FOR THE COMMUNITY OF LIFE

1. Respect Earth and life in all its diversity.

a. Recognize that all beings are interdependent and every form of life has value regardless of its worth to human beings.

b. Affirm faith in the inherent dignity of all human beings and in the intellectual, artistic, ethical, and spiritual potential of humanity.

2. Care for the community of life with understanding, compassion, and love.

a. Accept that with the right to own, manage, and use natural resources comes the duty to prevent environmental harm and to protect the rights of people.

b. Affirm that with increased freedom, knowledge, and power comes increased responsibility to promote the common good.

3. Build democratic societies that are just, participatory, sustainable, and peaceful.

a. Ensure that communities at all levels guarantee human rights and fundamental freedoms and provide everyone an opportunity to realize his or her full potential.

b. Promote social and economic justice, enabling all to achieve a secure and meaningful livelihood that is ecologically responsible.

4. Secure Earth's bounty and beauty for present and future generations.

a. Recognize that the freedom of action of each generation is qualified by the needs of future generations.

b. Transmit to future generations values, traditions, and institutions that support the long-term flourishing of Earth's human and ecological communities. In order to fulfill these four broad commitments, it is necessary to:

II. ECOLOGICAL INTEGRITY

5. Protect and restore the integrity of Earth's ecological systems, with special concern for biological diversity and the natural processes that sustain life.

a. Adopt at all levels sustainable development plans and regulations that make environmental conservation and rehabilitation integral to all development initiatives.

b. Establish and safeguard viable nature and biosphere reserves, including wild lands and marine areas, to protect Earth's life support systems, maintain biodiversity, and preserve our natural heritage.

c. Promote the recovery of endangered species and ecosystems.

d. Control and eradicate non-native or genetically modified organisms harmful to native species and the environment, and prevent introduction of such harmful organisms.

e. Manage the use of renewable resources such as water, soil, forest products, and marine life in ways that do not exceed rates of regeneration and that protect the health of ecosystems.

f. Manage the extraction and use of non-renewable resources such as minerals and fossil fuels in ways that minimize depletion and cause no serious environmental damage.

6. Prevent harm as the best method of environmental protection and, when knowledge is limited, apply a precautionary approach.

a. Take action to avoid the possibility of serious or irreversible environmental harm even when scientific knowledge is incomplete or inconclusive.

b. Place the burden of proof on those who argue that a proposed activity will not cause significant harm, and make the responsible parties liable for environmental harm.

c. Ensure that decision making addresses the cumulative, long-term, indirect, long distance, and global consequences of human activities.

d. Prevent pollution of any part of the environment and allow no build-up of radioactive, toxic, or other hazardous substances.

e. Avoid military activities damaging to the environment.

7. Adopt patterns of production, consumption, and reproduction that safeguard Earth's regenerative capacities, human rights, and community well-being.

a. Reduce, reuse, and recycle the materials used in production and consumption systems, and ensure that residual waste can be assimilated by ecological systems.

b. Act with restraint and efficiency when using energy, and rely increasingly on renewable energy sources such as solar and wind.

c. Promote the development, adoption, and equitable transfer of environmentally sound technologies.

d. Internalize the full environmental and social costs of goods and services in the selling price, and enable consumers to identify products that meet the highest social and environmental standards.

e. Ensure universal access to health care that fosters reproductive health and responsible reproduction.

f. Adopt lifestyles that emphasize the quality of life and material sufficiency in a finite world.

8. Advance the study of ecological sustainability and promote the open exchange and wide application of the knowledge acquired.

a. Support international scientific and technical cooperation on sustainability, with special attention to the needs of developing nations.

b. Recognize and preserve the traditional knowledge and spiritual wisdom in all cultures that contribute to environmental protection and human well-being.

c. Ensure that information of vital importance to human health and environmental protection, including genetic information, remains available in the public domain.

III. SOCIAL AND ECONOMIC JUSTICE

9. Eradicate poverty as an ethical, social, and environmental imperative.

a. Guarantee the right to potable water, clean air, food security, uncontaminated soil, shelter, and safe sanitation, allocating the national and international resources required.

b. Empower every human being with the education and resources to secure a sustainable livelihood, and provide social security and safety nets for those who are unable to support themselves.

c. Recognize the ignored, protect the vulnerable, serve those who suffer, and enable them to develop their capacities and to pursue their aspirations.

10. Ensure that economic activities and institutions at all levels promote human development in an equitable and sustainable manner.

a. Promote the equitable distribution of wealth within nations and among nations.

b. Enhance the intellectual, financial, technical, and social resources of developing nations, and relieve them of onerous international debt.

c. Ensure that all trade supports sustainable resource use, environmental protection, and progressive labor standards.

d. Require multinational corporations and international financial organizations to act transparently in the public good, and hold them accountable for the consequences of their activities.

11. Affirm gender equality and equity as prerequisites to sustainable development and ensure universal access to education, health care, and economic opportunity.

a. Secure the human rights of women and girls and end all violence against them.

b. Promote the active participation of women in all aspects of economic, political, civil, social, and cultural life as full and equal partners, decision makers, leaders, and beneficiaries.

c. Strengthen families and ensure the safety and loving nurture of all family members.

12. Uphold the right of all, without discrimination, to a natural and social environment supportive of human dignity, bodily health, and spiritual well-being, with special attention to the rights of indigenous peoples and minorities.

a. Eliminate discrimination in all its forms, such as that based on race, color, sex, sexual orientation, religion, language, and national, ethnic or social origin.

b. Affirm the right of indigenous peoples to their spirituality, knowledge, lands and resources and to their related practice of sustainable livelihoods.

c. Honor and support the young people of our communities, enabling them to fulfill their essential role in creating sustainable societies.

d. Protect and restore outstanding places of cultural and spiritual significance.

IV. DEMOCRACY, NONVIOLENCE, AND PEACE

13. Strengthen democratic institutions at all levels, and provide transparency and accountability in governance, inclusive participation in decision making, and access to justice.

a. Uphold the right of everyone to receive clear and timely information on environmental matters and all development plans and activities which are likely to affect them or in which they have an interest.

b. Support local, regional and global civil society, and promote the meaningful participation of all interested individuals and organizations in decision making.

c. Protect the rights to freedom of opinion, expression, peaceful assembly, association, and dissent.

d. Institute effective and efficient access to administrative and independent judicial procedures, including remedies and redress for environmental harm and the threat of such harm.

e. Eliminate corruption in all public and private institutions.

f. Strengthen local communities, enabling them to care for their environments, and assign environmental responsibilities to the levels of government where they can be carried out most effectively.

14. Integrate into formal education and life-long learning the knowledge, values, and skills needed for a sustainable way of life.

a. Provide all, especially children and youth, with educational opportunities that empower them to contribute actively to sustainable development.

b. Promote the contribution of the arts and humanities as well as the sciences in sustainability education.

c. Enhance the role of the mass media in raising awareness of ecological and social challenges.

d. Recognize the importance of moral and spiritual education for sustainable living.

15. Treat all living beings with respect and consideration.

a. Prevent cruelty to animals kept in human societies and protect them from suffering.

b. Protect wild animals from methods of hunting, trapping, and fishing that cause extreme, prolonged, or avoidable suffering.

c. Avoid or eliminate to the full extent possible the taking or destruction of non-targeted species.

16. Promote a culture of tolerance, nonviolence, and peace.

a. Encourage and support mutual understanding, solidarity, and cooperation among all peoples and within and among nations.

b. Implement comprehensive strategies to prevent violent conflict and use collaborative problem solving to manage and resolve environmental conflicts and other disputes.

c. Demilitarize national security systems to the level of a non-provocative defense posture, and convert military resources to peaceful purposes, including ecological restoration.

d. Eliminate nuclear, biological, and toxic weapons and other weapons of mass destruction.

e. Ensure that the use of orbital and outer space supports environmental protection and peace.

f. Recognize that peace is the wholeness created by right relationships with oneself, other persons, other cultures, other life, Earth, and the larger whole of which all are a part.

THE WAY FORWARD

As never before in history, common destiny beckons us to seek a new beginning. Such renewal is the promise of these Earth Charter principles. To fulfill this promise, we must commit ourselves to adopt and promote the values and objectives of the Charter.

This requires a change of mind and heart. It requires a new sense of global interdependence and universal responsibility. We must imaginatively develop and apply the vision of a sustainable way of life locally, nationally, regionally, and globally. Our cultural diversity is a precious heritage and different cultures will find their own distinctive ways to realize the vision. We must deepen and expand the global dialogue that generated the Earth Charter, for we have much to learn from the ongoing collaborative search for truth and wisdom.

Life often involves tensions between important values. This can mean difficult choices. However, we must find ways to harmonize diversity with unity, the exercise of freedom with the common good, short-term objectives with long-term goals. Every individual, family, organization, and community has a vital role to play. The arts, sciences, religions, educational institutions, media, businesses, nongovernmental organizations, and governments are all called to offer creative leadership. The partnership of government, civil society, and business is essential for effective governance.

In order to build a sustainable global community, the nations of the world must renew their commitment to the United Nations, fulfill their obligations under existing international agreements, and support the implementation of Earth Charter principles with an international legally binding instrument on environment and development.

Let ours be a time remembered for the awakening of a new reverence for life, the firm resolve to achieve sustainability, the quickening of the struggle for justice and peace, and the joyful celebration of life.

Annotated Bibliography

Compiled by Holmes Rolston III

1. REFERENCE WORKS

The most comprehensive bibliography is that of the International Society for Environmental Ethics, with more than 15,000 entries, updated annually. The bibliography is searchable and available from http://www.cep.unt.edu/bib.

Brennan, Andrew, ed. 1995. *The Ethics of the Environment.* Brookfield, VT: Dartmouth Publishing Company. International Research Library of Philosophy. *A large, single-volume collection of about three dozen basic and classic papers through 1995.*

Callicott, J. Baird, and Clare Palmer, eds. 2005. *Environmental Philosophy: Critical Concepts in the Environment.* London; New York: Routledge. *This is the most comprehensive collection in a single multivolume work. Nearly a hundred of the now-classic and important articles in the field are reprinted in five volumes.*

Jamieson, Dale, ed. 2001. *A Companion to Environmental Philosophy.* Malden, MA: Blackwell. *This is a major reference work with three dozen articles covering various aspects of environmental ethics. Topics covered include classical concepts of nature in philosophy and religion, and contemporary environmental ethics, not only in philosophy but also in literature, aesthetics, and economics. The volume deals with wilderness, population, sustainability, global warming, environmental justice, and related subjects.*

2. SYSTEMATIC OVERVIEWS

Attfield, Robin. 1992. *The Ethics of Environmental Concern.* 2nd edition. Athens: University of Georgia Press. *The first edition, one of the early systematic works in the field by a British philosopher, was published by Columbia University Press and Blackwell, Oxford, UK, in 1983.*

Brennan, Andrew. 1988. *Thinking about Nature: An Investigation of Nature, Value, and Ecology.* Athens: University of Georgia Press. *An effort to think ecologically about value and ethics by a then-British philosopher who later moved to Australia.*

Callicott, J. Baird. 1994. *Earth's Insights: A Survey of Ecological Ethics from the Mediterranean Basin to the Australian Outback.* Berkeley: University of California Press. *An exercise in comparative environmental philosophy. It explores and critically evaluates environmental ethics grounded in all the world's major religious traditions (Judaism, Christianity, Islam, Hinduism, South Asian and East Asian Buddhism, Daoism, and Confucianism) and representative indigenous traditions (from Polynesia, North America, South America, Africa, and Australia) and tests their ecological merits against the Leopold land ethic, which is recommended as the international gold standard for environmental ethics.*

Des Jardins, Joseph R. 2001. *Environmental Ethics: An Introduction to Environmental Philosophy.* 3rd edition. Belmont, CA: Wadsworth/Thomson Learning, 2001. *Third edition of an introduction addressed to those previously unacquainted with the field.*

Devall, Bill, and George Sessions. 1985. *Deep Ecology: Living as if Nature Mattered.* Salt Lake City, UT: Peregrine Smith. *Long a standard introduction to Deep Ecology, a philosophical position that seeks to raise ecological consciousnesss and reveal the unity of humanity and nature, a consciousness thought of as an enlarged ecological self.*

Ehrenfeld, David. 1978. *The Arrogance of Humanism.* New York: Oxford, UK: Oxford University Press. *Ehrenfeld, a biologist, not a philosopher, proved quite influential in awakening scientists to the anthropocentrism in their science and opening up the larger question of intrinsic values in nonhuman nature.*

Hargrove, Eugene C. 1996. *Foundations of Environmental Ethics.* Denton, TX: Environmental Ethics Books. *A far-ranging investigation of the intellectual history of environmental attitudes, with a focus on aesthetic arguments as a historical and contemporary foundation of environmental ethics.*

Johnson, Lawrence E. 1990. *A Morally Deep World: An Essay on Moral Significance and Environmental Ethics.* Cambridge: Cambridge University Press. *A rights-based theory of environmental ethics, extending rights to the kinds of things typically thought incapable of possessing them. Nonhuman animals and ecosystems are viewed as morally significant beings with interests and rights. Written for general readers.*

Kohák, Erazim. 2000. *The Green Halo: A Bird's-Eye View of Ecological Ethics.* Chicago: Open Court. *Originally written for students in a Czech university. The author fled Czechoslovakia with the coming of the pro-Soviet regime, long taught philosophy in the United States, and returned after the Soviet collapse. His life in multiple worlds gives him facility with Soviet ideology, continental philosophy (especially phenomenology), Central and Eastern European thought, as well as British and American philosopy, all brought to bear on environmental ethics.*

Mathews, Freya. 1991. *The Ecological Self.* London: Routledge. *A metaphysics of interconnectedness, based on the fundamental ecological intuition that humans are in some sense "one with" nature and that everything is connected to everything else; this work rejects the dominant atomistic metaphysics implicit in European and North American philosophy.*

Naess, Arne. 1989. *Ecology, Community, and Lifestyle: Outline of an Ecosophy.* New York: Cambridge University Press. *Translated and revised by David Rothenberg from* Okologi, Samfunn, og Livsstil, *published in Norwegian in 1976. The original Naess article envisioning a Deep Ecology is "The Shallow and the Deep, Long-Range Ecology Movements: A Summary,"* Inquiry *16 (1973): 95–100.*

Norton, Bryan G. 1991. *Toward Unity among Environmentalists.* New York: Oxford University Press. *Norton seeks to unite environmentalists in the common cause of environmental protection and appreciation despite their multiple and varied value systems. Notwithstanding these diverse worldviews, he believes that there can be converging policies (his "convergence hypothesis"). Norton illustrates his thesis using Muir, Pinchot, and Leopold, and applies it to growth, pollution, biodiversity, and land use.*

Palmer, Clare. 1997. *Environmental Ethics.* Santa Barbara, California: ABC-CLIO. *This reference work is especially good as a basic resource guide to materials, chronology, major figures, and principal issues.*

Passmore, John. 1974. *Man's Responsibility for Nature.* New York: Scribners. *One of the earliest works in the field, by a prominent Australian philosopher. Passmore argues that classical humanistic ethics can be applied to new environmental problems, a view challenged by many who hold that environmental ethics has many novel, nonanthropocentric dimensions.*

Plumwood, Val. 2003. *Feminism and the Mastery of Nature.* London: Routledge. *A magisterial critique of dualism as an ingrained habit of thinking, including male-female and human-nature instances of dualistic thinking.*

Rolston III, Holmes. 1988. *Environmental Ethics: Duties to and Values in the Natural World.* Philadelphia: Temple University Press. *One of the earliest systematic works in environmental ethics, ranging across animals, plants, endangered species, ecosystems, environmental policy and business, and a personal environmental ethic. Rolston throughout claims there are intrinsic values in nature that humans ought to respect, in addition to considerations about how humans are helped or hurt by the condition of their environment. (A critique of Rolston's work is found in Preston, Christopher J., and Wayne Ouderkirk, eds. 2006.* Nature, Value, Duty: Life on Earth with Holmes Rolston III. *Dordrecht, The Netherlands: Springer.)*

Rolston III, Holmes. 1994. *Conserving Natural Value.* New York: Columbia University Press. *This survey is written for use in introductory college classes on biological and natural-resource conservation and environmental philosophy, ethics, and policy. There is extensive use of cases to provoke thought, and Rolston also applies his ethics using a number of axioms designed to help those who confront practical decisions.*

Stone, Christopher F. 1987. *Earth and Other Ethics: The Case for Moral Pluralism.* New York: Harper and Row. *This book, by a lawyer, introduces a view of normative ethics that is pluralistic regarding the entities and situations that are morally relevant, foreshadowing later focus on a pragmatic environmental ethics. Different moral systems, he argues, must be used depending on levels of concern and relevant conditions of decision.*

Sylvan, Richard, and David Bennett. 1994. *The Greening of Ethics: From Human Chauvinism to Deep-Green*

Theory. Cambridge, UK: White Horse Press. *Environmental ethics from "down under" (Australia), claiming that the European/North American worldview is topsy-turvy. The authors set out a course for Australia's independent national development. They find environmental ethics in shallow, intermediate, and deep forms, and the authors delineate their deep-green theory.*

Taylor, Paul. 1986. *Respect for Nature: A Theory of Environmental Ethics*. Princeton: Princeton University Press. *The classic defense of biocentrism. All living organisms seek their own good and are centers of inherent worth that warrant respect. The biocentric outlook denies human superiority, in theory at least, and calls for a radical bioegalitarianism, although Taylor recognizes situations in which humans, based on the principle of self-defense, can sacrifice the basic interests of wildlife to further their own basic interests.*

Wenz, Peter S. 2001. *Environmental Ethics Today*. New York: Oxford University Press. *Wide ranging: overpopulation, free markets, human rights, future generations, global warming, animal liberation, medical research with animals, species diversity, the land ethic, hunting as a conservation strategy, aesthetics, and conservation. Wenz argues that a synergy can and ought to exist between respect for people and respect for nature. He contends that simultaneous respect for people and nature improves outcomes for both.*

3. COLLECTED ESSAYS OF A SINGLE AUTHOR

Callicott, J. Baird. 1989. *In Defense of the Land Ethic: Essays in Environmental Philosophy*. Albany, NY: SUNY Press. *A collection of previously published essays by the leading philosophical interpreter of Aldo Leopold. (A critique of Callicott's work is found in* Land, Value, Community: Callicott and Environmental Philosophy, *eds. Wayne Ouderkirk and Jim Hill. 2002. Albany, NY: SUNY Press.)*

Callicott, J. Baird. 1999. *Beyond the Land Ethic: More Essays in Environmental Philosophy*. Albany, NY: SUNY Press. *A second collection of Callicott's essays, most of them written between 1989 and 1999 and previously published.*

Rolston III, Holmes. 1986. *Philosophy Gone Wild*. Buffalo, NY: Prometheus Books. *A collection of fifteen essays articulating and justifying values in nature, generally progressing from the more theoretical to the more personal. Values in nature, following nature, subjective versus objective values, endangered species, nature and human emotions; immediate personal experience of nature.*

Sagoff, Mark. 1988. *The Economy of the Earth: Philosophy, Law, and the Environment*. New York: Cambridge University Press. *In this collection of his previously published essays, Sagoff concentrates on the interconnections between environmental policy, law, economics, and environmental ethics. There is a systematic attack on the basic assumptions of welfare economics and cost-benefit analysis as a basis for environmental policy. Values are community-based, intersubjective goals that evolve throughout the history of the community, state, or nation and cannot be reduced to consumer preferences.*

Sagoff, Mark. 2004. *Price, Principle, and the Environment*. New York: Cambridge University Press. *A collection of nine previously published essays further developing his critique of a purely economic approach to environmental concerns, particularly the claim that all values are preferences subject to expression in a monetary metric.*

4. ANTHOLOGY OVERVIEWS, COLLECTED ESSAYS BY MULTIPLE AUTHORS, TEXTBOOKS

Armstrong, Susan J., and Richard G. Botzler, eds. 2003. *Environmental Ethics: Divergence and Convergence*. 3rd edition. New York: McGraw-Hill. *Third edition of an anthology that has proven a classic anthology in previous editions. More than sixty articles. Comprehensive, but coverage is often limited.*

Attfield, Robin, and Andrew Belsey, eds. 1994. *Philosophy and the Natural Environment*. Cambridge, UK: Cambridge University Press. *Values in nature, restoration, awe in nature, order and disorder in nature, global environmental justice, genetic engineering, persons in nature, anthropocentrism, and more. This collection originated from the Royal Institute of Philosophy Conference, "Philosophy and the Natural Environment," held at the University of Wales in Cardiff in 1993.*

Chappell, Timothy D. J., ed. 1997. *Respecting Nature: Environmental Thinking in the Light of Philosophical Theory*. New York: Columbia University Press. *Features theory in environmental ethics in relation to classical philosophy. Plato and environmental ethics, nature as a social construct, aesthetics of environment, sustainability, animal welfare, whaling, zoos.*

Elliot, Robert, ed. *Environmental Ethics*. 1995. New York: Oxford University Press. *This collection focuses on philosophically seminal articles rather than seeking more comprehensive coverage by incorporating extracts from several dozen articles. Values in nature, anthropocentrism in environmental ethics, animal welfare, restoration, stability in natural systems, ecofeminism.*

Foltz, Bruce V., and Robert Frodeman, eds. 2004. *Rethinking Nature: Essays in Environmental Philosophy*. Bloomington: Indiana University Press. *Emphasizes continental philosophy. Aesthetics, ontology, phenomenology, gender and the environment, and the role of science and technology in forming knowledge about the natural world.*

Frodeman, Robert, ed. 2000. *Earth Matters: The Earth Sciences, Philosophy, and the Claims of Community.* Upper Saddle River, NJ: Prentice-Hall. *Brings together fifteen essays on environmental matters from a multidisciplinary group of authors, including scientists, policy analysts, and philosophers.*

Gruen, Lori, and Dale Jamieson, eds. 1994. *Reflecting on Nature: Readings in Environmental Philosophy.* Oxford. *Highlights the problems of environmental justice and sustainable development from a multicultural perspective; features feminist and minority scholars and scholars from developing countries. Biodiversity loss, the significance of wilderness, population and overconsumption, and the human use of animals.*

Light, Andrew, and Rolston III, Holmes, eds. 2003. *Environmental Ethics: An Anthology.* Oxford: Blackwell. *Forty classic and recent full-length articles in environmental ethics organized for classroom use. What is environmental ethics? Who counts morally? Intrinsic value in nature, environmental pluralism, Deep Ecology, ecofeminism, restoration, wilderness, sustainability, social choices, and environmental values.*

List, Peter C., ed. 1993. *Radical Environmentalism: Philosophy and Tactics.* Belmont, CA: Wadsworth. *Radical activism in environmental ethics critically examined.* The Monkey Wrench Gang *(Edward Abbey). Greenpeace; Earth First!; the Sea Shepard Society (Paul Watson); civil disobedience and tree spiking; activist protests against the destruction and pollution of natural systems.*

List, Peter, ed. 2000. *Environmental Ethics and Forestry: A Reader.* Philadelphia: Temple University Press. *Both forestry and philosophy have been rethinking their foundations; each needs the other. John Muir versus Gifford Pinchot; Leopold's land ethic; values in forests, both instrumental and intrinsic; aesthetic experience in forests; global forests; foresters as advocates. A particular feature is examination of codes of ethics as formulated by foresters.*

Pierce, Christine, and Donald VanDeVeer, eds. 1995. *People, Penguins, and Plastic Trees.* 2nd edition. Belmont, CA: Wadsworth. *Long a best-selling text, first published in 1986 and widely regarded as the easiest text to use with freshmen and sophomores. Ecofeminism, Deep Ecology, Native American land ethics, critiques of industrialized nations by those in less-industrialized nations, environmental racism, sustainability, biocentric views, intrinsic value, biodiversity, animal liberation, land ethics.*

Pojman, Louis, P., and Paul Pojman, eds. 2008. *Environmental Ethics: Readings in Theory and Application.* 5th edition. Belmont, CA: Thomson/ Wadsworth. *A perennially popular anthology that has gone through five editions since 1994. The Pojmans include articles on both sides of issues. The historical roots of our ecological crisis, animal rights, biocentrism, the land ethic, Deep Ecology, intrinsic natural value, ecofeminism, the Gaia hypothesis, biodiversity, obligations to future generations, Asian concepts of nature, world population, hunger, sustainable development.*

Schmidtz, David, and Elizabeth Willott, eds. 2002. *Environmental Ethics: What Really Matters, What Really Works.* New York: Oxford University Press. *Sixty-two selections, addressing the principal areas of inquiry in the field. Value in nature, the land ethic, animal liberation, environmental holism, rights in nature, wilderness, biodiversity, sustainability, poverty, cost-benefit analysis, and more.*

VanDeVeer, Donald, and Christine Pierce, eds. 2003. *The Environmental Ethics and Policy Book: Philosophy, Ecology, Economics.* 3rd edition. *Environmental ethics with a focus on how it affects public policy. Future generations, sustainability, corporate responsibility, population, consumption, marine environmental ethics, genetically modified foods, transgenic organisms, the impact of fast food production, patenting life. Jewish, Christian environmental ethics, aboriginal ecological knowledge. One feature is an Internet environmental resources section.*

Weston, Anthony, ed. 1999. *An Invitation to Environmental Philosophy.* New York: Oxford University Press. *Offered as an alternative to heavy academic anthologies, this compact anthology features five original essays by prominent environmental philosophers; intended as a first invitation to environmental philosophy.*

Zimmerman, Michael E., et al., eds. 2005. *Environmental Philosophy: From Animal Rights to Radical Ecology.* 4th edition. Upper Saddle River, NJ: Pearson/Prentice Hall. *Fourth edition of a time-tested and popular anthology. Animal welfare, biocentrism, the land ethic, ecofeminism, continental environmental philosophy, ecophenomenology, ecofascism, free-market versus political environmentalism, sustainability, social ecology.*

5. CASE STUDIES

Derr, Patrick G., and Edward M. McNamara. 2003. *Case Studies in Environmental Ethics.* Lanham, MD: Rowman and Littlefield. *More than forty cases, typically three or four pages each: includes Hawaiian feral pigs, oil and the Arctic National Wildlife Refuge (ANWR), golden rice, Bhopal, monkeywrenching, great apes, the Delhi Sands fly, and a host of other topics.*

Gudorf, Christine E., James E. Huchingson. 2003. *Boundaries: A Casebook in Environmental Ethics.* Washington, DC: Georgetown University Press. *The Everglades, Java forests, endangered ecosystems and endangered cultures in Madagascar, nuclear waste, coral reefs, hydropower versus free-flowing rivers, genetically modified foods, hunting in India, xenotransplants.*

Newton, Lisa H., and Catherine K. Dillingham, eds. 1997. *Watersheds: Classic Cases in Environmental Ethics.* 2nd edition. Belmont, CA: Wadsworth. *Impressive detail and documentation of dozens of specific cases in environmental ethics combined with insightful ethical analysis.*

6. ANIMALS AND ENVIRONMENTAL ETHICS

Hargrove, Eugene C. 1992. *The Animal Rights/ Environmental Ethics Debate: The Environmental Perspective.* Albany: State University of New York Press. *A collection of essays by a number of environmental philosophers offering criticism of and various alternatives to nonanthropocentric ethics limited to animals and excluding other nonhuman natural entities such as plants; higher levels of biological organization, such as species and ecosystems; and nature as a whole.*

Regan, Tom. 2004. *The Case for Animal Rights.* Berkeley: University of California Press. *A philosophically rigorous argument that animals have rights. First published in 1983, this book complemented Peter Singer's utilitarian* Animal Liberation*; these were the two most influential books concerned with animal ethics in the second half of the last century.*

Singer, Peter. 2002. *Animal Liberation.* 2nd edition. New York: Ecco (HarperCollins). *The book that launched contemporary ethical concern for animals, first published in 1975. Singer argues from a utilitarian viewpoint that humans are morally obligated to minimize animal suffering.*

Sterba, James P., ed. 1995. *Earth Ethics: Environmental Ethics, Animal Rights, and Practical Applications.* Upper Saddle River, NJ: Prentice-Hall. *Three dozen contributors analyze animal liberation, animal rights, their reconciliation with environmental ethics, anthropocentrism versus nonanthropocentrism, Deep Ecology, ecofeminism, biodiversity, climate change, economics, and environmental quality.*

Waldau, Paul, and Kimberley C. Patton, eds. 2006. *A Communion of Subjects: Animals in Religion, Science, and Ethics.* New York: Columbia University Press. *Animals are subjects who experience the world and have been pervasively incorporated into human belief systems, myths, and rituals, traditions that can serve as a basis for contemporary respect and conservation.*

7. BIODIVERSITY, WILDERNESS, RESTORATION, AESTHETICS

Callicott, J. Baird, and Michael P. Nelson, eds. 1998. *The Great New Wilderness Debate.* Athens: University of Georgia Press. *A large anthology on wilderness, covering the spectrum of views about the character and importance of wilderness conservation. Some contributors argue that wilderness is a European and North American idea, socially constructed. Others argue that indigenous peoples had so managed wilderness that primeval nature seldom continues in present wilderness landscapes. Others find substantial tracts of spontaneous wild nature, where ecosystemic processes are the dominant determinants, and the effect of humans is minimal.*

Carlson, Allen, and Sheila Lintott. 2008. *Nature, Aesthetics, and Environmentalism: From Beauty to Duty.* New York: Columbia University Press. *This collection combines the most important historical essays on nature appreciation and the best contemporary research in the field. Aesthetic of nature in relation to art and science; positive aesthetics, the view that all wild landscapes are beautiful; moral duties deriving from the aesthetics of nature.*

Elliot, Robert. 1997. *Faking Nature: The Ethics of Environmental Restoration.* London and New York: Routledge. *Natural value cannot be restored because original naturalness is a basis for intrinsic value in nature. Restored nature, however desirable, is second best because uninterrupted historical genesis cannot be restored; it is a faked nature because of this lost value.*

Nash, Roderick. 2002. *Wilderness and the American Mind.* 4th edition. New Haven, CT: Yale University Press. *A classic study, first published in 1967, by an environmental historian of changing ideas about wilderness in American thought.*

Nelson, Michael P., and J. Baird Callicott, eds. 2008. *The Wilderness Debate Rages On.* Athens: University of Georgia Press. *Organized into four parts, the first of which documents a little-known history of wilderness-preservation advocacy by ecologists that, had it been able to influence national policy, would have resulted in a very different system of wilderness preserves, focused nonanthropocentrically on critical habitat for threatened species and representative ecosystem types rather than on anthropocentric recreation. Also includes more non-European and liminal critiques of the wilderness idea, philosophical debate about the wilderness idea, and alternatives to the wilderness idea.*

Norton, Bryan G., ed. 1986. *Preservation of Species: The Value of Biological Diversity.* Princeton, NJ: Princeton University Press. *Scientific and social dimensions of extinction, management decisions regarding species preservation, ethical justification of species preservation, instrumental (such as economic) reasons versus the intrinsic value of species, aesthetic values in species preservation.*

Oelschlaeger, Max. 1991. *The Idea of Wilderness from Prehistory to the Present*. New Haven, CT: Yale University Press. *An intellectual history drawing on evidence from philosophy, anthropology, theology, literature, ecology, cultural geography, and archaeology.*

Rolston III, Holmes. 1985. "Duties to Endangered Species." *BioScience* 35: 718–726. *Duties to humans concerning endangered species, although important, must be complemented by duties directly to species. This requires an account, biologically, of what species are, and, ethically, of why species are morally considerable. Species are dynamic natural kinds, historical life lineages, that humans ought to respect. Another author in this special issue of* BioScience *is Edward O. Wilson.*

Throop, William, ed. 2000. *Environmental Restoration: Ethics, Theory, and Practice*. Amherst, NY: Humanity Books/Prometheus Press. *This anthology examines whether restoring nature is viable, legitimate, and practical.*

Willers, William B., ed. 1999. *Unmanaged Landscapes: Voices for Untamed Nature*. Washington, DC: Island Press. *Unmanaged landscapes are the focus of the struggle to protect and restore wildness, the autonomy of nature, and to allow for its preservation and return on a grand scale.*

8. ENVIRONMENTAL JUSTICE, ENVIRONMENTAL VIRTUE ETHICS

Attfield, Robin, and Barry Wilkins, eds. 1992. *International Justice and the Third World: Essays in the Philosophy of Development*. London: Routledge. *The contributors ask about justice among societies of unequal power and worry that development efforts, resulting in indebtedness of the developing world, are often exploitative. What are the relations between just development and environmental conservation?*

Bullard, Robert D., ed. 2005. *The Quest for Environmental Justice: Human Rights and the Politics of Pollution*. San Francisco: Sierra Club Books. *An anthology by a sociologist, one of the first people to become deeply concerned about the way in which the poor disproportionately bear the burdens of environmental degradation.*

Sandler, Ronald. 2007. *Character and Environment: A Virtue-Oriented Approach to Environmental Ethics*. New York: Columbia University Press. *Any ethic of character can and should be informed by many environmental considerations. A pluralist, virtue-oriented environmental ethic accommodates the richness and complexity of human relationships with the natural environment and provides effective and nuanced guidance on environmental issues.*

Sandler, Ronald, and Philip Cafaro, eds. 2005. *Environmental Virtue Ethics*. Lanham, MD: Rowman and Littlefield. *Contributors discuss the role that virtue and character have traditionally played in environmental discourse and reflect upon the role that it should play in the future. Environmental virtue ethics theory, particular environmental virtues and vices, and applying environmental virtue ethics to particular environmental issues.*

Shrader-Frechette, Kristin. 2002. *Environmental Justice: Creating Equity, Reclaiming Democracy*. New York: Oxford University Press. *Fundamental ethical concepts such as equality, property rights, procedural justice, free informed consent, intergenerational equity, and just compensation have been compromised for a large segment of the global population, among them Appalachians, African Americans, workers in hazardous jobs, and indigenous people in developing nations. Burdens like pollution and resource depletion need to be apportioned more equally.*

Wenz, Peter S. 1988. *Environmental Justice*. Albany, NY: SUNY Press. *Competing principles of distributive justice as they might guide environmental decisions: libertarian theory, laissez-faire economics, human rights, utilitarian theory, cost-benefit analysis, virtue ethics, John Rawls's theory of justice. Wenz offers concentric-circle theory of environmental justice.*

Westra, Laura, and Peter S. Wenz, eds. 1995. *Faces of Environmental Racism: Confronting Issues of Global Justice*. Lanham, MD: Rowman and Littlefield. *Racial minorities in the United States are disproportionately exposed to toxic wastes and other environmental hazards. Internationally, wealthy countries of the north increasingly ship hazardous wastes to poorer countries of the south. These authors argue that environmentalism and concern for human beings and justice can be entirely compatible.*

9. RELIGION AND NATURE

Foltz, Richard C., ed. 2003. *Worldviews, Religion, and the Environment: A Global Anthology*. Belmont, CA: Wadsworth/Thomson Learning. *First peoples, Buddhism, Chinese traditions, Japanese traditions, Judaism, new cosmologies, globalization, ecojustice. More than sixty contributors.*

Gottlieb, Roger S. 2006. *A Greener Faith: Religious Environmentalism and Our Planet's Future*. New York: Oxford University Press. *Theologians are recovering nature-honoring elements of traditional religions and forging bold new theologies connecting devotion to God and spiritual truth with love for God's creation and care for the earth.*

Northcott, Michael S. 1996. *The Environment and Christian Ethics.* Cambridge, UK: Cambridge University Press. *Environmental ethics from a perspective of Christian ethics, written by a theological ethicist with a thorough familiarity with the philosophical literature. The resolution of the environmental crisis requires the rediscovery of value and moral significance in the nonhuman natural world, an independence located in divine beneficence. Christians have often been the cause of environmental degradation, but the primal Hebrew vision and early Christians both had great respect for creation.*

Oelschlaeger, Max. 1994. *Caring for Creation: An Ecumenical Approach to the Environmental Crisis.* New Haven, CT: Yale University Press. *Only the churches, as the repository of moral values that lie outside the economic paradigm, can provide the social and political leadership and power to move our society to ecological sustainability. All faiths have an emphasis on caring for creation on which we can draw, and religion is necessary if we are to solve the environmental crisis politically.*

Rasmussen, Larry L. 1996. *Earth Community Earth Ethics.* Maryknoll, NY: Orbis. *An insightful analysis, from a theological perspective, of social justice and ecological concerns. Underlying themes are "justice, peace, and the integrity of creation" (World Council of Churches), areas in which Rasmussen has been influential. Humans have sought arrogant dominion over nature, denying the wholeness of creation. There is need now for symbols that effect a reenchantment of the world.*

Taylor, Bron, ed. 2005. *Encyclopedia of Religion and Nature.* 2 vols. London: Thoemmes Continuum. *An encyclopedia that is chronologically, geographically, and theoretically comprehensive, with a thousand entries from more than 500 contributors.*

Tucker, Mary Evelyn, and John Grim, eds. 1997–2002. *Religions of the World and Ecology.* 10 vols. Cambridge, MA: Harvard University Press. *Ten volumes, each on a major world religion.*

10. ECOFEMINISM

Clayton, Patti H. 1998. *Connection on the Ice: Environmental Ethics in Theory and Practice.* Philadelphia: Temple University Press. *Ecofeminist environmental ethics compared with other major types of environmental philosophy, taking as a critical case the rescue of three whales trapped in ice in Alaska. The real world displays quite multifaceted human-nonhuman relationships.*

Plumwood, Val. 2002. *Environmental Culture.* New York: Routledge. *A detailed and passionate argument for forms of culture that are logically and pragmatically superior to those cultures built on the rationalism, idealism, and empiricism that encourage moral distance. Humans are dependent on nature, men are dependent on women, and those with economic and decision-making power are dependent on the disempowerment of others. Sustainable cultures must care for creation.*

Ruether, Rosemary Radford. 1994. *Gaia and God: An Ecofeminist Theology of Earth Healing.* San Francisco: HarperOne. *European and North American theology often has a patriarchal tradition of dominance, but the classical Christian traditions also struggled with injustice and sin and sought to create just and loving relations between people in their relations with the living earth (Gaia). Christians today can use this heritage, enlarging it for a better vision of an abundant life on a sustainable earth.*

Warren, Karen, ed. 1994. *Ecological Feminism.* New York: Routledge. *The conceptual underpinnings of women-nature connections and the importance of seeing sexism and the exploitation of the environment as parallel forms of domination. Ecofeminism and the reconstruction of environmental ethics.*

Warren, Karen J. 2000. *Ecofeminist Philosophy: A Western Perspective on What It Is and Why It Matters.* Lanham, MD: Rowman and Littlefield. *Ecofeminism and animal welfare, vegetarianism, ecosystem ecology, Leopold's land ethic, ecojustice, patriarchy, spirituality.*

11. SUSTAINABILITY, FUTURE GENERATIONS

Burkhardt, Jeffrey. 1989. "The Morality behind Sustainability." *Journal of Agricultural Ethics* 2: 113–128. *Obligations to future generations entail more than sustaining sufficient food production or an adequate resource base; they extend to a continuing tradition of care and community.*

Daly, Herman E., and John B. Cobb Jr. 1999. *For the Common Good: Redirecting the Economy toward Community, the Environment, and a Sustainable Future.* 2nd edition. Boston: Beacon Press. *A steady-state economist and a theologian combine for a searching evaluation of whether and how far the global economy contributes to the common good, both social and environmental.*

Millennium Ecosystem Assessment. 2005. *Living Beyond Our Means: Natural Assets and Human Well-Being: Statement from the Board.* Available from *http://www.millenniumassessment.org/en/index.aspx. This is a summary document of a huge project sponsored by the United Nations and a host of organizations and corporations and involving more than 1,300 experts worldwide. There are multiple volumes, both in print and online. The focus is scientific, but there is a sustained effort to apply these results toward a humane environmental policy.*

National Commission on the Environment. 1993. *Choosing a Sustainable Future: The Report of the National Commission on the Environment.* Washington, DC: Island Press. *A private-sector initiative convened by the World Wildlife Fund that concludes that the natural processes that support life on earth are increasingly at risk.*

Norton, Bryan G. 2005. *Sustainability: A Philosophy of Adaptive Ecosystem Management.* Chicago: University of Chicago Press. *Sustainability ought to be the cornerstone of environmental policy and requires shared, multidisciplinary deliberation over environmental goals and policy. Such communication is now fragmented by disciplines and ideologies. Norton offers a vision of a nonideological vocabulary that can accommodate the scientific and evaluative environmental discourse.*

Partridge, Ernest, ed. 1981. *Responsibilities to Future Generations: Environmental Ethics.* Buffalo, NY. *What do humans owe to posterity? Two dozen contributors seek an answer. Concern for future generations is a vital dimension of the ecological crisis, essential to sustainability. Although humans' ability to affect the future is immense, their ability to foresee the result of their environmental interventions is incomplete. This poses challenging moral questions and novel responsibilities.*

12. GLOBAL ENVIRONMENTAL ETHICS, CLIMATE CHANGE

Adger, W. Neil, Jouni Paavola, Seleemul Huq, and M. J. Mace, eds. 2006. *Fairness in Adaptation to Climate Change.* Cambridge, MA: The MIT Press. *All countries will be endangered by climate-change risks from flood, drought, and other extreme weather events, but developing countries are more dependent on climate-sensitive livelihoods such as farming and fishing and hence are more vulnerable. Nevertheless, the concerns of developing countries are marginalized in climate-policy decisions.*

Attfield, Robin. 2003. *Environmental Ethics: An Overview for the Twenty-First Century.* Cambridge, UK: Polity Press. *A survey and synthesis of the enormous range of challenging issues: local and global environmental problems; theories of value, stewardship, anthropocentrism and biocentrism; sustainable development; population; global citizenship. Attfield advocates what he calls biocentric consequentialism.*

Dallmeyer, Dorinda, and Albert Ike, eds. 1998. *Environmental Ethics and the Global Marketplace.* Athens: University of Georgia Press. *Contributors present arguments for creating global business practices that work in harmony with the environment; discussions of free trade, private ownership, sustainability, environmental justice.*

Engel, J. Ronald, and Joan Gibb Engel, eds. 1990. *Ethics of Environment and Development: Global Challenge and International Response.* Tucson: University of Arizona Press. *This anthology, published in association with the International Union for the Conservation of Nature and Natural Resources, contains more than twenty articles with an international focus on forms of development that are compatible with wildlife conservation.*

Northcott, Michael S. 2007. *A Moral Climate: The Ethics of Global Warming.* London: Darton, Longman and Todd. *Response to the challenge of global warming requires learning to put the common good ahead of selfish interests, weaving together the physical climate and the moral climate. Relieving climate change opens opportunities for solving other problems: world poverty, the rich/poor divide, the overuse of resources, and the appreciation and conservation of nonhuman creation.*

Pojman, Louis P. 2000. *Global Environmental Ethics.* Mountain View, CA: Mayfield. *Classical ethical theories are challenged by both the global scale and the environmental dimensions of contemporary problems. Discussions of greenhouse effects, ozone depletion, population, world hunger, energy use, animal welfare, endangered species, wilderness, sustainability.*

Index

Bold *volume and page numbers (e.g.* ***1:1–3****) refer to the main entry on the subject. Page numbers in italics refer to illustrations, figures, and tables. Page numbers followed by the letter t indicate a table within the article.*

A

B

D

E

F

H

J

M

N

O

T

V

X

Y

Z